D1582992

SECOND EDITION

GRAZING MANAGEMENT

SECOND EDITION

GRAZING

MANAGEMENT

JOHN F. VALLENTINE

Emeritus Professor of Range Science
Brigham Young University
Provo, Utah

ACADEMIC PRESS

A Harcourt Science and Technology Company

San Diego   San Francisco   New York   Boston   London   Sydney   Tokyo

Academic Press
A Harcourt Science and Technology Company
525 B Street, Suite 1900, San Diego, California 92101-4495, USA
http://www.academicpress.com

Academic Press
Harcourt Place, 32 Jamestown Road, London NW1 7BY, UK
http://www.academicpress.com

Library of Congress Catalog Card Number: 00-104371

International Standard Book Number: 0-12-710001-6

PRINTED IN THE UNITED STATES OF AMERICA
00 01 02 03 04 05   MM 9 8 7 6 5 4 3 2 1

CONTENTS

3

ANIMAL NUTRITIONAL BALANCE

4

GRAZING SEASONS AND SEASONAL BALANCE

5

GRAZING EFFECTS ON PLANTS AND SOIL

6

GRAZING ACTIVITIES/BEHAVIOR

7

SPATIAL PATTERNS IN GRAZING

8

MANIPULATING GRAZING DISTRIBUTION

9

PLANT SELECTION IN GRAZING

10

KIND AND MIX OF GRAZING ANIMALS

11
GRAZING ANIMAL INTAKE AND EQUIVALENCE

12
GRAZING CAPACITY INVENTORY

13
GRAZING INTENSITY

14
GRAZING METHODS

15

GRAZING SYSTEMS

16

GRAZING AS AN ENVIRONMENTAL TOOL

PREFACE TO
SECOND EDITION

Grazing Management attempts to integrate the concepts, principles, and management techniques that apply to all grazing lands and to all grazing animals. The kinds of grazing lands grazed by these ungulate herbivores range from native and seeded rangelands through several kinds of perennial pasture to temporary and crop aftermath pasture. An attempt has been made to cover big game herbivores with an intensity equivalent to domestic livestock, but only in some areas was this fully accomplished.

Many academic disciplines for over half a century have focused attention on grazing management, and it is readily apparent that the subject must remain highly interdisciplinary. Although it seemed proper to compile a book about it, and now a second edition, the unknowns surrounding grazing management are readily acknowledged. It also is readily apparent that grazing management must remain dynmamic to accommodate and build upon new principles and applicatations as they become more fuly substantiated and documented. To accomodate the new information, concepts, and application during the past 10 years alone (since publication of the first edition), this second edition has been increased in size and coverage by about 50%.

New insights into grazing management may require grazing managers and scientists to modify substantially their present as well as future concepts and practices. New emphasis on grazing management in this second edition has included: (1) the role of post-ingestive feedback in diet selection, (2) spatial foraging decisions made by grazing animals, (3) the utility of the grazing animal as both an environmental enhancement tool as well as a forage harvester and converter, and (4) an expansion of the concept of grazing as a natural ecosystem component.

Recent research, evaluation, and use experience probably justify (5) the current

---

trend away from a universal application of complex grazing systems while re-serving their use for special situations, (6) greater attention to differentiating be-tween complex grazing systems and simpler grazing methods and a greater appli-cation of the latter, and (7) a general return to the basics of grazing management. These new recent developments in grazing management have required (8) the fur-ther refinement of grazing management terminology and (9) the development of decision support systems to aid the grazing manager (the latter acknowledged but included only minimally in this second edition of *Grazing Management*).

The principles and practices emphasized in *Grazing Management* should apply throughout the world, even though social customs, agricultural policy, and land ownership may restrict their unlimited application. Even in North America cul-tural, social, and political philosophies and agendas continually invade and even detour the application of grazing management science and technology. A majori-ty of the examples of grazing management provided and much of the literature cited herein are of North American origin. Since grazing management is a rapid-ly developing science, extensive documentation seemed justified and has been pro-vided. Figure legends have also been utilized to summarize some of the salient points made in each chapter.

Grazing Management is recommended both as a textbook and as a reference manual. As a textbook, its prime audience is expected to be university students in upper division and graduate levels with a major in range science, animal science, big game management, or agronomy. However, it should also serve as a compre-hensive reference manual for extension and research personnel, grazing land man-agers, innovative ranchers and farmers, agribusiness personnel, and conservation-ists generally. Although this book is not primarily designed as a field manual, the concepts and principles of grazing management have been tied to their application throughout.

As this second edition of *Grazing Management* is a synthesis of the concepts, research data, and application experiences contributed directly and indirectly by many people, it is sincerely hoped that it will properly represent their conclusions.

John F. Vallentine
Springfield, Utah

1

INTRODUCTION TO GRAZING

I. THE ROLE OF GRAZING MANAGEMENT

A. GRAZING MANAGEMENT OBJECTIVES

Grazing management is the manipulation of animal grazing to achieve desired results based on animal, plant, land, or economic responses, but the continuing immediate goal is to supply the quantity and quality of forage needed by the grazing animal for it to achieve the production function intended. Grazing management is important because this is where theory is put into practice (Walker, 1995). The **grazier** is the person who manages the grazing animals, i.e., the **grazers** (including browsers). These terms have a common stem in the verb **graze,**

1

which specifies the consumption of standing forage by ungulate herbivores. Grazing of standing forage on range and pasture is the counterpart of mechanical (machine) harvesting of harvested forage crops (Fig. 1.1), except that the grazing animal is both the consumer and the converter, as well as the harvester, of grazed forage.

Ranch management is the manipulation of all **ranch** resources—including not only the grazing resources but also all the financial, personnel, and physical resources of the ranch—to accomplish the specific management objectives set for the ranch. In this regard, grazing management is only one aspect, albeit a very important one, of total ranch management. Where animal grazing is a prominent component of a farm operation and involves substantial use of lands for grazing, grazing management will necessarily play a similar role in farm management. Emphasis on grazing management is well deserved on both farm and ranch because of its relatively low cost and potentially great returns per unit of management input (Lewis and Volesky, 1988).

Grazing land management—which principally integrates both range and pasture management—is the art and science of planning and directing the development, maintenance, and use of grazing lands to obtain optimum, sustained returns based on management objectives. The effective grazing planner/manager must inventory all sources of available grazing capacity and integrate them into the best animal production system. The management interrelationships of different kinds of grazing lands, when used in a single production system, are great and must not be overlooked or ignored.

The use of rangeland is generally comingled with the use of other types of grazing land, and most range livestock and many big game animals use multiple sources of grazing capacity to meet their annual requirements (Vallentine, 1978). Grazing management is also a major component in the multiple-use management of public lands, whether the grazing is by domestic livestock or big game animals or a combination of both.

The positive manipulation of the soil-forage plant-grazing animal complex is a

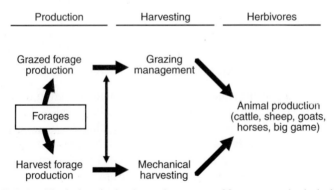

FIGURE 1.1   The dual production-harvesting avenues of forage conversion by herbivores.

central role of the grazing manager. Opportunities to enhance the energy efficiency of the soil–forage–ruminant complex exist in three principal areas:

1. Increase conversion of radiant solar energy to usable form through photosynthesis during growth of the forage plants, i.e., by enhancing the quantity and quality of forage produced on the site.
2. Increase consumption of the energy fixed by forage plants by the grazing animal (i.e., livestock or big game) through optimal management of grazing and reducing forage waste and nonproductive consumption.
3. Increase conversion of the ingested energy by the grazing animal into products directly usable by humans through improved animal genetics, nutrition, and health.

Savory (1987) has recognized two additional links for completing the grazing resource management and economics chain: (1) the market link, involving the active marketing of the "packaged product" and conversion into "solar" dollars of new wealth; and (2) the reinvest link, involving the return of "solar" dollars through reinvestment to strengthen the chain at its weakest link in a constant growth/stability process.

B. GRAZING MANAGEMENT PRINCIPLES

Managing and manipulating the grazing animal–forage plant–soil complex to obtain specific objectives are accomplished by blending ecological, economic, and animal management requirements. The principles of grazing management remain the same regardless of the kind of grazing land: (1) optimal stocking rate (i.e., number of animals), (2) optimal season(s) of use (i.e., timing of grazing), (3) optimal kind or mix of large herbivore species, and (4) optimal grazing distribution. Nevertheless, their application and the relative emphasis on cultural treatments may vary considerably depending upon the kind of grazing land, the management objectives, and the economic implications. And, finally, after these four principles have been made operative, it is appropriate to select and implement special management techniques referred to as grazing methods and grazing systems (Hart et al., 1993).

Common to the management of all grazing lands must be forage plant considerations such as plant growth requirements, providing for plant vigor and reproduction, defoliation and other animal impacts, seasonality and fluctuations in forage production, and ecological status of the plants. But equally high in priority are animal considerations including animal performance, animal behavior, nutrient requirements and intake levels, forage quality relative to specific animal needs, and forage palatability and animal preference. Opportunities exist to manipulate the forage resource to better fit the specific kind or class of grazing animal. Still another approach is the manipulation of the grazing animal (species of animal, stage of production, and calendar events in the animal production cycle) to better fit the forage resource, or an even better alternative will generally be the integration and simultaneous application of both approaches.

Heady (1974) has listed nine ways in which the grazing animal influences its habitat: (1) intensity of defoliation, (2) frequency of defoliation, (3) selectivity of defoliation, (4) seasonality of defoliation, (5) distribution of animals, (6) distribution of plants, (7) distribution of minerals, (8) physical effects, and (9) cycling of minerals. Each of these factors, in turn, affects the productivity and welfare of the grazing animal. Grazing management has the challenge of recognizing and beneficially enhancing the positive impacts of these factors or reducing any negative impacts on the plant–animal–soil complex. These factors will receive detailed attention in subsequent chapters.

C. GRAZING MANAGEMENT DECISIONS

Grazing management, since it is both a science and an art, should be based on both the knowledge of science and the wisdom of practical experience and application. Because of the intricacies and variability of the animal–forage–land biological system, the management of grazing animals on grazing lands may require as much art as science to make continual adjustments as needed. The concept of grazing management implies decision making (Fig. 1.2). Profitable decision making requires knowledge about forage plant species, animal responses desired relative to the market, and the forage plant–animal–land interactions (Matches and Burns, 1985).

FIGURE 1.2   Grazing management requires decision making based on both the knowledge of science and the wisdom of practical experience. (University of Arizona photo.)

The grazing land manager is frequently called upon to solve complex plant–animal problems associated with the management of grazed ecosystems, and the rancher additionally must meet the challenge of adding financial and personnel resources into the solution. "Decision support systems (DSS) . . . bring an integrated approach to bear on complex problems. . . . Decision support systems can encompass methodologies as diverse as hard-core computer models, expert systems, geographic information systems, discussion groups, and (even) structured thought processes" (Stuth and Smith, 1993). DSS can appear as computer programs but also on paper or videos or can emanate from human beings themselves in speech or writing.

Of the numerous computer models now being used or developed as decision support systems, only two are mentioned here: (1) the Grazingland Alternative Analysis Tool (GAAT) and the Grazing Lands Application (GLA). GAAT was developed at Texas A&M University to estimate the economic efficiency of a wide range of grazing land production systems. Systems that can be analyzed, either individually or in combination, include livestock, wildlife, leased grazing, grain and forage crops, nonforage crops, and even investment in grazing land development (Kreuter *et al.,* 1996). GLA is a computer planning tool for use on grazing lands that is based on an initial model developed by Texas A&M and is now used by USDA's Natural Resources Conservation Service in ranch and farm planning (Stine, 1998).

Nevertheless, there are caveats that must be considered in using computerized decision support systems to make grazing management and related decisions. Computerized grazing models must include not only a biological but also a managerial counterpart (Christian, 1981). Grazing lands cannot be managed by computers alone; qualified, trained individuals make good grazing management decisions (Rittenhouse, 1984). Until the data pool on the focal subject becomes large enough to provide reliable numbers for computer input, a reasoned expert opinion may be more appropriate for many difficult management choices. A precise number with insufficient or inadequate formulation is seldom better than an informed guess and may be even less reliable (Van Dyne *et al.,* 1984a).

Limitations of computer support systems have been pointed out by Stuth and Smith (1993): "Some decisions are clearly too trivial for DSS; others have so many sources of uncertainty that it is probably safer to leave them to the integrative powers of the human brain rather than give results an unwarranted aura of reliability." While there are circumstances where personal decisions are likely to operate more reliably than decisions by computer, some decisions are technologically inappropriate for answering by computer. Nevertheless, current computerized DSS represent a major breakthrough in providing a useful tool to aid in making grazing management decisions.

Effective grazing management requires a comprehensive plan to secure the best feasible use of the forage resources. Such a plan must provide for the daily, seasonal, and annual grazing capacity needs of the livestock and/or big game; it must also seek to match the quantity and quality of grazing animal unit months (AUMs) produced on the ranch or grazing land unit with the AUM needs of the grazing animals associated with it. However, grazing management is much more complex than even this; the grazing management plan and subsequent operational decisions must include a wide variety of economic, managerial, and biological considerations along with their many interactions.

The objectives of the grazing plan should include long-term stability as well as immediate profitability or other economic criteria of production (Morley, 1981). The planning time frame for individual grazing land units must be as long as the projected life of the forage stand, commonly a few weeks to several months for annual pasture, 25–50 years for some perennial pastures, or of unlimited continuation for native and seeded rangelands. For the longer term grazing lands, management including stand maintenance must provide for a continuation of desired forage and animal production on a sustained yield basis.

D. TERMINOLOGY OF FORAGES

The terminology related to forages and other feedstuffs available for animal consumption must be consistently used. The following are suggested:

Feed. Any noninjurious, edible material, including forage, having nutritive value for animals when ingested; synonym, **feedstuff.**

Forage. Edible (i.e., acceptable for animal consumption) parts of plants, other than separated grain, that can be consumed by grazing animals or mechanically harvested for feeding; includes the edible portion of herbage, also browse and mast.

Herbage. The biomass of herbaceous plants, generally above ground but may be specified to include edible roots and tubers; generally includes some plant material not edible or accessible to ungulate herbivores.

Browse. Leaf and twig growth of shrubs, woody vines, and trees acceptable (edible) for animal consumption.

Mast. Fruits and seed of shrubs, woody vines, trees, cacti, and other non-herbaceous plants acceptable (edible) for animal consumption.

Grazed forage. Forage consumed directly by the grazing animal from the standing crop; synonym, **pasturage.**

Harvested forage. Forage mechanically harvested before being fed to animals in the form of hay, haylage, fodder, stover, silage, green chop, etc.

Forage crop. A crop consisting entirely or mostly of cultivated plants or plant parts produced to be grazed (**grazed forage**) or mechanically harvested for use as feed for animals (**harvested forage**).

Roughage. Plant materials and other feedstuffs high in crude fiber (20% or more) and low in total digestible nutrients (TDN) (60% or less), usually bulky and coarse; synonymous with **forage** only in part.

Concentrate feed. Grains or their products or other processed feeds that contain a high proportion of nutrients (over 60% TDN) relative to bulk and are low in crude fiber (under about 20%).

II. KINDS OF GRAZING LANDS

Grazing lands include all lands vegetated by plants, either native or intro-duced, which are grazed or have the potential to be grazed by animals. An array of kinds of grazing lands results from a continuum of soil/site factors, vegetation/forage stands, and management projections and applications.

Use of the terms **wild, tame, artificial, synthetic,** and **natural** when applied to grazing lands is unreliable in projecting site adaptation, longevity, and even usefulness of forage plant species and the combinations or variety of forage stands they comprise. The development of improved cultivars of both native and introduced species seems to make such terms nearly redundant. Except where required for correlation of concepts, such terms have been avoided in this pub-lication. The terms "improved" and "unimproved," when applied to grazing land, can also be misleading unless precisely defined or described and consis-tently used thereafter. The case against using **annual** and **perennial** to describe grazing lands may also be misleading; annual forage species may be capable of reseeding themselves for a few years, and perennials can variously be used for a single year, limited to a specified number of years, or managed for unlimited continuation.

The following classification (outlined in Fig. 1.3) has the objective of describ-ing and correlating the major categories of grazing lands for use in their planning and management. It arranges all grazing lands into broad categories by blending ecological site factors with intended land use and management objectives (Val-lentine, 1988). Since grazing lands differ greatly in the following characteristics, these have been used as the principal basis of classification: (1) projected grazing use longevity, (2) climax orientation of the vegetation and its management, (3) ara-bility of the land, (4) land capability and potential productivity, and (5) relative emphasis on cultural treatments. However, boundaries between the categories as discussed are artificially abrupt rather than being naturally gradual and transition-

I. Long-term grazing lands
    A. Native range
    B. Seeded range (native species)
    C. Seeded range (introduced species)

II. Medium-term grazing lands
    A. Transitory pasture / range
    B. Permanent pasture

III. Short-term grazing lands
    A. Crop-rotation pasture
    B. Temporary pasture
    C. Crop aftermath/residue pasture

FIGURE 1.3   Classification of grazing lands (arrows indicate increasing levels of emphasis).

al, additional local adaptation may be required, and the kind of grazing land may be altered over time.

A. LONG-TERM GRAZING LANDS

Long-term grazing lands (synonym **range**) are mostly nonarable lands on which the present forage stand is projected for unlimited continuation. Ecological principles provide the management basis, but grazing manipulation and cultural treatment inputs may be used to manipulate the forage stand. Cultural treatments may be limited by low site potential and/or cost–benefit considerations but are not excluded. Grazing management levels may vary from extensive to intermediate, but fencing and stockwater developments are generally given high priority. Long-term grazing lands, often environmentally severe, consist primarily of land capability classes[1] IV through VIII but are not limited thereto (Fig. 1.4).

1. Native Range

Native range consists of natural vegetation of predominantly grasses, grasslike plants, forbs, and shrubs; tree overstory may be present or absent. Climax vegetation or a high seral stage is often the objective of management. No substantive artificial reseeding has occurred in the past, and useful, introduced species are minimal. Maintenance treatments will be provided when urgent but are currently planned as minimal.

2. Seeded Range (Native Species)

Seeded range (native species) has been treated to enhance or reestablish the natural vegetation by reseeding local strains or new cultivars of native species. Such grazing lands may include interseedings and/or marginal cropland or reclamation sites restored to long-term grazing. Minimal future cultural treatment is projected.

3. Seeded Range (Introduced Species)

Seeded range (introduced species) has been fully or partially converted by seeding to long-lived, adapted, introduced species. Such grazing lands may include interseedings and/or marginal cropland or reclamation sites restored to long-term grazing. The duration of the newly established stand under grazing is projected to exceed 50 years. Minimum to moderate maintenance treatment is projected, but some treatment may be required periodically to prevent natural succession.

B. MEDIUM-TERM GRAZING LANDS

These include both arable and nonarable lands on which the projected long-term grazing tenure is extended but uncertain. The levels of past and projected future cultural treatment are highly variable, depending upon site potential and own-

---

[1]Land capability classes I through IV are suited to cultivation, with I having few limitations and IV the most. Classes V through VIII are not suited to cultivation. Classes V through VII are generally suited to livestock but VIII only to minimal wildlife use (Klingebiel and Montgomery, 1961).

FIGURE 1.4   Examples of long-term grazing lands: (A) native range in the Uinta Mountains, UT; (B) seeded range (native species) consisting of warm-season grasses in Nebraska.

ership objectives. The establishment of new forage stands is not planned within the next 10 years since grazing may be only an interim use. Land capability is highly variable but commonly within intermediate classes (Fig. 1.5).

FIGURE 1.5   Examples of medium-term grazing lands: (A) transitory range/pasture showing grass recovery and lodgepole pine (*Pinus contorta*) and spruce (*Abies* spp.) seedlings on clearcut area in Uinta Mountains, Utah; (B) permanent pasture principally of Kentucky bluegrass (*Poa pratensis*) being grazed by horses near Lexington, KY.

1. Transitory Pasture/Range

Transitory pasture/range provides grazing capacity during an interim period of uncertain duration. The present plant species components in the forage stand are highly variable; the vegetation is generally undeveloped but substantially modified from the original vegetation either by design or by default. Minimal or no restoration has been provided, and major cultural treatment is not anticipated. This category includes go-back farmlands, timber clearings, burn areas in timber lands, semi-waste sites, sites reclaimed/stabilized with grazing secondary, pine plantations, and predevelopment lands. If originally forested, some interim practices may be used to maintain an herbaceous component to prevent reversion back to a forest-dominated plant community.

2. Permanent Pasture

Permanent pasture (replaces term **tame pasture,** in part) consists of forage stands principally of perennial grasses and legumes and/or self-seeding annuals (Rohweder and Van Keuren, 1985). Present forage stands have commonly resulted from a prior seeding or from the aggressive spreading of forage plants onto formerly cultivated sites or into indigenous stands. Grazing is given first priority, and this is to continue indefinitely; the long-term tenure under grazing is uncertain, but the planning horizon as pasture extends out to 50 years. Medium levels of treatment and grazing management are generally projected, but major manipulation of the forage stand in the future is not excluded. This category includes formerly cultivated lands removed from cultivation and provided with "permanent" cover for conservation reasons, but with varying degrees of potential to be returned to cultivation.

C. SHORT-TERM GRAZING LANDS
(CROPLAND PASTURE)

These are arable lands on which grazing is currently being realized but under limited duration. Introduced forage species are mostly utilized, but native species responsive to high management and cultural inputs may be included. High levels of development, maintenance, and management are common. Land capability classes are primarily I through IV (Fig. 1.6).

1. Crop-Rotation Pasture

Crop-rotation pasture (replaces **tame pasture,** in part) provides grazing for a period of 3–10 years in a predesigned crop-rotation cycle. Perennial forage species are mostly utilized. Intensive cultural treatments are generally required to be economically competitive with alternative cash crops. Such treatments include forage stand establishment, fertilization, pest control (weeds, insects, rodents, diseases), and irrigation, if necessary. Grazing is given top priority, but the forage stand may yield harvested forage or seed as a secondary crop.

FIGURE 1.6   Examples of short-term grazing lands: (A) crop-rotation pasture consisting of irrigated smooth brome (*Bromus inermis*) and orchardgrass (*Dactylis glomerat*) at the University of Nebraska, North Platte; (B) crop aftermath/residue pasture with potential for grazing stubble, volunteer, and weeds in a Kansas wheat field following wheat harvest.

2. Temporary Pasture

Temporary pasture (includes and mostly replaces terms **annual** and **emergency pasture**) consists of a forage plant stand established for grazing during a single year or part of year, or annual tillage and reestablishment is projected for perpetuation. Annual forage plants are most commonly utilized, and intensive cultural treatments are provided. Short-season grazing can be provided by emergency or catch-crop plantings or the **graze-out** of a failed cash crop, or it may comprise double cropping when interseeded into or following harvest of the primary crop for fall grazing, winter cover/grazing, or spring grazing. Although grazing is given top priority, harvested forage may also be taken from the stand.

3. Crop Aftermath/Residue Pasture

Crop aftermath/residue pasture provides grazing as a secondary product and is carried out after (or sometimes before) the primary crop is produced and harvested. The income or production derived is supplemental to the main crop (i.e., hay, row crops, small grains, horticultural crops). Grazable herbage includes regrowth of the primary crop (i.e., **crop aftermath**), excess foliage yield on small grain crops removed preharvest (i.e., **foremath**), and windrowed or bunched forages fed/grazed on the site where produced; it also includes **crop residue** (stubble, chaff, lost grain, weeds, and volunteer herbage remaining after primary crop harvest).

Improved pasture has been defined by Rohweder and Van Keuren (1985) as grazing land renovated or treated with one or more cultural practices such as clearing, weed and brush control, fertilization and liming, and irrigation or drainage. However, such practices cross over many categories of grazing lands, at least to some extent. In providing a useful term for collectively referring to grazing lands receiving substantial cultural inputs for development and/or renovation for improving quantity and quality of forage, improved pasture would exclude native and seeded range, transitory pasture/range, and crop residue pasture but include crop aftermath pasture provided by meadows and most other haylands. While implying enhanced productivity, it also assumes the potential to respond to intensive management including cultural treatments in the future.

III. GRAZING MANAGEMENT APPLICATIONS

A. MANAGEMENT INTENSITY AND FLEXIBILITY

Two contrasting levels of grazing management are sometimes recognized: extensive and intensive. **Extensive grazing management** utilizes relatively large land area per animal and a relatively low input level of labor, resources, or capital; this level of management often relies on cost-cutting measures in an attempt to maintain profits and is commonly but not exclusively applied to range. **Inten-**

sive grazing management utilizes relatively high input levels of labor, resources, or capital in attempting to increase quantity and/or quality of forage and thus animal production; this higher level of management is more commonly applied to improved pasture.

Management-intensive grazing (MiG) refers to an intensive pasture management system developed in Missouri but extended to other areas (Moore, 1999). This system has been described as maximizing the application of cultural treatments to pasture along with pasture subdivision, stockwater development, and use of rotational grazing. This system requires high management skills and targets backgrounding and other enterprises (not cow-calf or other breeding enterprises) where putting weight on animals is the primary goal, in order to improve gains and profits per acre.

Flexibility is one of the keys to effective grazing management. Blaisdell *et al.* (1982), while emphasizing federal grazing lands, proposed guidelines applicable to all grazing lands: "There are many ways to reach the desired objectives [of grazing management], and flexibility should not only be allowed but encouraged. Admittedly, uniformity in opening and closing the grazing season, in allowable utilization, in kind or class of livestock, in methods of salting, or in type of grazing system makes for easier administration of public rangeland, but it does not necessarily mean the best management. Early grazing can be tolerated and may be desirable if livestock are removed in time to allow adequate regrowth; heavy use can be allowed if sufficient rest is subsequently provided; change in season [of grazing] can be a useful management tool; and certainly no one grazing system is the best for all situations."

The successful grazing manager is one who avoids crisis in his management operation by anticipating the changes that might become necessary and making the needed changes in a timely manner as soon as they become needed (Heitschmidt and Stuth, 1991). Effective managers identify circumstances whereby desirable changes can be facilitated and undesirable changes avoided. Seizing opportunities and avoiding hazards depend on a philosophy based on timing and flexibility rather than a rigid plan or policy.

B. PASTORAL SYSTEMS

Williams (1981) has listed and described four systems of pastoralism (i.e., grazing management systems) used throughout the world: nomadism, semi-sedentary, transhumance, and sedentary. **Nomadism** is increasingly uncommon but still found in Africa, in the Middle East, and in Central Asia where rainfall is both sparse and unreliable. It is characterized by no main home base; herds, flocks, people, and belongings move together, following the rains and seasonal availability of forage but within no set annual pattern; and the people and livestock are found only temporarily in rural centers for rest or grazing livestock on crop residues. **Semi-sedentary** differs in utilizing a built village or common home base where the women and children permanently reside, but from which the men and boys with their herds are absent for extended periods.

Transhumance is associated with cyclical, annual movement of livestock between distinctive seasonal ranges; movable tents or mobile homes are utilized by shepherds or herders when accompanying the livestock. This system is found over many parts of the world, including the western United States. The **sedentary** system now includes the bulk of the world's livestock. The village-centered adaptation found in Europe and parts of Africa consists of taking livestock out daily to graze and then returning them to confinement each night. The open range or ranching adaptation is the most common in the western United States, South America, southern Africa, Australia, and New Zealand. Livestock remain continuously on grazing lands or feedgrounds and are not confined or provided housing at night; the livestock are mostly controlled by fencing and water provision. The transhumance and sedentary systems offer the greatest opportunity for effective grazing management.

C. RANGE VS. PASTURE

While **range** is defined as supporting mostly native or indigenous vegetation, **pasture** consists mostly of introduced plant species. Love and Eckert (1985) have concluded that ecological principles are not the sole domain of any given kind of grazing land and that the management of both range and pasture involves overcoming difficulties inherent in the interrelationships of climate, soil, plants, and forage harvesting. Nevertheless, an essential difference in management orientation is that range (even when consisting in part or entirely of long-lived, well-adapted introduced forage plant species) is managed as a natural ecosystem. By contrast, pasture must generally be managed to arrest natural plant succession in order to maintain the desired original seeding and prevent the encroachment of native vegetation or of brush and weedy grasses and forbs. On range, deferment or longer rest or dormant season grazing may be used to promote plant succession to a higher successional stage; on pasture, such practices are seldom applied and may only defeat grazing objectives.

Grazing management designed for pasture (particularly improved pasture) can generally target the maximization of animal production in the short-term; such pastures are mostly limited to growing season utilization during rapid growth stages, are generally characterized by prolonged growth periods, are often composed of forage species relatively tolerant of grazing, and can generally justify occasional total renovation or reestablishment. In contrast, rangelands compared to pasturelands are generally composed of more complex plant species mixtures, have more limited growth periods, have slower and less reliable regrowth potential, receive greater emphasis on the grazing resource, are more commonly used for grazing during vegetation dormancy, and must be maintained in productive condition for the long-term.

McMurphy *et al.* (1990) have listed additional ways in which optimal grazing management on improved pasture would logically differ from that on range. Improved pasture is often devoted to maintaining a monoculture or very simple mix of forage plant species of high quality and quantity to target specific needs in livestock enterprises; the management of range more commonly targets a multispecies stand to accommodate an assortment of livestock, wildlife, and other resource needs and uses with emphasis on improving persistence of desirable species. Improved pasture provides other advantages: (1) it is more adapted to higher utilization rates, (2) it offers greater opportunity for grazing to intercept maturity and thus prolong the growing season, and (3) it is more apt to repeat growth cycles and be more responsive to rotational grazing.

Cultural treatments (seeding, weed and pest control, fertilization, physical land treatments, irrigation, etc.) are not the sole domain of pasture development and management but are tools sometimes appropriate to range. Such tools should be considered in the overall grazing land management planning process and then be accepted and implemented or rejected after careful evaluation on the basis of feasibility, cost-effectiveness, duration, and ecological impact.

Many grazing management practices are additive. Contrary to general expectations, Shoop and McIlvain (1971a) found that four livestock management practices, when used in combination, increased yearlong gain per steer as much as the sum of the practices used alone. In their study, a base group of calves averaging 470 lb at weaning was grazed on sandhills range in Oklahoma from November 10 to the following October 1 under heavy stocking rates (6 acres per steer). Individual treatments increased animal gains per head during the $11\frac{1}{2}$-month grazing period over the base group as follows: (1) 14 lb from moderate grazing (9 acres/ steer), (2) 18 lb from feeding an additional 1.5 lb of cottonseed cake from November 10 to April 10, (3) 11 lb from feeding cottonseed cake at 1.5 lb daily from July 20 to October 1, and (4) 46 lb from stilbestrol implants made on November 1 and again on May 1. The total of the separate advantages was 89 lb compared to 92 lb when all factors were combined in the same cattle. It was concluded that the steers receiving the combination practices had still not reached their genetic potential.

IV. GRAZING AS AN ECOSYSTEM COMPONENT

A. PLANT EVOLUTION UNDER HERBIVORY

Grazing in most natural ecosystems is as much a part of the system as is the need for forage by grazing animals (Fig. 1.7). Most native rangeland evolved under animals grazing plants and plants tolerating grazing; i.e., the evolution of the herbivores and edible plants was simultaneous. The selective grazing pressures over thousands of years favored plants that developed resistance to browsing, grazing, and trampling. Grasses that evolved with large numbers of herbivores have been characterized as being small in stature with a large proportion of basal meristems; having minimum supportive tissue, high shoot density, and rapid leaf

FIGURE 1.7   Grazing in most natural ecosystems is as much a part of the system as is the need for forage by grazing animals; bison (buffalo) grazing in the Black Hills, SD.

turnover; and able to reproduce vegetatively (Miller *et al.,* 1994; Archer and Tiezen, 1986). In contrast, plants that evolved under reduced grazing pressure developed fewer morphological, physiological, and biochemical mechanisms to make them grazing tolerant and competitive under grazing in their environments.

It is widely held that past (pre-European) grazing had an important positive effect on the ability of tallgrass, midgrass, and shortgrass ecosystems of the Great Plains to withstand the introduction of and grazing by domestic livestock. Here, the previous grazing pressures by bison as the dominant herbivore, and to a lesser extent by pronghorn, jackrabbits, prairie dogs, and even elk, were merely replaced by the grazing of cattle, horses, and sheep. The result has been that the change from wild ungulates to properly managed domestic livestock has had minimal if any detrimental ecological impacts. The original prairie has been pictured as a vast region of wandering herds of grazers, having a long summer growing season and grasses maturing favorably for winter grazing, and with forage quantity and predation the main incentives to herd movement (Burkhardt, 1996).

Lauenroth *et al.* (1994) concluded that plant communities that have co-evolved with large herbivores for thousands of years before domestic grazers were introduced, such as the shortgrass steppe, are more likely to have a negative response to the removal of grazing than to a conversion to domestic livestock. Lack of subsequent grazing in a community that has evolved under grazing might well be considered a disturbance factor (Milchunas and Lauenroth, 1988). The corollary offered by Lauenroth *et al.* (1994) was that plant communities with a short evolutionary history of grazing were more likely to change under the introduction of domestic livestock grazing.

It seems probable that grazing played a greater role in the development of some ecosystems than in others. According to Platou and Tueller (1985), regular grazing pressures presumably played much less of a role in the evolution of the sagebrush steppe ecosystems of the Great Basin than in the evolution of the Great Plains grasslands and prairies. Pieper (1994) concluded that the introduction of domestic livestock probably had a much larger impact on vegetation in the Intermountain Region, Great Basin, and the Southwest than in the Great Plains. Miller *et al.* (1994) projected that herbivory by native herbivores was probably light historically in the Intermountain sagebrush steppe, with heavy grazing by native ungulates limited to localized areas, but was still an important process in the development of this ecosystem.

Burkhardt (1996) challenged the underlying assumption that large herbivore grazing was an unnatural impact on the plant communities of the Intermountain West. He noted that radiocarbon dating indicated that many large herbivore species and their associated predators became extinct between 12,000 and 10,000 years ago but that bison survived these extinctions and continued to populate the shrub–steppe landscapes of the entire Intermountain Region until the late 1700s or early 1800s.

Burkhardt (1996) rejected the hypothesis that biotic conditions and relationships at the time of European contact in the 1800s represented the pristine, stable-state ecology of the region; his belief was that human predation had played a large role in the Pleistocene extinction of the original large herbivores and that intensive Indian hunting pressure was largely responsible for the low numbers of bison and other herbivores noted in the shrub–steppe zone at the time European contact began in the early 1800s. Miller *et al.* (1994) concluded that both a combination of environmental conditions and hunting pressure by Indians appeared to have kept populations of large herbivores in the Intermountain sagebrush region at low levels.

Native herbivores in the Intermountain Region, in contrast to those in the Great Plains, according to Burkhardt (1996), had been required to develop seasonal grazing strategies during the co-evolutionary period and extend the 6-week green feed period at low elevations by seasonally migrating up into the mountains, by seeking riparian areas as the summers progressed, and by browsing on the numerous woody plants back down at lower elevations in fall and winter. In contrast, cattle and horse grazing in the area has been largely season-long and often under heavy stocking rates, which has provided the native vegetation little opportunity for seasonal recovery. Perhaps even more significant, the removal of the standing crop of fine fuel to carry fires greatly favored the increase of woody plants by fireproofing the range. This led to the conclusion that it was the intense stocking levels and shift of foraging patterns from seasonal to season-long use along with the reduction of fire that expedited development of the sagebrush monocultures and created safe havens for juniper seedlings and the invasion of aggressive annual grasses.

B. SUSTAINABILITY OF LIVESTOCK GRAZING

Livestock grazing per se has sometimes been equated with excessive utilization and deterioration of the environment. Early-day range research was largely involved with studying the impacts of livestock on rangeland vegetation when relatively little control of grazing was applied (Holechek, 1981). Even today, all grazing is sometimes improperly equated with improper, destructive grazing. "The current penchant for describing the bad effects of overgrazing far over-shadows descriptions of successful . . . grazing programs and the good results from proper grazing" (Heady, 1984). Even scientists, sometimes knowingly but probably more often unknowingly, report comparisons of the impact of "no livestock grazing" with "livestock grazing," when often all that was compared to no grazing was severe livestock use much beyond the pale of proper use but with no qualification made as to this aspect. Holechek (1991) noted that failure to distinguish properly controlled from poorly controlled livestock grazing is the major reason for the conflict between ranchers and environmental groups.

"Some avidly promote that grazing by livestock has caused and is continuing to cause, among other things, diminished biodiversity; deteriorated range condition; increased soil erosion; desertification; depleted watersheds and riparian areas; . . . impoverished wildlife habitat; declining wildlife population; and decreased recreational opportunities and experiences" (Laycock et al., 1996). For example, "Virtually all undesirable changes in the plant communities of the Intermountain Region are considered [by some] the result of livestock grazing in an environment not adapted to large herbivores" (Burkhardt, 1996). Vavra et al. (1994) concluded, in referring more particularly to public lands, that scientific evidence and other information indicate that, although public rangelands are being degraded in localized areas, current livestock-grazing practices are not degrading rangelands on a large scale. "In fact, with a few exceptions, U.S. rangelands are in their best condition [of] this century" (Laycock et al., 1996).

Substantial historical and present-day evidence shows unmanaged livestock grazing has been and can still be very destructive of soil, plant, water, and wildlife resources (Holechek, 1991). Pieper (1994) concluded that livestock grazing has played a role in reducing the amount of fuel for wildfires, altering nutrient distribution, and disrupting crytogamic crusts while acting to create patchiness at landscape levels in the environment; he further concluded that range deterioration associated with livestock grazing was mostly historical but still occurs on some rangelands.

Much of this destructive grazing occurred prior to World War I before the importance of placing grazing lands under proper management became widely known. In fact, prior to about 1890, there was essentially no compelling evidence among scientists, practitioners, or government agents as to what proper management of livestock on grazing lands consisted of. In commenting on the "Seven Popular Myths about Livestock Grazing on Public Lands," Mosley et al. (1990) challenged the "myth" that livestock grazing on public lands is widely causing these lands to deteriorate. While acknowledging that public lands were often damaged in the late 1800s due to improper livestock grazing and nonadapted government land policies, they concluded that soil and vegetation were still recovering from these past abuses in some areas. Today, as summarized by Holechek (1991), "Many long term studies are available that show controlled livestock grazing using sound range management principles will sustain and in many cases improve [range resources]." "Domestic livestock grazing at conservative levels appears to be sustainable, even on sensitive western rangelands" (Pieper, 1994).

Kothmann (1984) has concluded that, "The most important concerns [in grazing rangeland ecosystems] should be the stability of the range and its productivity with respect to the desired products which may include forage for livestock and/ or wildlife, water, recreation, or other [uses]." The current policy statement on "Livestock Grazing on Rangeland" of the Society for Range Management (1998) states: "The Society supports appropriately planned and monitored livestock grazing based on scientific principles that meet management goals and societal needs."

The more comprehensive position statement of the Society for Range Management (1998) follows in its entirety:

> Properly managed livestock grazing is a sustainable form of agriculture and is compatible with a wide array of other sustainable uses of rangeland. The Society recognizes the cultural and economic importance of livestock grazing especially to rural communities. Livestock grazing is an efficient method for converting low quality forages to high quality agricultural products that supply human needs worldwide. Managed grazing may be used for expediting desired changes in the structure and function of rangeland ecosystems. Livestock grazing can be complementary and [even] synergistic with other rangeland restoration technologies. Livestock grazing may not be appropriate on certain fragile and highly erodible lands; [but] the removal of livestock grazing on other lands may be of no benefit.

The concept of **sustained yield,** a pivotal principle in the management of renewable natural resources, opts for the continuation of desired animal or forage production or yield of other related rangeland resources. The application of sustained yield to rangelands in a long-time planning horizon should ensure that options in rangeland use are maintained for future generations. Managed livestock grazing, now the norm on most grazing lands, appears capable of sustaining or improving rangeland resources.

C. THE ECOLOGY OF GRAZING DISCONTINUATION

The discontinuation of grazing on rangelands, particularly on public lands, has been sought by some political action groups. This has often been based on the belief that the reduction or removal of livestock will solve any existing rangeland problems and rapidly return these lands to near pristine conditions. Rather than actually doing this, Pieper (1994) concluded that,

> In removing all livestock grazing, the changes in most cases would be subtle and in the long run might even be negative in terms of biodiversity and other desirable characteristics. . . . The idea that recovery of pristine conditions can be restored simply by removing livestock

is much too simplistic in light of other changes that have occurred such as introduction of alien species, changes in fire regimes, etc. Livestock constitute only one component of rangeland ecosystems, and many extrinsic factors, especially weather variations, are instrumental in altering ecosystem components.

The effects of grazing animals, within a particular ecosystem, are a function of the intensity, frequency, seasonal timing, and duration of grazing. Among ecosystems, a particular regime can result in very different responses depending on climatic conditions and structural, functional, and historical attributes (Lauenroth *et al.*, 1994). The pathways of change following removal or relaxation of grazing (on deteriorated range) may differ substantially from former pathways of retrogression, and the probability of ecosystem recovery to previous states may be greatly reduced or nonexistent. While vegetational changes are continuous, they are seldom reversible and consistent (Heady, 1994). Unless other ecological factors are modified, the new steady states—these often are very different from the original community—may continue indefinitely whether grazing is discontinued or continued under proper grazing.

Long-term experiments widely testify to the failure of plant communities to revert to their former pristine status merely by removing grazing. Exclusion of livestock over periods of many years from desert and semidesert rangelands has commonly had little beneficial effects on the vegetation (Atwood and Beck, 1987; Holechek and Stephenson, 1983; Hughes, 1983). Conservative livestock grazing or no livestock grazing has not prevented the change from sagebrush-bunchgrass to sagebrush-annual grass to annual grassland in the Intermountain Region (Burkhardt, 1996). Since grazing is an important process in many ecosystems, West (1993) concluded that the removal of grazing might destabilize some ecosystems.

"It appears that if plant communities have not crossed a threshold into another steady state that a return to good ecological condition can occur both under protection or light to moderate grazing by domestic herbivores. If a community has entered a new steady state, removal of livestock will most often not return this ecosystem to near pristine conditions" (Miller *et al.* 1994). Its return to the "presettlement" state may be impossible (at least in a time frame meaningful to management) or may require more management input (i.e., cultural treatment) than merely manipulation or reduction of grazing (Laycock *et al.*, 1996).

D. CLIMAX AS AN OBJECTIVE

Different philosophies have prevailed as to the optimal orientation the management of grazing, particularly on rangelands, should take with regard to **climax** or pristine conditions (Love and Eckert, 1985; Dyksterhuis, 1986). Van Dyne *et al.* (1984a) concluded:

> Where objectives of management and the desirability of climax coincide, such as in the case of native perennial grasslands, there is no conflict between the range condition classes [these ranked from poor to excellent based on similarity to climax] and range management. Where

woody species, shrubs, and trees make up a greater proportion of the climax vegetation, the
relationship between management objective(s) and the range condition concept often weak-
ens . . . . To manage for a less than excellent range condition appears incomprehensible, but
it is in reality often the desired management goal.

Emphasis has often been given to classical plant ecological concepts and range
conditions, according to Launchbaugh *et al.* (1978), while ignoring or slighting
grazing animal requirements, thus relegating grazing animals to a mechanical role
of harvesting the forage crop and converting it into meat and fiber.

The objective of rangeland management, according to Wilson (1986), is not an
ungrazed "climax" vegetation but rather grazing land that is productive and re-
silient, i.e., "a metastable ecosystem or disclimax that is maintained by management." He concluded that forage plant species are not superior or inferior because
of their place on a successional scale in relation to climax but rather differ in production characteristics and must finally be judged on their ability to support animal production and soil stability. Vogel *et al.* (1985) have warned: "The philosophy that native climax vegetation is optimal may have contributed to the plateauing
of rangeland productivity and limited productivity to that level. This doctrine also
has tended to discourage creative research on rangelands because research has
been focused on management towards climax, and suggestions that other concepts
and research approaches may be required have been viewed as heresy."

"Examples of alternative steady states, abrupt thresholds, and discontinuous
and irreversible transitions are becoming increasingly abundant for both succession and retrogression. When one group of plants has been displaced by another
as a result of altered climate-grazing-fire interactions, the new assemblage may be
long-lived and persistent, despite progressive grazing management practices"
(Heitschmidt and Stuth, 1991, p. 137). Laycock *et al.,* (1996) noted that, for most
rangelands, the greatest biodiversity occurs at the midseral stage, i.e., under high
fair to low good range condition.

As a result of the recognition that the **desired plant community** is often not
the climax community, the concepts and procedures for inventorying both rangelands and pasturelands are currently being modified. For example, the range condition concept makes no allowance for a management objective other than achieving and maintaining the climax plant community. In contrast, a **resource value
rating** has the flexibility of adapting not only to grazing as a land use but also to
other range resource uses. The 1997 edition of USDA-National Resources Conservation Service's *Natural Range and Pasture Handbook* has included many of
these new concepts and procedures for inventorying grazing lands.

V. THE IMPORTANCE OF GRAZING

On a worldwide basis, range is the largest land resource, encompassing about
50% of the land area of the earth (Anonymous, 1985; Busby, 1987). Additional
acreage, possibly 5–10%, comprises cropland pasture and permanent pasture (Fig.

1.8). In addition to the United States, other countries or areas of the world with large expanses of rangeland are Canada, Mexico, South America, the Middle East, Africa, Australia, Russia, and China. Substantial areas of nonrange pasture are found in the British Isles, western Europe, and New Zealand.

Approximately half (50.6%) of the land area of the United States, based on 1986 calculations, is grazing land (Table 1.1). Total rangeland, both open and forested, constitutes 41.9% of the total land area of the United States, while pasture excluding rangeland composes 8.7%. Over 50% of the open range (i.e., nonforested) and nearly 40% of the forested range (also referred to as grazed forest land) are federally owned, while most of the nonrange pasture is privately owned. Grazing lands constitute 64% of the total land area in the West, 48% in the Southeast, and 33% in the North Central area but only 12% in the Northeast. Range constitutes 97% of the total grazing lands in the West, 62% and 57% in the Southeast and North Central areas, respectively, but only 14% of the total grazing land in New England. Of the total forest and rangeland in the United States, 65% is grazed by livestock (USDA, Forest Service, 1981).

Grazing lands, based on 1975 data, contribute an estimated 40% of the feed consumed by livestock in the United States, harvested forages about 20%, and concentrated feed, including grains and protein supplements, the remaining 40% (Allen and Devers, 1975). Livestock products provide the major economic return

FIGURE 1.8   Rangelands are estimated to comprise about 50% of the land area of the world, with cropland pasture and permanent pasture an additional 5 to 10%. (U.S. Forest Service photo.)

TABLE 1.1   Grazing Lands of the United States by Category and Region[a]

Regions[b]	Total land area	Cropland pasture[c]	Perennial pasture	Open range	Forested range	Total grazing land	Percentage grazing lands
Northeast	111.7	2.4	8.8	0.0	1.7	13.0	12
North Central	481.3	23.5	43.4	78.0	11.9	156.8	33
Southeast	538.0	31.0	66.2	116.3	44.0	257.4	48
West	1120.9	7.4	13.1	592.2	99.7	712.3	64
Total	2252.0	64.3	131.5	786.5	157.2	1139.6	51
	(100%)	(2.9%)	(5.8%)	(34.9%)	(7.0%)	(50.6%)	—

[a]Data in million acres. Compiled by John L. Artz and Daniel L. Merkel (USDA, Ext. Serv., 1986).
[b]Northeast: Conn., Del., Me., Md., Mass., N.H., N.J., N.Y., Penn., R.I., Vt., W. Va.; North Central: Ill., Ind., Iowa, Kan., Mich., Minn., Mo., Neb., N. D., Ohio, S.D., Wisc.; Southeast: Ala., Ark., Fla., Ga., Ky., La., Miss., N.C., Okla., S.C., Tenn., Texas, Va., Caribbean; West: Aka., Ariz., Cal., Colo., Ha., Ida., Mon., Nev., N.M., Ore., Utah, Wash., Wyo.
[c]Cropland pasture is cropland reseeded and used for pasture at varying intervals.

from most range and pasture lands. Compared with harvested or purchased feeds, ranges and pastures provide a relatively inexpensive and energy-efficient feed source for livestock production. The National Research Council, Committee on Animal Nutrition (1987), has summarized that both beef cattle and sheep (goats would be a logical addition) are raised primarily as a means of marketing forages, especially those forages that have limited alternative markets. The combined value of cattle, sheep, and goats to the U.S. national economy, based on 1981 data, was approximately $25 billion annually; this includes the contributions made by forages. By comparison, $90 billion was the annual value of all other crops, including forest products (USDA, Forest Service, 1981).

The proportion of their total feed obtained from grazing varies with different kinds of livestock: sheep and goats, 80%; beef cattle, 74%; horses, 51%; dairy cattle other than lactating cows, 43%; and milking cows, 18% (Allen and Devers, 1975). Feed for maintenance of breeding herds of beef cattle, sheep, and goats and the production of their offspring comes primarily from grazing lands. Rangelands, together with forestlands, are the largest and most productive habitats for big game animals.

Another popular myth about livestock grazing on public lands, according to Mosley et al. (1990), is that public lands play an insignificant role in U.S. cattle and sheep production and that livestock grazing on public lands makes an insignificant contribution to the U.S. economy and the western livestock industry. At one time or another during the year, domestic cattle and sheep reportedly graze on about half of the federal lands in the adjoining 11 western states (Public Land Law Review Commission, 1970). In these states, public lands supply about 12% of the total forage consumed, the individual state with the high of 49% being Nevada.

VI. GRAZING AND LAND USE PLANNING

"Grazing lands are the physical, biological, cultural, and sensory environment where many [people] live, work, and enjoy recreational activities. Soil productivity, as well as water and air quality, is better maintained under the permanent vegetative cover of well-managed grazing lands than virtually any other land-use system. . . . Additional economic returns, as well as social and environmental values, occur through sound grazing land management systems" (USDA, Extension Service, 1986). Grazing lands play a prominent role in the conservation ethics comprising **grassland agriculture** (Fig. 1.9), a land management system emphasizing cultivated forage crops, pasture, and rangelands for livestock production and soil stability (Barnes, 1982). Rohweder and Van Keuren (1985) have recognized the widespread practice of permitting abandoned or unproductive cropland to go without improvement as the waste of a natural resource; they estimated that 55% of such lands require treatment through improvement or establishment of vegetative cover for site stabilization and conservation use as grazing lands.

FIGURE 1.9   Grazing lands such as this native range administered by the U.S. Forest Service in Montana play a prominent role in the conservation ethics comprising grassland agriculture. (Forest Service Collection, National Agricultural Library.)

The benefits of grazing, particularly on range, from the standpoint of national, regional, and even firm levels, have been highlighted by the USDA, Inter-Agency Work Group on Range Production (1974). These benefits include (1) low fossil fuel expenditure per pound of livestock weight gain, (2) reduction or elimination of need for commercial fertilizers, and (3) the production of red meat and fiber for export with improved balance of trade. Additional benefits listed were (4) increased rural income, (5) the release of feed grains for human consumption and/or export (i.e., one AUM of grazing being equivalent to 8 bushels of corn), (6) provision for the utilization of otherwise mostly idle or nonproductive land, and (7) producing animal protein necessary for meeting the nutritional needs of people around the world. Some intensive animal grazing programs, particularly some based on improved pasture, now use relatively high amounts of fossil fuels; however, extensive grazing programs carried out principally on native rangelands seem fully sustainable without heavy demands for fossil fuels (Heitschmidt *et al.*, 1996).

Technology exists to greatly increase grazing capacity on most kinds of grazing lands, particularly on intensively managed pasture but also on range, as the need develops and when economic conditions are favorable. Based on a study made by USDA, Forest Service (1972), it was concluded that grazing capacity of U.S. rangelands alone could be increased 49% by the year 2000 through extensive management alone. By maximizing livestock management systems and range improvements, but still maintaining environmental quality, it was concluded that grazing capacity could be increased by 166%.

The future demand for livestock grazing must be incorporated into land use planning. The demand for red meat production from rangelands will apparently persist because of (1) dietary demand for quality protein, (2) inherent dispersal of animal waste and nutrient cycling with range operations, (3) lack of direct or more efficient means for humans to utilize range forage, and (4) unrelenting growth in demand for food (Vavra *et al.*, 1994). Gee *et al.* (1992) summarized the consumption of grazed forages by cattle and sheep from all sources in the United States for the year 1985 and projected the increased demand for the year 2030. For the base year of 1985, 431 million AUMs of grazed forage was consumed by cattle and sheep; deeded nonirrigated lands provided 85.8%; public lands, 6.8%; crop residue, 5.2%; and irrigated lands, 2.2% of this total.

Gee *et al.* (1992) projected demand for livestock grazing in the United States at 637 million AUMs by the year 2030, an increase of 48%. After considering the current trend of decreasing per capita consumption of beef and lamb/mutton, partly counterbalanced by increased population in the United States and other factors, they set the minimal probable demand at 523 million AUMs by 2030. They projected that the AUMs coming from irrigated grazing land and from crop residue would hold constant, that the amount provided by public land grazing would fall by 37%, and that the remainder would be made up from improvements to and increased utilization rates on nonirrigated deeded (private) lands (i.e., both range- and pasturelands). However, in their later evaluation of these projections, Laycock *et al.* (1996) noted that the sharp drop projected in public land grazing had not occurred and that the large increase projected from nonirrigated deeded lands was neither probable nor ecologically feasible.

The following viewpoints of grazing management into the 21st century have been abstracted from Walker (1995):

1. Successful grazing management will be based on the ability to accomplish three objectives: (a) control what animals graze, (b) control where they graze, and (c) monitor the impact on both the environment and the animal.

2. Grazing management will place greater emphasis on the manipulation of plant communities. While continuing the production goal of maximization of long-term economic returns, the challenge will be in making an orderly transition from single-objective livestock production to dual-objective livestock production.

3. Though rangeland will continue to be important for livestock production, it will not be the growth area in grazing capacity into the 21st century; rather, the principal growth will be technology driven and will center on improved pasture and harvested forages produced on irrigated lands and non-irrigated lands in the more humid regions of the United States.

4. Societal demands for increased environmental quality and greater demand for open space values and recreation will impact public lands and probably also private lands by requiring greater attention to this area by the grazing manager.

5. New technology must be developed in manipulating diet selection and foraging efficiency, in manipulating rumen microbes to digest forages more efficiently, in risk management and enhancing economic returns to grazing, and in improving the efficiency of converting forage into livestock products, including the selection or genetic manipulation of grazing animals for high biological efficiency for converting forages.

6. Greater technology must be directed to determining and achieving proper stocking rates, the most important variable in grazing management, including the capability of meeting the challenge of grazing capacity variations over time.

2

GRAZING HERBIVORE
NUTRITION

I. RUMINANTS VS. CECAL HERBIVORES

Ungulate herbivores have two basic types of digestive systems, the ruminant system and the cecal digestive system (Fig. 2.1). Each system enables the animal to digest plant fiber, high in plant cell walls, by microbial fermentation. The fermentation processes by bacteria and protozoa are similar in both systems, but the anatomy of the respective systems is substantially different.

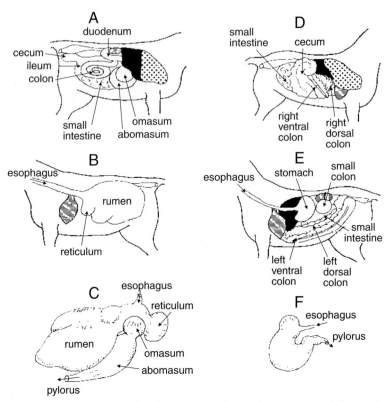

FIGURE 2.1   Comparison of digestive systems of the ruminant and the cecal digester. Ruminant (cow): A, right view; B, left view; C, stomach only. Cecal digester (horse): D, right view; E, left view; F, stomach only (Janis, 1976).

A. THE RUMINANT SYSTEM

A **ruminant** is any even-toed, hoofed mammal that chews the cud and has a four-chambered stomach, i.e., a member of the *Ruminantia* animal division. Included in *Ruminantia* is the large family Bovidae, which includes cattle, sheep, goat, yak, buffalo, bison, eland, gazelle, musk ox, and antelope. Other families within *Ruminantia* are Antilocarpridae (includes pronghorn antelope), Giraffidae (which includes the giraffe and okapi), Cervidae (includes caribou, reindeer, deer, elk, and moose), and the Camelidae (includes camel, llama, vicuna, and alpaca) (Fitzhugh *et al.,* 1978).

The first three chambers of the ruminant stomach are the rumen, reticulum, and omasum (Fig. 2.1); these are sometimes referred to collectively as the paunch. The **rumen** in combination with the smaller **reticulum,** commonly referred to as the **reticulo-rumen,** comprises the anterior large compartment of the ruminant stomach. It functions as a holding tank where fermentation can occur and from which the **ingesta**—the nutritive materials consumed by the animal—is regurgitated for rumination (rechewing). Here the symbiotic breakdown of cellulose and similar compounds occurs through fermentation. Extensive absorption of the resulting volatile fatty acids occurs in the reticulo-rumen and continues as the ingesta flows through the omasum into the fourth chamber, the abomasum. The **abomasum** comprises the true stomach and provides the site for digestive processes similar to that found in the non-ruminant stomach.

The reticulo-rumen provides a favorable environment for microbial populations; muscular contractions increase the contact between microbes and food particles, and the by-products of fermentation, principally volatile fatty acids, are reused so that fermentation continues (Demment and Van Soest, 1983). Selective delay in the passage of ingesta through the reticulo-rumen results, and the probability of passage is tied to particle size. Large particles that are recently ingested have a low probability of escape; the probability of passage increases as retention time increases and particle size is reduced. Retarding the flow of the plant tissues from the reticulo-rumen is a means of extending the period of time available for chemical and physical degradation of such plant tissues. The mean particle size escaping the rumen and appearing in the feces is remarkably constant across *ad libitum* fed or grazed forage diets (Ellis *et al.*, 1987).

Particle size reduction is a critical process determining digesta volume, rates of passage, and digestion of the food particles (Ellis *et al.*, 1987). These, in turn, largely determine the rate of forage intake by ruminants. Forage maturity and selective grazing affect voluntary intake, dry matter digestibility, and the rate at which large particles are reduced to a size capable of escaping the reticulo-rumen. The increase in digestibility of the ingested forage resulting from delayed passage may not be a net benefit to the ruminant; due to the finite capacity of the reticulo-rumen to harbor undigested forage residues and remove such residues by means of fermentation and passage, such retarded flow may limit the level of forage intake.

The digestibility of a forage is a function of the digestion rate acting on a particle for the duration of its retention time within the digestive system. Digestibility, then, is not only a function of the quality of the feed but is also related to retention time (i.e., the longer the retention, the more complete the fermentation and then digestion). **Apparent digestibility** is the balance of nutrients in the ingesta minus that in the feces; **true digestibility** requires additionally that the metabolic products added back into the intestinal tract during body processes be accounted for and be subtracted out of apparent nutrient losses through the feces (Van Soest, 1982), thereby elevating true digestibility over apparent.

The ability to masticate and remasticate their feed in order to reduce particle size in combination with high salivary secretion sets ruminants apart from other herbivores (Ellis *et al.*, 1987). Ingested forages are fragmented into various sizes of particles as the result of **ingestive mastication** (initial chewing) and **ruminative mastication** (rechewing the cud after regurgitation); further disintegration by digesta movements and microbial and chemical digestion aid slightly in further particle size reduction (Pond *et al.,* 1987). Ingestive mastication reduces the ingesta to sizes that can be incorporated into a bolus and swallowed. Ruminative mastication results in further particle size reduction and exposure to microbial attack.

Rumination appears to be the major factor in decreasing the size of forage particles in the rumen (Chai *et al.,* 1988). This was demonstrated in a study in which a ryegrass-alfalfa forage mix harvested in the vegetative stage was fed to cattle by McLeod and Minson (1988). The breakdown of large particles (greater than 1.18 mm.) to small particles was 25% by primary mastication (eating), 50% by rumination, and 17% by digestion and rubbing inside the rumen, the remaining 8% being fecal loss of large particles. The level of voluntary intake of the forage had no apparent effect on these ratios. The importance of rumination was demonstrated by Chai *et al.* (1988) by muzzling sheep during the nonfeeding portion of the day to inhibit and largely prevent rumination. Muzzling when animals were not feeding markedly limited subsequent intake of hay and increased rumen retention time from 7.4 to 17.4 hr, but it did result in the sheep chewing more often and in reduced particle size in the boluses while eating without the muzzles.

Particle size reduction by mastication, such that more fragment ends and a larger surface are exposed to microbial attack and passage along the intestinal tract is facilitated, plays a very important role in forage digestion. However, of additional importance are the crushing and crimping of plant tissues that take place without a change in particle size, thereby releasing the soluble cell contents for microbial access (Pond *et al.,* 1987). The disruption of "barrier" tissues—the cuticle and vascular tissue within the blade and stem fragments—allows entry of the microflora. The main effect of mastication is probably the exposure of more potentially digestible tissues previously encompassed within indigestible "barrier" tissues.

B. ADVANTAGES OF THE RUMINANT

In addition to the major advantage of greater efficiency of cell-wall digestion in the ruminant over simple-stomach animals, other advantages also result from the symbiotic relationships of microorganisms in the rumen. Microbial synthesis in the functioning rumen can supply the full complement of most, if not all, required amino acids and B vitamins (Demment and Van Soest, 1983). The rumen also apparently provides greater potential for detoxifying toxic agents in the diet than is found in animal species having simple stomachs.

Microbes passing from the rumen into the abomasum and small intestines are readily digested and absorbed, providing the ruminant with an expanded, if not the major, source of protein. Bacterial crude protein commonly provides 50% to essentially all of the metabolizable protein requirement of beef cattle (NRC, 1996). Rumen-degradable protein meets the microbial amino acid needs and thus ruminant protein requirements for maintenance and modest levels of production, in most cases (Kansas State University, 1995).

Attention is now being given to ensuring that some rumen-escape protein is made available for absorption from the small intestines to enhance the production of animals with high genetic potential. Achieving maximum ruminant productivity, such as during rapid growth, will generally require providing additional escape protein, with a favorable amino acid profile, to augment microbial protein synthesis (Heitschmidt and Stuth, 1991). In contrast, the metabolizable protein required for wintering gestating beef cows in Nebraska studies was met by the microbial protein flow to the small intestine (Hollingsworth-Jenkins *et al.,* 1996); thus, the cost of supplementing range cows during winter could be reduced by providing a highly rumen-degradable protein source and supplementing only to meet the rumen-degradable protein needs of the gestating cows.

More complete nitrogen conservation and recycling through the saliva take place in the ruminant, thus reducing dietary nitrogen intake needed and ameliorating short-term dietary protein deficiency periods. The ruminant, additionally, has the ability to more effectively use non-protein nitrogen sources for microbial protein synthesis (Owens, 1988), but this capability may be limited with low-quality, high-forage diets low in readily available energy.

The ability of the ruminant to convert organic substances not usable by humans and other monogastric animals into human food of high quality and desirability is a truly great natural phenomenon and benefit to mankind (Fig. 2.2). There are far greater tonnages of biological material in the world that the ruminant must convert for human use than of the materials that humans can consume directly. In addition to pasturage, large amounts of high-fiber by-products of agriculture, forestry, and industry can be converted by the ruminant into meat, milk, wool and mohair, and hides. These fibrous materials—inedible to humans and often of little value per se and even creating expensive disposal problems otherwise—would mostly be wasted were it not for utilization by cattle, sheep, and big game animals.

Animal agriculture has come under attack from some people with the simplistic assumption (this a misconception) that each pound of meat is produced at a cost of 4 to 10 lb of grain that could have gone directly into human consumption. However, this ignores the premier role played by the ruminant in converting forages and other fibrous waste materials into edible and highly nutritious human foods and other useful products. Fitzhugh *et al.* (1978) estimated that ruminants consume diets of about 90% roughage on a world average and about 70% in the U.S. More recent estimates (Reber 1987), however, raise these figures to 95% and 80%, respectively. It seems reasonable to anticipate that the ruminant in the future will be used primarily to convert low-quality biomass into useful production, but ruminant rations may continue to include some low-quality grain, grains bred specifically as "feed grains," and even food grains in surplus of market demands. Also, small amounts of grain or short concentrate feeding periods immediately prior to slaughter greatly improve carcass quality.

The ruminant animal has a relatively low conversion efficiency—even steers on finishing rations typically require 7.5 lb dry matter per pound of gain vs. 2.0 lb dry matter per pound liveweight gain with broilers. Rumen fermentation converts

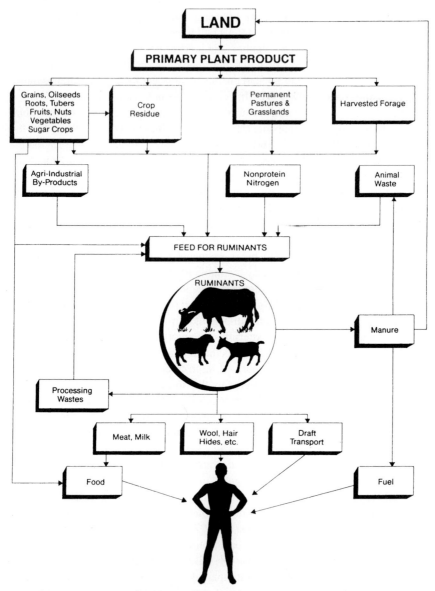

FIGURE 2.2   A graphic representation of land-ruminant-man relationships (Fitzhugh *et al.,* 1978).

much of the cell wall materials, not otherwise usable, and most of the soluble cellular contents into fermentation products before absorption occurs. Therefore, ruminants must rely almost entirely on the production of microbial volatile fatty acids for energy. This, in part, accounts for the reduced energy efficiency in ruminants. Ruminants on high roughage diets will have even lower conversion efficiency. Nevertheless, pasturage and other forage rations generally provide satisfactory, lowest-cost rations for maintenance and production by ruminants. This requires that abundant forages and roughages be available at appropriate periods to sustain profitable ruminant production (Blaser *et al.*, 1974).

Wild ruminants, like domestic ruminants, have microflora capable of digesting hay, as well as browse, and converting it to volatile fatty acids (Nagy *et al.*, 1967; Urness, 1980). Rumen bacteria capable of digesting both roughage and concentrate diets survive long periods of deer starvation. Thus, it seems certain that the buildup of rumen microfauna capable of effectively using harvested feeds is rapid, and the transition period from low intake of restricted, poor quality browse to such feeds is less critical than once thought. However, by the time that starving deer or elk find access to a hay supply, rumen function may already be largely halted, the microflora completely dead, or energy reserves too low to respond, and the animals die of starvation before recovery can begin.

C. THE CECAL DIGESTIVE SYSTEM

The principal cecal-digesting, non-ruminant, ungulate herbivore is the horse. (An **herbivore** is any animal species, including many insects and rodents, that subsist principally or entirely on plants or plant materials; an **ungulate** is any hoofed animal, including ruminants but also horses, tapirs, elephants, and rhinoceroses). The single-compartment stomach of the horse is relatively small and functions mainly as storage and regulation of ingesta reaching the small intestine (Burke, 1987). Food moves rapidly through the stomach of the horse, and the digestive activity therein is limited. The small intestines are the primary site of digestion of soluble carbohydrates, fats, and proteins in the horse. The large intestine—comprised of the cecum, large colon, small colon, and rectum—is the most important segment of the equine digestive tract relating to the utilization of forages.

The **cecum** is a blind sac appended at the posterior end of the small intestines and forms the forepart of the large intestine; it comprises only about 10% of the digestive tract in the horse but typifies the digestive system of many non-ruminant, forage-consuming mammals. It has some functional similarities to the rumen; the operational difference between them is that the rumen functions like a filter that selectively delays food particles, whereas the cecum provides less selective retention and functions more like a perfect mixer (Demment and Van Soest, 1983).

The equine large intestine contains large populations of bacteria, and here fermentation of the fibrous portions of feeds takes place, the end products being volatile fatty acids as from the rumen of cattle and sheep (Burke, 1987). From the small intestines, the ingesta passes into the cecum where fermentation of the fibrous portions of the ingesta begins. Protein and vitamin B synthesis takes place as in the rumen, but their utilization by the horse is less efficient since the synthesis takes place beyond the stomach and small intestines. Ample evidence of nitrogen recycling exists for non-ruminants, but, except for the horse where the evidence is probable, there is no evidence of the ability to absorb amino acids from the colon (Demment and Van Soest, 1983). Except in ruminants and cecal digesters—at least in the horse and probably other cecal digestors—most essential amino acids have to be present in the diet. Although some amino acids are apparently synthesized in the liver of all animals, amino acid synthesis appears quite limited in range and utility except in ruminants and cecal digesters.

The horse can apparently subsist on lower quality diets than even large ruminant species by increasing rate and amount of intake of fibrous feedstuffs (Holechek, 1984; Hanley, 1982b). The reduced fiber digestion resulting from faster and less restrictive or selective passage from the cecum compared to the rumen is compensated for by ingesting greater amounts of forage (Janis, 1976). This permits the horse to be less selective and spend less time grazing (Hanley, 1982b). Walker (1994) has generalized that, "Where forage quantity is limiting, a ruminant digestive system is advantageous; whereas where forage quality is limiting, a cecal digestive system is advantageous."

II. NUTRIENT REQUIREMENTS

The nutrient balance of animals, whether grazing or penfed, is dependent upon four basic factors: (1) the animal's nutrient requirements, (2) nutrient content of the feedstuff(s) consumed, (3) digestibility of the feedstuff(s) consumed, and (4) how much the animal consumes. The nutrient requirement of animals, including grazing animals, is dependent upon a number of factors including metabolic rate, metabolic body size, body condition, physiological and reproductive state, and production levels, but also on ambient temperature, wind, and hide conditions (hair cover and dryness). Grazing and the related voluntary travel also require substantial increases in energy expenditure. (Factors that affect forage dry matter intake are discussed in detail in Chapter 11.)

The nutrient requirements of domestic livestock are provided in detail by the respective current National Research Council (NRC) publications: beef cattle (NRC, 1996), goats (NRC, 1981b), horses (NRC, 1989), and sheep (NRC, 1985). Some information on the ingestion, digestion, and assimilation of forages as related to the energy requirements of wild herbivores and their management implications has been given in *Bioenergetics of Wild Herbivores* (Hudson and White, 1985). Using the mature beef cow as an example, Table 2.1 contrasts the nutrient requirements of early gestation (7th month since calving), late gestation (12th month), and peak lactation (2nd month) for energy, protein, and phosphorus. Note that late gestation increased the requirements over early gestation (approximated maintenance) for net energy (57%), metabolizable protein (54%), and phosphorus (33%); peak lactation increased the requirements over maintenance for net energy (80%), metabolizable protein (93%), and phosphorus (85%).

TABLE 2.1 Nutrient Requirements of Beef Cows

	Months since calving[c]		
	7th	12th	2nd
Daily requirements[a]			
Net energy (Mcal)			
Maintenance	8.54	8.54	10.25
Lactation	0.00	0.00	5.74
Pregnancy	0.32	5.37	0.00
Total	8.87	13.91 (57%)[d]	15.99 (80%)[e]
Metabolizable protein, total (g)	436	672 (54%)[d]	840 (93%)[e]
Phosphorus, total (g)	13	18 (33%)[d]	24 (85%)[e]
Diet density requirements[b]			
Dry matter intake, daily (lb)	21.1	21.4	25.0
Total digestible nutrients (% of dry matter)	44.9	55.7	60.9
Net energy, total (mcal/lb)	0.37	0.54	0.62
Crude protein (% of dry matter)	5.98	8.67	11.18
Phosphorus (% of dry matter)	0.11	0.15	0.21

Note: Data adapted from National Research Council (1998).

[a]Assumes 1175-lb mature weight; 17.6-lb peak milk daily; Angus breed code.

[b]Assumes 1000-lb mature weight; 20-lb peak milk daily.

[c]Seventh month = first month following end of lactation; fourth month of gestation.

Twelfth month = last month of gestation; non-lactation.

Second month = peak lactation month; non-gestation.

[d]Percent increase in 12th month over 7th month; approximates maximum gestation requirements over maintenance.

[e]Percent increase in 2nd month over 7th month; represents maximum lactation requirements over maintenance.

A. GRAZING VS. PENFED ANIMALS

The NRC figures for cattle in *Nutrient Requirements of Beef Cattle* (NRC, 1984) assumed the cattle (1) were fed in a no-stress environment (Fox *et al.,* 1988) and (2) were developed from penfed animals where maintenance requirements are readily calculated and tend to vary only slightly within a given weight, sex, age, and physiological state (Wallace, 1984). Maintenance energy requirements include not only essential metabolic processes but also body temperature regulation and physical activity (NRC, 1996). Even though grazing activity is not a factor in dry-lot feeding, maintenance energy requirements are greatly increased under extreme environmental conditions (Fox *et al.,* 1988); these environmental effects have been summarized by the National Research Council (1981a).

Extending nutrient requirements from the penfed to the grazing animal has been met with considerable difficulty. Nutrient requirements of grazing animals, particularly range animals, are not well defined. Nutrient requirements, particularly energy, can be substantially altered by grazing activity, travel, and environmental stresses such as temperature extremes (Fig. 2.3) (Allison, 1985). For example, the maintenance requirements of free-ranging animals for energy were estimated by Van Soest (1982) to be from 140 to 170% of the requirements of stall-fed animals.

CSIRO (1990), from a review of available literature, estimated the additional maintenance energy expenditure for grazing under Australian conditions at 10 to 20% under optimal grazing conditions and about 50% on hilly terrain when animals have to walk considerable distances to reach preferred grazing areas and water. Lachica *et al.* (1997) found that goats substantially increased their energy requirements under grazing, principally locomotion, above maintenance as follows: 14.2% in summer traveling 3.6 miles daily, and 8.7% in autumn traveling 2.2 miles daily. Doughterty (1991) concluded that the additional energy cost of grazing and grazing-related activities on improved pastures under the very best conditions would run about 25% more than under penfed conditions but much more under less favorable conditions.

Osuji (1974) has estimated the increased maintenance energy requirements (i.e., energy expended to keep animals in energy equilibrium while maintaining normal body functions) of animals on range at 25 to 50% over conventionally housed animals. This results from range animals walking longer distances, climbing gradients, sometimes ingesting herbage of low dry matter content, and thus spending more time eating and foraging for food. In cool temperate and cold cli-

FIGURE 2.3   The energy requirements for grazing animals are substantially increased by grazing activity, travel, climbing, and environmental stresses; showing cattle on winter range at the Cottonwood Range Field Station, SD (South Dakota Agricultural Experiment Station photo).

matic zones, grazing animals are also more subject to periods of energy-draining climatic stress. Havstad *et al.* (1986b) estimated that cows grazing rangeland use 30% more energy than confined cows because of longer eating (grazing) time and longer travel distance. He attributed an additional energy requirement of 3% for each additional hour spent grazing.

Havstad and Malechek (1982), based on studies with beef heifers on crested wheatgrass range in Utah, suggested that the free-roaming condition contributes as much as 40% to the energy requirements of range cattle. They estimated that the daily mean energy expenditure of free-ranging heifers grazing crested wheat-grass was 161 kcal/kg BW$^{.75}$/day, i.e., 46% greater than the mean 110 kcal/kg BW$^{.75}$/day estimated for stall-fed heifers consuming similar forage. The greater maintenance energy expenditure was attributed to more time grazing/eating (9 vs. 2 hr daily) and the walking and searching activities associated with grazing the sparse forage plants. When grazing is being concentrated on selectively searching for green forage in short supply, distance traveled daily by grazing animals can be expected to increase even more (NRC, 1996).

The maintenance energy requirement of the ruminant increases linearly during cold weather but nonlinearly during heat stress (Ames and Ray, 1983). Research indicates reduced rates of performance by animals exposed to adverse thermal environments. In each case, maintenance requirement increases with less energy available for production. When exposed to heat stress, animals reduce intake while their maintenance requirement is also increased, the increased body temperature resulting in an increased tissue metabolic rate and extra "work" of dissipating heat, again leading to reduced performance (Ames and Ray, 1983; NRC, 1996).

When exposed to cold stress, the heat production from normal tissue metabolism and fermentation in ruminants is inadequate to maintain body temperature; and animal metabolism must increase—thereby utilizing additional maintenance energy—to meet this additional need (NRC, 1996). Under certain conditions, animals may increase intake during cold weather (this often inconsistent on winter range), but maintenance energy requirements usually increase more rapidly than the rate of voluntary energy intake. Based on a temperature range of 50 to 68°F (the zone of thermoneutrality) minimizing maintenance expenditures of energy, Ames (1985) calculated increased maintenance energy needed for both increasing and decreasing effective ambient temperatures as follows: 104°F, 32%; 86°F, 11%; 50 to 68°F, the zero base; 32°F, 10%; 14°F, 26%; and −4°F, 51%.

Ames (1985) concluded that range beef cows expend approximately 90% of all metabolizable energy intake for maintenance. He estimated the average effect of selected variables on increased maintenance energy requirements for range beef cows as follows: climate, 40%; activity, 40%; level of feeding, 40%; acclimatization, 30%; biological, 30%; and body condition, 15%. The potential combined effect of these variables on maintenance energy requirements and ultimately on production costs, even if not additive, is staggering. Based on data for cattle in the Northern Great Plains (Adams and Short, 1988), Fig. 2.4 shows the effects of pregnancy and cold stress on the cow's maintenance requirements in relation to ex-

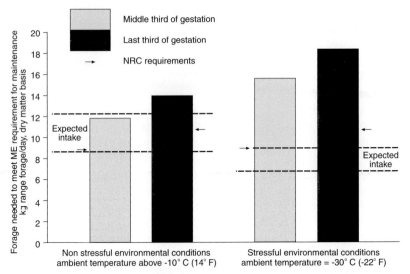

FIGURE 2.4   Effect of pregnancy and cold stress on the range forage required to meet the metabolizable energy requirements for maintenance of a 1200-pound cow (Adams and Short, 1988).

pected forage intake. This demonstrates the difficulty of maintaining body weight and condition on a winter range forage diet, particularly under conditions of cold stress.

The effects of environmental factors on the maintenance requirements of beef cattle have more recently become even better understood and have been more fully incorporated into the calculations of nutrient requirements for beef cattle in the 7th revised edition of *Nutrient Requirements for Beef Cattle* (NRC, 1996). In the table entitled "Maintenance Requirement Multipliers for Representative Environmental Conditions" (p. 219), the net energy requirement for maintenance was compared to a no-stress maintenance when cumulating the effects of selected individual environmental stress factors. When compared to a no-stress environment for wintering mature "Hereford type" cows, one scenario of cumulating multipliers for environmental stress factors was as follows: (1) 1.39, with lowering temperature to −10°F; (2) 1.69, with increasing wind speed from 1 to 10 mph; (3) 2.17, if hair coat was wet/matted rather than dry/clean; and (4) 2.39 if the cow was a thin-hide (Zebu) type rather than thick-hide (Hereford) type.

B. BODY STORAGE OF NUTRIENTS

The nutrients most apt to be deficient in range and pasture forages are energy (based on total digestible nutrients, metabolizable energy, or net energy), protein, phosphorus, and vitamin A. In some areas, micro-minerals and other macro-minerals should be considered. For example, both weanling calves and dry bred cows grazing native grassland winter range in Nebraska may require potassium supplementation to maintain adequate body weights (Karn and Clanton, 1977).

The ideal nutritional state in large herbivores is when all required nutrients are obtained in the same general time frame. However, since animals can withstand some variance from the normal average intake of nutrients, they need not maximize (optimize) intake of any particular nutrient or a specific mix of nutrients within each meal or even on a daily basis (Howery *et al.*, 1998a). Most nutrients required by grazing animals can be stored temporarily by the animal and remobilized when intake does not meet the animal's nutrient demands (Kothmann and Hinnant, 1987). Energy can be stored in the form of body lipids, but heavy demands against this energy source to meet daily needs will prove disruptive to animal performance (e.g., delaying estrus and conception in breeding females). Dietary requirements for energy should generally be met daily for energy and not less than weekly for protein.

In grazing livestock, phosphorus deficiency is considered the most prevalent mineral deficiency, not only in the U.S. but throughout the world, and the deficiency is often associated with phosphorus-deficient soils. Supplemental phosphorus in a mineral supplement is rather widely provided to grazing livestock, particularly on phosphorus-deficient soils or on cured pasturage. Although the skeleton provides a substantial reserve of phosphorus for meeting temporary deficiencies in diets of mature livestock, body reserves of phosphorus are able to meet only short-term dietary deficits without impairing animal productivity.

The only dietary vitamin requirement of major concern with grazing ruminants is vitamin A, which is obtained principally as carotene from green plants but also supplementally in the form of vitamin A in enriched feeds or by injection. Vitamin A and/or carotene can be stored in the liver in sufficient amounts to last considerable periods of time when carotene is low or even devoid in the diet, such as during winter dormancy or prolonged drought. High body-storage levels of carotene and/or vitamin A are considered adequate to last $3\frac{1}{2}$ to 5 months in mature, nonlactating cattle; 40 to 80 days in young, growing cattle; and intermediate periods for lactating females. Beef cattle exposed to drought, winter feeds (harvested or grazed) of low quality, or stresses such as high temperature or elevated nitrate intake are more prone to have carotene/Vitamin A storage deficiencies (NRC, 1996).

III. NUTRIENT LEVELS IN FORAGES AND INGESTA

When grazing animals are fed to appetite (free choice) and forage intake can be accurately predicted, nutrient requirements can be expressed as a percentage of the diet (NRC, 1987), i.e., as diet nutrient density requirements (NRC, 1996). Nutrient requirements thus stated can then be compared with forage listings in tables of average composition to imply the nutritional status of the diet being consumed. Tables of average composition, such as those available from NRC (1985 and 1996), are useful in initially providing general information on nutrient trends and differences between various kinds and species of forage plants.

General knowledge and even visual appraisal of such factors as stage of maturity, leafiness, green color, plant species and cultivars, and even palatability will be suggestive in predicting the adequacy of protein, phosphorus, carotene, and even energy content in pasturage. Forbes and Coleman (1993) concluded that diet digestibility of grazing animals was most influenced by the proportion of green leaves in the sward and in the diet.

The relative capability of grasses, forb, and browse at different growth stages to meet ruminant nutrient requirements is shown in Fig. 2.5. The nutrient content of grasses, forbs, and browse in relation to gestation requirements for winter grazing has been further generalized as follows by Cook (1972), assuming complete dormancy in grasses and forbs:

	Energy	*Protein*	*Phosphorus*	*Calcium*	*Carotene*
Grasses	Adequate	Low	Low	High	Low
Forbs	Low	Adequate/Low	Low	High	Adequate/Low
Browse	Low	Adequate	Adequate	High	High

Thus, maintaining a diverse plant community, containing not only desirable grasses but also palatable forbs and shrubs, should permit grazing animals to maintain a more nearly balanced and adequate level of intake of the key nutrients during periods of grass and forb dormancy.

A. NUTRIENT TRENDS IN HERBACEOUS PLANTS

A knowledge of generalized nutrient trends in the forage plants available to grazing animals will assist in achieving their most timely utilization, help predict nutrient deficiencies, and suggest supplementation needs. The stage of growth greatly affects the nutritive levels in forage plants. During rapid spring growth, herbaceous forage normally contains enough nutrients to promote growth, weight gains, improvement in body condition, and milk production. However, as these plants begin to mature, the levels of some nutrients drop sharply and dietary deficiencies may result.

Protein, phosphorus, and carotene follow similar patterns through the plant growth cycle, being high in fast-growing, herbaceous plants but low after maturity. Crude protein on a dry matter basis may be as high as 20% in new growth but as low as 2.5% in cured, weathered grasses. The digestibility of protein also decreases as plants mature. Digestibility of protein in new grass growth can be as high as 70 to 80% and 40 to 50% in good grass hay, but often as low as 25 to 30% in cured range grasses by midwinter. On blue grama rangeland in New Mexico, the forage consumed by cattle contained 12.8% crude protein during active growth compared to 6.5% at dormancy; the respective rumen ammonia concentrations were 12.6 and 3.5 mg/100 ml (Krysl *et al.*, 1987).

Phosphorus content commonly reaches a high of .35% in new growth but drops

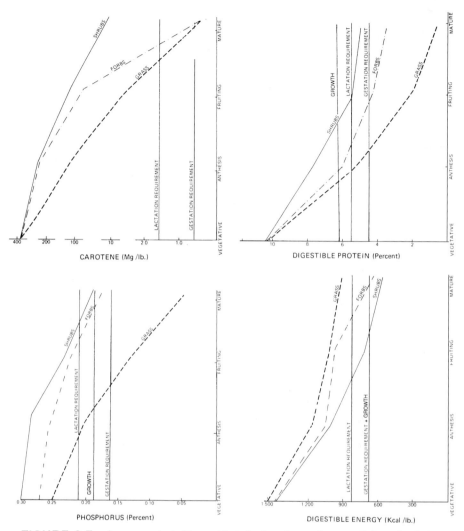

FIGURE 2.5   Average content of key nutrients for three forage plant growth forms in relation to plant phenological development and ruminant nutritional requirements (Cook, 1972).

below .10% in standing grass by midwinter. On soils greatly deficient in phosphorus, the phosphorus levels in the plants growing on them are further reduced. Calcium content in pasture and range grasses drops somewhat from early growth to maturity but remains above minimum animal requirements except on calcium-deficient soils. Under semitropical conditions in southeastern U.S., soils are heavily leached due to the high rainfall, resulting in frequent mineral deficiencies in the forage. The content of carotene (precursor of vitamin A) in herbaceous plants commonly drops from a high of 450 ppm in early spring growth to virtually zero in cured, weathered grass.

The usable energy in range grasses on a dry matter basis is relatively stable while plants are green and growing but drops somewhat nearing maturity and following maturity (Table 2.2). The stems of switchgrass (*Panicum virgatum*) were found by Twidwell *et. al.* (1987) not only to increase in dry-weight percentage with maturation but also to have large increases in acid detergent fiber and lignin and a large decline in *in vitro* dry matter digestibility. The leaf blade had the highest crude protein and *in vitro* dry matter digestibility at all harvest dates. Hendrickson *et al.* (1997) concluded that tissue aging was the major mechanism associated with reduction of leaf *in vitro* dry matter digestibility in sand bluestem (*Andropogon gerardii paucipilus*) and prairie sandreed (*Calamovilfa longifolia*); this reduction resulted principally in decreases in cell wall digestibility rather than from reductions in the cell soluble fractions. Leaf crude protein was affected more by year than by species but typically declined sharply early in the growing season and then leveled out abruptly in early July.

Grazing herbaceous plants in earlier growth stages—in contrast to near or following maturity—is nutritionally advantageous. Individual animal performance will be sacrificed on low-quality roughages produced one season but grazed after maturity in another. However, except in tropical and semitropical areas, green and growing plants are not available on a year-round basis, and compromises in nutritive levels are necessary in meeting long-season or 12-month grazing. Synchronizing the timing of the lower maintenance nutrient level requirements of the grazing animals with the period when the herbaceous vegetation is dormant is nutritionally optimal.

Small changes in the digestibility of ingesta in the diets of grazing ruminants are often under evaluated. An increase in forage energy digestibility from 50 to 55% represents more than a mere 10% increase in nutritive value, since the amount of energy available above maintenance may be increased as much as 200% (Malechek, 1984). Clearly, stand treatments including grazing management activities that delay forage maturity, one of the major factors influencing digestibility, have a potentially large and beneficial impact on animal performance.

Rainfall greatly influences the nutrient composition of individual forage plants but may also alter plant numbers and the relative yield of different plant species. Following dry periods, rainfall may result in rapid increase in protein, phosphorus, and carotene associated with accelerated growth when enabled by adequate temperature. On the other hand, drought is commonly accompanied by a decrease in these nutrients along with growth reduction and early maturity while maintaining moderate energy levels in grasses. If new growth is not seasonally initiated during a dry growing season, grazing animals may be forced to select older, more fibrous plant parts from the previous growing season.

In years of sub-normal rainfall when crested wheatgrass (*Agropyron cristatum* or *A. desertorum*) was forced into drought-induced dormancy before entering the reproductive stage, Malechek (1986) found the forage was prematurely "cured" in

TABLE 2.2 Nutrient and Digestibility Trends in Oregon Range Forage During the Grazing Season[a]

Date	Crude protein		Cellulose		Lignin content (%)	Dry matter digestibility (%)	Phosphorus content (%)
	Content (%)	Digestibility (%)	Content (%)	Digestibility (%)			
May 1	18.8	—	23.7	—	4.2	—	0.22
May 15	17.7	—	25.0	—	4.3	—	0.20
May 29	12.0	65.0	26.6	70.0	4.6	62.0	0.19
June 12	10.2	64.5	27.7	68.5	5.6	62.0	0.18
June 26	8.6	63.5	28.8	59.8	6.6	59.0	0.17
July 10	7.4	58.2	29.6	56.1	6.9	57.0	0.16
July 24	6.3	44.0	30.1	54.0	7.2	52.0	0.15
Aug. 7	5.3	36.5	30.4	53.2	7.4	49.0	0.14
Aug. 21	4.2	28.5	30.5	52.0	7.6	48.0	0.12
Sept. 4	3.3	26.0	30.4	52.0	—	48.0	0.11

[a]Digestibility data from steers and wethers on native forages and crested wheatgrass. Nutrient content and digestibility data averaged over several years. Grasses predominantly cool-season growers (Raleigh and Wallace, 1965).

a more nutritious, more immature stage and maintained higher levels of crude protein well into dormancy (i.e., late summer). Both protein and energy levels may be abnormally low in cured range grass produced during a growing season of high rainfall even though total forage yield may have been high, in part resulting from stemminess and high fiber content. From work with tropical grasses in Queensland, Wilson (1983) found that digestibility in plants subjected to limited water was higher than that in the well-watered controls because of the decreased proportion of stem and a higher digestibility of both leaf and stem.

Prolonged high nutritive levels associated with longer growth periods in temperate climates may be maintained on higher elevation sites where summer precipitation encourages continued growth, particularly when combined with grazing methods to delay maturity. Similar results at lower elevations may be obtained from continuing but not excessive rainfall or with supplemental irrigation.

Decomposition reduces the amount and quality of herbage available for grazing in winter and thereby reduces grazing efficiency. In a study of weathering losses on fescue grasslands, Willms et al. (1998) found that the degradation of standing litter was most rapid in late summer and tended to decline towards stability by December; while minimum crude protein and phosphorus levels were reached by December, leaf biomass of standing litter decreased from August to March. No evidence was found that leaf position in the standing biomass affected exposure and biomass losses, but wet fall-winter periods accelerated decomposition. Winter precipitation, whether in the form of rain or snow, typically further reduces the nutritive value of cured range forage by weathering and leaching, particularly where temperatures are intermittently above and below freezing but too low for new growth.

Herbaceous plants accumulate nonstructural carbohydrates, including sugars, during the day and use them up at night; this causes a diurnal cycling of forage sugars and overall quality. Mayland and Shewmaker (1999) found that late afternoon cutting of hay increased the feed value of the hay by 15%, this from increased energy values, but crude protein was apparently not changed. Tall fescue (*Festuca arundinacea*) hay mowed late in the day not only had higher nutritive value but was also higher in palatability, being consistently preferred by cattle, goats, and sheep over hay mowed early in the morning (Fisher et al., 1999). Cutting hay or green chop or even grazing in the afternoon would conceivably benefit from the higher afternoon energy values. However, the window of opportunity for taking advantage of the diurnal cycling of sugars in the forage is generally considered too short to be practical even with hay harvesting.

Forbs (non-woody, broadleaf plants) are commonly intermediate in nutrient content between grasses and browse by the time the plants mature. Legumes often play a prominent role in both irrigated and dryland pasture mixtures. Marten et al. (1987) noted that the forage quality of perennial weeds in improved pastures varied among species but was sometimes equal to or superior to that of alfalfa (*Medicago sativa*). This suggests that the decision as to whether to implement herbaceous weed control on grazing lands should consider the potential nutritive value of the resident forbs.

Even though grasses will remain the mainstay on rangelands for livestock, particularly cattle, the importance of forbs to grazing animals needs to be recognized (Fig. 2.6). Three primary reasons were offered by Cook (1983) as to why forbs should be recognized as important constituents of many native grasslands: (1) they may be present as either dominants or subdominants in the climax, (2) some forbs are readily eaten even by cattle (and other large herbivores), and (3) forbs contribute substantially to a higher animal nutritional level when found in mixture with grasses. Hoehne *et al.* (1968) found that forbs on Nebraska sandhills range during rapid growth sometimes contained higher nutritive levels than grasses and occasionally constituted up to 50% of the cattle diets for short periods.

However, most native forbs become unpalatable or largely disintegrate and disappear by winter or are at least reduced to unattractive stemmy remnants closer to mulch than forage. From their studies on the fescue grasslands of southern Canada, Willms *et al.* (1998) found that weathering losses in both quantity and quality were much greater for forbs than for grasses in the standing crop. Also, forbs are generally not as reliable as grasses for herbage production, many are unpalatable and pose weed problems, and some are toxic to grazing animal. Management systems to favor forbs or at least take advantage of them on rangelands are difficult to implement (Pieper and Beck, 1980).

Annual grasses and forbs are seldom considered as favorably as their perennial counterparts. Yet, some grow in cold periods when perennials are dormant and

FIGURE 2.6   Forbs often play important roles locally in meeting the nutritive needs of grazing animals; example shown above, scarlet globemallow (*Sphaeralcea coccinea*) growing luxuriantly during a high rainfall spring on southern Nevada range.

provide an important nutritional contribution to livestock as well as big game animals (Holechek, 1984).

B. NUTRIENT LEVELS IN SHRUBS

In contrast with grasses, the levels of phosphorus, carotene, and protein in browse from shrubs may decline only slightly as the growing season progresses (Fig. 2.5). Browse from non-deciduous shrubs shows the least loss of these nutrients with maturation. Even in deciduous shrubs, considerable amounts of carotene, protein, and phosphorus are retained in the buds and cambium (green bark) layers. Thus, diets of grazing animals on winter range that include a mixture of grasses and browse are apt to more nearly provide a balanced ration than are either straight grass or browse diets alone. Nevertheless, the woody materials from shrubs are generally highly lignified and low in nutritive value.

The presence or addition of high-protein browse to grass diets on winter range may well be a viable alternative to feeding costly high-protein concentrates. Based on studies in metabolism stalls in which sheep and beef cattle in New Mexico were fed a basal diet of low-quality blue grama (*Bouteloua gracilis*) hay, native shrubs elicited a nutritional response similar to alfalfa hay when fed as a supplement (Rafique *et al.,* 1988; Arthun *et al.,* 1988). The native shrubs influenced blue grama hay intake and nitrogen utilization in a favorable and similar manner to the alfalfa hay. The availability of desirable native shrubs for use during critical drought and winter periods can be an important stabilizing influence.

On fourwing saltbush-dominated sites at the Central Plains Experimental Range near Nunn, CO (Shoop *et al.,* 1985), cattle diets averaged 14% fourwing saltbush (*Atriplex canescens*) during the June to October period but 32% during the November to April period; the lowest intake was in June and July, the highest (55%) in March. It was calculated that a dry, pregnant cow on a 33% saltbush:67% blue grama diet on winter range would receive an adequate intake of carotene and metabolizable energy, about 90% of needed crude protein, and about 60% of needed phosphorus. By contrast a comparable cow eating only blue grama would receive about 25, 100, 50, and 60%, respectively, of her needs for these nutrients. When lambs in Texas grazed dry grass plus fourwing saltbush during the January to March period, the gain advantage over grazing dry grass alone was 11 and 8 lb per head during the first and second year of the study (Ueckert *et al.,* 1988).

Natural shrub-grass ranges are often quantitatively inadequate to support all livestock for which winter grazing is desired, but procedures for adding palatable shrubs to seeded and native grasslands are being developed (Fig. 2.7). In the Great Basin and the Southwest, the inclusion of palatable forbs and shrubs in seeding mixtures with grasses—or their later addition to grass stands—has greatly improved livestock as well as big game performance during grass dormancy (Holechek, 1984). Fourwing saltbush in Texas studies provided a source of supplemental nutrients, including protein, when grazed by sheep in conjunction with winter-dormant WW-Spar bluestem (*Bothriochloa ischaemum ischaemum*); year-

FIGURE 2.7   Including palatable shrubs in range seeding mixtures or adding to existing grass-lands has shown good potential in the Intermountain Region for improving the winter diets of livestock and big game; A, shrub mixture added to crested wheatgrass at Nephi Field Station, UT; B, forage kochia (*Kochia prostata*) planted in tilled strips in established crested wheatgrass at BYU Skaggs Research Ranch, Malta, ID.

ling Rambouillets lost an average of 9.3 lb/head during the 60-day winter grazing period on straight grass compared to gaining 9.3 lb per head on grass-saltbush mixtures (Ueckert *et al.,* 1990). Research in southern Idaho has shown that forage kochia (*Kochia prostrata*) seeded in mixture with crested wheatgrass, or later added to the grass stand, can make significant inputs to cattle diets not only during the winter but also during grass summer dormancy (Monsen *et al.* 1990).

In initial studies at the Nephi Field Station in central Utah, fourwing saltbush and winterfat (*Ceratoides lanata*) were more effective than rubber rabbitbrush (*Chrysothamnus nauseosus*) and big sagebrush (*Artemisia tridentata*) in providing supplemental protein to the base crested wheatgrass diet (Otsyina et al., 1982). In subsequent winter grazing trials with sheep, forage kochia also proved to be a useful supplement to the cured crested wheatgrass (Gade and Provenza, 1986). Sheep on the grass-shrub mix consumed forage about 50% higher in protein but similar in organic matter digestibility (i.e., energy potential), and organic matter intake was 18.5% greater on the grass-shrub mix. When snow levels were deep, the grass was less available and heavier utilization was made of the shrubs, thus reducing energy levels associated with the more lignified ingesta. The researchers concluded that supplemental protein obtained from the shrubs would be particularly beneficial in dry or cold winters when green vegetative growth was scarce or as an emergency feed when heavy snow covered the herbaceous vegetation.

C. ENHANCING NUTRIENT LEVELS IN FORAGE

On seeded pasture, using cultivars that are leafier, have a long growing period, prolong the vegetative stage, and make rapid regrowth is a means of maintaining high nutrient levels in forage plants. Concentrating grazing during the rapid growing period—such as with intensive early grazing—is another means of taking advantage of high nutrient levels early in the growing season. However, while the highest quality pasturage can be obtained by grazing only new shoots, restricting grazing to this period results in substantial sacrifice in forage dry matter yields and fails to provide forage for later growth stages and after maturity.

Since the highest quality forage is produced during the early stage of forage plant development (usually prior to entering the reproductive stage), interrupting the normal maturation processes of herbaceous plants is important in maintaining high nutrient levels in forage plants. Grazing practices that prolong the vegetative growth stage and delay or prevent the production of reproductive stems and inflorescences maintain higher nutritive levels in pasturage. Many grazing treatments— stocking rates, stocking densities, season of grazing, and mixing animal species— are directed towards encouraging regrowth and continued availability of nutritious forage. (Refer also to Chapter 16, Section III, "Manipulating Animal Habitat by Grazing.")

Moderate to heavy grazing early in the season will generally be required to effectively remove low-quality, senesced forage and to improve forage quality and availability of subsequent regrowth. Light stocking rates early in the season with cattle on old world bluestems (*Bothriochloa* spp.) in Oklahoma (Forbes and Coleman, 1993) resulted in tall, poor-quality swards late in the growing season; as a result of patchy grazing, the resulting herbage intake late in the season was often as low as on pastures continuously stocked under heavy grazing. However, in the long term, particularly on range, the quality of available forage under heavy stocking rates will depend upon the quality of replacement species resulting from changes in species composition caused by the cumulative effects of prolonged heavy grazing (Heitschmidt and Stuth, 1991).

Grazing early during rapid plant growth, then removing grazing in time to allow regrowth, thereby prolonging the period of high nutritive value, is an attainable objective only when good plant growing conditions continue. With improved pasture in moderate to high precipitation zones, optimal growing conditions may enable multiple regrowth periods, thus providing for the renewal of high-quality morphological units over an extended period of time. However, grazing as a preconditioning treatment to improve forage quality for later grazing may be quite unreliable in areas or on sites prone to dry years or short-term drought, such as is likely in arid and semi-arid regions.

Hart *et al.* (1993) found that regrowth was insignificant for most of the year on mixed grass steppe; grass tillers were seldom grazed more than once because regrowth was too minimal. In the Northern Great Basin, preconditioning by spring grazing failed to improve nutritive quality of forage for later grazing during dry years, growth being prevented by low levels of moisture, leaving only the remainder, more stemmy portion of the spring growth (Brandyberry *et al.,* 1993). Studies on rough fescue (*Festuca campestris*) grasslands in Alberta by Willms and Beauchemin (1991) confirmed the benefits of repeated grazing within a year to maintain high-quality forage by increasing crude protein, phosphorus, and *in vitro* dry-matter digestibility and reducing acid-detergent fiber and lignin. However, the study also demonstrated the need to limit the frequency of repeated grazing to avoid stand deterioration and to maximize nutrient yield.

Burning or mechanical treatments that interrupt maturation or remove old growth can prolong the green-growth period and make it more accessible by removing the old growth. The use of burning for this purpose has been particularly effective on sites where tallgrass predominates, such as on Midwest prairie, Gulf cordgrass (*Spartina spartinae*) vegetation (Angell *et al.,* 1986), and creeping bluestem (*Schizachyrium stoloniferum*) range (Long *et al.,* 1986; Sievers, 1985), but the effects seldom extend beyond the current growing season. Burning is also used for rejuvenating browse yields of sprouting shrubs in the southern Great Plains, Intermountain, and chaparral shrublands. Burning treatments are generally prescribed only periodically (every 2 to 5 years), rather than annually, and then only when the desired grasses, forbs, and shrubs are tolerant of fire. The short-term attraction of recently burned sites on tallgrass prairie in Oklahoma to bison was adjudged by Coppedge and Shaw (1998) to result from enhanced forage quality and palatability rather than forage quantity.

Nitrogen (N) or phosphorus (P) fertilizers, particularly when these elements are deficient in the soil, increase their levels in the forage. Combining the positive effects of fertilizers and burning has been used in the Midwest and southeastern U.S. to improve subsequent forage value. For example, burning and N-P fertilization, either individually or in combination, increased the dry matter digestibility and crude protein levels in big bluestem (*Andropogon gerardii*) while reducing levels of acid-detergent lignin (Mitchell *et al.,* 1991). Burning in May increased *in vitro* dry-matter digestibility (IVDMD) of forage harvested in June; fertilization increased crude protein content and, by a lesser degree, the IVDMD. The combination of mid-May burning and fertilization increased forage quality of big bluestem the most.

Herbicides applied at rates that "cure" rather than kill forage plants or otherwise enhance protein levels and even energy levels in standing forage include paraquat (Sneva, 1967, 1973; Kay and Torrell, 1970), tebuthiuron (Masters and Scifres, 1984; Biondini *et al.,* 1986), and simazine and atrazine (Houston and Van Der Sluijs, 1975). Even killing levels of 2,4-D or glyphosate (Kisserbeth *et al.,* 1986) may temporarily enhance the nutritive levels and/or palatability of affected plants. Spray-topping annual grass pasture with glyphosate at seedhead emergence stages reduced the loss of water-soluble carbohydrates and digestibility going into plant senescence; it also increased dry matter intake and sheep gains, particularly during the 8-week period following application, but slightly reduced forage dry matter pasture yields (Leury *et al.,* 1999).

Other plant growth regulators such as mefluidide, ethephon, and amidochlor appear even more promising in suppressing seedstalk production, delaying maturity, and increasing forage quality of grasses (DeRamus and Bagley, 1984; Roberts *et al.,* 1987; Wimer *et al.,* 1986; Haferkamp *et al.,* 1987; Slade and Reynolds, 1985; Turner *et al.,* 1990) and an alfalfa-grass mix (Fritz *et al.,* 1987). Mefluidide has generally enhanced forage quality of grasses under optimal conditions; however, on occasion, it has been ineffective during drought periods (McCaughey and Cohen, 1990), and it has a short period of effective application (White, 1991) and may substantially decrease forage yield (White, 1989). The use of plant growth regulators for inducing dormancy in forage plants is still mostly experimental, and promising chemicals may or may not be cleared for such use.

IV. ANTI-QUALITY AGENTS IN PLANTS

Anti-quality components of forage are any factors inherent in forage that limit the ability of the animal to reach its potential (Allen, 1999). Frequently encountered factors include natural toxins (including alkaloids, glycosides, and prussic acid), maladies associated with endophytes (including tall fescue toxicosis), frothy bloat associated with legume consumption, mineral disorders (including grass tetany and silica urinary calculi), nitrate toxicity, plant structural agents (lignin, tannins, spines), and many more. Anti-quality agents in plants often serve as defense mechanisms for the plants.

Anti-quality agents are numerous and widely distributed among forage plants. Grazing animals contend with many anti-quality agents by detecting and avoiding high concentrations of phytotoxins, adopting appropriate temporal patterns of toxin intake, and selecting a broad and varied diet (Provenza and Launchbaugh, 1999). Grazing animals also have the ability to tolerate and detoxify many of the anti-quality components in forages in the rumen; and the inheritance of systems to metabolize phytotoxins is probably widespread in herbivores (Launchbaugh *et al.,* 1999). An animal's ability to tolerate or detoxify anti-quality factors can be affected by the nutrient status of the grazing animal, intake of specific nutrients and supplements, familiarity with the toxic plant and the area where found, previous exposures to the phytotoxins, and intake of combinations of toxins. (Since anti-quality in forage plants is interrelated with anti-palatability, refer also to "Secondary Compounds as Anti-palatability Factors" in Chapter 9.)

A. PLANT POISONING OF LIVESTOCK

Since it appears that plant poisons are evolved defense mechanisms, it is probable that co-evolution has occurred in herbivores to prevent their being poisoned by plants either through avoiding the poisonous plants or by detoxifying plant poisons (Laycock, 1978). These adaptations appear to operate in both domestic and wild herbivores but are more prevalent in the latter. Since native herbivores apparently developed greater capacity than domestic animals to tolerate or detoxify toxins by co-evolving with the poisonous plants, the domestic animals are more subject to human management errors.

Animals have varying capabilities of reducing the risk of plant poisoning through negative post-ingestive feedback by (1) avoiding or reducing toxin intake through changes in diet selection, (2) selecting a mixed diet and diluting the toxin, or (3) consuming toxin in a cyclic or intermittent fashion. They also reduce the risk of poisoning by special handling of toxin once eaten by (1) ejecting the toxin; (2) complexing, degrading, and detoxifying the toxin; and (3) tolerating the toxin (Pfister, 1999). Aversion training of livestock against poisonous plants offers some hope for the future (see Chapter 9) but also rather formidable problems, such as the tendency of livestock to repeatedly sample highly poisonous plants that they have been conditioned against (Provenza *et al.,* 1987, 1988; Burritt and Provenza, 1989).

Poisonous plants cause significant direct livestock losses, not only in livestock deaths but also through deformities, abortions, lengthened calving and lambing intervals, decreased efficiency, and reduced weight gains (Fig. 2.8). Indirect costs often include medical costs, supplemental feeding, management alterations, additional herding or fencing, and loss of forage associated with efforts to prevent or minimize poisoning. Either thirst or hunger may result in poisoning that normally would not occur (Reid and James, 1985). Good forage and livestock management resulting in an adequate plane of nutrition is the first line of defense against poisonous plants. Hunger lowers the smell and taste rejection thresholds (Ralphs and

FIGURE 2.8   Locoweed (*Astragalus* spp.) growing in the Uintah Basin of Utah (A) caused birth
deformities in these lambs (B) when plant material was collected and fed to their dams at the USDA-
ARS Poisonous Plant Research Lab at Logan, UT.

Olsen, 1987), and grazing management that alleviates hunger is often helpful in
preventing livestock losses from poisonous plants.

   Livestock are more apt to eat enough poisonous plants to have harmful effects when there is a shortage of palatable, non-toxic forage. There are exceptions, however, of poisonous plants that are quite palatable at times to livestock—i.e., tall larkspur (*Delphinium barbeyi* or *D. occidentale*) to cattle in summer or green, growing locoweed to sheep on winter range. Stocking rate seems to have no consistent effect on tall larkspur consumption by cattle (Pfister *et al.,* 1988) but is involved in many other poisonous plant problem situations. In contrast with tall larkspur, cattle seldom eat toxic amounts of low larkspurs (*Delphinium* spp.) if sufficient other forage is available (Pfister and Gardner, 1999).

Prolonged drought and overgrazing sometimes force grazing livestock to eat harmful amounts of poisonous plants. On good condition grazing land, poisonous plants are subjected to intense competition from vigorous, high producing forage plants, and there is a greater variety of non-poisonous plant species available for selective grazing. However, even intensive grazing systems that incorporate high animal density and low alternative forage availability in the presence of poisonous plants can increase the likelihood of poisoning (Pfister, 1999).

There is not always a sharp distinction between poisonous and nonpoisonous plants. Most plants are poisonous only when eaten in large amounts (i.e., chokecherry [*Prunus virginiana*]) and may merely provide nutritious forage when consumed in smaller amounts mixed with other forages. Also, timber milkvetch (*Astragalus miser*) in British Columbia (Majak *et al.,* 1996) is a digestible and nutritious forage when taken in small amounts. On grassland sites, cattle preferred the associated grasses and consumed small amounts, while on forest sites the milkvetch was preferred over the associated pinegrass (*Calamagrostis rubescens*) and cattle consumed larger amounts of milkvetch. Cattle can tolerate substantial quantities of larkspur over many days if they do not consume excessive amounts at any one time, i.e., over 10 to 20% of diets during late summer (Ralphs *et al.,* 1989). With a few plant species (e.g., whorled milkweed [*Asclepias subverticillata*] and water hemlock [*Cicuta maculata*], the consumption of even small amounts can be lethal. Some plant toxins act very quickly; a dead animal may be the first symptom of trouble.

In most situations the only practical solution to poisonous plants is to manage grazing to avoid losses as much as possible. Prevention of livestock losses from poisonous plants is far more effective than treatment. Although antidotes are known for some poisonous plants, help often comes too late to save animals poisoned on grazing lands. Theoretically, feed supplements with antidotes added might provide protection in limited situations, such as supplementing with calcium to neutralize the effects of tannins on cattle (Ruyle *et al.,* 1986b) or to render oxalates insoluble in the digestive tracts of sheep feeding on halogeton (*Halogeton glomeratus*) (USDA, Agricultural Research Service, 1968). However, these have generally been ineffective under field conditions.

Most large losses of livestock from poisonous plants result from management mistakes (Ralphs and Olsen, 1987). Proper grazing management decisions require knowledge of poisonous plants and their toxins (Keeler and Laycock, 1987; Ralphs and Sharp, 1987). Information required to prevent or limit animal poisoning includes what poisonous plants are present, their relative palatability, the effects their toxins have on livestock, and seasonal and other trends in toxin concentration.

General management techniques to reduce losses of grazing livestock from poisonous plants include the following 16 tools (Vallentine *et al.,* 1984; McGinty, 1985):

1. Match livestock numbers with the grazing capacity; do not overstock; vary livestock numbers in a grazing unit to assure ample forage and forage selection; remove grazing animals when the desirable forage plants have been properly grazed.

2. Maintain forage stands in grazing units in good condition; many poisonous plants are increasers or invaders and become more prevalent when vegetation condition lowers over time.

3. Do not turn out hungry animals onto poisonous plant-infested areas; hungry animals often forego much of their normal selective grazing behavior. Hungry livestock unloaded from trucks or corrals may quickly graze poisonous plants they would otherwise ignore.

4. Avoid poisonous plant areas as much as possible, particularly during high-risk periods; check corrals and holding pastures carefully for the presence of poisonous plants. Do not provide salt or other supplements or place drinking water or bed livestock where poisonous plants are abundant.

5. If poisonous plant areas cannot be avoided temporarily such as in trailing, fill up animals with feed ahead of time; trail slowly but continuously through infested areas so that animals have opportunity to select against poisonous plants. Animals with full stomachs are less apt to consume poisonous herbage, and what they do eat will be more diluted with nonpoisonous forage or other feed.

6. Graze the kind of livestock that seldom graze or are more resistant to the poisonous plants present. Many plants are poisonous only to one kind of livestock; cattle are six times more susceptible to tall larkspur than are sheep (Ralphs and Olsen, 1989). Other poisonous plant species may be potentially poisonous to all kinds of livestock but may be eaten normally only by one kind; both cattle and sheep are subject to oxalate poisoning, but cattle generally avoid halogeton and are seldom poisoned.

7. Avoid grazing a class of livestock when they are most affected by the plant toxin. For example, ewes in late gestation or lactation are most susceptible to bitterweed (*Hymenoxys odorata*) poisoning (Taylor and Merrill, 1986). Keeping pregnant cows from grazing broom snakeweed (*Gutierrezia sarothrae*) will prevent this cause of abortion (Ralphs, 1985). Deformed calves result from their dams eating lupine (*Lupinus* spp.) only during the 40th to 70th days of gestation (Keeler, 1983). Deformed lambs result from their dams eating falsehellebore or cow cabbage (*Veratrum californicum*) on or about the 14th day of pregnancy (USDA, Agricultural Research Service, 1968). Pregnant cows nearing parturition should be kept away from Ponderosa pine (*Pinus ponderosa*) trees to prevent needle consumption, particularly during inclement winter weather.

8. Graze during the time of year or plant growth stage when the poisonous plant

is the least troublesome (i.e., least palatable or least toxic). When poisonous plants are troublesome in early spring, for example, death camas and low larkspur do not turn livestock out before desirable forage plants have achieved ample growth to sustain selective grazing. For example, early spring (May and June) has been shown to be the most troublesome period for the consumption and subsequent poisoning by woolly locoweed (*Astragalus molissimus*) and white loco (*Oxytropis sericea*) in northeastern New Mexico (Ralphs *et al.,* 1993). At this time of year, the associated warm-season grasses are still dormant, and cattle were apparently actively seeking green growth such as was available from the actively growing loco plants. Since the problem quickly abated once green grass became abundant, the solution recommended was to create or maintain a locoweed-free pasture for spring grazing. In contrast, cattle may avoid eating low larkspurs before flowering (Pfister and Gardner, 1999). From the elevation of flowering stems (i.e., early flower to full flower stage) and until pods begin to dry out, a combination of palatability and toxicity make this a high-risk period for grazing tall larkspur sites with cattle (Pfister *et al.,* 1997b). Even though the alkaloid levels are high in immature larkspur, the risk of cattle losses from grazing tall larkspur areas in the spring is low because of low palatability then. Thus, a window of opportunity exists for about 4 to 5 weeks in the spring on Intermountain ranges to graze cattle in tall larkspur areas without undue risk, or after larkspur pods dry out and shatter and toxicity levels are low by late summer.

9. Rotate grazing animals between areas contaminated and areas free of poisonous plants when poisoning effects result from cumulative, low-level intake over time, i.e., in milkvetches (*Astragalus* spp.) (James, 1983a), locoweeds (James, 1983b), or selenium-accumulating plants.

10. "Flash graze" high densities of susceptible livestock during periods too short to allow consumption of toxic levels, for example, sheep on bitterweed in Texas (McGinty, 1985) or cattle on locoweed in the Intermountain (Ralphs, 1987). Short-duration grazing readily provides opportunity for flash grazing or skipping a particular problem pasture during a critical time but can promote nonselective grazing and shift animals to poisonous plants if not rotated soon enough (McGinty, 1985).

11. Be aware of unusual environmental conditions that may restrict animal movement or change diet selection, such as drought, unseasonal frost, summer rains, or unseasonal snow storms. Research has verified the frequent reports that cattle deaths from tall larkspur increase during stormy periods, when the consumption of the plant increases, even to the animals becoming glutinous (Pfister *et al.,* 1988; Ralphs *et al.,* 1994a). Possible explanations for the sudden increase in palatability or attractiveness of the tall larkspur associated with the cold, damp summer storms include a temporary reduction or alteration of alkaloids in the plants, washing off of the bitter-tasting wax that builds up on the leaves, increased sugar levels in the leaves, or stimulated hunger in the animals. Another example is that pregnant cows have been shown to materially increase the consumption of Ponderosa pine needles, a cause of premature births and abortion, during winter as snow depths increase, the amount of alternative forage is reduced, and cold am-

bient temperatures reduce times of normal grazing (Pfister and Adams, 1993; Pfister *et al.,* 1998).

12. Remove livestock from poisonous plant areas as soon as unexpected problems develop; provide ample feed and water to poisoned livestock, keep them as quiet as possible, and administer any known effective antidote.

13. Provide drinking water in adequate amounts and frequency to maintain an adequate and constant intake of forage and flush toxic agents through the animals. When water intake is severely restricted, forage consumption often declines rapidly. After water is subsequently provided, hunger may be quickly evidenced and animals are prone to graze whatever is immediately available, including poisonous plants. For example, most sheep losses from halogeton have been associated with this scenario (James *et al.,* 1970).

14. Provide deficient minerals through supplements to prevent abnormal or depraved appetites and to maintain selective grazing. Make salt available on a free-choice basis; also provide supplemental phosphorus when it is deficient and correct any other mineral deficiencies known to be related to the consumption of specific poisonous plants.

15. Animals from other geographic areas may be attracted to unfamiliar poisonous plants as a novelty and may initially consume harmful amounts of them. Locally raised animals familiar with them may consume only low, nontoxic levels or may have developed an aversion or resistance to them. Familiarity with chokecherry apparently stabilizes intake levels, but Ralphs *et al.* (1987) found no evidence that inexperienced, introduced heifers were more inclined to consume white locoweed than were their long-term resident counterparts.

16. Train cattle to avoid eating poisonous plants using aversive training. (Refer to "Post-ingestive Feedback Aversion" in Chapter 9 for further details.)

17. Guard against herbicide effects in increasing toxicity or palatability of poisonous plants (see below).

Herbicides such as 2,4-D and related compounds 2,4,5-T and silvex (the latter two are now no longer marketed) are known to temporarily increase the palatability of many plants, including poisonous plants. A general recommendation is that areas with poisonous plants should not be grazed until after affected plants begin to dry and lose their palatability, generally three weeks or more after herbicide application. For example, levels of miserotoxin, a toxic glucoside in timber milkvetch, decreased rapidly after being sprayed with 2,4,5-T or silvex and were only one-third of controls after 4 weeks. A thoroughly bleached condition of timber milkvetch was recommended before allowing grazing to be resumed. It is also known that 2,4-D can increase levels of nitrate and cyanide (prussic acid) in plants that accumulate these toxic agents (Schneider, 1999).

The herbicides silvex and 2,4,5-T have also increased the alkaloid content of leaves and stems of tall larkspur, doubling the levels in some years (Williams and Cronin, 1966). Similar treatments increased the alkaloid levels on false hellebore by 2,4,5-T but not silvex. Metsulfuron application also increased the toxicity of tall larkspur; applications of glyphosate and picloram did not affect the absolute amount of toxic alkaloids, but they did not reduce larkspur toxic alkaloids (Ralphs *et al.,* 1998). With all three herbicides the risk of poisoning remained until plants had desiccated. Waiting until tall larkspur plants are dead, desiccated, and wilted has been recommended following spraying with metsulfuron, glyphosate, and picloram (Ralphs *et al.,* 1998).

Losses from poisonous plants have been most severe under heavy stocking rates, with fewer losses occurring at lighter stocking rates. Records kept at the Sonora Research Station in Texas have shown that grazing practices are related to livestock losses from poisonous plants (Merrill and Schuster, 1978; McGinty, 1985; Taylor and Ralphs, 1992). Moderate but not heavy stocking with cattle, sheep, or goats under the Merrill deferred-rotation system over a 20-year period prevented livestock poisoning by bitterweed, oaks (*Quercus* spp.), and sacahuista (*Nolina texana*); the same resulted from grazing a mix of livestock under light continuous stocking. These results were attributed to better range conditions and forage variety under the Merrill moderate rotation and light continuous treatments. Heavier grazing treatments increased animal poisoning; using a mixture of livestock reduced the incidence of bitterweed and sacahuista poisoning but not oak poisoning. Grazing combinations of cattle, sheep, and goats reduced bitterweed losses by decreasing spot grazing which limited the invasion of bitterweed.

It is seldom practical to remove widespread infestations of poisonous plants by chemical or mechanical means. However, biological plant control by nonsusceptible classes or kinds of livestock is sometimes effective in reducing levels of plants hazardous to other classes or kinds of animals. Only when the poisonous plants are concentrated in smaller patches is herbicidal or mechanical removal apt to be practical. Fencing off or removing poisonous plants growing in dense patches at normal concentration points such as wells, waterholes, meadows, trails, and corrals is particularly suggested. It should be noted that the palatability of toxic plants may be increased by recent burning. Also, livestock poisoning can sometimes result from feeding hay containing herbage of certain poisonous plants (Horrocks and Vallentine, 1999).

B. FESCUE TOXICOSIS

The deleterious effects of a tall fescue toxicity problem on animal production has seriously impacted animal production in the southeastern and midwestern U.S. (Stuedemann and Hoveland, 1988). The primary toxicity problem has been shown to be associated with a fungal endophyte contaminating tall fescue plants, and it is known that this endophyte produces an ergot alkaloid, the apparent cause of fescue toxicosis (Bacon, 1995). Three animal impact syndromes have been recognized: (1) "fescue foot," a gangrenous condition of the feet; (2) bovine fat necrosis; and (3) fescue toxicosis, or "summer slump." The last is widely associated with tall fescue and is characterized by low gains or even loss of weight, rough hair coat, general unthriftiness, low milk production, and impaired reproduction. The symptoms of tall fescue toxicosis are most severe in grazing animals but can also be a problem with feeding cured hay or haylage.

The fungal endophyte does not appear to be harmful to the tall fescue plants and is, in fact, symbiotic with it (Bacon, 1995). The initial solution to the problem lies in destroying contaminated plant stands and replacing by using fungus-free seed; plant breeding for resistance to the fungus also appears promising (Pedersen and Sleper, 1988). Endophyte-free tall fescue is more palatable and more readily consumed by grazing animals but is more difficult to establish and is less tolerant of environmental stress than is endophyte-infected tall fescue—apparently because of the loss of symbiotic benefits—and may require more careful management, including less severe defoliation (Hoveland *et al.,* 1990).

C. LEGUME BLOAT

Legumes are commonly used for grazing on subhumid and mesic sites because of their high grazing capacity and nutritive value. However, when legumes such as alfalfa, the clovers (*Trifolium* spp.), or sweetclover (*Melilotus* spp.) are grazed, there is the potential for subacute or acute frothy bloat. This condition results in formation of a frothy stable foam in the rumen, a retention of gas produced in normal rumen function, and an inhibition of the eructation (belching) mechanism (Reid and James, 1985). Stable foam production in bloating animals is due to a complex interaction of animal, plant, and microbiological factors. The direct cause of frothy bloat is the production of carbon dioxide and methane in the reticulorumen which results in a foam that is stabilized by legume leaf proteins (Lowe, 1998).

An effective synthetic compound for preventing frothy bloat is poloxalene (trade name Bloat Guard), a water-soluble, detergent-type chemical. This antifoaming agent is effective when daily intake is assured and feeding is begun 2 to 5 days before turning animals onto bloating legumes (Corah and Bartley, 1985). Poloxalene acts by destabilizing the frothy foam, thus allowing its escape from the animal's digestive tract.

Poloxalene can be provided in a liquid molasses-based supplement, mixed with dry energy or mineral supplements (possibly restricting intake by salt), or included in a supplement block. However, the supplement as a poloxalene carrier may be an expense item not otherwise justified, and the desired level and regularity of supplement intake may vary substantially between animals. While poloxalene has been shown to be 100% effective when given intraruminally at the prescribed dose, under practical conditions poloxalene can only be offered free choice, and total protection from bloat cannot be guaranteed (Majak and McAllister, 1999; Majak *et al.,* 1995).

Placing poloxalene in the drinking water, when this is restricted to a single, treated source, shows great promise for low-cost, uniform consumption of poloxalene but has not yet been approved in the U.S. A similar class of surfactant compounds, pluronics, is used for bloat control in New Zealand and Australia; they are normally administered by drenching or by addition to the water supply (Reid and James, 1985). A water-soluble polymer, trade name Blocare, when used in the water supply has proven to be 100% effective in bloat prevention but also has not yet been registered in North America (Majak and McAllister, 1999). In Australia, an anti-bloat gelatin capsule is also available which remains in the rumen and releases a foam-dispersing detergent for a period of up to 24 days (Walton, 1983). While alieving sub-acute bloat symptoms, poloxalene has also been found to reduce dry-matter intake rates; however, increases in grazing time tend to offset reduced intake rates, thus moderating ingestive behavior within grazing meals when grazing immature alfalfa (Dougherty *et al.,* 1992).

Monensin has also proven effective in controlling frothy bloat on alfalfa and white clover (*Trifolium repens*) pasture. Monensin provided in controlled-release capsules not only reduced bloat death losses but also the deleterious effects of sub-lethal bloat on animal performance, this in addition to the direct beneficial effects monensin has on efficiency of rumen metabolism (Lowe, 1998). The capsules used delivered 170 mg of monensin daily for 150 days for younger cattle and 300 mg daily for 100 days for older cattle. Monensin reduces the populations of bacteria that produce carbon dioxide and methane and increases populations of bacteria that produce more propionic acid. Drenching with a monensin solution was also effective in reducing frothy bloat but was considered mostly impractical with beef cattle grazing larger pastures.

Many management practices are helpful in reducing frothy bloat on legume or grass-legume pasture (Fig. 2.9). Although their effectiveness varies greatly from region to region and from pasture to pasture, management practices are generally effective in reducing the incidence of bloat but not totally preventing it. Practices that are useful and generally recommended for reducing bloat on legume pasture include:

1. Manage alfalfa and clovers in grass-legume mixtures not to exceed 30 to 40% legumes.

2. Delay initial turnout until legumes have reached the late bud to early bloom stage of maturity, since very immature legume growth is highly bloat promoting. Bloat potency is highest at the vegetative or prebud stage, decreasing progressively as the plant grows and matures to full flower (Howarth *et al.,* 1991)

3. When first turning out on legume pasture or after nightly lockup, feed animals with grass hay or other roughage before turn out.

4. Once accustomed to legume pasture, leave animals continually on pasture even at night, if possible.

5. Keep some dry roughage available to grazing animals at all times, or mow a strip 3 to 4 days before grazing.

6. Avoid advancing livestock to the next fresh pasture unit when they are very hungry, as they are then prone to overeat and consume large amounts of alfalfa.

7. Grazing under a strip or short-duration system to reduce selectivity for bloating components in a grass-legume sward has been suggested, but continuous grazing of pure legume or high legume stands is apparently superior to rotation graz-

FIGURE 2.9   Good management practices help alleviate legume bloat by cattle, as shown here grazing an alfalfa–intermediate wheatgrass stand in central Utah, but the feeding of poloxalene is a more sure preventative

ing in that it does not require frequent readjustment by the grazing animals (Majak and McAllister, 1999; Majak *et al.*, 1995).

8. Initiating grazing of alfalfa at midday or moving into a new pasture unit then, rather than early morning, has reduced the incidence of frothy bloat, this apparently associated with waiting until the dew is off the plants (Majak *et al.*, 1995).

9. Frost does not render alfalfa bloat-safe, and management precautions should continue even after the first killing frost (Majak *et al.*, 1995).

10. Provide good fencing to prevent cattle from accessing lush alfalfa or clovers by straying or getting out, particularly when beyond access to poloxalene.

11. Check livestock frequently; note chronic bloaters as trouble predictors but remove affected animals (or even all animals) before their condition becomes serious. Marked differences exist between individual animals in susceptibility to alfalfa pasture, and since bloat susceptibility is considered to be a heritable trait, it is a good practice to cull known bloat-susceptible animals from a breeding herd (Howarth *et al.*, 1991).

12. Use nonbloating legumes such as sainfoin (*Onobrychis viciafolia*), cicer milkvetch (*Astragalus cicer*), birdsfoot trefoil (*Lotus corniculatus*), or crownvetch (*Coronilla varia*). However, yield performance may be substantially less than for alfalfa or the clovers; bloat-causing legumes are digested rapidly while bloat-safe species and cultivars are digested more slowly. The consumption of a sainfoin-alfalfa diet compared to a straight alfalfa diet appears to reduce the incidence of bloat while reducing ruminal proteolysis and increasing levels of rumen-escape protein

(McMahon *et al.,* 1999). Co-feeding sainfoin at only 10% of the fresh alfalfa intake has substantially reduced the incidence of bloat (McMahon *et al.,* 1999); this beneficial effect of the sainfoin was attributed to the low levels of condensed tannin it added to the diet, which reduced degradation of forage protein without affecting the digestibility of the non-protein fraction.

13. Use low-bloat cultivars of alfalfa or clovers. A new cultivar of alfalfa (i.e., AC Grazeland), selected for a low initial rate of digestion—not for low total digestibility—is being released in Canada. It has been found to reduce the incidence of bloat by 62% in grazing trials (Majak and McAllister, 1999).

D. GRASS TETANY

Grass tetany (also known as grass staggers or wheat pasture poisoning) can be a major anti-quality factor when grazing on spring grain forage, crested wheatgrass, and other cool-season grasses in lush growth stages. Grass tetany is characterized by low blood serum magnesium concentrations (hypomagnesemia); this condition results from a simple magnesium deficiency in the diet or more often from reduced availability and absorption of forage magnesium being converted in the digestive system to an insoluble form. The complete causal relationships of the latter are only partly understood (Mayland, 1986; Greene, 1986). However, the malady is most prevalent in forage that is marginal or deficient in magnesium, calcium, and carbohydrates and high in potassium (Asay *et al.,* 1996).

The incidence of the problem is increased by growth reduction by cool or dry weather followed by a rapid flush of growth and the development of washy forage. High dietary concentrations of potassium decrease the availability of magnesium in spring forages, thereby increasing cattle susceptibility to grass tetany. Low dietary calcium levels also reportedly contribute to the onset of grass tetany. High non-protein nitrogen and low carbohydrate levels in the forage may also be involved. The development of the HiMag cultivar of tall fescue has increased magnesium levels in the forage; in areas where tall fescue is adapted and widely used, such as eastern U.S., this may provide a means of reducing the incidence of grass tetany during periods when risk is high (Crawford *et al.,* 1998). For the northcentral and western U.S., the selection for reduced grass tetany potential in crested wheatgrass appears practical and would also likely be accompanied by improved forage quality (Asay *et al.,* 1996).

Grass tetany occurs most often in older, lactating cows recently turned onto cool-season pasture in the spring. Growing yearling heifers and steers can also be affected then. Symptoms include nervousness, muscular incoordination, staggering, and paralysis; death usually occurs within 2 to 6 hr if affected animals are left untreated.

Grass tetany in cattle can be prevented or treated by one or more of the following practices (Grunes and Mayland, 1984; Greene, 1986):

1. Prevent a magnesium deficiency by feeding a minimum of $\frac{1}{2}$ oz (14 g) of supplemental magnesium per head daily; provide the supplemental mag-

nesium by feeding magnesium oxide or magnesium sulfate in a dry supplement, mixed in a molasses liquid, mixed with salt, added to the drinking water, or as a rumen bolus.

2. On intensively managed improved pasture, assuring adequate levels of magnesium and calcium by soil fertilization while keeping potassium at the lower recommended levels are suggested (Robinson *et al.,* 1989)

3. Graze native range with large admixtures of cured herbage or delay grazing on seeded pasture beyond the flush growth stage; these practices may be somewhat helpful but may thwart the other advantages of early cool-season pasture.

4. Give intravenous injections of magnesium sulfate or calcium-magnesium gluconate to animals that have already exhibited symptoms of grass tetany. (This treatment is also recommended for treating the similar symptoms of prussic acid poisoning.)

E. SILICA URINARY CALCULI

Silica can comprise up to 10% of mature or cured grasses on a dry matter basis, and at high levels it can substantially reduce forage digestibility. Silica, particularly in conjunction with low water intake, is also responsible for the development of silica kidney stones and the condition known as silica urinary calculi or urolithiasis. Mayland (1986) concluded that, when in combination with low water intake, a forage silica content greater than 2% can be expected to cause urinary calculi in susceptible animals.

This malady occurs in both cattle and sheep, especially in castrated males. Silica stones collect in the urethra, thereby interfering with urine flow; in advanced stages, the bladder may rupture and urine collect in the abdominal cavity, giving rise to an extended abdomen referred to as "water belly." Other symptoms include tail twitching, uneasiness, kicking at abdomen, and straining in an attempt to urinate. Prevention includes encouraging high water intake for diluting silicic acid and other interacting minerals in the urine by providing adequate supplies of clean water and even warming water on cold days.

Force feeding high levels of common salt or $\frac{1}{10}$ lb of ammonium chloride daily in the diet will materially increase water intake (Emerick, 1987). Both ammonium chloride and phosphorus supplements aid in acidifying the urine and reducing the formation of silica stones. Extending the green grass growing period into the fall for grazing by the use of crested wheatgrass or Russian wildrye (*Psathrostachys juncea*) in western U.S. and Canada (Bailey and Lawson, 1987) or feeding good quality legume hay on dry grass pasture has also been useful in reducing urinary calculi.

F. SELENIUM TOXICITY

Selenium is a naturally occurring mineral required in trace amounts in animal diets, but its presence in excessive amounts in forages and grains is apt to cause animal poisoning (Anderson *et al.,* 1961). Soils of specific parent materials in the central and northern Great Plains and other local areas of the western U.S., those receiving less than 25 in. of precipitation annually, are labeled seleniferous if they contain hazardous levels of selenium of 0.5 to 100 ppm or more. Animals consuming forage grown on these soils may be poisoned from consuming excessive levels of high-selenium forage in their diets.

Some native plant species growing on seleniferous soils actively accumulate selenium in their tissues at levels of 50 to 3000 ppm, and acute symptoms, including death, may result in grazing animals. Most grasses and other forage species passively develop lower but potentially toxic levels of 5 to 40 ppm. Animals consuming pasturage or hays and silages containing these lower levels of selenium over a period of several weeks slowly become poisoned and develop the malady referred to as **alkali disease.** Symptoms of this chronic illness include emaciation, lack of vigor, stiffness of the joints, rough hair coats, loss of long hairs, and cracking of the hooves, resulting in tender feet (Anderson *et al.,* 1961).

It appears that cattle have the ability to select against plants containing high levels of selenium but not against plants containing low levels; thus, lighter stocking rates providing greater opportunities to be selective should be followed. At lower levels of selenium in the soil, immature forage is generally higher in selenium levels than is more mature forage; this suggests that fall or winter grazing should be followed where possible (Minyard, 1961). Because alkali disease is a chronic form of poisoning resulting from accumulation of selenium in animal bodies over time, rotating animals biweekly between seleniferous and non-seleniferous pasturage may be a useful practice. All domestic livestock, and presumably big game animals as well, can be affected by selenium toxicity.

G. PLANT STRUCTURAL ANTI-QUALITY AGENTS

As plant tillers attain more advanced growth stages and become taller/longer and heavier, forage plants become more structurally complex. This process is accompanied by an increase in structural, anti-quality compounds such as lignin and a decline in crude protein, dry matter digestibility, and palatability factors (Northup and Nichols, 1998). With advancing growth and maturity, forage cells insert a noncarbohydrate material known as **lignin** into the cell walls. This complex compound gives additional tensile strength and rigidity to the plant but has negative nutritional consequences. Not only is the lignin mostly indigestible, but its presence also inhibits the availability of the associated cellulose and hemicellulose (Horrocks and Vallentine, 1999). Theories on the role of lignin in limiting fiber digestibility include the following: (1) interference with cell wall degrading enzymes, (2) direct toxicity to rumen microbes, (3) inhibition of microbial attachment, and (4) blockage of microbial access to potentially digestible cell wall tissue (Moore and Jung, 1999). (Since anti-palatability factors are also interrelated with anti-health factors, the reader is referred to treatment of anti-palatability factors in Chapter 9.)

Structural anti-quality factors in plants can be particularly detrimental to the ingestive abilities of the herbivore by reducing bite mass, bite rate, chewing efficiency, and chewing rate as well as depressing digestibility of associated nutrients. Plant structural anti-quality factors are related to higher tensile strength (can reduce bite area), modified canopy structure (can decrease intake associated with increased residence time in rumen), plant fibrousness (requires increased chews per unit of intake), leaf anatomy (vascular bundles and epidermis can be restraints on intake and digestion), and stemminess (increases time of prehension and reduces intake rate). Aspects of canopy structure that reduce intake rates include short stature, high density of ramets, and interspersion of palatable with unpalatable barriers (Laca *et al.,* 1999).

Tannins (proanthocyanidins), which function also as anti-palatability agents, can greatly reduce the digestion of fibrous materials. Their effects on fiber digestion include the inhibition of microbial enzymes that degrade fibrous polysaccharides, toxicity to fiber-degrading microorganisms, and formation of indigestible complexes with fibrous and proteinaceous substrates (Reed, 1999). Tannins occur in plants as a result of natural physiological processes but are more deleterious when associated with some plant species than others.

3

ANIMAL NUTRITIONAL BALANCE

I. FORAGE-ANIMAL PLANS

Forage-animal plans are combined forage and management practices directed to meeting the nutritional needs of ungulate herbivores in specific production phases or throughout a production cycle (Matches and Burns, 1985). A priority objective is to match forages with animal nutritional needs. In fact, nutritional requirements of the grazing animals should be given first consideration in planning a forage program. Unless forage can be managed to meet these needs, it is of little use to the livestock producer. Adams *et al.* (1996) concluded that when the cow and the range resource are well matched, the cow should receive most its needed nutrients from grazed forages. Even though matching animal requirements and nutrient supply from pasturage is often imperfect in practice, compensatory growth allows some deviation from a perfect fit (Riewe, 1981).

Based on beef cow production records for southern and southeastern New Mexico, Foster (1982) concluded that 50% more forage was required per cow in 1978 compared to 1925. This increase was based on increased size of cows, size of calves, and percent calf crop. Marked increase in potential growth and productivity of livestock has continued and has resulted from larger mature sizes, more rapid development, advanced growth and weaning weights, increased milk production, expanded use of exogenous growth stimulants, and accelerated reproduction; new techniques such as multiple births in beef females and genetically engineering high-gaining animals are also being developed (Bellows, 1985, 1988).

Biotechnology, including advanced techniques in animal genetics and reproduction, have made great strides in recent years. However, since most animal seedstocks have been selected largely on the offspring's performance in feedlot environments, these same seedstock are generally not highly efficient in converting grazable forages and other low-quality roughages (Heitschmidt *et al.*, 1996). These authors concluded that more attention to breeding and selection must be directed to converting grazable forages efficiently. The concept that livestock are merely large generalist herbivores should be replaced by concentrated efforts to genetically manipulate livestock to select diets that are most appropriate for the environment and management goals of the grazier (Walker, 1995). Burkhardt (1996) suggested that paying too little attention to "rangeability" has created "sedentary welfare cattle"; while such breed development and associated husbandry practices may have an immediate economic advantage, environmental sustainability as regards rangeland grazing is considered questionable.

The technology exists, according to Eller (1985), to breed and manage beef cattle to reach the following levels and goals: (1) a cowherd in which cows weigh 1100 lb, give birth to 70-lb calves, and wean calves at 6 months weighing 700 lb; (2) calves grazed for 3 months to achieve 900 lb at 9 months of age; (3) male calves carried through a finishing period to weigh 1200 lb at $11\frac{1}{2}$ months of age. Other achievable objectives listed include: (4) all carcasses at slaughter having ideal fat-to-lean ratios, having high cutability and quality, and being very uniform because their parents were full siblings (or otherwise closely related); (5) a designated 90% (or other desired proportion) of the cowherd producing only male calves and the other 10% (or other desired proportion) only females, each portion of the cowherd best equipped genetically to achieve their respective roles; and (6) breeding females, calving first at 18 months of age, producing a calf crop exceeding 100% (up to at least 150% with induced twinning) on a 350-day calving interval (or 300 day-interval with early weaning).

A. MATCHING FORAGES AND ANIMAL NEEDS

Advancements in livestock production potential carry with them increased nutrient demands, most of which are dependent on increased productivity and utilization efficiency of grazing resources, now commonly the limiting factor in this scenario (Fig. 3.1). The more productive grazing environments show more potential than many arid and semi-arid range environments to accommodate the increased maintenance and production requirements of high producing animals. To

FIGURE 3.1 With marked increase in the potential growth and productivity of livestock in recent years, grazing resources are now commonly the limiting factor in achieving these potentials; showing cattle used in grazing studies at the Texas Experimental Ranch, Vernon, TX. (Texas Agricultural Experiment Station photo by Rodney K. Heitschmidt.)

achieve these production goals, extensive use of improved pasture will generally be required to complement and supplement rangelands. (The use of complemental and supplemental pastures is elaborated upon further in Chapter 4.)

A key principle in developing forage-livestock systems is to utilize advantageously the inherent differences among forages in their pattern of seasonal production and nutritive levels (Matches and Burns, 1985). It is important to match the nutrient requirements of different kinds and classes of livestock with the nutritive value of the different sources of forage available for use. Successful grazing management must consider the type of livestock and their nutritive needs in relation to the seasonal quality of the forages.

Forage plant species that can be maintained high in digestibility or that offer plant parts, particularly leaves, of high digestibility through selective grazing are good choices for animal responses requiring high energy intake. Higher yielding forages that are lower in digestibility may be better choices where animal responses are less demanding (Matches and Burns, 1985). Paying an extra premium for the highest quality pasture, such as irrigated grass-legume pasture, will probably not be economical when only a maintenance ration is needed. Most forage plants adequately meet the nutrient requirements of some kind and class of livestock; however, for those classes of livestock having high nutrient requirements, fewer forage plants meet the needs. "Junk" feeds can be important in forage-livestock systems; these can include quackgrass (*Agropyron repens*) areas, stackyards with hay mats, weeds in wintering grounds, and corn fields previously harvested for grain (Salzman, 1983). However, junk feeds will often provide only maintenance rations and may even result in temporary weight loss in gestation cows and ewes.

Forage-livestock plans must be adapted to the changing nutritive requirements of animals as they move into different phases of production. The cyclic nature of reproduction in ruminant females and in the corresponding nutrient requirements (Fig. 3.2) results in the following critical periods: (1) development of the replacement females, (2) breeding and conception, (3) the last trimester of gestation, and (4) the postpartum period, including lactation, particularly for first-calving heifers (Bellows, 1985). Breeding beef cattle to calve first at 24 months of age is now widely considered the most economical (as compared to 3 or even 2.5 years of age), but enabling the 2-year-old cow to rebreed within 80 days after calving and to make continued growth requires careful nutritional monitoring (NRC, 1996). High-quality pasture will be required during these periods but also for young livestock at weaning and early post-weaning or when being finished for slaughter. The addition of energy and protein concentrates and harvested roughages may not only be required but also fully economical if strategically provided.

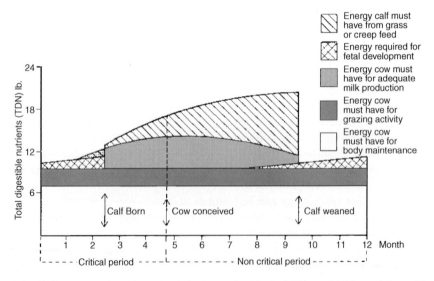

FIGURE 3.2   Estimated energy requirements of a mature, 1000-pound beef cow during a 12-month reproductive cycle, based on a 90-day calving season and 500-pound calf weaned at 7 months of age (Yates, 1980).

Cowherd management must be integrated with plant development, consistent with regional constraints (calving weather, pathogens and parasites, etc.), for best reproductive efficiency, minimum supplementation, and maximum nutritional advantage for cows and calves. In addition to females needing new grass for improved recycling, conception rate, and milk production, suckling calves need high-quality forage as they develop sufficient rumen function (Launchbaugh *et al.,* 1978). Immature animals, in particular, gain most on new growth, gain less as plants mature, and may lose weight thereafter. Creep feeding of nursing calves, or creep grazing or forward creep grazing, might be used to maintain or assure high gain by suckling calves.

Sims (1993) compared the common practice of spring calving and grazing cows year-round on native range with a fall calving system using a combination of native range and complemental pasture and concluded both systems were useful for the Southern Plains mixed-grass prairie. The traditional system utilized 20 acres of native range per cow-calf unit, while the combination system used 12 acres of native range and 1.5 acres of cropland double cropped with winter wheat and/or rye and warm-season annuals including forage sorghum (*Sorghum bicolor*), sudangrass (*Sorghum halepense*), or pearl millet (*Setaria spp.*). While reducing the land area required by 40%, the combination system provided nearly year-round green forage and a somewhat more consistent forage supply. The combination system also increased cow weight by 84 lb, calf crop by 6.7% (93.9 vs. 87.2%), and reduced the interval from breeding to estrus (295 vs. 304 days).

The alternative to changing the forage quality to meet the animal's changing needs is to change the livestock program to better coordinate with the changes that naturally occur in forage quality. The latter may require changing breeding and calving seasons, changing weaning ages (including early weaning of calves), or shortening the breeding season, but supplemental feeding during critical, high-nutrient demand periods may still be required (Vavra and Raleigh, 1976). Dormant vegetation in the fall or winter will fail to meet lactation requirements of fall-calving cows unless heavily supplemented but may adequately meet the requirements of dry, pregnant, spring-calving cows, except during periods of deep snows, drought, or other harsh weather (Fig. 3.3). Increased nutrient intake should be provided during the last trimester of gestation, and a combination of adequate body reserves and nutrient intake must carry through early lactation and breeding (Bellows, 1985).

Shortgrass range alone in eastern Colorado resulted in inadequate growth and development of replacement heifers for first calving at 24 months of age (Shoop and Hyder, 1976). Weaned heifer calves gained only .4 lb daily during their first winter on range. The March to May period on range was found often inadequate for reproduction and early lactation because of inadequate new herbage, poor quality of old growth, and excessive energy expenditures made in search of scarce green herbage. It was concluded the criteria for successful 24-month calving in the area should include: (1) increasing daily gains to 1.25 lb daily from weaning to 12 months of age, and (2) placing all 2-year-old heifers with first calf on gain-promoting forage and/or harvested feed during the March to May period.

FIGURE 3.3 Dormant vegetation on grass range in the fall and winter is incapable of meeting the nutrient needs of the cow-calf pair (shown above) but may those of the dry pregnant cow with minimum protein supplementation.

Adams *et al.* (1996) noted that 75% of the Nebraska Sandhills ranchers typically calved cows before March 10. Since this matches the highest nutrient requirements of the cows with the lowest nutrient value of the standing forage, this early calving has required significant inputs of harvested forages and concentrates. Assuming range is ready for grazing in early May, beginning calving from 2 weeks before (late April) to one month after (early June) the turnout date should allow reducing the hay requirement by one ton per cow compared to February-March calving. Delaying the calving date may also offer more opportunities to grow calves on a forage diet through their first winter and to graze as yearlings on range the next growing season.

Adams *et al.* (1996) also considered early weaning (September compared to November) to be an alternative for reducing the nutrient requirements of cows going into the fall when nutrient density is low in available range forage and improving body condition of the cows. Presumably, an additional alternative might be to set a more intermediate calving date and provide early cool-season pasture such as crested wheatgrass, Russian wildrye, or common rye for 30 to 45 days before moving the cowherd to native range.

From case studies of ranches recently shifting from late winter to mid-spring in Wyoming, May *et al.* (1999) concluded that mid-spring calving, even with lighter weaning weights (500 lb compared to 650 lb), was nevertheless generally more profitable. Factors favoring the mid-spring calving over late winter were: (1) a closer match between cow nutrient requirements and forage nutrient availability during late winter and early spring, (2) saving about one ton of hay per cow during this period by requiring the non-lactating cows to survive winters on minimal hay or purchased feed, (3) fewer health problems along with reduced veterinary and labor costs, (4) greater flexibility in choosing areas suitable for calving, and (5) a better price per pound generally received for the lighter spring calves.

While confirming that early weaning (September) of calves from late spring-calving cows in the Nebraska Sandhills was a viable practice, Lamb *et al.* (1997) found that grazing cow-calf pairs on subirrigated meadow regrowth for an additional 2 months (7 September to 7 November) before weaning was yet another alternative. When compared to cow-calf pairs grazed on native range during this 2-month extension period, the cows on subirrigated meadows gained an additional 62 lb, thereby improving body condition, and their calves gained an additional 76 lb. The diets of cows during this 2-month period was 12.3 and 7.6% crude protein and 71.1 and 55.1% *in vitro* dry matter digestibility, respectively, for cows grazed on meadow regrowth or on range. Where there was not enough subirrigated meadow regrowth or comparable high-quality forage to support both cows and calves, it was suggested that early weaning the calves and then grazing them on meadows while returning the dry cow to range would not only maintain calf gains but improve body condition of the cow as well.

B. FLUSHING

The practice of improving the nutrition of female breeding animals prior to and during breeding as a means of stimulating ovulation and reproduction is referred to as **flushing** and is commonly sought through feeding energy concentrates. Providing lush pasture during the post-calving and breeding period may be an effective alternative when used in a program of spring calving. However, neither approach to flushing has been universally helpful, suggesting that maintaining body condition at higher levels before parturition is equally or possibly even more important than flushing (Dziuk and Bellows, 1983).

Earlier return to estrus, improved conception and percent calf crop, concentrating calving in the forepart of the calving season and thus increasing average age and weight at a fixed weaning date, reducing length of breeding season needed down to 45 to 60 days, and enhancing milk flow and thus calf gains are some of the benefits attributed to flushing/breeding on high-quality pasture (Wiltbank, 1964; Houston and Urick, 1972; Hedrick, 1967; and Clanton *et al.,* 1971). Seeded, high-producing pasture of early growing, introduced grasses such as crested wheatgrass and Russian wildrye or irrigated grass-legume mixtures are means of providing an abundance of nutritious early spring forage, this often well in advance of many native grasses, particularly warm-season grasses (Fig. 3.4).

The advantages of special pastures for flushing will undoubtedly depend upon the available alternatives. From a study in Montana, utilizing seeded cool-season pastures from parturition through breeding did not improve the reproductive efficiency of beef cattle over those on native ranges including a large component of

FIGURE 3.4   Russian wildrye pasture being grazed by beef cows following calving for flushing at the University of Nebraska, North Platte Station.

cool-season grasses (Adams *et al.,* 1989); the seeded pastures gave no advantage in terms of calving date, occurrence of initial estrus, or fall pregnancy. It was concluded that native ranges in the Northern Great Plains in good condition because of not being overstocked or otherwise mismanaged and with a high component of native cool-season grasses are capable of producing forage of ample quality for achieving good reproduction. This suggests that timing the calving/pre-breeding season to begin with early rapid growth of native grasses can be a rewarding practice. Providing supplemental feed concentrates for purposes of flushing failed to improve the reproductive performance of Angora goat does in Texas when given access to high-quality range forage as an alternative (Hunt *et al.,* 1987).

Providing lush pasture for flushing fall-calving cows—pasture required December 15 to March 15 for August 15 to October 15 calving—is not possible in temperate climatic zones but might be approached under semitropical conditions. Difficulties are met in pasture flushing ewes on pasture in the fall for early spring lambing, the schedule to which sheep are mostly genetically or hormonally restricted. Subclover (*Trifolium subterraneum*) and hardinggrass (*Phalaris tuberosa stenoptera*) pasture mown prior to maturity and left in swaths to maintain good protein and energy levels has been successfully utilized in California for flushing ewes (beginning August 27, or 17 days prior to breeding and lasting through the first 17 days of breeding) (Torell *et al.,* 1972). This practice was similar to the flushing effects of concentrate feeding in drylot for increasing lambing percentage; it increased lambing percentage to 138% compared to 110% on dry, annual grass range. Similar or even improved results could be expected in areas where temperature and moisture permitted lush pasture to continue during late summer and fall.

C. PASTURAGE FOR FINISHING

An increasing demand for leaner meat produced at a lower cost per pound seems to assure that more pasture and harvested roughages and less grain and other feed concentrates will be used in the future than in the past in preparing cattle, and probably also sheep and goats, for slaughter. High-quality improved pasture has frequently been used in cattle and sheep finishing for slaughter. In some programs, finishing on pasture has been accompanied by increasing levels of supplementation until market weights have been reached. In other programs, growing and initial phases of finishing only have taken place on pasture, with the final phases of finishing (i.e., the last 30 to 100 days) taking place in drylot. Whenever pasture is utilized for high gains during growing-finishing preparation for slaughter, only pasture of the highest quality has generally been successful.

From studies in Colorado, Cook *et al.* (1981, 1983, 1984) concluded that acceptable beef could be produced directly from range if calves were kept on a growing diet of nutritious forage with minimum supplement until they were 18 months of age. However, animals on short-term, terminal drylot feeding systems following pasture made the most efficient gains and graded higher. They concluded that yearlings grazing native range plus crested wheatgrass in the spring, grazing forage sorghums in summer and fall without grain, and then being fed in drylot for the final 66-day period was the most efficient alternative in their study. Gains during the drylot period apparently benefited from compensatory gain after pasture. These cattle were more efficient in producing lean meat than those in the longer 97-day finishing period, produced carcasses grading mostly good or better (83%), required less fat trimming, and provided highly acceptable beef.

II. MONITORING GRAZING ANIMAL NUTRITION

Measuring the nutritional status of the grazing animal is a complex problem for both researchers and managers, and a rapid, cost-effective method to measure it is needed (Kothmann and Hinnant, 1987). Variation in both total daily ingestion and the nutritional content of ingested material is particularly high for free-grazing animals (Rittenhouse and Bailey, 1996). Information on the grazing animal's nutritional status has typically been determined by monitoring: (1) the chemical composition of the standing crop available to the grazing animal, (2) the intake and nutrient composition of the diet, (3) the grazing animal's physical measurements and performance, and (4) levels of nutrients stored in the animal's body (Anderson, 1987). Combinations of the above approaches may be required to make optimal grazing management decisions that affect the nutritional status of grazing animals.

A. SAMPLING THE FORAGE

While the total amount of edible forage available will primarily determine grazing capacity, the quality of the forage will greatly determine its effectiveness in promoting animal performance, providing the quantity available and corresponding intake are not limiting. Probably the best combination of quality measurement of forage is crude protein, energy value, and minerals of possible or probable deficiency. During extended dry periods such as drought, levels of carotene (precursor of vitamin A) may become urgent.

Methods of measuring dietary composition and nutrient intake have been reviewed by Holechek *et al.* (1982b) and Cook and Stubbendieck (1986). However, most of these procedures are experimental in nature and of limited or no direct use in applied grazing land management. As discussed previously, generalized nutrient levels and trends in forages provide some help in predicting animal nutritional status. The following discussion is geared more to providing a background rather than precise management procedures for sampling the forage and nutrient intake of grazing animals.

When samples for nutrient evaluation are taken from the forage stand being grazed rather than from the actual ingesta, forage samples must be taken that are representative of what the grazing animals are eating rather than what is available to them. Dougherty (1991) concluded that when the herbage in the grazing horizon from which grazing animals are removing forage is quite uniform in quality, which is often found in intensively managed improved pastures under rotational grazing, the metabolizable energy content of the ingesta will probably be very close to the metabolizable energy content of the available herbage. However, in less intensive grazing systems (i.e., rangelands), the metabolizable energy content of the ingesta will likely be considerably higher than the average of the sward.

Grazing animals, particularly in heterogenous vegetation, must be carefully observed to determine what plants and plant parts are currently being eaten if clipped samples are to be reliable measures of the grazing animal's diet. Bulk sampling by entire plant clipping or mower strips will seldom yield the required information. Better techniques that simulate what is being selected by the grazing animal include: (1) collection of new growth from selected plant species, or (2) plucking samples believed to represent the grazing animal's diet. However, each of these techniques is only an indirect measure of the forage actually being consumed at any given time.

The concentration of digestible nutrients in the diet is nearly always greater than the average of the standing crop of forage. Grazing animals are highly selective when given the opportunity (Launchbaugh *et al.,* 1978), consuming only certain plants and certain portions of the plants available to them. Since grazing animals select the greener, finer, leafier, and thus more nutritious plants and plant parts, total clipping of standing forage plants to ground level will underestimate the nutritive content of the grazing animal's diet. The result is that ingesta can be expected to be higher in both digestible protein and energy but also generally in essential minerals (Gengelbach *et al.,* 1990).

Range and most pasture vegetation is highly heterogeneous and dynamic across space and time, and grazing animals select diets much different from the average of what is available to them (Kothmann and Hinnant, 1987). Estimates of available forage by plant species, the consumption by the animal, and the contribution of the forage to the animal's diet must be synchronized with each other in the same time frame (Currie, 1987). Measuring the quality of standing forage available immediately prior to and during the grazing period will provide the most useful information. Repeated sampling at periodic intervals to reference frequent dietary changes is required. However, the lag time between sample collection and return of analyzed results for the laboratory must be short (probably under 7 days when plant nutrient levels and/or diet are rapidly changing) for any practical decision making (Holechek and Herbel, 1986).

Data collected during a $2\frac{1}{2}$-year period on semi-desert grassland range near Tucson (Cable and Shumway, 1966) showed that rumen protein varied from 1.53 to 2.91 times that of the clipped whole-grass samples. This higher protein content in the rumen was attributed to selective grazing by the steers for green parts of grasses rather than the whole plants and for high protein browse and annual forbs when they were available. Selectivity also enhances the nutrient intake from intensively managed, improved pasture. Botanical composition of first-day samples of irrigated, alfalfa-orchardgrass pasture was satisfactorily measured by both hand-clipped and esophageal fistula samples early in the growing season, but great disparity was found between the two methods near the end of the grazing period (Heinemann and Russell, 1969). It was concluded that hand-clipping forage samples—to about 2.75 in.—provided good estimates of available forage but only the esophageal samples measured the forage being selected by the grazing animal.

Esophageal fistula collection is considered the standard for diet analysis for grazing animals (Kothmann and Hinnant, 1987). The use of an esophageal or rumen fistula permits direct sampling of the diet but is available only for research purposes and establishing nutritional relationships; the procedures are too complex for routine monitoring of animal diets (Fig. 3.5). Although the esophageal fistula provides reliable information on the nutritive content of ingesta, it has some technical limitations. Salivary contamination of fistula samples, particularly in respect to phosphorus, is a problem that cannot be avoided, but statistical procedures are available for correcting the data. Steps must be taken to avoid rumen contents contamination of the fistula sample through regurgitation and when obtaining representative samples over large grazing units. Diet samples taken from a rumen fistula is another means of determining quality of ingesta shortly after grazing, but the samples are always subject to contamination from other rumen contents unless total evacuation of rumen solids and even fluids is done in advance of collection. Using an artificial rumen (providing a fermentation-digestive environment including actual rumen inocula) simulates actual rumen function and can be used to measure energy value.

Evaluating forages by proximate analysis—i.e., dry matter content, crude protein, ether extract, ash, crude fiber, nitrogen-free extract, and total digestible nutrients (TDN)— has been shown to have serious limitations, particularly in regard

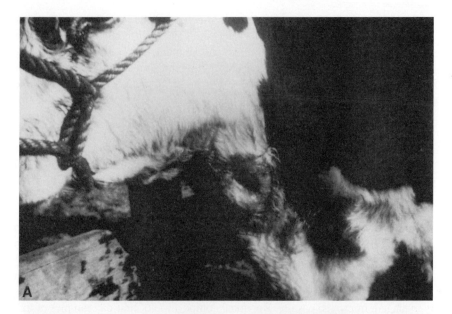

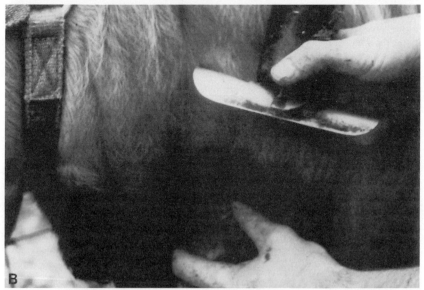

FIGURE 3.5   The esophageal fistula is a useful research tool in measuring the nutritional and botanical composition of the grazing animal's diet; A, fistula (opening) surgically placed in animal's throat; B, preparing to insert the canula and plug to close the fistula when not collecting; C, experimental animal prepared for forage collection by removing the canula and plug and attaching a collection bag under neck.

FIGURE 3.5   *(Continued)*

to energy value (Horrocks and Vallentine, 1999). However, differentiating be-tween the energy components of cell contents and cell walls aids materially in de-termining the relative energy value of forages and other feeds as well.

Cell contents are highly digestible to both ruminants and nonruminants (Table 3.1) and comprise most of the protein, starch, sugars, lipids, organic acids, and sol-uble ash of forages. The sugars, starch, pectin, and other soluble carbohydrates are almost completely digestible to all animals. The proteins, non-protein nitrogen, lipids (fats), and other solubles have high digestibility to all animals.

In contrast, cell walls are the less digestible portion of the plant cell. Making up a large part of forage (40 to 80%), cell walls are a complex matrix of cellulose and hemicellulose, lignin, some protein lignified nitrogenous substances, waxes, cutin, and minerals that resist normal digestive processes such as silica (Van Soest, 1982; Hatfield, 1989). Cellulose and hemicellulose are partially digestible to ru-minants and horses but have low digestibility to most other nonruminants. Heat-damaged protein, lignin, and silica are mostly indigestible to ruminants and non-ruminants alike. Neither ruminants nor the horse produce the enzymes necessary to digest cellulose and hemicellulose in forages per se, but microbial populations within their digestive systems are able to break down these components through fermentation so that normal digestion can then occur.

Basic differences in structure and chemical composition exist between the cell walls of grass leaves, particularly as maturation develops, and those of forbs and browse. Grass leaves have a thicker cell wall containing potentially digestible structural carbohydrates such as cellulose. The thicker more fibrous cell walls of

TABLE 3.1 Classification of Forage Fractions Using the Van Soest Method

Fraction	Components included	Nutritional availability	
		Ruminant	Nonruminant
Cell contents	Sugars, starch, pectin	Complete	Complete
	Soluble carbohydrates	Complete	Complete
	Protein, non-protein N	High	High
	Lipids (fats)	High	High
	Other solubles	High	High
Cell wall (NDF)	Hemicellulose	Partial	Low
	Cellulose	Partial	Low
	Heat-damaged protein	Indigestible	Indigestible
	Lignin	Indigestible	Indigestible

(After P. J. Van Soest, 1967. Development of a Comprehensive System of Feed Analyses and Its Application to Forage. *J. Anim. Sci.* **26**(1):119–128).

grasses make grass more difficult and energy expensive to fracture (bite and chew) than the more fragile leaves of browse and forbs (Shipley, 1999). In contrast, the leaves of forbs and leaves and stems of many woody plants have thinner cell walls and more cell contents. However, the thinner cell walls of browse contain more indigestible fibers such as lignin which interfere with digestibility but presumably permit a more rapid flow of indigestible food particles through the rumen of smaller browsing animals while promoting higher forage intake.

In order to differentiate cell contents and cell-wall fractions of forages and the components of the cell wall and thereby more accurately estimate energy values, a wet chemistry method referred to as the **detergent method** or the **Van Soest method** was developed by USDA (Van Soest, 1967). A functional comparison of the proximate and detergent systems is provided in Fig. 3.6. In the detergent method of analysis, cell contents are referred to as **neutral detergent solubles (NDS),** while the remaining insoluble portion is referred to as **neutral detergent fiber (NDF).** After removing the hemicellulose from NDF, the remainder consists of cellulose and lignin (also silica unless ashed) and is referred to as **acid detergent fiber (ADF).** Removing the cellulose leaves only indigestible **lignin** and silica (unless previously removed by ashing). Energy values, estimates of digestibility, and relative feed values reported on laboratory analyses are calculated using the ADF and NDF content of the forage (Horrocks and Vallentine, 1999).

Daily feed intake on a dry matter basis must be measured or accurately predicted to determine total intake of the various nutrients. Complete recovery of ingesta with an esophageal fistula to determine daily dry matter intake is impractical; collection periods must generally be limited to not over 30 minutes and even during such short collection periods some ingesta will bypass the fistula and not be collected. Dry matter intake during grazing sessions can be determined by ru-

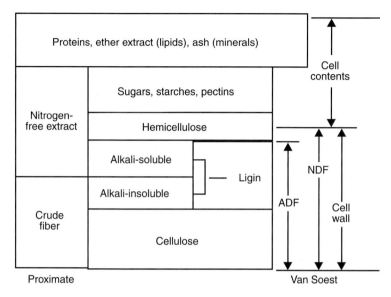

FIGURE 3.6 Forage analysis showing proximate (*left*) vs. Van Soest (*right*). (Redrawn from Holland *et al.* (eds.), 1990).

men evacuation techniques. However, this technique, as well as those that combine the measurement or prediction of total fecal output with a determination of ration indigestibility to estimate total dry matter intake, are experimental techniques not generally available for making management decisions.

Fecal nitrogen is directly related to both intake and digestibility of the diet and is generally correlated with dietary nitrogen. Kothmann and Hinnant (1987) have concluded that fecal nitrogen shows great promise as a reliable, rapid, inexpensive indicator for monitoring the nutrient intake of the grazing animal. If so, it could be used with either domestic or wild animals; however, fecal nitrogen must first show consistent reliability, and formulae or models for direct management application must be developed. Technology has also been developed using near-infrared reflectance spectroscopy to directly estimate nitrogen levels in feces and nitrogen, fiber, and digestibility in forage on site (Holechek and Herbel, 1986).

B. MEASURING ANIMAL PERFORMANCE

Since the animal itself is the final integrator of dietary effects, it has been suggested that the complete expression of adequate forage quantity and quality for the grazing animal should come from evaluating the animal itself (Anderson, 1987). Weight gain/losses; yields of milk, meat, and wool; reproductive rates; and blood values have been used as indirect indicators of the nutritional status of grazing animals (Cook and Stubbendieck, 1986); all are live animal measures and do not require destruction of the animal. However, these indicators measure primarily recent/past performance, thereby reflecting recent/past nutrition, and may not be indicative of the current and projected nutrient intake of grazing animals (Kothmann and Hinnant, 1987).

Free-grazing animals, particularly on extensively managed rangelands, experience periods of under-nutrition, over-nutrition, and compensatory nutrition in the process of attempting to meet requirements during their life cycle (Rittenhouse and Bailey, 1996). Prior nutrition of animals often has substantial impact on how they later gain or produce. **Compensatory gains** are subsequent gains that are enhanced (or depressed) as a result of gains during a prior period. When animals previously deprived nutritionally, but otherwise healthy, are placed on higher quality or quantity of feed, their subsequent gains and gain efficiency are generally greater than had they previously been fed on a higher plane of nutrition.

The National Research Council (1996) suggested that a reduction in maintenance energy requirement for a compensating bovine was a reasonable generalization and that this improved energy efficiency would typically last 60 to 90 days. For example, a practical application of compensatory gains is in restricting drylot gains during an overwintering period so that gains and gain efficiency are enhanced when growing animals are subsequently placed on high-quality pasture. However, these higher pasture gains may be largely lost if growing animals are later returned to near maintenance rations.

Compensatory gains of growing cattle under grazing are evident in many grazing studies. In one Kansas study, steers wintered on bluestem range gained 111 lb, while equivalent steers fed on a higher plane of nutrition in drylot gained 147 lb (Smith, 1981). However, their respective summer gains were 266 vs. 232 lb respectively, thereby narrowing the end of summer advantage of the steers wintered in drylot. Steers fed to gain 252, 146, and 42 lb during an overwinter period gained 153, 202, and 243 lb, respectively, during the subsequent summer when grazed on a moderately stocked midgrass range near Fort Hayes, KS (Launchbaugh, 1957). Steers from the high overwintering ration gained only 85 lb during the summer under heavy stocking rates; steers from the low level of overwintering gained 258 lb during the summer under light stocking. Thus, the combination of level of overwintering and subsequent summer stocking rates accounted for a maximum difference of 173 lb gain per head on summer pasture.

Similar results were obtained from orchardgrass-Ladino clover (*Trifolium repens*) irrigated pasture in California (Hull *et al.,* 1965). Steers overwintered at a low rate of gain (0.77 lb daily) gained 1.68 and 0.84 lb daily, respectively, under light and heavy stocking rates during the following summer. Steers overwintered at a high rate of gain (1.75 lb daily) gained only 1.15 and 0.45 lb daily, respectively, under light and heavy summer stocking rates. The low-winter to high-summer group during summer grazing converted forage to animal gain at a ratio of 11.0:1, while the high-winter to low-summer group converted at a ratio of 28.3:1. When short yearlings were overwintered in Nebraska in drylot at average daily gain levels ranging from 0.62 to 1.09 lb, it was found that for each additional pound of winter gain there was a compensatory gain reduction on summer grass pasture of 0.81 lb (Lewis *et al.,* 1988); this suggested the 0.62 lb daily gain was the optimal winter gain level. It is apparent that moderately low energy intake by growing cattle previous to grazing allows more latitude in stocking rates on pasture the following grazing season.

A principle in the applied nutrition of growing-finishing animals is continual improvement in dietary quality once the drive towards market condition begins, and this begins at birth in some systems. In contrast, compensatory gains and losses tend to net out to zero in the annual weight/condition cycle of the mature reproductive animal; this allows the loss of weight and body condition in non-critical reproductive periods to be restored in other periods. Mature, pregnant cows in good condition can lose 10% of body weight during winter and still produce 90% calf crops or more if they can gain weight after calving (Holechek and Herbel, 1986). In fact, it is generally accepted that the most economically efficient beef cow systems are those that allow cows to lose weight and condition over winter when feeding is expensive and regain weight and condition while grazing relatively cheap pasturage during the plant-growing season.

Compensatory gains are not unique merely to all species of domestic livestock but to big game species as well. Big game species in the wild naturally lose body condition and weight over winter only to regain their winter losses by grazing new plant growth beginning in the spring. Compensatory gains were readily demonstrated in an elk ranching study in Alberta with weaned males calves (Wairimu and Hudson, 1993). One group of elk calves was overwintered at a low nutritional level (native range plus medium quality hay) and compared to a second group overwintered at a high nutritional level (native range plus hay and alfalfa-barley pellets). The low-level elk calves averaged 33 lb lighter than the high-level calves at the end of the overwintering period in April. After being combined into a single herd and grazed on rapidly growing spring-summer pasture, compensatory gains resulted in the average weight of calves in both groups equalizing at about 420 lb by July.

Body condition scores correlate well with stored energy in the animal body and also with reproductive efficiency and milk production. A review of the historical background of body scoring with cattle and a comparison of the several scoring systems proposed or then in use were made by Anderson (1987). Techniques for body scoring based on visual appraisal or external palpation for subcutaneous fat have been developed and have become effective and rather widely used by both the producer and researcher. Prepartum condition in the reproductive female can be used to effectively manage fat and thin animals according to their nutritional needs for optimum sustained production.

A body condition scoring system recommended by Herd and Sprott (1986) utilized nine categories: 1–3, thin condition; 4, borderline; 5–7, optimum condition; and 8–9, fat condition. Guidelines were provided for achieving and maintaining adequate body condition through nutrition during the last third of pregnancy and lactation for Texas conditions based on utilizing maximum forages and minimal supplements. Drought and overstocking pastures were given as common causes of poor body condition and reproductive failure. NRC (1996) has adopted the cow body condition scoring of Herd and Sprott with modification and has generally accepted a body condition score of 5 ("moderate") as adequate at conception. Bennett and Wiedmeier (1992) have concluded that body condition scores for the range cow are a somewhat more sensitive measure of dietary nutritional change than body weight, are much easier to obtain, and are much more sensitive than the suckling rate of gain of calves, as the calf's gain is somewhat buffered by milk consumption from the cow.

Techniques for measuring live animal performance in wild ungulates as related to nutritional status are more difficult and more limited since they are usually not under direct control (Kie, 1987); tamed animals are exceptions but are applicable only to research, except possibly for game ranching. Determining condition in wild big game animals is mostly limited to the use of live weights, reproductive rates measured by non-destructive techniques, and blood values. Measures of performance can also include indicators of condition and health such as fat reserves and parasite loads. Anderson (1987) included hair analysis, blood analysis, urine analysis, bone analysis, milk analysis, and body water as a fat estimate as indirect measures of the nutritional status of grazing animals, either big game or livestock, but some of these measures are performed on the carcass rather than live animal.

Leckenby and Adams (1986) have suggested weather indices be used in projecting the performance of free-ranging domestic and wild ruminants. They utilized a weighted index of temperature, wind, and snow cover to reflect episodes of positive and negative energy balances of free-ranging deer to aid scheduling of feeding programs and planning cover-forage manipulations.

III. SUPPLEMENTATION PRACTICES

A. OBJECTIVES OF SUPPLEMENTATION

Additional feedstuffs beyond grazable forage are fed to grazing animals for various reasons—improve forage utilization, provide supplemental nutrients, improve animal performance, provide additional carrying capacity (substitution for forage), or stretch forage supplies (Lusby and Wagner, 1987). Other reasons for feeding feedstuffs beyond pasturage to grazing animals are to provide a carrier for growth promotives, to aid in preventing or treating certain health problems (i.e., poloxolene or other medicants), to enhance cattle management for gathering for checking or moving, and to teach calves to eat supplements on pasture before weaning and preconditioning.

Since energy expenditure associated with the physical work of grazing can be substantial, Caton and Dhuyvetter (1996) have suggested that "supplemental" feeding of grazing animals in some cases may reduce the amount of time spent grazing, often without reducing forage intake, thus reducing maintenance energy requirements. However, the effect that providing harvested feeds and concentrates has on forage utilization and the performance of grazing animals is complex and largely determined by supplement composition, the quantity of supplement fed, and the quality and quantity of the forage being supplemented.

The term **supplement** more precisely refers to feedstuffs high in specific nutrients (protein, energy, phosphorus, salt, or other nutrients) intended to remedy deficiencies in the grazing animal's diet or other basal ration, thereby balancing animal diets. Maximum efficiency of diet utilization results from providing nutritionally balanced diets, and performance is limited to that which is supported by the first-limiting nutrient (NRC, 1996). For example, when energy is first limiting, protein, minerals, and vitamins are not efficiently utilized; supplemental protein, in this case, will primarily be used for energy until energy and protein are equally limiting. If protein is first limiting, providing additional energy will not improve performance, may substitute for part of the normal forage intake, and may even depress performance. Since the first-limiting nutrient in low-quality forage (less than 7% crude protein) is often protein, the best approach may be to supplement with rumen-degradable protein for increasing both total protein and the potential energy supply (Kansas State University, 1995), with the limitation that protein may be an expensive source of energy.

Supplements are generally concentrates or less commonly nutrient-rich harvested roughages such as alfalfa hay or even pasturage of exceptional quality (i.e., **supplemental pasture**) grazed simultaneously with low-quality pasturage. When nutrient levels are marginal in grazed forage, any reduction in forage intake associated with low palatability, digestibility, or availability may cause dietary deficiencies not otherwise encountered. Thus, a feedstuff such as alfalfa hay fed on dry pasture may bring up dry matter intake while providing supplemental protein.

Feeding long-stem or pelleted alfalfa (typically 20% crude protein or higher) on a daily or alternate day basis to provide supplemental protein has been similar to a protein concentrate in its effect on performance, grazing behavior, forage intake, or diet digestibility of beef cattle grazing winter range in eastern Oregon (Brandyberry *et al.,* 1992). Also, high-quality meadowgrass hay, typically 16% crude protein, such as produced in subirrigated meadows in the Nebraska Sandhills, was equally effective as soybean meal-based concentrates for supplementing gestating beef cows grazing native winter range (Villalobos *et al.,* 1992, 1997). Still another consideration is that fall regrowth of native range consisting of a high component of cool-season grasses may provide a high-protein forage for winter grazing, complementing if not supplementing the standing dead forage component. It was concluded that providing a protein supplement to mature cattle on winter range comprised of bluebunch wheatgrass (*Agropyron spicatum*) and Idaho fescue (*Festuca idahoensis*) may be of little economic benefit in mild winters when there is substantial fall growth (Houseal and Olson, 1996).

On the other hand, it is common practice to replace part or all of the grazing resource with harvested forages (or less commonly concentrates) on a regular basis for livestock as part of a year-round forage program or when the standing crop is inadequate in supply (preferably referred to as **maintenance feeding**). A related term, **emergency feeding,** refers to supplying such feedstuffs when the available standing forage crop is insufficient because of heavy storms, fires, severe drought, or other emergencies. However, maintenance and emergency feeds are mostly basal roughages fed to increase carrying capacity rather than enrich the basal diet. Special drought practices related to providing supplemental and emergency feeds are suggested in "Meeting the Lows" in Chapter 13.

The benefits from additional feedstuffs fed to grazing animals may be substantial during significant stress periods (e.g., severe winter, drought, extreme weed infestation, or heavy stocking rates), but under more favorable conditions the usefulness of supplementation of protein and energy should be closely monitored and questioned (Cochran *et al.,* 1986). Supplemental feeding of protein and energy represents one of the major variable cash costs in livestock production, and sometimes even in game ranching, and must be carefully controlled; unnecessary supplements or unnecessary additives added to the needed supplement often only increase costs. Torrell and Torrell (1996) concluded that the economics of supplemental feeding for added weight gain and growth has been variable and requires continual re-evaluation as range and pasture conditions, type of feed, beef prices, and supplemental feed prices change. In contrast, the economics of supplemental feeding of the cowherd to maintain body condition and reproductive potential was more obvious, and the economic risk was considered often too great not to supplement.

The following four rules should govern the supplementation of grazing animals: (1) supplement for proven or probable deficiencies in the diet only, (2) use only supplements that are profitable or otherwise meet priority objectives, (3) provide supplements so each animal in the herd or flock gets its share, and (4) use supplemental feeding methods that keep the grazing animals rustling (see Chapter 6) and well distributed (see Chapter 8).

Winter feeding of big game animals—often emergency feeding rather than supplemental feeding—is generally recommended only when absolutely necessary to prevent massive die-off and not as a general practice. Reasons for not providing supplemental feed to big game routinely during winter, in addition to excessive cost, according to Olson and Lewis (1994), include: (1) adverse animal physiological problems (i.e., too late to help); (2) increases in disease transmission, such as brucellosis, from animal concentrations; (3) impact on concentrating animals at feeding sites and on adjacent natural habitat; (4) interference with the natural selection process (i.e., genetically inferior animals survive); and (5) stress associated with the feeding process (e.g., human contacts, dogs, transportation, or reduced shelter).

Urness (1980) has favored winter feeding of big game animals only under the following conditions: (1) when necessary to reduce land-use conflicts that cannot be resolved any other way, i.e., keep animals off private property where not wanted; (2) in unusually severe winter weather such as deep snow, but only in situations of a limited and temporary nature; and (3) as a substitute for lost winter range resulting from such circumstances as urban sprawl and not habitat deterioration per se. Developing existing winter habitat and improving forage production or reducing the herd to habitat capacity are apt to be better expenditures of funds earmarked for big game winter feeding (Urness, 1980; Olson and Lewis (1994). However, supplementation and maintenance feeding practices similar to those in livestock production may be realistic under intensive game ranching practices.

Olson and Lewis (1994) suggested that winter feeding of big game animals be based on body fat reserves. They considered winter feeding to be best for animals with marginal fat reserves going into winter; if conditions become severe enough, survival may depend upon winter feeding but it must be started soon enough. The authors found that animals entering winter with high fat reserves were likely to survive regardless of winter severity or winter feeding, and animals entering winter with extremely low fat reserves would probably not have survived anyway. These authors concluded that the condition and quality of summer/fall range, which directly affects the amount of stored fat reserves available to supplement forage intake during winter, may have as much influence on winter survival as does the quality of the winter range. Further, these authors noted that both mule deer and elk typically lose substantial body weight on winter range regardless of forage conditions, their winter survival depending on: (1) amount of stored fat reserves, (2) rate of fat reserve used as influenced by available forage, and (3) the degree of stress from cold temperatures and human disturbance.

B. UNIFORMITY OF INTAKE

Protein and energy (grain) supplements cannot be rationed precisely to individual grazing animals unless complex and costly equipment is provided. The amount of supplement consumed by different animals, in both even-aged and uneven-aged herds, will generally vary greatly and often inversely to animal needs unless special precautions are taken (Morley, 1981; Allden, 1981). Based on an extensive review of literature, Bowman and Sowell (1997) concluded that variation in individual supplement intake exists for cattle and sheep almost regardless of the supplement form or method of delivery.

Livestock inexperienced in taking supplement, particularly young animals, must be encouraged or trained to take supplement; training should be done using a palatable, limit-fed, low salt-content meal for a week or so. Even then a few may refuse to take supplement, either from neophobia to feed or feed-delivery devices or in response to more aggressive animals in the group. Wagnon (1965) found that the average response for mixed-age cows to take supplement was 83.9% when called, but it rose to 94.8% when the cows were called and started or driven towards the supplement. The percentage coming to supplement was less (1) when there was green regrowth available in the standing crop of range forage; (2) when the cows were 2 to 3 years old, in contrast with older cows; and (3) during and after calving for several days.

Although strong social dominance is seldom shown by female sheep when group supplemented, it is readily observed with cattle taking salt or supplement or even when receiving palatable roughage such as alfalfa hay. The younger, smaller, less robust, and/or less aggressive animals in a cattle herd—or less often in a sheep flock—are apt to be deprived of supplement by the others, particularly when small amounts of supplemental feeds are group fed each time.

Wagnon et al. (1959) and Wagnon (1965) found serious social dominance as related to supplemental feed intake in range cattle in California. Hand feeding supplements to a mixed-aged cowherd resulted in many 2- and 3-year-old cows being driven from the feeders before they had an opportunity to eat supplement. As a consequence, the younger cows, because of their lower dominance, suffered greater weight losses than similar animals of the same age pastured and supplemented separately from the older cows. Dominance was favored by older age cows, heavier weight, more aggressiveness, more agility, and less timidity (less afraid of other cows). Competition from cows and even from their own dams may largely deny the calves access to supplement when fed together. In contrast, little expression of dominance was found within weaned calf-yearling groups supplemented together on pasture.

When supplements are handfed, providing adequate feeder space is very important. Also, grouping livestock first by species and then into age classes aids in each animal more nearly getting its fair share (i.e., the calculated average need of supplement). A suggested division for cattle is: (1) weanling calves; (2) yearlings, by possibly growing steers separate from replacement heifers if their supplemental needs are different; (3) young cows calving as 2 and 3 year olds; (4) older cows; and (5) bulls, unless during the breeding season.

Other practices suggested for improving uniformity of supplement intake between individuals in group feeding are using feeding intervals less frequent than daily, using a salt-meal mix to limit daily supplement intake when fed free choice, using a lick-wheel feeder containing liquid supplement, using hardness in a supplement block to reduce supplement intake, or spaced placement on the ground (Fig. 3.7). A condensed molasses block (32% crude protein) proved to be an effective method of limiting daily supplement consumption with beef calves grazing bermudagrass (*Cynodon dactylon*) pasture during the summer in Texas (Grigsby et al., 1988); the consistency of the blocks was not adversely affected by the high temperature or unseasonably high rainfall.

C. EXTENDED FEEDING INTERVALS

Less frequent feeding intervals, ranging from every other day to weekly, have proved to be an effective practice for obtaining uniform intake of high-protein supplements with grazing cattle or sheep. This is accomplished by group feeding a 2-day allotment of supplement every other day, a 3-day allotment every third day, or a weekly allotment every seventh day. This practice has been verified by extensive research (Adams, 1986; Duval, 1969; Huston et al., 1997, 1999; McIlvain and Shoop, 1962b; Pearson and Whitaker, 1972; Pope et al., 1963; Smith, 1981; Thomas and Kott, 1995; Thomas et al., 1992; Wallace et al., 1988) and has be-

FIGURE 3.7   Equipment developed by the Fort Keogh Livestock and Range Research Laboratory, Miles City, MT, to improve uniformity of supplemental intake by range cattle by spaced placement on the ground includes: (1) a cube drop unit mounted on a pickup, and (2) a mobile unit (shown in background) for transporting large bales or small stacks of alfalfa hay and slicing into segments for dispensing.

come widely used in practice, even to the extent of use as a research tool to maximize uniformity of supplement intake (Cook and Stubbendieck, 1986).

Using extended feeding intervals with high protein supplements has caused no digestive disturbances, has been nutritionally effective, has resulted in similar to improved performance compared to daily feeding with all ages and classes of cattle, has not affected forage consumption, and has had positive effects on grazing activity. The practice has also saved labor, commonly 40 to 60% over daily feeding. The advantages of feeding protein supplements—either protein concentrates or alfalfa hay—over extended intervals has resulted from prolonged or even multiple feeding bouts in which socially dominant individuals have been largely unable to dominate access to the supplement. The practice could presumably be adapted to other species of grazing ruminants as well as cattle and sheep. Even with less frequent feeding, the amount of supplement consumed by individuals probably still varies considerably (Morley, 1981), and other practices such as gathering the animals together in advance, encouraging reluctant ones to eat, assuring adequate feeder space or drop-spacing on the ground of individual animal allowances, and separation by species and class are encouraged.

When substantial amounts of non-protein nitrogen (NPN) are substituted for true protein to provide nitrogen equivalency in a supplement, extended feeding intervals should be used with caution. Foster *et al.* (1971) found that a barley-biuret supplement, fed at a daily equivalent of 2 lb, was more effective in promoting gains of growing cattle when fed daily to growing cattle on crested wheatgrass compared to alternate days or every fourth day. In fact, the value of NPN as a substitute for "natural" protein in low-protein, high-forage ruminant diets has been questioned by Clanton (1979), and NRC (1996) has advised caution in using NPN in such diets. It is known that lipids provide little if any energy for ruminal protein synthesis, and the energy obtained from protein degradation is minimal for this purpose. Kansas State University (1995) has recommended not using NPN such as urea for over 15% of the total supplemental crude protein for range livestock.

Feeding high-energy, low-protein supplements less frequently than daily to grazing animals has not been successful. A grain cube fed twice a week to yearling range heifers on dormant range in New Mexico rather than daily was less effective in growth rate and reproductive performance (Wallace *et al.,* 1988). On fall-winter range in Montana, the supplemental feeding of cracked grain to range cows at the equivalent rate of 0.3 lb/100 lb liveweight daily was compared under daily and alternate-day feeding (Kartchner and Adams, 1982; Adams, 1986). During the 70-day study period, the cows fed 3.3 lb daily gained 142 lb and tended to improve in body condition, whereas the cows fed 6.6 lb on alternate days gained only 69 lb and tended to decrease in body condition. Since the time spent grazing did not differ between the two groups of cows (8 hr/day average), the decreased performance of alternate-day grain supplementation was attributed to less favorable rumen conditions for fiber digestion.

D. SALT-MEAL MIXES

Both high-protein and high-energy concentrates may be self-fed to grazing cattle by mixing in adequate amounts of loose salt (30 to 50% of the total mix) to limit intake. This permits self-feeding either in open or covered feeders. The mix can be put out biweekly or weekly, and the necessity of bunching animals is eliminated since they can come in individually to feed. Plus, it provides a means of adding phosphorus, vitamin A, or other feed additives. Self-feeding a salt-meal mix also provides minimal opportunity for socially dominant individuals to control the feeders and prevent timid animals from eating. Idling "boss" cows in California range studies (Wagnon, 1965) were sometimes observed keeping other cows from coming to the self-feeder for short periods of time but would eventually return to grazing, allowing the others to move up to the feeders. Providing several feeders may further reduce the potential for dominance being expressed.

Cattle consuming salt-meal mixes commonly ingest 0.25 to 1.25 lb of salt daily in contrast to normal consumption of 2 to 4 lb per month. These levels of salt have seldom proven toxic to livestock, and then almost always in limited situations when animals were prevented free and unlimited access to drinking water (e.g., water tanks frozen over, watering places gone dry, animals accidentally locked away from water, or animals isolated by blizzard conditions near salt-meal feeders and away from water). An abundance of drinking water is necessary for eliminating the excess salt from the animal's body. Lusby (1983) recommended the following safeguards in using salt-meal mixes: (1) do not allow salt-hungry animals sudden, ready access to salt-meal mix; (2) do not force cattle to eat large quantities of the mix with an inadequate water supply; and (3) do not force cattle to drink water with high salt concentration in addition to consuming the salt-meal mix.

As a rule of thumb, cattle on salt-meal mix drink 50 to 75% more water than normal, or approximately 5 gal of additional water daily for each pound of salt consumed (Rich *et al.,* 1976). This increased water intake is disadvantageous when water supplies are limited or must be hauled or during extremely cold weather when the large intake of cold water must be warmed by an extra expenditure of body energy for heat production. The higher water intake associated with force feeding salt, however, provides a means of flushing out the urinary tract of male cattle or sheep and reducing the incidence of urinary calculi.

Salt is an economic regulator of meal intake similar to the use of extended feeding intervals but additionally includes the cost of the salt. Salt is not a precise regulator of intake since certain animals tolerate more salt than others; also, the salt level in the mix must be varied as required to regulate meal intake at desired levels. Animals should be familiar with eating supplement or be trained before being allowed access to salt-meal mix. The possibility of ill effects from sudden heavy consumption of salt-meal mix by inexperienced animals can be minimized by increasing the proportion of salt from a sprinkling at the outset with limit feeding to the quantity of salt required to regulate self-feeding, starting the practice by daily hand-feeding and gradually working into self-feeding. Coarse-ground, white salt and not a trace-mineralized salt should be used.

Lusby (1983) concluded that high salt intake will generally have no deleterious effects on fertility, calf crop percentage, weaning weights, or bloom on animals when water is readily available. Smith (1981) concluded that salt may be used satisfactorily to limit supplement intake on summer pasture with growing animals with little reduction in performance. Also, self-feeding supplement during the summer by salt regulation with yearling steers on native range in Nebraska (Berger and Clanton, 1979) and in Oklahoma (McIlvain *et al.,* 1955) was equal to hand-feeding.

Slightly reduced gains on winter range, however, can be expected from using salt to limit intake of meal by young cattle, according to Smith (1981). In his studies, the salt-soybean oil meal mixture—in which .63 lb of salt per steer daily was required to restrict intake of meal to 2 lb daily—reduced the winter gains on grass winter range to 23 lb/steer compared to 58 lb for daily or alternate-day feeding of straight meal. A 16-lb winter gain reduction was also experienced with growing steers on grass winter range in Oklahoma due to salt regulation of meal intake in young cattle (McIlvain *et al.,* 1955). Although generally considered a satisfactory feeding method when used with mature cattle, Duvall (1969) did not recommend the use of salt as a regulator with either mature or young cattle based on his research on pine-bluestem range in Louisiana. The general consensus seems to be that salt limiting of meal intake generally works, but its use has greatly declined in recent years, particularly in regulating the intake of high-protein supplements, because of the reliability of extended feeding intervals.

E. LIQUID SUPPLEMENTS

Liquid supplements dispensed through a lick-wheel feeder containing liquid supplement is another alternative for dispensing particularly high-protein equivalent and mineral supplements. This approach utilizes molasses or other sweetening agent to attract animals and uses slow dispensing, time required for licking, and satiety as possible limiting factors in consumption. However, as is true of salt-meal mixes, considerable variation in individual and average animal consumption results from using liquid supplements. Tank proximity to water and preferred grazing areas, supplement formulation and palatability, and time available for consumption influence liquid supplement consumption.

A computer-controlled lickwheel feeder for dispensing liquid protein supplements (limiting animal consumption to 2.2 lb daily) was compared to the typical lickwheel feeder (*ad libitum*) in grazing studies with cows on winter range in Montana (Daniels *et al.,* 1998). Across treatments, liquid supplement daily intake was lowest for 2-year-old cows (1.76 lb), intermediate for 3-year olds (2.64 lb), and highest for 4- to 6-year olds (3.3 lb); daily consumption of liquid supplement averaged 4.2 lb for *ad libitum* dispensing and 1.5 lb for computer-regulated dispensing. Forage intake was increased by 15% when cows had *ad libitum* access to liquid supplement but by 47% with computer-limited dispensing, compared to unsupplemented cows. The results of this study suggest that the forage consumption stimulation resulting from protein supplementation was retarded by high *ad libitum* intake levels of the liquid supplement.

4

GRAZING SEASONS AND
SEASONAL BALANCE

Determining when to harvest the standing forage crop with grazing animals must consider: (1) plant factors, (2) physical site factors, (3) animal factors, and (4) economic and management factors. While some forage stands can be utilized any season of the year, many are adapted to grazing only when grazing is confined to a specific season of the year. If grazing animals are to be grazed beyond a single season, the grazing plan must consider how best to coordinate the multiseason demand for grazing capacity with seasonal forage supplies. Except on yearlong grazing lands, this will require the complemental use of various sources of grazing capacity to provide seasonal balance. This chapter will describe the forage production year-grazing year complex and suggest alternative solutions for synchronizing seasonal grazing needs with seasonal forage production.

I. THE FORAGE GROWTH CYCLE

The annual forage production cycle in temperate climates includes both a forage quality and a forage quantity cycle, and grazing animals under a set season-long stocking rate face a forage supply that is constantly changing both in quantity and quality. The forage quality cycle was covered in some detail in Chapter 2. A summary of the forage quality cycle is that herbaceous plant foliage during rapid growth is high in protein, phosphorus, and carotene (precursor of vitamin A), but all three components decline rapidly as the plants mature. However, cured herbage in the standing crop maintains moderate levels of energy as maturity and dormancy are reached.

It is optimal to have grazing animals on green, growing forage as much of the year as possible. In tropical and semitropical areas, plant growth is more or less continuous and subject mostly to only precipitation and soil moisture. Soil moisture limitations can be overcome in sub-humid to arid areas under irrigation, but atmospheric temperature in temperate zones restricts the green growth period of forage plants even under irrigation. Even though many areas along the Gulf Coast in southeastern U.S. have the potential for plant growth during 12 months out of the year through the use of small grains, annual ryegrass, and cool-season legumes for winter pasture (Rohweder and Van Keuren, 1985), cool weather and low light intensity often result in substantial reduction in growth rates of forage plants during winter. Coastal bermudagrass and kleingrass (*Panicum coloratum*), both introduced, warm-season perennial grasses, have been found feasible in central and south Texas for providing year-round grazing programs for cows (Conrad and Holt, 1983).

In temperate and cold latitudes, the forage production year is distinctly cyclic and plant growth is concentrated in a limited **growing season,** during which time temperature and soil moisture are usually conducive to plant growth. This results in one distinct forage supply cycle (or sometimes more) during the year consisting of the following phases (Heady, 1975):

1. Initial growth (slow)
2. Flush growth (rapid)
3. Reduced growth (slow)
4. Maturity-early dormancy (no growth)
5. Post-maturity (herbage loss and deterioration)

During the growing season, animals are faced with increasing supplies of forage (phase 1 into phase 2), and forage production often exceeds consumption. Entry into phase 3 (the "summer slump") may require adjustments in animal numbers to meet the slower forage growth rate. Two alternatives to meeting the slow-rapid-slow forage production cycle have been proposed for managing tall fescue pastures in West Virginia. One alternative is to increase livestock density by increasing livestock numbers per unit land area and selling the heaviest animals as herbage growth slows and becomes limiting. The second alternative is to decrease the size of land area grazed at a fixed density of livestock—possibly by temporary fencing—and harvesting the excess herbage in the fenced-out area for stored feed and later grazing the regrowth as needed.

During dormant periods (phases 4 and 5), growth is halted, and the forage supply in the standing crop then declines due to consumption, wastage, and natural weathering. These forage supply-animal demand relationships for a cow-calf enterprise grazing yearlong on range are shown in Fig. 4.1. This suggests that a program of spring calving or lambing and summer or fall weaning best fits the normal herbage supply cycle on western rangelands (see Chapter 3). However, year-to-year variations in precipitation, often as high as 150% in some arid areas, exert an overriding influence on such relationships (Malechek, 1984).

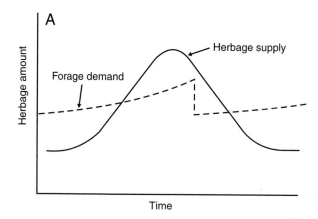

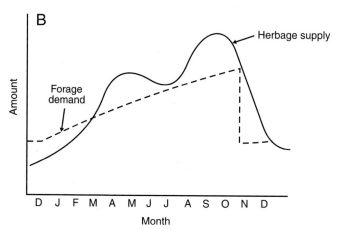

FIGURE 4.1   Forage supply-animal demand relationships for a cow-calf enterprise: (A) under a single growing period; (B) under a bimodal growing period on southwestern ranges (Malechck, 1984).

Peak growth periods differ substantially among various forage plant species. Forage plant species, particularly grasses, are commonly classified as **cool-season** (the major portion of growth occurs during late fall, winter, and early spring, depending on location) and **warm-season** (most or all of their growth occurs during late spring to early fall). Crested wheatgrass has been widely used in range seeding in western U.S. and southwestern Canada, in large part because of its abundance of early spring forage production. In southern Alberta it produces 90% of its total annual yield by July 1; in contrast, 65% of the total production of native prairie in the area, comprised predominantly of warm-season grasses, is made after June 1 (Lodge *et al.*, 1972)

The season of grazing of crested wheatgrass by cattle in central Utah has been shown to substantially affect forage yield and availability (Harris *et al.*, 1968). When based on 65% utilization of the crested wheatgrass standing crop being full proper use, the season of grazing affected on grazing capacity as follows:

Early spring (April 25–May 24): 11.5 AUD/acre (limited growth)
Late spring (May 24–June 21): 18.5 AUD/acre
Early summer (June 21–August 8): 21.0 AUD/acre (maximum herbage)
Late summer (August 8–September 16): 17.5 AUD/acre
Early fall (September 16–October 31): 16.5 AUD/acre
Late fall (October 31–December 9): 11.0 AUD/acre (limited herbage)
Early spring and regraze in early fall: 19.5 AUD/acre (high herbage)

Grazing capacity of the crested wheatgrass was least when grazed only in early spring (limited growth) or late fall (reduced availability) but was maximized when grazed in early summer or a combination of early spring and early fall. When a crested wheatgrass foothill range in Utah was grazed by sheep, an average turn-out date of April 14 compared to April 27 produced 20 less sheep-days per acre but increased the gains of suckling lambs by 0.08 lb/day (Bleak and Plummer, 1954); the grazing period in both treatments ended on May 25 average.

Under Midwestern conditions, Alta tall fescue, a cool-season introduced grass, was found to produce 58% of its yield by June 1 and 42% thereafter; by contrast Caucasian bluestem (*Bothriochloa caucasica*), an introduced warm-season grass, produced nearly all of its annual growth between June 1 and August 31 (Roundtree *et al.*, 1974). Although annual yield between the two species was not greatly different, Caucasian bluestem provided virtually no green-growth grazing capacity prior to June 1 but two to three times that of tall fescue from June through August. It was concluded by Forwood *et al.* (1988) that a mid-June defoliation, with further defoliation up to late summer, would optimize Caucasian bluestem's potential for livestock grazing systems in the Midwest; digestible dry matter yield increased throughout the season until it declined in late September. From pasture grazing studies in West Virginia, Turner *et al.* (1996) concluded that using herbage earlier in the spring before maturation may compromise total herbage productivity but should sustain a greater number of grazing days and livestock productivity for a given land area.

Peak standing crop and plant cover are inadequate to predict livestock management needs because they do not provide any assessment of forage quality and overlook the basic nutrient needs and responses of the animal (Currie, 1987). Management must consider the interdependence of feed efficiency, compensatory gain principles, and economic functions, instead of only maximum plant growth (Launchbaugh *et al.,* 1978). Nutrient conversion efficiency and production rate per animal are greatest when forage plants are grazed when growing and thereby providing high-quality forage. (Refer to Chapter 3, section I, "Forage-Animal Plans," for further discussion on priority use of high-quality forages in animal production systems.)

The levels of digestible and/or available nutrients in herbaceous plants are generally inversely related to dry matter yields. Delaying the grazing or mechanical harvest of forage plants until near the end of the plant growth cycle (multiple growth cycles per year may be realized under some situations) generally maximizes dry matter yield while greatly reducing nutrient content and often even digestible protein and total digestible nutrients (TDN) yield per acre. Kalmbacher *et al.* (1986) recognized the quandary as to when to graze creeping bluestem range in the Southeast. Grazing only in winter resulted in high yield but poor quality; grazing year after year only in summer resulted in loss of vigor, lower yields, and weaker stands but better forage quality. In order to provide less mature but vigorous stands when fall-winter grazing begins, one recommendation made was to defer range from grazing from June to September every other year; another was to extend winter grazing into the spring.

White and Wight (1980) combined dry matter yield, digestibility, and dry matter intake for several forage plant species in calculating the optimal time to harvest. Peak potential weight gains per acre under dryland conditions when grazing 665-lb steers were estimated as follows:

1. When Russian wildrye, crested wheatgrass, meadow brome (*Bromus biebersteinii*), green needlegrass (*Stipa viridula*), and reed canarygrass (*Phalaris aruninaria*) have produced less than half of their peak dry matter yield
2. When pubescent wheatgrass (*Agropyron trichophorum*) has produced 55% of peak dry matter
3. When cicer milkvetch and altai wildrye (*Elymus angustus*) have produced 75% peak dry matter
4. When alfalfa has produced 90% of peak dry matter

Peak weight gains per acre were projected for most grasses at 2 to 12 days before first inflorescence appeared, for alfalfa at 10% bloom, and for cicer milkvetch 8 days after 10% bloom.

Defoliation management can be used to modify the seasonal distribution of forage production (Matches and Burns, 1985). For example, partial spring defoliation during a short grazing period during early to mid-June not only provided some very high-quality herbage from switchgrass but also delayed the major growth period (George and Obermann, 1989). Not only did this practice make more forage available when most needed in midsummer, it did so without serious reduction in herbage quality. Grazing during rapid growth often lowers the peak growth rate due to added stress on the plant, but this can stimulate regrowth and increase late spring and summer forage production of many forage plant species. Although total forage production may be decreased under dryland conditions, the forage available for grazing during the later period should be more nutritious and palatable than if not grazed previously. Under extended optimal soil moisture conditions, such as in humid areas or under irrigation, prolonging the rapid growth phase of herbaceous plants under grazing should result in substantially greater total annual forage production.

Early spring grazing of crested wheatgrass on seeded Intermountain and Rocky Mountain rangelands has allowed further growth prior to fall grazing. Spring-fall grazing of crested wheatgrass at Manitou, CO, produced more forage, provided more days of grazing, and produced higher weight gains per acre than either spring or fall grazing alone (Currie, 1970). Average annual data during the 10-year study were as follows when grazed to a 1-in. stubble height during the respective seasons:

	Spring-fall	Spring only	Fall only
Forage yield (lb)	1152	910	1045
Grazing capacity (heifer days/acre)	65	40	48
Daily gains (lb)	.80	1.10	.10
Gain/acre (lb)	54	43	3

Crested wheatgrass pastures grazed in the spring and again in the fall have provided slight additional gains in the fall compared to fall-grazed only (Harris *et al.,* 1968; USDA, 1961b). Close grazing just prior to the boot stage has favored subsequent regrowth. Late spring grazing ending in late June has generally been followed by no regrowth, and the remaining forage has been less valuable for fall grazing. Full use by late June leaves insufficient forage for fall grazing except in those unusual years when summer rainfall permits regrowth in late summer and fall. The highest forage yield of crested wheatgrass has been provided by grazing from the time seedheads are in sheath until flowering. The longest spring grazing period has come from early spring grazing of one pasture, rotating to a pasture unit grazed in mid-spring for the most feed, and then back to the first pasture. (Refer to the two crop-one crop grazing system in Chapter 15.)

II. DETERMINANTS OF THE GRAZING SEASON

Season of grazing, a principle of grazing related to the timing of grazing, is generally considered second only to stocking rate in impacting natural plant communities (Thompson *et al.,* 1999). The **grazing season,** by definition, is the time period during which grazing is normally possible or practical each year; it may be the entire year or a short portion of the year. Plant, soil, climatic, and animal factors combine to dictate if grazing of the vegetation on a given site can occur anytime of the year (i.e., on **yearlong grazing lands**) or whether grazing must be restricted to a given season of the year (i.e., **seasonal grazing lands**). However, the management plan for any given year may specify a reduced time period of grazing within the overall grazing season in order to optimize: (1) the quantity and/or quality of forage produced, (2) its utilization by grazing animals, and/or (3) its utility in meeting the nutritional needs of grazing animals.

Grazing lands with restricted feasible grazing seasons are appropriately referred to as spring, fall, spring-fall, summer, or winter range or pasture. Livestock may be managerially confined to a particular kind of range or pasture during a given season, or big game and less commonly livestock under no management restriction may graze progressively in a sequence of moves from one seasonal range to another as vegetation develops.

The factor, or combination of factors, that determines the grazing season varies greatly from area to area and between different vegetation types. Not all of the limiting factors are constants; many vary from year to year. Such limiting factors include:

1. Extremely hot temperatures, frigid temperatures, and deep snow or icing over the top of vegetation
2. Declining plant palatability associated with plant maturity
3. Low nutritive quality or availability of vegetation after curing and/or subsequent weathering
4. Unstable soils or slopes when wet
5. Seasonal danger periods with poisonous plants
6. Seasonal drought often associated with deficits of both stockwater and forage supply
7. Plant species differences in tolerating defoliation in different seasons or growth stages

A. RANGE READINESS

A concept developed by range managers for seasonal native rangelands and later extended to seeded rangelands is that of **range readiness** (Fig. 4.2). It has been defined as the stage of plant development at which grazing may begin without permanent damage to the vegetation or soil. It has been directed primarily to the earliest date in the spring at which grazing can begin on mountainous spring and summer range, but the concept can also be applied to the time by which grazing should be terminated on winter range, sometimes set at the beginning of plant growth. Range readiness is often the stipulated turnout date on federal lands; it is similarly meaningful for both livestock and big game grazing but can seldom be enforced with big game except possibly in the case of game ranching.

When it was based on low larkspur (*Delphinium* spp.) being in full bloom and

FIGURE 4.2   Range readiness is a concept originally developed to describe when livestock grazing could safely be begun on mountainous spring and summer range; photo shows sheep beginning the summer grazing season on the Manti-La Sal National Forest, UT. (U.S. Forest Service photo.)

the appearance of Idaho fescue seedstalks, Mueggler (1983) found from a study on western Montana mountain grasslands that the calendar date of range readiness varied considerably between years. He found it differed by as much as 5 weeks over a 10-year period but could be expected to be within 2 weeks of the long-term mean in 2 out of 3 years. Early growth appeared to be most closely associated with May and June temperatures. Range readiness varies with the kind of grazing land and elevation. In the Intermountain Region, grazing is commonly begun on crested wheatgrass in the lower valleys and foothills between April 15 and May 1 but May 15 to June 1 in the upper foothills and mountains. Range readiness on native mountain range is often June 1 to June 15 but on western U.S. alpine range is often July 10 to 15 (Thilenius, 1975).

As discussed further under "Grazing Effects on Soil" in Chapter 5, range readiness is probably more pertinent to soil and site limitations than plant limitations even though it is commonly based on plant growth criteria. Wet mountain meadows have been particularly susceptible to trampling damage if grazed before seasonally becoming stabilized; recommendations for the Sierra Nevada mountains have included grazing the meadows as early as herbage is sufficient for livestock but not until the soils have sufficiently dried such that normal animal movements will not cut the sod (Ratliff *et al.,* 1987).

If turnout dates are too early, livestock may suffer from an inadequate forage supply and from susceptibility to poisonous plants; otherwise, utilizing early spring growth is beneficial to animal performance as far as taking advantage of the high nutrient levels. Where soil and site factors are not restrictive, routinely delaying turnout dates may result in the optimal forage quality stages being missed, and this is seldom fully compensated by increased forage availability later in the grazing season when forage quality is lower.

Under the concept of range readiness, the delay of grazing until several inches of new growth are available was intended to prevent grazing throughout the growing season on seasonal-long spring and summer ranges, particularly in the mountain and Intermountain western U.S. (Bawtree, 1989; Miller 1989). The idea was to defer grazing as long as possible to shorten the impact of a long season of grazing; this concept led to the misunderstanding that early grazing was necessarily harmful to the range. However, since most seasonal ranges are not grazed season-long today, following the old rule of thumb for range readiness is probably often unnecessary and may actually be damaging under today's grazing practices. In fact, timing the cessation of grazing may be more important than initiation of grazing for plant vigor (Burkhardt, 1996).

Economics, animal nutrition, and range research all suggest consideration be given to grazing seasonal spring or spring-summer ranges earlier than followed under the range readiness concept (Bawtree, 1989). Many studies have shown that grazing bunchgrasses from the boot stage to early flowering has greater impact on plant vigor than at any other stage of growth; this has application to perennial grasses such as crested wheatgrass, bluebunch wheatgrass, Thurber's needlegrass (*Stipa thurberiana*), and Idaho fescue. In mountainous areas, the solution may be to graze early for a 2- or 4-week period, then rest through the flowering stage, and rotate the early grazing into another unit the following year. Burkhardt (1996) concluded that this agreed with the concept that Intermountain bunchgrasses evolved under the selective pressure of early spring and dormant season herbivory and that the concept of range readiness ran counter to the instinctive nature of both native and domestic ungulates. Grazing for a short period in early spring will help ensure regrowth is obtained for fall grazing by livestock, where this is practiced, or for winter grazing by big game animals. (Refer to "Manipulating Animal Habitat by Grazing" in Chapter 16.)

III. SEASONAL TOLERANCE OF DEFOLIATION

Defoliation during the early stages of plant growth in the spring was formerly presumed to be invariably the most detrimental time; this assumption was based on root total available carbohydrate (TAC) reserves being the lowest at that point and on regrowth requiring a major drawdown of TAC reserves (Waller *et al.,* 1985). However, vigorous plants have a great capacity to replenish TAC reserves during the period of peak growth. Early season defoliation of native plants followed by nongrazing during the remainder of the growing season has often had less impact than severe defoliation late in the growing season only.

Late growing season grazing is a critical period for many perennial forage plants, and adequate time is optimally provided them after grazing and before dormancy for TAC accumulation and bud development. For example, TAC reserves of fourwing saltbush and antelope bitterbrush (*Purshia tridentata*) were most sensitive to a single defoliation (90%) at the seed-shattering stage (Menke and Trlica, 1983). Environmental conditions late in the growing season seldom favor the burst of growth observed in early season. On occasion, it may be necessary to intensively graze native range late in the growing season. If the grass has been properly managed in previous years, it should recover from this late season grazing; however, the same grazing unit should not be the last one grazed the following year (Waller *et al.,* 1985).

Defoliation after maturity during late fall and winter generally has the least effect, either detrimental or beneficial, on subsequent growth or TAC levels in either grasses (Heady, 1984) or shrubs (Garrison, 1972). Both clipping and grazing studies generally reveal that perennial forage species can withstand much more defoliation during periods of dormancy than during periods of active vegetative growth (Cook, 1971). Fall or winter grazing at lower elevations in the sagebrush steppe tends to incorporate litter, break surface crusts, and disperse plant seeds (Burkhardt, 1996) as well as provide an alternative to expensive winter feeding of livestock.

On flatwood range in southeastern U.S. dominated by wiregrass (*Aristida stricta*), commonly managed for cattle production by biennially burning and continuous grazing, was found not highly responsive to either grazing or deferment from grazing (Kalmbacker *et al.,* 1994). Resting wiregrass range December–March, April–July, or August–November did not affect the frequency of desirable grasses nor annual biomass production, and the advantages over the cost of more intensive grazing management was questioned.

It has long been recognized that full to heavy grazing of rapid spring growth and fall regrowth of cool-season grasses favors the warm-season grass component in the standing crop. By contrast, full to heavy grazing during the summer growth period of warm-season grasses favors the cool-season component. The effects are sufficiently consistent that early spring or summer grazing, respectively, can be used to enhance the warm-season or cool-season grass component in the standing crop of mixed grass stands or to maintain a balance of both types (Waller *et al.,* 1985; Jameson, 1991). However, these differential seasonal effects probably result as much from seasonal grazing selectivity as from tolerance or susceptibility to defoliation per se.

A. GREAT PLAINS AND SOUTHWEST

In order to learn more about the seasonal susceptibility or resistance of native plant species to grazing on shortgrass range, repeated heavy grazing treatments in individual months were applied annually over a period of 7 years at the Central Plains Experimental Range, Nunn, CO (Hyder *et al.,* 1975). Hereford heifers were stocked at the first of each month on the assigned paddock with the estimated number to utilize all but 100 lb/acre dry weight of forage in 4 weeks (i.e., very heavy rate). The stocking rates were heavy enough that animal gains were substantially reduced compared with moderate stocking under continuous grazing.

Heavy grazing in April, May, and June was considered the most unfavorable because of its detrimental effects on the cool-season vegetation component and corresponding reduction of grazing capacity. Heavy grazing in September favored the cool-season perennials such as western wheatgrass (*Agropyron smithii*) but disfavored warm-season grasses such as blue grama and sand dropseed (*Sporobolus cryptandrus*). The best forage quality, as interpreted from animal gains, was obtained in June and July; the greatest herbage production, as interpreted from animal days of grazing, was obtained by grazing in August and September and the least in June and July. The reduced grazing capacity in May, June, and July was attributed to the least standing crop, but energy intake per animal unit day was greatest at that time also.

On Oklahoma range, where warm-season grasses such as the bluestems and blue grama dominate the plant composition, it was suggested that grazing plans avoid heavy fall grazing (Shoop and McIlvain, 1972). In the northwestern Oklahoma studies, predominantly blue grama ranges were heavily grazed for 3 consecutive years only at designated 2-month periods and then rested during the fourth year. Yields following growing season rest in the fourth year were 470 lb under the July–August grazing treatment but only 300 lb from the September–October treatment. Yields for the April–May treatment were similar to the July–August treatment, and the May–June treatment was intermediate. Heavy early fall grazing or mowing before dormancy also reduced vigor and increased winterkill of weeping lovegrass (*Eragrostis curvula*), an introduced warm-season grass (Shoop and McIlvain, 1972).

Mowing Flinthills tallgrass range, previously grazed during the growing season, at a 2-in. height anytime from October to April had no effect on total nonstructural carbohydrates in big bluestem rhizomes or on herbage production the following season (Auen and Owensby, 1988). It was concluded that range units grazed closely in the spring could be safely restocked after dormancy (October 1) to remove the regrowth. From studies on native, tallgrass haylands in Kansas, it was concluded that late summer-early fall grazing lowered meadow productivity the next growing season; thus, livestock should not be allowed to graze the regrowth until after frost (Kansas State University, 1995).

It was concluded in Nebraska studies that intensive grazing or haying tall prairie grasses should preferably occur prior to mid-July to permit regrowth before maturity; intensive defoliation after late July was considered capable of weakening stands and reducing plant vigor (Horn and Anderson, 1988). Heavily grazing Nebraska Sandhills range in late April prior to prairie sandreed tiller emergence benefited this warm-season species by defoliating and reducing cool-season grasses in the forage stand (Reece *et al.,* 1999). Excluding livestock from these stands from the time of prairie sandreed tiller emergence to the end of the growing season was suggested as a means of more fully benefiting this species. It has also been noted that dormant season grazing (October) in the Nebraska Sandhills was similar to total rest for both prairie sandreed and sand bluestem during a 4-year study (Reece *et al.,* 1996).

The spring growth of semi-desert grasses in the Southwest is apparently critical to plant vigor and maintenance, even though they produce little forage at that time. When November–April, May–October, and yearlong grazing at the Santa Rita Experimental Range near Tucson was compared over a 10-year period, grazing during the November–April period was least favorable to the perennial warm-season grasses (Martin and Cable, 1974; Martin, 1975b). Because green forage was scarce and was especially sought by cattle in early spring, heavy defoliation at that time was the apparent cause of reducing the number of summer culms, reducing diameter growth of established plants, and a reduction of forage yields during the main summer growing season. In contrast, fall grazing of big sacaton (*Sporobolus wrightii*) in Arizona was the most deleterious because defoliation at that time exposed the plants to below freezing temperatures, and crown damage resulted (Cox *et al.,* 1989).

Rough fescue prairie in southern Alberta was greatly reduced in vigor by heavy continuous spring-through-fall grazing or heavy grazing restricted to June only (Horton and Bailey, 1987; Willms *et al.,* 1998); late fall grazing was found best for maintaining the maximum vigor of rough fescue. Dormant season grazing was concluded to have no negative effects on forage yield on the more favorable sites occupied by fescue prairie and may enhance plant vigor by stimulating tillering in the grasses (Willms *et al.,* 1986b). However, on the more arid sites occupied by the mixed prairie, heavy winter defoliation over a 3-year period reduced yields by 43% compared to the controls. The negative effects on the mixed prairie sites were attributed to accelerated evaporation resulting from the removal of standing plant litter and mulch. This resulted in an induced soil moisture deficit, shallower infiltration, and concentration of roots near the soil surface. On the more mesic fescue grassland, the higher levels of plant residues left by grazing inhibited excessive losses in soil moisture.

B. ROCKY MOUNTAIN AND INTERMOUNTAIN

Pinegrass in the Douglas fir (*Pseudotsuga menziesii*) zone of British Columbia was most sensitive to defoliation during the last half of July and early August when growth was slowing down and summer dormancy was setting in (Stout *et al.,* 1980). It was suggested that pinegrass be grazed for a short time while it was actively growing early in June and then later after mid-summer dormancy is well achieved by late August to maintain its vigor. When pinegrass had to be grazed during July, it was suggested it receive nongrazing during July the following year.

Slimstem muhly (*Muhlenbergia filiculmis*), a warm-season grass on mountain grasslands in the Colorado Front Range, was greatly reduced by repeated annual late growth period harvest (summer), reduced by late harvest in alternate years, and slightly promoted by early harvest of the companion cool-season grass, Parry oatgrass (*Danthonia parryi*). Compared to the traditional grazing sequence of spring and fall on the native grasslands of interior British Columbia, fall grazing over a 16-year study period resulted in a plant community similar to the ungrazed exclosure while concentrating grazing in the spring had the greatest impact (Thompson *et al.*, 1999); the latter reduced the dominant bluebunch wheatgrass and increased the sub-dominants junegrass (*Koeleria cristata* or *pyramidatus*) and needle-and-thread (*Stipa comata*).

Fall-only heavy grazing (43 sheep days per acre average) and a more moderate level of spring and fall grazing (19 plus 10 equals 29 sheep days per acre average) was compared on native sagebrush-grass range at the U.S. Sheep Experiment Station near Dubois, ID (Mueggler, 1950). After 25 years of treatment the fall-only pasture remained in good condition while the range grazed both in spring and fall had declined to poor condition and lost two-thirds of its grazing capacity. During a subsequent 14-year study at the same station, heavy spring grazing was compared to heavy fall grazing (Laycock, 1970). The heavy spring grazing by sheep damaged native sagebrush-grass range in good condition, and the production of herbaceous plants in the understory was reduced by 50% while big sagebrush *Artemisia tridentata*) production was increased by 78%. In contrast, heavy fall grazing maintained range condition associated with high production of herbaceous perennials and an open rather than dense stand of sagebrush.

The fall heavy grazing treatment with sheep was so successful in increasing forage production of the grasses and forbs while reducing big sagebrush that it was considered an effective method of sagebrush control (Laycock, 1970). The cool-season grasses (predominantly bluebunch wheatgrass and Idaho fescue) and the forbs were susceptible to grazing during spring growth but were dormant by fall when sagebrush was still growing and subject to vigor reduction by fall grazing. It was concluded by Laycock (1979) that spring and early summer grazing of sagebrush-grass range requires careful selection of grazing intensity and system because defoliation coincided with the growing season of the herbaceous understory, while fall or winter grazing was generally less critical. An 80% clipping treatment was found by Wright (1970) to reduce yields of big sagebrush most when applied during July, moderately when applied during spring, and least when applied during late summer through the winter months. In the Idaho studies, the selective grazing of big sagebrush by sheep in late fall apparently overcame its relatively higher tolerance of defoliation at that time.

The interactions of season of grazing, stocking rates, plant life cycles, and climatic patterns have been studied on salt-desert shrub range at the Desert Experimental Range in western Utah beginning in the 1930s. These grazing studies with sheep were carried out during the winter through early spring grazing season typical in the area. The effect of season of grazing did not become apparent during the first dozen years of the study and so were not reported by Hutchings and Stewart (1953). After more than 30 years of treatment, however, the most striking differences in vegetation resulted from season of use (Holmgren and Hutchings, 1972). During early winter and mid-winter (late November to late February), the moderately desirable native grasses and the desirable winterfat and bud sage

(*Artemisia spinescens*) are tolerant of moderate to heavy grazing because it occurs during plant dormancy. However, the desirable grasses and shrubs were reduced in vigor at these same grazing intensities by late winter grazing (March into early April), during which period they break dormancy and begin growth. Cook (1971) concluded that this desert vegetation could tolerate 50% utilization under dormant season use but only about 25% during spring and early summer growth.

From their evaluation of the long-term studies on this same area, Clary and Holmgren (1982) made the following conclusions: (1) There is an intensity of grazing that at any time during winter is harmful to the desirable plants and results in an increase of less useful species. (2) Stocking at even a light rate is damaging if repeated year after year in late winter-early spring (March–April). (3) Range can improve in quantity and quality of forage production when grazed under a moderate annual stocking rate in early and middle winter. Whisenant and Wagstaff (1991) also concluded that annual March–April grazing was an important cause of retrogression in the salt-desert shrub ecosystem. While late fall or early winter grazing during dormancy had effects differing little from nongrazing, the greater impact of late winter-early spring grazing coincided with what was generally the only reliable growing period of the year.

C. LEGUMES

Alfalfa is a C_3 plant (cool-season grower), is vigorous, grows rapidly, and produces a high volume of forage under optimal growing conditions (Allen *et al.,* 1986a,b). It is more tolerant of grazing and competes better in a grass-legume mixture and against weeds when grazed in the spring than in the summer, particularly when summer conditions are dry and/or hot. Delaying grazing until alfalfa has reached 1/10 bloom is commonly recommended, both for maintaining vigor and reducing the bloat hazard; extended grazing duration prior to early bloom is particularly damaging during the summer. However, alfalfa, in pure stands or in mixture with one or two cool-season grasses, can be maintained for 5 to 8 years under full growing season grazing in temperate or cool summer areas when: (1) soil fertility is maintained at optimal levels, (2) ample soil moisture is assured during summer, (3) stocking is regulated to prevent overgrazing, and (4) ample nongrazing periods are provided for regrowth.

In West Virginia studies, grazing alfalfa only during a 3- to 4-week period in the spring did not reduce total annual yields; this permitted the flexibility of grazing alfalfa in early spring for balancing seasonal grazing capacity (Wolf and Blaser, 1981). The spring grazing delayed the first hay cutting by about 3 weeks, thereby foregoing only about one-half cutting of hay for the season. During a 5-year study in Nevada with alfalfa grown under irrigation, dormant season grazing of aftermath during November, January–February, or April did not reduce yields of the first hay cutting made the following early May (Jensen *et al.,* 1981). The dormant season grazing treatments did not significantly affect the number of plants per unit area or increase the incidence and severity of root and crown diseases; it did provide an additional half ton of forage when grazed in the fall or about half that much if not grazed until winter. When little or no growth occurred from January to March on coastal California perennial ryegrass (*Lolium perenne*)-white clover pasture, short, intensive grazing periods during the winter had little effect on growing season yields (Jaindl and Sharrow, 1987).

The control of early fall defoliation and a gradual decrease in fall temperatures has long been held important in permitting TAC storage and ensuring winter hardiness in alfalfa cut for hay; however, the warmer the climate, the less necessity of high storage of carbohydrate reserves in the roots for adequate winter survival and high yields the next year (Tesar and Yager, 1985). Fall harvest treatments in Oklahoma did not affect total forage production during the lifetime of alfalfa stands, root carbohydrate concentrations, or the uniformity of harvests (Sholar *et al.,* 1988); the key to preventing fall harvest damage was concluded to be using adapted cultivars and growing on properly fertilized soils.

In the northern portion of the U.S., the common recommendation has been to avoid cutting alfalfa in September and early October or during the 4- to 6-week period preceding the first killing frost. Current fall management recommendations for alfalfa in Montana have been to schedule the next to last harvest during the growing season at least 30 days, and preferably 45 days, before the first 32°F frost based on long-time averages, and that the final harvest of the year should not occur before a 25°F frost has occurred (Welty *et al.,* 1988). Horrocks and Zaifnejad (1997) concluded from their studies in Utah that the critical period for avoiding cutting alfalfa was the 2 to 6 weeks prior to the mean first-killing frost date. While younger stands were less affected, harvesting older stands of alfalfa during this period allowed insufficient replenishment of root carbohydrates, resulting in yield losses ranging from 0.45 T/acre the first year to 1.12 T/acre the second year.

Tesar and Yager (1985) concluded that it is the interval of time between the last two growing season cuttings that is critical, rather than the interval between the last growing season harvest and the first killing frost. For the northern U.S., where three cuttings is the most popular management system, they concluded from studies in Michigan and Minnesota that making the third cutting in September or early October was not harmful as long as there was adequate time for replenishment of carbohydrate reserves (this indicated by at least 10% blossoming) between the second and third cuttings. This conclusion was based on the finding that mid-December TAC levels were similar regardless of the date of the last cutting when the proper interval was maintained between the last two cuttings prior to frost.

IV. BALANCING SEASONAL SUPPLY AND DEMAND

A. GUIDELINES AND PROCEDURES

A ranch or livestock-forage enterprise on a general farm may produce the required number of AUMs on an annual basis and still have a serious forage deficiency during one or more seasons of the year. Size of the breeding herd in a straight cow-calf enterprise will be limited by the carrying capacity during the season or month of lowest supply. Where total yearlong carrying capacity is comprised of many diverse sources and grazing lands are generally highly seasonal, such as found typically in the Rocky Mountain and Intermountain regions, achieving the seasonal balance desired between forage production and forage needs is both urgent and challenging.

Seasonal balance is achieved by modifying seasonal inventories of livestock, by modifying seasonal carrying capacity produced, or meeting seasonal deficits with off-ranch purchases of carrying capacity. Seasonal imbalances in grazing capacity in operations depending primarily on grazing may suggest modifications of existing livestock enterprises or adding or deleting one or more enterprises in order to bring seasonal needs in line with available seasonal grazing capacity. Seasonal excesses in grazing capacity can sometimes be profitably marketed to other livestock producers, but the value may be less than grazing capacity seasonally of limited availability in the vicinity.

A balanced livestock operation requires sufficient quality and quantity of grazed and harvested forages and other feedstuffs to promote continuous satisfactory maintenance and production of the livestock. A comprehensive plan to secure the best practicable use of forage resources is a key management step in ruminant animal production enterprises. This includes providing the day-to-day carrying capacity from the combination of available sources to best match the quantitative and qualitative requirements of the animals (Fig. 4.3).

Gains of grazing animals reflect not only the source of grazing but also stage of plant growth. The average daily gains of cattle and sheep by month on various sources of grazing are given in Table 4.1. These gains are not intended to represent target gains but merely to compare monthly differences in gains within each study. Additional seasonal animal gains, as related to grazing intensity, are included in Table 11.3. Animal gains are highest during the spring flush of growth, continue at moderate levels during late spring and summer, decline rapidly as plant dormancy is approached, and approach zero or become negative during dormancy. When continued good gains are desired with growing livestock, their removal from pasture prior to advanced maturity and dormancy and being provided with good growing rations consisting of harvested forages and concentrate feeds will generally be required.

The nature of the ruminant animal enterprise will determine whether a single season, multiple season, or year-round plan is required. A similar plan will be required for successful big game ranching enterprises, except that the carrying capacity will be mostly limited to grazing resources. Livestock-carrying capacity balance evaluations for specific case studies can be accomplished manually through the following more or less distinct steps:

1. Designate and locate individual grazing units and associated cropland field units.

FIGURE 4.3   Developing a plan for providing the seasonal and monthly carrying capacity for each ruminant animal enterprise is a key management step; showing development of the plan for an Idaho ranch by a Brigham Young University class.

2. Determine acreage by land use and productivity within each unit, and estimate yields in terms of AUMs of grazing capacity, tonnage and/or AUM equivalents of harvested roughages, and tonnage of other feedstuffs.
3. Designate feasible use by season and month of all projected yields within each grazing and cropland field unit, and summarize available seasonal and monthly carrying capacity for the entire ranch or individual livestock-forage enterprises where carrying capacity sources are assigned by enterprise.
4. Develop a continuous monthly livestock inventory by enterprise and class of animals and determine associated carrying capacity demands.
5. Complete a carrying capacity balance between what is produced and what is required by month, season, and year (by total ranch or individual enterprise); show the planned use of the available carrying capacity by source and month; utilize whatever seasonal flexibility exists in livestock practices and use of carrying capacity in deriving the best monthly balances.
6. Document carrying capacity deficits and surpluses that result from case farm or ranch production, and indicate plans for obtaining deficit carrying capacity while utilizing any surpluses.

Use AUMs for the basis of carrying capacity comparisons when capacity is taken exclusively from grazed forages, use AUMs or a combination of AUMs and

TABLE 4.1   Average Daily Gains of Livestock on Grazing Lands

State	Kind of of livestock	Class of livestock	Kind of pasture	No. years	Jan.	Feb.	Mar.	Apr.	May
Alberta	Sheep	Ewes	Mixed grass prairie	19	0.17	0.17	0.17	1.07	0.1
Alberta	Sheep	Lambs	Mixed grass prairie	19					0.67
Colorado	Cattle	Yearlings	Shortgrass	10					2.31
Colorado	Cattle	Yearlings	Shortgrass		−0.9	−0.6	−0.4	−1.0	0.4
Kansas	Cattle	Yearlings	Shortgrass	11					2.0
Kansas	Cattle	Yearlings	Tallgrass	17					1.85
Kansas	Cattle	Yearlings	Tallgrass	16					1.8
Kansas	Cattle	Yearlings	Tallgrass	16					2.45
Kansas	Cattle	Yearlings	Tallgrass						2.28
Montana	Cattle	Calves	Midgrass-shrubs	7				2.4	1.9
Montana	Cattle	Calves	Foothill grassland	10					1.8
Nebraska	Cattle	Yearlings	Sandhill rangeland	10					2.14
Oklahoma	Cattle	Weaners	Sandy grassland	10	0.43	0.47	0.47	0.59	2.31
Oregon	Cattle	Calves	Flood meadows	4					1.92
Oregon	Cattle	Calves	Sagebrush-grass	4					1.68
Texas	Cattle	Calves	Clover-grass				2.9	1.4	1.8
Texas	Cattle	Calves	Clover-grass				3.2	2.4	2.2
Utah	Cattle	Yearlings	Crested wheatgrass	4					2.6
Utah	Cattle	Calves	Crested wheatgrass	4					1.7
Utah	Sheep	Ewes	Salt-desert shrub	4–5	0.071	0.071	0.120	0.120	

TABLE 4.1   (*continued*)

June	July	Aug.	Sep.	Oct.	Nov.	Dec.	Average	Explanatory information	Reference
0.17	0.07	0.07	0.07	0.03	0.13	0.0		Grazed May 1–Nov. 1; fed hay Jan. 1–April 1	Smoliak (1974)
0.67	0.5	0.43	0.40	0.17			0.47	Lambing to weaning	Smoliak (1974)
2.13	1.94	1.71	1 15	0.26			1.6	Heifers; 12–18 mo.; moderate	Klipple and Costello (1960)
1.2	1.6	1.1	0.7	−0.5	−1.0	−1.0		Heifers; Oct–Sept.: heavy grazing	Hyder *et al.* (1975)
1.4	1.2	1.2	0.7	−0.3			1.0	Steers, 12–18 mo.; moderate	Launchbaugh (1957)
1.75	1.6	1.25	1.45				1.6	Steers; moderate	Owensby *et al.* (1973)
1.75	1.6	1.25	1.2				1.5	Unburned	Anderson *et al.* (1970)
2.0	1.7	1.3	1.5				1.8	Burned early spring	Anderson *et al.* (1970)
1.93	1.64	1.23	1.29				1.7	Steers	Smith (1981)
1.7	1.8	1.9	1.8	1.2			1.8	Birth to weaning; cows range year-round; moderate	Woolfolk and Knapp (1949)
2.1	2.3	2.1	1.9	1.7			2.0	Suckling	Church *et al.* (1986)
2.04	1.76	1.4	0.4				1.6	Steers; 30-day gains from mid-month; moderate	Burzlaff and Harris (1969)
2.19	2.01	1.39	1.2	0.24	0.37	0.57	1.1	Oct–Sept.; 7–19 months; moderate	McIlvain and Shoop (1961)
1.92	1.86	1.86	1.3	0.83			1.6	Suckling	Cooper *et al.* (1957)
1.68	1.42	1.42	1.0	0.68			1.6	Suckling	Cooper *et al.* (1957)
1.5	1.3	1.0	0.8				1.5	Weaned; moderate	Roth *et al.* (1986b)
2.2	2.2	1.9	1.5				2.2	Suckling; moderate	Roth *et al.* (1986b)
2.0	1.5	0.7	0.7	0.7	−0.7		1.1	Unsupplemented	Harris *et al.* (1968)
1.5	1.5	1.5	1.2	0.7			1.35	Unsupplemented	Harris *et al.* (1968)
					0.052	0.052	0.079	Dry ewes; winter range; moderate	Hutchings and Stewart (1953)

tonnage where both grazed and harvested forages are included, and use both AUMs and tonnage where grazed forages, harvested forages, and other feedstuffs are all involved in providing carrying capacity. (Refer to Chapter 11 for further guidelines for applying the animal unit and animal unit month concepts.) Detailed procedures and worksheets for making a comprehensive livestock-carrying capacity balance evaluation can be found in various workbooks such as Waller *et al.* (1986). Although seasonal carrying capacity balance evaluations can be done manually, computerized decision-support computer software is available for carrying out this analysis and comparing various alternative solutions. The livestock-carrying capacity balance process is given a prominent role in conservation farm and ranch plans developed through the USDA, Natural Resources Conservation Service (formerly Soil Conservation Service).

When balancing the annual and seasonal carrying capacity with livestock needs, the grazing lands should generally be considered first. On most ranches carrying capacity comes primarily from grazing, and efficiency in the use of the forage produced on grazing lands directly influences livestock performance and thus net income from the ranch. The seasonal utilization of rangeland and permanent, dryland pasture is generally the least flexible, except on yearlong range found mostly in the Great Plains and Southwest. Croplands provide more flexibility through crop-rotation pasture, temporary pasture, and crop aftermath or residue pasture in meeting seasonal grazing capacity deficits; if not needed for pasture, such lands can often be utilized in producing harvested forages. And, finally, the production of harvested forages and grain can be used to fill carrying capacity gaps that cannot be filled by the optimum combination of grazing resources.

B. STOCKPILING

When the grazing season is extended beyond or begins prior to the plant growing season, grazing animals are required to graze matured, dormant vegetation. This will impact nutrient intake levels and must be coordinated with animal nutrient requirements. An alternative to grazing dormant standing forage, of course, is the feeding of harvested forages to livestock and less commonly big game herbivores. Allowing standing forage to accumulate during the rapid growth stage in the forage production cycle for grazing at a later period has been called **stockpiling.** Although originally coined in conjunction with the grazing use of improved pasture, the term is now commonly applied to rangeland, including the delayed grazing inherent in deferred grazing (see Chapter 14).

The standing forage stockpiled or "saved" during periods of rapid growth is then grazed during periods of reduced growth rate, maturity and early dormancy, or even full dormancy. The earlier into the growing season that stockpiling is started and the longer it is continued through the growing season the greater the accumulation of potential forage dry matter. However, stockpiling generally occurs at the expense of decreased quality of the stockpiled forage because of advancing stages of forage maturity (Fribourg and Bell, 1984; Matches and Burns, 1985). The

FIGURE 4.4   Stockpiling refers to allowing standing forage to accumulate during growth stages for consumption in a later period; greatly delaying utilization of the stockpiled forage can result in high dry matter and nutrient losses to weathering as in this intermediate wheatgrass (*Agropyron intermedium*) pasture in Nebraska.

effects of stockpiling on forage quality may be minimal if it is of short duration during rapid growth but will be substantial if it is continued to maturity or dormancy. Also, the greater the delay after growth ceases that grazing begins the greater will be the dry matter and nutrient losses to weathering (Fig. 4.4). The length of the accumulation period during forage plant growth can be used as a management tool for forage yield, but shorter accumulation periods will ensure higher forage quality where required for higher livestock performance.

Optimal temperature and rainfall for late summer growth will enhance stockpiling of herbage for grazing during dormancy. Winter grazing of dormant vegetation will be most effective with forage species that cure well and where dry climate after cessation of growth reduces the rate of weathering and the probability of the standing crop being buried under snow. On shortgrass range in the Great Plains (Hyder *et al.*, 1975) and semi-desert grasslands in the Southwest (Holechek *et al.*, 1989), herbage decay and disappearance is seldom of great consequence until spring. But, in areas where soils remain wet into the fall and winter, trampling losses as well as deterioration from weathering are apt to be severe. The remainder of the herbaceous standing crop not utilized will eventually deteriorate to mulch (usually completed by early in the subsequent growing season) and have no further potential as forage. Besides having the advantage of a greater component remaining above the snow, browse is more resistant to over-winter deterioration than herbaceous vegetation, and the twigs and even the foliage of non-deciduous species often remain available and edible through one or more subsequent growing seasons.

C. SUPPLEMENTAL VS. COMPLEMENTAL FORAGES

When their use is integrated with a primary forage resource base, such as rangeland or medium- to long-term pasture, secondary sources of a different kind or mix of forages, either grazed or harvested, can function as a **supplemental forage** (with the principal role of filling in nutrient deficiencies) or as a **complemental forage** (with the principal role of providing additional or alternative carrying capacity). When the secondary forage sources are limited to grazed forages confined in grazing units, their counterparts are **supplementary pasture** and **complementary pasture** (McIlvain, 1976).

While a supplementary pasture unit is grazed simultaneously with the base grazing unit, a complementary pasture unit is grazed in chronological order (or sometimes simultaneously) with the base grazing unit as a means of balancing and extending individual grazing seasons or the total annual grazing season. However, as complementary grazing units are increased in number, expanded in scale, and/or extended in longevity over years, their relation to the base grazing unit or units marks gradual transition into **sequence grazing** (synonymous with repeated seasonal grazing). (Further discussion of sequence grazing as a grazing method is found in Chapter 14.)

A supplementary pasture of enhanced nutrient quality is matched with a base grazing unit (either pasture or range) that is quantitatively adequate but nutritionally inadequate; while the base grazing land unit provides the primary source of grazing capacity, the supplementary pasture serves to correct nutrient deficiencies in grazing animal diets and achieve a balanced daily ration, much the same as does a supplemental feed, but will provide some additional grazing capacity as well. An example might be limiting time access of yearling steers to a high-protein irrigated pasture during a period when the base diet is being provided by dry rangeland (Fig. 4.5). Also, a creep pasture will serve as a supplementary pasture to the suckling offspring.

Supplementary pastures have played prominent roles in the southern U.S. Small acreages of perennial ryegrass pasture have been found effective in meeting the winter nutritional needs of cattle grazing southern pine-bluestem range while replacing the need for protein feed supplements (Pearson and Rollins, 1986). Seeded grass fuelbreaks, when combined with rotational grazing and burning of southern forests, has also been a viable alternative to protein supplements for cattle wintering on southern native forest range (Linnartz and Carpenter, 1979). Combining the use of improved supplementary pasture during the spring-summer grazing period with native pine-wiregrass range in Georgia (0.6 plus 8 to 12 acres per cow, respectively) increased average weaning weights by 115 lb (370 vs. 485 lb) (Lewis and McCormick, 1971). Limit-grazing Angora does being overwintered on a basal ration of bermudagrass hay to 2 hr daily access to wheat-ryegrass pasture,

FIGURE 4.5   High-quality improved pasture (foreground) available for supplementing California foothill range (background) in summer.

compared to being fed 1 lb daily of a 16% crude protein supplement, improved animal performance while reducing required amounts of supplement and reducing hay consumption by 50% (Hart and Sahlu, 1995).

V. COMPLEMENTARY GRAZING PROGRAMS

A complementary grazing program should describe the recommended order of grazing the different kinds of forage resources. Limitations of forage quality and timeliness of production are often inherent in rangeland or pasture land that can be compensated by adding other forage sources. Three major considerations determine the feasibility of using complementary forages (i.e., this concept expanded to include harvested forage crops as well as pasturage): (1) increased production per unit of land, (2) improved forage quality for better animal performance, and (3) reduced overall production costs or a more cost-effective means of filling a need, thereby increasing profitability of the farm or ranch (Nichols and Clanton, 1987).

The number of possible complementary grazing programs are almost infinite. Each should be custom fitted to the respective case situation (Fig. 4.6). Each will

FIGURE 4.6   The excess foliage produced by winter wheat, more properly referred to as foremath than aftermath, provides urgently needed complementary pasture for fall, winter, and early spring
use in the central and southern Great Plains; stocker steers grazing wheat pasture in Clark County, KS.

be determined by the combination of specific grazing land properties being included, the number and size of animal enterprises, and the management practices
followed in each. As a result, no two programs will be exactly the same even
though common designs will be apparent. Different kinds of complemental forages should generally be fenced under separate land units so that stocking levels
and distribution as well as season of grazing can be fully regulated.

Seeded range as well as temporary pasture complement native range by: (1) extending the green growing period, (2) fitting green forages into dry periods and
serving as a buffer against forage production fluctuations, (3) providing special use
grazing such as flushing, breeding, calving or lambing, or pasture finishing for
market, and (4) increasing grazing capacity per average acre allotted under the
plan. Using seeded pastures in conjunction with native range in the Northern Great
Plains was referred to by Lodge (1970) as "complementary grazing systems."
However, current thinking and definition reserve the term "grazing system" for describing the grazing and nongrazing period interrelationships between units managed under rotation grazing.

A. YEAR-ROUND PROGRAMS

On ranges of the southern Great Plains and the Southwest that are grazed year-
round, carrying capacity for all seasons comes from a single source or kind of

range and sometimes from a single grazing unit. Although this theoretically mostly circumvents the need for any planning for complemental grazing, grazing units for special seasonal needs or uses are commonly provided for. However, related planning must be directed to meeting problems of seasonal fluctuations in forage availability, and drought emergency plans must be kept current. (Refer to "Managing Forage Production Fluctuations" in Chapter 13.)

Ocumpaugh and Matches (1977) have recommended a year-round forage program for beef cattle in the Midwest that includes stockpiling based solely on tall fescue (Fig. 4.7). In this program, a combination of grazing and round baling of the forage is carried out during spring and summer. From August 10 through October, part of the pasturage is grazed and part is left ungrazed for stockpiling. The stockpiled forage is then grazed during November and December, and the round bales are then fed/grazed during the winter until green grass growth, thereby completing the year-round forage program.

A grazing program utilizing a 1:3 ratio of cropland pasture acreage to native range acreage has been recommended by Launchbaugh (1987) in Kansas for reproductive beef cattle operations. The cropland, ideally managed under double cropping in a warm-season and cool-season pasture sequence, was related to the native range in both a complemental and a supplemental-complemental relationship as follows:

FIGURE 4.7   Tall fescue, as shown growing in northern Tennessee, can provide a year-round forage program for beef cattle in the Midwest from a combination of grazing sequentially the growing plants, the stockpiled plants, and forage left on site in round bales.

June 1–July 15: Graze range only; sorghum-sudangrass hybrid cropland pasture being established.

July 15–August 10: Graze both range and cropland pasture (mostly complemental).

August 10–September 1: Graze cropland pasture only for graze-out; range benefits from late season nongrazing.

September 1–December 15: Graze range only; winter cereal crop being established on cropland.

December 15–May 1: Graze both range and cropland pasture (supplemental-complemental).

May 1–June 1: Graze cropland pasture only for graze-out; range benefits from early-season nongrazing.

This grazing program ensures additional grazing capacity over range alone; enhanced weaning weights are also anticipated from the higher quality forage mix provided by the combination of sources. The pasture double-cropping on cropland was considered ideal under the mesic conditions of central and eastern Kansas but also possible in semi-arid western Kansas if each pasture crop was grazed out early for temporary fallow to save moisture before the next was planted; another possibility would be supplemental irrigation of the cropland pasture. Some compaction problems might be anticipated on very heavy soils from the double-cropping and grazing.

B. GROWING SEASON PROGRAMS

Three complementary grazing programs built around native range have been reported for northwestern Oklahoma (McIlvain and Shoop, 1973). One plan utilizing native sandhills range (90%) and weeping lovegrass (10%) increased grazing capacity by 82% and gain per acre by 73% over native range alone. A second program used 75% range and 25% double-cropped wheat (*Triticum aestivum*) and sudangrass; grazing capacity was increased 89% and gain per acre over 100% compared to native range alone when averaged over 6 years. The greatest gain advantage under this system occurred during August when steers were on sudangrass. Expanding this plan to include weeping lovegrass in conjunction with wheat, sudangrass, and native range comprised a third program; this increased grazing capacity 360% and gains per acre by 335% compared to native range alone. Still another successful program involved the substitution of double-cropped millet for the winter wheat-sudangrass double-cropping in conjunction with native range (Sims, 1989)

When based 20% on crested wheatgrass pasture, 50% on native range, and 30% on Russian wildrye pasture, the acreage required per animal unit for a 7.5-month grazing season in the Northern Great Plains was only 11.4 acres compared to 24.8 acres when native range alone was used (Lodge, 1970). In Wyoming, a combination of seeded pastures (crested wheatgrass for spring, intermediate wheatgrass

[*Agropyron intermedium*] for summer, and Russian wildrye for fall) gave daily gains similar to those of native range when grazed with yearling cattle but had grazing capacity and gains per acre from 2 to 3 times those from the native range alone during the same time period (Lang and Landers, 1960). Manske and Conlon (1986) have described a complementary grazing plan for South Dakota using seeded range in conjunction with native range that extended the traditional grazing season for native range alone from 6 to 7.1 months with the capability to extend further to 8.4 months. Hart *et al.* (1988b) developed economic models for combining optimal acreages of dryland crested wheatgrass pasture or irrigated smooth brome-alfalfa pasture with native range in eastern Wyoming.

Cool-season grass seedings are often able to provides high-quality forage during early spring and late fall when associated warm-season range is dormant or of low quality. The addition of seeded range for grazing in the spring and again in the fall was compared at Manitou, CO, to year-round grazing cow-calf pairs on native bunchgrass foothill range and meadow (Currie, 1969). The seeded ranges used included Russian wildrye (April 15 to May 15), crested wheatgrass (May 15 to June 15), and Sherman big bluegrass (*Poa ampla*) (October 15 to December 15). The combination seeded pastures and native range increased weaning weights by 33 lb (451 vs. 418 lb) over the control treatment. This advantage, at least in part, resulted from increasing the adequate protein intake period from 8 to 10 months (Malechek, 1966).

Crested wheatgrass in the Intermountain Region has filled a critical role of providing a grazable forage that will meet the nutritional requirements for lactating livestock (and often big game as well) during early spring. While filling this critical gap in many complementary grazing programs, it has greatly reduced dependence upon harvested forages and supplements during the spring and permitted delayed entry onto native range. In central Utah, cows and calves gained an average of 1.73 and 2.02 lb per day, respectively, on crested wheatgrass foothill range but only 1.02 and 1.37 lb per day on native sagebrush-grass range during the same 5-week spring grazing period (Cook, 1966a).

It has been found that crested wheatgrass can be grazed satisfactorily from April to December (Harris *et al.,* 1968). Fall grazing can be continued into December when snow depth permits; even summer grazing has been satisfactory. At the end of a 3-month summer grazing season, calf weights from cows grazing seeded crested wheatgrass foothill range were similar to those from cows grazing adjoining mountain forest range, but the cows summered on crested wheatgrass weighed 20 lb less by the end of summer.

Irrigated pasture in areas of low rainfall or subject to drought can be used to provide reliable growing season balance in the forage supply. Nichols and Lesoing (1980) concluded that irrigated pasture can fulfill two prime roles in a complementary grazing program for beef cattle in Nebraska: (1) provide high-quality forage for yearling steers in late summer when native grasses are declining in forage value, and (2) provide early spring pasture for beef cows after calving to improve conception rates. The non-emergency use of irrigated pasture in cattle programs is best oriented to the production and sale of beef (Gomm and Turner, 1976). Therefore, the most productive gains from an acre of irrigated pasture will be from weaned stocker calves and yearlings rather than from cow-calf pairs. The early-weaned, fall-born calf is a prime candidate for irrigated pasture the following spring and summer; another is the 12-month-old yearling born in the spring and following overwintering at a light rate of gain.

Irrigated pastures capable of carrying 2 to 2.5 cow-calf pairs or 4 to 5 yearling steers per acre during the growing season are common in the Intermountain Region. Irrigated grass-legume pastures in Utah have provided up to 12.5 AUMs per acre during a May to September grazing season and produced from 1320 to 1650 pounds of beef per acre when grazed by steers (Harris *et al.*, 1958). However, these levels are obtained only under ideal soil conditions and intensive management of soil fertility, irrigation, and grazing. When optimal levels of management are applied and used strategically in livestock programs, well-managed irrigated pastures on high-quality tillable lands can compare favorably with other high-value cash crops.

Spring grazing of native hay meadows may be a practical means of increasing spring grazing without seriously affecting subsequent wild hay production. Continuation of grazing into mid-spring on hay meadows near Big Piney, WY, did not affect yield of the typical single cutting, and continuation into late spring depressed hay yield only slightly (Stewart and Clark, 1944):

Grazing discontinuation	AUD/acre grazing	Hay yield, (T/acre)	Total (T/acre)
Early (May 3 ave.)	10.8	1.66	1.79
Mid (May 26 ave.)	15.4	1.67	1.89
Late (June 8 ave.)	38.7	1.46	1.93

C. OVERWINTERING PROGRAMS

Young and Evans (1984) have concluded that the labor and capital demands of hay production for winter feeding of livestock are responsible for the resurgence of interest in winter grazing in the Intermountain Region. Forage reserves for winter grazing must be set aside from spring and summer growth, and there must be sufficient total grazing capacity to accommodate extending the total annual grazing season into winter. The decision to graze open native or seeded range during winter must consider probable and potential snow depths, snowfall frequency, and duration of snowpack. The remoteness of many range areas conducive to winter grazing raises concern for the care of livestock during winter storm emergencies; this is a problem particularly with cattle since they are not handled by herding as are sheep in the Intermountain Region. Emergency winter reserves must be available, and a means must be ensured of getting such to the livestock when urgently needed. Also, greater opportunity for individual nutrient deficiencies must be guarded against.

Salt-desert shrub ranges of the western U.S. in good condition are well adapted for winter grazing and maintenance of pregnant sheep and even beef cows in gestation, adjoin other range types mostly unsuited to winter grazing, and should generally be conserved for this use (Cook, 1971; Holmgren and Hutchings, 1972). However, this source of winter grazing capacity is much too limited to handle all of the demand (Fig. 4.8). On lower elevation range, cattle have been wintered satisfactorily on crested wheatgrass and other introduced wheatgrasses.

In an early Montana study, yearling steers were satisfactorily winter grazed on crested wheatgrass during a 140-day period when fed 3 lb of a 15% protein supplement daily, but gains were limited to 0.34 lb daily (Williams *et al.*, 1942). One rancher near Snowville, UT successfully wintered beef cows on range supporting a mixture of crested wheatgrass and native bluebunch wheatgrass for some 17 years (Fig. 4.9) (Malechek, 1986; Gade and Johnson, 1986). Hay feeding during this period was required only in three of the 17 winters when deep snow prevented access to the standing grass crop. When following the practice of early spring calving and weaning in mid-November, this overwintering practice resulted in a 93% average calf crop and 450- to 500-lb weaning weights. Native range was left interspersed with part of the crested wheatgrass seedings to provide shelter and forage during severe storms and for calving grounds.

In areas of substantial winter snowfall, fall regrowth is apt to be accessible only if part of the range blows free of snow; otherwise, plants with abundant standing dead material are more accessible to cows than forage trapped beneath the snow

FIGURE 4.8   Salt-desert shrub ranges in good condition, as shown here at the Desert Range Experiment Station in western Utah, are well adapted to winter grazing by cattle and sheep but are too limited to handle the full demand.

FIGURE 4.9 Crested wheatgrass is widely used in the northern Great Plains and the Intermountain Region to provide spring and fall grazing but can also be used for winter range when properly supplemented; showing crested wheatgrass winter range near Snowville, UT.

even if less nutritious (Houseal and Olson (1996). Seeded pastures of high-stature
forage grasses that protrude through the snow or are able to remain erect under
heavy snow loads are an alternative to using native range. One approach suggested for the Northern Great Plains and northern Intermountain Region has been to
remove the poorest haylands from production and establish pastures such as crested wheatgrass, tall wheatgrass (*Agropyron elongatum*), basin wildrye (*Elymus
cinereus*), or Altai wildrye (Majerus, 1992). Other possibilities include taller cultivars of Russian wildrye, intermediate or pubescent wheatgrass, or reed canarygrass.

Wheatgrasses and wildryes and their hybrids are being investigated for winter
grazing potential, with height, aboveground dry matter, carbohydrate (energy) levels, and total nitrogen and phosphorus content being examined (Johnson *et al.*,
1988). Both crested wheatgrass and Russian wildrye provide good-quality winter
forage but concentrate their productivity in the lower part of the plants; this low
stature is more subject to being covered by snow than taller growing grasses. Also,
the finer-stem cultivars of crested wheatgrass, even though often more palatable,
are subject to winter breakage and added loss of availability. Altai and basin
wildrye appear to maintain acceptable nutritional levels while remaining available
above deeper snow. Tall wheatgrass has in the past found considerable use in winter grazing. Reference has already been made to the potential of seeding grass-
shrub mixtures to provide nutritionally balanced forage for winter grazing.

Willms (1992) found that Altai wildrye was readily consumed by cattle in winter and was superior to crested wheatgrass and Russian wildrye in resisting weathering, being tall, and remaining erect after maturity; nitrogen fertilizer further increased forage yield and height of growth. Basin wildrye has also been found adapted to providing winter forage, shelter, and bedding for both cattle and big game animals (Majerus, 1992); the susceptibility of basin wildrye to having elevated growing points removed by spring and summer grazing, along with associated reductions in plant vigor and longevity, is mostly avoided under winter grazing.

D. ALTERNATIVE FORAGES FOR OVERWINTERING

Complementary grazing/harvested forage programs developed for west central Nebraska for use in conjunction with sandhills range (Nichols and Clanton, 1987) have included: (1) cornstalks as a fall and winter grazing resource in lieu of grazing winter range, and (2) the use of harvested forages as a protein supplement for winter range. The production or purchase of high-protein forages such as alfalfa hay for feeding, in conjunction with winter grass range, fills the role both of a complemental and a supplemental forage.

Similar to stockpiling is the routine practice of grazing off the aftermath residue on hayfields provided by herbaceous perennial plants after dormancy (Fig. 4.10). Fall or winter harvesting by grazing animals of the aftermath has generally resulted in no hay yield reduction the following year but rather has permitted utilizing herbage that would otherwise be wasted. After stockpiled forage or aftermath is depleted, livestock may remain on the meadows and be fed on the sod, providing deep treading and associated damage to the perennial plant rootcrowns are avoided. Frequently rotating the feeding site aids in the distribution of manure and urine, improves sanitary conditions, and lessens animal damage to the living sod (Taylor and Templeton, 1976; Baker et al., 1988).

In addition to the use of stockpiling for meshing the animal demand and forage production cycles, cutting for hay or haylage during flush growth or varying animal numbers in response to growth rates can be used on intensively managed, high production pastures (Blaser et al., 1974). Leaving the last cutting of hay as round bales in the field for grazing along with the aftermath is practiced in many areas of the Midwest (Fig. 6.9), particularly for beef cows (Wedin and Klopfenstein, 1985). Although such bales may be evenly dispersed for more uniform animal access and availability, limiting access by temporary fencing may be required for optimal utilization.

Rake-bunched meadow hay (last cutting) was found in Oregon to be a cost-effective strategy of overwintering pregnant beef cows (Angell et al., 1987); rake-bunched hay was more nutritional than equivalent standing crop, and cows readily opened up the windrows. Grazing windrowed hay, compared to baling hay, reduces labor and fuel costs. However, windrowed hay cannot be stored and carried over and must be used during the first fall/winter period, and the hay meadows cannot be irrigated in the fall after cutting and windrowing. Where big game ani-

FIGURE 4.10   Hay meadow aftermath can provide valuable grazing during fall and early winter, and sometimes even early spring as foremath, thereby helping to balance the year-round carrying capacity needs, such as on this Rocky Mountain cattle ranch.

mals are present in substantial to high numbers, losses of the windrowed hay through consumption, scattering, and contamination may be serious (May *et al.,* 1999).

Stockpiled perennial forages (regrowth following last cutting of hay) and corn crop residues for winter grazing by mid-term pregnant cows were compared in Iowa to those kept in drylot (Hitz and Russell, 1998). Grazing stockpiled tall fescue-alfalfa, smooth bromegrass-red clover (*Trifolium pratense*), and corn residues provided 85, 83, and 57 days of grazing per acre, respectively, before beginning hay feeding. During the total overwintering period, cows in the three grazing treatments, compared to cows confined to hay feeding in drylot, required 2352 lb, 2268 lb, and 1380 lb less hay dry matter to maintain a body condition score of 5. Grazing was begun following the first killing frost, the perennial pasture aftermath was grazed with strip grazing, and cows in the three grazing treatments were offered baled hay only when deemed necessary.

Secondary sources of carrying capacity such as stubble, crop residues, grain lost in harvesting, weed and volunteer herbage, and excess foliage yields on small grains should not be overlooked; such can often play critical roles in filling gaps in fall and winter carrying capacity. Even go-back farmland and range in poor condition invaded by cheatgrass brome (*Bromus tectorum*) may provide palatable and nutritious forage to fill a serious short-term forage gap in early spring (DeFlon, 1986). Added to the list of potential sources for winter grazing in Saskatchewan were frozen safflower and other failed grain crops, chaff piles or straw-chaff rows, matured grain-hybrid corn or cornstalk refuse, or cereal crops such as common oats (*Avena sativa*), common barley (*Hordeum vulgare*), or rye planted in June and swathed at late-milk to early-dough stage for winter grazing (Klein, 1994).

In corn producing areas such as the Midwest, cornstalk fields are widely available for grazing in late fall and winter (Clanton, 1988). This forage resource has been used primarily for maintaining breeding herds during the winter or putting weight on cull cows prior to sale; it can also be used for calves going subsequently to summer pasture if animals are supplemented and provided with adequate shelter during the winter. Shortened, high-density grazing periods are recommended for increasing grazing capacity by causing less selective grazing, the consumption of more low-quality material, and less trampling. Since cattle will tramp the leaves, husks, and grain and even the stalks into mud, it is suggested that cattle be removed from cornstalk fields when they become muddy or covered with snow and returned when the field dries or the ground freezes.

5

GRAZING EFFECTS ON
PLANTS AND SOIL

I. GRAZING EFFECTS IN PERSPECTIVE

Grazing animals affect individual plants and plant communities in several interrelated ways (Balph and Malechek, 1985), including: (1) plant defoliation, with effects on both plant morphology and physiology; (2) mechanical impacts on soil and plant materials through trampling; and (3) nutrient removal and redistribution through excreta. The short-term or immediate effects of grazing on a plant can: (1) be detrimental (i.e., reduce plant vigor or even kill it), (2) be beneficial (e.g., increase size or growth rate), or (3) have no apparent beneficial or negative effect on it. Most, if not all, plants can withstand some loss of foliage and still maintain their position in the plant community. Grazing by large herbivores in the short-term often is of little importance in the process of vegetation change, unless grazing is so excessive that the grazed plants cannot restore themselves (Dwyer *et al.,* 1984).

The perspective that grazing can potentially benefit plant growth and function has substantially altered scientific approach to plant-animal interactions (Briske and Richards, 1994). Several novel hypotheses have been proposed to explain the potentially beneficial effects of herbivory. These hypotheses have often been related to the following responses observed in various forage plant species following defoliation: (1) compensatory or increased photosynthetic rates by residual foliage, (2) accelerated growth rates, (3) modified carbohydrate allocation patterns, and (4) enhanced nutrient absorption (Briske and Richards, 1994).

The grazing optimization hypothesis states that primary production increases with an increasing intensity of grazing to an optimal level and then decreases with greater grazing intensity (Vavra *et al.,* 1994); however, this is only a hypothesis and not completely validated. If grazing optimization does occur, Pieper and Heitschmidt (1988) found it more plausible on mesic grazing lands and less likely to occur on arid and semi-arid rangelands. Plant growth stimulation by grazing seems most likely to occur when plants have high regrowth potential and are favored by optimal amount and timing of precipitation.

While it is possible that plants increase their growth rate following defoliation, this may only partially compensate for the total amount of biomass removed (Briske and Richards, 1994). Oesterheld and McNaughton (1991) concluded that defoliation of grasses by grazing, in decreasing order of likelihood, may stimulate: (1) photosynthetic rates, (2) relative growth rates, (3) yield to grazers, (4) live production or absolute growth rate, (5) total production, (6) final live biomass, and (6) litter or standing dead accumulation.

Oesterheld and McNaughton (1991) suggested that defoliation can release grass plants from the growth limitation imposed by their own accumulation of old and dead tissue and that this release may override the negative effects of the loss of biomass. They concluded that the compensatory response of grasses to grazing depends on the type and level of self-imposed stress-limiting growth. This was based on their finding that growth response to defoliation went from partial compensation when plants were growing at high relative growth rates to overcompensation when plants were more self-stressed and growing at low relative growth rates.

The long-term effects of grazing on plants will largely depend not only on their inherent morphological and physiological characteristics but also on their adaptation to local environmental factors and on the relative effects of grazing on associated plants and plant species. The extent to which a plant is under competitive pressure from other plants will determine in large part its tolerance of defoliation and trampling by grazing animals (Caldwell, 1984). Plants grazed less severely, capable of regrowing rapidly following defoliation, or possessing a combination of these resistance components realize a competitive advantage within the plant community (Briske and Richards, 1994). Recovery from grazing necessarily involves both the re-establishment of photosynthetic tissues and the ability to retain a competitive position in the plant community, with the latter being in part a consequence of the former (Caldwell and Richards, 1986).

Modifications of competitive interactions caused by herbivory may potentially constrain growth to a greater extent than the direct effects of biomass removal (Briske and Richards, 1994). Intense competition by surrounding plants may be more suppressive of foliage yield than severe defoliation, but the two in combination are additive in their negative effects on yield and survivability. Excessive defoliation reduces both root system activity and leaf area and may limit the plant's capacity to compete for and utilize soil moisture and nutrients. When this coincides with the comparatively short period of time when soil moisture and nutrients are available in semi-arid and arid environments, a forfeiture of these resources to competing plants can result (Caldwell, 1984).

II. DEFOLIATION AND PLANT MORPHOLOGY

A. FORAGE PLANT STRUCTURAL ORGANIZATION

Plant growth is an irreversible, quantitative increase in size that is accompanied by changes in plant form, structure, and general state of complexity of the plant (Dahl and Hyder, 1977). Growth is localized in meristems capable of cell division and in the subsequent enlargement of the existing cells. Cell division is mostly restricted to the buds and immature growth stages of leaves and twigs. Meristems constitute a very small part of each plant but give rise to and regulate all growth, differentiation of functional and structural tissues, and the development of form.

Meristems are named according to position on the plant, i.e., principally apical, intercalary, and lateral (Dahl and Hyder, 1977). **Apical** meristems occur universally at the tips of roots and stems and for a time at the leaf tips of vascular plants. The plant shoot's apical meristem along with the embryonic plant parts derived from it are known as a **shoot apex** or **growing point. Intercalary** meristems, soon separated from the apical meristem by nonmeristematic tissue, are located in narrow bands at the base of leaf blades, leaf sheaths, and internodes of monocots and petiole bases of dicot leaves. **Lateral** meristems are situated laterally in a plant organ (i.e., at leaf or twig axils or the cambium of trees and shrubs).

The modular unit of growth in plants is the **phytomer,** which consists of a leaf, an internode, an axillary bud or potential bud, and a node (Fig. 5.1). The phytomer develops from the apical meristem or growing point and is the basic unit of the **shoot,** which is a collective term applied to a growing stem and its leaves or any young growing branch or twig. Phytomers in graminoids are organized into **tillers**—young vegetative lateral shoots growing upward within the enveloping leaf sheath, the phytomers being initiated from the apical meristem from the base upwards (Briske, 1986). Tillers are further organized into anatomically attached groups forming the complex graminoid plant (Fig. 5.2). A grass tiller is composed

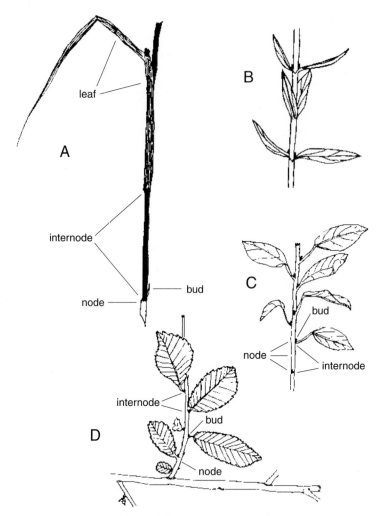

FIGURE 5.1   The phytomer (basic unit of the shoot) consists of a leaf, an internode, an axillary bud or potential bud, and a node; (A) grass, (B) forb (opposite leaves), (C) forb (alternate leaves), (D) shrub (Dahl and Hyder, 1977). (Society for Range Management drawing.)

of a single growing point (apical meristem), a stem, leaves, roots, nodes (joints), and dormant buds and the potentiality of producing a seedhead (Waller *et al.,* 1985) (Figs. 5.3 and 5.4).

The grass seedling is comprised of the primary tiller. The expansion and continuation of the young grass plant, however, is eventually dependent upon the original tiller producing buds from which develop the replacement tillers, i.e., secondary tillers, tertiary tillers, etc. (Fig. 5.5). The dormant or inactive buds have the potential to produce the new tillers (shoots), each with a new growing point. The

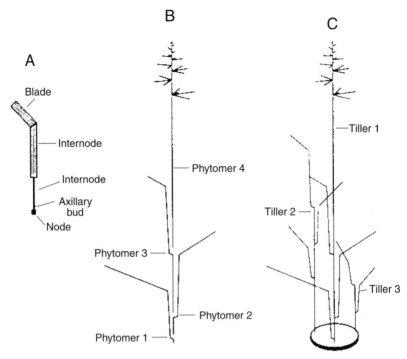

FIGURE 5.2   Structure of the graminoid life form at the (A) phytomer, (B) tiller, and (C) plant levels of organization (Briske, 1986). (Society for Range Management drawing.)

supply of buds is the origin of new tillers as the growing point of the old tiller is removed or dies, and these buds are what permit perennial plants to live from year to year. As long as the tiller is vegetative, it has the potential to produce multiple leaves. However, intercalary meristem development in the tiller ceases when the apical meristem dies or develops an inflorescence.

All grass tillers are initiated by a growing point developing from axillary buds of previous tiller generations located basally on the plant or at the nodes of stolons or rhizomes. These buds are at or below ground level until elevated, as when triggered to become reproductive. In effect, the tillers of most perennial grasses act as annuals or biennials. The tillers of a few graminoids, such as Arizona cottontop (*Digitaria californica*) (Cable, 1982) and sedges (*Carex* spp.) (White, 1973), may live three years or more. However, the leaves on the grass tillers generally live less than one year (often 2 to 4 months) in temperate climates.

Cool-season perennial grasses as well as winter annual grasses start growth in the fall, commonly maintain some basal green leaf material throughout the winter, and resume growth in the spring (White and Wight, 1973). Cool-season pasture grasses such as timothy (*Phleum pratense*), reed canarygrass, tall fescue, and smooth brome require tiller initiation in the fall and subsequent exposure to cold

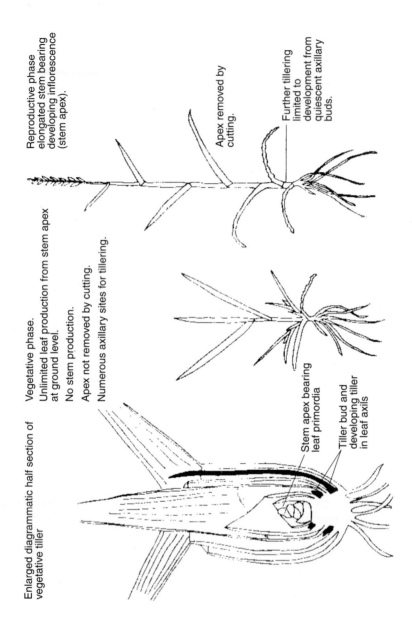

Enlarged diagrammatic half section of vegetative tiller

Vegetative phase.
Unlimited leaf production from stem apex at ground level.
No stem production.
Apex not removed by cutting.
Numerous axillary sites for tillering.

Reproductive phase elongated stem bearing developing inflorescence (stem apex).

Apex removed by cutting.

Further tillering limited to development from quiescent axillary buds.

Stem apex bearing leaf primordia

Tiller bud and developing tiller in leaf axils

FIGURE 5.3   The morphology of a temperate region perennial grass (Jewiss, 1972). (Society for Range Management drawing.)

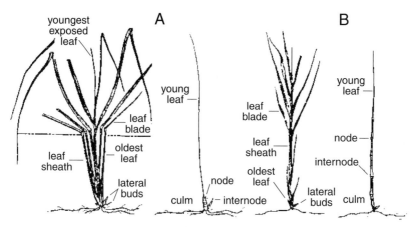

FIGURE 5.4 Vegetative shoots of (A) sea oats producing primarily short shoots and (B) bitter panicum producing primarily long shoots (Dahl and Hyder, 1977). (Society for Range Management drawing.)

temperature for formation of reproductive structures the following spring; tillers of perennial grasses initiated in the spring will generally not flower. In the case of indiangrass (*Sorgastrum nutans*) (Waller *et al.,* 1985) and blue grama (Sims *et al.,* 1973), both warm-season grasses, tillers develop in late summer and early fall, die back at frost, and become next year's shoots (Waller *et al.,* 1985). By contrast, all shoots of sand bluestem were found to be annual structures, dying back to the base at the end of the first growing season (Sims *et al.,* 1973).

The overwintering meristems in conjunction with developing axillary buds provide for initial biomass production each spring in perennial and winter annual forage plants. Morphological constraints influence growth form by determining individual tiller development and tiller arrangement in the complex grass plant. The capacity of plants to replace leaf tissue following defoliation is also subject to morphological constraints. Growth proceeds as phytomers are differentiated from the apical meristem associated with each tiller.

Following defoliation of grasses, the most rapid regrowth occurs from the previously differentiated cells of the intercalary meristems located at the base of leaf blades, leaf sheaths, and internodes of existing phytomers. This is followed by growth comprising the differentiation and development of additional phytomers from the extant tiller. Growth from axillary buds of tillers, rhizomes, or stolons is delayed until the bud is activated and the new tillers and their phytomers are differentiated. Tillers from aerially located axillary buds are the slowest and least productive with grasses but comprise a major source of growth for most forbs and shrubs (Briske, 1986; Waller *et al.,* 1985). Perennial forbs and many shrubs also have axillary buds located basally on stems or roots capable of producing new

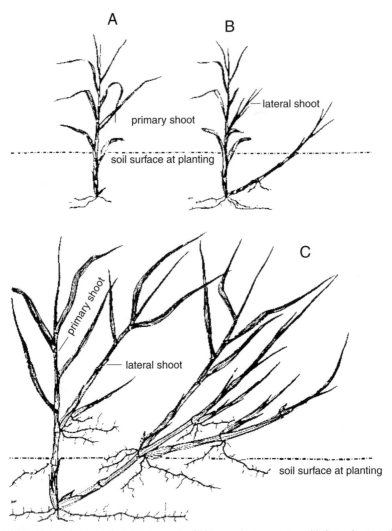

FIGURE 5.5   Lateral shoot development of bitter panicum, a grass with long shoots and a spreading growth habit. (A) a single shoot at time of planting, (B) lateral shoot development on the same shoot after a few weeks, (C) lateral shoot development at the end of the first growing season (Dahl and Hyder, 1977). (Society for Range Management drawing.)

shoots. When the apical meristem of the grass tiller assumes reproductive status or is removed by grazing, leaf replacement is dependent upon new tillers being formed from axillary buds and requires the greatest time interval following defoliation (Briske, 1986; Waller *et al.*, 1985; Briske, 1991).

Dead centers are a common characteristic of many bunchgrass species, but this is principally a natural morphological development rather than a negative response to grazing. Mean tiller replacement and percent tillering in crested wheatgrass is higher on the perimeter of the bunches, whether in the presence or absence of grazing. This differential replacement of tillers—primarily at the perimeter rather than in the core—is the apparent process leading to dead center development and breakup of crested wheatgrass clumps, particularly since it occurs with or without grazing (Olson and Richards, 1987). The hollow crowns of native bunchgrasses such as little bluestem (*Schizachyrium scoparium*) have been explained on the same basis (Briske and Brumfield, 1987). The latter authors concluded the dead centers provide physical barriers for tiller development, thus resulting in mostly peripheral development; however, it was also noted that heavy grazing did reduce tiller production on the periphery.

In many plant species the apical meristems located at the tips of stems and upper branches exert **apical dominance** over basal buds. Defoliation by grazing is one means by which the dominating apical buds are removed and the basal buds allowed to break dormancy and produce new shoots. Weak apical dominance is considered a valuable characteristic of range grasses—and possibly other forage species as well—since it permits the development of axillary shoots without removal of the growing points, thereby increasing herbage production compared to that of species with strong apical dominance (Cable, 1982).

B. DEFOLIATION BY GRAZING

Defoliation refers specifically to the removal of leaves from a plant, but the term is commonly extended to include the removal of edible stems, twigs, and inflorescences. Although primary attention here is directed to defoliation by grazing and browsing ungulates, defoliation can also result from insects and rodents, fire, hail, mowing, and contact herbicides. Since all forage plants can be safely and properly defoliated to some degree, the challenge in grazing management is to apply the maximum level that will still maintain optimum, sustained plant productivity. The component parts of optimum defoliation are

1. How much (intensity of defoliation; refer to Chapters 12 and 13 for more details)
2. When (seasonality of defoliation; refer to Chapter 4 for details)
3. How often and how long (frequency and duration of defoliation; refer to Chapter 15 for details)

The response of a forage plant to defoliation will depend upon: (1) availability of meristematic tissue and the developmental stage of new tiller buds, (2) carbon reserves and carbon balance, (3) remaining leaf area, (4) light interception, (5) time of year and physiological growth stage, (6) root area and root growth factors, and (7) the physical effects of grazing animals on plants and soils. However, these must be considered in light of the current environmental conditions and the constraints the latter may impose upon the forage plants.

The effects of defoliation on a grass tiller in the vegetative stage will depend upon which portions of the shoot are removed (Walton, 1983). If most of the leaf blades only are removed, the effect on growth rate of vigorous plants will be minimal if growing conditions are good. If a substantial part of the leaf sheaths of most leaves are removed, regrowth will be delayed, as most of the photosynthetically active parts of the plant will have been removed. If the apical dome along with all active photosynthetic areas is removed, the tiller will die, as the growing point will have been removed. Ideally, it will be replaced by the activation of one or more basal/lateral buds giving rise to new tillers. However, under most grazing conditions, in contrast with mowing, the degree of defoliation will be highly variable even between tillers on the same plant.

Some defoliation often promotes greater plant vigor than no defoliation at all (Heady, 1984). Defoliation may be advantageous in rejuvenating forage plants by: (1) reducing excess accumulations of standing dead vegetation and mulch that may chemically and physically inhibit new growth, (2) pruning away senescing and excess live biomass that may negatively affect net carbohydrate fixation, and (3) removing apical dominance that is more or less continuous in some species or exhibited in other species when inflorescences are developing.

From their studies on the Edwards Plateau of Texas, Reardon and Merrill (1976) suggested that even decreaser plants need some grazing in order to remain vigorous and productive. Zhang and Romo (1995) found that northern (or thickspike) wheatgrass (*Agropyron dasystachyum*) responded to repeated defoliation by increased tillering and longevity of tillers, this presumably enabling the species to increase herbage production in a relatively short time when grazing pressure is relieved. Fall tillers of crested wheatgrass were stimulated in growth rate by early spring grazing, were unaffected or reduced by mid-spring grazing, and were reduced by late spring grazing (Olson and Richards, 1988); mid-spring defoliation also stimulated the emergence of axillary spring tillers, but they contributed little additional forage because growth was limited by summer drought in this cold-desert ecosystem in Utah.

Briske and Richards (1994) have noted that physiological and morphological mechanisms potentially identified as capable of increasing plant growth following defoliation include compensatory photosynthesis, resource allocation, nutrient absorption, and shoot growth. While these compensatory mechanisms may frequently prevent plant growth from being suppressed by defoliation, these scientists concluded that only infrequently does defoliation increase total growth beyond that of undefoliated plants. Mitchell (1983) concluded that defoliation cannot be expected to increase the net primary production of cool-season native bunchgrasses, especially in xeric environments. However, Busso *et al.* (1990), in studies with crested wheatgrass and bluebunch wheatgrass, found that plants, after being exposed to prolonged periods of drought plus defoliation, may have rapid initial regrowth upon alleviation of these stresses; this phenomenon was attributed to high amounts of available carbohydrates being accumulated in their storage organs during stress.

While the commonly held belief that light to moderate defoliation stimulates tillering in bunchgrasses may not hold for all species, tillering is the method of replacing tillers removed by grazing or rendered nonfunctional by senescence. Although bunchgrasses are relatively long lived, tillers are relatively short lived (mostly 1 or occasionally 2 years) (Murphy and Briske, 1992). Thus, tiller initiation must occur annually to offset tiller mortality in order to maintain plant productivity, size, and competitive ability.

Prolific tiller production enables crested wheatgrass to tolerate grazing well. Richards *et al.* (1987) found that crested wheatgrass produced up to 18 times more new tillers immediately following a severe defoliation than did bluebunch wheatgrass. Moderate grazing, in contrast to heavy grazing, after the beginning of internode elongation did not reduce tiller numbers per plant the following year. The reduction in tiller numbers under heavy grazing was due to higher overwinter mortality rather than an inadequate number of tillers emerging in the fall. Tiller numbers in April were suggested as a measure of potential to produce forage for the year. Grazing of needle-and-thread and blue grama in western Nebraska increased the number of tillers per plant but reduced total organic reserves of both species (Reece *et al.*, 1988).

As the elongated grass tiller approaches the heading stage, it may be desirable to remove the young seedhead since the tiller has expanded most of its leaf area, and the quality of the tiller will begin to drop sharply (Moser, 1986). The palatability and quality of forage produced for later use in late summer or winter can be improved by defoliation if a large percentage of maturing growing points is removed from grasses and environmental conditions for regrowth are present. This practice is most effective when done early in the growing season, after growing points are elevated but before seed heads emerge (Reece, 1986).

The probable inhibition of the replacement tiller buds by the developing inflorescences is suggested as a reason for the poor growth during mid-summer of plants in improved pasture even in favorable environments (Jewiss, 1972). However, when growth through new leaf production slows or ceases because of a combination of phenological state and environmental limitations as commonly found on rangelands, different defoliation regimes often fail to enhance regrowth until after a rest period and/or a return to favorable environmental conditions (Brown and Stuth, 1986).

Tiller initiation in grasses and the development of sprouts from stem bases and roots of shrubs and forbs—or at least their acceleration—have traditionally been attributed to the removal of apical dominance. However, the concepts of an auxin produced in apical meristem directly inhibiting tiller initiation or indirectly inhibiting lateral bud growth—by either the production of a secondary inhibitor or the regulation of resource allocation to regions other than lateral buds—fail to explain the many inconsistencies. Apical meristem removal does not always promote tiller initiation in grasses, and tillering may occur in plants with intact apical meristems (Murphy and Briske, 1992).

Murphy and Briske (1992) further concluded that numerous environmental variables are known to exert strong influences on the timing and magnitude of tiller initiation in grasses and may play an equal or greater role than the physiological mechanisms responsible for apical dominance. Not only is the frequency and intensity of defoliation apt to affect tiller initiation, other potentially intervening factors include plant competition, resource availability (temperature and soil moisture and radiation quality and quantity), species-specific responses, and state of phenological development. In spite of a brief period of increased tiller recruitment immediately following defoliation, defoliation may not produce a greater number of tillers when evaluated over one or more growing seasons (Briske and Richards, 1994).

Browsing or even mechanical topping of desirable shrubs tends to remove the apical buds, often resulting in an increase in twig numbers due to the activation of lateral buds and associated lateral branching. Tueller and Tower (1979) found vegetation stagnation of shrubs fully protected in exclosures, wherein nonuse resulted in an average reduction in productivity of 70% in bitterbrush and 36% in big sagebrush. Mechanical topping of overmature bitterbrush shrubs in the northern Intermountain Region averaging 50% canopy removal every 6 years has resulted in a flush of new twig growth and a several-fold increase in browse yield (Ferguson, 1972). Similar results were reported with other shrub species by Garrison (1953).

Bilbrough and Richards (1988, 1993) reported that clipping stimulated bud elongation and growth in bitterbrush but not in big sagebrush. The greater grazing tolerance in bitterbrush was evidenced, in part, by increased node production and increased frequency and size of new twigs in the clipped plants, resulting in biomass production equal to or greater than production of unclipped plants. In New Mexico studies, browsing stimulated growth of fourwing saltbush but only when proper rest sequences were provided during the following growing season (Price *et al.*, 1989).

C. MORPHOLOGICAL RESISTANCE TO GRAZING

The ability of a plant to survive and maintain its abundance and productivity within a plant community subjected to grazing is dependent upon both an avoidance and tolerance component of grazing resistance (Briske 1986; Briske *et al.*, 1996):

Avoidance mechanisms are those that reduce the probability or severity of defoliation (i.e., plants are grazed lightly or infrequently); biochemical compounds that cause low palatability may be involved.

Tolerance mechanisms are those that facilitate regrowth following defoliation (i.e., plants readily replace photosynthetic tissues following defoliation).

Both mechanisms include morphological and physiological factors. These plant factors, when combined with various phenological and chemical factors, comprise the plant's defense against grazing.

Briske (1996) proposed the hypothesis that late-successional dominants, in contrast to mid- or early-seral dominants, rely primarily on tolerance mechanisms for grazing resistance. After the species relying on tolerance mechanisms have declined, herbivores will begin to successively graze species with progressively greater expression of avoidance mechanisms. This hypothesis continues that the species with the most well-developed avoidance mechanisms (mid- to low-seral dominants) will be grazed to a lesser extent, and this results in their relative abundance to remain constant or increase in response to continued intensive herbivory.

Morphological characteristics that provide resistance (tolerance or avoidance) to grazing in forage plants are commonly listed to include the following:

1. Low growing points
2. Delayed elevation of growing points
3. Elevation of growing points and foliage growth beyond reach of grazing/ browsing animals
4. Low, decumbent, or even prostrate foliage growth associated with low availability to herbivory, including protection by taller growing associated plant species
5. Predominance of vegetative only over reproductive shoots
6. An abundance or even variety of vegetative reproduction potential
7. Mechanical deterrent characteristics
8. Deep and expansive root systems, particularly for added drought tolerance and acquisition of minerals in short supply
9. Reduced length, increased tensile strength, and greater silication of leaves
10. Rapid completion of annual growth cycles, particularly characteristic of annuals

Vegetative tillers that do not elevate or delay elevation of their growing points (apical meristems) or produce a large proportion of vegetative only tillers are more resistant to grazing than those that do otherwise (Branson, 1953; Briske, 1986; Rechenthin, 1956). Early in the growing season all grasses and forbs have their perennating or growing points close to or at ground level. Since grazing animals cannot physically graze any closer than about one inch from the ground surface, there is no danger of removing the growing points at this time. Even when such tillers are grazed, they can continue to produce new leaf material. Grass species with a predominance of vegetative shoots generally have high photosynthetic potential for post-grazing recovery.

Little bluestem and sideoats grama (*Bouteloua curtipendula*) produce multiple, short, telescoped, basal internodes which remain basal until just before seedheads are produced (Rechenthin, 1956). On the other hand, both produce an abundance of inflorescence-oriented tillers (Branson, 1953), thus are intermediate in grazing resistance. Some grasses such as indiangrass and switchgrass have only 2 to 4 short basal nodes, and the growing points are elevated above ground level soon after growth starts and are within reach of grazing animals. By contrast, the vegetative growing points of Kentucky bluegrass, buffalograss (*Buchloe dactyloides*), blue grama, and white clover are low to the ground. Their low growth form also aids in avoiding severe defoliation.

The growing points of upright legume species are at the tips of elongated stems, which are frequently elevated; consequently, when tillers of these legume species are grazed, the growing points are removed and growth from them is terminated (Walton, 1983). Grass species with few vegetative but mostly inflorescence-oriented tillers are Canada wildrye (*Elymus canadensis*), switchgrass, plains muhly (*Muhlenbergia cuspidata*), marsh muhly (*M. racemosa*), little bluestem, and crested wheatgrass (Branson, 1953; Rechenthin, 1956). All except the latter two have rather low grazing resistance. The moderate grazing resistance of little bluestem is attributed, in part, to delayed elevation of its growing points. The high tolerance of defoliation by crested wheatgrass comes primarily from its physiological characteristics.

Low foliage accessibility and thus avoidance of grazing can result from low stature (culmless, decumbent, or prostrate growth habit), very short leaves, an accumulation of unattractive culms, or elevation above the reach of defoliation, such as for many large shrubs and trees. Mechanical deterrents as grazing avoidance mechanisms include thorns, sharp twig ends, and inflorescence awns. Epidermal characteristics such as pubescence, prickles, waxes, and silication reduce the probability of defoliation, as do high tensile strength and prominent vascular bundles in the leaves (Briske, 1986).

In contrast with many herbaceous plants and some half-shrubs, the perennating or annual growth points of woody plants are elevated for ready access by browsing unless protected by other characteristics. One defensive mechanism of trees and tall shrubs is growing above the reach of grazing animals, often resulting in a highline at the upper level of reach of the specific animal species (Fig. 5.6A). Another defensive mechanism against continuing heavy browsing by shrubs is developing a low, compact, and tightly hedged form (Fig. 5.6B), which is observed in such plants as antelope bitterbrush, true mountain mahogany (*Cercocarpus montanus*), blackbrush (*Coleogyne ramosissima*), fourwing saltbush, and rubber rabbitbrush (Plummer, 1975; McConnell and Smith, 1977). The hedged form permits the presence of leaf growth within the protection of short, profuse lateral branches not readily penetrated by most browsing animals. New growth outside the hedge line will be readily accessed by browsers.

Although the hedged form permitted secondary leaders to develop on fourwing saltbush in New Mexico (Price *et al.,* 1989), few were able to develop under continuous heavy cattle browsing because of low plant vigor. While lightly to moderately browsed plants of jojoba (*Simmondsia chinensis*) in Arizona had greater twig growth compared to ungrazed plants (attributed to compensatory growth), biomass yields were similar (Roundy and Ruyle, 1989). The increased twig growth was negated by the reduction in canopy size, the diminutive size minimizing the base from which leaders could develop. Periodic spring rest—the season of greatest browsing impact—or control of browsing intensity was recommended to maintain individual shrub size and thus maintain or increase total production. Teague (1992)

FIGURE 5.6 Morphological adaptations in shrubs that provide defensive mechanisms against browsing include (A) elevation of growing points above the browsing level (curlleaf mountainmahogany (*Cercocarpus ledifolia*)) and (B) developing a low, compact, and tightly hedged form (bitterbrush).

concluded that in the short-term, frequent defoliation tends to increase production of many shrubs but that such treatment, if continued over 3 or more years, can be expected to cause the plants to decline in vigor and size.

Shifting from a taller, upright growth form to a shorter, more spreading growth form as a defensive response to long-term heavy grazing has also been observed in a number of grass species, including crested wheatgrass, western wheatgrass, needle-and-thread, and the gramas. The shorter growth, often accompanied by an increase in total number of plants, is more resistant to grazing because livestock are less able to physically remove as much of the foliage as from the larger, more upright growth form. Hickey (1961) found that crested wheatgrass was more spreading on dry than on moist sites and where 70% or more of the herbage was removed annually; however, the past intensity of cattle grazing had a greater effect upon growth form than did the site.

A shift by little bluestem from a low density of large plants to a high density of small plants in response to defoliation appears to increase herbivory tolerance through increased tillering ability (Anderson and Briske, 1989). Carman and Briske (1986) found that little bluestem plants with a history of grazing had a lower growth form than those without recent past grazing history. The grazed plants also had an increased number of tillers, thus an increased number of growing points from which growth could occur following defoliation. While the little bluestem plants with a grazing history had lower mean tiller weight, they had equivalent or greater herbage production and greater basal area expansion (Briske and Anderson, 1992).

Being unable to demonstrate with little bluestem that contrasting growth form was a genetic selection response, Carman and Briske (1986) suggested these differences might have been an environmental response to different microsites. However, further study of little bluestem led to the revised conclusion that long-term herbivory had selected against genotypes possessing an erect canopy architecture (Briske and Anderson, 1992; Briske and Richards, 1994). The decumbent genotypes possessed greater grazing avoidance because of greater amount of meristematic and photosynthetic tissues protected from removal by grazing animals to facilitate growth following defoliation. Based on the little bluestem study, it was concluded that herbivory by domestic cattle can function as a selection pressure to induce architectural variation in grass populations within an ecological time frame equal to or even less than 25 years (Briske and Anderson, 1992).

Based on the little bluestem study, it was assumed that avoidance as a component of grazing resistance can be affected by grazing-induced selection as well as can the grazing tolerance component (Briske and Richards, 1994). In a comparison of creeping-rooted and tap-rooted alfalfa cultivars on dryland near Cheyenne, WY (Gdara et al., 1991), it seemed apparent that the greater persistence of the creeping-rooted cultivars under grazing did not result from greater intrinsic productivity or more rapid physiological recovery from defoliation. Rather, the more rapid recovery of the creeping-rooted cultivars apparently resulted from the open stands formed which increased the probability that some stems would escape defoliation at each grazing. Brummer and Bouton (1992) concluded that criteria for selecting grazing-tolerant alfalfa cultivars should include the genetic potential to store large amounts of carbohydrates (tolerance), the decumbent or semi-decumbent growth habit (avoidance), and the production of many stems and crown buds (tolerance); they also concluded that stubble leaf area was important in allowing maintenance of total nonstructural carbohydrates under frequent defoliation.

Bunchgrasses may be eliminated from a site by intensive, long-term grazing. The process of declining range trend in bunchgrass communities is probably initiated by the decrease of basal areas of individual plants (Butler and Briske, 1986). The fragmentation of large plants into smaller plants may initially increase total basal area as well as number of tillers. With continued intensive grazing, however, plant basal areas are eventually reduced below a critical size, and plant mortality increases in response to extreme environmental conditions and to competition from neighboring plants that are grazed less severely.

Root patterns under nongrazing or light grazing are generally dense, heavily branched, spreading, and deeply penetrating. Under progressively heavier grazing, roots have progressively fewer branches and are sparser, shorter, and more concentrated in the top portion of the soil profile (Fig. 5.7). Svejcar and Christiansen (1987b) found that continuous, heavy grazing reduced both the root mass and the root length of Caucasian bluestem but reduced the leaf area even more. This improvement in the ratio of root surface area to leaf area and associated improved water status may have been partly compensatory to the intense defoliation.

D. GRAZING AND PLANT REPRODUCTION

Herbivory may adversely or even beneficially affect reproduction by seed in existing forage stands but often has minimal effect, depending on the circumstances involved (Sindelar, 1988). Perennial grass-dominated range communities at higher levels of succession, such as in the Northern Great Plains, have minimal recruitment requirements (Sindelar, 1988). On excellent condition range, plant mortality is so low and competitive dominance is so complete that seedlings of climax grass species are rarely found. Where ranges are in low condition, recruitment of additional individuals is typically desirable but is relatively slow if required from seed alone, unless the vegetation exists as an open community or is made so by cultural treatment. Major obstacles to recruitment, particularly from seed, are weather variables and seed depredation as well as competition from nearby plants.

Hyder *et al.* (1975) concluded that the role of seed is most substantial and profound in the earliest stages of secondary plant succession on long-term grazing lands. Seed production in a grazing program was found important for perpetuating annuals and pioneering-type perennials, but increases by dominant and subdominant perennial species were primarily related to physiological, morphological, and climatological factors. Recruitment of new grass plants via rhizomes or stolons can proceed much faster and is apt to be more amenable to range improvement treatments (i.e., short of complete seedbed preparation and reseeding). Scholl and Kinucan (1996) concluded that grazing favored curly mesquite (*Hilaria belangeri*) as a rapid colonizer because of its stoloniferous growth habit, but its

FIGURE 5.7   Excessive levels of defoliation deleteriously affect forage plants, including reduction of the mass, spread, branching, and penetration depth of the roots; (*left to right*) light, moderate, and heavy defoliation levels.

capacity as a prolific seed producer and its rapid response to soil moisture added to its competitive advantage.

The "seed theory" of managing rangeland and perennial pastures attempts to explain and justify certain grazing practices (e.g., deferred grazing and rest-rotation grazing) in terms of seed production, planting of seed by trampling, and subsequent seedling establishment (Hyder *et al.,* 1975). Emphasis on promoting seed reproduction of perennial grasses for the maintenance or improvement of many ranges is probably not justified. While reproduction from seed is an important survival strategy for perennial grasses, vegetative reproduction is clearly more efficient and reliable (Sindelar, 1988). Nevertheless, reproductive potency may be a reasonable indicator of a favorable carbon and nutrient budget and thus a useful indicator of vigor of perennial grasses (Caldwell, 1984).

From his review of the role of reproductive efficiency in perennial grasses, Sindelar (1988) came to the following conclusions:

1. In most perennial grasses, reproduction from seed is relatively infrequent; vegetative reproduction plays the major role in perpetuation of perennial grasses.
2. Grazing practices directed at improving reproduction from seed are less important than is commonly believed.
3. Long-term range rest may not favor grass reproduction from seed in many rangeland environments.
4. Grazing management practices to improve rangeland should encourage vegetative reproduction with secondary consideration for reproduction from seed.

Rhizomes, stolons, basal stem or root sprouting, tillering, and layering are all vegetative methods of reproduction that complement or even replace the need for seed production in established stands of many perennial forage plant species. For example, black grama (*Bouteloua eriopoda*) depends on the production of an adequate number of stolons for recovery, while reproduction from seed, even when produced, is limited by the naturally droughty climate (Valentine, 1970). Briske and Stuth (1986) determined that a portion of the tillers on multi-tillered plants within a pasture—this presumably being true also for many single-stemmed plant species—generally remain ungrazed even under heavy grazing. Although this results in an inefficient harvest of available tillers on plants, it does permit substantial seed production under all except extreme grazing pressures.

Annuals, biennials, and some perennials have minimal capability for vegetative reproduction and must rely on seed production. Grazing methods and intensities are less important with annuals than with perennials since individual plant longevity is not a problem with annuals; annuals mostly evade the effects of grazing by rapid and prolific seed production. Many desirable forbs and shrubs on rangelands depend upon seed production for plant recruitment. For such species, the production of adequate fruiting shoots and adequate seed yield is important.

The removal of aftermath forage by judicious grazing following seed harvest of perennial, herbaceous forage species grown primarily for seed is a widely recommended practice in increasing the seed crop the following year. When winter wheat grows too rank, its extra foliage growth may be reduced to advantage by grazing. Grazing winter annual cereal crops grown primarily for grain, particularly wheat and barley in the central and southerly latitudes in the U.S., also permits the beneficial use of excess foliage produced in the fall and late winter. This can often be accomplished without reducing grain production and may even improve grain yield. Grazing in the central and southern Great Plains tends to reduce the number of heads per plant but increases the number of fertile florets in the remaining heads and helps prevent lodging; grazing also delays maturity slightly (Dunphy *et al.,* 1982; Harwell *et al.,* 1976; Swanson and Anderson, 1951; Winter and Thompson, 1987).

It is apparently not the amount of forage removed by grazing, at least prior to jointing or bolting (stage at which the growing points begin to elevate above ground level), but rather the length of time the grazing period is extended into the spring that primarily reduces grain yield of wheat and barley. Grazing animals must be removed prior to jointing to maximize grain yields. The negative effects of defoliation on grain yield during the jointing stage have been attributed to severe reduction in photosynthate production resulting from reduced leaf area during a high demand period rather than change (reduction or increase) in number of tillers (Dunphy *et al.,* 1982; Sharrow and Motazedian, 1987; Winter and Thompson, 1987; Winter, 1994). A contributing factor is the removal of the growing points by late grazing. Redmon *et al.* (1996) concluded that the first-hollow-stem stage was the best time for removing grazing in Oklahoma for maximizing net return for a combination of forage and grain yield; they also suggested that monitoring vegetative development be done in ungrazed patches since grazing tended to delay the indicator criteria.

In more northerly climates, the production of excess foliage, particularly with spring planting, is often insufficient to justify complementary grazing of wheat and barley grown for grain production. Also, the short, high yielding cultivars of wheat have been found in Texas studies to be more sensitive to excessive spring grazing than are the older, taller cultivars (Winter, 1994); the conclusion was that the use of the taller varieties and these planted early in the fall provided the optimal conditions for grazing and grain yield combinations. Nevertheless, if grain yield prospects become minimal because of adverse weather, disease, or insect damage, grazing can be continued as the sole product for the year. Also, at most latitudes and elevations receiving 15 in. average annual precipitation or more, cereals can be considered for use solely as a forage crop.

III. DEFOLIATION AND PLANT PHYSIOLOGY

A. CARBOHYDRATE RESERVES

The soluble carbon compounds found in cell contents—primarily sugars and starches—largely comprise the reserve food material in plants. Proteins, fats, and oils comprise lesser reserve substances; however, the complex, structural carbohydrates of cell walls such as cellulose are unavailable for reserve use. The pool of nonstructural carbohydrates is collectively referred to as total available carbohydrates (TAC) or sometimes total nonstructural carbohydrates (TNC). From this labile pool, the plant can draw material to offset both major and minor fluctuations in the levels of simple sugars needed for maintenance respiration, initial growth, and many other routine and emergency needs. The concentration of TAC in various parts of the plant changes with the production of photosynthetic products and their translocation to various parts of the plant. The rate of TAC use and export from storage to other areas of the plant also changes TAC concentration. Thus, any factor affecting the rate of photosynthesis or utilization of soluble carbohydrates will affect the TAC plant reserves.

TAC reserve levels have commonly been used as a key indicator of plant vigor and an index of the consequences of grazing (Caldwell, 1984). Cook (1971) concluded that the fall level of TAC in a forage plant is a good index of treatment severity during the previous growing season, but climate can also influence TAC storage levels and the effects must be carefully distinguished. The advantages of adequate TAC food reserves in perennial forage plants—moderate levels may be as advantageous as very high levels—are commonly summarized as follows:

1. Maintain overall high plant vigor.
2. Support perennial plant function during dormancy, principally respiration.
3. Enable earlier and more rapid regrowth following dormancy or severe defoliation.
4. Promote extensive root and rhizome growth.
5. Increase both vegetative reproduction and seed production.
6. Provide higher drought, frost, and heat tolerance.
7. Maintain high resistance to insect and disease injury.
8. Promote root nodulation in legumes; lack of TAC causes root nodules to cease functioning and be shed into the soil (Walton, 1983).

If a perennial plant is kept at depleted TAC levels for an extended time by overgrazing or in combination with severe restriction of environmental resources required for growth, the following sequential set of events can occur (Moser, 1986): (1) reduced root growth, (2) reduced tiller bud development, (3) reduced rhizome development, (4) reduced forage yield, and (5) even death, particularly when triggered by some adverse environmental factor. This downward spiral in plant vigor relates largely to the lack of initiation and survival of new tillers, which in turn relate to inadequate TAC levels.

Hyder and Sneva (1959) attributed the following TAC-related advantages to crested wheatgrass (*A. desertorum*): (1) rapid accumulation of abundant leafage (photosynthetic potential), (2) moderately early accumulation of TAC reserves, and (3) relatively high storage levels. However, Richards *et al.* (1987) and Richards (1984) concluded that it was morphological-developmental flexibility, including resource partitioning between above and below ground plant parts, rather than large amounts of stored TAC per se, that accounted for the rapid recovery of crested wheatgrass plants following grazing. Other factors adding to high grazing tolerance in crested wheatgrass are apparently: (1) delay and then rapid elevation of flowering culms, (2) short basal internodes, (3) a short pre-flowering period, and (4) a low degree of apical dominance.

The concentration of TAC reserves follows rather consistent annual trends in perennial plant species. Maximum concentrations usually occur immediately before fall dormancy, after which these levels gradually decrease over the dormancy period, followed by a period of rapid depletion during the initiation of spring growth. As the leaf is first developing, it may be in a negative carbon balance even during the daylight period (Caldwell, 1984). Photosynthetic competence increases to a maximum at about the time of full-leaf expansion. Thereafter, there is a slow decline of photosynthetic capacity until leaf senescence is complete.

Carbohydrate reserve levels in perennial plants generally follow a V-shaped or U-shaped cycle beginning with the breaking of spring dormancy (Trlica, 1977). In plants that exhibit the V-shaped seasonal cycle in TAC reserve levels, a rapid drawdown of reserves accompanies initiation of spring growth, but this is followed by a rapid accumulation of reserves after the low point in reserve levels has been reached. An example is crested wheatgrass, which experiences an early rapid recovery during the growing season, a moderate decrease during flowering, but a final recovery to its pre-heading high (Hyder and Sneva, 1959). In plants with a U-shaped TAC reserve cycle, low levels normally remain during active growth, with reserve stores being replenished only after growth rates decline as plants approach maturity or seed shatter.

The seasonal variation of TAC reserves differs among species. For many species the lowest reserve levels occur during early growth, but in others the lowest level is found at seed ripening. The highest TAC levels are generally found as senescence approaches, but an additional drawdown in reserves may accompany fall regrowth. In addition to the stage of plant development, other factors that can drastically change the reserve level are temperature, water stress, nitrogen fertilization, and defoliation (White, 1973).

Stored carbohydrates (TAC) are the only source available for winter survival and initial spring growth of deciduous perennials. Also, following intensive defoliation by heavy grazing or cutting for hay, developing buds must depend on storage TAC if no residual leaf material remains for photosynthesis. Once some leaf material is restored, photosynthesis occurs, and soluble carbohydrates are again manufactured and made available. However, growth and seed production have priority over storage for carbohydrate use (Waller et al., 1985). Carbohydrate storage increases primarily after leaf area has expanded and the growth rate slows.

The dependence of regrowth upon TAC following defoliation will be minimal and of short duration if adequate leaf area remains for continuing photosynthesis (Fig. 5.8). With adequate leaf area remaining, regrowth following defoliation may depend on stored TAC for energy for as little as 2 to 4 days (Caldwell, 1984; Richards and Caldwell, 1985). However, dependence on organic reserve com-

FIGURE 5.8 Rapid recovery of forage plants following defoliation resulting in regrowth depends on both adequate remaining leaf area and available carbohydrate (TAC) reserves.

pounds, both protein and carbohydrates, will be magnified and prolonged as severity of defoliation increases, particularly when regrowth is dependent upon replacement by new tillers (Vickery, 1981). Defoliation intensive enough to remove the apical meristems in switchgrass resulted in a substantially greater decline in TAC than lighter levels that did not remove the apexes (Anderson *et al.*, 1989), suggesting that a lengthy recovery period will be required when most apical meristems and leaf area are removed.

White (1973) concluded that TAC reserves in grass stem bases affect regrowth rate for 2 to 7 days following herbage removal, while other factors such as leaf area and nutrient uptake subsequently become of primary importance. However, Richards *et al.* (1987) concluded that within a few hours to at maximum two days following intensive defoliation, root growth and root respiratory activity are significantly depressed through lack of readily mobilized TAC, this being quickly utilized by continuing plant respiration and growth processes following defoliation. High TAC will not prolong growth for long if meristematic activity is not high.

Caldwell (1984) concluded that the labile carbon (TAC) pools in forage plants should be thought of as a small buffer, rather than as a large reservoir, which nevertheless plays a critical role in regrowth of foliage when the plant is unable to support regrowth directly from new photosynthesis. The absolute size of the TAC pool varies considerably from species to species, but only a relatively small amount of labile carbon may be all that is necessary to bring on recovery. The total TAC level in the plant provides a more meaningful assessment of carbohydrate reserves than does the concentrations of these compounds in selected plant parts or individual storage organs (Caldwell, 1984).

While perennial forage plants are influenced by conditions in current and preceding years which affect their TAC reserves and spring regrowth, annual forage plants are not. While annual forage plants have regrowth potential when defoliated during their single growth season, they are primarily dependent upon remaining leaf area rather than storage to provide the plant nutrients needed for regrowth. Annual rangelands in California are comprised mostly of annual rather than perennial forage grasses. These annual plants depend upon an adequate but not excessive seed bank in the soil and respond favorably to fall weather that promotes plant germination and establishment, to high water-holding capacity of the soil, and to optimal amounts of residue in the form of mulch and litter (Clawson *et al.,* 1982). The emphasis on annual rangelands, then, properly shifts away from level and frequency of defoliation to leaving adequate dry matter residue to provide adequate soil organic matter and water-holding capacity for favorable microenvironments needed for early seedling growth while preventing or containing erosion.

B. CARBOHYDRATE STORAGE AND TRANSFER

Nonstructural reserve carbohydrates may be stored temporarily in most perennating plant parts and even leaves (Trlica, 1977). Storage organs for carbohydrate reserve substances occur both above and below ground. These organs include roots (e.g., alfalfa and other forage legumes), rhizomes (e.g., smooth brome, reed canarygrass, and western wheatgrass), stolons (e.g., Ladino white clover, buffalograss, and bermudagrass), stem bases (orchardgrass and big and little bluestem), seeds (particularly annuals), leaf bases, and tubers, bulbs, and corms. Twigs of woody species are also important in TAC storage (Cook, 1966c). In grasses, TAC reserves are stored mostly in the lower regions of the stems, but nonstructural carbohydrates located in the fine roots are probably unavailable for use by the shoot system pool for regrowth of herbage (Caldwell, 1984; White, 1973).

TAC storage in shrubs occurs primarily in the rather small diameter portions of the roots and secondarily in the older materials of the canopy tops (Garrison, 1972). The twigs of woody species are important storage locations. Menke and Trlica (1983) concluded that root reserves in range plant species were not totally effective in maintaining aboveground stem reserves in the absence of aboveground contributions. TAC storage in plant parts of grasses and shrubs not generally preferred and defoliated by large herbivores adds materially to their tolerance of grazing.

The efficiency with which a species directs available carbon to aboveground growing points and utilizes it for synthesis of new foliage appears to be a key physiological feature which determines the plant's ability to tolerate defoliation by large grazing animals (Richards and Caldwell, 1985). Bilbrough and Richards (1993) concluded that the greater tolerance of bitterbrush than big sagebrush to herbivory was due to its greater ability to activate buds and then allocate TAC to their development as long shoots rather than higher storage levels of TAC or higher bud availability per se.

The relatively greater tolerance of crested wheatgrass to defoliation compared to bluebunch wheatgrass has been attributed to allocation of carbohydrates (Richards, 1984). A 50% reduction in root growth in crested wheatgrass following defoliation correlated with the allocation of relatively more TAC resources to aboveground regrowth, thus aiding re-establishment of a shoot-root balance. In contrast, root growth of bluebunch wheatgrass continued relatively unabated during the 90-day recovery period. In both species, carbohydrates necessary for continued root growth following defoliation were supplied primarily by photosynthesis during regrowth rather than by carbohydrates synthesized before defoliation.

Sinks refer to those areas in a plant where consumption of carbohydrates occurs and into which carbohydrates are channeled from other parts of the plant. The stronger a sink is and the closer it is to the source, the more carbohydrate it receives (Moser, 1986). The strongest sink is an actively growing area in the shoot; a seedhead is also a strong sink. Generally, root growth does not occur at the same time as active tiller growth because the growing areas in the tillers are strong sinks located relatively close to the source of carbohydrate manufacture. Root sinks are generally weaker sinks in most species, with the weakest sinks being the storage of reserve carbohydrates.

TAC pools within root systems are primarily used for root growth and respiration and are not readily remobilized for subsequent use in shoots following defoliation (Briske and Richards, 1994). Therefore, a decrease in TAC pools in root systems following defoliation results more from a reduction of current photosynthate allocation to the root system and its continued utilization in root respiration than to remobilization and allocation to the shoot system. Carbohydrate pools play an important role in initiating plant growth when photosynthetic capacity is severely limited, but the limited amount of carbohydrate stored in tiller bases of grasses limits their use as an effective index of shoot regrowth in perennial grasses and potentially other growth forms as well.

Also to be considered is the inaccessibility of root carbohydrates to support shoot growth and the poor correlation between shoot growth and carbohydrate concentrations or pools limits. The reduction in the amount of TAC available to active tillers following defoliation will mostly be compensated by remobilization and export from remaining photosynthetically active leaves to the regrowing shoots rather than to the roots (Briske and Richards, 1994). Herbivory tolerance among warm-season perennial grasses studied in Texas was associated with the capacity to rapidly reprioritize TAC allocation to the shoots rather than to the roots, but the relative ability of some species, such as purple threeawn (*Aristida purpurea*), to avoid herbivory was considered possibly more important in herbivory resistance (Briske *et al.*, 1996).

Both carbohydrate and nitrogen were shown by Welker and Briske (1986) to be shared among tillers of individual little bluestem plants. Both resources were rapidly transported from parent to daughter tillers, and the transport continued even after the newly initiated tillers had developed several roots and leaves and attained a height of 4 to 12 in. Nitrogen was observed to be readily shared among three generations of attached tillers within the complex, multi-tillered grass plants. Following random tiller defoliation to a height of 2 in, carbohydrate import began to increase within 30 min, and the rate of both carbohydrate and nitrogen import approximately doubled but slowed as leaf tissue on the defoliated tiller began to regrow. It was also demonstrated by Stout and Brooke (1985b) that clipped sod patches of pinegrass were provided carbohydrates through connecting rhizomes from intact tillers.

Continued carbohydrate transport among tillers provides a mechanism for rapid TAC re-allocation within the complex plant (Briske, 1986). Anatomical connections allow daughter tillers to be nurtured until becoming self-sufficient; any tiller following defoliation can once again import available growth resources from other connecting tillers. The capacity for rapid resource allocation among structural units within the same plant substantially increases survival and competitive ability in response to a variety of stresses within the environmental complex. Thus, grazing can be expected to be less detrimental than uniformly close defoliation by clipping. Except when excessive, grazing typically leaves some ungrazed tillers on a plant while removing others, thus allowing for the transfer and import of carbohydrates and nitrogen.

The time of year and phenological stage in which the plant is subjected to defoliation will have a great bearing on the severity of defoliation the plant can tolerate (Caldwell, 1984). Although grazing during any growth period will reduce TAC reserves, the impact is short-lived if defoliation is not complete and continuous and environmental conditions are adequate for recovery. For surviving the winter, beginning growth in the spring, and recovering after complete defoliation, Waller *et al.* (1985) have recommended the following practices to maintain adequate TAC levels in perennial grasses:

1. Delay initial spring defoliation or keep early defoliation periods short.
2. Allow adequate leaf area to remain at the conclusion of a defoliation period.
3. Allow adequate time between defoliations to permit leaf area and TAC reserves to accumulate.
4. Allow adequate residual leaf area and enough time late in the growing season to permit TAC build-up and bud development prior to dormancy.

The effects of grazing upon TAC levels by the end of the growing season depend upon the number of times the plant is grazed and the proportion of the photosynthetic tissue allowed to remain after each defoliation (Cook, 1966c). The more frequent and more intense grazing treatments reduce the number of roots and rhizomes and the level of food reserves remaining. Kothmann (1984) concluded that the stems and seedheads of reproductive tillers in grasses represent a major use of energy and that grazing during rapid plant growth will optimally divert some of the energy to growth of new vegetative tillers and roots. On the other hand, desert plants, particularly shrubs, will not tolerate heavy and continuous spring use because they do not have an opportunity for regrowth and TAC replenishment prior to and during normally dry summers. Shrub defoliation either during rapid growth (about May 1 in desert environments) or at maturity (about July 1) greatly reduced fall TAC reserves (Cook, 1971), while defoliation in early spring (about April 1) or during quiescence in the fall or winter had the least effect.

The management of grasses in late summer and fall (the period of maximum TAC storage rate) not only affects winter survival, but also has an impact on bud initiation (Waller *et al.,* 1985). The lack of spring vigor in a grass stand may be caused by a lack of development of tiller buds the previous year due to low TAC levels. Severe defoliation late in the growing season not only has a negative effect on stored TAC, but also removes nearly all of the insulation and protection of buds at the root crown or newly developed tillers against frost damage. While adequate TAC levels going into fall dormancy are necessary to prevent winterkill in northerly climates, ample reserves are equally important for surviving summer dormancy in more southerly hot summer climates.

C. LEAF AREA

Apart from stored labile carbon pools in the plant, recovery from defoliation depends on the quantity of the remaining foliage and its photosynthetic capacity; also important is the rate of development of new foliage and the photosynthetic capacity of the new leaves (Caldwell, 1984). If the available carbon buffer is limited, as is likely the case for plants under many grazing conditions, the photosynthetic capacity of the remaining plant canopy is of critical importance. The degree of defoliation during the growing season should be designed to allow enough leaf area to remain to provide carbohydrates for regrowth rather than prolonged dependence upon stored TAC (Burns, 1984; Waller *et al.,* 1985). If adequate leaf area remains after defoliation, the plant can regrow with minimal demand upon storage TAC. Consequently, the remaining leaf area after grazing plays an important role in regrowth during the remainder of the grazing season and in replenishing TAC reserves. The possibility of compensatory carbohydrate production in the intact leaves and tillers also exists.

Adequate leaf area remaining after defoliation for light interception and photosynthesis is important in promoting regrowth. An index commonly used is the **leaf area index (LAI)**—the ratio of the total upper surface leaf area of the plant community to the corresponding ground area. For well-watered and fertilized pasture swards, an optimum LAI can be designated to prevent defoliation from removing excessive amounts of leaf area and thereby minimizing wastage of the light energy available for pasture growth. An LAI which will prevent all but 5% of the light from reaching the soil surface has been considered optimum for dense swards (Walton, 1983).

Rates of photosynthesis and gross biomass yield are close to maximum on swards maintained at moderate LAI, but this requires an adequate and substantial proportion of the leaves produced to remain in the sward to contribute to photosynthesis (Parsons and Johnson, 1986). Excess levels of standing biomass in dense swards inevitably give rise to shading, reduced photosynthetic efficiency, and a high rate of foliage loss due to senescence and death. In swards maintained at a low LAI, a greater proportion of the leaf tissue is removed and utilized but photosynthesis and gross biomass production are substantially reduced. While an LAI at which virtually all of the incident light is intercepted by grass and grass-clover swards is usually in the range of 4 to 6 (Hodgson, 1990), an intermediate LAI—often set at an LAI of 1 to 2 in dense swards—provides the best compromise between gross biomass yield, herbage intake, and foliage death.

Allen *et al.* (1986a) recommended that sufficient foliage to maintain an LAI of about 1 be left on alfalfa under early spring grazing, but Wolf and Blaser (1981) indicated that very early alfalfa grazing should leave about half the potential leaf area, or an LAI of about 0.5. The application of the LAI concept to rangelands has not been widely used, and Vickery (1981) has questioned whether a heterogenous, semi-arid grazing area will have an optimum LAI. Further suggestions have been provided by Walton (1983) and Parsons and Johnson (1986) for the management of grazing of dense swards on the basis of LAI.

Managing for rapid regrowth utilizing LAI guidelines requires that ample new leaf tissue as well as adequate TAC levels remain after grazing (Burns, 1984). While defoliation that greatly reduces total photosynthetic area of the plant can dramatically reduce photosynthetic capacity, it is also well established that the relative proportion and age of foliage elements (i.e., blade, sheath, and culm) remaining after defoliation are important determinants of photosynthetic capacity following defoliation (Briske and Richards, 1994). Removal of young foliage elements, which attain maximum photosynthetic capacity at about the time of complete expansion, reduces canopy photosynthesis more than equivalent removal of older foliage.

The greater photosynthesis of the plant tussocks of crested wheatgrass after old leaf defoliation in mid-May, when the growing points are still basal, has been shown to permit greater light interception by the new tillers and more rapid growth rate (Gold and Caldwell, 1989, 1990). However, after the growing points of active tillers have elevated, the growth of new leaves may largely be found at the top of the plant. Defoliation at this time is apt to result in removal of the most active meristem. Since grazing generally proceeds from the younger top leaves to the lower older leaves in dense swards, progressive levels of defoliation tend to reduce photosynthetic efficiency by increasing the proportion of older leaves in the stand. Although the shading of lower by upper leaves in legumes plays an important part in causing a rapid decline in photosynthesis, the older, shaded leaves apparently do not parasitize the plant (Walton, 1983).

IV. PHYSICAL EFFECTS ON PLANTS

Clipping, shearing, or mowing are commonly used to simulate the effects of grazing on plants; however, the eating and trampling action of grazing affects forage plants in several ways not simulated by mere defoliation alone. Most of these effects are deleterious to the plants affected but will be tolerated unless they become excessive. The eating action involves biting, pulling, and breaking off plant parts at random heights. This often results in pulling from the plant unpalatable plant parts that will subsequently be discarded rather than being ingested. Entire plants, particularly seedlings, may be pulled out of the ground if not well rooted. For this reason, new forage plant seedlings must be allowed to become firmly rooted before being grazed (Vallentine, 1989).

The inadvertent pulling up of stolons or shallow rhizomes during grazing can also be a problem. Since black grama recovery from drought or prior defoliation depends upon the production of an adequate number of stolons, careful grazing management or even temporary exclusion of grazing and trampling will be required for the establishment of new offset plants from the outer buds on these stolons (Valentine, 1970). Stout and Brooke (1985a) found that more tillers of pinegrass are normally uprooted (i.e., a tuft of tillers) than are torn off (i.e., individual tillers torn out of the tuft) by grazing animals. The organic matter of the forest floor was found ineffective in holding the shallow rhizomes and roots of pinegrass against the pulling action of the grazing animal. An average of 32% of the distant tillers were removed during the first grazing pass and an average of 62% during several passes.

Another impact of grazing not applied by clipping is the trampling and treading of both the plants and the soil by the hooves of grazing animals. Grazing results in some plants being crushed, severed, or bruised by the hooves. Trampling losses of forage, resulting in a direct addition to the mulch component, may become excessive in dense forage stands and may be a major factor contributing to utilization inefficiency. Trampling damage and loss of forage on arid and semi-arid rangelands is generally much less than on mesic sites because of the lower density of animals.

Trampling losses were studied on native mountainous, forb-grass range by Laycock and Harniss (1974). When 18% of the herbage in paddocks was consumed by sheep in late summer, an additional 17% was damaged by trampling. When herded sheep in late summer on open range consumed 23%, an additional 27% was damaged. When cattle grazing in paddocks all summer consumed 50%, an additional 13% was damaged. Because they were succulent and easily broken, forbs suffered disproportional high trampling losses. While making up less than 20% of the diets of cattle, 66% of the missing material (i.e., trampling damaged or converted to litter) was forbs.

Trampling losses by cattle on sandhills range in northeastern Colorado varied from about 1% (20 lb/acre) under light grazing, to 2% (37 lb/acre) under moderate grazing, to 5% (60 lb/acre) under heavy stocking (Quinn and Hervey, 1970). Using before-and-after sampling techniques in paddocks grazed during short time periods in the summer, blue grama proved less susceptible to trampling than the mid and tall grasses but was more susceptible after maturity in September than in July. When trampling was simulated in a greenhouse study using native shortgrass species, more total vegetation was detached under continuous than short duration grazing, but about the same amount of live plant biomass was detached (Abdel-Magid *et al.,* 1987a).

Additional mechanical effects of foraging on woody plants include intentional or inadvertent breaking off of limbs and bark wounding by horning, rubbing, feeding, or hooves. Both elk and moose exert substantial physical damage to shrubs and small trees by these methods. In tallgrass prairie in Oklahoma, the horning and rubbing by bison, particularly bulls, has caused mechanical damage to only a few tress but has significantly affected shrubs and saplings, especially willows (Shaw, 1996; Coppedge and Shaw, 1997). Their horning and rubbing has been attributed to defense against insects, shedding hair, or activities associated with the rut. It was speculated that bison numbers during the pre-settlement period may have been enough to limit sapling encroachment into grass lands, at least in some areas, by mechanical impact.

The covering or otherwise fouling of vegetation with feces and urine is yet another deleterious effect of herbivory and must also be considered in grazing management. (Refer to "Patch Grazing: Problem or Benefit" in Chapter 9 for a discussion of the effects of excreta on forage growth, palatability, and utilization.)

V. GRAZING EFFECTS ON SOIL

A. TREADING AND TRAMPLING

All grazing land receives treading to a greater or lesser extent as a natural consequence of grazing. Treading of soil by grazing animals has the potential of being deleterious to soil in the following ways: (1) compacting the soil, (2) penetrating and disrupting the soil surface, (3) reducing infiltration, (4) vertical displacement of soil on steep slopes, (5) developing animal trails, and (6) increasing erosion. The interaction of many site, soil, weather, and vegetation factors will determine the severity of hoof action on the soil; the effects will range from inconsequential, or less commonly beneficial, to very destructive.

Livestock grazing affects watershed hydrologic properties by potentially removing protective vegetation as well as causing trampling disturbances. Reductions in the vegetation cover may: (1) increase the impact of raindrops, (2) decrease soil organic matter and soil aggregates, (3) increase surface soil crusting, and (4) decrease water infiltration rates (Blackburn, 1983, 1984). These effects may cause increased runoff, reduced soil water content, and increased erosion. Abusive grazing can severely damage both range and forest watersheds, but many watershed grazing studies historically have compared only heavy or uncontrolled grazing with no grazing, leading to an erroneous conclusion that livestock grazing is necessarily synonymous with heavy damage to watersheds. Van Dyne *et al.* (1984a) have emphasized that geologic erosion on rangeland must be more carefully differentiated from accelerated erosion than in the past.

Soil compaction, usually measured by soil bulk density, is a universal process associated with use or activity occurring on the soil surface (Stephenson and Veigel, 1987). The major effect is reduction of pore space through which water moves into and through the soil, thereby reducing infiltration and percolation, increasing runoff, and encouraging erosion. The extent to which a soil is compacted is determined by a complex interaction of the compacting force and soil water content, texture, and porosity. Studies on such diverse vegetation types as salt-desert shrub range in western Colorado (Thompson, 1968) and grasslands of south Texas (Blackburn, 1983) suggest that seasonal changes affect surface soil characteristics as much as or more than grazing.

Soil as well as the vegetation growing thereon has substantial resiliency that permits it to overcome many short-term effects of trampling (Abdel-Magid *et al.*, 1987b; Stephenson and Veigel, 1987). Soil compaction from grazing often disappears or decreases after seasonal wetting and drying or freezing and thawing, and, although related to infiltration rates, increased compaction does not necessarily result in lower soil water because the effects of grazing on reducing evapo-transpiration may be even greater (Lauenroth *et al.*, 1994).

Available information on the hydrologic impacts of grazing strongly suggests there are few hydrologic differences between perennial pastures and rangelands continuously grazed lightly or moderately. Studies in humid regions commonly report some increase in bulk density under grazing, but many studies in semi-arid ranges have failed to show a difference in soil loss, infiltration rates, or soil bulk density among light, moderate, and ungrazed pastures. Watershed research data strongly suggest that watershed condition can be maintained or improved under moderate grazing intensity (Blackburn, 1983, 1984; Laycock and Conrad, 1967). Thus, there appears to be no hydrologic advantage to grazing a watershed lightly at 30 to 40% utilization rather than moderately at 45 to 55% (Dwyer *et al.*, 1984).

Stocking rate seems consistently to be a more important influence on infiltration rate and bulk density than does the type of grazing system (Abdel-Magid *et al.*, 1987b; Blackburn, 1983, 1984; Weltz and Wood, 1986; Wood and Blackburn, 1981). In Texas studies, pastures grazed under the Merrill grazing system (continuous grazing plus deferment) had hydrologic characteristics similar to the livestock exclosures (Blackburn, 1983); at the same grazing intensities, the hydrological effects of continuous grazing and high-intensity systems were similar. Excessive soil loss along with reduced infiltration resulted from heavy grazing, regardless of grazing system, in part because of the reduction of the midgrasses (McCalla *et al.*, 1984ab).

However, infiltration on New Mexico rangelands was higher in shrub canopy areas than grass interspaces and least in shortgrass interspaces (Wood and Blackburn, 1981). Based on their central Texas research, Pluhar *et al.* (1987) concluded that differences among grazing treatments are directly related to their effect on amount of bare ground; grazing treatments which cause a reduction in vegetative cover and standing crop, with a corresponding increase in bare ground, tend to reduce water infiltration rates and concurrently enhance sediment production.

Concern has been expressed about the effects that trampling by grazing animals may have on cryptogamic crusts composed of nonvascular plants of algae, lichens, mosses, and diatoms. Such crusts are common in arid and semi-arid ecosystems, appearing in the interspaces between vascular plants on what would otherwise be mostly rock pavement or bare ground. The benefits of cryptogamic crusts have been attributed to enhancing soil stability by reducing water and wind erosion, increasing water infiltration into soil, improving seedling establishment and survival, and sometimes fixing nitrogen. Cattle grazing during winter at light to moderates rates have had the least effects, particularly on frozen ground; continued repetitive summer and especially spring grazing have the potential to do the greatest damage (Memmott *et al.,* 1998). Offsetting the damage to cryptogamic crusts by treading is the potential recovery during nongrazing periods, particularly long nongrazing periods.

Warren *et al.* (1986a) found that infiltration rates declined and sediment production increased following the short-term intensive grazing periods inherent in the short-duration grazing system during drought and winter dormancy but not during periods of active growth; however, some recovery was evident by the middle of the intervening rest period. In their study, soil conditions suggested that lower stocking rates and/or longer rest periods were required during winter dormancy or during periods of drought.

High-density grazing periods within short-duration grazing on semi-desert grasslands in New Mexico induced low infiltration rates, but recovery was made during the intervening rest periods (Weltz and Wood, 1986). Soil benefits from increasing the number of pastures beyond the minimum number required to qualify as short-duration grazing (i.e., more than about six) have not been found (Pluhar *et al.,* 1987; Warren *et al.,* 1986b); in the short term, the highest stocking density tends to produce the lowest infiltration rates and the greatest sediment loss.

Soil treading generally has a much greater impact on wet, heavy soils than on dry, sandy soils (Fig. 5.9). Fine-textured soils are more at risk than coarse-textured soils, except on dry sandy soils where hoof impact may encourage wind erosion; and soils covered by dense sod are less susceptible to deep treading damage. Trampling of fine-textured pasture soils even when at field capacity rather than saturated may greatly affect forage yields due to compaction (Tanner and Mamaril, 1959). When clay or even loam soils are very wet following rains or irrigation or during periods of high water table, hooves are apt to deeply penetrate and disrupt the soil surface. This deep treading in grazing has been referred to as **poaching** in the British Isles (Wilkins and Garwood, 1986). Not only does this high-impact hoof action disrupt soil structure and soil surface, the shearing action may destroy foliage, growing points, and roots of the plants. Mud may also be deposited on the remaining herbage, rendering it unavailable or less acceptable to grazing animals.

Frequent, light sprinkler irrigations of pasture on loamy to sandy soils may not require livestock to be removed from the pasture during irrigation (Nichols and Clanton, 1985), but on medium- to heavy-textured soils higher water application rates or rainfall may require animals to be temporarily removed from the pasture

FIGURE 5.9   Deep soil treading on wet, heavy soils disrupts soil structure and the soil surface and destroys foliage, growing points, and even roots of forage plants. (Forest Service Collection, National Agricultural Library.)

until the soil becomes capable of supporting animals without severe impact. A coordinated schedule of irrigation, fertilization, and grazing under a rotation grazing program is often the best plan. Particular care should be exercised in grazing meadows and subirrigated or irrigated sites during the first year or two after planting. Older stands and particularly those providing fibrous, supporting ground cover will be at lower risk (Wilkins and Garwood, 1986).

The best approach to reducing deep treading in high-risk situations such as wetlands is to reduce the density or duration of grazing of livestock or big game or remove them altogether during the most sensitive periods. Special grazing practices and even drainage may be required on wetlands, or grazing or even mechanical harvesting may have to be delayed until the soil has dried and firmed sufficiently to support ungulates or harvesting equipment. Special holding pastures with firm soil can often be provided during interim periods of wet pasture conditions. The concept of range readiness, as applied to medium to high elevation rangelands in the West, appears much more applicable to soil and site conditions than to vegetation development per se.

High density of livestock such as in a rotation system should be avoided during high-risk periods on susceptible sites. Also, livestock should not be held on high-risk pasture sites even during non-growing seasons if deep treading is apt to occur. Sheep seem to cause less damage by deep treading than do heavier species such as cattle, elk, and moose. From studies on northwestern pine-bunchgrass range in Oregon, Skovlin *et al.* (1976) noted that big-game trampling exerted as much or more compaction than that by cattle, because seasonal migration patterns placed heaviest big game use during the periods of wet and saturated soils. Although both moose and elk inhabit riparian zones, moose seldom cause the problems in deep treading that elk do (Skovlin, 1984).

The concentrated, heavy treading and grazing of large ungulates on intermixed riparian and aquatic zones can degrade all four components that make up the stream and lake fisheries habitat: (1) streamside vegetation, (2) stream channel condition, (3) shape and quality of the water column, and (4) the structure of the soil portion of the streambank (Platts, 1986). Heavy, concentrated grazing can affect the streamside environment by changing, reducing, or eliminating vegetation bordering the stream (Platts, 1981a); abundant and vigorous vegetation will minimize but not always totally prevent high water flow impacts on streambanks. Livestock or elk can trample and shear streambanks, causing them to slough off and thereby causing outsloping of the streambanks and eliminating natural overhang banks, further exposing banks to accelerated soil erosion. Channel morphology can be changed by sedimentation, alteration of channel substrate, disruption of the relation of pools to riffles, and making the channel wider and shallower. The water column can be deleteriously altered by increasing water temperature, nutrient levels, suspended sediment, and bacterial populations and affecting the timing and volume of streamflow.

Severe effects on riparian and aquatic zones, however, result primarily from uncontrolled animal grazing and trampling rather than from moderate, regulated use. In their northeastern Oregon study Buckhouse *et al.* (1981) found that no significant patterns of accelerated streambank deterioration occurred under moderate livestock grazing. In mountainous watersheds the combination of streambank erosion, bank cutting, channel scouring, and silt and rock deposition often relate more directly to natural hydrologic phenomena such as high water flows and ice shearing than to animal trampling (Buckhouse *et al.*, 1981; Dwyer *et al.*, 1984). (Management recommendations for controlling animal impact on riparian and aquatic sites are included in Chapter 8, "Special Problem: Riparian Zones.")

The natural habit of livestock and big game animals is to tread repeatedly along the same path; this consequence of hoof action results in the formation of trails. Trails or walkways are less commonly established by managers to facilitate efficient movement of livestock about the landscape (Ganskopp and Cruz, 1999). Although the problem of trailing on soil is more serious with cattle, unherded sheep are more prone to form trails than herded sheep. Cattle trails generally connect favored grazing, resting, and watering areas and can be an important factor contributing to soil erosion. Trails usually form along routes of least resistance such as the crest of ridges, in valleys, or parallel to contour lines. In developing trails by cattle in rugged terrain in northeastern Oregon, more moderate grades were sought (Ganskopp and Cruz, 1999). While the mean slope of the study area was 14%, the mean slope of the areas traversed by trails was 8%, while the mean slope of the actual trails was reduced to 5% by the selection of cross-slope routes.

The following hypotheses were concluded to be correct from research at the Texas Experimental Ranch in north Texas (Walker and Heitschmidt, 1986b): (1) the density of cattle trails and proportion of heavily used trails increased close to water, (2) the number of cattle trails increased under short-duration grazing as the number of pastures was increased, and (3) the number of cattle trails per unit area of land was greater under heavy short-duration grazing than under heavy continuous, moderate continuous, or moderate deferred-rotation grazing. Cattle trail density was similar under the latter three grazing systems in the various distance zones from water. The greater trail density under short-duration grazing was attributed to the high stock density and the pasture shape under the cell center arrangement of the pastures.

Terracettes, the name applied to parallel, contour patterns found on steep slopes in the Pacific Northwest, were concluded to be fundamentally of natural origin but livestock do use these natural walkways and probably accentuate them (Buckhouse and Krueger, 1981). Certain areas were found to have geological and climatic conditions that favored terracette development. The authors noted terracettes that ended flush with an emerging rock face, with livestock trails cutting across the terracettes at these points.

B. HERD EFFECT

Although grazing management has been generally based on the timing and amount of forage removed, a grazing program based prominently on beneficial manipulation of the soil surface by hoof impact has been widely promoted by Savory (1987). This program—referred to originally as the Savory Grazing Method—incorporates short-duration grazing (multiple pastures grazed in rapid rotation under high density stocking) and packages several grazing impacts on soil considered beneficial under a concept of **herd effect** (Fig. 5.10).

The concept of a large herd of ungulate herbivores managed under high animal concentration to achieve this herd effect is described as follows (Savory, 1987): "In an excited state, herding-type grazing animals no longer place their hooves carefully. As a result of this behavior, soil surfaces tend to be 'chipped,' dust is raised, plants are trampled and thus more material is laid on the ground, and steep soil banks are broken." The two prime objectives of herd effect were to "change the nature of the trampling to better break exposed soil surfaces and lay old plant material as soil cover. . . . Animal impact is just a tool . . . to cause breaks and irregularities in exposed soil surfaces, compact soil underground, lay dead plant ma-

FIGURE 5.10   Beneficially manipulating the soil surface through hoof impact of grazing animals, packaged under a concept of "herd effect," has been widely discounted and opposed by range and pasture scientists; showing heifers managed under short-duration grazing at the Tintic Experimental Pastures, Utah State University.

terial, break solid mats of algae, lichen or moss, result in dung and urine reaching the soil and a few other things of a physical nature. . . . Herd effect is generally required less in non-brittle environments than in brittle environments to control successional communities, but must be provided in brittle environments to prevent desertification."

The concept of herd effect became highly controversial in scientific circles, with some practitioners supporting the concept while range and pasture scientists generally discounted the concept and often vigorously opposed its application. The consensus has largely consolidated around the concept that trampling on a frequent basis cannot benefit most rangeland ecosystems and that this applies under short-duration grazing just as well as under any other grazing scheme (Pieper and Heitschmidt, 1988; Taylor, 1988). After summarizing the application of the concept in Africa, the locale in which the concept was originally developed and from which it was later extended to U.S. rangelands, Skovlin (1987) has labeled the proposed benefits of herd effect as mostly myth; he concluded that soil compaction with reduced infiltration and increased sediment was more apt rather than less apt to occur under its application.

Balph and Malecheck (1985), from studies in central Utah, found that cattle even under short-duration grazing deliberately avoided stepping on crested wheatgrass plants (principally *A. desertorum*), which grow in slightly elevated tussocks, but rather stepped in the mostly bare interspaces between the plants. Their conclusion was that the hoof action was minimal in breaking up the standing dead vegetation and mixing it with the surface soil. A response to this finding (Savory, 1987) has been that only under ultra-high stock densities and when under continual milling and agitation will livestock place their feet in the careless manner required to achieve herd effect! It was conceded by Balph and Malecheck (1985) that when first seeded the crested wheatgrass would not yet have concentrated growth into slightly elevated tussocks and cattle would have been more likely to step at random.

After further study of cattle grazing crested wheatgrass, Balph *et al.* (1989) rejected the hypothesis that cattle were avoiding dark areas (tussocks) by stepping on light areas (interstices) but rather were avoiding stepping on the tussocks because they present an uneven surface upon which to walk. When tussock height was reduced, the trampling frequency increased. When the tussocks were totally clipped and no vegetation was visible, the cattle continued to select against stepping on the mound itself. This indicated that the mound itself was a cue governing hoof placement. Only near salt or water where animal use was high and where jostling among animals would prevent them from avoiding tussocks might severe trampling of bunchgrass tussocks be expected.

Soil crusts that commonly develop on rangelands are characterized by low organic matter, high silt content, and low aggregate stability. Such crusts generally have low infiltration rate and are a prime factor associated with runoff and erosion. Livestock trampling may loosen and pulverize the soil when dry and initially incorporate mulch into the surface soil, providing the stepping is at random; but Blackburn (1983) has concluded that the "churned soil" does not remain beyond the initial impact of falling raindrops, which effectively destroy the modified surface and interrupt any increase in infiltration rates. It was further concluded that livestock grazing practices which promote plant and mulch cover will reduce soil crusts the most. Based on their application to fescue grasslands in Alberta, Dormaar *et al.* (1989) concluded that high hoof action associated with high herd density was unable to negate the effects of high utilization and thereby improve range condition. Instead, it reduced soil moisture, increased bulk density, did not significantly incorporate litter into the soil, and decreased fungal biomass.

Each soil type can be expected to respond differently to herd effect, and the balance between positive and negative effects will differ greatly between seasons of the year. Burleson and Leininger (1988) have emphasized the importance of planning carefully as to when, where, how much, and even if herd effect is wanted and that too much herd effect can be self-defeating. Although they suggested there may be some utility in loosening up capped soils and breaking up clubmoss mats to improve conditions for seed germination, they emphasized that its use must not lead to soil compaction or soil surface movement resulting in watershed damage.

The survival of natural seedlings of crested wheatgrass in old stands in Utah during the establishment year was compared (1) under protection from grazing, and (2) under hoof action of trampling under short-duration grazing (Saliki and Norton, 1987). By the following spring—12 months after emergence—survival was 11.6% under no trampling compared to only .4% under trampling/grazing.

The intensive grazing/trampling treatment did not enhance but rather greatly reduced establishment; the negative impact was greater during the second grazing period when the seedlings were older.

C. REDISTRIBUTION OF SOIL NUTRIENTS

Freely grazing animals inefficiently distribute excreta, both manure and urine; excreta is deposited most heavily where animals spend the most time rather than where the forage is produced and consumed. Thus, forage producing parts of the grazing unit become progressively more deficient in soil nutrients—removal being greatest on sites most selected for grazing—while animal concentration areas near water, salt, feeding areas, bedgrounds, shade, and selected level areas are enhanced with soil nutrients.

On Intermountain sagebrush steppe Miller *et al.* (1994) estimated that 35% of the excreta may be redistributed to 10% of the grazing unit area. Urine is particularly involved in the redistribution of nitrogen but also potassium, magnesium, and sulfur, while a large assortment of minerals including phosphorus and potassium are passed through the manure (Gerrish *et al.*, 1995b). The fertilizer effects are primarily found on the immediate area covered by feces and urine with lesser effects out to 2 to 3 times this area (Petersen *et al.*, 1956).

Grazing by large herbivores increases nutrient cycling rates by reducing particle size but also by accelerating the rate of nutrient conversion from an organic to inorganic form available to plants. Fecal nitrogen is largely insoluble and becomes available to plants only after incorporation into the soil by soil fauna and mineralization by microorganisms; the nitrogen in urine is readily available or rapidly becomes so (Simpson and Stobbs, 1981), the proteins and amino acids having been converted to nitrate and ammonium.

Nutrients consumed, digested, and deposited in feces and urine return to the soil more rapidly than through senescence-decomposition pathways, but this also introduces the potential for greater nutrient losses (Lauenroth *et al.*, 1994). Around 75% of the nitrogen and phosphorus and from 80 to 90% of the potassium normally passes through the animal, but the losses are both irregular and substantial. While only minimal amounts of nitrogen and other nutrients are exported from the site as animal tissue, high stocking rates and forage utilization efficiency can gradually deplete soil nutrients. Even higher nutrient losses may result from nitrogen volatilization, nutrient redistribution to unproductive sites, and water transport through leaching and soil erosion from accumulation sites. However, the general conclusion is that grazing does not seriously increase nutrient losses from grazed ecosystems, particularly rangelands, when atmospheric inputs of nitrogen and the buffering effects of soil parent material and soil organic matter are considered, and the greater losses of soil nutrients from grazing improved pasture are readily corrected by the use of fertilizers.

The management goal should be to keep manure evenly distributed over the grazing land unit to maintain uniform soil fertility. On rangeland, deposition areas usually comprise less than 5% of the land area, and the effects are generally only temporary except where animals congregate and large accumulations result (Heady and Child, 1994). Maximizing animal grazing distribution and preventing local overgrazing and long-term livestock concentration will minimize nutrient redistribution. In small pasture studies in Missouri, manure distribution was more uniform with smaller pasture size, with higher stocking density and frequent rotation, with minimal landscape variation within the paddock, and with water placed and made readily accessible in each paddock (Gerrish *et al.*, 1995b; Peterson and Gerrish (1995).

Excreta also accumulates in lanes provided to access water or transfer animals between paddocks. Concentrating animal feeding or feed placement on range or pasture concentrates not only excreta and urine but also feed residues. This may be more serious if deposition areas are located on riparian sites or other areas where groundwater or stream flow can become contaminated with high bacterial or excess mineral levels. Areas receiving excess excreta often receive excess trampling as well; while providing extra fertility, the combined effects may be to dramatically alter vegetation composition and permit the entry of undesirable weedy vegetation.

6

GRAZING ACTIVITIES/ BEHAVIOR

I. INGESTIVE BEHAVIOR

A. MECHANICS OF INGESTION (FEEDING)

Before considering the extent to which grazing animals are selective in their feeding habits and how they make selection, it is necessary to consider the mechanics of grazing. Grazing is a complex activity and (1) involves searching for and selecting suitable forage, after which (2) the forage is prehended (grasped) and defoliated from the plant, and (3) taken into the mouth. The forage is then (4) chewed and mixed with saliva, manipulated and formed into a bolus, and then swallowed and ejected with some force into the anterior part of the rumen (Arnold and Dudzinski, 1978; Van Soest, 1982). Variable amounts of time are spent on each phase of feeding activity by the grazing animal. Jaw activity during grazing is also complex, since it involves initial movements to arrange the herbage in the mouth, gripping the herbage with mouth parts, severing it from the plant by biting or jerking the head, and masticating and arranging the herbage for swallowing (Leaver, 1982).

The anatomy of the jaw, teeth, and other mouth parts results in differences between animal species in how the herbage is grasped and severed from the plant and, in part, what plants they eat. Cattle depend on their mobile tongue to encircle a mouthful of forage and draw it into the mouth, unless the vegetation is very short. The forage is then gripped between the upper and lower molars or between the incisor teeth in the lower jaw and the muscular pad in the upper jaw and severed from the plant by a backwards jerk of the head (Ellis and Travis, 1975; Heinemann, 1969). The horizontal movement of the grazing animal's head results in a mower effect, with the tops of the plant being "trimmed" off. However, this is a simple mechanical action and does not comprise selective grazing per se (Arnold and Dudzinski, 1978).

Bison ingest forage in a manner similar to cattle but seldom prehend forage with their tongues in a horizontal plane (Hudson and Frank, 1987). If the forage plants or plant parts are very short, attempts may be made to bite or break the forage directly from the plant even though this is not highly efficient for cattle or bison because of the lack of incisor teeth in the upper jaw. The large, flat muzzles of cattle and bison allow relatively large clumps of vegetation to be drawn into the mouth at one time (Fig. 6.1). The associated forage consumption rate is high, but more old tissue is consumed along with the current annual growth than by grazers with narrow mouth parts (Hanley, 1982a).

Sheep bite the foliage off the plant or it is broken as they grip it and jerk their heads backwards or less commonly forwards (the latter about 20% of the time) (Arnold and Dudzinski, 1978). Although sheep do not use the tongue to prehend forage as do cattle, similar results are accomplished by motions of the head and lips (Laca *et al.*, 1992). Sheep are similar to cattle in having only molar teeth in the upper jaw, the incisors being replaced by a muscular pad. In contrast with cattle, however, sheep have a cleft upper lip that permits close grazing if they so choose. Cattle seldom graze closer than about 2 inches from the ground unless forced to do so to obtain forage (Heinemann, 1969). Horses are able to bite close to the ground, having the advantage over sheep and cattle of both upper and lower incisors. All three species move with their muzzles in a horizontal plane as they graze and select forage in a vertical plane (Arnold and Dudzinski, 1978). Because sheep have smaller mouths, they can take smaller bites and so are able to be more selective of plant species and plant parts if they wish. However, all three species are able to vary their methods of harvesting forage somewhat according to the structure of the vegetation (Arnold and Dudzinski, 1978).

Goats have mouth parts and ingestive techniques similar to sheep but are noted for mobile upper lips and prehensile tongues that permit them to eat tiny leaves of browse even from thorny species (Fig. 6.2), which most other domestic livestock cannot readily or normally consume (Martin and Huss, 1981). Camels are like

FIGURE 6.1   The large flat muzzle of the cow allows relatively large clumps of vegetation to be drawn into the mouth at one time, this with the aid of its mobile tongue. (Texas Agricultural Experiment Station photo by Robert Moen and Charles A. Taylor, Jr.)

sheep and goats in having mouth parts adapted for browsing. The muzzle of the pronghorn is long and narrow, its mouth is small, and it has a cleft upper lip like sheep, giving it a great deal of manipulative ability (Ellis and Travis, 1975). The anatomy of the deer mouth is similar, and forage is either gripped between the molars and severed by biting action or seized between the incisors and upper dental pad and sheared off with an upward or downward jerking action (Willms, 1978). Elk forage by using their lips, dental pad, and lower incisors to grasp and break the forage instead of using the tongue to sweep and prehend forage (Jiang and Hudson, 1994).

Herbivores exhibit considerable plasticity in feeding behavior, and this is necessary to be able to feed on plants that may vary greatly in structure (Arnold, 1985). The mechanics required to remove chosen plant parts differ with the plant and plant parts being eaten, the size of bite, rate of biting, and total time spent biting; these are all varied as the animal attempts to achieve its intake potential.

B. MOVING AND SEARCHING

Grazing has been considered to have both an exploration phase and a subsequent daily routine phase (Arnold and Dudzinski, 1978). When introduced to a

FIGURE 6.2   Narrow mouth parts, a mobile upper lip, and a prehensile tongue permit goats to select individual leaves even from armed plants. (Texas Agricultural Experiment Station photo by Robert Moen and Charles A. Taylor, Jr.)

new paddock, both sheep and cattle explore it. Initially, they move around the boundaries and follow this by moving farther into the paddock. This process will be rapidly completed in small, flat paddocks, and use of all of the area will be quickly begun, but the process can be substantially prolonged by physical or visual constraints, particularly in large grazing units. One factor causing poor grazing distribution under continuous grazing, in contrast to grazing in smaller rotation paddocks, may be the extended amount of time required for animals to search the entire area for forage (Kothmann, 1984).

Cattle are taller and have a clearer field of view than do sheep, and this is reflected in their exploration and search behavior (Arnold and Dudzinski, 1978). Unherded sheep appear to use the fence lines for orientation. In large paddocks, sheep tracks usually run parallel to fences, providing there are no obstructions, for some distance before branching out into various directions. On the other hand, cattle tracks rarely run close to fences, unless confined by terrain, but instead run in nearly straight lines between locations, for example, from grazing to resting and/or watering points.

Among domestic livestock the goat is unique in its ability to browse taller shrubs and small trees in not just two but three canopy strata: a lower strata that can be reached in a quadrupedal stance, a middle strata reached in a bipedal stance (standing on hind legs), and a top strata generally considered above the goat's reach (about 59 in.) (Owens *et al.,* 1992). In one study, the degree of use of guajillo (*Acacia berlandieri*) was 79% in the middle strata and 63% in the lower strata; with blackbrush (*Acacia rigidula*), the degree of use was 39% in the middle strata and 27% in the lower canopy strata. Surprisingly, degree of use in the top strata, achieved only by the goats climbing up into large shrubs and small trees, was 28% on guajillo and 9% for blackbrush.

The typical activity of a grazing animal can be described as interrupted forward movement with the head swinging from side to side in front of the forelegs (Hodgson, 1986). Foraging behavior has two components—the feedings and the moving intervals between feedings. At intervals the foraging animal walks a number of steps in search of desired forage and then pauses to feed at the new location. This pause is referred to as the **feeding station interval** and the location as the **feeding station** (Fig. 6.3), the latter more fully defined as the area available in a half-circle shape in front of and to each side of the grazing animal while its front feet are temporarily stationary (Ruyle and Dwyer, 1985). Then, as animals reach away from their forefeet, the shift in balance may trigger locomotion leading to the selection of a new feeding station (Bailey *et al.,* 1996a).

The grazing animal "probably re-evaluates its environment after each feeding station. In head-up position, it sequentially: senses its degree of comfort, updates angle and distance from navigational cues, scans the potential pathway for food

FIGURE 6.3 The feeding station is the area available in a half-circle shape in front of and to each side of the grazing animal for foraging while its front feet are stationary; showing grazing scene on irrigated pasture in Nebraska.

abundance cues, evaluates effort vs. potential reward, and moves and establishes the next feeding station" (Kidunda *et al.*, 1993). From where they are at any point in time, grazing animals are attracted to a new feeding station of high nutrient resource availability; animals move shorter distances when food is abundant and longer distances when food is scarce (Kidunda *et al.*, 1993).

Besides interrupting its biting activity to move to a new feeding station, a grazing animal may be forced to pause to move out of the way of another animal or in response to any one of a number of disturbance factors. Interruptions of this kind tend to be more frequent and of longer duration at the beginning and end of a grazing period than in the middle (Leaver, 1982). Dwyer (1961) noted that during the intensive early morning and late afternoon grazing periods, cows ate almost frantically and were not easily distracted; several bites of forage were consumed between each step. During other grazing periods, the cows walked several steps between bites and were more easily distracted. It was reported by Smith *et al.* (1986) that cattle grazed more avidly in early morning than during other times of the day; the lower diet quality associated with early morning grazing was attributed to less selective grazing. The reduced bite rate of beef cattle grazing winter wheat at 90 minutes after turnout compared to 30 minutes was attributed principally to rumen fill (Kanyama-Phiri and Conrad, 1986).

Feeding station intervals are normally short, seldom more than a few seconds, unless animals are feeding selectively on a large plant such as a shrub. Spacing as well as time intervals between feeding stations are normally also very short in dense swards. In grazing studies in Texas shrublands, searching time and number of steps taken between feeding stations were greatest in seasons of active herbaceous growth, when selectivity was high (Mastel *et al.*, 1987). The number of steps between feeding stations was highest in communities characterized by small, well-defined patches of vegetation.

Sheep studied in grazing trials on high-elevation summer range in southwestern Utah reduced the time spent per feeding station as the amount of desirable forage declined (Ruyle and Dwyer, 1985). Grazing animals move more slowly through areas with greater nutrient abundance (higher quality and/or quantity of forage) because they spend more time biting than moving, and it may take them longer to process the more abundant forage (Laca *et al.*, 1994). In contrast, animals finding limited nutrient resources increase their forward movement velocity, this commonly noted on semi-desert range with sheep when they refuse to settle into a slow grazing mode and are prone to run. This suggests that an animal's grazing behavior may be a more sensitive indicator of range forage quantity and quality than are direct measurements of the vegetation. Monitoring animal behavior during feeding periods may allow the grazier to recognize limitations in the available forage and adjust management strategies accordingly.

Coleman (1992) concluded that livestock grazing uniform, seeded swards use different strategies for ingesting forage than those grazing diverse, indigenous ecosystems. Animals grazing heterogenous, indigenous standing crops spend more time and energy seeking feeding stations and searching within the feeding station.

When grazing seeded, uniform swards, feeding stations receive lower priority and the grazing animals are driven predominantly by their ability to prehend and sever leaves and succulent stems.

Deep snow impairs the search capability, mobility, and even selectivity of grazing animals. Ranchers have often observed that when 6 inches or more of snow covers the forage, cattle stop grazing. Horses are more adept at obtaining forage from beneath snow and willingly paw down through the snow to the forage (Salter and Hudson, 1979). In Canadian studies, cattalo did not paw to uncover grass as horses do but burrowed with their muzzles through the snow to secure feed (Smoliak and Peters, 1955). Rocky Mountain bighorn sheep are willing to nose or paw to reach herbaceous forage under snow cover (Goodson *et al.*, 1991). Heavily crusted snow or icing over will prevent any form of foraging for low-growing plant species (Severson and Medina, 1983). Thus, the animal survival advantage during winter with plants such as shrubs that remain exposed above deep snow is obvious.

C. TRAVEL DISTANCE

Walking locomotion is an inherent part of foraging by the grazing animal. For free-ranging deer and elk Hanley (1982a) determined that traveling occurred at the same time as foraging, and only occasionally did animals travel without foraging. This is also a general observation with domestic livestock. When the forage stand is heterogenous and grazing animals are being highly selective, additional travel time and distance may result. Travel distance for elk and deer in the Cascades of Washington was found to increase from May through August, seemingly resulting from increased plant selectivity (Hanley, 1982a). Thereafter, travel distance decreased as a result of grazing mostly in selected microhabitat patches.

Daily travel distances may be greatly increased when grazing animals must travel longer distances for adequate food and water (Fig. 6.4). Additional energy expenditures associated with travel can also result from livestock being driven, chased by predators or insect attacks, or responding to outside disruption such as noise or hunting and from a general lack of being contented or unusual weather conditions. The energy cost of this additional movement varies, depending largely upon the additional travel distance required and the slope and difficulty of terrain to be traversed. Cook (1970) reported that the energy cost to walk downslope was approximately the same as walking on a horizontal plain, while Clapperton (1961) and Christopherson and Young (1972) concluded that walking up a steep gradient may be up to 10 times as energy demanding as walking on the level.

When cow-calf pairs are maintained in confinement, cows and calves have been found to travel as little as .4 and .1 mile daily, respectively (Schake and Riggs, 1972). Research cattle in a Texas grazing study walked 2.2 miles per day when restricted to small range paddocks (10 acres each) in a rotational grazing treatment (high-intensity/low-frequency system using 21-day grazing intervals) but walked 3.2 miles per day when grazed continuously in a 50-acre unit (Anderson and Kothmann, 1980). In a related study, daily travel distances increased as the frequency

FIGURE 6.4   Daily travel distances of free-ranging herbivores may be greatly increased when forced to travel longer distances for either food or water or from various kinds of disturbance; this additional travel is an extra energy expenditure; showing cows on grassy foothill ranges in California. (Forest Service Collection, National Agricultural Library.)

of rotation was increased from non-rotated continuous grazing to 14- and 42-day short-duration rotation (3.6, 4.1, and 5.1 miles average daily) (Walker and Heitschmidt, 1986a). However, most of the increased travel was caused by travel associated with rotating between grazing units. Cattle on African grassland range traveled less per day under continuous than under short-duration grazing when all cattle were confined to 30-acre paddocks (Gammon and Roberts, 1980a).

Yearling cattle in lightly grazed 50-acre sandhills range units in Colorado averaged 1.5 miles of travel per day, while travel distance increased to about 2.0 miles daily under moderate and heavy stocking rates (Quinn and Hervey, 1970). Beef cows on prairie grasslands in Oklahoma traveled an average of 3.13 miles daily within a 1500-acre range unit, in spite of the fact that salt, water, and good forage were uniformly found throughout the unit (Dwyer, 1961). The average travel distance of lactating cows on Montana foothill range was 2.9 miles per day; the level of milk production, body weight, calf weight, and age of calf had no measurable effect on the distance the cows traveled daily (Havstad *et al.*, 1986b).

While cows traveled an average of 4.8 miles daily on semi-desert range in New Mexico (Rouda *et al.*, 1990), neither lactation nor feeding cottonseed pellets (self-fed about twice weekly) affected daily travel distance. (All cattle in the study grazed together in the 5000-acre range unit except when temporarily taking supplement.) The suckling calves did not limit dam travel in the large range unit but were left behind when the cows traveled to the water source. The latter was in agreement with the report by Arnold and Dudzinski (1978) that cows typically leave their calves in charge of a "guard cow" when water location or forage availability requires the dams to cover excessive distances. However, cows in the New Mexico study traveled 34% farther between May 29 and July 8 than they did between July 21 and August 18; possible explanations for greater travel in the spring were exploratory behavior when placed in new pasture, more selective grazing, longer day length, lower temperatures, and the recent cessation of supplementation.

On extensive rangelands in Australia, cattle normally walk from 4 to 9 miles daily but up to 12 miles in more severe situations such as forage being located at long distance from water or during severe drought (Squires, 1981). Travel distances of 7.5–10 miles per day are common for flocks in inland Australia. It was calculated that sheep walking 7.5 miles per day would expend more than eight times the energy per day for travel than when confined to a pen and almost twice as much as when placed on a small pasture with abundant forage (Squires, 1981). Goats have been noted to travel longer distances in search of preferred forage than other domestic ruminants (Taylor, 1986b); however, average daily travel distance on mixed brush range in Texas was only 2.8 miles per day (Askins and Turner, 1972).

Neither daily travel distance nor time spent grazing on Montana foothill grassland was related to beef cattle genotype (i.e., no differences between Angus, Hereford, Simmental, and their crosses) (Funston et al., 1987). In a related Montana study, the breed type with the smallest energy requirement (Hereford × Hereford) spent the same amount of time grazing, traveled the same distances, and covered the same area of the range unit as the breed type with the highest energy requirement (75% Simmental-25% Hereford). However, it is generally accepted that the Brahma, Santa Gertrudis, and Africander cattle will range farther than British breeds of cattle in hot climates (Arnold and Dudzinski, 1978). Santa Gertrudis cattle on semi-desert range in New Mexico spent more time walking (12.1% vs. 6.5% of the 24-hr day) and walked farther (7.8 vs. 4.9 miles) than did Hereford cattle (Nelson and Herbel, 1966). In an eastern Oregon study Brahma × Hereford cows walked 1.4 miles farther daily than did Hereford cows (Sneva, 1970); distances traveled by crossbred steers averaged slightly more than for the straightbred steers of the two species.

Differences in daily travel distance were also found between sheep breeds grazed but unherded on mountain summer range in southern Utah (Bowns, 1971); Rambouillets traveled an average of 2.9 miles daily compared to 2.4 miles for Targhees and 1.9 miles for Columbias. Antelope in Wyoming's Red Desert tended to move about more than free-ranging sheep, covering about $1\frac{1}{2}$ times the distance in equal time periods (Severson et al., 1968).

The distance livestock travel is influenced by weather factors such as temperature, wind, and storminess (Anderson and Kothmann, 1980). Cattle are more restless, graze less intensely, and cover more ground while grazing in stormy and unsettled weather (Culley, 1938; Arnold and Dudzinski, 1978). The distance cattle were willing to travel during mid-winter on northern Utah range was highly and inversely correlated ($-.90$) with average wind velocities (Malechek and Smith, 1976). Daily travel distances averaged 3.6 miles but decreased to 1.5 miles when wind speeds exceeded 2 mph (Malechek and Smith, 1974). When very cold temperatures are combined with high wind speeds (10 mph or more), grazing animals may increase travel distance by drifting downwind in the absence of protective shelter or travel barriers.

D. SOCIALITY IN GRAZING

Grazing behavior of animals of the same species is affected by a conflict in choice between group and individual activities (Balph and Balph, 1986). While in a group their behavior is governed by two desires: (1) gregariousness (the desire to be with friends), and (2) social facilitation (the desire to mimic the activity of friends). Individuals in the group also act to satisfy their own needs, such as drinking when thirsty. A genetic inclination to gregariousness exists within sheep breeds, with Merino and Rambouillets exhibiting a high degree and Cheviots and Southdowns at the other extreme exhibiting a low degree of gregariousness. Gregariousness in sheep, besides functioning as an anti-predator strategy, affects foraging since animals tend to forage in similar areas, in similar diurnal periods, and at similar foraging rates (Fig. 6.5) (Provenza and Balph, 1988).

Movements in a group of sheep appear initiated by individuals that are both less gregarious and more independent, frequently grazing with their backs to other sheep or at a greater distance from other sheep than is usual (Arnold and Dudzinski, 1978). The movements of these "leader" sheep are then followed by others, but leadership is not exhibited in any positive way. Free-ranging or loosely herded sheep will aggregate in large groups or as a whole flock when resting or when drinking and then gradually split up into smaller and smaller groups as they graze away from water or bedgrounds (Arnold and Dudzinski, 1978). Stress caused by cold or wet weather causes sheep to graze in a more compact flock (Campbell *et al.*, 1969).

Cattle exhibit the strongest social facilitation when traveling in trail formation and coming into water as a group (Fig. 6.6). In large range units in Oklahoma stocked with numerous cattle, cows tended to graze in groups of 20–30 each on the easier topography but in smaller groups of 6–10 when grazing rougher topography (Dwyer, 1961). Individual cattle within two subgroups on Oregon summer range showed considerable uniformity in timing of activity and movement, although the home range and the time when an activity occurred often were different between the two subgroups (Roath and Krueger, 1982b). On New Mexico desert range, Santa Gertrudis cattle were noted to stay together more and were easier to round up than were Herefords (Herbel and Nelson, 1966a). Bison tend to re-

FIGURE 6.5   Gregariousness in sheep induces animals to forage in similar areas, in similar diurnal periods, and at similar foraging rates; showing sheep at Shasta National Forest, CA. (Forest Service Collection, National Agricultural Library.)

main in one herd when grazing or resting (Hudson and Frank, 1987). In Wyoming's
Red Desert, antelope were found to be much less gregarious than free-ranging
sheep (Severson *et al.,* 1968).

Beef heifers on crested wheatgrass in Utah exhibited distinctly different behavior in grazing units ranging from 2.5–20 acres in size and from 3–24 head of
heifers per grazing unit (Hacker *et al.,* 1988). Animals in the small unit—thus few
in number—always remained in close proximity and moved as a tightly knit unit.
Animals in the larger herds were more dispersed, indicating diminishing group cohesion and greater individual independence, expressed in watering behavior and
in sporadic grazing activity during non-peak periods. In Australia, cattle were
noted to graze more widely when the supply of forage was limited; their spacing
within groups was also greater and the size of sub-groups was smaller (Squires,
1981). This again suggested that the social structure or dispersion of a herd might
be used to predict the forage conditions. Smith *et al.* (1986) also reported that cattle tend to graze more as a herd when feed is ample but as individuals when feed
is short or animals are very hungry.

While social dominance is shown by cattle in their order of walking, neither

FIGURE 6.6   Cattle exhibit the strongest social facilitation when traveling in trial formation and coming to water in a group, this having the potential for inducing erosion on some sites; showing cattle in the central Great Plains. (Soil Conservation Service photo.)

sheep nor cattle show evidence of appreciable dominance while grazing (Arnold and Dudzinski, 1978; Squires, 1981; Dwyer, 1961). However, Greenwood and Rittenhouse (1997) noted the existence of "leaders" and "followers" in groups of grazing cattle, and that these roles, once established, seldom changed. A possible explanation of reduced grazing time by cattle under short-duration compared to continuous grazing in Texas studies was that under high livestock density animals may be inhibited from expressing a drive for more search time because of the proximity of other animals and the potential for conflict caused by the intrusion into another animal's individual space (Walker and Heitschmidt, 1986a).

Mosley (1999) concluded that both interspecific and intraspecific social competition are largely passive processes in which subordinates avoid conflict with dominants even though dominant animals make few overt attempts to supplant subordinates during grazing. Subordinates appear to monitor their spatial relationships relative to the dominants and, as they get closer, may reduce their bite rate, stop feeding, or move away. However, the grazing behavior of the dominants appears largely unaffected by the proximity of subordinates, which permits dominants greater freedom in habitat selection while potentially restricting the amount of forage resources available to the subordinates.

Social dominance is exhibited in both cattle and sheep by males during mating and when the cattle or sheep are provided supplementary feed in a restricted space or at the water trough, particularly by very thirsty animals. The reduced performance of beef heifers under the high density of short-duration grazing during the breeding season in Utah studies on crested wheatgrass (Utah State University, unpublished data) apparently resulted from the continuous agitation of heifers in estrous and the interruption of bull service. No substantial differences in spring grazing behavior of steers on Kansas Flint Hills were found between normal stocking density and the 3× density under intensive-early stocking (Lugenja *et al.,* 1983).

Concern has been expressed about the effect that mixing strange animals of the same breed and even age will have on grazing behavior. Except for breeding males, particularly those not having opportunity to become familiar with each other in advance, this does not seem to have a lasting effect on grazing behavior or performance. It may take several weeks for two groups of sheep to become completely integrated (Arnold and Dudzinski, 1978), but no antagonism should be anticipated between individuals of the different groups. A minimal amount of sparring and fence walking occurred when steers on Oklahoma range were routinely mixed and moved to new pastures every month for 11 months following weaning, but this lasted for only 1–2 days (McIlvain and Shoop, 1971b). Compensatory gain during the last three weeks of the monthly grazing period offset most of the slightly smaller gain that occurred during the first week after the steers were moved and mixed. Although the 11-month gains favored the continuous grazing over the monthly moving and mixing, this difference was smaller than anticipated and was concluded not to be an important factor at least with steers.

E. RATE OF INGESTION

The mechanical task that is presented to large grazing herbivores in biting off their daily requirements of green forage (154–209 lb for mature cattle) appears almost formidable (Fig. 6.7). Where pasture conditions are optimum (i.e., a dense stand of easily harvested vegetation), Walton (1983) calculated that a cow must take about 80 bites per minute through an 8-hr grazing day to harvest 198 lb of green material. Biting rates of 30–50 bites per minute appear common in both cattle and sheep. Bites per day for adult cattle have ranged from 12,000–36,000 (Freer, 1981), but bite weight, which interacts with biting rate, has varied greatly, ranging from 0.05–8 grams of organic matter per bite (Burns, 1984).

Bite weight (usually expressed as dry matter equivalent weight), biting rate (i.e., number of bites per unit of time), and time spent grazing determine forage intake. Some new direction has been given to the analysis of daily herbage consumption as the product of these three components (Forbes, 1988; Erlinger *et al.,* 1990). This has been formularized for clarification as follows (Kothmann, 1984; Leaver, 1982), when weight is expressed either as total dry matter or limited to organic dry matter:

$$\text{Forage intake (g/day)} = \text{bite weight (g/bite)} \times \text{biting rate (bites/min)} \times \text{grazing time (min/day)}$$

Although this formula is useful in showing relationships, it is impractical to determine forage intake this way. Many problems exist in determining precise values for each of these three factors because of variability as well as differentiating

FIGURE 6.7   Collecting the daily diet by the grazing cow hypothetically requires an 8-hr work day, 12,000 to 36,000 bites, and 0.05 to 8 g per bite; all three factors interrelate in determining forage intake. (Texas Agricultural Experiment Station photo by Robert Moen and Charles A. Taylor, Jr.)

between prehensive bites and mastication bites (Freer, 1981). (Daily forage dry matter intake, the factors that determine it, and its relation to grazing capacity are discussed in greater detail in Chapter 11.)

The interrelations between grazing time, bites per minute, and amount per bite vary with the physical structure of the forage stand, its bulk, its density, and its height (Squires, 1981). Grazing animals vary bite size (or weight), biting rate, and grazing time to deal with a variable and changing environment, but their ability to effectively do so is limited. As available forage increases, bite size usually increases also (Burns, 1984). Biting rate usually declines as sward height or herbage mass increase and as intake per bite increases, principally because the ratio of manipulative to biting jaw movements increases as intake per bite and the size of individual plant components prehended increase (Hodgson, 1986). A critical bite size below which intake is suppressed is apparent for all ruminants and has been reported at about 0.3 g (dry weight) of organic matter for dairy cows (Minson, 1990).

Reciprocal changes in intake per bite and bite rate may balance to maintain a roughly constant rate of intake on relatively tall (or abundant) forage stands, but on shorter (i.e., limiting) stands any increase in biting rate is inadequate to balance the decline in intake per bite, and rate of intake declines. Under very limiting forage conditions, animals often stop searching and grazing even though time appears not to be a limiting factor. Coleman (1992) concluded that the grazer may compensate for reduced bite size by increasing biting rate or time spent grazing or both, but such compensation is limited to about 15%. This is in agreement with the conclusion of Minson (1990) that as a sward is grazed down there are large reductions in bite size but changes in grazing time and biting rate are small.

Several interrelationships between ingestion rate, biting rate, and forage availability have been revealed in studies with beef heifers grazing crested wheatgrass in central Utah. Biting rate was found to increase as the standing crop decreased through utilization—from 54–63 per minute as the standing crop decreased from 372–181 lb dry matter per acre during the first year of the study, and from 37–50 bites per minute as standing crop was reduced from 819–128 lb dry matter per acre the second year (Scarnecchia *et al.,* 1985). Ingestion rate decreased with decreasing forage availability, while both biting rate and grazing time increased (Olson *et al.,* 1989), but both biting rate and ingestion rate declined as the nutritional quality of the sward declined. The number of bites per feeding station also decreased as the season progressed, suggesting that animals were intensifying their selective strategies, particularly later in the season, as protein content in the grass declined and cell wall percentage increased (Flores and Malechek, 1983). The heifers also increased their step rate as the season progressed, but steps between stations and time spent at each station were not affected appreciably.

Forage intake rate is a product of bite weight and biting rate, total grazing time being the other component of forage intake. Of these three components, bite size has the greatest influence on forage intake, with rate of biting and grazing time being compensatory variables (Forbes, 1988; Minson, 1990; Forbes and Coleman, 1985). From their studies on grazing crested wheatgrass with cattle, Olson and Malechek (1988) concluded that forage intake was largely a function of ingestion rate, with the compensatory response of increasing grazing time having little effect on the decline in forage intake as ingestion rate declined.

Bite size was confirmed by Erlinger *et al.* (1990) to be a major determiner of forage intake, which was directly influenced by forage availability, and the compensating effect of longer grazing time with smaller bite size was demonstrated as a regulator of intake. However, there was no evidence of a compensatory role for rate of biting when intake per bite decreased under cattle grazing seeded Asiatic bluestem (*Bothriocloa* spp.) swards in Oklahoma. Confirming that maximizing bite size and minimizing grazing time, thereby reducing the energy expenditure of grazing, results in enhanced performance, Erlinger *et al.* (1990) concluded that livestock could be selected for these desirable grazing traits.

Bite weight is the product of bite volume and the bulk density of the grazed stratum, and the rate of intake at any given herbage mass is largely determined by bite weights which, in turn, largely depend on the spatial organization of the herbage in the biting plane (Ruyle and Rice, 1996). With taller swards, surface height (operating primarily through bite depth and hence bite volume) was found by Burlison *et al.* (1991) to have the dominant influence on bite weight, with grazed stratum bulk density having only a minor effect. On shorter swards with a reduced range of surface height, grazed stratum bulk density had more influence than surface height on bite weight.

Arias *et al.* (1990) found on tall fescue swards that rate of intake, principally resulting from larger bite size, increased with surface height of sward and availability of herbage. Steers grazed on legume or grass swards garnered heavier bites on tall sparse swards than on short dense ones with the same herbage mass per unit area (Laca *et al.,* 1992). Alfalfa swards were found by Dougherty *et al.* (1989a) to permit cows to prehend larger bites, albeit at slower rates, than did tall fescue swards.

Bite size is reduced by low forage mass per acre, low allowance per animal, and the presence in the sward of plant material rejected or selected against by the grazing animal. Minson (1990) concluded that the reduction in forage intake below a forage mass of 2000 kg of dry matter per hectare (782 lb per acre) in seeded swards resulted in part from the shorter stature of the standing crop and smaller associated bite size. When herbage availability on alfalfa-grass swards in Saskatchewan was not limiting, daily herbage intake by cattle was not affected by grazing intensity (Popp *et al.,* 1996). While animals in lightly stocked pastures compared to moderately stocked pastures had a higher intake rate, they spent less time grazing, the latter possibly the cause of higher gains.

Both wild and domestic large grazers may attain higher maximum forage intake rates than small grazers but require relatively higher forage biomass to do so (Hudson and Frank, 1987). These authors attributed the high rate of feeding of bison to both large bite size and rapid bite rate, both factors benefiting from high available forage biomass.

In an elk ranching study near Edmonton, Alberta, bite size of yearling elk males decreased but biting rate increased as biomass increased as the spring-summer grazing season progressed, this presumably associated with greater time and effort required to assemble the larger bites (Wairimu and Hudson, 1993). The feeding rate increased from 9–15 g per minute as forage biomass increased in the summer and grazing time declined from 12 hr in April to 8 hr in June. Jiang and Hudson (1994) found that the sacrifice of bite size in maturing tall summer and fall swards was compensated by diet quality maintained at about 14% protein by elk through selective grazing. However, when elk were confined to grass openings in aspen (*Populus tremuloides*) boreal forests during a 7-day period in late summer under high stocking density, both intake rate and bite size declined linearly with sharply declining biomass availability and plant height; foraging rates (bites per minute) increased but only in partial compensation (Hudson and Nietfield, 1985).

Ease of prehension also affects forage intake rates. Exceedingly long or very short leaves or plant parts that are stiff or have high shear strength delay grasping, severing, and ingesting by the animal and generally reduce intake rate. Larger bites and thus increase in intake rate resulted from sheep eating the large leaves of shrub liveoak (*Quercus turbinella*) compared to eating the small leaves of blackbrush

(Ortega-Reyes and Provenza, 1993). In improved pasture studies by Burns (1984), forage intake appeared maximized at an extended plant height of around 16 inches. At taller heights the ratio of manipulative to ingestive jaw movements increased.

In taller, denser standing crops, grazing was noticeably concentrated in the upper horizons of the sward where forage collection was more efficient (Minson, 1990). Burns (1984) noted defoliation mostly removed forage from the top half of improved pasture swards, and little penetration occurred into horizons containing dead material even when intake was being severely limited. Arias *et al.* (1990) found that the grazing horizon for cattle grazing endophyte-free tall fescue was about 4 in., below which were the pseudostems along with more senescent and dead material. **Pseudostems** are described as concentrically arranged sheaths of fully expanded leaves that surround the immature growing leaves, tillers, and growing points (Dougherty, 1991); they are generally recognized as a distinct barrier to prehension and biting depth. Flores *et al.* (1993) noted that leaves growing from the bottom of plants were protected from grazing by residual stem stubble on mown swards or range with abundant residual stems from the previous year's growth. However, using artificially hand-constructed sword of dallisgrass (*Paspalum dilatatum*), these authors concluded that stems of current year's tillers but not their pseudostems restricted bite depth and thus bite size.

In grazing studies on Lehmann lovegrass (*Eragrostis lehmanniana*) in Arizona, green tiller heights and amounts of standing dead material interacted to influence cattle grazing patterns and ingestive behavior (Abu-Zanat *et al.,* 1988). The presence of residual stems in lightly grazed patches increased handling time per bite approximately 0.5 sec over bites in previously grazed patches characterized by regrowth and minimal residual old growth. As green tiller heights increased, biting rates also increased in both previously grazed and ungrazed patches but remained higher in previously grazed patches throughout the year. It was apparent that the necessity of sorting the green material from standing dead material reduced biting rate. Biting rates on Lehmann lovegrass were the lowest during winter dormancy and under the heaviest stocking rates, the herbage mostly having been previously grazed but without regrowth (Ruyle and Rice, 1996).

The presence of thorns and spines have been shown to restrict bite size, particularly for larger animal species, with bites often being limited to individual leaves or leaf clusters (Cooper and Owen-Smith, 1986). Grazing animals are unable to increase their biting rates under these circumstances to compensate for the smaller bite size. In particular, the hooked thorns of certain species tend to slow down biting rates by catching on the lips and tongues or ears of feeding animals.

F. SATIETY AND HUNGER

Satiety, a feeling of satisfaction from rumen fill resulting from ingesting adequate or optimal kinds and amounts of forage, is expected to reduce the forage intake rate of grazing animals and eventually terminate the grazing bout (Dougherty, 1991). Owen-Smith and Novellie (1982) have predicted that bites per minute and per feeding station will decrease and steps between feeding stations will increase as satiety increases. Jung and Koong (1985) found that the rate of intake by grazing sheep decreased as amount of feed eaten before grazing increased from 0–30% of daily intake. Steers on perennial ryegrass were found capable of increasing their intake rate without decreasing diet quality after fasting (Greenwood and Demment, 1988); however, this was at the cost of a lower mastication rate with the implication of larger ingested particle sizes and therefore slower dry matter turnover rates in the rumen. Based on their studies with steers on stockpiled orchardgrass, Dougherty *et al.* (1988) accepted the current concepts that herbage intake of grazing animals is determined by (1) hunger-satiety status, (2) the forage harvesting capacity of the mouth and tongue, and (3) the properties of the sward.

When grazing alfalfa in 3-hr sessions—each session begun on a new forage plot—cattle consumed 47, 29, and 24% of their intake during the first, second, and third hours, respectively, forage plant utilization was 51, 32, and 27% for the corresponding periods (Dougherty *et al.,* 1987). The mean rates of biting were 26, 21, and 19 bites per minute for the same periods, while dry matter intake per bite declined linearly from 1.96 to 1.54 to 1.36 g. Declines in the rate of biting and bite size as the grazing sessions progressed may have resulted from essentially nonselective grazing at the beginning, with the animals becoming more selective as hunger was alleviated and the available forage became reduced in volume and uniformity. After being fasted for 16 hr, cows were found by Dougherty (1991) to initially graze vegetative tall fescue swards about 50% faster than their normal rate, by taking larger bites at a faster rate.

Weaned elk calves, overwintered at a lower nutrient level resulting in 30 lb lower spring weights compared to a higher level overwintering group, spent more time foraging and foraged faster when grazed together on spring-summer pasture (Wairimu and Hudson, 1993). While consuming more forage daily, resulting from a 10% higher cropping rate, the low-winter group were more dedicated to grazing and spent less time in energetically expensive activities. However, both groups selected spring-summer diets of equal quality and grazed similar habitats.

However, Bond *et al.* (1976) found no increase in rate of intake by grazing steers after satiety was reduced by fasts of 12–48 hr. Similarly, Freeman and Hart (1989) found that hay feeding prior to grazing by steers had little effect on feeding station behavior of steers (e.g., bites per minute, bites per station, or steps per station). It was suggested that the effects of satiety or time already spent grazing may require more time to appear than provided in their study; it was also suggested that increasing fatigue rather than increasing satiety may reduce biting rate and bites per station.

Short-term fasts—induced by lengthening nongrazing intervals—resulted in higher rates of intake of alfalfa by beef cows during the following grazing session but had no effect on ingestive rates the next day (Dougherty *et al.,* 1989a). In contrast, short-term fasting did not affect the corresponding rate of intake for cows on tall fescue swards during the following grazing session but depressed rates of intake 24 hr later. The delayed behavioral responses were attributed to differences in flow characteristics of ingesta of alfalfa (minimal retention) and tall fescue (de-layed retention). Dougherty *et al.* (1989b) found that accessing fresh herbage by lengthening tethers or turning grazing animals into new plots nearly doubled the rate of biting initially of non-starved cattle, but this effect was too transitory and did not last long enough to significantly affect rates of herbage intake. These au-thors predicted that the stimulation of biting rate generated by the availability of ungrazed swards would be enhanced by hunger but diminished by satiety.

II. DAILY ACTIVITIES: TIME SPENT

The grazing ruminant divides its day among three main activities: grazing, ru-minating, and idling. By themselves, these activities may be of greater academic interest than of practical value. But when considered with other criteria, their con-sideration and also deviation from the normal may signal stress factors and sug-gest management changes (Campbell *et al.,* 1969).

A. TIME SPENT GRAZING

Domestic livestock commonly spend 7–12 hr per day grazing, including time spent searching for as well as consuming forage (Burns, 1984; Arnold and Dudzin-ski, 1978; Walton, 1983). Representative grazing time per day from various stud-ies are as follows for beef cattle: steers on improved pasture in West Virginia, 7.3 hr (Sheppard *et al.,* 1957); cows with calves on native grass range in Oklahoma, 9.7 hr (Dwyer, 1961); heifers on Ozark ranges, 10.6 hr (Bjugstad and Dalrymple, 1968); steers on sagebrush-grass range in Oregon, 9.5 hr (Sneva, 1970); pregnant cows on Montana grassland winter range, 8.3 hr (Adams *et al.,* 1986); and cattle on Australian rangelands, 10 hr (Squires, 1981).

When cattle were grazed in adjoining paddocks of rangeland in Texas, Cory (1927) obtained daily grazing times of 7.7 hr for cattle, 6.6 hr for sheep, and 5.9 hr for goats. Angora goats on mixed brush range in Texas spent 2.9 hr daily trav-eling and 7.3 hr daily grazing and browsing (Askins and Turner, 1972). Both cat-tle and pronghorn antelope on rolling shortgrass range in northeastern Colorado spent about 12 hr per day grazing and traveling (Ellis and Travis, 1975). Elk on summer range commonly forage for about 9 hr per day and rarely extend grazing time beyond 13 hr (Hudson and Nietfeld, 1985).

Differences in grazing time per day between European breeds of beef cattle and their crosses have been quite small (Arnold and Dudzinski, 1978; Funston *et al.,* 1987). Kropp *et al.* (1973) concluded that Hereford, Holstein × Hereford, and Hol-stein heifers on Oklahoma native range were generally similar in range behavior-al activities such as grazing, ruminating, idling, drinking, walking, and sleeping. When compared while cattle grazed kleingrass in south Texas, daily grazing times for Angus, Brahman, Angus × Brahman, and Tuli × Brahman were not different (Forbes *et al.,* 1998). However, substantial difference has been found between

Hereford and Santa Gertrudis cattle on New Mexico range (Nelson and Herbel, 1966), with the Santa Gertrudis spending 1.4 hr more per day grazing.

Based on grazing behavior of beef cows on Montana rangeland, the level of milk production was concluded to be a driving force to increase the amount of time spent grazing (Lathrop et al., 1985; Havstad et al., 1986b). It was found the greater the milk production, the more time the cow spent grazing, and the older the calf was, the less time the cow spent grazing. Grazing time on Montana winter range by cows was not affected by animal size or body condition (Adams et al., 1986). Pregnant ewes were found to spend the same amount of time grazing as non-pregnant ewes, but lactating ewes spent 12% more time grazing than dry ewes at all but the highest level of forage availability (Arnold and Dudzinski, 1978). Calves at 4 to 6 months of age spent 18% less time grazing than their dams but grazed at least as long as the mature cows when reaching a year of age (Arnold and Dudzinski, 1978). Dwyer (1961) also noted that beginning at about 4–5 months of age most grazing activities of the calves began to closely mimic that of their dams.

B. FORAGE AVAILABILITY AND GRAZING TIME

As grazing time increases, more energy is used for activity and less for production; thus, the minimum grazing time that results in adequate dry matter intake is considered optimum (Fig. 6.8). Grazing time depends on ease of ingesting, which varies with accessibility of plant parts, availability of total forage, and quality of the consumed diet (Burns, 1984). Grazing time is generally lowest when forage is abundant and of good quality and highest when forage is of low quality or availability is limiting. Intake can be maintained for a time when forage and thus bite size are limiting by compensation in increased grazing time and number of bites per minute. However, this compensation is seldom adequate to prevent a reduction in daily intake once the short-term rate of intake starts to decline. Grazing time may fall when grazing animals are in a severe caloric deficit and forage availability is severely limited (they give up on further searching), thus contributing even further to decline in forage intake (Hodgson, 1986).

Campbell et al. (1969) concluded that cattle spend more time grazing and less time resting (1) as stocking rates increase, (2) when pastures are naturally short or eaten down, (3) as the nutritive value of a pasture declines, and (4) on swards comprised of species of different quality or growth habit (i.e., more heterogeneous plant stand). Under one or more of these stresses cattle that normally graze 8 hr per day may graze up to 12 hr per day. Campbell et al. (1969) further concluded that sheep in temperate environments commonly graze for 8 to 9 hours daily on good pasture but up to 12 hr when a pasture is overstocked or herbage is otherwise short.

Heifers grazing crested wheatgrass in central Utah compensated for decreasing forage supply by increasing grazing time and increasing biting rate (Scarnecchia et al., 1985). During the first year of study, grazing time increased from 7.6–10 hr daily as the standing crop decreased from 372–181 lb dry matter per acre; during the second year, grazing time increased from 6.3–10.9 hr daily when the standing

FIGURE 6.8 Grazing time (ranging from 7 to 12 hr per day for domestic livestock) and associated energy expended are reduced when forage is abundant, of good quality, and readily accessible; cattle grazing high condition range in Cherry County, NE. (Soil Conservation Service photo.)

crop decreased from 819–128 lb dry matter per acre. Since there were no significant differences in daily organic matter intake, it was concluded that individual animal intake may have reached a plateau and could not increase due to low digestibility, longer retention time, and slow rates of ingesta passage.

Lactating cows on rugged Montana foothill terrain spent 8.2 hr per day grazing in July but increased to 10 hr when available forage began to decline (Havstad *et al.,* 1986b). However, when herbage availability is greatly restricted, reduced herbage intake of both cows and calves may be associated with reduced grazing and rumination and increased idling (Baker *et al.,* 1981). Deer have been found to spend more time feeding under heavy cattle grazing than under light or no grazing by cattle (Kie *et al.,* 1986).

Cattle grazing time in Texas studies was one hour per day longer under moderate continuous stocking than heavy short-duration grazing (10.9 vs. 9.8 hr), but grazing time was similar between 14- and 42-pasture short-duration grazing (Walker and Heitschmidt, 1986a). The shorter grazing time under short-duration grazing may have resulted from less time spent in selecting habitats, feeding stations within a habitat, or bites within a feeding station. Both grazing time and intake were apparently controlled in this study by bulk fill of the rumen. Thus, cattle consumed forage until they were full and then stopped grazing. Grazing time was longer in the spring when forage quality was highest and shortest in the winter when forage quality was the lowest. The longer grazing times associated with seasons or days within rotation cycles when forage quality was high apparently resulted from the additional forage that could be consumed and processed.

Cattle on African grassland range grazed under continuous grazing averaged one hour more grazing time daily than those on short-duration grazing (7.4 vs. 6.2 hr), with the grazing time under the latter system increasing from the first to the last day in each paddock (Gammon and Roberts, 1980b). Steers on improved pasture in West Virginia spent similar amount of grazing time for rotational and continuous grazing (Sheppard *et al.,* 1957), presumably because of uniformity and availability of the standing forage.

On mountain range in southern Utah, unherded sheep increased their grazing time during daylight hours (i.e., from leaving the bedground to bedding down) from 67.1% during the first 3 months to 79.7% during the last 3 months of the summer grazing season (Bowns, 1971). Bison in boreal habitats in British Columbia increased their grazing time from 8.7 hr per day in the summer to 10.7 hr per day in the fall (Hudson and Frank,1987).

High selectivity for limited green material results in difficulty in harvesting enough forage; sheep may spend 12 hr a day selecting out small green shoots from a bulk of dry pasture and still have reduced intake (Arnold and Dudzinski, 1978). On lightly grazed foothill range in California, cows grazed 8.9 hr daily on abundant mature growth but 13.9 hr on new growth in short supply; on closely grazed range, cows grazed 6.7 hr on scant dry forage but 13.2 hr on scant new forage (Wagnon, 1963). Under both abundant and limited biomass availability, searching time increased when the palatable new green growth was present but in short supply. The time spent grazing by cattle on intermediate wheatgrass was inversely related to the mean number of green leaves per tiller (Pierson and Scarnecchia, 1987); this suggested that the cattle were selectively grazing green leaves and were spending more time searching for them as they became more limited.

From their work at the Central Plains Experimental Range in northeastern Colorado, Shoop and Hyder (1976) recommended two contrasting management practices. When the herbage was cured and plentiful, it was recommended that cattle be concentrated to reduce wandering in search of green herbage. But, when plant growth was beginning and the grass was in short supply, it was recommended that cattle be scattered as widely as possible. Based on their study of sheep grazing on improved pasture, Gluesing and Balph (1980) concluded that sheep graze inefficiently when searching for a few preferred plants; they recommended that the time and energy spent in the search might be reduced by confining the sheep to a small portion of the larger grazing unit.

C. WEATHER AND GRAZING TIME

Ruminants respond to short-term thermal stress, either hot or cold, by reducing grazing activity but subsequently readjust to running mean temperatures. The time required for livestock to re-acclimate to lasting changes in ambient temperature over time was 10 to 11 days with horses (Senft and Rittenhouse, 1985) and 10 to 14 days with cattle (Senft *et al.,* 1982). Cows on prairie range in Oklahoma spent about 2 hr less time grazing on hot days—6.7 hr when daytime temperature averaged above 85°F vs. 8.7 hr when the average was below 80°F (Dwyer, 1961). Heifers on Missouri Ozark range were reluctant to graze when high temperature was combined with high humidity, a situation commonly occurring during the summer months (Bjugstad and Dalrymple, 1968); both high temperature and high humidity suppressed grazing time, and the decreased grazing time required extra good forage conditions for satisfactory production. Following days of very hot weather, grazing time may be increased during intervening cool, cloudy days in summer by both sheep (Springfield, 1962) and cattle (Springfield and Reynolds, 1951).

It was concluded from the Missouri research that the U.S. Weather Bureau's temperature-humidity index (THI)—a comfort index used for people—was closely related to the time beef cattle spend grazing (Ehrenreich and Bjugstad, 1966). The multiple correlation between the THI index and time spent grazing was 0.968. It was noted that a breeze reduced the restriction of a high THI.

Morning temperatures were negatively related to percentage of cattle grazing during the summer in the Rio Grande Plains (Shaw and Dodd, 1979); however, summer daytime vapor pressure deficit (VPD) showed a better correlation with grazing habits. As VPD increased, the number of cattle grazing decreased; conversely, as VPD decreased, cattle grazing increased. Cattle on Utah winter range grazed and ruminated for longer time periods following changes in atmospheric pressure (Malechek and Smith, 1976); but these changes appeared short lived and were compensated for later. Rittenhouse and Senft (1982) concluded that low to intermediate rates of change in barometric pressure resulted in spurts of increased grazing time, while higher rates of barometric change led to a gradual decline in grazing time back to the 12-hr basis.

Adverse winter weather often reduces both grazing activity and forage intake of cattle on rangelands (Fig. 6.9) (Adams *et al.,* 1986). In studies on Montana winter range, cold temperatures reduced the grazing time of pregnant cows, but wind speed, relative humidity, and barometric pressure seemed not to affect grazing time (Beverlin *et al.,* 1987). Although the average daily grazing time was 7.2 hr on southeastern Montana winter range, individual cow grazing time varied from 0.5–11.6 hr per day (Adams *et al.,* 1986); 6-year-old cows were found to begin grazing earlier and grazed longer on cold winter days than 3-year-old cows.

Prescott *et al.* (1989) found that sudden temperature changes reduced grazing time in the fall more than in the winter; they concluded that cows responded to rapid temperature changes in the fall as being novel but by winter had become familiarized with their environment and became more insensitive to temperature changes. Because of more rapid acclimation to winter temperature changes, it was concluded that consistently cold temperatures had less effect on both total daily grazing time and daily forage intake than did other environmental conditions, es-

FIGURE 6.9   Adverse winter weather often reduces both grazing activity and forage intake of livestock on rangelands; on this northern Great Plains prairie range cattle diets are being provided by a combination of standing forage and round bales of prairie hay left in place.

pecially qualitative and quantitative aspects of the available forage (Prescott *et al.*, 1994). Hereford cows in Utah spent less time grazing (and undoubtedly consumed less forage) on cold days than on warm days (Malechek and Smith, 1976), but the greater time spent standing on cold days presumably conserved energy. Between the extremes of 32 and $-40°F$ in daily temperatures, total daily grazing time was reduced by about 50%. The cattle further reduced their travel distances on cold days when wind velocities were high.

The availability of protected sites in a grazing unit may allow cows to continue grazing, thus maintaining intake even when severe winter weather might otherwise cause them to reduce or temporarily stop grazing. In their studies on foothill range in Montana, Houseal and Olson (1995) found that cattle sought available moderate microclimate sites for both grazing and resting to avoid high winds and cold temperatures; they tended to remain in such sites but readily grazed there if forage was available. However, when winter weather conditions were mild, cows tended to graze on exposed upper slopes where periodic strong winds kept ridges and windward slopes free of snow, making forage more available to cattle and travel easier.

In Alberta bison and cattalo (bison × cattle) were found more willing to graze on more days under very cold conditions than European breeds of cattle (Smoliak and Peters, 1955), but both cattle and cattalo further decreased their grazing time during winter as wind speed increased (Arnold and Dudzinski, 1978). The net result of these alterations in behavioral patterns during periods of winter weather stress is a reduction in energy expenditures for physical activities, which compensates only in part for reduced forage consumption. In the Front Range of Colorado, the grazing time of mule deer during winter was not materially affected by wind and low temperature but was reduced by snow depth over 14 in. (Kufeld *et al.*, 1988).

D. WINTER MANAGEMENT AND GRAZING TIME

Range animals can apparently endure great stress from cold and related winter weather factors with reasonably good forage and available water. However, high maintenance requirements for grazing and additional energy required for thermoregulation in very cold weather, along with reduced intake of standing forage (of lower digestibility) can result in a large negative energy balance and loss of body weight. Winter management strategies with grazing ruminants should therefore include consideration for increased energy requirements and reduced levels of forage intake as air temperatures become very cold.

Management strategies that complement or enhance grazing activity are desirable during periods of cold weather. Such practices include feeding supplements that enhance forage intake and digestibility and timing supplementation so that grazing activities are not disrupted and range forage intake reduced (Adams *et al.*, 1986). Sorting cows by age and/or nutritional need and selection of winter grazing sites that provide feeding and resting protected from the wind should also be advantageous, particularly with younger animals. In Montana winter range studies, young (3-year-old), inexperienced cows were found less efficient than the older (7- to 8-year-old) cattle with experience on the site in using the available forage and thermal resources (Beaver and Olson, 1997). The younger cows used unprotected areas more frequently, were more often exposed to lower critical temperatures, and were presumably cold-stressed more often.

Grazing time is substantially reduced when supplements are fed in amounts resulting in their partial substitution of herbage (Allden, 1981; Arnold and Dudzinski, 1978). Feeding large amounts of high-energy supplement (i.e., 4 lb or more daily to cattle) greatly reduces both grazing time and forage intake (Cook and Harris, 1968b; Rittenhouse *et al.*, 1970; Clanton *et al.*, 1971). After being fed 8.5 lb grain daily in the corral before being turned out to graze during the winter, Montana range cows were observed to lie down for periods of up to 4 hr before starting to graze, thus resulting in a total period of about 6 hr away from grazing (Bellows and Thomas, 1976). High-energy supplements (23% protein) fed to sheep on New Mexico winter range decreased both forage intake and grazing time (309 vs. 242 min daily) (Hatfield *et al.*, 1990). However, the supplemented ewes gained less over the winter than the non-supplemented ewes, indicating that any advantage from feeding the low-protein supplement was diminished by its effects on herbivory.

From a review of literature, Krysl and Hess (1993) concluded that, when grazed on dormant winter range, supplementing cattle with a high-protein concentrate reduced grazing time by approximately 1.5 hr per day compared to non-supplemented cattle. While protein supplementation increased forage harvest efficiency (i.e., grams of forage intake per kilogram of body weight per minute spent grazing), high-starch supplements either did not alter or decreased harvest efficiency. Cattle supplemented with protein concentrates when wintering on intermediate wheatgrass spent less time grazing, but their forage intake was similar to that of non-supplemented cattle, thus increasing their forage harvesting efficiency (Hess et al., 1994). Under winter grazing on foothill range in California, supplemented cows (3 lb of 31% protein supplement daily) traveled slightly less (1.25 vs. 1.6 miles daily) and spent less time grazing (10.6 vs. 13.4 hr) than non-supplemented cows (Wagnon, 1963).

Less frequent feeding than daily of equivalent daily amounts of a protein supplement interferes less with grazing, reduces time spent at feed grounds, and encourages at least ample grazing time. Feeding a daily equivalent of 1.5 lb of cottonseed meal but on alternate days to Herefords on mixed prairie winter range did not affect grazing time (Box et al., 1965), but the supplemented cows walked less (1.9 vs. 4.1 miles), spent less time walking, and were easier to handle and appeared more contented. In Oklahoma studies, cattle hand-fed supplement every third day did not wait at feed bunks but were easy to call in (McIlvain and Shoop, 1962b). When fed daily, they waited at the bunks before and after feeding; when fed only weekly, they occasionally had to be gathered for feeding. Similarly, cows on Texas range supplemented daily and thrice weekly came to bunks readily when called, but cows fed only once per week did not respond quickly to being called. Cattle arriving first at the feed bunks under weekly supplementation would depart while feed remained; cattle supplemented daily were subject to more turbulence at the feed bunks as dominant cows attempted to keep others away.

E. TIME SPENT RUMINATING, RESTING, ETC.

Rumination is the second most time-consuming activity, after foraging, of free-ranging livestock (ruminants). Cattle commonly ruminate for 4–8 hr daily (Campbell et al., 1969), but time spent chewing their cud can vary from 1.5–10.5 hr a day, depending on the quantity and quality of the food eaten and the amount of grinding it requires (Squires, 1981). The time ruminants spend chewing their cud is proportional to cell wall intake, with a limit of around 10 hr daily, according to Van Soest (1982). Rumination time for cows grazing California foothill range averaged 7.7 hr daily, varying from 6 hr on new forage growth in short supply to 10.3 hr when the forage supply was abundant but had reached maturity (Wagnon, 1963). Sheep masticate more than cattle before swallowing and commonly ruminate only about 3.5 hr daily when on good pasture (Campbell et al., 1969).

In studies with deer and elk in the Cascades of Washington, periods of both food harvesting and rumination were short during June, when forage quality was at its peak (Hanley, 1982a). As forage quality decreased as the season advanced, longer periods were required to reach rumen fill and in rumination. The increased harvesting time was attributed to increasing selectivity for plant parts or individual plants, the increased ruminating time to decreased diet quality and rumen turnover rate.

The remaining 10% to over 50% of the 24-hr day, after foraging and ruminating, is spent by cattle in idling and resting. Idling is considered important in being a low-energy activity; longer free-choice idling time reduces stress on both the sward and the grazing animals (Campbell *et al.,* 1969).

III. DAILY ACTIVITIES: TIME OF DAY

A. DAILY GRAZING ACTIVITIES

Grazing animals exhibit a daily grazing cycle that is remarkably consistent and recurs each day with minimal change. By contrast, there seems to be no distinct diurnal pattern in rumination as there is in grazing (Squires, 1981). Most studies in temperate environments have shown that a major grazing period begins at about dawn and another in late afternoon, with shorter, less regular, and more casual periods during mid-day and at night (Arnold and Dudzinski, 1978; Campbell *et al.,* 1969). Cattle on mixed prairie during summer in Kansas began grazing between 5 and 6 a.m., depending on the time of sunrise, and continued unabated for about 3 hr; the evening grazing period was well under way by 5 p.m. and lasted until 8 p.m. or later when the cattle bedded down (Moorefield and Hopkins, 1951). The middle part of the day was marked by alternate periods of resting and feeding during the first part of the summer, but by a distinct resting period in late summer.

Similar patterns of intensive grazing in early morning and late afternoon and of shorter, more irregular grazing bouts in between have been shown by cattle on California foothill range (Wagnon, 1963), on forested range in the Blue Mountains of Oregon (Roath and Krueger, 1982b), in Canada (Campbell *et al.,* 1969), on Zimbabwe range (Gammon and Roberts, 1980a), on New Mexico desert range (Herbel and Nelson, 1966a), on south Texas range (Shaw and Dodd, 1979), on Ozark range (Bjugstad and Dalrymple, 1968), on northern Oklahoma prairie (Dwyer, 1961), on shortgrass range in southeastern Wyoming (Hepworth *et al.,* 1991), on Lehmann lovegrass in Arizona (Ruyle and Rice, 1996), and on improved pasture in West Virginia (Sheppard *et al.,* 1957).

Diurnal grazing patterns have been shown for sheep in Australia (Squires, 1981), in Wyoming desert areas (Severson *et al.,* 1968), on crested wheatgrass in New Mexico (Springfield, 1962), in the mountains of Utah (Bowns, 1971), for Angora goats in Texas (Askins and Turner, 1972), and in New Mexico (Velez *et al.,* 1991). Intensive early morning and late evening feeding periods have also been reported for most big game animals in temperate climates including antelope (Ellis and Travis, 1975) and deer and elk (Hanley, 1982a). Mule deer in the Front Range of

Colorado fed most during sunset, night, and sunrise periods and least during the day (Kufeld *et al.,* 1988). The period of lowest deer activity, except in winter, is during daytime, this time used primarily for resting, rumination, and escape.

Adams (1985, 1986) found that yearling steers fall grazing Russian wildrye gained 0.4 lb more per day (1.8 vs. 1.4 lb) when supplemented with 2 lb of cracked corn in the early afternoon than when fed in early morning or fed no supplement at all. This was attributed to the morning supplementation—the traditional time— falling during a major grazing period and disrupting normal grazing activity (Fig. 6.10). The morning feeding of supplement resulted in reduced forage intake and apparently in increased energy expenditure for maintenance. In contrast, early afternoon supplementation coincided with a time when cattle were engaged in mostly nongrazing activities.

Smith *et al.* (1986) also suggested that feeding in the afternoon interrupted grazing less and that the feed was less of a forage substitute than when fed in the morning. When grazing steers on dormant tall wheatgrass, Barton *et al.* (1989) found that animals fed protein supplement in early morning grazed less daily (368 min) than either the afternoon-fed steers (388 min) or the unsupplemented steers (442 min); however, forage intake was not affected. In a later study with cattle grazing dormant intermediate wheatgrass, the time of day cattle were fed protein supplement did not affect forage intake, digestion, or digesta kinetics, but supplementa-

FIGURE 6.10   Since the most consistent and avid grazing periods by most grazing animal species begin around sunrise (shown above) and again in late afternoon, protein or energy supplementation in early afternoon interferes least with grazing; showing steers grazing native range in central Utah.

tion did reduce grazing time by 1.5 hr daily (Barton *et al.*, 1992); these authors suggested that time of day for feeding limited amounts of high-protein supplements to cattle grazing dormant rangeland should not be of great concern. However, time of day supplementation is probably more important when low-protein supplements are being fed or supplement is fed in higher amounts.

Comparison was made with beef cows grazing fall native range in northeastern Colorado of supplementing with alfalfa hay at 8 a.m. and 4 p.m., both presumably normally intensive periods (Yelich *et al.*, 1988). Cows fed at 8 a.m. compared to 4 p.m. grazed slightly more (9.4 vs. 8.7 hr daily) and gained more (.82 lb vs. .55 lb daily); the non-supplemented cows grazed 10.1 hr daily (Yelich *et al.*, 1988).

B. WEATHER AND GRAZING ACTIVITIES

Current weather conditions may modify the time when animals graze during the 24-hr day. During hot summer days, livestock reduce or eliminate mid-day grazing bouts, seek shade or water sites, and spend time remaining idle or ruminating. During very cold weather, livestock may limit early morning and evening grazing and concentrate grazing during mid-day and afternoons. If possible, periods of combined sunshine and reduced wind velocity are sought in very cold weather. Livestock commonly search for breezy points in a pasture on hot days while seeking protection from the wind on very cold days. Cattle will graze and stand with their flanks to the wind on hot days for cooling (Dwyer, 1961) but with their heads downwind during periods of cold wind or rain or snow. They often assume a body position at right angles to the sun during periods of sunshine on cold days to absorb the maximum radiation (Malechek and Smith, 1976).

Temperatures above the thermoneutral zone ($30-85°F$) most commonly occur between 9 a.m. and 4 p.m. (Campbell *et al.*, 1969). Dwyer (1961) observed that cattle grazing rangeland moved as they grazed in the morning from one watering point to another and appeared to time their arrival at the second watering point earlier on hotter days. During hot days on the Rio Grande Plains, cattle grazed intensively for $3-5$ hr in early morning before moving to water; a lengthy midday rest period followed, lasting about 5 hr, during which time only about 20% of the cows grazed (Shaw and Dodd, 1979). Arnold and Dudzinski (1978) reported that sheep could seemingly predict the hotter days by starting their grazing earlier in the morning.

Hot, humid weather is more stressful than hot, dry conditions (Squires 1981), and high summer humidity may depress livestock gains more than the high summer temperature alone. When grazing yearling Hereford steers on Oklahoma range, McIlvain and Shoop (1971a) found that the combined effects of humidity above 45% and temperature above 85°F were especially harmful; each hot, muggy day was found to reduce summer-long steer gains by about 1 lb. The availability of shade during hot weather can reduce livestock heat stress and improve productivity, but natural shade is often more effective than artificial shade.

Cattle show heat stress by panting, reduced rumination time, frequent drinking, excessive milling around, standing instead of lying, and general irritation; sheep show heat stress by open grazing, spreading of hind legs, greater water intake, and rapid breathing (Campbell *et al.,* 1969). Brahman cattle or Brahman crosses (including Santa Gertrudis) are more tolerant of hot and hot/humid weather than European breeds and spend more time grazing and less time hunting shade under such conditions (Ittner *et al.,* 1954; Herbel and Nelson, 1966a).

It appears that European breeds of cattle, to compensate for reduced daytime grazing in hot daytime temperatures, crowd more of their grazing time into darkness or near-darkness hours. Cattle grazed on native range in southeast Wyoming an average of 10.2 hr daily; of the 3.5 hr classified as night grazing (7 p.m. to 7 a.m.), 2 hr were continuations with late evening and early morning grazing periods and 1.5 hr occurred in late night (Hepworth *et al.,* 1991). About 25% of the summer grazing time of cattle on California foothill range occurred at night, this usually around midnight (Wagnon, 1963); the presence or absence of a moon had no effect on time spent grazing at night. On hot summer days, cattle in Kansas were found to do most of their grazing in early morning and in the evening but more than usual at night (Weaver and Tomanek, 1951).

When compared during hot, humid weather in south Texas, Angus (representative of northern European breeds) had shorter grazing periods, spent less time grazing in daytime, and spent much more time in the shade during the day than did Brahman, Brahman-Angus, or Tuli-Angus cross (Forbes *et al.,* 1996, 1998). The Tuli and Brahman crosses appeared equally adapted to heat, and both much more so than Angus. While total 24-hr grazing time was similar between breeds, Angus spent more time grazing at night to compensate for reduced daytime grazing.

Opportunity for night grazing during hot weather should be provided livestock unless protection from predation requires otherwise. In the tropics and subtropics and during prolonged periods of hot weather in temperate zones, Campbell *et al.* (1969) reported night grazing may account for up to 80% of the total grazing time by cattle. However, the evidence is mounting that cattle rely heavily on vision to move about in their environment, so in temperate zones grazing in total darkness is expected to be much less than this (Heitschmidt and Stuth, 1991).

Little or no night grazing was found by cattle on Oregon range (Sneva, 1970), on Ozark range (Bjugstad and Dalrymple, 1968), or on Oklahoma range (Dwyer, 1961). Askins and Turner (1972) reported that Angora goats in Texas did not normally graze at night, but Velez *et al.* (1991) found that goats grazed heavily at night in New Mexico when daytime temperatures were high. Herded sheep are normally kept on the bedgrounds at night, but unherded sheep on pasture or range may graze minimally at night (Campbell *et al.,* 1969; Squires, 1981). Bison have been found to do a substantial amount of grazing at night during summer (Hudson and Frank, 1987).

C. DRINKING, SALTING ACTIVITIES

Cattle commonly graze for a period in early morning during summer before moving to water. This move to water occurred after about 3 hr in the mixed prairie of Kansas (Moorefield and Hopkins, 1951), 3–5 hr in the Rio Grande Plains (Shaw and Dodd, 1979), and as soon as 1.5 hr but peaking at 2.5–4 hr in South Texas (Prasad and Guthery, 1986). Cattle commonly do some grazing along the way as they work towards water. On sagebrush-grass summer range in Oregon, some drinking occurred in mid-morning, but 53% occurred between noon and 4 p.m.; three-fourths of the total travel time but only 20% of the total grazing time occurred around the time of watering (Sneva, 1970). In central Australia, two distinct peaks in time were noted when the majority of cattle came in to drink; one was in the morning after grazing had ceased and the other in the afternoon before the evening grazing commenced (Squires, 1981).

Individual cattle in Texas usually spent only 4–20 min at water in summer before moving away (Prasad and Guthery, 1986); white-tailed deer spent only 1 or 2 min at water, taking water shortly after sunrise and again before sundown. After reaching water in Kansas, some cattle drank immediately, while others ruminated a short time before drinking (Moorefield and Hopkins, 1951). When grazing in Wyoming's Red Desert, cattle usually remained close to water but without grazing for an extended time period, laying down in summer but standing in winter (Plumb et al., 1984); the heavy trampling losses within the 0 to 50-ft zone from water was attributed to cattle remaining close to water following drinking.

Feral horses in Wyoming remained only briefly at water before leaving as a group upon finishing drinking but did materially increase the trampling impact over cattle alone out to about 100-ft, particularly in summer (Plumb et al., 1984). Feral horses on sagebrush-bunchgrass steppe in Oregon preferred to drink during the first period of daylight (46% of watering) and last period of daylight (33% of watering) (Ganskopp and Vavra, 1986); watering events averaged 16 min in duration. Groups of horses typically moved rapidly to and from water, with very few feeding or loafing near water.

Beef cattle on rangelands in the U.S. generally drink water between one and three times per day during both summer and winter (Box et al., 1965; Dwyer, 1961; Herbel and Nelson, 1966a; Sneva, 1970). Wagnon (1965) noted considerable variation in frequency of watering by beef cattle on California foothill range: 4% of the cows watered four times daily, 18% did three times, 57% did twice, and 27% watered once daily. In the Red Desert in Wyoming, cattle generally traveled to water three or four times a day during summer but only one or two times each day in winter (Plumb et al., 1984); feral horses watered five to seven times daily in summer and three or four times daily in winter.

The frequency of watering will depend on a complex interaction of weather factors, feed conditions, size of grazing units, and accessibility of water (Arnold and Dudzinski, 1978; Squires, 1978, 1981). Factors that increase the frequency of watering are hot temperature, dry and/or salty feed, small grazing units, short distances to water, and multiple watering points (Fig. 6.11).

In small paddocks, drinking frequency may be high, but as the grazing area becomes larger and the travel time and distance to water become greater, drinking frequency generally declines. When forage is green and abundant, sheep and cattle may drink infrequently (even less than daily) but will need to drink regularly

FIGURE 6.11   Frequency of drinking by cattle ranges from several times per day in small paddocks to every other day when required to travel long distances from foraging areas; cattle commonly graze, drink water, take salt, and return to grazing in that order; Nebraska Sandhills scene.

while on dry feed. On extensive rangelands in Australia (Squires, 1981), cattle watered at least daily even when forage was high in water if water points were nearby, but half watered only on alternate days when water was available only at longer distances. Cattle watered an average of every other day in winter.

Frequency of drinking is reduced when the distance traveled to water is increased, and the grazing distribution pattern is altered (Squires, 1978, 1981). As drinking frequency is reduced, a greater area of land can be served by each watering point and the use of rangeland made more efficient. For example, sheep drinking once daily can probably range over an area of up to 6.5 miles from water, while those requiring twice daily drinking can only range out to 1.5–1.75 miles from water (Squires, 1981). Under extensive rangeland grazing, animals often have to compromise between preferred frequency of drinking and the distance to travel from water to reach less heavily grazed areas; this interaction is of considerable significance in conservation of soil and forage resources (Arnold and Dudzinski, 1978).

Allowing mature dry cows on Oregon high desert range consisting of seeded crested wheatgrass and native range to drink only every other day did not adversely affect them (Sneva *et al.,* 1977). However, suckling calves were found to be more susceptible to water stress through reduced milk production; suckling calves without direct access to water gained 0.4 lb less daily over the 60-day study period. It was concluded that watering daily rather than every other day should be done if the herd consists of lactating cows with calves.

Yearling heifers in the Oregon study that were watered every 72 hr during July lost weight, but the weight losses were compensated for after returning to normal watering schedules. Reduced water intake by pregnant heifers forced to travel considerable distances to water reduced gains but did not seem to affect calving date, weight of calf at birth, or ability of heifers to reproduce. There was evidence that range cattle could be conditioned to withstand some additional water stress.

Withholding water for 96 hr in the Oregon study began to subject cattle to water intoxication after water was restored. It was suggested that dehydrated cattle should be watched closely for a period exceeding 4 hr after access to water is restored to prevent water intoxication from over consumption.

Reducing watering frequency with Zebu steers in Kenya from daily to once every 2 to 3 days was recommended as a means of enhancing utilization of extensive range and saving on cost of providing water more frequently (Mushimba *et al.,* 1987); watering frequencies from daily to every third day did not adversely influence steer performance. A comparison of 24, 48, and 72-hr watering intervals for sheep and goats in the arid north coast region of Peru resulted in similar gains (Pfister *et al.,* 1987).

The practice of trailing sheep 3–5 miles to permanent snowbanks every second to third day during the winter used to be common on desert range in the Intermountain Region. During a 40-day winter trial at the Desert Range Station in western Utah, dry ewes watered every day gained 3.4 lb, those watered every second day gained 0.8 lb, but those watered every third day lost 6.0 lb (Hutchings, 1958). It was also noted that a better percent lamb crop and less abandoned lambs were associated with more frequent watering. In more recent studies in the same area it was concluded that watering sheep on alternate days may be an acceptable management practice during periods of cool temperature (below 40°F) or as an emergency measure up to 78°F (Choi and Butcher, 1961).

A number of observations have shown that cattle ordinarily lick salt after drinking rather than before, and that when salt is placed away from water cattle usually leave the salt station to graze rather than go directly to water (Martin, 1975b). Sheep on mountain summer range in southern Utah generally watered and took salt in the morning and in that order (Bowns, 1971). Steers on sagebrush-grass range salted most often around watering points (Sneva, 1970), taking salt only every second or third day. Wagnon (1965) noted relatively heavy traffic by range cattle between the water supply and salt-meal mix feeder; some cows took meal before water, others took water before the meal; cows readily returned to grazing from either meal or water. (Refer to Chapters 7 and 8, respectively, for discussions on [1] the relationships of distance from water and site selection for grazing, and [2] drinking water requirements and water as a tool for distributing grazing.)

7

SPATIAL PATTERNS
IN GRAZING

I. SPATIAL FORAGING DECISIONS

Foraging (i.e., the search for forage) requires numerous decisions to be made by large herbivores as to where to graze (the spatial choice emphasized in this chapter) as well as what to graze (the species, plants, and plant parts chosen, as emphasized in Chapter 9). Sites for grazing and the associated grazing distribution patterns result principally from numerous decisions and processes made by grazing animals at a variety of spatial and temporal scales (Bailey *et al.,* 1996a). The grazing animal must integrate these decisions into specific procedures as it initiates a **grazing bout.** However, as the grazing bout is continued and eventually terminated, the grazing animal frequently adjusts its movements both spatially and temporally to achieve adequate forage and nutrient intake. Movement allows animals to adjust and take advantage of heterogeneous forage resources and compensate for depletion of nutrients in an area. (Refer to "Nutritional Wisdom and Optimal Foraging" in Chapter 9 for a discussion of whether the drive for nutritional intake by large herbivores results from nutritional wisdom or other motivational aspects.)

According to Senft *et al.* (1987), grazing herbivores encounter forage resources in an ecological hierarchy at the following scales: region, landscape, plant community (or large patch), and feeding station (or small patch). Although associated with diminishing size, the boundaries of each scale according to these authors must ultimately be defined by animal perceptions and foraging responses. Similarly, Stuth (1991) viewed the "diet selection process" of large herbivores as occurring at the following levels: landscape, plant community, patch, feeding station, and plant.

Six spatial levels, each with an associated temporal level (interval between decisions), for describing large-herbivore foraging have been defined by Laca and Ortega (1996) based on characteristic behaviors: home range (or landscape), camp, feeding site, patch, feeding station, and bite. These levels (or scales) are based on functional definitions rather than corresponding to soil types, plant communities, and geomorphic features, in order to focus more directly on grazing mechanisms and foraging decisions. Each spatial level is described and functionally defined, with additional refinement by Bailey *et al.* (1996a) in Table 7.1. Note that in some home ranges (landscapes) there may be only one camp.

These spatial levels are associated with different units of space that vary in absolute dimensions with the body size and foraging strategy of the herbivore. A regional scale, as such, was not included in the six functional spatial levels. This seems appropriate since the location of a large herbivore species at this level results basically not from animal decision or choice but rather from environmental adaptation, evolution, continental barriers, or management decisions (transhumance or nomadism). Grazing animals apparently extract information from their environment at these various levels to help meet the following goals or needs: (1) locating high-quality food, (2) minimizing intake of low-quality food, and (3) avoiding toxic food; this enables them to maintain nutrient intake at a level equal to or greater than their requirements (Rittenhouse and Bailey, 1996).

Foraging decisions made at higher spatial levels (landscape or home range, camp, or feeding site) constrain decisions that can be made at lower spatial levels (patch or feeding station), particularly if the home range or grazing unit is large in size. This results in distant plants and patches not being available during the current bout because of geographic isolation; distant vegetation may not be visible, and animals would incur energy costs for travel to other feeding sites. However, the energy costs of moving between feeding stations are much smaller than the requirement of choosing new feeding sites or even new patches during the grazing bout (Bailey *et al.*, 1996a). It is apparent that the consequences of decisions made at smaller spatial levels can be integrated and used to develop expectations and make decisions between alternatives at higher levels (Bailey *et al.*, 1996a).

Large herbivores usually allocate time spent in different areas of a grazing unit or habitat in relation to the nutrient resources found there. When nutrient retrieval at a feeding station becomes unrewarding, this apparently triggers a move to a new feeding station, a new patch, or even a new feeding site. Intake rate decreases and movement rate increases if forage availability is reduced appreciably (Bailey *et al.*, 1996a).

TABLE 7.1 Spatial Foraging Levels for Large Herbivore Decisions

Spatial foraging level	Spatial level description	Defining behaviors	Temporal level, i.e., interval between decisions	Entity involved in movement	Potential selection criteria	Motivation to move	Mechanisms that may affect grazing distribution patterns
Home range (landscape)	A collection of camps; defined by fences, barriers, extent of migration, transhumance	Dispersal or migration	1–12 months or longer	Population	Water availability; forage abundance; plant phenology; competition; thermoregulation	Sociality; reproduction; plant phenology; competition; water; thermoregulation	Migration; dispersal; transhumance
Camp	A set of feeding sites sharing a common foci for drinking, resting, seeking cover	Integrated feeding, drinking, resting	1–4 weeks	Herd	Water availability; forage abundance; plant phenology; cover; competition; thermoregulation	Plant phenology; water; cover; forage depletion and regrowth	Transhumance; migration; frequency of selection (spatial memory)
Feeding site	A collection of patches in a contiguous spatial area that animals graze during a foraging bout	Feeding bout	1–4 hrs	Sub-herds	Topography; distance to water; forage quality; forage abundance; plant phenology; predation	Forage depletion; intake rate; digestion rate	Frequency of selection
Patch	A cluster of feeding stations separated from others by a break in the foraging sequence when animals reorient to a new location	Reorientation of animals to a new location with break in foraging	1–30 min	Few individuals	Forage abundance; forage quality; plant species; social interactions; topography	Forage depletion; intake rate; species composition; social interactions; visual stimuli	Transit rate; turning frequency; intake rate; optimal foraging; frequency of selection
Feeding station	An array of plants available to a herbivore without moving its front feet	Pausing, with front feet in place	5–100 sec	Individual	Forage abundance; forage quality; plant species; social interactions	Forage abundance or depletion; diet selection; mouthful enabling	Transit rate; intake rate; turning frequency
Bite	Forage ingested as defined by a sequence of prehension, gripping, and severance motions	Jaw, tongue, and head or neck movements	1–2 sec	Individual (head)	Nutrient concentration; toxins and secondary compounds; plant size; palatability; plant structure	Forage depletion; diet selection; touch, taste, and smell stimuli	Intake rate; diet selection; post-ingestive consequences

Adapted from Laca and Ortega (1996) and Bailey *et al.* (1996a).

When an animal begins a grazing bout, it first selects a location (patch) and then begins its search for desirable forage by lowering its head and establishing the first feeding station along its foraging path (Kothmann, 1984). The most palatable plants and plant parts within reach are selected until palatability of the remaining forage within the feeding station decreases to a minimal acceptable level. The animal then moves to a new feeding station. Distance traveled in a feeding bout is also determined by the size of forage plants and their spatial distribution; when forages are sparsely distributed, the size of a feeding site must be increased as animals move more quickly between feeding stations (Bailey *et al.,* 1996a).

Bailey and Rittenhouse (1989) have suggested that the herbivore's grazing pathway is constrained by factors such as mobility, barriers, and topography. The daily search area may comprise most or all of the landscape area or may be restricted to only a portion thereof by size of area, shape of area, topography, or distance from water. When turned into new grazing units, livestock commonly range more widely initially in exploring their new environment before making more lasting site selections. After full utilization has been made of the selected sites by the grazing animal, the search for new sites may be accelerated. Collins and Urness (1983) concluded that both deer and elk are innately motivated to explore their environments for alternate food sources.

The transition on each following day is generally to a nearby area rather than a more distant part of the unit, thus gradually moving around the grazing unit. When grazing crested wheatgrass near Fort Collins, cattle were rarely observed grazing in the same area of the pasture on two successive mornings, and cattle grazing midgrass range in Texas seldom grazed in the same area for more than two successive mornings (Bailey *et al.,* 1990). In both locales, nearly homogeneous grazing units, cattle generally began grazing in an adjacent area on the following morning.

Bailey (1995) found that cattle in a heterogeneous range area did not return to the feeding site with lower forage quality for about 21 days but alternated among the remaining two feeding sites with higher quality forage. On heterogeneous areas, animals continue to visit selected nutrient-rich sites until they are no longer able to find nutrient-rich forage, then leave or give up and move toward the next best site. On homogeneous areas, animals alternate among foraging sites. Returning more frequently to nutrient-rich patches and feeding sites, and thus spending proportionally more time on them, is apparently an important factor in determining grazing distribution patterns (Bailey *et al.,* 1996b). "While it is known that grazing animals can perceive differences among feeding stations and small patches, it is not clear whether they are able to directly perceive larger units of spatial selection" (Bailey *et al.,* 1996a).

Problems in grazing management, such as overgrazing, habitat deterioration, and riparian area degradation, are more related to larger scale grazing patterns (i.e., large patches to landscape levels) than those that occur at smaller scales (small patches to individual plants). "Except for stocking rate, most range management practices probably have the greatest impacts on patch, feeding site, and camp selection. Subdividing pastures and implementing rotational and other intensive grazing systems can, in some cases, reduce and possibly eliminate feeding site selections by large herbivores. If a pasture is very small, animals may graze throughout the entire pasture during a bout. Fencing, water development, riding, and other management practices have little, if any, effect on diet selection or feeding station processes (Bailey *et al.,* 1996a).

II. MEMORY, PAST EXPERIENCE, AND TRAINING

The location that grazing herbivores select for grazing, resting, and bedding results from a complex interaction of memory, past experience, and training in addition to environmental factors (Fig. 7.1). It is apparent that large herbivores have accurate spatial memory and use this in their foraging decisions to improve foraging efficiency. They can remember and avoid locations with little or no nutrient resources as well as remember patches that have been recently depleted (Rittenhouse and Bailey, 1996; Bailey *et al.,* 1996a,b). Memory permits grazing animals to return to nutrient-rich sites more frequently than to nutrient-poor sites. The fre-

FIGURE 7.1   Site selection by grazing animals for grazing, resting, and bedding results from a complex interaction of memory, past experience, and training; showing cattle on the Texas Experimental Ranch, Vernon. (Texas Agricultural Experiment Station photo by Rodney Heitschmidt.)

quent return to a grazing area (patch and/or feeding site) after a period of days or a few weeks apparently requires long-term memory; also, it is assumed that short-term (working) memory covering at least a few hours allows grazing animals to avoid recently depleted patches or to return following a grazing bout. However, the following mechanisms were concluded by Bailey *et al.* (1996a) not to require large herbivores to use memory during foraging and require little judgment from the animal: **foraging velocity** (the rate at which herbivores transit different portions of a landscape), turning frequency and angles, intake rate, and neck angle/head placement.

Storing information in memory and obtaining new information are not mutually exclusive; as stored information is used in foraging, new information is gathered simultaneously as foraging proceeds (Laca and Ortega, 1996). Bailey and Rittenhouse (1989) concluded that the decision of where to graze is based on perception, knowledge, and memory of potential choices. They noted that cattle quickly explore a new grazing unit and appear to develop long-term memory of the spatial relationships among vegetation patches and plant communities.

Bailey and Sims (1998) proposed that memory decay may be the reason cattle eventually return to a feeding site with poor nutrient resources, or animals may purposely sample areas in their home range on a periodic basis to reassess available food resources. Laca (1998) has suggested that impeding spatial memory might improve grazing patterns; animals that were unable to establish and use spatial memory were less efficient in harvesting but established a more systematic search pattern, thereby interacting with a larger available area.

Knowing that large herbivores use previous experiences to decide where to graze should enable new techniques to be developed to modify grazing patterns. Past experience appears to play a prominent role in which plants and plant parts individual grazing animals select and in the sites they choose to graze. A foundation cattle herd established over long periods of time on a particular type of range becomes adapted/conditioned to that type of range; performance may suffer if animals are moved to a greatly contrasting type of range (Rittenhouse, 1984). Livestock that forage efficiently in the environment where they were raised may spend more time foraging but still ingest less forage in a new environment (Provenza and Balph, 1988).

The best rustlers among cattle have been reported from ranches where cattle are fed on the range and are accustomed to rugged topography (Skovlin, 1957). As summarized by Provenza (1990), inexperienced animals generally spend more time foraging but eat less forage, spend more time walking and walk greater distances in search of preferred foods, and suffer more predation, malnutrition, and the harmful effects of poisonous plants than do animals that know the terrain. In outperfoming naive animals, experienced animals may use nutrient-rich portions of the landscape more frequently because the expectations from these areas are more developed (Bailey *et al.,* 1996a).

Some individual cattle range farther than others; some readily range into upland habitats, while other individuals seem content to utilize riparian habitats (Rittenhouse, 1984). Two distinct home range groups of cattle were identified by Roath and Krueger (1982b) on the same mountain allotment in Oregon; this was determined through examination of quality and patterns of forage use, cattle distribution, herd social structure, and cattle activities. Both groups were semi-independent of each other; one group grazed almost exclusively on upland, and the other spent much more time on lowlands and riparian sites. Both groups occupied the same home range area year after year.

One Nevada rancher (Zimmerman, 1980) has concluded that desert-raised cattle can become just as much at home in their natural habitat as native wild herbivore species. He has characterized his cattle as traveling and grazing in small bunches of not over 20 head, ranging widely in rough terrain, going up to 8 miles to water, utilizing primarily several species of browse, and adapting and surviving on the open range year-round. His prescription for success is to wean the steer calves but leave the heifer calves with their mothers to learn how to survive on the desert—to select a variety of desert plants, seek protection from storms, find forage during drought periods, or find water or eat snow when there is a shortage of water. This reportedly prepares the young heifer, when she is moved from the fields back out to range as a bred heifer, to take care of herself and raise her calf at the same time.

Differences in grazing distribution between different classes of the same breed of beef cattle have not been consistent; however, this inconsistency may have been contributed to by prior experience. Bryant (1982) reported that cows on mountain range in Oregon distributed themselves over the range better than yearlings. Slopes less than 35% were preferred by both cows and yearlings, but cows made more use of steeper slopes than yearlings. However, Hickey and Garcia (1964) reported that yearling cattle utilized grasses more uniformly over variable terrain than did either cows with calves or mixed classes of cattle. Cows with calves tended to utilize areas around water more heavily than did yearlings, but younger cows with calves used the open grasslands more than old cows and entered the edges of the more inaccessible terrain. Skovlin (1965) concluded that cows with calves on mountain range do best on gentle terrain, while yearlings foraged well on the rougher range. Finding that cows with calves are more reluctant to graze steep slopes or travel as far from water than yearlings or cows without calves, Bailey (1999) suggested better grazing distribution may be obtained in large, rugged range units if managers graze yearlings or dry cows.

Naive (inexperienced) cattle were demonstrated by Greenwood and Rittenhouse (1997) to follow and mimic cattle trained to know the location of high-quality feed resources. They concluded that trained leaders can influence the feeding area selected by the herd and consequently influence where the herd begins a grazing bout following a foci event such as watering. It does appear that knowledge of food quality and location can be socially transmitted by cattle. Interactions with social models help young animals learn about the kinds and locations of foods, sources of water, and nature of hazards in their environment (Provenza and Launchbaugh, 1999). Howery *et al.* (1999) concluded that their research supported the hypothesis that: (1) animals can transfer, through their actions, information to naive followers regarding the relationship of visual cues and food qualities/locations, and (2) "familiar leaders" (penmates) are more efficient in transmitting foraging information to naive followers than "unfamiliar leaders" (non-penmates).

In studies with sheep held in relatively small pastures, Scott *et al.* (1995) noted that individuals often occur in subgroups that differ in their choice of forage or habitat, even within the same environment. Conversely, social interactions and individual food preferences both influence the choice of foraging location. As lambs age, they interact increasingly less with their dams and more with their peers, thus affecting each other's behavior (Ralphs and Provenza, 1999, as cited by Provenza and Launchbaugh, 1999).

It may be best to begin new grazing management practices with young animals; Malechek (1984) noted that effective implementation of a specialized grazing system can be thwarted when old cows who are accustomed to long-standing patterns of grazing, watering, and travel suddenly have new management restrictions imposed on them. However, Howery *et al.* (1998a) found that young cattle returning to mountain range in Idaho as yearlings tended to use similar locations and habitats as those used by their dams with calves at side. Environmental (drought) and social factors (peer activities) tended to modify the habits set as calves; in dealing with environmental and social vagaries, the yearlings continually altered their behavior.

A combination of training when young, breeding, and culling out offending individuals may offer opportunities of adapting livestock to specific site, terrain, and forage conditions. Culling individuals based on behavior might alleviate specific management problems (Rittenhouse, 1984; Roath and Krueger, 1982b). Noting that the cattle run by Zimmerman (1980) were Hereford with some Brahman blood, Platou and Tueller (1985) suggested that changing animal behavior by breeding for selective foraging and for terrain adaptation might be effective. However, beef cattle of different European breeds and breed crosses on rugged foothill terrain in Montana evidenced no difference in distribution of grazing nor in time spent grazing (Havstad *et al.,* 1986b).

In creating cattle herds that disperse more and utilize riparian areas less, Howery *et al.* (1998b) recommended: (1) culling animals with undesirable habitat use characteristics, and (2) implementing practices that foster a predictable social environment (e.g., separating young animals with desirable distribution patterns from young animals with undesirable distribution patters). However, simultaneous use of management application training, including (3) herding routinely to change distribution patterns, and (4) developing water and shade in upland areas to attract the grazing animals, were also strongly recommended.

Bailey (1999) suggested managers should consider selection among animal species, livestock breeds, ages, and nursing status and perhaps cull and breed individual animals as a means of improving grazing distribution. He noted that a few animals may be the leaders in a cattle herd; if the leaders have bad distribution habits, it may be effective to cull them from the herd. While animal breeding, culling, and intensive habit training to improve grazing distribution seem to show considerable potential, their widescale application to management situations requires further research and technique development.

Adult mule deer have demonstrated strong fidelity to seasonal movement patterns, i.e., being creatures of habit, returning to essentially the identical locales occupied on summer and winter ranges in previous years (Garrott *et al.,* 1987). Their apparent lack of dispersal seems to retard their discovery and occupation of new areas, even including those enhanced by habitat manipulation.

III. DETERMINANTS OF GRAZING DISTRIBUTION

Animals are not dispersed randomly in any environment, and free-ranging wild or domestic animals may exhibit extreme non-randomness in the use of the nutritional resources of the environment (Arnold and Dudzinski, 1978). Site preference (e.g., habitat selection for grazing, resting, bedding, etc.) results from complex interactions of both abiotic and biotic factors, including human intervention. The determinants of grazing distribution are grouped here for convenience as abiotic and biotic factors (Skiles, 1984; Squires, 1981):

Abiotic: weather (temperature, precipitation, wind, storminess, etc.), soil characteristics, topography and landform features, elevation, aspect, water availability, salt availability, and fencing.

Biotic: plant communities, botanical composition, quantity of forage, quality/palatability of plants, shade and shelter, escape cover, brush or tree barriers, inter- and intra-specific animal behavior, insect pests, and human activity.

Abiotic factors such as distance to water, steepness of slope, and other physiographic complexity are primary determinants of grazing distribution and act as constraints in which foraging mechanisms based on forage characteristics may operate (Senft *et al.,*1987; Pinchak *et al.,* 1991). The movement of grazing animals between camps and home ranges may be motivated by the need to find a new water source or avoid adverse climatic conditions. Microsite characteristics, such as the presence or absence of shade and wind, will affect both where animals rest as well as where they graze (Bailey *et al.,* 1996a).

Grazing distribution patterns on heterogeneous grazing lands are difficult to predict with useful precision, but abiotic constraints must be combined with responses resulting from biotic factors such as forage quantity and quality to adequately predict grazing distribution patterns (Senft *et al.,* 1987). Site selection is influenced by a complex of factors, including animal social behavior; and each factor interrelates and exerts its influence in a complicated manner (Cook, 1966b; Gillen *et al.,* 1984; Krueger, 1983). Anything that induces grazing animals to forage radially from some more or less fixed attraction point (water, salt, shade, bedding area, etc.) results in a heavily exploited zone nearest that point and a gradient of decreasing resource exploitation that diminishes with distance from that point (Squires, 1981).

The major factors believed to affect the spatial grazing use of mountain summer rangeland in northern Utah by cattle were studied by Cook (1966b). No single factor proved reliable as an index in predicting use. Including all of the 21 selected factors in the analyses of distribution accounted for only 37–55% of the variability in grazing utilization. The average utilization of grasses by cattle on the mountain slopes was 19.4% but varied from 5–55%. It was concluded that actual use obtained under good management was the most accurate method of determining the utilization obtainable on a particular terrain.

In mixed brush savannah with grass understory near Uvalde, TX, where topographic range site differences were minimal, the actual amount of grass and brush abundance were the major opposing factors affecting utilization by cattle (Owens et al., 1991). When forage was abundant, the major factors affecting distribution of utilization were plant related; green herbage availability, grass quantity, brush abundance, remoteness from roads, and water availability accounted for 70% of the variation. When forage was limited, green herbage availability was less important; brush abundance, grass quantity, green forb frequency, road location, fence proximity, and water availability accounted for 70% of the variation.

When grazing animals are confined to paddocks with a single type of vegetation, every square meter may be visited by the animals daily (Arnold and Dudzinski, 1978). As heterogeneity of vegetation and topography increases, so does the variation in use of the area by grazing animals. When environmental resources are heterogeneous and patchy, both spatially and temporally, animals are likely to strongly select against some sites and congregate on others. High animal densities and heavy grazing pressures are commonly believed to reduce site selectivity. However, from a study of sheep grazing mountainous terrain in southern Utah it was concluded that large changes in animal density may be required to significantly alter the relative attractiveness of natural plant communities (Senft, 1986). (Plant community preference for grazing—or some other activity—is based on the percentage of the total grazing time in a given 24-hour period that is allocated to that particular plant community.)

IV. FORAGE FACTORS IN SITE SELECTION

Forage factors play a prominent role in grazing site selection by grazing animals. As a result, contiguous plant communities on rangeland are normally subjected to substantially different grazing pressure. Differences between attractiveness of the various plant communities is often the dominant factor in cattle grazing distribution (Senft, 1986).

An abundance of palatable plants attracts grazing animals into the communities in which the plants are found; nevertheless, animal preferences for plant species and plant communities are interrelated. Animal preferences for plant species greatly influence selection of the grazing site, but the site being grazed then influences the plant composition of the diet (Skiles, 1984). Although animal attraction and grazing management attention may be directed to the most preferred sites, the less preferred sites can play an important role in providing a reserve forage supply.

A. UPLAND SITES

The main influence on site selection for grazing by cattle on shortgrass range in Colorado was concluded to be dietary preferences, but this was modified by constraints such as the location of water (Senft *et al.,* 1982, 1985a). During the growing season, cattle frequented the lowlands because of the frequency of a highly preferred cool-season grass found there, western wheatgrass. During the dormant grazing season, cattle selected the uplands because of the relative abundance of the dominant winter dietary component, blue grama. Drainageways passing through grasslands or other seasonally dry sites often receive additional run-in water from surrounding slopes; the resulting supply of palatable, regrowth forage during the typically prolonged growing season often attracts cattle and can lead to overutilization in the drainageways much as on true riparian sites (Engel and Schimmel, 1984).

Relative community preference often changes during the grazing period due to changing levels of forage availability and quality among plant community types. Since many forage species are only seasonally preferred (i.e., their relative palatability changes substantially with the seasons), a corresponding seasonal vegetative type preference often occurs. Cattle and sheep in Texas frequent woody plant sites in late summer because of the continuing presence of green growth of Texas wintergrass (*Stipa leucotricha*) after grasses on adjoining vegetation types have dried. In the aspen parklands of Canada, cattle typically make greater use of the forage in the grassy parks than under aspen, but seasonal drought greatly increases grazing of the greener forage under canopy (Hilton and Bailey, 1972). The variety of forbs and grasses found on aspen sites in good condition are preferred by sheep for spring and summer grazing (Fig. 7.2).

Both deer and elk in the Cascade Range of west-central Washington shifted in foraging emphasis from individual plants and/or parts to microhabitat patches as the season advanced (Hanley, 1982a). In June, when forage quality and quantity was relatively high in all microhabitats, dietary selection was directed to individual plants across an array of sites. However, by October green forage remained only on microsites on which favorable slope, exposure, shading, and soil moisture had delayed phenology, and these pockets of green herbage became highly favored. This shift took place earlier for elk, primarily a grazer, than for deer. Aspen groves within mixed conifer forests in northern Arizona received higher deer and cattle use (about six times) than adjacent conifer forest; this higher use was associated with the greater abundance of attractive understory vegetation under aspen than under conifers (Reynolds, 1969).

FIGURE 7.2   The variety of forbs and grasses growing under aspen, as shown above in the mountains of central Utah, are preferred by sheep for spring and summer grazing.

Cattle consistently graze nontimbered grassland and open parks over adjoining forested sites and rarely graze densely timbered sites (Pickford and Reid, 1948; Johnson, 1953). In the Blue Mountains of Oregon, cattle grazed nearly 100% of the riparian zone and adjoining sagebrush-sandberg bluegrass (*Poa secunda*) vegetation, about 75% of the pine uplands, but only 20% of the Douglas-fir type (Roath and Krueger, 1982b). Low use of the Douglas-fir type apparently resulted from associated steep slopes, closed overstory canopy, north and west aspects, and sparse vegetation understory. Utilizing a plant community preference index based on time spent grazing, Gillen *et al.* (1984) rated plant communities in mixed vegetation areas in Oregon as follows: meadow (9.4), grassland (3.0), open ponderosa pine-Douglas-fir (0.9), mixed conifer (0.5), and grand fir (0.1). Logged forest areas had an index greater than 1. Clary (1975) recommended that special efforts be made in Arizona ponderosa pine forests to distribute cattle grazing since cattle are attracted to forest openings and/or repelled by dense timber stands. Heifers grazing Ozark ranges preferred open woods, open glades, old fields, brushy glades, and closed woods (in that order) for grazing (Bjugstad and Dalrymple, 1968).

Deer on Oregon range consistently preferred the forested areas over grassland openings, but elk showed no special preference (Skovlin *et al.,* 1968). In related studies (Edgerton and Smith, 1971), elk and mule deer made highest use of open forest from spring through fall; grassland ranked second over dense forest only in the summer. Dense forest placed second in summer and fall for time spent therein by mule deer and elk, but its use was more important as cover than as a source of food. In the Sierra Nevada Mountains of California deer preferred aspen habitat all summer (Loft *et al.,* 1986). In northern Utah summer range studies, mule deer, and elk exhibited strong grazing preference for open habitat over dense forest subunits (Collins and Urness, 1983). Elk most preferred the highly productive meadow bottoms, whereas deer most preferred less productive clear-cut lodgepole pine and aspen forest. On non-forested range in Colorado's Front Range, female mule deer preferred the grassland types for feeding and resting at night but the mountain mahogany type for both activities during the day (Kufeld *et al.,* 1988).

The seasonal migration of mule deer is, in part, a response to seasonal changes in forage quality, as well as quantity, between different vegetation types (Garrott *et al.,* 1987; Willms, 1978). Forage growth begins earliest at lower altitudes and progressively later at higher altitudes; hence, palatability increases with increased altitude, but availability of current production decreases. Therefore, movements upward from winter through spring to summer range and then back down to fall range appears related to the profitability of occupying a particular niche along the forage continuum. Mule deer generally move to lower elevations before significant snow cover and spend winters on lower elevations, gradually shifting from use of northerly to southerly aspects, but may be forced to move off traditional winter range following severe winter weather.

B. RIPARIAN SITES

Heavy stocking rates, lack of alternative forage sources, and season-long or even yearlong concentration of large herbivores can greatly damage riparian areas. Riparian sites are favored areas for many species of grazing animals, including cattle, moose, elk, and sometimes sheep, horses, and deer (Fig. 7.3). As with livestock, wild herbivores may congregate there and be a factor in site deterioration. Moose are perhaps the best known inhabitants of riparian zones and associated mountain meadows, but elk also have a high preference for forage associated with riparian vegetation and moist meadowland. The effects of browsing and compaction by large, wild ungulates in riparian zones takes place primarily in spring and early summer before livestock grazing begins (Skovlin, 1984).

A major attractant to grazing animals is the relatively high palatability, quality, and variety of forage on most riparian sites. Because of improved moisture conditions and the resulting species composition, meadowlands and riparian sites stay green and succulent longer, often until the first killing frost; their attractiveness relative to uplands becomes even greater and they are sought after as the dry season progresses (Skovlin, 1984). The contrasting attractiveness of riparian sites over dry to xeric upland sites is magnified in arid and semi-arid regions, and high and disproportionate levels of cattle use may result on the riparian sites. Nevertheless, cattle in the Blue Mountains of Oregon favored the upland vegetation late in the summer during years when it remained green (Bryant, 1982).

In addition to forage factors, other major attributes of riparian sites that attract and hold grazing animals are the availability of water, shade, and thermal cover

FIGURE 7.3   Grazing animals of many species, including cattle shown here, are attracted to riparian sites for grazing because of relatively high palatability, quality, and variety of forage found there.

and ease of accessibility (Kauffman and Krueger, 1984). Shade, wind movement in drainages, and evaporation all have a cooling effect over adjacent upland conditions that attract livestock to riparian areas in hot weather (Skovlin, 1984). In contrast, livestock were noted to avoid wet meadows in Oregon because of boggy conditions (Korpela and Krueger, 1987).

From a survey made of nine sites in the Intermountain Region, Platts and Nelson (1985a) found that average forage use was 25% higher by livestock on streambanks than on uplands but varied up to 60% more. These differences were reduced by attractiveness of the adjoining upland sites (e.g., plants growing and succulent rather than dry and stemmy), by extreme wetness of the riparian sites, and when grazing early in the season. On rangelands in the Blue Mountains of Oregon characterized by steep slopes and erratic distribution of watering areas away from the creek, Roath and Krueger (1982a) noted that the riparian zone characterized by bluegrass bottoms, covering only about 2% of the grazing allotment, accounted for 71–81% of the total herbaceous vegetation removed by cattle.

Cattle grazing in Montana foothills was primarily in the uplands in June to mid-July but then gradually shifted more towards the riparian zone as the upland forages matured and became less palatable (Marlow and Pogacnik, 1986); the riparian zones became the favorite resting areas for cattle during hot summer afternoons. Cattle increased their grazing time on riparian zones from about 10% in early July to about 50% by mid-August to about 70% by early October (Marlow, 1985). During drought the use of the riparian zone was commenced earlier and was heavier than when forage growth was abundant in the uplands (Marlow and Pogacnik, 1986).

In eastern Oregon's high desert, meadow ecosystems produce high summer gains of cattle as well as high grazing capacity; an acre of meadow produced 244 lb of yearling beef annually (Cooper *et al.,* 1957). Adjacent sagebrush-bunchgrass range produced comparable per head rates of gain until early July but only about 50% of the daily gains thereafter compared to meadow. Even horses are attracted to the grasses and sedges found in riparian areas. In one mountainous study area in central Wyoming, the riparian habitat comprised only about 1% of the study area but received 21% of the summer grazing use by feral horses.

Sheep generally do not prefer wet or marshy grazing areas but rather drier areas and uplands, and properly managed sheep will have minimal impact on riparian habitats. When sheep are managed by herding, an added advantage is that the herder can control how much time sheep spend in grazing riparian areas. However, sheep can seriously impact riparian areas when herders select such sites for camp sites and bedding grounds or when riparian areas are used for excessive periods as holding areas. An exception to sheep not preferring riparian areas are during hot summer weather when the only shade is found along streams or late in the growing season when green herbage and woody plants found in riparian areas are the most palatable plants available (Glimp and Swanson, 1994). (Refer to "Special Problem: Riparian Zones" in Chapter 8 for a discussion of management practices to control distribution of grazing on riparian sites.)

V. NON-FORAGE VEGETATION FACTORS

Non-forage vegetation factors—often directly or indirectly expressed through forage factors as well—also influence where animals select to graze, rest, and bed down (Fig. 7.4). Timber harvesting methods affect site selection by grazing animals; domestic livestock generally prefer timber clearcuts and areas where the logging slash has been removed. The greatly increased midstory browse and understory herbage associated with the clearcuts may be similarly attractive to big game animals. Sheep utilized 2.5 times more forage from logged lodgepole sites in Oregon than unlogged areas, while cattle used only the fringes of the unlogged areas and greatly preferred areas where slash concentration was minimal (Stuth and Winward, 1977). Where little forage was produced in unlogged lodgepole communities, even deer harvested 7–10 times more forage in adjoining logged areas. Sheep are reluctant to penetrate dense vegetation higher than their line of vision, which may result in spot grazing in areas with excess foliage production (Glimp and Swanson, 1994).

Clearcutting in northern Utah lodgepole pine has greatly increased both deer

FIGURE 7.4   Nonforage vegetation factors that influence where animals graze, rest, and bed down include timber harvesting, edge effects, shade, and shelter. Cattle and deer range on Pine Ridge in northwestern Nebraska.

and elk grazing use, but aspen clearcuts have been used at about the same level as uncut aspen (Collins and Urness, 1983). Both elk and deer in Montana preferred clearcuts with cover in the openings except where such cover inhibited forage growth (Lyon and Jensen, 1980). Both preferred openings in which logging slash was not a barrier to movement. The overall use by elk increased 350% in small clearcuts in the Blue Mountains of Oregon compared to the uncut prior to logging, but the use in the clearcuts declined to near normal levels by the fifth year. Patch clearcut logging resulted in greater elk use than partial cutting. In New Mexico, cattle and elk preferred areas with less dead and downed slash, while deer seemed to tolerate or even prefer moderate amounts of slash (Severson and Medina, 1983).

Extensive clearings of pinyons (*Pinus* spp.) and junipers (*Juniperus* spp.) (up to several hundred acres) may greatly increase forage production and be highly attractive to livestock, but small patch cuttings within conifer woodlands (25 acres or less) are often more readily used by deer and elk than larger cuttings (Short *et al.*, 1977). Clearcutting Colorado lodgepole pine and spruce-fir forest in strips 66, 132, 198, and 264 feet wide, with alternating uncut strips of the same widths, doubled mule deer use of the area 10 years after logging (Wallmo, 1969). The increase in use was in the cut strips, where mean pellet-group densities were three times those on uncut strips and on adjacent virgin forest.

An **edge** is the place where plant communities meet or where structural conditions within plant communities come together (Thomas *et al.,* 1979). The area influenced by the transition between two plant communities or conditions, referred to as an **ecotone,** is often richer in forage plants and often provides forage and cover in close proximity. Such areas are particularly attractive to big game animals, and special consideration is often given to creating or maintaining edges in management plans. On non-forested Front Range habitat in Colorado, deer during daylight showed preference for the ecotones between grassland and shrub patches which offered escape cover, but no such preferences were observed at night (Kufeld *et al.,* 1988). White-tailed deer in Georgia used closed vegetation types during daytime but more commonly open types at dusk, dawn, and during the night (Beier and McCullough, 1990).

The willingness of big game animals to enter openings is influenced by a desire for security during the feeding period but is locally modified by the past experience of the animals in the available environment. Elk are more tolerant than deer of large openings, particularly where natural openings are already present in the environment. On open foothill winter range in southwestern Montana, elk sought areas where grass was relatively abundant, and deer the areas where more sagebrush was present (Wambolt and McNeal, 1987). The elk apparently selected feeding sites where the relationship of food intake to energy expenditure was optimized, while deer selected feeding sites where forage availability, security, and thermal cover were optimized. Based on studies of elk use of sagebrush-grass habitat in Washington, McCorquodale *et al.* (1986) suggested elk can be successful in habitats with limited thermal or security cover, even in severe climates, under conditions of infrequent disturbance and adequate forage. Elk were found to enlarge their home ranges on treeless, arid shrub-steppe to compensate for low food densities; that is, the quantity of area and forage compensated for lower forage quality to provide successful animal performance (McCorquodale *et al.,* 1989).

The natural cover of trees and large shrubs provides shade in very hot weather, thermal protection in very cold weather, and possibly security from real or imagined enemies for grazing herbivores. Such cover attracts grazing animals for idling and resting but may induce animals to graze and trample vegetation under or in the immediate proximity to the cover (e.g., around saltbushes) (Squires, 1981) or under mesquite canopy (Gamoughoun *et al.,* 1987). Natural shade can be either a help or a hindrance in livestock grazing units depending upon where it is located—good if located in otherwise less preferred areas but bad if located in overused areas (McIlvain and Shoop, 1971a).

Cattle and sheep consistently seek shade during hot summer days, primarily during mid-day periods and afternoons when the temperature exceeds the comfort zone (i.e., 85°F). Where shade is limited, livestock may travel considerable distances to reach shade on hot days. Livestock, particularly cattle, are more apt to loiter around water if shade is not available elsewhere (Arnold and Dudzinski,

1978; Bjugstad and Dalrymple, 1968; McIlvain and Shoop, 1971a; Weaver and Tomanek, 1951). Cattle with Brahma breeding are less prone to seek shade during hot mid-day periods but rest more in the open than do European breeds (Tanner *et al.*, 1984). During mild weather, livestock are more apt at mid-day to rest at the point where the morning grazing stops; temperatures of up to 80°F in aspen parkland did not induce increased grazing in shade (Hilton and Bailey, 1972). Sheep are less prone to rest and idle near water than cattle (Springfield, 1962). However, sheep on mountain range in southern Utah concentrated grazing on hot summer days in the bottoms of the canyons, where water and shade were both located (Bowns, 1971).

Walker *et al.* (1987) concluded that cattle show greater selectivity for loafing sites than for grazing sites. In contrast with grazing being closely correlated with vegetation, Senft *et al.* (1985b) concluded from work with cattle on shortgrass range in eastern Colorado that resting behavior was correlated primarily with topographic variables. During June through August, daytime resting was correlated with (1) draws and lowlands, (2) fencelines, and (3) stockwater areas, in that order. The order of preference was similar during the September through May period, except that south-facing slopes became important in winter. Nighttime resting occurred very little in stockwater areas but rather was scattered elsewhere except on ridgetops in both summer and winter.

Grazing distribution is affected by the interaction of inclement weather in site selection. During cold, stormy weather livestock and big game seek shelter under natural tree cover, in tall grass or brush, behind ridges or planted tree windbreaks, in gullies and other depressions, or selected manmade structures. Sites in direct sunshine are sought for warmth on cold, stormless days. Both cattle and sheep graze or travel downwind during cold winds, blizzards, or rain squalls; thus, shelter will be ideally located in the downwind side of the grazing unit (Arnold and Dudzinski, 1978). Horses during driving rain are more apt to stop grazing and stand with their heads downwind. On hot days both cattle and sheep graze principally into the wind and utilize parts of a grazing unit which receive the most wind (Dwyer, 1961; Squires, 1981).

Winter distribution of grazing is governed by available forage as modified by snow depth. Winds cause snow drifting on leeward slopes, depressions, and narrow arroyos but also remove snow from other sites, such as ridgetops and upper portions of windward slopes. Snow also tends to melt faster on some sites such as south-facing slopes. A majority of the grazing/browsing is then concentrated on limited sites with minimal or no snow accumulation. These concentration areas are often so small and used so intensely and frequently that the vegetation has little opportunity for recovery (Severson and Medina, 1983). During two years of a 3-year study in Middle Park, CO, over 90% of the winter range was excluded from mule deer use because of snow depth (Gilbert *et al.*, 1970), the remaining 10% being heavily impacted. Even Rocky Mountain bighorn sheep, during periods of snow cover, shift from feeding in open sites to areas of shrub cover, this change enhancing forage accessibility (Goodson *et al.*, 1991).

Mule deer movements are impeded by snow depths of 10 inches and upwards, and depths of over 20 inches immobilize deer and preclude their use of an area. Snow depths of 16–18 in. impede elk movements, and depths in excess of 2 feet, except for space under conifer cover, have prohibited use by elk (Severson and Medina, 1983; Leege and Hickey, 1977). Moose are apparently even more tolerant than elk of deep snow (Telfer, 1978a). Snow conditions, such as crusting and high density, lower these critical limits even more. When these snow depth tolerance levels are reached, big game are forced to lower elevations and/or more xeric exposures.

VI. SLOPE AND RELATED PHYSICAL FACTORS

Slope is an important factor affecting grazing distribution in hilly or mountainous country, but its effect varies greatly between kinds of grazing animals (see also Table 10.1 for additional comparison by animal species of terrain adaptation and tolerance). Cattle prefer accessible areas such as flatlands and rolling lands, valley bottoms, low saddles between drainages, level benches, and mesas (Fig. 7.5). Phillips (1965) has reported that the degree of utilization of key forage species by cattle may reach 75 to 80% on gently sloping drainages in the Intermountain area while steep slopes 500 feet away may receive only 5% use or less.

On rolling hills to mountainous topography on Bureau of Land Management (BLM) lands near Vale, OR, slopes between 0 and 19% supported 94% of cattle

FIGURE 7.5 Thurber fescue site in LaSal Mountains of eastern Utah with ample forage production but not being grazed by cattle because of steepness of slope.

use, 79% of feral horse use, 66% of mule deer use, and 25% of bighorn sheep use (Ganskopp and Vavra, 1987). Cattle, horses, and deer demonstrated negative curvilinear responses to increasing slope with initial site avoidance exhibited on 20, 30, and 40% slopes, respectively. Where large expanses of level topography were available, cattle and horses made less use of steep slopes than their counterparts inhabiting more rugged terrain. Cattle appeared more willing to utilize slopes in early spring and late fall than during the warmer summer months. This suggests that cattle, and possibly also horses and bison, are more unwilling than incapable of grazing steeper slopes.

In a study on mountainous range in southwestern Montana, Mueggler (1965) used cow chips as an index of cattle distribution. Included in his study were 38 slopes ranging from 0–78% and accessible to cattle only from the bottom of the slope. The influence of slope steepness and distance upslope on relative cow use is shown in Fig. 7.6. On 10% slopes, for example, 75% of the utilization occurred within 810 yards of the foot of the slope, while 75% of the utilization occurred within 35 yards on 60% slopes. On a Southwest semi-desert study area, slopes of

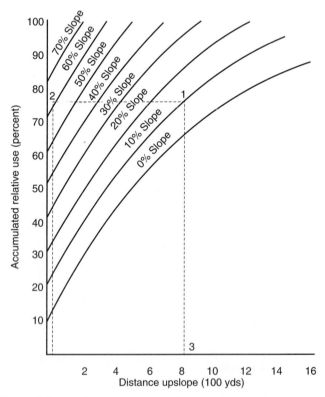

FIGURE 7.6   Influence of slope steepness and distance upslope on relative cow use. (Mueggler, 1965).

25–40% and partial barriers to cattle movements markedly reduced forage utilization, and topography influenced utilization more strongly than did distance from water (Martin and Ward, 1970).

Based on animal days of use, 79% of cattle use on Wyoming foothill range was on slopes less than 7% (Pinchak *et al.,* 1991); the area on or surrounded by slopes greater than 10% comprised 35% of the total area but received only 7% of the observed use. Also from this study, stocking rate was found to be a factor affecting cattle grazing on slopes; at a very light stocking rate the mean slope grazed was 3.2%, but under a moderate stocking rate the mean percent slope grazed rose to 5.2% (Hart *et al.,* 1991).

Slope gradient was the only physical factor consistently associated with cattle grazing distribution on Oregon mountain range (Gillen *et al.,* 1984). Using a site preference index (1.0 = neutral slope effect), high preference (index 3 to 5) was shown for the 0–5 and 5–10% slope gradients, slight preference on the 11–15% slopes (index 1.3), slight negative preference on the 16–20% slopes (0.8), and mostly avoidance on slopes over 20% (index 0.1–0.4). Five of the eight most important factors increasing cattle utilization of mountainous terrain in northern Utah were slope factors (Cook, 1966b): (1) lower percent slope at site, (2) lower percent slope adjacent to water, (3) lower percent slope from the site to water, (4) lower percent maximum slope between site and water, and (5) lower percent slope from site to salt. (The other three major factors were reduced distance to water below the site, higher percentage of palatable plants on the site, and low brush levels over the site).

Cattle utilizing Idaho mountainous rangelands were found unwilling to walk up slopes much steeper than 30% but were willing to walk up gentle watercourses and then walk on the contour across slopes up to 70% (Patton, 1971). Level contour trails extending across slope faces made it possible for cattle to graze on steep slopes at distances of up to one mile from water (Fig. 7.7). Heavier utilization by cattle of locations along previously developed grazing routes was observed on hilly rangeland in Nebraska (Weaver and Tomanek, 1951). Cattle on forested range in Oregon made extensive use of old logging roads and skid trails in accessing particularly steep and broken country, but distribution from these access routes decreased dramatically with increase in slope above or below the roads (Roath and Krueger, 1982b). In contrast, roads in gentle terrain seemed not to be an important factor in distribution. Downed timber may effectively block access trails and contour trails, and its removal may be required to increase forage utilization beyond the blockage.

It was concluded from studies with cattle on 25% slopes in rough fescue grasslands in Alberta that the location on the slope had little effect on cattle distribution when water availability was not a factor (Willms, 1989). When grazing land units were fenced perpendicularly to the slope and water was made available at the bottom, middle, and top of the slope, cattle distributed themselves uniformly over the slope. Total water consumption was greatest (43%) at the middle and least (22%) at the bottom.

FIGURE 7.7   Cattle are generally unwilling to walk up steep slopes but are willing to walk up gentle watercourses and then out on the contour to graze on slopes. (Colorado State University photo.)

Sheep are less intimidated by steeper slopes than cattle and tend to prefer up-land grazing sites (Glimp and Swanson, 1994). Unherded sheep on mountainous, juniper-pinyon terrain in New Mexico favored mountain ridgetops and saddles for both grazing and bedgrounds (McDaniel and Tiedeman, 1981). Sheep used all slopes regardless of steepness, but when terrain was especially rough the animals mostly trailed through the area, making little use of the available forage. Sheep utilization was relatively uniform on all side slopes less than 45%, but utilization was reduced by 50–75% on the steeper slopes. The unherded sheep tended to use the same bedding grounds on the ridgetops with up to 70% forage removal but with significantly less forage use on the midslopes and bottomlands. Unherded sheep in Utah also chose high ground for bedding down, and the same selected spots on high ridges in the mountains were used throughout the summer; overgrazing on and in the vicinity of established bedgrounds was caused by animals grazing these areas in the evening prior to bedding down (Bowns, 1971).

Feral horses in Wyoming's Red Desert (Miller 1983) and in the sagebrush steppe of Oregon (Ganskopp and Vavra, 1986) preferred ridgetops for grazing. This preference for elevated terrain was attributed to a desire to view the sur-rounding areas for potential threats to their safety, but the availability of ungrazed forage and cover may have been contributing factors. Livestock were noted to avoid wet meadows in Oregon because of boggy conditions (Korpela and Krueger, 1987). In addition to very wet clay soils or bogs, some grazing animals decline use of stony or very sandy areas (Skiles, 1984).

VII. DISTANCE FROM WATER

The location and number of watering points on grazing lands are important in controlling the movement, distribution, and concentration of grazing animals. When other factors do not limit grazing distribution, distance from drinking water ultimately controls the limit of vegetation utilization (Roath and Krueger, 1982b). While forage factors play a major role in determining where grazing animals will actually graze (i.e., set the inner boundary), distance from water will set the outer boundary within which animals will graze (Rowland and Stuth, 1989). Abundant forage in large grazing units often is found at considerable distance from water; thus, knowing the distances different kinds and classes of livestock and big game will travel out from water and the utilization they will make of the standing forage at these greater distances are necessary for management planning. Type of vegetation, topography, season, and kind and even class and age of grazing animals can modify the relation of vegetation utilization to distance from water (Arnold and Dudzinski, 1978). When snow is available on the ground as a substitute for water, both cattle and sheep commonly range farther from open water (Blaisdell and Holmgren, 1984).

Water is the major focal point from which grazing activities radiate out (Fig. 7.8). Concentric rings of utilization are generally found around the water point on level terrain, with utilization decreasing as distance from the water point increases. Cattle, and sometimes other livestock and big game species as well, often heavily graze forage plants near water rather than traveling long distances to better forage. This results in deterioration of forage resources near the water supply and wastes forage at long distances from water. It is a common recommendation that water points be located no more than one mile from all forage supplies, but travel distances far beyond this are commonly reported. Thus, it is apparent that it is not the ability of the animals to travel which primarily restricts utilization away from water but rather their willingness to do so (Kothmann, 1984). When considerable time and energy are needed to travel between the forage and the drinking water, the land area that the animal can cover in a day's grazing, and often food intake as well, may be restricted, particularly in semi-arid grazing areas (Freer, 1981).

On gentle terrain near Las Cruces, NM, the average percent forage utilization was determined in half-mile-wide concentric zones centering on the single watering place (Valentine, 1947). The respective utilization in the 0–0.5, 0.5–1.0, 1.0–1.5, 1.5–2.0, and 2.0–2.5 mile zones were 50, 38, 26, 17, and 12%, respectively. Within these distance-from-water zones, the location of palatable plant species determined where cattle actually grazed. The grazing patterns of Hereford and Santa Gertrudis were compared in this same area and were found similar in grazing time spent in each zone (Herbel et al., 1967). However, indications are that in larger grazing units than those used in the study, the Santa Gertrudis would develop better grazing patterns than the Herefords because of their far-ranging tendencies (Ares, 1974).

On eastern Colorado shortgrass range of gentle terrain, distance from water and

FIGURE 7.8   On level to gently rolling terrain, distance from water is inversely related to level of forage utilization; showing ideal location of windmill in the Nebraska Sandhills. (Soil Conservation Service photo.)

the abundance of seasonally preferred plant species were the most important determinants of the pattern of year-round grazing use (Senft *et al.,* 1982). Distances up to 1 mile from water did not greatly reduce utilization by cattle on the relatively level, rock-free study area at Santa Rita near Tucson, AZ (Martin and Cable, 1974). Utilization $\frac{1}{4}$ mile from water averaged 48% compared to 44 and 43% at $\frac{5}{8}$ and 1 mile, respectively. The inverse relationship between distance from water and herbage yield was strongest for black grama and Arizona cottontop, two of the most palatable forage species.

On foothill range near Arlington, WY, 77% of cattle utilization, based on animal unit days of use, was within 1200 feet of water (Pinchak *et al.,* 1991); 65% of the land area was beyond 2375 feet of water but received only 12% of the observed grazing use. As a part of this same study, cattle grazed farther from water as stocking rate increased and as the summer grazing season progressed (Hart *et al.,* 1991). The mean distance cattle grazed from water was 885 feet under very light stocking rate but 1738 feet under moderate stocking rate.

In mountainous areas of northeastern Oregon, cattle preferred areas within 565 feet of water and mostly avoided areas beyond 2000 feet from drinking water (Gillen *et al.,* 1984); in the uplands, where numerous water points were available, water distribution had practically no association with grazing patterns. From other studies in the area it was observed that cattle were using slopes adjacent to the riparian zone on the contour; it was concluded that vertical distance above water was the most negative factor in reducing utilization on moderately steep slopes (Roath and Krueger, 1982b). Distance from water as a primary factor in determining cattle use of an area was magnified in drought years. It appeared that both distribution and utilization became zero at about 6200 feet from water.

In mountainous terrain in Idaho distance from water to site along the cattle contour trails was more closely related to cattle utilization than distance from water in a straight line to the site (Patton, 1971). Contour trails originating at watering locations and extending across slope faces induce cattle to graze on steep slopes at distances up to one mile from water. On New Mexico mountain range, distance from water out to 6500 feet did not limit utilization of the forage plants by free-ranging sheep; a slight reduction in utilization was noted up to 7850 feet (McDaniel and Tiedeman, 1981).

When feed conditions are good, cattle on Australian range generally concentrate their grazing within 2 miles of the water source and spend most of the time on their preferred plant communities (Squires, 1981). As feed conditions deteriorate or when preferred areas close to water are fully grazed, the cattle move as far out as 4 miles or more from water in search of better forage. Under extreme drought conditions, cattle range up to 15 miles from water in search of forage, but distances of more than 6 miles are considered unusual under normal conditions. Salt-induced thirst and demand for more frequent watering to flush the salt load in the diet has greatly reduced the grazing range of sheep on saltbush-dominated ranges (Squires, 1981).

Shape of the grazing unit as well as size can greatly affect grazing distribution. When 10-acre cool-season, grass-legume pastures in Missouri were square, all areas of the pasture were grazed uniformly and averaged 35% utilization for a single short-term (2–4 days) grazing period. In 10-acre rectangular (4:1 ratio) pastures, utilization was variable, ranging from 40–50% within 100–200 feet at the front end of the pasture (water source) to less than 20% when distance from water exceeded approximately 1100 feet. It was concluded that pasture cattle will likely stay closer to water as long as forage is abundant. In contrast, range cattle must travel greater distances to satisfy their forage intake needs, and there is opportunity for even greater utilization differences in relation to water sources.

A distinct temporal utilization pattern was found in grazing rectangular range units, $\frac{1}{4}$ mile wide and 2 miles long on rolling grasslands near Edmonton, Alberta. (Irving et al., 1995). In this study, three adjoining range units of equal size, with water available only at one end of each unit, were grazed rotationally under a single cycle of short-duration (5 days) stocking. The utilization could best "be described as a wave, with defoliation beginning near the water source located at one end [near water source] on day 1 of grazing and proceeding outward from water until the ends of the units were reached on day 5." Final utilization was uniform out to 1.5 miles (i.e., 50–60% by visual estimate) but averaged 40% from 1.5 miles to the far end of grazing units.

It was concluded by the authors of the above study that high levels of utilization had to be reached near water before utilization was achieved farther from the single water source. Selectivity for areas close to water was not removed by high stocking densities, but was masked by the speed at which the defoliation wave progressed. Two projections were made from the study: (1) If the grazing units had been large enough or grazing periods long enough to have permitted regrowth, cattle might have begun grazing regrowth near water before or instead of grazing the far ends of the units. (2) Temporal utilization patterns are probably not significant from a management point of view, at least under short-duration grazing, when final herbaceous utilization is relative uniform and falls within the moderate range, unless livestock performance is detrimentally affected by excessive travel distance.

In the mountains of northern California, the mean distance of mule deer from water was 1.5 miles; it was concluded that the preferred spacing between watering sources was 2 miles with a maximum spacing of 3 miles (Boroski and Mossman, 1996). Along the Missouri River Breaks in Montana, deer grazing decreased at distances over 1 mile from water, elk seldom grazed over a mile from water, and cattle grazing was mostly within $\frac{3}{4}$ mile from water (Mackie, 1970). Grazing distance from water was more restricted for all three species in summer and fall than in winter and spring. Feral horses in the Owyhee Breaks of Oregon also showed a slight increase in average grazing distance from water in winter (1.25 miles) compared to the warm, dry summer months (about 1 mile) (Ganskopp and Vavra, 1986).

Water distribution was concluded to have little effect on mule deer distribution in the central Intermountain Region as long as the forage remains green and succulent (Julander, 1966). However, during periods when the forage is dry, particularly in late summer and fall, deer may trail to water and tend to concentrate there. Water is apt to play an even greater role on semi-desert foothill ranges, particularly in drought years.

During the summer season, a high proportion of elk grazing in Oregon was found within $\frac{1}{2}$ mile and even $\frac{1}{4}$ mile from permanent water (Skovlin, 1984). Elk in Arizona preferred habitat within $\frac{1}{4}$ mile of permanent water sources during early spring (dry period) and within $\frac{1}{2}$ mile during late spring-summer (Delguidice and Rodiek, 1984). Desiccating weather conditions, the association of succulent forage with most of the permanent water sources in the upper elevations, and the occurrence of calving activities were concluded largely responsible for the greater elk use of areas near water.

Antelope densities are highest on range where water is available every 1–5 miles (Yoakum, 1978). Good distribution of antelope grazing during dry periods in western Utah has been dependent upon adequate water distribution (Beale and Smith, 1970). Spreading the distribution of pronghorn antelope and desert bighorn sheep and their successful introduction into new areas in the Intermountain Region have often required additional water developments, particularly during dry seasons and drought years (Scotter, 1980).

VIII. OUTSIDE DISRUPTION

Incidental human activity and transportation noise seldom materially affect the grazing activities of domestic livestock, and it appears that wild herbivores such as deer respond more intensely to unpredictable than predictable human activity (Stephenson *et al.,* 1996). However, human disturbance in northern California studies did not preclude or seriously impede mule deer from using traditional water sources; the deer responded by adjusting the amount of time spent at water sources, moving away from water to return later and/or increasing the frequency with which they drank. Nevertheless, timber harvesting, subsequent cleanup activities, road construction, intensive vehicular use of existing roads, military maneuvers, and intensive hunting and camping activities may displace elk and deer from preferred sites (Lyon, 1979; Lyon and Jensen, 1980).

Elk have been noted to retreat from logging areas as long as men and machinery are active, the distance moved appearing to be only as far as necessary to cross a topographic barrier, but they then return to the timber harvested areas once logging activities have ceased (Lyon, 1979; Ward, 1976). Irregular, light traffic on service roads has had little effect on elk activity, especially on areas more than 0.25 mile from the road. Deer on foothill range near Fort Carson in southeastern Colorado responded to military maneuvers not only by moving out of their normal home ranges during maneuvers but also by increasing their core home ranges within maneuver areas (Stephenson *et al.,* 1996); this was apparently a response to a combination of human harassment, alteration of security cover, and localized destruction of the forage base.

Access to favored vegetation types by big game animals may be intercepted by intensive developments such as major housing, business, and transportation areas; this can substantially alter normal migration routes and greatly limit access to critical seasonal range. Providing access through or around such obstructions to favored areas is often difficult. Equally serious problems result when the favored areas are actually replaced by intensive development.

Both mule deer and elk can be greatly disturbed by horsefly or other biting insect attacks, and this may substantially affect the summer distribution of these ungulates (Collins and Urness, 1982). Elk readily move into the forest from the preferred meadow bottoms during fly attacks; deer already grazing in the forest will be less affected. Elk may find relief from mosquitoes by seeking windy ridges, but wind seems not to lessen horsefly agitation.

Cattle commonly seek the protection of brushy sites or open water for protection against heel flies. During periods of intensive heel fly attacks, cattle in California (Wagnon, 1963) were noted to seek protection before the normal morning grazing period was over and remain there until late in the afternoon. Other insects such as horn flies and deer flies may also cause livestock to seek protection.

Minor activity in defense against insects such as head and ear movements, tail switching, stamping legs, and skin twitching require slight additional energy expenditure. Substantial head and front leg movements may interfere somewhat with prehension and biting while having only minor effects on forage intake. Cattle have been noted to graze deeper into the sward in an attempt to dislodge face flies from the muzzle. However, major evasive actions by free-ranging animals such as taking flight and sharply reducing foraging time and thus intake may have substantial energy consequences to the grazing animal. If insect attacks are severe, relief can be provided by spraying livestock with recommended insecticides.

Drought, large burns (temporarily removes forage but regrowth may be a strong attractant), heavy stocking, or inter-specific competition from other grazing animal species may force heavier use of less preferred sites by an animal species. Predation or antagonism from other animal species may also modify normal site preferences. (Refer to section on "Interspecific Sociality" in Chapter 10 for discussion of the effects of social interaction between different kinds of grazing animals on site selection for grazing.)

8

MANIPULATING GRAZING
DISTRIBUTION

The dispersion of grazing animals and associated forage utilization within a grazing unit or area is the fourth principle of grazing management and one of its more important facets. The goal is to obtain the maximum safe grazing use over as wide an area as possible without causing serious damage to any portion within it. Management plans must consider all natural, cultural, and management factors that affect site selection or avoidance by grazing animals, as elaborated upon in the previous chapter. Mountainous rangelands, for example, often exhibit complex combinations of topography, plant communities and successional stages, water distribution, and other habitat factors which create especially difficult grazing distribution problems (Gillen *et al.,* 1984).

When grazing animals are not kept well distributed, the areas grazed too heavily as well as those grazed too lightly expand in size while those areas receiving optimal use become smaller. Grazing distribution will have a direct effect on realizable grazing capacity. Many grazing land units appearing overstocked may merely be suffering from lack of uniform use. Even on properly stocked range there are generally areas where forage is wasted because of the great distance from water, difficulty of livestock access, or lack of manipulative, manmade attractants. There may also be localized heavily grazed areas close to water sources, main trails, and corrals on moderately or even lightly stocked grazing land units.

I. THE TOOLS OF DISTRIBUTION

There is no single prescription that will disperse grazing animals in every situation; every situation is unique, will be affected by different combinations of factors that affect grazing distribution, and requires a tailored plan for distributing grazing. Probably no aspect of grazing management is more of an art but at the same time more rewarding than concentrated efforts to more effectively distribute grazing. Using a combination of several tools for distributing grazing is generally the most effective approach.

To aid in selecting the appropriate tools of grazing distribution to fit each given situation, the following checklist is offered in two parts. The first set includes the primary tools that will have almost universal application; the second set includes the secondary tools that will primarily have application in special situations.

1. **Primary tools for grazing distribution:**
   1. Provide additional watering places where their present number or distribution is inadequate.
   2. Utilize properly located fences to provide more direct control of grazing.
   3. Utilize herding for dispersing animal concentrations and movement into undergrazed areas.
   4. Place salt and other supplements in areas where more grazing is desired, minimizing their location at water and other animal concentration areas while making sure that grazing animals know where the supplements have been placed.
   5. Assure accessibility by grazing animals into all areas to be grazed, giving consideration to needed stock trails and accessways over difficult terrain, browseways through dense browse fields, walkways across marsh range, and strategically located gates.
2. **Secondary tools for grazing distribution:**
   1. Utilize high-density, reduced-duration grazing (refer to Chapters 14 and 15 for further details).
   2. Use a mix of animal species having different site preferences (refer to Chapter 10 for further details).
   3. Provide shade or shelter at strategic locations where not currently available.

4. Cut herbage for hay or green chop, mow old grass, or spot burn to induce palatable regrowth to attract more grazing into undergrazed areas.
5. Apply nitrogen fertilizer or phenoxy herbicides to increase palatability of new growth if primary objectives for use are economically feasible.
6. Control insects that discourage livestock grazing in certain areas by rendering forage unattractive.
7. Plant less palatable, grazing-tolerant species in natural concentration areas and more palatable plants in normally undergrazed areas.
8. Apply artificial attractants to vegetation on microsites not being fully utilized and repellents where being overgrazed.

These tools are directed primarily towards the distribution of livestock grazing, but some of the techniques apply to big game animals also. After reviewing the literature on methodology and successes in improving the distribution of game animals, Scotter (1980) concluded that salting, fertilizer application, and water developments were often effective in attracting big game to graze underused areas.

Although a high degree of dispersion of grazing animals and associated uniformity of use is common on small units comprised of uniform vegetation and terrain, this is seldom achieved on expansive rangelands. Thus, the effectiveness of grazing distribution plans is often evaluated on the basis of results achieved on key forage plant species in key areas. However, the recognition of **critical areas** (i.e., areas on which special grazing safeguards are mandated) and the acceptance of sacrifice areas on the same grazing unit add materially to the complexity of executing and evaluating grazing management results.

A **sacrifice area,** by definition, is a small portion of a range or pasture unit on which overgrazing is willingly allowed in order to obtain efficient grazing use on the remaining majority of the unit. Such areas commonly develop at animal concentration areas such as watering places, handling facilities (corrals, lanes, corridors), feed grounds and other holding areas, trails, and limited shade. Stocking lightly enough to eliminate all such areas would undoubtedly result in unwarranted waste of forage (Valentine, 1947). In fact, failure to accept even minimal sacrifice area will mostly preclude grazing on many grazing lands by livestock and often big game animals as well.

Conversely, stocking heavily enough to make fullest possible use of outlying areas will often result in an unacceptable extension of the depleted sacrifice zone. While it is generally recognized that sacrifice areas around points of animal concentration such as permanent water are unavoidable, there is no clear answer as to what size they should be permitted to develop. A few acres may be unacceptable on highly productive grazing lands, particularly when these acres coincide with critical areas, but a quarter section may be acceptable on extensive desert grazing areas where grazing units are large (Valentine, 1947; Arnold and Dudzinski, 1978). Management objectives will obviously play a substantial role in determining the maximum acceptable size or percent of the grazing unit in sacrifice area, but a policy of realistic containment seems urgent. Holding sacrifice area to not more than 1–5% may be appropriate for most permanent grazing land units.

II. DRINKING WATER REQUIREMENTS

A. NEEDS BY LIVESTOCK

Providing adequate amounts of drinking water is important for grazing animals, both livestock and big game (Fig. 8.1). Free-choice consumption of water without attempts to limit water intake should normally be permitted. Since water is both a nutrient and a medium of metabolic functions in the body, is an important milk and tissue constituent, and provides the means of animal waste removal, adequate consumption is necessary for animal health and production. Daily animal gains are directly related to the amount and quality of feed consumed each day, but feed consumption can be sharply reduced by inadequate water intake. Restricting water intake by failing to provide adequate water supplies sharply reduces milk flow in lactating females, reduces gains in both suckling and weaned animals, and may contribute to or even cause death losses if severe (Vallentine, 1963).

Normally, 8–10 gallons of water per day is ample for mature cows, 0.75–1 gallon per day for adult sheep and goats, and 10–12 gallons per day for horses (Stoddart *et al.*, 1975; USDA, Soil Conservation Service, 1976). Mature cows on the

FIGURE 8.1   Ample drinking water for free-choice consumption should be provided both domestic livestock and big game animals; example shown, water piped from the mountains in the background to various outlets on a Bureau of Land Management grazing allotment in eastern Utah.

Jornada Experimental Range in New Mexico consumed an average of 15 gallons per day during the May–July period, with lactating cows consuming 24% more water than non-lactating cows (Rouda *et al.*, 1994). Lactating cows on dry summer range in Oregon drank 12.4–16.7 gallons per day, and 3½-month-old calves consumed about 1.1 gallons per day (Sneva *et al.*, 1977). In order to meet maximum free-choice water consumption and allow some evaporation, the U.S. Forest Service (USDA, Forest Service, 1969) has recommended that water development plans consider 12–15 gallons per day for cattle and horses and 1.0–1.5 gallons per day for ewe-lamb pairs.

The daily demand for water at a given watering point will, of course, depend directly on the number of animals watering there daily. Since cattle on a body-size equivalent require about 50% more water in a year than sheep, extra water must be made available when cattle are grazed (Squires, 1981). Water requirements per grazing unit will increase as stocking rates and stocking densities increase. Under short-duration grazing, where stocking densities are magnified several times over, proportionately more water will be required daily per unit area stocked during the grazing period compared to that under season-long continuous grazing (Reece, 1986).

B. WATER INTAKE FACTORS, RESTRICTIONS

Live weight and animal condition, stage of production, amount of activity, and environmental factors will affect the amount of water consumed by grazing animals. High temperatures, low humidity, high salt or protein content of the diet, dry feeds, increased feed intake, and high levels of activity all increase water consumption. Winchester and Morris (1956) estimated cattle will increase their water intake as the air temperature rises from 40–90°F as follows: cows nursing calves, from 11.4–16.2 gallons; 600-lb heifers and steers, from 5.3–12.7 gallons. Water intake estimates were based on 0.37 gallons of water per pound of dry matter intake at 40°F, increasing to 0.88 gallon of water per pound of dry matter intake at 90°F. Squires (1981) concluded that *Bos indicus* (Brahman) are more efficient water users than *Bos taurus* (European breeds), turning over body water at a slower rate, thus reducing water intake needs by as much as 40%.

Green succulent forage intake decreases water consumption by grazing animals. When forage on mountain summer range in southern Utah was very succulent or wet from rain or dew, sheep did not seem to want or require open water (Bowns, 1971). However, ewe-lamb pairs commonly drink about 1 gallon daily on summer range under typical vegetation conditions. On desert range in the Intermountain Region, ewes were found to drink daily about 1 gallon in the fall, 0.75 gallon in the winter, and 0.5 gallon on spring foothill range (Hutchings, 1958). On dry grass or shadscale (*Atriplex confertifolia*)-winterfat range in late winter, ewes drank 1.8–2.5 gallons daily, the latter amounts when the weather was warm. Sheep watered daily drank slightly less water on a weekly basis than those watered every other day. When forced to go 2 or 3 days without water, sheep normally drank 2–

3 gallons of water the following day. Squires (1981) provided the following guide to water intake in relation to moisture content of forage ingested by *B. taurus* cattle: 10% moisture content, 1 gal per lb dry matter; 40%, .8 gal per lb; 70%, 0.4 gal/lb.

When cattle grazing crested wheatgrass in Oregon were allowed to drink water only every other day or were required to trail daily 1 to 2 miles to water, water intake was reduced by 25 to 35% compared to cattle with unlimited access to nearby water (Sneva *et al.,* 1977). Lactating cows, when so stressed, still tended to gain weight but their calves showed reduced performance. The calves after $3\frac{1}{2}$ months of age showed a strong desire for water; when water was withheld, they performed poorly. A combination of every other day access to water during spring and summer and requiring trailing did not reduce water intake over either treatment alone but did permanently reduce weight of heifers due to calve in the fall. Water intake by pregnant yearlings averaged about 9 4 gallons daily with free access to water, 8.2 gallons when offered only every 24 hr, and 6.4 gallons daily when offered every 48 hr.

The suckling calf was found the most susceptible to water stress in the Oregon studies (Sneva *et al.,* 1977); suckling calves without direct access to water gained 0.4 lb less daily over the 60-day study period. On ranges where the trailing distance between forage and water is excessive, the reduced water intake of the dam is critical to calf performance in reducing milk production. This comes at an age when the calf is still unable to adequately handle range forage which is seasonally decreasing in quality. Faced with great trailing distances and increasing summer temperatures, the calf is less likely to travel to water. Thus, the calf is affected in two ways, and the condition of the cow is no indication of how well the calf is performing.

The following conclusions were also reached from the Oregon studies: (1) where water is hauled and provided every other day, restricting water intake by 25% compared to free-choice intake had no ill effects on performance; (2) limiting water intake may get animals through a drought period without serious death losses; and (3) unless water stress is severe, the resulting weight losses are compensated for after returning to normal water schedules.

Restricting water intake of 920-lb steers has been studied in Utah in drylot under controlled air temperatures of between 30 and 45°F (Butcher *et al.,* 1959). At the end of 26 days, steers with free access to water, 25% restriction, and 50% restriction weighed 940, 920, and 870 lb, respectively. All three groups were then allowed free access to water for a second 24-day period. By the end of the second period, both the free-access group and the 25% restricted group averaged 990 lb, but the 50% restricted group still weighed 18 lb less. It was concluded that a 25% restriction only reduced intestinal fill of water but that the 50% reduction reduced actual body weight.

Cattle with access to salt-meal mix consume an additional 1 to 2 lb of salt per day. For each pound of salt eaten in salt-meal mixes, an additional 5 gallons of water is commonly consumed and should be provided (Cardon *et al.,* 1951). When used with beef cattle on California foothill range, salt-meal mixes increased the number of visits to water as well as the intake of water (up to 20 gallons daily) (Wagnon, 1965). Salt-meal mixes can have the added advantage, in addition to regulation of meal intake, of providing an indirect method of controlling urinary calculi. The increased water intake associated with higher salt levels reduces the concentration of minerals such as silica in the urinary tract, flushing them away, and thereby reducing the incidence of urinary calculi. However, as noted subsequently, the consumption of the additional water, if cold, will comprise an additional energy expenditure.

Livestock may not drink enough water if compelled to drink it ice cold or may be prevented from drinking at all if tanks freeze over and the ice is not broken by hand. Plumb *et al.* (1984) found that horses on winter range in Wyoming would attempt to break ice over drinking water with their hooves but cattle would not. Heating drinking water for livestock up to 50°F is apt to increase animal productivity as well as eliminate the labor costs of hand-breaking ice; however, the latter is now commonly being circumvented by using frost-free watering devices. Drawing upon heat reserves of the animal body to warm cold water ingested failed to influence body temperatures of sheep in one Canadian study (Bailey *et al.,* 1962), but in very cold weather the ingestion of cold water may require additional conversion of productive energy to heat energy to maintain body warmth. Leckenby *et al.* (1982) calculated that a 150-lb mule deer doe would expend the same amount of energy (1) to walk 0.6 miles on level ground, (2) produce 7 oz of milk, or (3) to raise a liter of water from 32°F to body temperature. But, three times this amount of energy would be expended if water was not available and the doe consumed the water equivalent as snow.

C. SNOW AS WATER SUBSTITUTE

Where providing drinking water is costly, difficult, or impossible, snow can be substituted in some situations (Fig. 8.2). It has been traditional in the West Desert of Utah not to trail sheep to water when snow is available (Hutchings, 1958). Free-ranging sheep in the area often do not drink at all when snow is on the ground, and cattle come into water less frequently (Blaisdell and Holmgren, 1984). It was concluded from controlled studies in the area that soft, wet snow was as effective as open water for ewes on winter range (Choi and Butcher, 1961). However, caution was suggested in relying on frozen or grainy snow or allowing access to snow for less than 2 hr per day (Butcher, 1966). The availability of snow for consumption also appeared to reduce visits to watering places by both cattle and horses in Wyoming's Red Desert (Plumb *et al.,* 1984). However, snow as a substitute for water on New Mexico winter range was concluded to be more appropriate for sheep than for cattle (McDaniel and Tiedeman, 1981).

Lactating ewes housed outdoors in Alberta receiving snow plus free water were compared to those receiving only snow (Degen and Young, 1981). Ewes relying on snow as their only source of water reduced their total water intake by 35% without affecting their milk yield or total body water, the energy content of their milk,

FIGURE 8.2   Soft snow can be substituted for drinking water in some situations, such as shown here in northern Utah, particularly where providing water is costly, difficult, or impossible; but water in liquid form is generally the best management practice.

or liveweight of the ewes. In related trials both steers and pregnant cows consumed similar amounts of total water whether offered as liquid water, part snow and part water, or all snow (Degen *et al.,* 1979). In subsequent trials with cattle fed hay-grain rations, denying water in winter when snow was available resulted in no differences in body mass or subcutaneous fat depth of the cows nor in the birth or weaning body masses of their calves (Young and Degen, 1991).

D. WATER FOR GAME ANIMALS

The availability of drinking water is an important limiting factor for wild ungulate populations in the arid West (Scotter, 1980). Open water for drinking by big game animals should normally be provided on a year-round basis. Except when consuming very succulent forage or when snow is available, water in the free state for drinking is required by big game animals. Drinking water is also considered important to upland game birds such as sage grouse, quail, turkey, chukar, and mourning dove (Lamb and Pieper, 1971).

Water is an essential part of successful desert bighorn sheep management in southwestern U.S. (Halloran and Deming, 1956). The development of water lowers disease and predation potential as well as increasing available bighorn range, reducing pressure on other limited watering places, and allowing more uniform range utilization. During July when temperatures were high, desert mule deer in southern Arizona watered once a night; mature deer consumed from 1.6–6.4

quarts daily while lactating does in late summer drank an average of 4.4 quarts daily (Hervert and Krausman, 1986; Hazam and Krausman, 1988). When denied access to water, the does searched outside of their known home range for alternative sources. Nichol (1938) estimated that the average daily water consumption of mule deer in Arizona was 1–1.5 quarts per 100 lb liveweight in winter and twice that amount in the summer.

White-tailed deer in southern Texas were found susceptible to rapid and precipitous declines in dry matter intake and body weight during droughts when drinking water was restricted. During water restriction studies by Lautier *et al.* (1988), the deer drank 3.8 quarts daily when unrestricted, but water restriction by 33% of free-choice intake reduced both dry matter intake and body weight but not as severely as a 67% restriction. It was concluded that managers should give urgent consideration to the need for supplemental water sources for white-tailed deer in drought-prone areas of high ambient temperatures.

Drinking water located somewhere within the home range of antelope at all seasons of the year is considered important, and providing 1–4 quarts per head daily is suggested (Yoakum, 1975, 1978). Antelope water consumption also varies inversely with the quantity and succulence of preferred forage species. In western Utah, antelope did not normally drink water even if readily available when forbs were succulent and their moisture content was 75% or more (Beale and Smith, 1970). As vegetation lost succulence, water consumption began; during hot, dry periods antelope drank up to 3 quarts per day.

From a review of published and unpublished literature, Broyles (1995) concluded that desert wildlife in southwestern Arizona, including Sonoran pronghorn, desert mule deer, and desert bighorn sheep, have seemingly adapted to an existence without requiring water. While these big game species apparently use free water when available, it was suggested they were able to subsist and reproduce without it when not available. However, this is not a widely accepted principle among big game managers.

III. LOCATING DRINKING WATER

The movement, distribution, and concentration of grazing animals on medium to large grazing units are highly dependent upon the number and distribution of watering places (Vallentine, 1989). Watering points should not be too widely spaced and should be adequate in number. Parts of a grazing unit with less than optimal availability of drinking water are often undergrazed, thus wasting forage, while other areas with accessible water are overgrazed and tend to deteriorate (Fig. 8.3). Excessive travel associated with reaching grazable forage beyond optimal distance from the nearest watering point will cause inefficient harvesting of forage but often be harmful to parts of the grazing unit as well as to the grazing animals themselves.

More watering points are required on rough than on level or rolling terrain. Cat-

FIGURE 8.3   Adequate number, distribution, and accessibility of watering places are required to (A) foster good distribution of grazing, and (B) prevent heavy animal concentration and killing out the local vegetation; Nebraska scenes. (Soil Conservation Service photos.)

tle should not have to travel more than $\frac{1}{4}$ to $\frac{1}{2}$ mile from forage to water (i.e., $\frac{1}{2}$–1 mile between watering points) in steep, rough country, or more than 1 mile (2 miles between watering points) on level or gently rolling range (USDA, Forest Service, 1969; USDA, Soil Conservation Service, 1976; Vallentine, 1963). On highly productive grazing lands, it may be realistic to provide at least one watering place per section for cattle and possibly two if water can be developed at low cost. The spacing of watering points for sheep and horses can apparently be widened somewhat.

Another rule of thumb is that at least one watering facility should be provided for every 50–75 animal units for full plant growing season use. However, on arid rangelands in Australia up to 300 cattle is considered a reasonable number to have on one water point (Squires, 1981). In some areas only temporary water can be developed; these water sources are less reliable and may be dry when needed most. In such areas or when temporary sources are interspersed with inadequate sources of permanent water, livestock should be grazed in areas near temporary water when the water is available. This will allow nearby ranges with permanent water to be used effectively during more critical periods and relieve prolonged congestion around permanent water.

The availability, accessibility, and spacing of watering points are also important on big game range. Recommendations for game range improvement in New Mexico have included the following provisions for drinking water (Lamb and Pieper, 1971): (1) water should be available in all grazing units through all seasons for antelope, (2) at least one watering place per four sections should be provided for deer, and (3) elk and bighorn sheep should not be more than 1 mile from water. Antelope densities in the Intermountain Region are highest on range where watering places are available 1 mile or at most 2–3 miles apart (Yoakum, 1978).

Not only do lactating elk cows require free water sources but also in Arizona they were found to utilize areas primarily within $\frac{1}{4}$ to $\frac{1}{2}$ mile of water (Delguice and Rodiek, 1984). In arid mountainous country in the Southwest, a watering place every 5 miles in favorable habitat was considered minimum for desert bighorn sheep (Halloran and Deming, 1956). Development and better distribution of water sources on arid rangelands often permit yearlong use of ranges by big game animals that otherwise would be seasonally unusable. Most big game animals readily adapt to using manmade water developments such as reservoirs, troughs and tanks, and water catchments. Antelope in western Utah readily located and drank from partly buried shallow water tanks into which water had been hauled (Beale and Smith, 1970).

The location of watering points and the size of the grazing unit can be adjusted to some extent for the needs of the livestock and the grazing resources, but economic constraints often force a compromise between what is optimal and what is feasible. Where a single watering point is to serve a grazing unit, its location in the center of the unit is optimum, but special grazing unit arrangements, limited availability of water, and terrain factors may dictate otherwise. A single, central water facility is the focus of herd management under some short-duration grazing systems, and peripheral water facilities are sometimes fenced off or deactivated to force livestock visits to the cell center and facilitate herd movement.

A system of rotating access to permanent water sources has been used in the Southwest to enable systematic deferment of livestock grazing on unfenced ranges (Martin and Ward, 1970). This seasonal opening and closing of watering places was effective in reducing livestock grazing intensity near water sources if (1) the ranges were otherwise properly stocked, (2) the watering places were not too close together but reasonably well distributed, and (3) the grazing units were not too small. In this study, cattle had to be driven away from the closed water points only in the first fall-winter period. Within a year, the cattle had learned to move to and use range surrounding the open watering point. However, the use of access to water as a means of controlling livestock grazing in lieu of fencing or shutting down watering points when not being used by livestock can have adverse effects on big game distribution (Urness, 1976). When it is desirable to seasonally exclude livestock from a watering point, using a style of fencing that will exclude livestock but not big game animals has been suggested (Prasad and Guthery, 1986).

Hauling water into undergrazed areas may be resorted to where adequate watering points cannot be developed or when existing watering places are dry seasonally or during drought emergencies (Fig. 8.4). Providing the terrain permits, water can be hauled with a 1000-gallon tank truck directly to areas of ungrazed or undergrazed forage. The primary water supply should generally be within distances not to exceed 10–15 miles of where the livestock are to be watered. Lightweight, portable tanks that can be readily stacked and transported are suggested for watering bands of sheep but can be adapted for use with cattle also. A single

FIGURE 8.4   Hauling water to sheep on western Utah winter range; water is made available in lightweight, portable, plastic tanks in sufficient number that the entire band can drink simultaneously.

large tank mounted on skids for frequent moving to new areas can also be used with cattle.

Water hauling has been very effective with cattle in obtaining more uniform use of forage by enticing them to graze where they might not otherwise do so; it has also reduced travel and thus energy expenditures, has permitted grazing at the most appropriate time, and has reduced trailing damage to the range (Costello and Driscoll, 1957). Tanks should be placed not more than 1 mile apart on flat or undulating terrain, but spacing $\frac{1}{2}$ mile apart in conjunction with moving every few days is even preferable.

Best results have been obtained when water is routinely hauled to sheep daily and tanks are moved to a new location after each watering (Hutchings, 1958). The daily routes of herded sheep should be planned in advance and the tanks placed ahead of the flock and filled with water so that the flock can graze quietly to water, drink, and then move out onto fresh forage with a minimum of disturbance. Enough tanks should be used and spaced far enough apart so that the entire band can water simultaneously without crowding.

When additional watering points are added in minimally grazed areas in a grazing unit previously having only a single watering point, long-term benefits in overall forage utilization and livestock production are assured only with good subsequent management. Any increases in stocking rates must be carefully monitored to assure that degradation around the new watering points does not exceed the improvement around the old watering point. Noy-Meir (1996) has concluded that benefits will be only short term if associated increased stocking rates result only in increasing the area under heavy grazing and removing emergency forage resources; another concern expressed was the reduction or elimination of "corner refuges" for plants available for conserving biodiversity and genetic resources as a continuing source of natural regeneration of losses of perennial vegetation.

IV. FENCING

While the placement of watering, salting, and supplementation points are indirect techniques, fencing and herding are direct means of regulating the distribution of grazing. Fences on grazing lands have a multitude of objectives and uses, and an assortment of fence types and designs can be selected for the kind or mix of animal species involved (Vallentine, 1989). However, their function in controlling distribution of grazing is the focus here (Fig. 8.5). Once the fence style and design have been selected to best accomplish the objectives, the location of the fence takes on paramount importance.

Guidelines for locating fences to enhance grazing distribution include the following:

1. Separate grazing areas that will require different grazing management practices.

FIGURE 8.5   Fencing is a direct means of regulating the distribution of grazing; (A) a 5-wire range fence near Thermopolis, WY, and (B) a combination net wire-barb wire fence provided with gaps for antelope crossing. (Bureau of Land Management photo (bottom) by Raymond D. Mapston.)

2. Cross fence large grazing units to provide more direct control of grazing, provide greater forage homogeneity within units, or enable specialized grazing techniques.
3. Utilize existing or projected watering points to greatest advantage.
4. Assure that each fenced unit can or will be provided with adequate spacing and number of watering points.
5. Coordinate size of grazing units with the planned use including grazing capacity projected.
6. Provide for efficient access to and exit from the fenced unit.
7. Expedite the proper movement of grazing animals within the fenced unit while reducing trailing along fences or through erosive or hazardous areas.
8. Locate fences along crests of ridges and other natural division lines, when possible.
9. Determine the location (and type of fence) that expedites the proper grazing of multiple species when grazed in common (for example, livestock and big game or even sheep with cattle).

Separately fencing individual plant communities of different animal preference is often impractical due to their limited size and occurrence in a mosaic pattern; however, separately fencing large areas requiring different management such as seasonal grazing or special use is often justified. Highly productive areas such as riparian zones or irrigated sites may require being fenced separately for optimal management. Protection of new forage seedings during establishment and preventing differential grazing between new seedings and adjoining unseeded land may require fencing.

Fences are ideally constructed along the hogbacks or dividing lines between individual watersheds; cattle do not walk these fence lines, and utilization patterns are not deleteriously influenced by the presence of such fences (Hickey and Garcia, 1964). Location of fences along the crests of ridges also reduces the amount of time sheep spend on any one bedground or part of a bedground and prevents them from crossing from one drainage to another and greatly extending travel distance (Bowns, 1971).

V. HERDING AND HANDLING

Control can be exercised directly by the grazier as animals, principally livestock, are grazing or moving about over the grazing unit. **Herding** refers specifically to the control exercised by the grazier in assembling and keeping grazing animals together in a group, with reference generally to a single kind of animal. By contrast, **free ranging** or **loose running** (unherded) implies no direct control by the grazier on the movements of the grazing animals.

Close herding (tightly grouped) involves holding a herd or flock of animals in a closely bunched manner, restricting the natural spread of animals when grazing.

In North America this method is primarily applied to sheep on unfenced range and under the more or less constant control of the herder and his dogs; cattle and other domestic and semi-domesticated animals are similarly handled in some foreign countries. **Open herding** (loosely bunched) is a method in which the individuals in a herd are allowed to spread naturally for grazing under more relaxed control by the herder and dogs but are kept within a prescribed area.

Trail herding (or **driving**) implies close control of livestock by the grazier when being moved towards a specific destination. And, finally, **drifting** refers to slow urging or even leading by the grazier of animals in a certain direction, utilizing the natural movement of the animals as much as possible. In this regard "low-stress" handling of cattle has recently begun receiving widespread acclaim; its advantage is that cattle can be managed in a loose herd and their movements easily regulated while "high-stress" handling induces animals to leave the herd (Cole, 1999). In actual practice, the use of these animal control practices intermix and vary by grazier, time, locality, and immediate management objectives.

A. HERDING SHEEP

Increasing labor costs, lack of capable herders, and narrow profit margins have forced more attention to be given in western U.S. to herderless management of sheep under fence. The loose running of sheep on fenced, non-range situations has been a long-standing practice in the U.S. and Canada; the practice on fenced range began in eastern Wyoming and the southern Great Plains, and from there the concept and practice spread to other range areas. Major savings in labor costs must be compared with increased fence investments, and making the change to loose running of sheep requires that management assume part of the sheep supervision previously provided by herders (Roberts, 1961).

Ranchers who have converted from herding to herderless management of sheep under fence generally indicate they would not return to herding because of the benefits of fenced management (Roberts, 1961; Blankenship, 1969). However, the following reported benefits of loose running of sheep on fenced range are made more prominent when compared to close herding rather than open herding:

1. Increased grazing capacity and improved range condition
2. Improved distribution by spreading out grazing and allowing individual sheep to reach difficult spots
3. Permitting sheep to water singly whenever they desire, with the advantage of effective use of smaller, low-capacity watering sources
4. Improved condition of ewes and better mothering of twins because of less disturbance
5. Similar to slightly improved lamb weights
6. Cleaner wool
7. More breed choices with sheep since the herding instinct is of reduced importance
8. Easier to run a mix of sheep and cattle together in the same grazing unit

However, new problems that may be encountered, particularly on large range units, from changing to herderless management of sheep under fence include: (1) some trailing; (2) more winter and storm losses; (3) local concentration of grazing and trampling, particularly on bedgrounds; (4) interference of added fencing with big game migration; (5) reduced access to sheep for treatment; (6) more difficulty in gathering sheep at the end of the grazing season; and (7) increased predator losses and theft. Failure to adequately control predators in the area may prevent loose running of sheep from being a practical alternative to herding. Sheep reportedly can be herded part time and fenced part time, but an adjustment period is involved each time a change is made.

After reviewing sheep herding practices on high-elevation mountain range, Thilenius (1975) concluded that close herding often resulted in range deterioration, particularly when combined with continuous bedding in the same location, usually near water. Instead, open herding practices were recommended with unherded management as a possible alternative (Fig. 8.6). Specific practices under the open herding method were recommended as follows: (1) move herd steadily in one direction, (2) herd by guiding the movement of the lead animals rather than driving from the rear, (3) prevent excessive use of dogs, (4) plan to reach water once each day and graze quietly (drift) rather than drive to water, (5) graze each area only once, (6) follow one-night bedding when animals finish grazing for the day, and (7) provide salt at the bedding ground in movable containers. The latter is a permissible convenience since both the salt and the bedground will be moved nightly under herding management.

Open herding was compared to herderless grazing of sheep under fence on adjoining allotments on Wyoming alpine range during a 10-year study (Thilenius and Brown, 1987). It was concluded that neither system could be recommended over the other from the standpoint of the vegetation. Both systems provided acceptable levels of forage utilization, and there were no consistent differences in the weight of lambs produced under the two systems. It was concluded that other factors, such as the availability of herders or economics, should be used to select one of the alternative management system.

Suggestions for herding sheep on desert winter range have been similar to those for mountain range (Hutchings, 1954). Herding was considered a full-time job with the larger bands on winter range, and a good herder spends most of his time with the flock, directing the course of grazing and allowing the sheep to graze quietly throughout the day without excessive trailing. For good management, the following herding practices were recommended: (1) plan the daily route of grazing in advance to provide a variety of forage with some fresh ungrazed area each day, and graze rather than trail to new areas; (2) allow the sheep to spread out and graze quietly but do not allow them to trail back and forth across the range; and (3) bed the flock in a new location each night wherever nightfall overtakes them. The advantages of sheep being in a flock were noted when water was provided by truck hauling and when salt and other supplements were fed.

Optimal distribution of sheep does not result automatically from loose running

FIGURE 8.6   Open herding of sheep in the Stanley Basin of Idaho allowing them to spread for efficient grazing while being kept within a prescribed area. (Forest Service Collection, National Agricultural Library.)

of sheep. Although the sheep break up into smaller bunches, they continue to maintain site preferences for grazing, bedding, etc. and require fencing to augment natural barriers and greater effort to distribute water and salt (Thilenius, 1975). The potential also exists for grazing or bedding on critical areas that should be avoided, such as steep, easily eroded slopes, wet boggy areas below snowbanks, or riparian sites.

B. HERDING CATTLE

Cattle untended on mountainous or other heterogeneous areas tend to settle in one area and graze there indefinitely. Grazing associations on many public grazing lands are required to provide "cattle herders" or "cow riders" to tend the cattle. Their duties include salting, health care, dispersing bulls for adequate breeding service, and repairing fences and stockwater facilities. They also have the responsibility of gathering cattle from normally overgrazed parts of a range and establishing them on lightly or ungrazed parts, often improving not only range condition but cattle gains as well (Fig. 8.7). This is more properly referred to as drifting than herding, but the latter term remains in common use. A rider who knows the range and the cattle can keep 500 head well distributed over 50 square miles of mountain summer range, perhaps twice this many on highly productive foothill range, or only half this many on low productive brush or chaparral range (Skovlin, 1965).

Particularly when applied along with other tools, the use of intermittent drifting of cattle away from natural concentration areas can be a useful tool in improving animal distribution. In fact, adequate fences, water, salt, and access often fail to achieve proper use of all the suitable range without the aid of a rider (Skovlin, 1965). In an 8-year distribution study on mountain summer range in northern Utah, cattle permitted to remain along streams in the bottom grazed forage on slopes (up to 35%) only 7% compared to 27% when drifted two to four times per week onto the slopes (Cook, 1967). It was calculated this practice would increase grazing capacity 3.6 cow-days per acre; however, this practice was even more effective when salt was methodically located on the slopes to be grazed. A common-sense rule in riding to increase distribution is to move cattle to the forage supply only if there is water, and possibly necessary cover there also (Martin, 1975b). Cattle may have to be met daily and pushed back to the new location where

FIGURE 8.7   Gathering and drifting cattle from heavily grazed areas for establishment in lightly or ungrazed areas; winter range near Garrison, UT.

both grass and water are present for a week or more before they will accept the move (Howery *et al.,* 1999).

Improving cattle distribution on western mountain rangelands, according to Skovlin (1965), should begin when cattle are first placed in the range unit. Ideally, the cattle should be gathered, trailed, and released in small bunches at selected areas throughout the new range. In no case should the cattle be turned out through the boundary gate and left to seek their forage location at will. The help of extra riding at turn-on time will simplify the regular rider's job for the entire grazing season. Once full use is reached on various parts of the range, the rider should begin moving cattle to lightly grazed areas and get them established there. This is helped by moving entire "family-type groups" together, gathering them near water or salt during mid-day when cow-calf pairs are together and trailing them to new areas in the cool of late afternoon. Upon arrival in the new area they should be moved to the nearest water and then on to the salt ground, giving ample opportunity for calves to mother-up at both places.

Studies on forested summer range in Oregon have suggested herding cattle away from concentration areas is apt to be ineffective in making lasting effects once their grazing pattern has been set (Roath and Krueger, 1982b). It was concluded that great care must be taken to avoid initial concentration on riparian sites when cattle are first turned into an area; allowing access from a different point and fostering immediate dispersion was suggested. Cattle that had not grazed the area previously might be more effectively trained by bonding them to areas which had been previously undergrazed, given that water, forage, shade, and salt were available in that area. Animals that refuse to stay where they are placed might well be removed from the herd.

VI. SALTING AND SUPPLEMENTATION

Guidelines for the placement of salt and other supplements to improve the distribution of grazing commonly include the following:

1. Feed salt or protein and energy supplements in areas of ungrazed or undergrazed forage to encourage grazing in those areas.
2. Place salt methodically over the range and not less than $\frac{1}{3}$ mile from water on grazing units of one section or more in size or less than $\frac{1}{5}$ mile in smaller units.
3. Move creep feeders away from water or natural animal concentration areas as soon as calves learn to eat creep feed.
4. Move salt and supplement locations frequently; use movable bunks or feeders rather than permanently located facilities.

Moving salt away from sites having other attractants such as water, riparian forage, main roads, trails, and bedgrounds has long been considered a tool for improving livestock distribution (Fig. 8.8). Some research reports have indicated the

FIGURE 8.8   The methodical placement of salt on good forage sites and away from other attractants such as water, riparian forage, and main trails has long been used as a tool for improving grazing management; cattle in Hooker County, NE. (Soil Conservation Service photo.)

substantial favorable influence of salt placement on grazing distribution: Bjugstad and Dalrymple (1968), on Ozark range; Cook (1967), on Utah mountain range; Martin (1975b), on Southwest grass-shrub range; Patton (1971), on Idaho mountain range; and Roath and Krueger (1982b), on Oregon mountain range. Others have indicated little influence of salt placement on grazing distribution: Bryant (1982), on riparian zones; Gillen *et al.* (1984), on Oregon mountain range; and Wagnon (1968) on California foothill range.

It does not appear that salt placement is capable of circumventing or overriding all of the attraction that open water, favorite forage, favorable terrain, or protective cover or shade has for grazing animals. It also appears to be less useful when the vegetation is naturally salty or natural salt licks occur in the area (Martin, 1975b). Cook (1967) noted that salting is more effective when combined with other tools; while salting alone increased the grazing capacity for livestock by about 13% (up to 20% on some slopes), salting plus occasional herding increased grazing capacity by 21% (up to 30% on some slopes). Locating salt where it cannot be found or realistically reached will be ineffective. To be effective, salt should be placed on sites where animals do not naturally congregate but are easily accessed. Even though its individual effectiveness is open to question, in most range and pasture situations, salt location should still be considered as a usable tool when combined with other tools.

For mountainous terrain in northeastern Oregon, Skovlin (1957) recommended the following cattle salting practices: (1) use movable salt grounds, about one to every 300 acres; (2) move the salt grounds when forage in the immediate area has been grazed; (3) pick up small bunches of cattle around noon at watering points, drift them to the newly established saltgrounds, and hold them there for a few hours; (4) repeat the operation every several days, more frequently in mid- and late summer; and (5) pick up the unused salt near the end of the grazing season to encourage cattle to move into the bottoms for final gathering.

The placement of protein and energy supplements or salt-meal mix is apt to be even more effective in manipulating grazing patterns than salt alone (Fig. 8.9). Bailey and Welling (1999) found that cattle can be lured to under-utilized range areas by the strategic placement of dehydrated molasses supplement blocks, this being more effective in moderate terrain than in difficult terrain, while salt alone did not affect where cattle grazed. During a 2-year study with cattle during the November-June period on the Jornada Experimental Range in New Mexico, feeding a salt-meal mix away from water only ($\frac{1}{2}$–3 miles) was compared to feeding both at and away from water (Ares, 1953). The benefits of feeding the mix only away from water were listed as follows: (1) increased proper use zone by 84%, (2) reduced

FIGURE 8.9   Feeding protein supplements, as shown here on Nebraska winter range near North Platte, is a useful means of locating cattle where ample forage supplies are available. (University of Nebraska photo by Donald C. Clanton.)

heavy use zone by 52%, (3) almost eliminated the excessive use zone, and (4) reduced the light use zone by 29%. When the meal was provided both at and away from water, 80% of the meal consumed was at water.

Moving supplemental feeding locations away from water sources on annual grass-hardwood range at the San Joaquin Experimental Range, CA, drew cattle grazing into areas where high amounts of residual dry matter remained (McDougald *et al.*, 1989). This practice reduced heavy cattle grazing over approximately 50% to less than 1% of the riparian areas. However, based on their research on the Santa Rita Experimental Range near Tucson, Martin and Ward (1973) concluded that the placement of salt or salt-meal mix alone cannot be expected to cure serious distribution problems. In their studies, salt-meal mixes tended to pull livestock out into undergrazed areas but did not alleviate the heavy grazing and trampling at the watering points, the latter presumably inadequate in number and/or location.

Alternate day or less frequent feeding of a protein supplement compared to daily feeding has induced cattle to graze more widely over the grazing units (Melton and Riggs, 1964; Rothlisberger *et al.*, 1962). Since cattle tend to wait at permanently located bunks and graze out short distances only, it is suggested that pelleted or cubed supplements be fed on the ground or in movable bunks. When supplement is fed daily or even on alternate days, alternating the place and possibly also the time of supplement should reduce prior concentration of animals.

When native range or permanent pasture is utilized for prolonged periods as feedgrounds for feeding harvested roughages, great damage can result to the forage plants by heavy utilization and intensive trampling, not only just on the feedground but also throughout the fenced unit of which the feedground is a part (Fig. 8.10). As a result of the extensive damage such a practice can cause, federal land management agencies and many state agencies forbid feeding livestock other than mineral supplements on their lands.

When replacement rations of harvested forages are required during the winter, a practice common in many areas, the feed should be fed in drylot, on unseeded cropland, non-erosive wasteland, or on a restricted area of rangeland that can be accepted as a sacrifice area. Where livestock must be frequently bunched for specialized breeding programs or otherwise require frequent handling, providing high production, intensively managed holding/breeding pastures will alleviate stress both on the livestock and on adjoining rangelands.

Continuous winter feeding of hay under high animal density on improved pasture previously rotationally grazed in summer in Ohio resulted in increased runoff and erosion (Owens *et al.*, 1997). Damage resulted particularly in late winter and early spring, the deleterious effects noted principally in the subsequent grazing season. Since the impacts were substantially meliorated within a year, feeding only on pastures with less severe slopes and rotating winter feeding were suggested.

The practice of emergency feeding of big game animals during severe winters has become more common in recent years. Even though it may be required for sheer survival of the animals, this tends to have a negative influence on the grazing/browsing

FIGURE 8.10   Use as a winter feedground devastated this sagebrush-grass range in southern Idaho, formerly in good condition, within a 4-year period; although harvested feeds were fed on only a small, unfenced portion of this range unit, prolonged heavy grazing and intensive trampling continued throughout the entire range unit.

behavior of the animals. Once the big game animals learn that food can be found at the feeding stations in the valleys, this trains the animals, by default, to come to the valleys and even into the towns (Nielsen *et al.,* 1986). In subsequent years the big game animals are prone to spend much more time down in the valleys, sometimes even the entire winter, and some remain there even after winter has gone.

For supplemental or emergency feeding of big game animals, Olson and Lewis (1994) have recommended selecting feeding areas that (1) are easily accessible by animals without obstacles, (2) are near thermal cover (principally trees) or broken terrain, (3) avoid cold air inversions, and (4) minimize outside disruption; sufficient feeding areas of adequate distribution should be used to prevent large concentrations of animals. For feeding mule deer, Bryant and Morrison (1985) have recommended that one feeding station of 0.5–1.0 acre in size be established for each 500 acres, and that feeding stations be located within 0.25 mile of cover and away from areas of high human activity.

VII. OTHER SITE ATTRACTANTS

A. PROVIDING SHADE

Shade and cover often fail to give consistent results in improving distribution of grazing since the location where animals loaf and seek protection from weather may be poorly related to where they graze. Also, localized heavy grazing and trampling often develop around points of scattered shade. However, shade was found on Oklahoma range to be nearly as effective as water and supplemental feeding location as a tool to promote uniform cattle grazing on hot summer days (McIlvain and Shoop, 1971a). Living shade during hot weather or cover during inclement weather as an attractant to grazing animals can be provided by (1) utilizing natural tree and shrub stands, or (2) planting trees individually, in clusters, or in windrows. Constructed, artificial facilities can be developed more rapidly than the latter and can be made mobile but may be less useful.

The advantages of convertible units for providing shade in summer and shelter in winter were determined under yearlong grazing of yearling steers on treeless Oklahoma sandhills range (McIlvain and Shoop, 1971a). The artificial shades were consistently used every hot, sunny day in summer, and summer gains were increased an average of 19 lb. However, the units had no measurable effect on winter steer gains. Even though the protein supplement was fed in the shelters, the steers did not use them additionally during cold, windy days or during storms, preferring instead the protection of sand sagebrush (*Artemisia filifolia*) near the base of southeast-facing dune slopes.

During hot, humid weather in Louisiana in late spring and summer, Hereford and Angus cattle grazed more and were more attracted by natural than artificial shade (McDaniel and Roark, 1956). The gains of the calves but not the cows were improved by providing artificial shades made with roofs of hay, straw, or pasture clippings. Inexpensive shade provided to yearling bulls summer grazing bermudagrass in Oklahoma had no clear effects on time or patterns of grazing (Coleman *et al.,* 1984). These results agree with observations of Arnold and Dudzinski (1978) that artificial shelters must allow enough space for animals to keep their normal social distance when lying or standing and allow maximum air movement on hot days in order to attract animals. Providing they are constructed so that they attract livestock, artificial shades can be provided with skids to allow them to be moved to undergrazed portions of the grazing units.

B. ENHANCING FORAGE PALATABILITY

When applied to areas previously undergrazed, treatments that enhance palatability of most or all forage plants on an area may serve as tools in attracting grazing animals onto the treated sites and holding them there once placed. Such treatments include burning, chemical or mechanical treatments, or grazing to remove unpalatable plant species or remove old growth of otherwise palatable species and stimulate palatable regrowth. Another treatment is the application of nitrogen fertilizers.

Certain herbicides such as 2,4-D and related compounds temporarily increase the palatability of affected plants (i.e., initial wilting to early desiccation) and increase grazing on sites where affected plants are prominent. Herbicides at typical control rates in Texas studies increased grazing preference by cattle in the following order: (1) tebuthiuron, (2) picloram, and (3) 2,4-D (Scifres *et al.,* 1983). When followed from post-treatment day 1 through 21, glyphosate increased the attractiveness of forage for about one week; during the second week, there was no notable preference for treated plants, and during the third week there was an aversion to the affected herbage, the plants by then becoming dry and brown (Kisserberth *et al.,* 1986). In Texas studies (Tanner *et al.,* 1978), white-tailed deer tended initially to evacuate mixed-brush strips sprayed with phenoxy herbicides but were subsequently attracted by the succulent woody plant regrowth and forb recovery in greater than normal numbers.

Plants stimulated by added nitrogen from fertilizers or legume association are more palatable to livestock than unaffected plants. This effect is nearly universal, regardless of life form or whether the plants are recognized forage plants or weeds or brush or, unfortunately, even poisonous plants. For example, on desert grasslands in southern Arizona 25-, 50-, and 100-lb nitrogen per acre increased utilization of grasses by 300, 400, and 500%, respectively (Holt and Wilson, 1961). Phosphorus addition generally has had no significant effect on palatability of grasses grown without legumes (Cook, 1965), but an exception may be with grasses grown on extremely phosphorus-deficient soils.

From studies on aspen and sagebrush-grass range in northern Utah, Cook and Jefferies (1963) concluded that cattle will not seek out nitrogen-fertilized or 2,4-D-treated plants on slopes but will use these areas more readily after they have once grazed over it or are drifted onto it. In conjunction with drifting cattle onto the study sites, 2,4-D alone, nitrogen fertilizer alone, and both in combination resulted in an average utilization of grasses of 47, 34, and 50% on slopes compared to 27% on similar but untreated slopes. When combination treatment strips were placed perpendicular to the bottoms of the slopes, cattle without herding used the treated strips materially more than the untreated strips. During the first part of the summer grazing season, animals that drifted onto slopes stayed on both the treated and the untreated slopes only during the day, but during the latter part of the grazing season, animals stayed longer on the treated sites, and 45% of the drifted animals later returned to graze. This presumably resulted, at least in part, from the training the animals had received but possibly also from relatively greater availability of forage on the slope by then.

In a continuation of the Utah study, Hooper *et al.* (1969) noted that nitrogen fertilization not only increased utilization on treated areas but also increased utilization to a lesser extent on range adjoining the treated areas. Forage utilization by cattle on Wyoming mountain range was increased by two to five times by application of 67.5 lb nitrogen per acre (Smith and Lang, 1958). Cattle tended to graze untreated areas in the immediate vicinity of the treated plots. Hooper *et al.* (1969) concluded that (1) nitrogen fertilizer placement must be coordinated with herding, salting, and water development to be effective; (2) care must be taken not to fertilize areas (without fencing separately) where animals normally congregate; and (3) the areas should be made sufficiently large (perhaps 30 acres) so that concentrated heavy use is not experienced. If only productive, overgrazed areas within a grazing unit are nitrogen fertilized, even heavier localized grazing can be expected; nitrogen fertilizer can extend areas of excessive use if improperly used.

Nitrogen fertilization has also increased the palatability of forage grasses and browse plants such as bitterbrush and big sagebrush for big game (Anderson *et al.,* 1974; Bayoumi and Smith, 1976; Thomas *et al.,* 1964), but the benefits have often lasted only one year (Skovlin *et al.,* 1983). The effects of nitrogen fertilization have generally been insufficient to be used solely as a means of controlling movements and distribution of big game animals. However, Scotter (1980) has noted that nitrogen fertilization of state-operated hayfields in Washington has made them more attractive to elk and thereby reduced damage to adjacent private hayfields.

New regrowth on local areas recently prescribed burned affect grazing distribution by attracting animals into affected areas as a result of enhanced palatability and availability of green herbaceous regrowth or suckering or vegetative sprouting of woody plants. Burning is a site attractant to both livestock and large wild herbivores. From their studies with bison on prairie range in Oklahoma, Coppedge and Shaw (1998) concluded prescribed burning might even eliminate the need for multi-unit, rotational grazing systems in bison ranching, thereby reducing the need for extensive cross-fencing and associated maintenance requirements. Such a rotational burning/grazing practice has long been employed in southeastern U.S. to, in part, seasonally and annually redistribute grazing (see "Rotational Burning/ Grazing" in Chapter 14). The use of prescribed burning as well as fertilizer or herbicide is seldom cost effective for improving distribution of grazing alone, either for livestock or big game, and must be justified primarily on the basis of enhancement of forage quantity and quality.

VIII. SPECIAL PROBLEM: RIPARIAN ZONES

Since **riparian** zones are the focal points for many multiple-use values of grazing lands, their management and utilization are major sources of potential conflict between different land uses (Dwyer *et al.,* 1984). Not only do they attract livestock and many big game animals, but they also provide habitat for fish and waterfowl, water for agricultural, industrial, and domestic use, and opportunities for water recreation. Vavra *et al.* (1994) have concluded that maintaining a healthy riparian ecosystem does not require total exclusion of livestock but rather proper management.

Even under moderate grazing of adjacent upland areas, riparian zones often receive heavy grazing by domestic livestock, particularly cattle (Dwyer *et al.,* 1984). Proper grazing of streamside vegetation requires controlled animal distribution (Clary and Booth, 1993). In the absence of proper control of distribution of grazing, even light stocking rates applied to grazing units including riparian areas may not prevent excessive use of them by livestock or some big game species because of their attractiveness. Belsky *et al.* (1999) found that nearly all scientific studies record that livestock do not benefit stream and riparian communities, water quality, or hydrologic functions. However, the findings were that their damage can be reduced by improving grazing methods, herding or fencing cattle away from streams, reducing livestock numbers, or increasing the period of rest from grazing.

Conventional management strategies, tailored to extensive livestock production and forage maintenance on upland and/or large areas, may not be fully effective in achieving acceptable animal distribution and forage use in highly preferred riparian zones (Fig. 8.11) (Platts and Nelson, 1985a; Skovlin, 1984). From a review of literature, Larsen *et al.* (1998) found 428 articles were directly related to grazing impacts on riparian zones and fish habitat, but only 89 articles were classified as experimental where treatments were replicated and results statistically validated. Their generalizations from their literature review were that: (1) livestock and big game can and do co-exist within sustainable riparian systems, (2) vegetation responses are highly site specific, and (3) ecosystems are highly variable in space and time.

Nevertheless, special practices suggested for controlling the level and distribution of livestock impact on riparian areas and wetlands have commonly included the following (Skovlin, 1984; Dwyer *et al.,* 1984; Platts and Nelson, 1985a; Behnke and Raleigh, 1978; Vavra *et al.,* 1996; Chaney *et al.,* 1993):

1. Use all normal tools available for improving livestock distribution.
2. Avoid using riparian areas as driveways or holding livestock while awaiting shipment to avoid harm to the riparian and stream environment (Platts, 1981b).

FIGURE 8.11   Regulating grazing use on riparian areas often requires a combination of conventional and special practices because of the great attractiveness of such areas to both livestock and many big game species; scene near Panquitch, UT.

3. Implement specialized rotational grazing systems that reduce duration of grazing and provide adequate nongrazing recovery periods.
4. Prioritize livestock access to a grazing unit containing critical riparian areas, to the extent realistically possible, to periods (1) when soils of riparian areas are not wet or boggy but are firm and stable, and (2) when acceptable forage is available on non-riparian sites within the same grazing unit.
5. Manage riparian and floodplain areas, when their larger size and management independence justifies, by fencing into one or more "riparian pastures" separated from uplands for controlling timing and intensity of grazing.
6. Rest entire grazing unit or at least riparian zones for 2–5 years until target levels of recovery have been achieved.
7. As a last resort, fence out the most vulnerable streamside corridors for complete habitat preservation, while providing strategic access to drinking water for grazing animals where needed.

It is seldom practical or economically feasible to fence off all streamside corridors in an area to exclude grazing, and this would result in the loss of a large amount of forage (Platts and Nelson, 1985a,b). Fencing large riparian/floodplain areas into separate pasture units for controlled grazing appears promising, and combining separate fencing of riparian areas and grazing under rest-rotation or deferred-rotation grazing shows even more promise; but many riparian areas are too small, too irregular in shape, or too scattered to make fencing riparian areas entirely separate from upland areas being practical. Fencing an equivalent acreage of upland range in with the riparian area may be an acceptable compromise, and livestock have been noted to search for a variety of vegetation types, including use of adjoining uplands. When small, scattered riparian sites are intermixed with uplands, the only practical option may be to consider them as sacrifice areas and apply best management practices holistically to the entire grazing unit.

No particular season of grazing use is universally recognized for minimizing damage or maximizing improvement of streamside zones (Smith *et al.,* 1991). The optimal season for grazing areas including riparian sites should consider the elevation of the site, average annual precipitation, precipitation timing, forage growth on riparian and adjoining sites, soil wetness, and aspect. Early spring grazing and/or winter grazing appears best in many low and middle elevation areas (Vavra *et al.,* 1994). Grazing dormant vegetation can decrease the stress of herbage removal (Masters *et al.,* 1996; Glimp and Swanson, 1994; Platts and Nelson, 1985a).

Clary and Booth (1993) concluded that spring grazing should be favored in many areas because cattle have less tendency to concentrate along streams and wet bottoms during that season. From their study of cattle grazing during June in the mountains of central Idaho, these authors found that increasing stocking rates from light to medium levels induced cattle to concentrate most of their additional use on adjoining drier meadow rather than on riparian sites. Late spring through summer and into early fall on hot/dry ranges is generally the least desirable grazing season; at that time of year, the combination of green forage, shade, and drinking water on or in close proximity to riparian habitat multiplies the attraction to grazing animals. Deferred grazing seems only to further concentrate cattle in riparian areas (Clary *et al.,* 1992).

Cool, mesic ranges at higher elevations with riparian sites may respond best to summer grazing (Masters *et al.,* 1996). Roath and Krueger (1982a) noted in Oregon that turning in at mid-season as well as the accumulation of cold air pockets over the riparian zones tended to disperse animals, but delaying use until late summer, particularly in dry years, increased cattle browsing of shrubs over riparian herbaceous plants. However, less damage occurred on riparian sites in Montana foothills when grazed beginning in mid-summer than in late June and early July (Marlow, 1985); although relatively more grazing occurred on the riparian sites after mid-summer, the soil moisture content and level of damage to soil and streambanks were less at that time.

Reducing length of stay of grazing animals on riparian and aquatic areas may be as important as delaying livestock entry until streambank moisture content has lowered (Marlow and Aspie, 1988). Grazing practices that have reduced duration of grazing and increased duration of nongrazing while avoiding heavy impact during any period have shown some promise in restoring riparian habitats (Kauffman and Krueger, 1984). Based on a 5-year comparison of grazing systems on riparian soils, Bohn and Buckhouse (1985) concluded that rest-rotation but not deferred-rotation or season-long continuous grazing enhanced hydrologic expression, i.e., higher infiltration, lower sediment production, and lower bulk density. (Refer to "Rest-Rotation Grazing" in Chapter 15 for further details on this grazing system.)

Rest-rotation grazing provides both deferment and 12-month rests, but safeguards against excessive impact during full grazing years may be required. Masters *et al.* (1996) concluded that three-paddock rest-rotation grazing under moderate stocking rates is more apt to maintain than improve riparian sites included in range units. When only small amounts of riparian area are fenced in with extensive upland range, rest-rotation grazing may have minimal effect in reducing animal concentration on the riparian portion (Skovlin, 1984). Masters *et al.* (1996) concluded that four- or five-paddock rotation without rest may be more suitable to areas that require increased streambank vegetation by allowing for a shorter grazing periods and greater flexibility in rotation schedules. However, Clary (1995) has noted that riparian areas have not responded consistently to grazing systems.

Platts and Nelson (1985b) evaluated seven special management units in the Intermountain area that included riparian zones and adjoining upland sites within $\frac{1}{4}$ mile on each side of the stream in approximate 50:50 ratios. Livestock averaged 29% heavier utilization (72 vs. 43%) on the riparian portion; however, the differential utilization on small units of about 10 acres was reduced using rest-rotation grazing under early season use and almost eliminated under late season grazing.

Clary *et al.* (1996) in a 7-year study in Oregon found few differences in responses by either plants or animals from the following treatments on riparian sites: (1) no livestock grazing, (2) no livestock grazing and woody species planted,

(3) light to moderate fall grazing, (4) light to moderate spring grazing, and (5) heavy season-long grazing. This suggested even longer recovery or treatment periods. While plantings of herbaceous and woody species and reduced grazing pressure provided initial improvements, these tended to be removed by natural high flood streamflow.

Sheep managed under herding can be controlled to minimize the impact on riparian sites, but frequent herding of cattle—and possibly loose-running sheep, as well—away from riparian sites in conjunction with assuring water availability on adjoining upland can be successful in limiting animal numbers and time spent on stream bottoms. From grazing studies in mountainous terrain near Fairfield, ID, Howery *et al.* (1996), after finding a high degree of home range fidelity among individual cattle, suggested that selective culling may be required to effectively change cattle distribution and decrease use of riparian areas. Further suggestion was that a rider could note the eartags of animals habitually in riparian areas and move or cull those that were the worst offenders. In their study, these authors noted four sub-herds that remained moderately constant, with individuals and even groups expressing faulty distribution habits.

Over 99% of the time that water is contaminated by cattle, according to Miner *et al.* (1992), is by direct deposition of animal fecal matter into the stream rather than the washing in of fecal material during a runoff event. Under winter hay feeding conditions on a riparian site with a stream flowing through the center, these authors compared providing additional water in a tank located 300 feet from the stream to having water available only at the stream. Providing an alterative source of water away from the stream decreased the time (both drinking and loafing) spent at the stream by 90%. This held true whether the hay was fed beyond the tank or halfway between the tank and the stream, i.e., in comparison with stream-only water availability when feeding locations were at comparable distances from the stream.

When given the choice in another study (Sheffield *et al.,* 1997) between drinking from a stream or from a conveniently located water trough, grazing cattle were observed to drink from the water trough 92% of the time. Installation of the alternative water source reduced stream bank erosion by 77%. These authors concluded that off-stream water sources effectively reduced the loss of sediment and sediment-bound pollutants to adjacent streams without resorting to total stream bank fencing.

Spring sites surrounded by riparian areas should often be fenced and the water piped outside to more stable sites for livestock access. Whyte and Cain (1981) studied shorelines of manmade ponds in south Texas and concluded that carefully planned grazing including key rest and grazing periods controlled the impact of grazing on shoreline vegetation; fencing half of the shoreline from cattle use provided additional assurance of stable waterfowl habitat.

9

PLANT SELECTION
IN GRAZING

I. SELECTIVITY

Grazing animals are always **selective** in what they eat; that is, they choose or harvest plant species, individual plants, or plant parts differently from random removal or from the average of what is available. Herbivores range from generalists to specialists in their diet selection; however, there are no obligatory ungulate herbivores—restricted to a single species or genera of plants—as there are for insects, for example. Diet selection affects not only the grazing animal's nutrient status but also the successional processes in plant communities. Selecting for some plant species and against others, large herbivores have a profound effect on the competitive relationships of plants growing in mixed plant communities. In fact, the impact of livestock on rangelands occurs primarily because they selectively defoliate the available herbage rather than indiscriminately consuming herbage according to its availability (Walker, 1995).

Some grazing techniques have the objective of **nonselective grazing**—the utilization of forage by grazing animals so that all forage species, plants, and plant parts are grazed to a comparable degree. Selectivity is diminished by grazing techniques such as heavy stocking or high animal density; strip grazing at a high stocking rate will largely abolish selectivity (Van Soest, 1982). However, the use of traditional grazing systems does not appreciably affect selective foraging behavior (Walker, 1995). Nevertheless, selectivity should not be totally eliminated in consideration for the grazing animal. Production of high-producing animals, in particular, may suffer if they are compelled to mix in their diets substantial quantities of less preferred, low-quality forage.

Although palatability and preference are interrelated in their determination of selectivity in grazing, they are different terms (Skiles, 1984). Restricting the term "palatability" to plant characteristics or conditions and the term "preference" to the reactions of the animal to these differences provides a proper basis for evaluating the extent and causes of selectivity (Heady, 1964). **Palatability** refers to that combination of plant characteristics that stimulates animals to prefer one forage over another, or can be extended to any feedstuff. These differences in attractiveness or acceptability stimulate a selective consumption response by the herbivore.

Palatability can be applied collectively to a group of plant species or all plants of a single species or can be restricted to a single plant or individual parts of that plant. Individual plant characteristics affecting the relish shown for the plant by the grazing animal can be negative or positive. While defining palatability as being pleasant or acceptable to the taste, Provenza (1996b) concluded that the term was best understood as the interaction between taste and post-ingestive feedback. (See later section in this chapter on "Post-ingestive Feedback, Aversion.")

Great differences in palatability exist not only between plant species but also between subspecies and cultivars of the same plant species. Palatability can be selected for by the plant breeder; this is accomplished by relating observable plant characteristics to palatability and/or differentiation by actual grazing. Even selection for drought tolerance may improve palatability if associated with a longer period of green growth. When the utilization by sheep of 21 big sagebrush accessions was compared under winter grazing, great differences in palatability between accessions were found (Welch et al., 1987). The sheep tended to remove significant amounts (60–70%) of current growth from the more preferred big sagebrush accessions before removing even small amounts (15%) of less preferred accessions. Skiles (1984) noted that plant species with the C_3 photosynthesis pathway have in some cases been found to be more palatable than those with the C_4 pathway (i.e., mostly warm-season plants), but such comparisons may be confounded by differential plant growth stages.

   Preference refers to the selective response made by the animal to plant differences and is essentially behavioral. It presumes both initiative and opportunity to choose between alternatives and implies active selection of its diet by the animal. The breakdown of an animal's diet by plant species or forage classes such as grasses, forbs, and shrubs is not totally dependent on animal preference. Since climate and topography of a particular site dictates what species of plants can grow there, the ingestion of certain plants by animals over the area they graze is precluded if those plants are not found there.

   Selectivity by an animal may be influenced by the presence, either concurrently or previously, of one or more other animal species in the area, either by changing the short-term relative availability of the different plant species or differentially affecting the palatability of the remaining forage; even rodents and insects may affect food preference by the grazing animal (Wallace, 1984). In the short run, ungulates continuously "high-grade" the forage supply by eating the most preferred—and usually the best (i.e., most nutritious)—plant species and plant parts (Cooperrider and Bailey, 1984). As a result, the timing of when an animal utilizes the pasturage in relation to other animals of the same species is an important determinant of the quality of forage that will be available to it and thus determine its forage selection and probably nutritional intake. Providing there is substantial overlap in diets between animal species, a follower group of animals of a different species may be similarly disadvantaged.

   Great differences in preference are exhibited by various herbivore species but also by individuals within animal species, and this varies from place to place, from season to season, and even from year to year. Dietary selectivity appears especially sensitive to seasonal changes in forage plants, but body size and related nutritional-energetic demands may require grazing animals to shift to a less selective foraging strategy (Schwartz and Ellis, 1981). Grazing behavior is altered by the grazing animal to conform to the feed conditions available to it; these changes allow the animal to sustain its food intake under a wide range of rangeland or pasture conditions (Arnold and Dudzinski, 1978). Predicting a grazing animal's diet is complicated because selectivity may vary not only between animal species and even individual animals but also with stage of plant maturity, location, weather, and availability of plants (Holechek *et al.,* 1984; Malechek, 1984).

   Forage plant species that are commonly preferred by grazing and/or browsing animals and are selected first by choice are commonly labeled as **preferred species. Ice-cream plant** is a label commonly applied to exceptionally palatable plant species, these often physiologically over-utilized even under proper stocking rates when present in mixed stands. The designation of plant species as **decreasers, increasers,** and **invaders** in range condition classification partly implies decreasing levels of palatability, but exceptions result from differential tolerance or avoidance of grazing and other factors.

   Diet selection may be minimal when hungry cattle graze, but conscious selection of higher quality herbage increases as satiety slows the rate of herbage intake under liberal herbage allowance (Dougherty, 1991). Grazing animals make dietary

shifts as utilization increases, but the direction of the change depends on the botanical make-up of the particular plant community being grazed (Pieper *et al.,* 1959).
As grazing pressure increases, grazing animals will generally shift towards the less
palatable species. When herbaceous forage was abundant on brush-grass prairie
near Uvalde, TX, cattle maintained similar diets regardless of the upland range site
grazed (Launchbaugh *et al.,* 1990). However, as herbaceous forage became limited, cattle diets conformed more closely to the botanical composition available to
them.

Botanical composition of the diet alone is not an accurate index of palatability
in a typical grazing situation; it might be considered so only in a cafeteria setting
in which equal amounts of alternative forages are offered in a similar manner. In
fact, the composition of the diet may be quite unrelated to the proportions of various species or plant parts available to the animal. Only on grazing lands with relatively few species, and these of similar acceptability to the grazing animal, will
the diet closely reflect the relative availability of the species. Thus, the amount of
a forage plant species consumed in a particular grazing situation will depend on
both the relative abundance and palatability of the plant.

Difficulties are met in comparing the results of selectivity studies because of
the many and varied techniques that have been used to measure preference or
palatability. Six broad categories of measuring forage preferences by grazing animals as given by Skiles (1984) are (1) percent of the grazing time spent grazing
the species, (2) percent of individual plants of the species grazed, (3) animal presence or density (i.e., actually a preference measure of a site on which the species
predominates), (4) average percent utilization of the species (or utilization rank),
(5) cafeteria feeding (based on herbage removed or mouthfuls of forage taken),
and (6) relationship between botanical composition of the ingesta and occurrence
in the sward (i.e., selectivity ratio).

The **selectivity ratio** (or selectivity index) provides one means of balancing
availability and palatability as it is responded to by the grazing animal; it is the
proportion in the animal diet of any species, species group, or plant part relative
to its proportion in the available herbage. The selectivity ratio is determined as follows:

$$\text{Selectivity ratio} = \frac{\text{Proportion in the diet } (\%)}{\text{Proportion in the available herbage } (\%)}$$

Rosiere *et al.* (1975), while referring to the formula as a preference index, suggested the following categories for evaluational purposes: (1) 2.1 or greater, defi-
nite preference; (2) 1.4–2.0, some preference; (3) 0.7–1.3, same in diet as available; (4) 0.3–0.6, some avoidance; and (5) 0.2 or less, avoidance. Examples of the
selectivity ratio applied to specific situations are included in Table 9.1. Only when
forage is provided in amounts in excess of immediate ingestion needs will selec-

TABLE 9.1   Selectivity Ratios Exhibited by Several Animal Species for Specific Dietary Items[a]

Animal	Plant species or group	Selectivity ratio	Season	Location	Source
Cattle	Alkali sacaton	7.1	Yearlong	New Mexico	Herbel and Nelson
	Six-weeks grama	6.4			(1966b)
	Black grama	1.7			
	Russian thistle	1.3			
	Broom snakeweed	0.2			
	Burrograss	0.2			
Cattle	Hardinggrass	2.6	Summer	California	Van Dyne and Heady
	Purple needlegrass	2.2			(1965)
	Clovers	1.9			
	Bromes (annual)	1.0			
	Slender oats	0.6			
	Silver hairgrass	0.5			
Sheep	Hardinggrass	10.0	Summer	California	Van Dyne and Heady
	Purple needlegrass	3.1			(1965)
	Clovers	1.8			
	Bromes (annual)	0.9			
	Silver hairgrass	0.4			
	Slender oats	0.3			
Sheep	Grasses	1.5	Summer	Utah	Smith and Julander
	Forbs	1.7			(1953)
	Browse	0.5			
Angora goats	Grasses	0.7	Winter	Texas	Malechek and
	Forbs	—			Leinweber (1972)
	Browse	0.7			
Angora goats	Grasses	1.1	Summer	Texas	Malechek and
	Forbs	8.0			Leinweber (1972)
	Browse	0.7			
Mule deer	Grasses	0.3	Summer	Utah	Smith and Julander
	Forbs	3.3			(1953)
	Browse	0.4			
Mule deer	Grasses	5.9	Spring	Utah	Smith and Julander
	Forbs	2.4			(1953)
	Browse	0.3			
Pronghorn antelope	Grasses	0.2	Yearlong	Alberta	Mitchell and Smoliak
	Forbs	2.8			(1971)
	Browse	10.9			

[a]Adapted from Heady (1975).

tivity be clearly evidenced. Selectivity rankings will be different for different animal species and often for different groups of the same animal species based on past experience, degree of hunger, and other animal factors.

The **palatability factor,** formerly known as proper use factor, is another expression of palatability. It is defined as the percent utilization of a specific plant species when the plant species mixture as a whole is considered properly grazed, i.e., when percent utilization is set at what is physiologically optimal for one selected key forage plant species in the stand. At this reference point the percent utilization of any other species in the stand, generally either under or over physiological proper use for that plant species, is its palatability factor, i.e., the best management compromise on degree of utilization relative to all other plant species in the stand.

II. PALATABILITY (PLANT)

A. RELATIVE PALATABILITY

Both palatability and preference are always relative to the variety or alternatives offered for selection (Fig. 9.1). Palatability is subject to seasonal changes in the forage plant species and in associated plant species, and plant preference varies between animal species and even individual animals (Ivins, 1955). As much as 80% of the diet during a season may come from only 1% of the total forage available on rangelands (Arnold and Dudzinski, 1978). Animals often continue to graze on preferred species even after utilization has greatly decreased availability. If a choice is present, grazing animals may shift to forage species of otherwise lower palatability. If no alternative is offered, the animal may find the specific feed either acceptable or unacceptable; the palatability of singly offered feeds is extremely difficult to measure and interpret (Marten, 1978).

FIGURE 9.1   Palatability and preference are always relative to the variety or alternatives offered for selection; selectivity is limited on this brushy range south of San Antonio, TX, but experience gained early in life in an adverse habitat seems to help grazing animals adapt.

Some forage species are consumed well if they are provided as a sole choice but may be discriminated against or even rejected if offered with alternative forages. In a study at Kinsella, Alberta, six grasses and three legumes were seeded in sets of pure stands but grazed cafeteria style (Gesshe and Walton, 1981). The relative palatabilities were rated as follows (values greater than 1.0 indicate preference, less than 1.0 avoidance):

	Vegetative stage	Flowering stage	Seed set stage
Smooth brome	1.2	1.0	1.2
Creeping red fescue	1.1	0.6	0.0
Crested wheatgrass	0.8	0.2	0.0
Intermediate wheatgrass	1.2	0.2	0.0
Redtop	0.9	1.1	1.5
Russian wildrye	1.2	1.9	1.7
Alfalfa	1.5	1.5	1.3
Birdsfoot trefoil	0.9	1.8	1.9
Sainfoin	0.5	0.7	0.8

Although their relative palatability was different and this changed between different growth stages, it was concluded that all species would have been readily utilized by cattle had they been in pure stands without alternatives (Gesshe and Walton, 1981). In studies in northwestern Wyoming, elk and mule deer demonstrated distinct preference for certain sagebrush taxa over others (Wambolt, 1996). However, these big game species used even the least preferred taxon very heavily if the more palatable taxons were unavailable.

In southern Arizona, Lehmann lovegrass readily invades native grasslands and annually produces three to four times more green forage than the native grasses, but cattle greatly prefer the native grasses over Lehmann lovegrass. Because of the palatability differential, cattle selectively remove dormant native grass herbage from the standing crop before grazing green growth of Lehmann lovegrass, and the selective grazing pressure on the native grasses is particularly great under yearlong grazing (Cox *et al.,* 1990). Management recommendations were to fence pure Lehmann lovegrass areas from native grasslands mostly free of Lehmann lovegrass invasion and to graze separately. Since Lehmann lovegrass stays green longer into the late summer growing season, recommendations were to graze the Lehmann lovegrass units when nutrient levels, digestibility, and palatability peak in spring and summer. This allowed native range be rested until being grazed in fall and winter dormancy, thereby nearly equalizing the grazing pressure on Lehmann lovegrass and the native range grasses.

Where forage plant species of different palatabilities are combined into mixtures for reseeding, the relative palatability of the component species will largely determine the future of the stand (Gomm, 1969; Vallentine, 1989). Livestock concentrate on the most palatable species and graze more heavily. If these species are reduced in vigor or killed, the remaining less palatable species may be capable of replacing the most palatable species by spreading or reseeding. If this happens, the expense of including the most palatable species might well have been avoided in the original seeding mixture. Although selecting species of high palatability is generally desirable, species or cultivars of reduced palatability may be selected as a means of reducing plant utilization on critical sites.

The relative availability (i.e., percent botanical composition in a forage stand) appears to have an effect on relative palatability. In cattle grazing studies on foothill range in Colorado, the palatable Arizona fescue (*Festuca arizonica*) was more heavily grazed where it was relatively scarce in a dense stand of mountain muhly (*Muhlenbergia montana*) than where it was more abundant (Johnson, 1953). On native Utah desert ranges grazed by sheep in winter, a palatable species was found to be grazed very heavily if it constituted less than 10% of the total yield of vegetation even in pastures stocked at moderate rates (Hutchings and Stewart, 1953). When a desirable species was in short supply, it was sure to be severely grazed even under light stocking. Selective grazing of desirable species poorly represented on deteriorated sites may make their recovery particularly difficult.

In the desert range study above (Hutchings and Stewart, 1953), less palatable species were also taken to a greater degree where they made up a small proportion of the total vegetation than where they were abundant. This novelty attraction to grazing animals of scarce plants can be seen in occasional heavy grazing by cattle of widely scattered juniper trees on crested wheatgrass range. Such deviations from the expected norm may be incidental consumption associated with the animals sampling the environment as conditions change (Heitschmidt and Stuth 1991).

Where sacahuista (*Nolina texana*) was abundant on Edward Plateau ranges in Texas, it served as a non-preferred reserve forage through the year as the more palatable species were depleted (Ralphs *et al.,* 1986a). However, on sites were sacahuista was scarce, it was the preferred forage during fall and winter. Under very heavy stocking during winter, the former reserve status of sacahuista was shifted to pricklypear (*Opuntia* spp.).

Willms (1978) has arrived at a conclusion in opposition to the novelty theory. He has concluded that a low availability of both palatable and unpalatable plants can result in a low relative preference simply because the search effort required is not effective, and alternative species are utilized instead. He concluded that as the availability of the preferred forage increases so does use (relative preference) until a threshold is reached. From their work on the utilization of Utah desert range plants by sheep, Cook *et al.* (1962) found that the utilization of a palatable species increased as it increased in proportion to the other palatable species; however, as the quantity of an unpalatable species increased in proportion to the total palatable species, the utilization of it decreased. These findings should be balanced against more recent findings that grazing animals often show reluctance to initially sample novelty plants (i.e., plants with which they are not familiar).

The proximity of certain plant species may reduce the acceptability of nearby otherwise acceptable species. Yearling heifers in Colorado studies not only rejected sixweeks fescue (*Vulpia octoflora*) but also stayed away from blue grama in the infested areas, or they carefully grazed around sixweeks fescue when encountered

(Hyder and Bement, 1964). Greatest restriction was apparent in mid-season when the sixweeks fescue matured and turned brown. Tarweed (*Madia glomerata*) secretes a sticky exudate that emits a highly objectional odor; not only does this make the plant extremely unpalatable, but it also prevents the utilization of plants of otherwise palatable species near the tarweed plants (Bowns, 1989). The undesirability of cholla (*Opuntia* spp.), a tree-like cactus found in the Southwest, results from its impedance of livestock movements and rendering unavailable the desirable forage plants growing near it more than from competition with the desirable vegetation (Kunst *et al.,* 1988).

Although cattle avoid leafy spurge (*Euphorbia esula*), they often reduce their grazing on or even avoid other plants growing within leafy spurge infestations (Kirby and Lym, 1987; USDA, Agricultural Research Service, 1987), apparently because of the latex content of leafy spurge (Lym and Kirby, 1987). At a level of 10% leafy spurge canopy cover in Montana studies (Hein and Miller, 1992), cattle utilization of associated forage grasses was about 45%, but as leafy spurge canopy cover increased above 10%, forage utilization declined rapidly on associated species. From this it was concluded that to achieve 50% forage utilization by cattle of the associated native grasses, the level of leafy spurge canopy cover had to be kept at less than 10%. Based on research in North Dakota (Trammell and Butler, 1995), the reduction in use of spurge-infested sites by native ungulates (deer, bison, and elk) was attributed to both lower forage production on infested sites and simple avoidance of such sites.

B. PLANT PHYSICAL FACTORS

Physical or morphological factors that commonly improve or lower palatability include the following:

Improve palatability	*Lower palatability*
High succulence	Low succulence; high percent dry matter
High leaf:stem ratio	Low leaf:stem ratio
Seedstalks scarce	Seedstalks abundant
New growth/regrowth; long growing season	Old growth; dormancy
Leaves fine, tender; growth not rank	Leaves coarse, tough; growth rank
Twigs small, spaced	Twigs large-diameter, compacted, sharp-tipped
High accessibility	Low accessibility
Not thorny or spiny	Thorny, spine barriers, awns

The application of nitrogen fertilizer or certain herbicides, burning or mechanical defoliation, or natural or artificial sweeteners also affect palatability and have implications in distributing grazing; these are discussed elsewhere.

Succulence appears to be a major factor, if not the major plant characteristic, sought by grazing animals (Fig. 9.2) (Beale and Smith, 1970; Freer, 1981; Kothmann, 1984). The degree of selection exerted by grazing animals between the components of the standing crop increases with contrasts in plant maturity (Hodgson, 1986). When sheep in Utah were initially placed on crested wheatgrass during mid-

FIGURE 9.2   Succulence and general attractiveness to grazing animals is extended through the summer by subirrigation (dark area) on this combination hay and grazing meadow in Nebraska.

winter, 92% of the available herbage was mature stems and leaves and 8% was green vegetative growth (Gade and Provenza, 1986). During the experiment, the sheep consumed about 39% of the mature crested wheatgrass herbage but essentially all of the green growth.

Animals prefer young plants to older plants or the young and actively growing leaves and stems of older plants to their mature counterparts. A high preference for living over dead plant material is generally exhibited by grazing animals; an almost exclusive preference for green materials often persists down to very low proportions in the stand. When new green leaf material is not available, the order of preference is for older green leaves, green stems, dry leaves, and dry stems in that order (Wallace, 1984). Leaves that are tender and of low tensile strength are generally preferred. It has been suggested that ruminants select against forages with thick cell walls and, as a result, against forages of low digestibility and passage rate through the reticulo-rumen (Spalinger *et al.*, 1986). Soft-leaved, low-growing shrubs with small-diameter twigs are generally preferred over their opposites.

Because leaves are the readily grazed portions of most plants, Cook *et al.* (1948) recommended that leaves rather than the entire plant be used as the basis for utilization determination. They found that preferred parts of the plant, mostly leaves, will often be grazed 90–100% when the entire plant is utilized only 40–50%. Throughout their study on Utah mountainous summer range with sheep, leaves averaged 40% utilization compared to 9% for stems. Browse leaves produced 2.7 times as much forage as browse stems but comprised 4.5 times as much of the diet.

Forb leaves and stems were available in approximately equal amounts, but leaves contributed 7.5 times as much to the diet as did stems. Grass stems were 4 times as abundant as grass leaves but made up only 1.3 times as much of the diet. In dense, improved pastures, cattle seldom graze below the sward level where the pseudostems and more senescent and dead material are encountered (Arias *et al.*, 1990).

Sheep grazing grasslands in South Africa apparently balanced the unfavorable and favorable characteristics of the different plant species and then selected accordingly (O'Reagain, 1993). Preferred plant species were generally short and not stemmy and had leaves of low dry matter (i.e., succulent), low tensile strength, and high crude protein content. In contrast, avoided species tended to be tall and stemmy with a high grazing horizon and had leaves of high dry matter and tensile strength but low crude protein levels. Acceptability to sheep was therefore concluded to be determined by the interplay between plant structure and leaf quality attributes.

The relative use of browse varies widely with the season, the alternative vegetation, and the type of animal. Goats grazing shrublands in northwestern Mexico did not browse the different shrubs at the same phenological growth stages but had characteristic preferences for different plant parts and species at different periods of the year (Genin and Badan-Dangon, 1991). Browse is typically eaten in greatest quantity at the height of the dry season or in winter when green grasses and herbs are sparse, but browsing herbivores may make substantial use of the most palatable shrubs during their growing season even in the presence of palatable forbs and grasses. Less palatable shrubs are generally grazed most heavily in winter or other stress periods when herbaceous forage and more palatable browse are not available (Merrill, 1972). Also, because of seasonal and annual variation in the availability and palatability of the herbaceous understory, it is often not possible to predict from past observations the proportions of browse that will be eaten.

The ability of animals to select green material depends to some extent on the accessibility of the green material. Selective ability is enhanced when the green material is physically more separated from the dead material and depressed when the green material is growing up through an overburden of standing dead materials. When new growth becomes available in old grass bunches in the spring, it is often difficult for livestock to select out and they tend to leave the entire plant ungrazed. Defoliation by burning, chemical or mechanical treatments, or intensive grazing are methods of improving or prolonging palatability by delaying maturity, removal of old growth, stimulating regrowth, and effectively prolonging the green, growing period. Herbicides or other plant control treatments may be used to kill unpalatable plant species and increase availability of or improve accessibility to desirable species. However, removing the standing dormant herbage of Lehmann lovegrass by mowing did not improve the selective ability of grazing cattle grazing regrowth, and thus enhance diet nutrient levels, but presumably it did affect grazing efficiency (Rice *et al.*, 1990).

Cattle grazing crested wheatgrass in Utah exhibited a preference for moderate-sized plants over the smallest and the largest size plants (Fig. 9.3) (Norton and Johnson, 1983, 1986; Hacker *et al.,* 1988). Animals not only selected against the very large plants but also against those areas on which these robust plants predominated. Of the smallest size class of plants, 60% were not grazed at all; smaller plants also seemed more likely to be damaged by trampling than by defoliation. The largest size plants contributed relatively less to forage harvested and received some protection from grazing due to the physical impediment of standing stalks from old inflorescences; 20% of the large plants were not grazed at all. For those plants that did experience defoliation, severity of grazing was inversely related to plant size. Very small plants were grazed in a uniform manner similar to clipping; grazed plants of medium to large size basal areas experienced a loss of less than half their plant volume.

Any growth form or anatomical characteristic of plants that enhances avoidance of grazing will effectively reduce relative use. The barrier effect of thorns and spines of woody plants are readily recognized; the spines of pricklypear render an otherwise very succulent plant highly unattractive (unpalatable). Thick, stubby twigs, particularly when compacted and sharp-pointed at the ends, may also prove to be effective barriers against utilization of leaves within the protected area. Stiff, spiky leaves or feathery inflorescences or sticky foliage may have similar inhibiting effects. When the grazing horizon of vegetative tall fescue swards approached

FIGURE 9.3   Cattle grazing crested wheatgrass generally prefer medium-sized plants over either the smallest or the largest, most robust plants; robust plants on the left (shown at the Benmore Experimental Range in Utah) resulted from a thin stand and extra fertility caused from burning out brush windrow.

a plane represented by the tops of coarse **pseudostems,** the latter were found by Dougherty *et al.* (1992a) to serve as a barrier against defoliation by cows of any leaf blades below this level. These plant physical characteristics enhance the ability of otherwise palatable plants to avoid grazing; as a result, they also reduce the utilization and thus relative palatability of the protected foliage.

Prickles on leaves seem to be ineffective in deterring ungulate herbivore feeding, unless coupled with small leaf size (Cooper and Owen-Smith, 1986). Hairiness and stiff pubescence have often been attributed to reducing palatability over a glabrous condition, but the effects seem not to be universal. Lenssen *et al.* (1989) discovered that the presence of erect glandular hairs and their exudates on alfalfa did not negatively affect preference by sheep but did provide resistance against insects (e.g., alfalfa weevil and potato leafhopper).

Utilization of palatable plants can be sharply reduced when shielded by overstory plants as a result of low accessibility. Utilization will also be reduced when plants grow on steep slopes, rough terrain, or unstable soils of low accessibility to the grazing animal species.

C. ENVIRONMENTAL FACTORS

Environmental factors associated with weather and site and even management affect palatability of forage plants. These factors, in relation to their effect on palatability, are listed as follows:

Improve palatability	*Lower palatability*
Weather promotes current growth	Weather imposes dormancy
Prior growth normal	Prior growth rapid, coarse
Plant surface moist from dew, light rain	Plant surface dry
Shallow soil/unfavorable site	Deep soil/favorable site
High sunlight (except when maturation advanced)	Low sunlight (except when maturation delayed)
Plant surfaces clean	Plants dust, mud, or dung covered
Plants normal	Plants damaged by insects or disease
Initial herbicide response	Desiccated by herbicide

Although adequate soil moisture and temperature are necessary to promote current growth and associated high palatability, excessive growth rate early in the season may promote coarse, stemmy growth of both the primary forage crop and weeds. Slow growth, as long as growth remains active, is often associated with increased nutritive levels and palatability. For example, the first cutting of hay each season is generally coarser, less palatable, and less nutritious than subsequent cuttings. Although forage quantity may be higher in high rainfall years, forage quality in terms of nutritive value and steminess often suffers.

Moisture in the form of rain, dew, or light melting snow on semi-arid rangelands commonly increases palatability by "softening up" the matured forage (Mayland, 1986). Livestock may graze herbaceous forage plants less discriminately following rain showers (Springfield and Reynolds, 1951); this might result in more uniform utilization between plants and plant species but might also result in accelerated use of poisonous plants, such as tall larkspur. Weather conditions that force forage plants into rapid dormancy may preserve higher palatability, while those conditions that expedite rapid weathering following maturity will reduce palatability.

The forage of herbaceous plant species produced on localized unfavorable sites (shallow, gravelly, low fertility, low moisture) is commonly of greater palatability than when produced on adjoining but favorable sites. Sites on sagebrush-juniper range in central Utah found unfavorable from the standpoint of reduced biomass yield had more rocks and pebbles, lighter soil color (low organic matter), and shallow or exposed calcium carbonate horizon (Cook, 1959). However, seeded wheatgrass plants on favorable sites had about 50% more stems than leaves by weight, while plants on the poorer sites had about equal portions of leaves and stems. As a result of the higher palatability of the forage produced on the unfavorable sites, heavier grazing may cause these sites to deteriorate more rapidly when they occur intermixed with favorable sites. Tillers of native, warm-season grasses in Texas growing on shallow soils were also selected more often and were grazed more intensively than on deep sites (Hinant and Kothman, 1986).

The comparison and evaluation of palatability of herbage produced on shaded and adjoining unshaded sites is made difficult because of many interacting factors. In addition to amounts of radiant energy received, substantial differences are often found in soil moisture and temperature, botanical composition, plant growth stages, prevalence of disease and insects, etc. Woody plants growing in open clearcuts in direct sunlight in Alaska compared to those under dense conifer overstory were less light-limited, allocated carbon primarily to growth and maintenance, had higher levels of digestible protein, and were more palatable (Hanley *et al.,* 1987). The normal correlations between environmental factors and palatability may be negated by secondary plant compounds.

Snow cover decreases the availability of forage species from which animals can select; in the absence of alternative choices, the relative palatability of otherwise low palatability plants can be greatly enhanced. Deep snow can decrease animal mobility, further biasing palatability comparisons. When grazed by sheep during winter, 60% of the mature, dormant crested wheatgrass was compacted by snow and trampled under by the sheep (Gade and Provenza, 1986). Snow was found by Willms and Rode (1998) to affect the accessibility of associated grasses in fescue prairie and in shifting the grazing pressure towards rough fescue. The snow formed a dome over rough fescue plants that cattle targeted for cratering. Additionally, the wind redistributed the snow, commonly exposing the taller rough fescue plants and accumulating between them, thereby further covering the shorter plant species such as Parry oatgrass and Idaho fescue.

D. PLANT CHEMICAL FACTORS

The chemical composition of forage is a very important palatability factor, although the effective constituents are not well known (Willms, 1978), and their associations with palatability have sometimes only been situation-specific rather than universal (Marten, 1978). The following chemical factors generally affect the palatability of forage as indicated, but exceptions are probable:

Improve palatability	Lower palatability
High crude protein, sugars, fat, cellular contents	High fiber, lignin, silica; low magnesium, phosphorus
Low anti-palatability metabolites	High secondary plant metabolites
High digestibility	Low digestibility

The order of selection of plant parts by the grazing animal—leaves over stems and green herbage over mature herbage—is advantageous in that nutrient content and digestibility are higher in the herbage eaten than that offered but rejected. Enhanced levels in the ingesta of nitrogen, phosphorus, carotene (vitamin A precursor), and generally useful energy and lower levels of fiber, lignin, cell-wall materials, and silica result from this selectivity. This is readily evident when material taken from esophageal fistula samples is compared to samples clipped from the same pasture; this is generally true even when clipped samples have simulated ingested samples as nearly as possible. Nevertheless, the potential exists for anti-quality factors present in the forage to mask the normal correlations of favorable nutrient content with palatability.

Sugars and soluble carbohydrates are believed to be important in determining palatability, but some sugars give responses other than sweetness, and sweetness is not confined to sugars (Skiles, 1984). When cut for hay or used as pasturage, the higher sugar levels found in forage in the late afternoon compared to morning are strongly associated with increased palatability (Mayland and Shewmaker, 1999a). Less preferred plants commonly have lower protein, magnesium, phosphorus, and soluble carbohydrate concentrations or higher silica levels. Low forage palatability has commonly been attributed to specific mineral nutrient deficiencies in soils, but these effects may be only indirect in that they affect growth aspects of the plant rather than palatability directly through their deficiencies. Palatability is often seemingly unrelated to the proximate composition analysis of plants (Yabann et al., 1987).

Forage quality is commonly equated with crude nitrogen content, and high nitrogen levels are often correlated with high palatability but without necessarily demonstrating a cause and effect relationship. Crude protein content by itself is apparently not a strong determining factor in food selection when animal diets include plants of diverse growth forms. Other nutrients and repellent constituents produced in the plants have to be considered in order to account for effects of plant chemistry on diet selection (Genin and Badan-Dangon, 1991).

Unpalatable plants such as certain perennial legumes, broomsedge bluestem (Andropogon virginicus), and threeawn grasses (Aristida spp.) were readily eaten in Oklahoma studies when sprayed with natural sweeteners (in order of preference): blackstrap molasses, sorghum syrup, sugar, and corn syrup (Plice, 1952). Dilute solutions of synthetic sweeteners including saccharine rendered unpalatable

FIGURE 9.4   Alternating strands of curly mesquite (fine, low growth) and tobosagrass (coarse, rank growth) near Albuquerque, NM, provide a real challenge to the grazier in obtaining proper grazing utilization.

plants quite palatable as well. Saccharine aromatized with vinegar or anise oil was preferred to plain sugar and was equal to blackstrap molasses in causing low-quality forage to be relished by grazing animals.

Dry annual range forage on California foothill range of low palatability was completely utilized by weaned calves after spraying with cane molasses or a cane molasses-urea mixture, even when more desirable or nutritious fine-stemmed forage was abundant (Wagnon and Goss, 1961). Similar unsprayed forage was mostly left ungrazed. However, light dews diluted the molasses on the plant leaves, and light rain washed it off completely. Molasses spray has also been effective in enhancing the palatability of old growth of coarse perennial grasses such as tobosagrass (*Hilaria mutica*) (Fig. 9.4). The possibility of spraying a palatability enhancer on dense, low-palatable forage, while serving also as an animal supplement, may be a useful practice in limited situations.

E. SECONDARY COMPOUNDS AS ANTI-PALATABILITY FACTORS

Presumably any plant constituent that gives rise to or modifies the odor or taste response of the animal can change the relative palatability of the forage plant. Anti-palatability factors such as phenols, tannins, monoterpenes, and alkaloids are commonly associated with certain plant species. Although sometimes considered to be only waste products in plants, such toxic secondary compounds provide defense against insect and large animal herbivory. In addition to causing animal avoidance because of low palatability, such compounds produced in plants may serve as defensive mechanisms by causing aversive conditioning (the animal learns the plant makes it ill and so avoids it) or having extreme toxicity (kills outright and thus prevents further attack) (Laycock, 1978; Provenza *et al.*, 1988). (Because of the interrelationships between anti-palatability and anti-quality agents in plants, reference should also be made to "Anti-quality Agents in Plants" in Chapter 2.)

The effects of toxic secondary compounds range from individualistic to a composite of all secondary metabolites present (Bray *et al.*, 1987). Reichardt *et al.* (1987) have suggested that plant chemical defenses rely on the properties of individual compounds, that unpalatability is closely tied to exact molecular structure, and that no broad class of secondary metabolites contains individual compounds with uniform deterrent properties.

Levels of secondary metabolites in herbage are both environmentally as well as genetically induced. The plant carbon/nitrogen balance controls the expression of plant chemical defenses (Bryant *et al.*, 1987); nutrient stresses in woody plants tend to cause an increase in carbon-containing defenses such as phenolics but a decline in nitrogen-containing toxins and deterrents such as alkaloids and nitrates. Stresses resulting from inadequate light generally result in the opposite: carbon-based defenses decrease and, as the growth rate is reduced, nitrogen-containing deterrents increase. Bryant *et al.* (1983) concluded that woody plants adapted to growing on low-resource sites (with low nutrient levels) are often more dependent on evolved chemical defenses to counteract their inability to grow rapidly beyond the reach of most browing animals.

Many evergreen shrubs such as the oaks, sagebrushes, and junipers have high levels of tannins as well as other anti-palatability chemicals. Owen-Smith and Cooper (1987) found that condensed tannin level was the primary factor that negatively correlated with the palatability of woody plants (i.e., increasing levels with decreasing palatability) and that crude protein levels were only minimally related to palatability. Tannins bind up proteins and markedly reduce the availability and digestibility of protein in flowers and forb, tree, and shrub leaves and restrict microbial fermentation of structural carbohydrates. However, Robbins *et al.* (1987b) reported that ruminants that commonly consume tanniferous forages such as deer are less affected than are predominant grazers such as cattle and sheep. They concluded that deer defend against tannins by producing salivary proteins that bind tannins, thereby reducing the absorption of hydrolyzable tannins and the potential for tannin toxicity.

At low levels in the diet, oak was considered an important winter dietary constituent for cattle in Arizona (Ruyle *et al.*, 1986b), but as tannin levels reached high levels in green forage early in the spring, the reduction of protein and energy digestibility appeared to be the cause of decreased calf crops and calf weaning weights. The avoidance of tannin-containing plants and plant parts by herbivores may be advantageous if tannins decrease digestibility and lower animal fitness.

Tannin levels in blackbrush were found by Provenza and Malechek (1984) to be high in current-season twigs and leaves, lower in basal twigs, and least in old terminal twigs. However, high tannin levels were found to be associated with higher crude protein levels and increased digestibility by Provenza and Malechek (1984). When goats selected against the current-season twigs and leaves, presumably because of their high tannin concentrations and their location within the protected periphery of the hedged canopy, they were also selecting against a higher nutritional plane. Thus, goat nutrition was apparently affected more by the adverse effects tannins had on palatability than by the negative effects they had on digestibility.

Palatability, and possibly digestibility as well, is reduced by high monoterpenoid concentrations in plants such as the sagebrushes, junipers, and rabbitbrushes. Browsers such as deer, antelope, and goats are more willing to eat plants high in monoterpenoids than are grazers such as cattle, horses, and bison. Monoterpenoids at high levels have been shown in *in vitro* trials to inhibit rumen microorganisms. This has led to the presumption that high-monoterpenoid forage levels of 20–30% of the diet can be expected to decrease digestibility of the diet and, consequently, increase retention time of the ingesta in the rumen, thereby reducing dry matter intake (Schwartz *et al.*, 1980). Yet, ingesta *in vitro* digestibility in mule deer has apparently not been suppressed by high levels of sagebrush or juniper; and inocula from deer not previously exposed to high-monoterpenoid plants readily digest such browse (Pederson and Welch, 1982).

The solution to this dilemma may be found in the report by Cluff *et al.* (1982) that an 80% reduction in monoterpenoid level occurred in the rumen of mule deer. This loss of monoterpenoids was attributed to their release as gasses through chewing and rumination; browsers are known to chew their ingested food more thoroughly than do the grazers. The probability follows that this reduced the assimilation of the monoterpenoids and/or that they were excreted in the urine or metabolized in the liver after absorption.

Browsers, probably because of their narrow mouth parts, appear to select against plant parts high in total or selected monoterpenoids. From summer-fall sheep grazing trials on big sagebrush range, Yabann *et al.* (1987) found that sheep selected against high levels of monoterpenoids rather than for nutrient levels in the sagebrush plants. The sheep selected the older plants and plant parts, which are lower in monoterpenoids, rather than current season's growth. Rejected plants had 2.6 and 3.3 times the total monoterpenoid levels of accepted plants in summer and fall, respectively; corresponding levels in plant parts rejected vs. accepted were 5.9 and 3.3 times. In general, sheep refused to consume plant parts that contained more than 0.33% monoterpenoids.

It appears that individual monoterpenoids differ in their effect on forage palatability and digestibility. Total monoterpenoid content of 21 accessions of sagebrushes in a Utah study varied from 3.62–7.75% of dry matter (Welch *et al.*, 1983). However, when the accessions were made available to mule deer in a uniform shrub garden, total monoterpenoid content of the various accessions was not significantly related to deer preference. Wambolt *et al.* (1987) concluded that the monoterpenoid chemistry of sagebrush negatively affects digestibility but that the digestibility of an individual taxon was not directly related to total crude monoterpenoid content.

The alkaloid content of some species (i.e., reed canarygrass, lupines [*Lupinus* spp.], crotalaria [*Crotalaria* spp.], and grounsels (*Senecio* spp.]) is inversely related to palatability or has been shown to deter grazing (Ralphs and Olsen, 1987). However, some species high in total alkaloids (i.e., larkspurs and locoweeds) apparently remain moderately palatable to livestock; in such cases, other palatability factors appear to override the effects of the high alkaloid levels (Pfister *et al.,* 1988; Ralphs, 1987). (Refer to "Plant Poisoning of Livestock" in Chapter 2 for further discussion of the interrelationships between alkaloid levels, palatability, and toxicity, particularly in tall larkspur.)

Robbins *et al.* (1987) have hypothesized that soluble phenolics, which reduce herbage ingestion rates through their toxicity rather than inhibiting digestion, are more effective in defending plants against ruminants than are tannins. Levels of cyanogenic compounds reportedly have little influence on palatability.

III. PREFERENCE (ANIMAL)

Animal genetics sets rather broad but less than rigid neurological (instinct), morphological, and physiological constraints on foraging behavior of grazing animals, while learning fine tunes diet selection and harvesting ability to meet local necessity (Provenza and Balph, 1988). What grazing animals actually ingest is a complex phenomenon determined by the animal, by the plants offered to the animal, and by the environment in which the selection occurs (Marten, 1978). Walker (1994) concluded that diet selection is primarily a function of (1) post-ingestive consequences, (2) the animal's ability to discriminate between alternative plant species, and (3) the ability to physically select among alternative choices. Animal factors that influence food preference can be placed into four major categories: (1) initial use of the senses; (2) previous experience or adaptation of the animals; (3) variations between animal species, breeds, and individuals; and (4) post-ingestive feedback (Fig. 9.5).

A. USE OF THE SENSES

All five senses—taste, smell, touch, sight, and hearing—appear involved in forage preference behavior by grazing animals, but their interactions are complex, and no one sense seems to predominate in every situation (Arnold, 1966; Tribe, 1950). Least used is hearing, but even this has been implicated in the response by mast feeders to the sound of falling acorns striking the ground. The sense of touch is important in that grazing animals generally select against rough, harsh, or spiny material. The epidermis of the muzzles of most large herbivores are provided with sensitive nerve structure; discriminating against coarse materials results.

FIGURE 9.5   Animal factors that influence forage preferences of grazing animals fit into four categories: (1) initial use of the senses; (2) previous experience or adaptation of the animals; (3) variations between animal species, breeds, and individuals; and (4) post-ingestive feedback; showing sheep in west desert of Utah.

Both taste and smell can be prominent in selecting against, as well as for, certain plants. From working with tame black-tailed deer, Willms (1978) concluded that smell appeared to be involved in initial selection, while taste determined the duration and ingestion rate of a particular forage. Even after being positively sensed by receptors in the nose and throat, reappraisal by taste receptors of old or new flavors arising from the broken cells may be responsible for secondary rejection of some foods. Even after being prehended and severed from the plant, forage is commonly discarded by grazing animals before being swallowed.

In grazing studies with sheep, smell was reported by Arnold (1966) to be closely related to selection of plant parts or specific phenological stages of plants. Smell seems to be a mechanism by which sheep discriminate against high-alkaloid varieties of forage plants (Marten, 1978). Animals have also been noted to discriminate against high selenium levels in plants, presumably a result of smell. Cattle avoid forage growing vigorously near their own dung pats; smell is highly involved since harvesting and feeding the forage in another area, spraying the dung with a chemical that masks its smell, or anesthetizing the animal's olfactory organs readily restores acceptability of contaminated plants (Walton, 1983).

It is generally concluded that taste is the most important sense in determining forage selection and that the other senses serve primarily to supplement or modify its expression, and then primarily in the preliminary stages of discovery and selection. Taste is apparently the final discriminating sense used by the grazing animal. Sensitivity to and tolerance of different plant tastes are inherited but appear capable of being substantially shaped by the foraging environment (Provenza and Balph, 1988).

It has been hypothesized that large herbivores have minimal color discrimination; thus, sight was presumed relatively unimportant in distinguishing between color of forages (Walton, 1983). However, Kidunda *et al.* (1993) concluded from their work and that of others that cattle, sheep, and horses can distinguish colors and shapes and associate these cues with the locations of foods. In their study vision was apparently important in patch selection by cattle with color being one of the clues used in selecting a new patch. Arave *et al.* (1993) found that Holstein heifers were able to discriminate among all colors used in their trials, even though some colors were apparently more difficult to determine. They suggested the heifers were trichromatic, with color perception deficiency, much like partially color-blind humans.

It is apparent that sight aids grazing animals in spatial orientation in relation to vegetation (i.e., in the discovery of food areas), while other senses primarily determine actual plant selection (Marten, 1978). Bailey *et al.* (1996a) concluded that cattle may rely heavily on visual cues to select their grazing pathway once the foraging bout is initiated and they choose a feeding area. Sight also appears useful in distinguishing between widely differing plant growth forms (Tribe, 1950). In a spaced nursery study in which 4% of the plants were Spredor 2 alfalfa, sheep within hours recognized the presence of the highly relished alfalfa plants which were randomly spaced (Mayland and Shewmaker, 1999b). Several of the lead ewes were observed stetching their necks and scanning for other alfalfa plants; once sighted, the sheep walked and sometimes ran to eagerly graze the alfalfa plants they had located.

The chemical impairment of taste, smell, and touch and the physical obstruction of sight have been studied in relation to forage preferences of sheep on mountain summer range in Montana (Krueger *et al.,* 1974). Taste was the most influential sense in directing forage preference, although taste of some plants had no apparent influence on their palatability. Smell was of minor importance in selection. Touch and sight were related to such specific plant conditions as succulence and growth form. Sight was related to the selection of certain palatable plants such as sweetanise (*Osmorhiza occidentalis*). Even impairment of all four senses did not result in random selection of plants, but it did result in an increased preference for unpalatable plants and decreased preference for palatable ones.

Blindfolding steers in grazing studies at Miles City, MT, did not alter apparent selection behavior, sight being confirmed as playing primarily a spatial orientation role (Truscott and Currie, 1983). Blindfolded steers took fewer total bites per unit time, primarily resulting from the extra time required to find and select plants. The blindfolded animals actually had a higher bite rate per plant but took smaller bites.

It is also commonly noted that large herbivores are known to be effective grazers even in total darkness. Thus, doubts remain about whether color or other aspects of sight are substantially used to select between plants of similar form and shape at the feeding station level.

B. EXPERIENCE IN DIET SELECTION

Preference is the result of dynamic interplay between "nature and nurture" throughout the lifetime of the individual grazing animal. Natural selection provides each animal species and each individual within that species with a set of genetic instructions, but learning about herbivory from maternal observation, peer interaction, and post-ingestive consequences continually modifies food selection throughout the animal's lifetime (Provenza, 1995, 1996a). Foraging experiences early in life exert great influence on the grazing animal's later forage selection (Arnold and Dudzinski, 1978). Animals acquire preferences for familiar foods, first from their dams and second from their peers, but also by trial and error. The influence of their mothers' foraging habits apparently has a greater and more lasting effect on their lambs than the influence of other lambs (peers) (Thorhallsdottir et al., 1987). However, Mirza and Provenza (1990) concluded that lambs were influenced more by their mothers' dietary habits when 6 weeks of age than when 12 weeks of age.

Young animals have a remarkable ability to learn and to remember what they have learned; that ability is useful in molding dietary preferences (Balph and Balph, 1986). As a result of eating particular foods and not eating others, young animals acquire dietary habits. Learning or training in diet selection by lambs is optimal when they are 5–8 weeks of age (Thorhallsdottir et al., 1987; Squibb et al., 1987). At this age, lambs readily learn forage selection from their mothers, who are generally experienced foragers (Thorhallsdottir et al., 1987).

Previous experience of the grazing animal can render a particular forage species more or less acceptable (Squires, 1981; Walton, 1983), and it can result in preferences for or aversions to available plants and aid in acquiring the motor skills necessary to harvest and ingest the preferred forages (Provenza and Balph, 1988). Failure to utilize a newly offered forage species may result from lack of acceptance or lack of prehension skills or both. Experience is closely associated with palatability factors and is demonstrated by animals testing a forage and repeatedly returning to it (Willms, 1978).

Grazing animals are reluctant to eat novel forages and search for preferred foods in unfamiliar environments (Provenza, 1996a). In an unfamiliar environment, social interactions seem to have more influence than dietary preference on diet selection, while in familiar environments food preferences seem to be more influential in food selection (Scott et al., 1996). Ewes introduced into Montana pastures offering less preferred diets—i.e., alfalfa was missing from the forage stand—adjusted within a few days (Gluesing and Balph, 1980). Where alfalfa had been a major constituent of the previous pasture, the ewes in the new pasture initially spent much time searching for the nonexistent alfalfa.

Previous experience can influence both volume of intake and the efficiency with which grazing animals harvest forage; they show the greatest decrease in intake when they are faced with novel foods located in novel environments (Provenza, 1996a). Sheep reared to 3 years of age without any grazing experience were shown to obtain much less forage per hour of grazing than sheep reared on pasture (Arnold and Dudzinski, 1978). Lambs experienced in harvesting serviceberry (*Amelanchier alnifolia*) were found to be about 10% more efficient (amount ingested/unit of time) than inexperienced lambs (Flores *et al.,* 1987); the difference appeared to result from improved prehension skills of the experienced grazers.

Lambs without experience grazing crested wheatgrass lacked prehension skills and spent more time prehending and masticating forage, this resulting in reduced rate of forage intake (Flores *et al.,* 1989a). Grass plants in head were more difficult to ingest than when in vegetative stage, but these differences were mostly offset by prior experience by the lambs in foraging on grass.

Skills acquired by foraging one plant form may be largely specific to that plant form (Flores *et al.,* 1989b). Goats given repeated browsing experience on older growth blackbrush foraged more efficiently than naive goats, not only on older growth blackbrush but on shrub live oak as well (Reyes and Provenza, 1993). They concluded that animals moving from one range to another are likely to generalize foraging skills, providing the plants are similar in life form, resulting in a shorter period of adaptation and better foraging efficiency. Flores *et al.* (1989b) found that lambs seemed to learn the prehension skills (including head movements and orientations) appropriate for a particular plant form but noted that browsing experience by lambs was less useful in grazing grass than was grass grazing experience. They also found that grass-experienced lambs were more skilled at foraging on shrubs than were those experienced with shrubs at foraging on grass.

The influence of prior conditioning on the foraging habits of the grazing animal must be recognized. Researchers generally provide a preconditioning period prior to the collection of grazing data to allow animals to adjust to new pasture conditions. Short adjustment periods of 1 to 2 weeks tend to remove much of the effects of prior experience. Longer periods may be required to modify long-standing grazing habits. Sheep have adjusted to grazing leafy spurge, without any prior experience, within a 1- to 3-week adjustment period; grazing pressure exerted on the leafy spurge subsequently resulted in biological control without harmful effects to the sheep (Lacey *et al.,* 1984).

Bartmann and Carpenter (1982) observed that mule deer raised on pinyon-juniper range had different foraging habits than mule deer raised on sagebrush-steppe range; they also observed that tame mule deer maintained in pens subsequently had foraging habits different from those maintained on native range. The deer seemed to determine palatable and unpalatable species at first contact, but evidently needed time to adapt and develop new feeding habits.

From a study comparing adult tame-experienced, tame-naive, and wild mule deer, Olson-Rutz and Urness (1987) attributed behavioral differences between the groups to the effect of experience with new environments rather than experience with foraging per se; they concluded that being tamed had little influence on deer foraging behavior. Spalinger *et al.* (1997) concluded that forage selection by white-tailed deer is largely an innate behavior and that hand-reared deer are essentially the foraging equivalents of maternal-reared or wild animals. Following 30 days of preconditioning on similar pelleted and freshly cut browse, naive fawns in their study selected a diet similar to that of the experienced fawns and adults.

C. VARIATIONS BETWEEN INDIVIDUAL HERBIVORES

There are marked differences between individual members of the same species in their preferences for plant species (Walton, 1983; Marten, 1978). This is demonstrated by the use of esophageal fistula collections, wherein the botanical composition of the diet varies substantially between individual animals of the same species on the same day as well as for the same individual on different days. In contrast, a high degree of gregariousness may reduce differences between individuals by inducing animals to forage in similar areas, at similar times, and at similar foraging rates (Provenza and Balph, 1988).

The physiological condition of the animal alters forage preferences; animals with voracious appetites may discriminate less than those which may be finicky eaters with lower demands (Van Soest, 1982). Hunger lowers the threshold for forage acceptance, and this has particular significance when grazing animals are stressed. When livestock become very hungry, such as on drives or under severe drought conditions, they often feed indiscriminately. As pointed out previously, this can result in increased consumption of poisonous plants.

Individual animal variation in forage selection may be inherited as well as acquired; this appears to be true of both deer and cattle (Willms, 1978). Black-tailed deer of a nervous disposition have been noted to feed on taller vegetation, presumably to maintain an alert position. Particularly on windy days when their senses of hearing and sight are impaired by the noise of the wind and movement of shrubs, these more agitated individuals forego grazing on the palatable understory to feed only from the tops of the shrubs. Environmental factors acting directly on the grazing animal may also influence dietary selection. For example, during special winter hunts on Rocky Mountain foothill range in Alberta (Morgantini and Hudson, 1985), elk reduced their consumption of rough fescue from 87–34%, while browsing increased proportionately; following the end of the hunting season and associated animal stress, the undisturbed elk returned to their normal diets.

Comparisons of the diets of Herefords, Angus × Hereford crosses, and Charolais × Hereford crosses grazing sandhills range at the Eastern Colorado Range Station were made by Walker *et al.* (1981). It was concluded that variations in the botanical composition of the diets because of age and breed effects were probably inconsequential. Also, no apparent differences were found between Hereford and Santa Gertrudis cows on New Mexico range in the quantity of coarse plants consumed; differences in species preference between the two breeds were small (Herbel and Nelson, 1966b). Diet botanical composition in comparisons made on rangelands in the Chihuahuan desert were similar between Barzona, Brangus, and Beef-master, all apparently having similar adaptations to semi-desert environments (Becerra *et al.,* 1998). In this same area some dietary differences were found between Hereford and Brangus cattle, possible explanations being (1) the greater distances traveled by the Brangus, or (2) simply breed differences in plant species selection.

Although Heady (1964) suggested that forage preference may be related to pregnancy, fatness, and lactation, Arnold and Dudzinski (1978) concluded there was insufficient evidence to substantiate that food preference changes when the physiological status of the animal raises or lowers nutrient requirements. Both cattle and sheep on seeded pastures selected the same diet whether pregnant, in lactation, or dry (Arnold, 1966). In Montana studies, calves selected diets of higher quality in the spring than did mature steers, but no diet differences were noted during the September-November period (Grings *et al.,* 1995). Possible explanations for the higher grazing selectivity of calves in the spring were (1) higher forage quality in the spring allowed selective grazing, (2) low forage intake of calves still suckling may have allowed more time for selection, or (3) spring diets of calves were mostly exploratory in nature.

D. POST-INGESTIVE FEEDBACK, AVERSIONS

The concept that forage selection by the grazing animal is best understood as the functional relationship between taste and post-ingestive feedback (either negative or positive) has recently gained wide acceptance (Fig. 9.6). This concept suggests that animal preference for foods (and thus their palatability) results from involuntary processes that cause animals to associate the taste of a forage with its positive or negative post-ingestive feedback and form either conditioned preference or a conditioned aversion for that forage (Provenza, 1996b). If a forage causes **malaise** (nausea or unpleasant feelings of physical discomfort), animals acquire for it a conditioned taste **aversion** (mild to strong, depending on the amount ingested). On the other hand, if a forage causes pleasant feelings of satiety (the sensation of being full), animals acquire for it a conditioned taste preference (mild or strong). These cause-and-effect relationships apparently operate both between meals and within a meal; aversions are involuntary and are not the result of conscious decisions by the animal (Provenza, 1996a).

"Animals use their senses (smell and sight) to seek foods that cause positive feedback (i.e., nutritional well-being) and avoid foods that cause negative feedback (i.e., nutrient deficiencies and toxicosis)" (Howery *et al.,* 1998a). Thus, what makes a forage taste good or bad (thus, sought or avoided) is not taste per se, but rather results from nutritional benefits or deficits received from ingestion of that particular forage. The post-ingestive feedback from consuming the forage is sensed by animals and linked with the forage's taste.

Food aversions are considered beneficial to the extent the animal is induced to seek a more varied diet. This presumably results in greater sampling of foods and a more balanced diet, reduces ingestion of toxic food, optimizes foraging and rumina-

FIGURE 9.6   The concept that forage selection by the grazing animal is best understood as the functional relationship between taste and post-ingestive feedback (either negative or positive) has recently gained widespread acceptance.

tion times, and maintains a diverse microflora in the rumen. Aversions apparently diminish preference and cause animals to eat a variety of foods, but the malaise, rather than the benefits, is the cause of the varied diets (Provenza, 1995, 1996a). "Malaise may occur when the forage ingested contains excess nutrients (e.g., energy, protein, minerals), excess toxins (e.g., tannins, alkaloids), or inadequate nutrients" (Howery *et al.,* 1998a). These authors concluded that what constitutes excess and deficits in nutrients depends on the animal's age, size, type of digestive system, and physiological condition. They also reported that an animal can become confused and "blame" a novel food for negative feedback even when it is not responsible for the malaise.

There is poor correlation between plant toxins and either decreased or increased palatability, according to Molyneux and Ralphs (1992). Grazing animals generally do avoid plants that cause toxicosis, inhibit digestion, and cause malnutrition (Howery *et al.,* 1998a). Consumption followed by negative post-ingestive feedback (i.e., trial and error) is probably the most efficient and permanent way for grazing animals to learn about harmful forages. By this mechanism, grazing animals develop a conditioned aversion to harmful plants after first ingesting some threshold amount of it (Kronberg *et al.,* 1993). Lambs that observed their mothers during exposure to harmful foods ate significantly less than when they were exposed alone, indicating that they also learned to avoid the harmful from observing their mother avoid it (Thorhallsdottir *et al.,* 1990).

In contrast to long-standing tradition, grazing animals consuming locoweed do not develop physiological or psychological addiction to the toxic alkaloid swainsonine it contains but typically eat loco while searching for green plant growth (Ralphs *et al.*, 1990, 1991a; Pfister *et al.*, 1999). Sheep show no initial preference to locoweed, but individual animals may acquire a preference for it from repeated use of the plants. Most ewes preferred other feed if a choice was offered, even when severely intoxicated animals were returned to locoweed-infested rangeland; any seeking out of locoweed may have resulted from the unavailability of more succulent and palatable plants. Affected ("locoed") animals showed reduced ability to prehend and ingest adequate amounts of forage under range conditions, and the common involuntary seizures or trembling of the head were related to reduction in biting rate; these acquired traits are probably major contributors to the emaciated condition typical of animals previously poisoned by locoweed. Cattle showed no evidence of addiction to Columbia milkvetch (*Astragalus miser* ex Hook. var. *serotinus*) in British Columbia studies (Majak *et al.*, 1996), but poison hemlock (*Conium maculatum*) was concluded to be addictive (Pfister *et al.*, 1999).

Occasionally, both the post-ingestive aversion process and the capability of the large herbivore to bind, metabolize, or detoxify the toxicant breaks down, and the animals are killed by consuming poisonous plants. Possible reasons why these safeguards fail to perform and animals are deleteriously affected include (Howery *et al.*, 1998a):

1. Over-ingestion of certain toxins may not stimulate the emetic system, which is apparently required to produce a conditioned taste aversion (example, bloat on alfalfa).
2. When aversive feedback is delayed following consumption of plant toxins, poisoning may result before the effect takes place (example, various alkaloids).
3. Animals may experience difficulties in differentiating nutritious from toxic forages in unfamiliar environments because all forages may be new.
4. A change in environment may alter animal susceptibility to a toxin, including lack of alternative forage, differential relative palatability, social solicitation, changes in weather, or changes in animal physiology.
5. Subtle molecular changes in plant toxins increase their toxicity.
6. Animals have difficulty in associating aversion to a specific plant if two or more plant species contain the same toxicant, or when two or more compounds in different plants interact to cause toxicity.

Not all natural aversions are beneficial. The aversion of cattle to robust needlegrass (*Stipa robusta*) is more related to its endophyte infection than to the grass itself. However, livestock avoidance of the grass is a greater economic impact than the infrequent narcotic effects caused by the grass (Jones and Ralphs, 1999). Endophyte-free populations have proven to be much more palatable and are being increased to replace existing endophyte-infected stands. A similar problem exists with tall fescue, with a solution following similar avenues.

Diet training has potential uses and may eventually become a useful technique for making livestock production more efficient and rangelands more productive for livestock (Provenza and Balph, 1988). Diet training has been described by Provenza and Balph (1987) as the manipulation of livestock dietary habits to meet grazing management objectives. They proposed possible objectives of the application of behavioral concepts through the training of animals as follows: (1) to recognize, readily accept, and efficiently harvest forages they will encounter later in life; (2) to avoid undesirable or poisonous plants; and (3) to better accept plants that have low palatability but acceptable nutrient levels. Both sheep and calves develop and can be taught aversion to foodstuffs that have unpleasant gastronomic consequences (Balph and Balph, 1986).

Conditioned food aversion is a training tool with potential to prevent livestock from ingesting poisonous plants that are palatable, abundant, and cause persistent poisoning problems (Ralphs, 1992). This process functions by accelerating normal trial-and-error consequences by expediting the negative consequences associated with the consumption of poisonous plants (Walker, 1995). "The process is fairly simple: animals are offered a food, they smell it, eat it, and then are given an emetic (usually lithium chloride, LiCl) to induce nausea (malaise)." An association is made between the taste of the food and the induced illness, and the animal will subsequently refuse that food (Ralphs, 1992). However, a controlled condition is generally required where a single feed is ingested and associated with the induced illness. Animals that are already familiar with the plant will require reinforcement with repeated aversion treatment.

Social facilitation has been the major factor inhibiting the retention of conditioned aversions to poisonous plants. This problem has been observed with cattle on larkspur (Ralphs and Olsen, 1990; Lane et al., 1990; Ralphs, 1992), with cattle on locoweed (Ralphs et al., 1994b), and with sheep on locoweed (Pfister and Price, 1996). It is readily apparent that mixing averted animals with non-averted animals in the presence of the targeted poisonous plant must be avoided for conditioned aversion to work.

If developed for widespread practical application, training foraging animals by aversive conditioning, in addition to reducing losses from poisonous plants, might also be useful against browsing on tree seedlings in forest plantations or on shrub seedlings during establishment of rangelands (Olsen and Ralphs, 1986). Of note is the fact that many ranchers in locoweed-infested areas of New Mexico watch their cattle closely and remove those they see grazing locoweed; this reduces the incidence of social facilitation influencing other cows to eat locoweed while preventing the progression of locoweed intoxication (Ralphs et al., 1993, 1994b).

E. NUTRITIONAL WISDOM AND OPTIMAL FORAGING

The interrelated theories that grazing animals exert "nutritional wisdom" and "optimal foraging" remain controversial, and the evidence supporting each appears only circumstantial at best. The existence of generalized **nutritional wis-**

dom by grazing animals in the forage they consume has been rejected by many scientists (Marten, 1978; Wallace, 1984; Walton, 1983). However, a variable relationship does exist between the forages ruminants prefer and their nutritional value, and this has been explained by Provenza (1995) as follows: "Ruminants possess a degree of nutritional wisdom in the sense that they generally select foods that meet nutritional needs and avoid foods that cause toxicosis. There is little reason to believe that nutritional wisdom occurs because animals can directly taste or smell either nutrients or toxins in food." Rather, neurally mediated interactions between the senses (i.e., taste and smell) and the viscera apparently enable ruminants to sense the consequences or nutritious food ingestion, thereby subtly and profoundly affecting food selection and intake.

However, examples of fallibility in nutritional wisdom are numerous and suggest that the theory of nutritional wisdom is limited, at best. For example, the palatability of the different perennial weed species found in pasture are not directly associated with nutritive value at the time of grazing (Marten *et al.,* 1987). In spite of sub-maintenance crude protein contents of grasses in winter, undisturbed elk have preferred grazing to browsing (Morgantini and Hudson, 1985). While mule deer demonstrate a distinct order of preference for big sagebrush varieties, high palatability to mule deer does not correlate positively with high digestibility (Wambolt *et al.,* 1987). In fact, varieties with the highest levels of crude protein in Utah studies were the least palatable to mule deer (Welch and McArthur, 1979). While the new growth of blackbrush twigs contain more nitrogen and is more digestible than old growth, goats prefer the old growth to the current-season growth because the higher condensed tannin levels in the latter cause negative post-ingestive feedback (Howery *et al.,* 1998a).

The theory of **optimal foraging** is based on the assumption that grazing animals will optimize some objective function in their grazing, often assuming that the animal will maximize its energy intake per unit of time or effort expended. Wallace (1984) found erroneous the concept that both forage preference and forage intake are exercises carried out by a grazing animal to satisfy a given energy requirement. He concluded that forage consumption by grazing animals does not result directly from the grazing animal's specific energy demands, particularly when associated with forage diets of lower digestibility. Rather than consume more forage when digestibility is low, ruminants tend to consume substantially less forage when digestibility is noticeably reduced. Arnold and Dudzinski (1978) cautioned that care must be taken in judging if there is any nutritional motivation in selective grazing and suggested instead that selection by grazing animals results from inborn reactions in combination with environmental effects. Provenza (1990) noted the optimal foraging theory fails to explain how animals select nutritious diets and avoid toxins, and neither this theory nor the nutritional wisdom theory explains dietary blunders made by livestock, such as why some animals seemingly often ignore nutritious morsels but occasionally kill themselves by eating poisonous plants. Senft *et al.* (1987) *assumed* that foraging beef cattle attempt to maximize net energy (nutrient) capture when they presented their findings and conclusions. Willms (1978) has concluded that where chemical composition is corre-lated to palatability, it must be through related secondary compounds or physical properties. Noting that cattle seek out nutrient-rich patches to graze, Pinchak *et al.* (1991) concluded: "Though we do not propose that cattle have euphagic [nutri-tional] wisdom, we interpret these results to indicate that not only is absolute for-age standing crop an important characteristic for site preference but also the chem-ical and growth form composition of that standing crop as these affect caloric density of the diet." Bailey and Sims (1998) concluded that, "Cattle can remem-ber the *quality (or palatability)* of forage found at different spatial locations." Thus, the assumed desire by grazing animals to maximize energy or other nutrient intake may well be an incidental, but nevertheless highly beneficial, secondary re-sponse to a complex of animal preference, positive post-ingestive feedback (sati-ety), or other sensual achievement.

F. DETERMINING BOTANICAL COMPOSITION OF DIETS

Reliable information on the botanical composition of the grazing animal's diet is required for making many grazing management decisions. It is required in (1) selecting the kind or mix of animal species to best utilize a specific forage resource, (2) allocating forage to different kinds of herbivores, (3) predicting the manipula-tive effects of grazing on vegetation, (4) identifying key forage species on which to base management, and (5) applying intensive range improvements such as plant

FIGURE 9.7   Comparing direct observation (quantified by using bite counts or feeding minutes) with esophageal fistula collection for determining the botanical composition of the grazing animal's diet. (Utah State University photo.)

control and reseeding. However, the numerous tabulations by species list of diet composition data are not satisfactory for explaining observed diet differences between animal species (Hanley, 1982b).

Procedures for estimating the botanical composition of diets of grazing animals have been grouped and evaluated under five categories: (1) direct observation, (2) utilization (degree of use) studies, (3) fistula methods, (4) stomach contents, and (5) fecal analysis (Holechek *et al.*, 1982a, 1984; Theuer *et al.*, 1976). Direct observation (quantified by using bite counts or feeding minutes) requires minimal time and equipment inputs, but accuracy and precision are a problem, particularly with wild animals (Fig. 9.7). Utilization studies are generally unsuitable when plants are actively growing and more than one herbivore is using the area. The fistula methods are the most accurate measures of botanical composition of the diet, the esophageal fistula being more accurate than the rumen fistula. However, fistula methods are difficult to use with wild animals, are costly and time consuming, and are primarily research tools. Stomach analysis can provide useful data but generally involves sacrificing the study animals. Fecal analysis shows good promise for management application providing accuracy can be maintained at high levels.

IV. PATCH GRAZING: PROBLEM OR BENEFIT

A. THE PROBLEM

The problem of achieving uniform grazing occurs not only at the macrosite level discussed previously but also at the microsite to individual plant level. Even when substantial grazing has been achieved over all major geographical, terrain, and vegetation types in a grazing unit, this expedited by the use of the tools of grazing distribution, the problem commonly continues at the microlevel. **Patch grazing** (synonymous with **spot grazing**) is the close and often repeated grazing of small patches (or even individual plants) while adjacent but similar patches (or individual plants) of the same species are left ungrazed or only lightly grazed (Fig. 9.8).

Patch grazing is an inefficient utilization of forage since a significant portion of the major forage plants are not grazed or are grazed only after they have deteriorated from weathering, while others are damaged by repeated close grazing. The objectives of proper stocking rates can be thwarted if patchy grazing results in alternating patches of forage plant undergrazing and forage waste and patches of forage plant overgrazing and deterioration; this may require downward adjustment from initial inventories to achieve proper stocking rates. Romo (1994) concluded that as much as 55–65% of the forage in crested wheatgrass pastures may be located in the ungrazed patches; unused residual forage present in the wolf plants comprised an average of about 41% of the total standing crop of crested wheatgrass in this Canadian study (Romo *et al.*, 1997). Accessing this forage can substantially increase the real grazing capacity of this bunchgrass species.

The loss of quantity and quality of grazing capacity in the ungrazed patches and

FIGURE 9.8   Patch grazing of blue panicgrass (*Panicum antidotale*) in south Texas perpetuating adjoining areas of inefficient forage utilization and repeated, severe, heavy grazing. (Holt Machinery Co. photo.)

the corresponding ever increasing grazing pressures in the grazed patches leading to vegetation and habitat retrogression can have long-term consequences. Fuls (1992a) regarded widespread patch-selective grazing and subsequent patch overgrazing as the main cause for the continued retrogression of semi-arid and arid rangelands worldwide. He concluded that patch grazing not only reduces rangeland productivity but also enhances desertification and adversely affects rangeland stability, with low and erratic rainfall aggravating the situation. An additional conclusion by Fuls (1992b) from his work on climatic climax grasslands in South Africa was that their continued retrogression was not caused principally by widespread overstocking as generally believed, but rather was the result of the expansion and increase of small heavily grazed patches into large overgrazed and subsequently degraded areas.

B. DIFFERENTIAL DEFOLIATION

Patch grazing is particularly prevalent on sites of high plant productivity and with species of mediocre palatability; bunchgrasses that produce abundant seedstalks, particularly when grazing occurs only during advanced growth stages, are prone to the problem. Ungrazed patches of perennial forage plants in one year tend to be perpetuated as ungrazed patches the next year. However, under summer season-long grazing in the mixed prairie with steers, the development of ungrazed patches depended mostly on not being grazed at the beginning of the grazing season (Ring *et al.*, 1985).

Adjoining patches of grazed and ungrazed vegetation frequently develop, not because the animal cannot search the total area, but because of factors affecting preference for individual plants or clusters of plants over others (Kothmann, 1984) or initially from mere random selection. Herbivores contribute directly to the formation of grazed and ungrazed patches by (1) localized defoliation, (2) altering competitive interactions between plants, (3) plant trampling, (4) urine and dung deposition, and (5) transformation and redistribution of nutrients from animal waste.

Post-ingestive feedback presumably is a major determinant of whether or not a large herbivore will return to graze a previously grazed patch. When regrowth of plants is slow or absent, there is a negative feedback between present and future grazing, and grazed patches are unlikely to be grazed again (Hobbs, 1996); when regrowth is rapid, the feedback is positive and grazing enhances the conditions for future grazing. It then follows, according to this author, that positive feedback will increase landscape heterogeneity by heightening the contrast between grazed and ungrazed patches while negative feedback reduces or does not change landscape heterogeneity because grazing intensity tends to become more uniform across patches.

Patch grazing often occurs when forage supply exceeds livestock demand and grazing animals have the opportunity to graze selectivity and is more characteristic of season-long stocking. Regrazing of patches is most common under continuous grazing and is apparently reduced but not eliminated under rotation grazing (Willms and Beachemin, 1991). High-density grazing systems may largely overcome moderate patch grazing, but when high grazing pressure is achieved only by increasing the stocking rate, an unstable site situation frequently results (Kothmann, 1984). High density of steers under intensive-early stocking on native mixed-grass prairie in central Kansas effectively controlled patch grazing (Ring *et al.*, 1985).

Adjoining undergrazed and overgrazed patches on rough fescue grasslands in Alberta have been studied by Willms *et al.* (1988) and concluded to be caused by grazing rather than by pre-existing site factors. Under high grazing pressure, the patches were relatively unstable. Although the patch boundaries under low grazing pressure fluctuated somewhat from year to year, the patches were relatively stable between consecutive years and over the 4-year study period. Soil organic matter and depth of the A_h horizon were greater on the undergrazed patches, but NO_3N, NH_4, and available phosphorus were lower than on the overgrazed patches. On the overgrazed patches seral plant species were more prominent, and the climax grasses were 50% shorter and produced 35% less forage. While patches in this study persisted for 4 or more years, the use of grazed patches has been shown to decay over time, particularly in the presence of fire as a natural event or treatment (Hobbs et al., 1991).

Distinct patches of ungrazed vegetation surrounded by areas of grazed vegetation are even found on shortgrass range under both light and moderate grazing and to a lesser extent under heavy grazing (Klipple and Costello, 1960). Understanding the variability and patchiness within native stands of blue grama in eastern Colorado was further complicated by the finding that such can also result from a combination of plant genetic and edaphic factors (McGinnies et al., 1988); the mosaic of microenvironments and soil properties was evident and seemed to be correlated with a corresponding mosaic of genotypes and phenotypes of blue grama occupying the diverse ecological niches.

On buffalo-blue grama range in eastern Colorado that was homogeneous prior to introducing grazing, patterns that tended toward heterogeneity gradually developed as grazing occurred under both light as well as heavy grazing intensity (Bonham and Remington, 1991). On tall fescue pasture in Argentina cattle grazing created and maintained a mosaic of areas with different degrees of utilization (Cid and Brizuela, 1998); as grazing pressures increased, the percentage of the land surface occupied by highly utilized patches increased as did the degree of utilization of the patches. However, on native range near Woodward, OK, cattle in homogeneous habitats exhibited low plant community selectivity and did not develop preferences for patches (Bailey, 1995); in heterogeneous areas, cattle preferred and enhanced the grazed patches.

Cattle grazing patterns on Lehmann lovegrass has also resulted in uneven utilization with patchy areas of heavy use. After one year of grazing in an Arizona study (Ruyle et al., 1986a), over 75% of the total defoliation events the following year occurred in previously grazed patches. Regardless of stocking rate there were higher percentages of nondefoliated tillers in the previously ungrazed patches. In a similar study under year-long grazing of Lehmann lovegrass range in Arizona (Nascimento et al., 1987), utilization in the grazed patches averaged 95% under all stocking rates, and 60–70% of the overall area was not grazed at all. Stocking rates at moderate and heavy levels did not alleviate patch grazing. Even very heavy stocking resulted in use of the previously ungrazed patches only when forage in the grazed patches was exhausted and cattle were sharply declining in condition. Ungrazed patches had greater total and green biomass than grazed patches but a much lower percent of green biomass of the total vegetation.

Cattle appear to concentrate grazing on grazed patches of Lehmann lovegrass because of nutrient density and absence of old standing dead biomass which interferes with selection for green material even though green biomass is greater on ungrazed areas (Nascimento et al., 1989). At light stocking rates, new patch development and maintenance are minimal since animal nutrient needs may be satisfied by patch growth of Lehmann lovegrass (Ruyle and Rice, 1996). At very heavy stocking rates, cattle were forced into previously ungrazed areas where residual stems reduced grazing efficiencies and nutrient benefit was limited.

Naive cattle initially employed different grazing strategies than native cattle, but as they gained experience these differences decreased (Ruyle and Rice, 1996). Native cattle grazed almost exclusively in the previously grazed patches but directed their efforts to the new tillers at the base of Lehmann lovegrass plants either from the top, working their muzzles through the residual stems, or from the side with their head held at an angle. They grasped and pulled the tillers loose with incisors and dental pads, never employing tongue sweeps or even tongue anchorage in the mouth to prevent slippage. Naive cattle initially began biting previously ungrazed plants and used tongue-sweeping bites to graze the upper (old herbage) horizon of the sward. After gaining experience, the naive cattle first shifted to bites at the base of ungrazed plants and then to bites taken from the base of plants in grazed patches similar to the native cattle.

In seeded pastures continuously grazed and dominated by tall fescue in Argentina (Cid and Brizuela, 1998), cattle grazing created and maintained a mosaic of areas with different degrees of utilization over a broad range of stocking rates. As stocking rates increased, the percentage of land surface occupied by highly utilized patches increased as did the average degree of utilization. Stocking density principally affected the height of lightly used patches; the heavily utilized patches remained fairly constant at a height of about 2 in. Patch boundaries fluctuated throughout the year at all cattle stocking rates (Cid and Brizuela, 1998) but were more stable at the lower stocking rates.

Bringing down differential grazing from the patch to the individual plant level, it is commonly observed that some individual plants in a population of a given species are utilized much less intensively than others. **Wolf plant** is a term that refers to individual plants of a species generally considered palatable that remain ungrazed or minimally grazed when exposed to grazing. It is mostly a matter of chance that these individual plants have access to more soil resources or receive less utilization and develop into large, robust plants with a great deal of coarse foliage and/or seedstalks which subsequently deter utilization (Caldwell, 1984). Grazing animals tend to graze heavily on the same herbaceous perennial plants they have utilized the year before. Conversely, plants of the same species not grazed one year are less likely to be grazed the next year because of robust growth and the accumulation of remnant flower stalks and old growth (i.e., **thatch**).

In grasses used in range and pasture seedings, patchy grazing is less common with Russian wildrye, smooth brome, and intermediate wheatgrass, but species such as tall fescue, tall wheatgrass, weeping lovegrass, Lehmann lovegrass, blue panicgrass, and crested wheatgrass are prone to develop wolf plants. Because crested wheatgrass has been widely used for reseeding on western North American ranges, its efficiency of utilization by grazing animals has been widely studied (Fig. 9.9). The degree of use of crested wheatgrass plants growing side by side in central Utah has been observed to range from 15–80% for cattle and 5–90% for sheep (Cook, 1966a). In related studies, 25% of the crested wheatgrass plants were commonly ungrazed during typical spring grazing with cattle (Norton *et al.,* 1983).

Light grazing of crested wheatgrass has encouraged the development of wolf

FIGURE 9.9   Patch grazing of crested wheatgrass, resulting in severely overgrazed plants and ungrazed "wolf plants" growing side by side, is a serious problem that requires special grazing or other treatments.

plants more than moderate use (Currie and Smith, 1970; Sharp, 1970), and year-long rest followed by growing season grazing has stimulated wolf plant develop-ment even more. Such large bunchgrasses become progressively less attractive to cattle and eventually die out in the center; reversal of this trend with overall mod-erate levels of utilization becomes difficult (Norton *et al.,* 1983). The development of wolf plants, even under intensive grazing, is accelerated by delaying the start of grazing too long into the spring; yearling heifers appear less effective than mature cows in preventing seedstalk development because of being more selective and discriminating in their foraging habits (Hedrick *et al.,* 1969).

Cattle (also sheep and bison) have an aversion to large bunchgrass plants that carry an abundance of straw over winter, and the degree of defoliation declines as the plant size and the amount of straw increases. Cattle grazing crested wheatgrass were found by Ganskopp *et al.* (1993) to graze less plants and remove less mate-rial from plants as the density of cured stems increased. Under controlled condi-tions, plants with one, two, or three cured stems were, respectively, 8, 20, and 32% less likely to be grazed than control tussocks with no stems; also, 35, 39, and 60% less plant material, respectively, was removed by the cattle.

Even though presumably collecting additional snow during winter, the presence of wolf plants of crested wheatgrass were found by Romo (1994) to have no con-sistent effect on the water relations or on the growth of subordinate crested wheat-grass plants (i.e., those located nearby and receiving the heaviest grazing pressure).

Because they played no apparent beneficial role in the water status and productivity of subordinate plants, it was recommended that management techniques should be implemented to exploit the forage produced by the wolf plants.

C. SITE TREATMENT

Although 80% utilization annually in the spring removed wolf plants of crested wheatgrass, stands on central Utah foothill range deteriorated to about 30% of their potential after 7 years of such grazing treatment (Cook, 1966a). In contrast, wolf plants were effectively controlled by heavy spring grazing every third or fourth year while permitting recovery in following years from the occasional intensive, early spring grazing. During the special treatment years, tripling the normal stocking density and grazing for a few weeks in early spring after the new growth has reached about 3 in. long have been suggested. When set aside for intensive grazing during the boot stage in late spring of the previous year, that pasture would thus be prepared to sustain the main grazing pressure early the following spring (Norton et al., 1983). In contrast to the above emphasis on spring grazing, Ganskopp et al. (1992) concluded that heavy grazing with an objective of obtaining utilization of wolf plants was more successful after all forage had cured and cattle grazed less selectively.

Romo et al. (1997) concluded that abundant wolf plants developing in pastures of crested wheatgrass were indicators of both poor grazing management and potential economic loss. They concluded that management that encourages more uniform and complete use of the unused residual forage in wolf plants and ungrazed patches was economically beneficial in most situations. The order of decreasing profitability and sustainability of management practices in their study were

1. Periodic haying, water development, and cross-fencing
2. Periodic haying without cross-fencing or development of water along with unchanged grazing management
3. Burning without water development or new fences
4. No improvement in management

Swathing and baling the crested wheatgrass every 5 years was economically feasible when unused residual forage in wolf plants averaged about 200 kg per ha (178 lb per acre) (Romo et al., 1997).

Mechanical removal of the seedstalks has also resulted in substantial grazing of the wolf plants of crested wheatgrass (Norton and Johnson, 1986). In Nevada, clipping, crushing, or dragging effectively removed the coarse old growth from established crested wheatgrass plants (Artz and Hackett, 1971). Burning completely controlled wolf plants for two growing seasons; fertilizing with 40 lb nitrogen per acre reduced wolf plants, while heavy rates (160 lb nitrogen per acre) controlled the wolf plants completely. Winter grazing of dormant crested wheatgrass stands also offers the potential to control wolf plants (Young and Evans, 1984).

Weeping lovegrass is highly productive, but palatability drops off rapidly as it reaches maturity, and ungrazed patches become decadent and of little value for livestock grazing. Old, decadent stands in Texas (Cotter *et al.,* 1983) have been made productive and more effectively utilized by the following practices: (1) use high enough stock density to graze off the high-quality forage in a few days, (2) mow or shred to remove old growth and allow tillering from clumps, (3) burn to remove old growth (often failed to get into densely packed bases), (4) graze and trample by animals to break up dead centers, and (5) hay to prevent maturing and maintain active growth. Patchy grazing continued under cattle grazing even with high stocking rates. After a 14-day grazing period in the spring following winter burning in a related study (Klett *et al.,* 1971), cattle had grazed 52% of the new growth of weeping lovegrass in the burned area but only 8% in the unburned.

Fire was found effective in tallgrass prairie in removing grazing-induced patches (Hobbs *et al.,* 1991). Both fire and fall grazing of bluebunch wheatgrass near Kamloops, B.C., removed the barrier effect created by the old litter and improved spring palatability for both deer and cattle (Willms *et al.,* 1980b). The unburned dead stubble of small plants was a less effective barrier than that of large plants. Patch grazing can also be caused by the barrier effect of clumps or thickets of unpalatable species; these effects can be removed through plant control or even top removal of the offending species. Dense brush or the presence of cholla plants in southwestern U.S., or pricklypear in the Great Plains, or silverberry (*Elaeagnus commutata*) in the central Alberta parklands (Bailey, 1970) are examples. In eastern Colorado about 29% of the blue grama produced per acre was rendered unavailable because it grew within the protection of pricklypear clumps (Bement, 1968). Removal of the pricklypear, surprisingly enough, did not increase forage production but did make the previously protected plants available for grazing.

Treatments applied during plant dormancy to manipulate mulch accumulation on switchgrass and mixed grass sites that occurred under 10 years of the Conservation Reserve Program included: (1) October shredded and left on site, (2) October shredded and removed from the site, (3) October grazed, (4) March grazed, and (5) late April prescribed burn (Schacht *et al.,* 1998). In this Nebraska study, the grazing treatments were applied under high intensity to remove 80% of the standing crop within 6 days. All treatments were similarly effective in removing the standing dead material and enhancing utilization of forage by grazing animals during the following growing season, and no treatment increased the first-year growth rate of marked tillers or total yield compared to the control. High-intensity grazing, in addition to being as effective as the mowing and burning treatments in removing accumulated plant residue, also provided a means of utilizing the stockpiled herbage.

Patchy grazing can cause serious damage in new grass seedings even under very light stocking rates. Haying while guarding against severely close mowing is one means of harvesting early growth for beneficial use while preventing development of harmful grazing patterns before full plant establishment. Grazing subirrigated sandhills meadows in Nebraska during consecutive growing seasons has not been recommended because of deterioration of range conditions, resulting in part from severe patch grazing (Clanton and Burzlaff, 1966). However, grazing every third year or even every second year with cattle in conjunction with haying during the intervening years resulted in favorable response of both vegetation and grazing animals. Formerly only very early spring growth or aftermath was commonly grazed in the sandhills meadows, and growing season grazing was avoided.

Repellents have proven effective against the grazing of ornamental plants such as shrubs and tree seedlings; however, this approach to reduce grazing on localized overgrazed sites seems to hold little potential. In South Dakota a commercially available deer and elk repellent with an active ingredient of putrescent whole egg solids was applied to subirrigated range sites (Engle and Schimmel, 1984). However, there was no significant difference in cow chip numbers between treatment and control areas.

D. ANIMAL WASTE

Animal waste (dung) can be the cause of patch grazing. Based on their research in grass-clover swards, Gibb *et al.* (1989) concluded that when grazing pressure is high, the taller ungrazed patches arise mainly from avoidance of fecal deposits. Where grazing pressure is low, in addition to avoidance of fouled patches, areas of the sward infrequently grazed mature and are subsequently rejected by the animals in preference for younger growth patches.

Cattle select against or even reject herbage directly contaminated by their own feces and, of even greater total consequence, that growing around the dung pats (Fig. 9.10). The rejection of herbage growing around dung pats varies from 5–12 times the area covered by the dung pat itself (Wilkins and Garwood, 1986). This problem is minimal on arid and semi-arid rangelands, except in animal concentration areas at water, permanent supplemental feeding areas, and under shade, but can be severe on meadows and highly productive pasture. Under intensive pasture systems in temperate areas, where a grazing season may be 180 days, up to 45% of the area may be covered with herbage rejected by cattle by the end of the grazing season (Arnold and Dudzinski, 1978).

In one grazing study with cattle on coastal bermudagrass in Florida, halo spots left ungrazed and associated with fecal contamination comprised 35% of the area after 42 days but reached 70% after 98 days of continuous grazing (Brown *et al.,* 1961). On intensively managed pastures in another study (Simpson and Stobbs, 1981), 2–3% of the surface area had feces dropped on it by the end of the grazing season, but this caused incomplete consumption over 15–29% of the area.

In a rotational grazing study with dairy heifers and steers on brome and alfalfa-brome pasture, Marten and Donker (1964) found that 93% of the patches rejected by the livestock was caused by dung deposited 3–4 weeks earlier during the previous grazing period. The effect of dung often lasts from two to several months (Wilkins and Garwood, 1986). Thus, rotational grazing will not affect the degree of wastage from feces contamination because only by chance are feces dropped on sites where such already lie; however, any effect on grazing patterns is essential-

FIGURE 9.10   Cattle reject herbage growing around their dung pats, often resulting in wastage of 20% or more of the total usable herbage on highly productive pasture such as this irrigated grass–alfalfa stand in Nebraska.

ly eliminated by overwintering. The fact that horses and cattle do not graze close to their own droppings for several weeks does reduce the risk of infectious larvae that hatch from eggs in the feces and move onto nearby herbage (Arnold and Dudzinski, 1978).

The reduction in herbage intake and animal production associated with the fouling from dung appears greatest at intermediate grazing pressures but minimal at either very low or very high grazing pressure (Wilkins and Garwood, 1986). With very low grazing intensity, herbage intake would not be affected because there is plenty of unfouled forage. With very high grazing intensity, herbage intake would be minimally affected because intake of all animals is depressed by low herbage availability, and this overrides the tendency to reject herbage on affected spots.

Smell appears to be important in causing the rejection, initially from the fresh feces and then from decomposition products, but smell may later be succeeded by secondary effects including maturation and associated reduction of natural palatability (Wilkins and Garwood, 1986). Marten and Donker (1964) concluded that the rejection was not related to protein/carbon imbalance or sugar concentration in the affected forage plants. However, the application of nitrogen fertilizer greatly increased consumption of the affected plants, while phosphorus fertilizer had no significant effect.

Sheep show only minimal aversion to herbage around their own feces, except around sheep camps where it is often very dense (Arnold and Dudzinski, 1978).

Whereas cattle reject forage growing in proximity to cattle dung but will graze close to sheep dung, sheep will accept forage growing close to dung from either species (Forbes and Hodgson, 1985). Thus, grazing sheep in common with cattle or alternating animal species in a rotation grazing system offer major solutions to the problem on high-production pastures of cattle aversion to forage affected by their own dung. The mixed grazing also appears helpful in reducing parasite burdens in sheep by the parasites being consumed and destroyed by cattle as well as the dilution effect of replacing some sheep with cattle (Nolan and Connolly, 1977).

Dispersing cattle dung pats by harrowing have provided only moderate to minimal benefits. Mowing may help alleviate the secondary effects of maturation, and alternate grazing and harvest cutting may be economically justifiable (Wilkins and Garwood, 1986). Spraying with natural or artificial sweeteners has effectively prevented or overcome the problem (Plice, 1952).

Neither cow nor sheep urine appears to cause more than initial aversion by the same or alternate species, but subsequent preferential grazing of urine-patch areas often follows because of enhanced green growth (Marten and Donker, 1964; Wilkins and Garwood, 1986). Day and Detling (1990) found that bison urine deposition resulted in vegetation patches that were preferentially grazed by bison. When natural urine patches covered only 2% of the surface area, higher feeding preference for the urine patches led to these patches contributing 7–14% of the aboveground biomass and nitrogen, respectively, during the June through August grazing period. The urine patches in both little bluestem and Kentucky bluegrass had higher aboveground biomass, higher root mass, lower root-shoot ratios, and higher nitrogen concentration in herbage (effect on Kentucky bluegrass greater than with little bluestem); initiated growth earlier in the season; and delayed senescence.

E. THE BENEFITS

Grazing animals need to be able to select significant quantities of immature forage in order to meet high production requirements (Taylor, 1986a). Allowing animals to select the most nutritious part of the available forage increases nutrient intake and improves animal production. The most obvious foraging strategy of grazing animals, particularly ruminants, is the selection of green material above dried, mature, or weathered forage. The green forage will be high in protein and phosphorus, very high in carotene, and a good source of digestible energy.

At low and moderate stocking densities on mixed swards dominated by tall fescue in Argentina (Cid and Brizuela, 1998), cattle were found to be nutritionally benefited by patch grazing. Live biomass of the highly utilized patches had higher nitrogen concentration and density than that of the lightly utilized patches. Such findings can lead to the concept that properly stocked ranges are optimally grazed in small to large patches rather than uniformly throughout in order to give livestock opportunity to be selective in choosing their diets (Kansas State University, 1995).

However, the objective of increasing evenness of utilization (and thus higher forage harvesting efficiency often enabling higher stocking levels) conflicts with the objective of maintaining high-quality diets for herbivores. Elevating grazing pressures to the extent that grazing animals are restricted to forage they would not otherwise select is a major cause for reduced stocker gains, reproductive performance, and weaning weights. The desire to improve rangeland condition, resulting from more even utilization, must be balanced with the need to improve livestock performance. The key to maintaining optimal foraging conditions would seem to be maintaining choices between a variety of patches on the landscape and/ or choices of plants and plant parts within patches (Rittenhouse and Bailey, 1996).

Rittenhouse and Bailey (1996) have concluded: "The value of nutritional heterogeneity on rangelands is to dampen the amplitude of temporal changes in food availability and quality and to extend the period that high-quality nutrients are available. Whenever we do things that create more homogeneous conditions (e.g., uniform grazing), either within or among patches, we increase the risk of creating a nutritional bottleneck for the animals." This assumes that management will be unable otherwise to make optimal quality and quantity of forage continually available to the grazing animals. This condition more appropriately describes the grazing environments of rangelands characterized by great heterogeneity in plant species composition, phenology and maturation, and relative palatability than improved pastures managed under intensive grazing.

Potential secondary advantages of patch grazing listed by Willms *et al.* (1998) include: (1) providing a more diverse habitat, (2) contributing to the survival of climax species in the undergrowth, and (3) carrying over emergency forage into drought years. Herbage in mostly ungrazed patches as well as that available from lesser selected species in the vegetation mix provides emergency forage and may allow animals to survive in subsisting situations. Spatial heterogeneity created by non-uniform grazing may be valuable in enabling recolonization of desirable plant species. Noy-Meir (1996) suggested recolonization will be faster if many small "plant refuges" remain in the grazing parts of the landscape rather than in only remote or inaccessible areas. Laycock *et al.* (1996) even equated patchiness with habitat diversity.

10

KIND AND MIX OF
GRAZING ANIMALS

I. CHOICE OF ANIMAL SPECIES

The traditional domestic livestock produced on grazing lands in the U.S. have been cattle, sheep, goats, and sometimes horses. The principal choices have been European or Brahman (Zebu) cattle or their crosses, mutton and wool breeds or crosses for sheep, and Spanish or Angora goats. The use of the genetic background provided by the Africander, the Longhorn, and continental, multi-purpose breeds has added further diversity and potential in beef cattle production. Grazing lands can play a prominent role in the cow-calf, growing, and finishing phases of beef cattle production, but high-quality grazing lands are also ideal for dairy cattle. Ewe-lamb and doe-kid enterprises can make extensive use of grazing resources. Grazing should normally also play a prominent role in the diets of pleasure and working horses during the plant growing season.

Interest is substantial and growing relative to commercially producing native big game animals including bison, deer, elk, pronghorn antelope, and moose. These less traditional kinds of grazing animals have recently received more attention by North American ranchers due to the opportunity to enhance ranch income and utilize certain types of forage plant species and land types more effectively. A prime example is the American bison (buffalo), which is naturally adapted to native grasslands and is readily produced under domestication. Although its cross with domestic cattle, referred to as the "Beefalo," has been researched, particularly in Canada, it has met with production problems, including low fertility. The introduced reindeer, as well as the native caribou and muskox, have proved well adapted for meat production in arctic and subarctic environments of North America.

Selecting the animal enterprises and associated grazing management practices that will most efficiently use the available grazing lands is a major step in ranch planning. The financial feasibility of each alternative animal enterprise should be given major consideration. However, priority considerations should also be given to the quality and quantity of grazing lands, the availability of harvested forages and concentrates, available financing and markets, ranch facilities, prevalence of predation and specific animal diseases, kind and amount of care and attention required, and the experience and preference of management and ownership. Choosing an alternative or additional class or species of grazing animal often requires different investment requirements and new production and marketing techniques.

The kind or mix of grazing animals to be grazed will generally be the first grazing management decision required of the grazier. This decision will often determine season of use and grazing methods and will also affect stocking rate (Lewis and Volesky, 1988). The effect of choosing the optimum kind and mixture of grazing animals may be as great or greater than the effect of the grazing system employed (Lewis, 1986; Lewis and Volesky, 1988). Although some grazing lands may be suitable only to certain species of grazing animals and not to others, most grazing lands have some flexibility relative to the choice of grazing animal species; grazing a mix of animal species may also be desirable. Opportunities exist to adapt the kind and class of grazing animals to the available forage resource, or to modify, manipulate, and even convert the forage resource to better fit the selected grazing animals, or as a combination of both approaches. The urgency of matching forage sources with animal requirements has been emphasized in Chapter 3.

II. KIND OF ANIMAL

A. BODY SIZE AND RUMEN CAPACITY

The greater efficiency in energy digestion of fibrous material and the lower basal metabolism requirement per unit of body size favors the larger ruminant species (Hanley, 1982b; Willms, 1978). Reticulo-rumen volume relative to the size of the animal materially affects the type of forage each ruminant species is efficient in processing. High reticulo-rumen volume to body size (or body volume) (0.25–0.35 in cattle and bison and 0.25 in sheep) is considered an adaptation to exploiting thick cell-walled, high-cellulose diets (i.e., graminoids). High reticulo-rumen volume is associated with higher consumption rate, more time spent in rumination, and longer retention time.

By contrast, low reticulo-rumen volume to body size (0.1 in deer and pronghorn antelope) is considered an adaptation to exploiting high cell-soluble and thin but lignified cell-walled diets (e.g., browse and often forbs); this is associated with lower consumption rates, more selective grazing, less time spent in rumination, and faster ingesta passage (Hanley, 1982b). Pronghorn appear to engage in frequent, relatively short alternating bouts of foraging and resting (including ruminating), while cattle engage in longer alternating bouts of feeding and resting (Ellis and Travis, 1975); the shorter foraging/resting cycles of the pronghorn have been attributed to its relatively small rumen and its selection and ingestion of a higher quality diet that does not require a long rumen retention time.

Animal species with very small rumen capacity relative to weight have been referred to as "concentrate selectors" based on their relying more on fruits, forb leaves, and tree and shrub foliage to provide higher levels of crude protein, phosphorus, and energy (Hanley, 1982b; Hofmann, 1988). Grasses are noted for having lower fiber content in their stems than trees and shrubs but higher cell wall and lignin fractions in their leaves, particularly in advanced growth stages. However, selective foraging on individual plants or specific plant parts may not always be advantageous to the animal in nutritional terms because of additional time and energy costs of foraging (Hodgson, 1986; Hanley, 1982b). Animals which are actively discriminating between alternative forage stand components are likely to take smaller bites at lower bite rates than those which are not. Thus, any advantages in terms of the nutrient concentration in ingested herbage may be offset by the disadvantages of a reduction in the rate of herbage ingestion.

Demment and Van Soest (1983) concluded similarly that small ruminants with a high metabolic requirement:digestive tract or gut capacity (MR:GC) ratio must eat food composed largely of rapidly digestible fractions. In the small ruminants, increased intake cannot compensate for depression in digestibility associated with ingesta of slower digestibility or greater indigestible fraction. Small ruminants must compensate for their proportionally higher MR:GC ratio by higher rates of energy production per unit volume of the rumen. In ruminant species of larger body size, a reduced MR:GC ratio can be compensated for by slower rates of passage of ingesta, longer retention time resulting in greater digestibility of the slowly digesting fraction of the forage ingested, and relatively large reticulo-rumen capacity.

Demment and Van Soest (1983) suggested that the upper limits of ruminant body size are influenced by the ability of ruminant herbivores to maintain adequate intake of low-quality forages. Within ruminants, feeding time decreases with larger size while rumination time increases. If rumination is considered feeding activity, then small and large ruminant feeding times may be about equal. Larger ruminant species are apt to be eating more easily available, higher-fiber diets that are rapidly harvested but slowly ruminated.

Hanley (1982b) concluded the following hypothesis: (1) large ruminant species are more limited by time (combining both ingestion and rumination) than are smaller animals, spend less time per nutrient unit consumed, and thus cannot be as selective; and (2) small body size is advantageous to the ruminant if forage quantity is limiting, while large body size is advantageous if forage quality is limiting. However, small ruminants being selective does not necessarily imply the use of a reduced number of forage plant species; examples are deer and antelope which typically include small amounts of many plant species in their diets. Bailey *et al.* (1996a) concluded that when vegetative conditions restrict maximum bite size, larger herbivores will invariably be more affected than smaller animals.

Bailey *et al.* (1996a) related body size and degree of nutrient selectivity by herbivores to distribution of grazing as follows: "Spatial distribution of high-quality forage will more strongly influence the spatial patterns of foraging by selective feeders, and forage availability as measured by short-term intake rate will determine patterns of foraging by bulk feeders." The result is that when high-quality forages are limited, smaller herbivores are predicted to feed in areas where they can maximize diet quality, while larger herbivores are predicted to feed in areas where they maximize intake rates.

B. HERBIVORE DIETS BY FORAGE CLASS

Substantial differences in forage preference are exhibited among animal species groups and even among related species. Generalized diet and terrain/site preferences useful in comparing the adaptation of the major North American grazing animal species are provided in Table 10.1. The terrain/site adaptation has been summarized from Chapter 7; the dietary adaptation has been summarized from the selected examples of botanical composition of diets of the principal North American ungulate herbivores in Table 10.2. The dietary information in Table 10.2 has been categorized by grasses (including grass-like plants), forbs, and browse and has been taken from published references that have, in most cases, included the four seasons of the year. However, a breakdown of diets by forage classes is probably never totally dependent on animal preference since availability by forage class will also affect grazing animal diets.

In addition to body size and reticulo-rumen capacity, inherited differences in animal anatomy (teeth, lips, and mouth structure), animal prehensive and grazing ability, animal agility, and digestive systems may account for some of the differences in forage preferences among kinds of animals. Mouth size directly affects the degree of selectivity that is mechanically possible for the foraging animal to exhibit; ruminants with small mouth parts such as goats, deer, and pronghorn, in contrast with cattle, horses, and elk, can more effectively utilize many shrubs while selecting against woody material (Holechek, 1984). Narrow mouth parts and thus the ability to be highly selective may permit such animals to be efficient browsers and survive on greatly deteriorated brushlands (Schwartz and Ellis, 1981). Animal species with broad mouth parts may be inhibited from feeding on spinescent species when coupled with especially small leaf size (Cooper and Owen-Smith, 1986).

TABLE 10.1   Generalized Dietary and Terrain/Site Preference Characteristics of the Principal
North American Ungulate Herbivores

Species	Diet preferences[a]	Terrain/site preferences
Bison	Grazer: mostly grasses, minimal forbs and browse	Grasslands and semi-deserts; prefers level to rolling terrain
Horse	Grazer: mostly grasses, minor forbs and browse	Widely adapted to plains and semideserts
Cattle	Grazer: mostly grasses, some seasonal use of forbs and browse	Prefers level to rolling land; capable of but often unwilling to graze steep or rocky areas
Elk	Grazer to intermediate feeder: also considerable forbs and browse; highly versatile	Prefers meadows, parks, bottoms, and lower slopes; grazing often concentrated
Domestic sheep	Intermediate feeder: high use of forbs, but also use large volume of grass and browse	Better adapted to steep lands and rough terrain than cattle
Domestic goat	Browser to intermediate feeder: high forb use, but can utilize large amounts of browse and grass; highly versatile	Adapted to a wide variety of terrain and vegetation types
White-tailed deer	Browser: typically high browse; forbs and even graminoids locally important	Adapted to a wide variety of terrain and vegetation types
Mule deer	Browser: *spring* and *summer*— high forb use, also green grass and browse; *fall* and *winter*— mostly browse, forbs and grasses beginning late winter	Range widely and use large amount of range inaccessible or unused by domestic livestock, particularly cattle
Pronghorn antelope	Browser: wide assortment of of browse and forbs, minimal grasses	Found principally in dry plains and deserts; prefer rolling terrain
Moose	Browser: mostly browse use, including bark; some use of forbs and aquatics in spring and summer	Frequent mountain stream beds, adjoining slopes, and timbered areas

[a]Arranged in order of dietary preference (graminoids to browse).

Ungulate herbivores have been divided into three groups by Van Soest (1982):
(1) bulk and roughage eaters, (2) concentrate selectors, and (3) intermediate feeders (Fig. 10.1). These three groups are mostly equivalent to those in a corresponding classification: (1) grazers, (2) browsers, and (3) intermediate feeders, respectively (Holechek, 1984). The former classification has been described by
Hofmann (1988) as an evolutionary system of morphophysiological feeding types;
the latter attempts to more directly describe herbivore diets by forage class—
graminoids (grasses and grasslike plants) and shrubs and forbs—and has been
mostly followed here and in subsequent discussions.

TABLE 10.2 Botanical Composition of Diets of the Principal North American Ungulate Herbivores

Herbivore	Spring Gr	Spring Fo	Spring Br	Summer Gr	Summer Fo	Summer Br	Fall Gr	Fall Fo	Fall Br	Winter Gr	Winter Fo	Winter Br	Annual Gr	Annual Fo	Annual Br	State, vegetation type, number of years, method[b]	Reference
GRAZERS																	
Cattle	35	40	25	71	23	6	50	41	9	50	27	23	51	33	16	Jornada, NM; desert range; 3 years; DO	Herbel and Nelson (1966b)
Cattle	58	42	0	78	14	8	76	12	12	85	11	4	75	19	6	Sinton, TX; mixed vegetation; 1 year; UM	Drawe and Box (1968)
Cattle	91	9	1	89	11	0	82	1	17	88	4	8	87	6	7	Edwards Plateau, TX; oak–juniper savannah; 1 year; BC	McMahan (1964)
Cattle							59	37	4	64	33	3	57	26	16	CO: shortgrass prairie; 5 years; EF	Kautz and Van Dyne (1978)
Cattle	47	41	11	77	20	3				58	33	9	62	33	5	Ft. Stanton, NM; pinyon–juniper grasslands; 1 year; EF	Thetford et al. (1971)
Cattle	45	55	0	79	21	0	80	20	0				68	32	0	Manitou, CO; pine–bunchgrass; 1 year; RF	Currie (1977)
Cattle	19	71	10	46	52	2	68	26	6				48	45	7	University Ranch, NM; semi-desert grass–shrub; 1 year; EF	Rosiere et al. (1975)
Cattle	86	14	0	90	10	0	96	4	0	99	1	0	93	7	0	CO: plains sandhills; 1 year; EF	Wallace et al. (1972)
Cattle	97	0	3	84 (early summer)	0	16				76	0	24	86	0	14	Santa Rita, AZ; desert grassland; RF	Galt et al. (1982)
Cattle	67	0	33	71	15	14	64	10	26	39	26	35	57	18	25	NM; north-central rolling plains and mountains; FA	Holechek et al. (1986)
Cattle	54	22	24	92	7	1	64	12	24	32	25	43	63	13	24	LA; pine–bluestem range; 3 years; BC	Thill and Martin (1986)
Cattle	63	9	28	71	28	1	86	9	1	98	2	0	83	11	6	CO: shortgrass; 1 year; BC	Peden et al. (1974)
Cattle	78	4	18	91	3	6	93	4	6				93	4	3	CA–NV; brushland foothills; 1 year; FA	Hanley and Hanley (1982)
Cattle	94	4	2	48	16	36				66	1	33	57	9	34	Red Desert, WY; sagebrush–grass; 1 year; FA	Krysl et al. (1984)
Horse				95	4	1	95	2	3	81	7	12	89	6	5	CA–NV; brushland foothills; 1 year; FA	Hanley and Hanley (1982)
Horse	86	10	4	70	3	27				60	1	39	65	2	33	Red Desert, WY; sagebrush–grass; 1 year; FA	Krysl et al. (1984)

Botanical composition in diet (%)[a]

Animal	set 1 (G, F, B)	set 2 (G, F, B)	set 3 (G, F, B)	set 4 (G, F, B)	set 5 (G, F, B)	Location; duration; technique	Reference
Horse	99, 1, 0	—, —, —	94, 5, 1	85, 9, 6	89, 7, 4	Alberta; foothills; 1 year; FA	Salter and Hudson (1980)
Bison	62, 4, 34	58, 20, 22	—, —, —	—, —, —	89, 6, 4	CO; shortgrass prairie; 5 years; EF	Kautz and Van Dyne (1978)
Bison	93, 7, 0	—, —, —	93, 1, 0	93, 7, 0	96, 4, 0	CO; shortgrass; 1 year; EF	Peden et al. (1974)
Elk	74, 3, 23	—, —, —	58, 19, 23	76, 2, 22	63, 12, 25	CA; northwestern, redwood–hardwood mountains; 1 year; FM	Harper et al. (1967)
Elk	71, 3, 26	—, —, —	83, 6, 11	71, 1, 28	75, 2, 23	CO; Sangre de Cristo foothills, 1 year; FA	Hansen and Reid (1975)
Elk	73, 25, 2	—, —, —	78, 21, 1	42, 56, 2	57, 38, 5	Black Hills, SD; grass–pine woodlands; 1 year; RF	Wydeven and Dahlgren (1983)
Elk	79, 4, 17	—, —, —	75, 4, 21	31, 1, 68	61, 11, 28	CO; subalpine forest and foothills; RF	Boyd (1970)
Sheep, mtn. bighorn	57, 10, 33	—, —, —	65, 6, 29	23, 11, 66	50, 7, 43	Saguache Co., CO; mesas and mountains; 1½ years; FA	Todd (1975)
Sheep, mtn. bighorn	56, 5, 39	—, —, —	29, 7, 64	63, 3, 34	52, 5, 43	WA; open forests to alpine; 1 year; not stated	Johnson (1983)
Sheep, mtn. bighorn	78[c], —, 22	—, —, —	86[c], —, 14	73[c], —, 27	78[c], —, 22	ID; Salmon River range; 3 years; DO	Smith (1954)

INTERMEDIATE FEEDERS

Animal	set 1 (G, F, B)	set 2 (G, F, B)	set 3 (G, F, B)	set 4 (G, F, B)	set 5 (G, F, B)	Location; duration; technique	Reference
Sheep, domestic	37, 47, 16	61, 8, 31	68, 3, 28	82, 1, 17	62, 15, 23	Edwards Plateau, TX; oak–juniper savannah; 1 year; BC	McMahan (1964)
Sheep, domestic	27, 67, 6	56, 14, 30	—, —, —	—, —, —	47, 22, 30	CO; shortgrass prairie; 5 years; EF	Kautz and Van Dyne (1978)
Sheep, domestic	23, 72, 5	56, 14, 30	30, 69, 1	31, 69, 0	36, 62, 2	Ft. Stanton, NM; pinyon–juniper grasslands; 1 year; EF	Thetford et al. (1971)
Sheep, domestic	72, 5, 23	75, 14, 11	92, 4, 4	88, 11, 1	84, 10, 6	Hopland Field Sta., CA; foothill grassland–shrubland; 1 year; RF	Longhurst et al. (1979)
Sheep, domestic	—, —, —	25, 0, 75	42, 30, 28	25, 0, 75	—, —, —	UT; northern desert shrub and northern mountains; 2 years; B & A	Cook and Harris (1950)
Sheep, domestic	25, 62, 13	2, 3, 95	65, 6, 29	50, 2, 48	70, 3, 27	Red Desert, WY; 2 years; RF	Severson et al. (1968)
Sheep, domestic	62, 13, 25	—, —, —	29, 14, 57	46, 24, 30	41, 30, 29	NM; north-central rolling plains and mountains; FA	Holechek et al. (1986)
Sheep, domestic	51, 7, 42	38, 42, 20	70, 28, 2	—, —, —	—, —, —	CO; shortgrass; 1 year; EF	Peden et al. (1974)
Sheep, domestic	47, 12, 41	68, 22, 10	47, 12, 41	—, —, —	—, —, —	CA-NV; brushland foothills; 1 year; FA	Hanley and Hanley (1982)
Sheep, domestic	68, 29, 3	26, 7, 67	49, 14, 37	58, 5, 37	60, 18, 22	Sonora, TX; grass–oak savannah; 1 year; EF	Bryant et al. (1979)

(continues)

TABLE 10.2 (Continued)

Botanical composition in diet (%)[a]

Herbivore	Spring Gr	Fo	Br	Summer Gr	Fo	Br	Fall Gr	Fo	Br	Winter Gr	Fo	Br	Annual Gr	Fo	Br	State vegetation type, number of years, method[b]	Reference
Burro	25	54	21	33	14	53	40	8	52	6	36	58	26	28	46	AZ; desert canyons and mountains; FA	Seegmiller and Ohmart (1981)
Sheep, desert bighorn	4	32	64	21	53	26	29	52	19							NM; semi-desert mountains; DO and FA	Howard and DeLorenzo (1975)
Sheep, desert bighorn	6	55	39	16	22	62	16	15	69	3	21	76	10	28	62	AZ; desert canyons and mountains; FA	Seegmiller and Ohmart (1981)
Goat, mtn.	72	14	14	72	23	3	76	21	1	58	16	25	70	19	11	MT; alpine and subalpine; FM & RF	Saunders (1955)
Goat, mtn.	46	9	45	43	20	37	48	13	39	31	5	64	42	12	46	WA; open forests to alpine; 3 years; FA	Campbell and Johnson (1983)
Caribou	67	32	1	1	2	97	1	98		3	96	1	18	57	25	Northern Yukon; 1 year; FA	Thompson and McCourt (1981)
Caribou				28	33	39				13	47	40	21	40	39	N. Canada; DO	Kelsall (1968)
BROWSERS																	
Goat, domestic	40	25	35	65	8	27	47	12	41	47	4	49	50	12	38	Sonora, TX; moderate grazing; 1 year; EF	Malechek and Leinweber (1972)
Goat, domestic	26	12	62	10	2	88	18	5	76	13	0	87	17	5	78	Edwards Plateau, TX; oak–juniper savannah; 1 year; BC	McMahan (1964)
Goat, domestic	54	6	40	35	12	53	18	12	70	40	8	52	36	10	54	TX; mixed grass–shrub; 1 year; FA	Warren et al. (1984b)
Goat, domestic	53	17	30	53	22	25	44	7	47	41	4	55	48	12	40	Sonora, TX; grass–oak savannah; 1 year; EF	Bryant et al. (1979)
Goat, domestic (Angora)	51	19	30	59	15	26	37	9	54	39	4	57	45	13	42	Sonora, TX; grass–oak savannah; 1 year; EF	Bryant et al. (1979)
Goat, domestic (Spanish)	5	35	60	4	44	52	6	8	86	2	4	94	4	23	73	NM; southwestern brushlands; few years; RA	Boecker et al. (1972)
Deer, mule	6	85	9	42	16	42	10	1	89	7	1	92	16	26	58	CO; Sangre de Cristo foothills; 1 year; FA	Hansen and Reid (1975)
Deer, mule	33	37	30	9	57	34	19	45	36							Manitou, CO; pine–bunchgrass; 1 year; BC	Currie et al. (1977)
Deer, mule	12	6	82	16	18	66	11	6	83	17	15	68	14	11	75	WA; open forest to alpine; 3 years; FA	Campbell and Johnson (1983)

Animal	Gr[a]	Fo	Br	Gr	Fo	Br	Gr	Fo	Br	Gr	Fo	Br	Gr	Fo	Br	Location; vegetation; duration; method[b]	Reference
Deer, mule	9	6	85	8	4	88	4	4	92	3	3	94	6	4	90	CA–NV; brushland foothills; 1 year; FA	Hanley and Hanley (1982)
Deer, mule	10	5	85	6	18	76	0	8	92	3	3	94	5	8	87	Medicine Bow Mtns. WY; mixed shrub; 1 year; FA	Goodwin (1975)
Deer, mule	26	19	55	3	23	74	3	8	89	4	8	88	9	15	76	Ruby Butte, NV; sagebrush–grass; 4 years; RF	Tueller (1979)
Deer, white-tailed	34	65	1	5	71	24	27	66	7	37	59	4	18	69	13	Sinton, TX; mixed vegetation; 1 year; UM	Drawe and Box (1968)
Deer, white-tailed	38	18	44	1	54	45	2	17	81	6	29	65	12	29	59	MT; river flood-plain/breaks; 1 year; RF	Allen (1968)
Deer, white-tailed	11	42	48	6	6	88	6	8	87	0	0	35	6	14	80	Edwards Plateau, TX; oak–juniper savannah; 1 year; BC	McMahan (1964)
Deer, white-tailed	5	59	36	7	11	82	8	30	62	10	30	60	8	32	60	TX; southern rolling brushland; 2 years; RF	Arnold and Drawe (1979)
Deer, white-tailed	7	38	55	3	28	69	10	14	76	15	20	65	9	25	66	LA; pine–bluestem range; 3 years; BC	Thill and Martin (1986)
Deer, white-tailed	8	41	51	10	54	36	6	25	69	5	6	89	7	32	61	Sonora, TX; grass–oak savannah; 1 year; FM	Bryant et al. (1979)
Antelope, pronghorn	25	57	18	13	62	25	13	37	50	9	47	43	13	51	35	Alberta; rolling grasslands with some shrubs; 4 years; RF	Mitchell and Smoliak (1971)
Antelope, pronghorn	16	60	24	2	22	76	2	10	88	0	4	96	5	24	71	UT; semi-desert shrub; 7 years; UM and RF	Beale and Smith (1970)
Antelope, pronghorn													46	45	9	CO; shortgrass prairie; 5 years; BC	Kautz and Van Dyne (1978)
Antelope, pronghorn	8	22	70	3	65	32	3	44	53	2	1	97	4	33	63	MT; sagebrush–grasslands and croplands; CO and RF	Cole (1965)
Antelope, pronghorn	47	39	14	5	93	2	45	55	0	90	9	1	47	49	4	KS; western mixed grasslands and cropland; 1 year; FA	Sexson et al. (1981)
Antelope, pronghorn	2	16	82	5	27	68	2	11	87	1	8	91	3	15	82	CA–NV; brushland foothills; 1 year; FA	Hanley and Hanley (1982)
Antelope, pronghorn				7	5	88	6	9	85	3	1	96	5	5	90	Red Desert, WY; 2 years; SC	Severson et al. (1968)
Moose				1	70	29	1	8	91	0	3	97				Gravelly Mountains, MT; 1 year; DO and RF	Knowlton (1960)
Moose, shiras	6	21	73	4	26	70	2	7	91	5	2	93	4	14	82	CO; Rocky Mountain, N.P.; NP	Stevens (1974)

[a] Gr, grass; Fo, forbs; Br, browse.

[b] Method used: EF, esophageal fistula; RM, rumen fistula; FA, fecal analysis; DO, direct observation; BC, bite count; FM, feeding minutes; B & A, before and after; UM, utilization method; SC, stomach contents; NP, not provided.

[c] Percentage for grass and forbs combined.

FIGURE 10.1   All three groups of ungulate herbivores, based on reliance on different types of forage plants, are represented in this grazing scene from the Edwards Plateau of Texas: the bulk and roughage eaters (cattle), the concentrate selectors (goats), and the intermediate feeders (sheep). These are roughly comparable to the more commonly used categories of grazers, browsers, and intermediate feeders. (Texas Agricultural Experiment Station photo by Robert Moen and Charles A. Taylor, Jr)

Both classifications have obvious limitations, such as failure to adequately treat forbs and implying invariable rigidity in dietary selection. Hofmann (1988) considered the term "browser" as being too narrow and misleading and preferred the term "concentrate selector." However, the concept of concentrate selection seemingly ignores the high concentration and digestibility of nutrients in immature grass forage and the preference shown it by "concentrate selectors" (browsers) in late winter and early spring.

Gordon and Illius (1994) considered incorrect the assumption that browsers were "concentrate selectors" and the implication that the higher levels of cell solubles in browse corresponds necessarily to higher relative amounts of digestible protein, fat, or carbohydrate in the diets of "concentrate selectors" (browsers). Considering that tree and shrub leaves commonly contain soluble secondary compounds that are non-nutritive and lead to an overestimate of their nutritional value, Robbins *et al.* (1995) concluded against tree and shrub leaves, in contrast to the leaves of graminoids, being considered "concentrates."

Classification of an ungulate herbivore as a grazer or as a browser rather than as an intermediate feeder properly suggests forage preference, but it tends to ignore seasonal differences and implies a rigidity in dietary selection that is often not experienced. Care must be taken so that the versatility of the individual grazer or browser species is not underestimated; the concept of forage preference being relative to what is available applies here as well. Also, interspecific and intraspecific animal competition can greatly affect what an animal species or individual is actually ingesting at a given time (Taylor, 1986a). Management practices that control grazing pressure and modify the mix of animal species also influence animal foraging and diet selection (Baker, 1985).

Along with the perception of plant group preferences according to the classifications of a grazer, browser, or intermediate feeder, there is also the connotation that browsers are invariably more selective than the grazers in their feeding behavior. However, Demment and Van Soest (1983) attributed this view to a "temperate zone bias" where cattle (grazers) are less selective than deer (browsers). They pointed out that in tropical systems there is often a great differentiation of the nutritive value of grasses, and some grazers such as the oribi can be selective, while some browsers such as the giraffe are apt to be much less selective. Their conclusions were that genetic body size across the array of ungulate herbivore species was inversely related to degree of forage selectivity but that an array of body sizes occur relative to grass vs. browse emphasis in diets. From studies with cattle grazing Lehmann lovegrass, Ruyle and Rice (1996) concluded that classifying cattle as nonselective was erroneous; they found cattle were very selective and adapted to forage characteristics yielding the best return—considering both volume and nutrient density—for effort expended.

Based on feeding mountain meadow grass hay, Baker and Hansen (1985) concluded that elk were better adapted than mule deer for digesting non-lignified fiber (grass) diets. The digestion of dry matter and energy in the hay was 8% higher for elk (larger in size, classed either as an intermediate feeder or sometimes grazer) than for mule deer (smaller in size, classed as a browser); the digestion of neutral detergent fiber and acid detergent fiber was 13 and 18% higher for elk, respectively, apparently the result of increased retention time of the ingesta. However, Robbins *et al.* (1995) found that fiber digestion is not significantly different between browsers and grazers per se but does increase as body weight increases. These authors, in contrast to Hofmann (1988), concluded that fiber digestion was not inherently lower in browsers than in grazers.

When the animals are fed similar diets, Robbins *et al.* (1995) also concluded there was little evidence that browers have inherently faster passage rates through the rumen than grazers. They were in agreement with the hypothesis of Spalinger *et al.* (1986) that the more fragile leaves of the browser's diet can be chewed and ruminated to smaller particles and thus passed from the rumen faster than the grass of a grazer's diet. However, this was attributed not to differences in ruminant anatomy but rather to the chemical and physical characteristics of the respective diets.

C. THE GRAZERS

The **grazers** include bison (*Bison bison*), muskoxen (*Ovibos moschatus*), horses (*Equus caballas*), cattle (*Bos* spp.), mountain bighorn sheep (*Ovis canadensis*), and Dalles sheep (*Ovis dalli*). Grazers (1) mostly consume diets dominated by grasses and grass-like plants and (2) may locally consume substantial amounts of forbs and shrubs (with the possible exception of bison and muskoxen), particularly when graminoids are not available. They also (3) show a strong avoidance of browse high in volatile oils and appear to lack mechanisms to reduce the toxic effects of these substances (Holechek, 1984). (Note: because of diet versatility, elk are included under intermediate feeders.) The bison is the most dedicated consumer of graminoids, followed by the muskox and the horse. Graminoids comprise the principal plant group in the diets of cattle, mountain bighorn sheep, and Dalles sheep, but greater acceptance of forbs and/or browse is seasonally common.

Bison eat grass and rarely consume forbs or shrubs (Fig. 10.2) (Skiles, 1984; Plumb and Dodd, 1994). In tallgrass prairie in Oklahoma, graminoids comprised at least 98% of their diet across all seasons (Coppedge *et al.*, 1998). Bison on the shortgrass plains of Colorado demonstrated a high preference for warm-season grasses (Peden *et al.*, 1974). Except during late spring and early summer, when bison made substantial use of cool-season grasses, warm-season grasses comprised about 80%

FIGURE 10.2   Bison demonstrate a high preference for graminoids; they also tend to concentrate their grazing on selected sites and can cause serious overgrazing, as demonstrated above in Yellowstone National Park, unless numbers are closely regulated.

of the diet. Unlike sheep and cattle, with which bison were compared, bison consistently selected less than 13% forbs in their diet. It was also concluded that bison had greater digestive power than cattle when consuming low-protein, poor-quality forage and consumed a greater quantity of forage when of low quality than did cattle.

When bison and cattle were grazed on a shrub-grass range in the Henry Mountains of Utah, grasses and sedges predominated in the diets of both—99% for bison and 95% for cattle (Van Vuren, 1984). Even when foraging in boreal habitats in British Columbia, bison highly preferred graminoids and spent the most time foraging in grassy upland meadows and the least time in poplar forests (Hudson and Frank, 1987).

Horses primarily consume graminoids (generally comprising 85% or more of their diets), and improved pastures developed for domestic horses largely consist of grasses but sometimes with a smaller component of palatable clovers. Of the six feral horse food habit studies on western rangelands evaluated by Skiles (1984), only one reported shrub use of greater than 5%, and forbs were generally a minor diet component. However, wild horses on shrub-grass range in Wyoming's Red Desert consumed 27% browse in their summer diets and 39% browse in their winter diets, suggesting some acclimation to utilizing shrubs (Krysl et al., 1984). This agrees with the conclusion of Wagner (1978) that feral horses gone wild on western U.S. rangelands readily adapt to browse when grass is not available.

The diets of free-ranging beef cattle have probably been studied more thoroughly than any other domestic or wild herbivore. The annual diets of range cattle commonly include 60–90% graminoids (Table 10.2). However, grasses should seldom be used as the sole criteria for estimating forage production or grazing capacity for cattle since they show some versatility in adapting to current forage resources. Cattle may switch to browse if the dry matter intake of grass is severely restricted or to forbs when temporal flushes of palatable species occur (Heitschmidt and Stuth, 1991).

Forbs may be important contributors to cattle diets (up to 50%) early in the growing season or after summer rains on grasslands ranging from sandhills prairie in Nebraska (Hoehne et al., 1968) to semi-desert grasslands in New Mexico (Rosiere et al., 1975). Cattle effectively utilized forb-dominated high mountain rangelands in summer (Ralphs and Pfister, 1992); forbs comprised only 11–32% of diets in grass-dominated plant communities but 46–83% of cattle diets in forb-dominated plant communities. However, the seasonality of most forb species in most plant communities and the variability in forage production of forbs from year to year prohibit major dependency upon them on most cattle ranges. Shrub use is generally limited (see Table 10.2), but cattle sometimes consume considerable browse on desert or semi-desert shrub ranges (Skiles, 1984). Substantial cattle use of palatable shrubs has also been observed by late summer after grasses under southern pine canopy have become coarse (Thill and Martin, 1986) or after full grass utilization under western conifers (Mitchell and Rodgers, 1985).

FIGURE 10.3   The cow primarily selects graminoids in its diet but makes some seasonal use of forbs and shrubs; it prefers level to rolling grazing lands. Photo of cattle on preferred range in the Nebraska Sandhills.

D. THE INTERMEDIATE FEEDERS

The **intermediate feeders** include domestic sheep (*Ovis aries*), elk or wapiti (*Cervus elaphus canadensis*), burros (*Equus asimus*), caribou (*Rangifer tarandus*), desert bighorn sheep (*Ovis canadensis*), and mountain goats (*Oreamnos americanus*). They (1) use large amounts of grasses, forbs, and shrubs and (2) have substantial capability to adjust their feeding habits to whatever forage is available. Because of the versatility in their diets, domestic goats (forb and browse preferring) could also be included here; their diet has been discussed under "browsers" because of its unique propensity for browse consumption.

Hanley and Hanley (1982) proposed that sheep are advantaged in having the time and ability to be highly selective foragers as well as being physically able to exploit the high-cellulose forage resources (graminoids). Sheep were considered able to selectively harvest the most palatable, most nutritive portions of the grasses while supplementing their diet with high cell-soluble forbs and browse (Fig. 10.4). Sheep are generally thought to consume substantial amounts of forbs and lesser amounts of grass, but Skiles (1984) concluded from a review of literature that sheep, depending on season and pasture treatment, often consume roughly equal amounts of forbs and grass.

High diversity in the diets of range sheep is implied by a comparison of the results of selected studies summarized in Table 10.2. Of the seven studies providing botanical composition on an annual basis, browse ranged from 2–30%, forbs from

3–62%, and grasses from 36–70%. Based on a 6-year study on northern Utah mountain summer range comprised of parks and open aspen stands, Cook (1983) emphasized the importance of forbs in the diet of sheep. Of the total quantity of forage consumed, sheep diets included 70% forbs compared to 37% by cattle. The weighted utilization (amount consumed:amount available) of grasses, forbs, and shrubs for sheep was 18, 42, and 8%, respectively. By comparison, corresponding figures for cattle were 37, 17, and 9%. However, sheep proved capable monocot grazers in a big sagebrush-grassland area in southeastern Montana (Alexander *et al.,* 1983); the summer food habits of sheep were grasses and sedges, 96%; forbs, 1%; and shrubs, 3%. The capability of sheep for consuming large amounts of grasses apparently led Hofmann (1988) to classify them as grass/roughage eaters, but their versatility and ability to utilize high proportions of forbs and browse as well suggests better placement as intermediate feeders.

While elk commonly include large amounts of grasses in their diets, suggesting they might well be classified as grazers, they are listed here as intermediate feeders because of the great diet versatility shown. Elk appear capable of handling both woody material and high-cellulose graminoids. Although graminoids ranged

FIGURE 10.4 Because of great diversity in their diets (forbs, grasses, and browse), sheep are readily classified as intermediate feeders; they are considered as being advantaged in being highly selective foragers, also able to utilize high cellulose forages such as the graminoids. (Texas Agricultural Experiment Station photo by Robert Moen and Charles A. Taylor, Jr.)

from 57–75% of the annual diets of elk in studies cited in Table 10.2, elk are noted for being versatile due to their ability to utilize substantial amounts of forbs and shrubs (Fig. 10.5). In northern Utah, forb use by elk was often locally greater than grass use (Collins and Urness, 1983). On Black Hills range in South Dakota, elk consumed diets of 37% forbs and 49% grass during summer where sweetclover was prominent in the standing forage, but where sweetclover (*Melilotus* spp.) was absent, their diet consisted of 11% and 78% forbs and graminoids, respectively (Wydeven and Dahlgren, 1983). When available, sweetclover was important in fall and winter as well as summer elk diets. However, on winter range in northwestern Colorado, elk consumed diets consisting of 68% browse (Boyd, 1970). Winter feeding by elk on grass or browse may be dependent on availability, with a low grass component in the stand or grass being buried deeply under snow resulting in heavy browsing.

Burros apparently eat whatever plants are in bloom or in season, the diets of feral burrows on desert range varying from high shrub and forb consumption in some areas to high grass consumption in other areas (Skiles, 1984). The annual diet of barren-ground caribou includes a mixture of graminoids, forbs, and shrubs

FIGURE 10.5 Although graminoids commonly range from 55 to 75% in their annual diets, elk are known as being very versatile in being able and willing to utilize substantial amounts of forbs and shrubs; they are also more commonly found grazing in larger openings than are mule deer. (U.S. Forest Service photo.)

(Table 10.2), but lichens and mosses commonly comprise a majority of the forb category. Desert bighorn sheep typically include a greatly reduced graminoid component in their diets compared to mountain bighorn sheep; this difference probably results primarily from a lack of availability of grasses, but past experience and even genetics may play a role. Since the diet of the mountain goat varies greatly depending on season and location, it is classified here as an intermediate feeder.

E. THE BROWSERS

The **browsers** include the domestic goat (*Capra aegragus*), mule deer (*Odocoileus hemionus*), black-tailed deer (*Odocoileus hemionus columbianus*), white-tailed deer (*Odocoileus virginicus*), pronghorn antelope (*Antilocapra americana*), and moose (*Alces alces*). Browsers (1) are noted primarily for consuming large amounts of forbs and shrubs, (2) commonly consume substantial amounts of green grass during rapid growth stages but avoid dry, mature grass, and (3) often experience digestive upsets if forced to consume diets dominated by mature grass, particularly when unaccustomed to such. As discussed previously, the smaller browsers are better able to metabolize plants high in volatile oils such as the monoterpenes.

The fact that goats consume many shrubs is an attribute that can be beneficially exploited, and goats can survive even after other vegetation has been destroyed by other kinds of livestock (Martin and Huss, 1981). Both physical characteristics and foraging skills allow goats to select preferred forages even at excessive grazing pressures (Taylor and Kothmann, 1989). A mobile upper lip and a very prehensile tongue permit the goat to eat short grass and browse and feed in areas that offer no other choice (Huss, 1972). Since the goat prefers leaves and tender twigs and does not normally consume tough, woody growth, it is capable of consuming young tender growth of many otherwise undesirable woody species when it has no other choice. Nevertheless, goats respond to good grazing management with higher production as do other species of livestock even though they can survive on deteriorated shrublands.

Most studies agree that goats are primarily browsers but will readily consume grasses and forbs, thus allowing them to maintain a high-quality diet under adverse conditions (Huss, 1972; Owens *et al.*, 1992). Coblentz (1977) proposed that goats are not primarily browsers by preference but are opportunistic generalists and tend to consume the most palatable vegetation available. This versatility in their eating habits could qualify them as "intermediate feeders" as well as "browsers" (Fig. 10.6). After finding that Angora goats at the Sonora Station in Texas consumed 50% grasses, 12% forbs, and 38% browse in their annual diets, it was suggested that they might sometimes even qualify as "grazers" (Malechek and Leinweber, 1972). Nevertheless, Spanish goats in Utah consumed large amounts of Gambel oak by choice in summer (Riggs *et al.*, 1988). It is commonly acknowledged that goats are most effective in biological control of shrubs by grazing when stocked

FIGURE 10.6. Although the domestic goat is widely recognized for its ability to consume browse, it is also capable and often willing to consume large amounts of forbs and graminoids, suggesting it might well qualify as an intermediate feeder as well as a browser. (Texas Agricultural Experiment Station photo by Robert Moen and Charles A. Taylor, Jr.)

at high densities for short periods and when the palatability of shrub leaves and twigs is highest relative to alternative vegetation planned for increase in the stand.

The willingness of Spanish goats to consume browse has increased the efficiency of forage utilization on mixed-brush rangelands in south Texas as well as making them an important method of brush control (Warren *et al.*, 1984b). (Refer also to "Biological Control by Goats" in Chapter 16.) Although Bryant *et al.* (1979) found great similarity between the diets of Angora goats and Spanish goats, it is commonly held that Spanish goats make the greater use of browse. This was borne out by research on the Edwards Plateau of Texas by Warren *et al.* (1984a). In one trial running from November to August, Spanish goats consumed 25% grass, 5% forbs, and 70% browse while Angora goats consumed 53, 6, and 41%, respectively. In a February-November comparison, Spanish goats consumed 28% browse in their diets while Angora goats consumed 19%.

Mule deer are noted for consuming foliage and twigs from shrubs and small trees. Of the six yearlong dietary studies of mule deer summarized in Table 10.2, 83–92% of the fall diets and 68–94% of the winter diets consisted of browse. Even though herbaceous vegetation became important in these studies during spring and summer—in some cases their consumption exceeding browse for a short time—58–90% of the annual diets consisted of browse (Fig. 10.7). Forbs and browse were found by Krausman *et al.* (1997) to make up over 90% of the diets of desert mule deer in southern Arizona, and grasses and succulents were generally less than 5% of their diets.

Browse is an indispensable component of deer winter diets when snow depths prevent access to short forb and grass forages (Bartman, 1983; Urness, 1986). At snow depths of about 8 in., deer essentially cease to paw through snow for low-growing grasses and forbs (Austin and Urness, 1983). On seeded big sagebrush-grass range in Utah, Austin and Urness (1983) found that as snow cover increased from 23–98% the percent grass in the diet decreased from 51–2%. Nevertheless, on many mule deer wintering areas, snow-cover periods often alternate with snow-free periods on south and west-facing slopes, permitting access to palatable herbaceous species (Urness, 1986). As snow decreases and deer move to south aspects in late winter and early spring, a shift from nearly complete browse intake to large amounts of new forbs and grasses may occur rapidly.

The green growth of exotic grasses has been shown in some cases to be very important from fall to mid-spring in mule deer diets in the Intermountain Region, supplementing browse diets of frequently modest or low value until new forb growth in spring (Urness *et al.,* 1983; Urness, 1986). Mule deer in northern Utah preferred both green grass and cured forbs over shrubs in the spring as long as they

FIGURE 10.7 Mule deer are noted for consuming large amounts of browse but may consume high levels of graminoids and forbs when green and succulent; they range widely and use large areas of range minimally grazed by domestic livestock, particularly cattle.

were available (Smith *et al.*, 1979). When available, forbs and grasses are important in easing browsing pressure on shrubs during the critical mid- and late-win-
ter periods and helping to maintain diet quality (Bartman, 1983). However, fall regrowth of annual and perennial grasses may be sparse or absent in drought years,
making shrub resources more critical during the dormant winter season.

Where winter grain crops are being produced, such as triticale, barley, rye, and
wheat in the Texas Panhandle (Wiggers *et al.*, 1984), their foliage may figure
prominently in the winter diets of mule deer. In southeastern Montana, mule and
white-tailed deer as well as antelope were also attracted to non-range agricultural
lands and utilized an array of alfalfa fields, winter wheat, and conservation reserve
lands (CRP), the latter providing both forage and cover, but made minimal use of
wheat stubble lands (Selting and Irby, 1997).

Skiles (1984) has noted in many mule deer studies the long lists of food items
making up less than 5% of the diet. This led to the conclusion that mule deer were
less highly selective for restricted plant species than generally assumed but rather
sampled a large number of forage species when available.

White-tailed deer are found in a variety of habitats across the U.S., and dietary
differences resulting from plant species availability and probably experience as
well are not surprising (Table 10.2). On a southern Texas shrub type, browse comprised 65, 84, 81, and 95% of their diets in spring, summer, fall, and winter, respectively (Varner and Blankenship, 1987). However, based on their 3-year study
in a grassland-brushland complex, also in south Texas, Chamrad and Box (1968)
found that white-tailed deer were primarily grazers rather than browsers during the
winter-spring period, when only 5% of the diet was browse. In a 1-year study, also
at the Welder Foundation near Sinton, browse comprised only 1, 24, 7, 4, and 13%
of deer diets in spring, summer, fall, winter, and annually, respectively (Drawe and
Box, 1968).

Browse, particularly from halfshrubs, is generally the preferred food item for
pronghorn antelope; but forbs may comprise 25–50% of their diet in the spring
and summer when plentiful (Fig. 10.8). At the Desert Range Experiment Station
in western Utah, pronghorns varied their diets between seasons and years (Beale
and Smith, 1970; Smith and Beale, 1980). Browse typically amounted to 85% or
more of antelope diets during the November-March period, but was also high in
summer and fall during dry spring-summer seasons when green herbaceous foliage was minimal (Beale and Smith, 1970). Forb use was normally high only during the April-June period, extended into the summer during wet spring-summer
periods, or rose again in the fall following forb production resulting from summer
rains. Grass intake was negligible except during early spring.

According to Yoakum (1978), pronghorns thrive best on rangelands with a diversity of vegetation, an abundance of plants with high succulence, and height
growth between 15 and 24 in. The pronghorn appears prone to sample any palatable plant. In common with mule deer and white-tailed deer, pronghorn antelope
consume many food items in their diets which make up less than 5% of the total

FIGURE 10.8   Browse is generally the preferred forage for pronghorn antelope, but forbs may comprise 25 to 50% of the diet when plentiful; antelope are principally found in dry plains and semi-desert areas.

intake (Skiles, 1984). Pronghorn were found able to overwinter satisfactorily in western Kansas with minimal amounts of browse by substituting winter wheat and alfalfa (Sexson *et al.,* 1981). By consuming native forbs in late spring and summer and cropland plants the remainder of the year, pronghorns were able to adapt to mixed cropland and rangeland areas. The annual diets of pronghorns on short-grass range in southeastern New Mexico were dominated by forbs (51–99%), compared to 40–50% for sheep, thus demonstrating reduced dependence by pronghorns on shrubs when ample forbs are available (Beasom *et al.,* 1981).

Moose are capable of utilizing more woody material from shrubs and trees (even including bark) than deer and pronghorn antelope (Holechek, 1984). In southwest Montana the diet of moose was found to consist of over 95% browse in both winter and summer (Dorn, 1970). As the principal taiga range animal in Canada, moose were noted to depend on mostly browse from deciduous shrubs and deciduous tree saplings with moderate amounts from yew and pine trees (Telfer, 1978b). However, although browse will normally consist of 90% or more of the fall and winter diets of moose, forbs and aquatics may be of high local significance in spring and summer diets where available and palatable.

III. MIXED SPECIES GRAZING

Mixed grazing (synonymous with **multispecies grazing** or **common use**) is the practice of grazing the current year's forage production by two (**dual use**) or more kinds of ungulate herbivores, domestic or wild, either at the same time or at different seasons of the year. The concept of mixed grazing applies equally to a combination of livestock species, or of big game species, or a mixture of livestock and big game species. The importance of kind and proportion of managed grazing animals has been greatly underestimated by most grazing land managers; indeed, such management provides an opportunity to achieve uniform plant utilization under moderate stocking rates even with continuous grazing (Merrill *et al.,* 1957a). Mixed grazing seeks to promote stability in the botanical composition of vegetation while preventing trends in vegetation.

Erroneous traditions, prejudice, conflict of interests between land-user groups, and even administrative edict have restricted the use of this principle of grazing management. Animosity between growers of different livestock species has historically contributed to the problem; such disagreements along with greater managerial/administrative requirements have led some landlords to view multispecies grazing with disfavor. Also, some wildlife conservationists have favored eliminating livestock grazing from public lands in the belief that livestock grazing is too competitive with big game or other wildlife species. However, multispecies grazing is the norm for wild ungulates (Walker, 1994). Nevertheless, Smith (1965) argued for situations in which the best-suited animal alone might provide maximum grazing capacity, providing grazing distribution problems were not a factor, but gave no verified example.

Grazing research and producer/grazing manager experience suggest the following should be considered as advantages and limitations of mixed or multispecies grazing (Glimp, 1988; Baker, 1985; Taylor, 1986a; Walker, 1994):

Advantages

1. Complementarity due to differences in forage plant and terrain preferences
2. Maintaining a desired balance between forage species
3. Providing stability in grazing land ecosystems
4. Providing diversity of income and more uniform cash flow
5. Aiding in the control of internal parasites (i.e., for sheep when cattle present)
6. Developing mutually beneficial interrelationships between animal species
7. Maximizing yield of animal products through greater biological efficiency
8. Sheep utilization of forage affected by cattle feces

Limitations

1. Increased facility costs, such as fencing, watering, and handling facilities
2. Reduced scale of enterprise resulting in reduced technological efficiency
3. Conflicts in labor needs
4. Need for increased management skills and knowledge
5. Greater predator problems (adding sheep or goats to cattle)
6. Marketing made more complex
7. Antisocial behavior between animal species in limited situations
8. Differential suitability of climates to different animal species
9. Required proper stocking ratios between animal species

9. Elevating some plant species from weed to forage status
10. Increasing grazing capacity
11. Reducing predator problems (adding cattle with sheep)

10. Trend toward larger and more specialized agricultural enterprises

Changing from single to mixed species grazing may be prevented by predation (e.g., adding sheep or goats to a cattle enterprise) unless high predation can be controlled. Although mixed grazing may result in more even use of forage plant species within plant communities, spatial use of pastures by the respective animal species may not be readily changed (Senft *et al.,* 1986). While grazing in common on large summer range units in southwestern Utah mountains, cattle and sheep commonly separated on a topographic basis (Ruyle and Bowns, 1985). However, when stocking densities were increased, cattle and sheep congregated together more often, thus reducing the spread of topographic usage. It is commonly observed that adding sheep or goats or deer to traditional cattle range can spread the grazing impact by their negotiating steeper terrain and making better use of poorly watered areas than cattle.

The overriding principle favoring mixed grazing, according to Walker (1994), is that intraspecific competition (between individuals of the same species) is always greater than interspecific competition (between different species). This ecological principle conceivably applies to every situation where grazable vegetation grows (Huston, 1975), even on homogenous improved pasture, but limitations including that of management, human social status, and targeting goals other than profit maximization (or sometimes lack of economic incentive) have made the application of mixed grazing uncommon outside of a few areas (Walker, 1994).

Combination grazing by two or more animal species having different diet and/or site and terrain preferences and habitat requirements provides the greatest opportunities for mixed grazing. Cook *et al.* (1967) compared the degree of use made of the primary herbage species on Utah mountain summer range under common-use grazing by cattle and sheep with single use by either animal species alone. When equivalent levels of forage availability were held constant, utilization of the primary herbage species was reduced by common use compared to single use in most instances. It was concluded that if utilization of the primary forage species were allowed to reach the equivalent level under the highest single use, additional animal days of grazing per acre would be realized with mixed grazing.

The advantage of mixing kinds of grazing animals increases as the diversity of vegetation and site and terrain within a grazing unit increases. Also, the more kinds of grazing animals grazed in common the more likely that more plant species will be utilized and a greater portion of the grazing unit will be thoroughly covered. This relationship is widely recognized in the Edwards Plateau area of Texas where cattle, sheep, goats, and white-tailed deer are frequently grazed in common (Fig. 10.9). Grazing four wild ungulates in common on north-central New Mexico ranges—mule deer, pronghorn antelope, elk, and wild horses—also showed the advantages of mixed grazing (Stephenson *et al.,* 1985). However, increasing the

FIGURE 10.9   The advantage of multispecies grazing has been widely recognized in Texas; research on multispecies grazing management has long been given high priority at the Texas A&M University research station at Sonora shown above. (Texas Agricultural Experiment Station photo by Robert Moen and Charles A. Taylor, Jr.)

number of animal species will increase management requirements and the urgency of proper stocking rates and ratios.

An early study of common use by cattle and sheep was made in Utah by Cook (1954) on mountain summer range consisting of a combination of grassland, aspen, and meadow types. Grazing capacities and exchange ratios for cattle and sheep were based on forage factors derived from grazing on separate but adjoining allotments. It was calculated that the 2800 acres of land included in the study would provide 560 AUMs under cattle use alone, 306 AUMs from sheep use alone, but 652 AUMs from common use under an optimum 65:35 cattle to sheep animal equivalent ratio. This was equivalent to mixed grazing of the area being 2.3 times more effective than sheep alone and 1.16 times more than cattle alone.

When cattle and sheep were grazed in common on mountain summer range in southwestern Utah (Bowns and Matthews, 1983), total grazing capacity was increased, more efficient and uniform use was made of the range area, and improved range conditions resulted. Sheep alone made the greatest use of forbs and snowberry (*Symphoricarpos* spp.) but the least use of grasses; the reverse resulted from grazing cattle alone. Essentially no differences were found in mean daily gains or weight changes between groups of animals grazed alone and those in mixed grazing groups. In a continuation of this study (Matthews and Foote, 1987; Bowns, 1989; Olson *et al.,* 1999), sheep production improved when they were grazed with cattle, suggesting that sheep compete more successfully with cattle than among themselves. However, the advantages to cattle being grazed in common with sheep were less consistent and were often nonexistent to slightly negative for calf gains.

After many years of sheep grazing only on high elevation summer range in southwestern Utah, Letterman needlegrass (*Stipa lettermani*) increased in abundance while forbs decreased (Ruyle and Bowns, 1985; Bowns and Bagley, 1986). By contrast, on cattle range shrubs greatly increased. Thus, either animal species grazed singly caused a profound directional trend in botanical composition of the native vegetation. It was projected that common use would have the advantage of enabling the vegetation to maintain a stable composition at higher levels of use than does single species stocking. It was further projected that common use by both cattle and sheep should result in more efficient range management and improved range condition by balancing forage demand and preventing or minimizing directional trends resulting from plant species selection during grazing.

Under controlled management of pasture at Beltsville, MD, when only orchardgrass was available for grazing, there was no advantage in total gain/acre to co-grazing of steers and sheep at a ratio of 1:5 over cattle alone (Reynolds *et al.,* 1971). However, multiple plant species, including weeds, are generally available for selection even in improved pastures. While dual grazing promoted a stable plant mix of Kentucky bluegrass and white clover, single species grazing by cattle alone reduced the bluegrass while sheep alone reduced the white clover (Abaye *et al.,* 1994).

Another consideration is that with mixed grazing of improved, intensively managed pasture, a greater proportion of the herbage is normally available to sheep than to cattle since cattle reject grass growing in proximity to cattle dung while sheep will accept forage growing close to dung from either species (Forbes and Hodgson, 1985). This may account for cattle appearing more disadvantaged than sheep when grazed in common under heavy grazing. When improved pasture was stocked at a 1:3 or 1:6 steer-to-sheep number ratio, sheep grazing pressure was adequate to consume the rank herbage developed around cattle dung pats (Nolan and Connolly, 1977).

When evaluated across 14 independent grazing studies using cattle and sheep, Walker (1994) found that dual species grazing consistently increased meat production per unit area compared to cattle only and by an average 24%. Although the benefits of dual grazing averaged 9% compared to sheep only across all studies, in some studies sheep-only stocking resulted in the higher meal yields per unit area. This occasional reversal resulted from the higher relative growth rate of lambs compared to calves and ewes being more prolific than cows. In mountain summer range studies in southwestern Utah (Bowns, 1989), lb/acre of total progeny at equivalent stocking rates were 16.6, 10.1, and 16.2 for lambs alone, calves alone, and lambs with calves. Across all treatments in the study, the lb gain per 100 lb of dam weight were 56.5 for sheep and 22.5 for cattle, an increase of 151% for sheep over cattle when multiple births were included, or 47% more when adjusted to single births.

In the 14 study comparisons made by Walker (1994), dual grazing benefited individual sheep more than cattle. Sheep grazed in combination with cattle averaged 30% higher (range: 12–126%) per head performance; cattle grazed in combination with sheep averaged 6% (range: −3–21%) higher performance. The depressed cattle performance in some studies was attributed to sheep being more competitive for forage when availability is low by grazing closer to the soil surface and more selectively from the total standing crop (Walker, 1994). Based on grazing research made on Kentucky bluegrass-white clover pasture, Abaye *et al.* (1994) concluded that dual grazing favored earlier weaning, increased lamb performance (daily gain, total gain, and weaning weights), and improved ewe condition at the start of the breeding season.

At equivalent stocking rates, sheep in Australia did better—in rate of pregnancy, size of lamb crop weaned, and in weaning weights—when run with cattle than by themselves (Squires, 1981). Merrill and Young (1954) reported a beneficial response from either cattle or sheep grazed in combination with goats in central Texas, but grazing dually or alone apparently had no effect on body weights or the mohair production of the Angora goats.

In the absence of empirical research for each major vegetation type and case situation, Walker (1994) suggested increasing stocking rates using the following rules of thumb, these initial points subject to later adjustments from experience, in changing from single to mixed species grazing: (1) 10% increase from single cattle or sheep use to dual cattle-sheep use on improved pasture; (2) 25% increase from similar single use to dual use on native range; and (3) even greater increase from single cattle or sheep use to three species (goats-sheep-cattle) stocking (i.e., as high as 70% in diverse shrub-forb-grass communities). However, it should be noted that the grazing capacity advantage associated with mixed grazing will be specific to each site and situation.

IV. ALTERNATIVE LARGE HERBIVORES

Maintaining game animals in a wild state, while harvesting them primarily to keep populations in check and to reduce cycle extremes in numbers, has been traditional in North America; this has commonly been referred to as **game cropping.** In contrast, under **game ranching** either native or exotic game animals are maintained under semi-domestication (Teer, 1975), and animal management is provided to enhance breeding, health, nutrition, and production and marketing as a ranch earning enterprise (Fig. 10.10). A variant of game ranching, sometimes referred to as **game farming,** follows even more intensified management practices similar to those used with domesticated animals and often includes the use of improved pastures, harvested forages, and feed concentrates.

Game ranching has come into vogue since about World War II, beginning par-

FIGURE 10.10   A game ranching enterprise, such as raising white-tailed deer on this ranch on the Edwards Plateau of Texas, can be a highly profitable program when operated in conjunction with traditional livestock enterprises. (Texas Parks and Wildlife Dept. photo.)

ticularly in Texas and from there spreading into the southeast and throughout much of the west. Initially applied on privately owned lands, it has more recently been applied to mixtures of private and public lands. Attention presently is geared toward the big game animals, more from the standpoint of economic returns through hunting and recreation potential than from their meat potential, although the latter should not be ignored. Terrill (1975) has recommended attention be directed primarily to domestic livestock, where genetics and management have made great strides, for meat and fiber production while maintaining and managing wild species for their economic potential through hunting and associated recreation. Nevertheless, in his comprehensive treatise, *Big Game Ranching in the United States,* White (1987) extends the concept of game ranching to meat production as well.

A wide array of grazing and/or browsing species can be considered in game ranching and can be complementary with domestic livestock production. The concept of game ranching by means of fee hunting and fishing or land leasing extends to upland game birds and even small upland game mammals as well as big game and fish. Game ranching offers additional opportunities for diversified ranching, alternative or supplemental sources of ranch income and thus more stable cash flow, and a way of compensating private landowners for the forage consumed or damaged on private lands by state-owned big game animals (Nielsen *et al.,* 1986).

Native big game animals in the U.S. and Canadian provinces are typically owned by the individual states and provinces but often graze substantially or, in some areas, almost exclusively on privately owned lands. Stringently enforced human trespass laws have allowed the game management system based on private property rights to emerge in North America. Under this system, access to the land for hunting rather than the actual animal is the product sold and bartered by the private landowner. Virtually every aspect of big game ranching is subject to numerous laws and regulations, both federal and state/provincial, with the latter often differing widely among states and provinces (White, 1987). Even when individual states and provinces permit the introduction and private ownership of exotic game animals, such activities are tightly controlled by laws.

Reimbursing private landowners for forage consumed or damaged by publicly owned big game animals on their lands is economically important to landowners and vital to maintaining populations of many native big game species. For example, the production of white-tailed deer for recreational and economic purposes is dependent upon their protection, production, and management on private lands over widespread areas (Teer, 1996). Strategies for reimbursing landowners vary widely depending on state game laws and local situations. Although such strategies must be accomplished within existing hunting seasons, numbers (bag limits), and sex quotas and by hunters with valid hunting licenses, variances are sometimes issued to the landowner. Incentives to the landowners may include one or a combination of the following:

1. Fee hunting charges hunters a daily, weekly, or seasonal access fee to hunt on private lands.
2. A block of private land is leased to individuals, hunting clubs, or outfitters or hunting brokers for specified hunting/recreational purposes.
3. Landowners buy a specified number of hunting tags or permits from the state and sell them to hunters as they wish (Long, 1996).

4. Landowners get longer seasons and/or more liberal bag limits but are required to do definitive habitat work (Long, 1996).
5. Coupon attached to each big game license is redeemable by the landowner for a cash payment by the state if the animal is killed on private property (Dana et al., 1985; Van Tassell et al., 1995).
6. Payments are made to landowners for damage or other substantial use by state-owned big game.
7. Block hunting easements for public hunting access are made by federal, state, or private organizations.
8. Conservation easements are made by governmental or private organizations that provide for game habitat along with other agricultural activities.

Fee hunting and land leasing have become the principal means of providing income to private landowners for state-owned big game raised entirely or in part on their private lands (Dill et al., 1983). A popular approach to hunting has been for landowners to lease to clubs or outfitters rather than selling access permits directly to hunters (Jordan and Workman, 1988). This permits landowners to assert more control over hunter numbers and behavior on private lands and reduce or at least be compensated for property damage. Rules can be set and help obtained from club representatives in getting members to abide by the rules while also providing help in patroling to prevent unauthorized hunting. Property damage is minimized by retaining the option of refusing permission to hunt during future years.

Game ranching under land-lease or fee-hunting systems offers incentives for environmental thought and consideration as follows (Dill et al., 1983; Renecker and Kozak, 1987): (1) It provides an opportunity to reverse the transformation of wildlands into other exclusive land uses. (2) It provides for selecting animals adapted to fragile habitats and marginal agricultural lands. (3) It provides an opportunity to diversify rather than replace wildlife management concepts. (4) It provides an incentive to improve wildlife habitat on private lands. (5) It provides an economic basis for wildlife to compete successfully with other agricultural enterprises and an incentive to nurture wildlife as carefully as the other agricultural commodities. (6) It improves hunting success or even live animal sales while maintaining and supporting optimal wildlife populations. Game ranching offers "ecologically gentle and socially acceptable means for the support of man" (Mossman, 1975).

In most states and Canadian provinces, the American bison on private land are legally treated as private property (Shaw, 1996). This established tradition of private ownership readily adapts bison to intensive big game ranching. Bison can tolerate both heat and cold extremes with minimal shelter (White, (1987); appear better adapted to poor quality and less accessible sites, low quality forage, and limited availability of water than cattle; and respond to good pasture and range management (Shaw, 1996). Bison can be managed successfully in non-supplemented herds grazed yearlong under extensive management on Great Plains grasslands (Hamilton, 1999; Steuter, 1999), while being capable of producing 80–85% weaned calf crops from 3- to 10-year-old cows without calving assistance, with bull and heifer calves weighing 310–350 lb and 350–400 lb, respectively, at weaning.

Elk are another ruminant native to North America that is well adapted to game ranching or even game farming (Friedel and Hudson, 1994; Klein, 1997). Being intermediate feeders, elk can readily adapt to a wide variety of vegetation profiles; and because elk are virtually bloat-free, legumes are readily grazed by them.

In some areas, exotic big game herbivores introduced principally from Africa and the Indian subcontinent have become common, but the laws in the various states and provinces vary widely in permitting the introduction of exotics. Exotic big game species offer new opportunities for mixed species grazing. In the state of Texas, ranchers have the opportunity to select not only from the domestic live-stock species but from seven species of native ungulates and 68 species of exotic ungulates (Nelle, 1992); other states are less liberal and some do not allow the introduction of big game species.

Not all exotic big game species introduced into North America are browsers; rather, a wide array of diet preferences are represented. Although diet preferences for most exotic introductions have not been intensively studied in their North American habitats, tentative listings of some of the exotic big game species are as follows (Mungall and Sheffield, 1994; Nelle, 1992; White, 1987):

Grazers: barasingha (*Cervus duvauceli*), blackbuck antelope (*Antilope cervicarpra*), blesbock (*Damaliscus dorcas*), gaur (*Bos gaurus*), gazelle (*Gazella* spp.), gemsbok (*Oryx gazella*), hartebeest (*Alcelaphus* spp.), mouflon (*Ovis oreintalis*), oribi, reedbuck (*Redunca* spp.), waterbuck (*Kobus ellipsiprymnus*), wildebeest (*Connochaetes* spp.), European wisent (*Bison bonasus*), and zebra (*Equus* spp.).

Intermediate feeders: auodad or Barbary sheep (*Ammotragus lervia*), axis deer (*Axis axis*), common eland (*Taurotragus oryx*), fallow deer (*Dama dama*), impala (*Aepyceros melampus*), nilgai antelope (*Boselaphas tragocamelus*), red deer (*Cervus elaphus elaphus*), sika deer (*Cervus nippon*), and springbok (*Antidorcas marsupialis*).

Browsers: bushbuck (*Tragelaphus scriptus*), grey duiker (*Sylvicapra grimmia*), giraffe (*Giraffa camelopardalis*), greater kudu (*Tragelaphus strepsiceros*), and nyala (*Tragelaphus angasi*).

When compared to white-tailed deer in central Texas, the exotic blackbuck antelope, fallow deer, axis deer, and sika deer were found to have greater rumen capacity and capability of digesting a grass diet (Henke *et al.*, 1988). Grasses comprised about 95% of the late spring diets of the blackbuck antelope, fallow deer, and axis deer; the diet of the sika deer was comprised of 40% grass (plus 48% forbs), but the diet of the white-tailed deer was only 1% grass (plus 91% forbs). Of the 68 exotic big game species censused in Texas in 1988, 87% of the total number were axis deer, nilgai antelope, blackbuck antelope, auodad sheep, fallow deer, and sika deer.

In the states and Canadian provinces where exotic game animals are legal, they are generally considered private property and not subject to regulated hunting seasons and bag limits. Thus, they may enter into trophy hunting; live animal sales, including breeding animals; sale of meat, antlers, and other by-products; or nonconsumptive uses such as for aesthetic pleasure and photo safaris. Although most states have laws to control, monitor, and reduce the escape of exotic big game animals from game ranches, escapees are not uncommon.

The introduction and establishment of exotic big game in North America remains controversial. According to Demarais *et al.* (1990), the positive aspects of exotic big game are (1) year-round income to the landowner, (2) increased opportunities for hunters, (3) preservation of endangered species through private increase, (4) filling of open ecological niches, and (5) aesthetic value. However, these same authors summarized the negative aspects of exotic big game as (1) competition for niches with native big game species, (2) uncontrolled spread with the potential of becoming pests, (3) disease complications with native big game species and domestic livestock, and (4) and interbreeding with similar native big game animals.

Game ranching can effectively complement domestic livestock production, but it can also be competitive. Grazing multiple species, whether livestock or big game, is always more complex and requires greater planning, coordination, and management of grazing. The principles of grazing management apply equally to big game herbivores as to domestic livestock, and the potential impacts of large herbivorous wildlife species on plant growth and development, ecological succession, and watershed condition are similar to those of livestock (Heitschmidt and Stuth, 1991).

Special fences and facilities are generally required to properly confine and handle big game ungulates, which add substantially to animal production costs. New markets must be developed for big game products, and initial investments may be high. Big game enterprises must generally complement rather than replace traditional livestock enterprises. One strategy in big game ranching has been to target the use of arid or semi-arid areas or other marginal lands that often fail to meet the minimal forage requirements for domestic livestock (White, 1987).

V. INTERSPECIFIC SOCIALITY

With mixed grazing there are social interactions between the different species of grazing animals. These can be negative when two or more species come together, particularly under conditions of crowding such as at watering places or where supplemental feed is made available. When they come together, horses dominate cattle and sheep, and cattle dominate sheep and goats. It is usual to see sheep and cattle grazing apart when in the same grazing unit, but cattle and horses may often intermingle (Squires, 1981). Sheep may use the same area of the pasture as cattle or horses but at different times of the day. Negative interactions are seldom observed between sheep and goats.

Does and kids and ewes and lambs may be susceptible to injury by cattle and horses. However, adverse interactions between large and small livestock species will normally occur only during confinement or crowding or at times of commotion; under extensive grazing conditions, sheep grazing, watering, and lambing commonly occur in the presence of cattle without apparent difficulties (Walker, 1994). It may be advantageous to separate ewes and does around lambing and kidding time from cattle or horses under intensive management; placing them in separate pens or small paddocks will facilitate providing help at parturition when needed, in mothering of the newborn, and reducing predation during this critical period.

Where both large and small livestock utilize the same watering places at points of concentration, such as at the cell center of a specialized grazing system, providing separate troughs or tanks for the sheep and goats that are fenced off separately with access only through creep gates to exclude cattle and horses has been suggested (Taylor, 1986c). Spatial separation, either by animal choice or through use of separate facilities, may be needed during periods of supplemental feeding as well as during watering.

Individual cattle or horses that become unduly aggressive, either playfully or from intent to harm, may have to be removed from the presence of sheep and goats. Greater success has been expressed in combining cow-calf pairs than yearling cattle with sheep and goats; yearling cattle tend to injure lambs and kids more, while cows may be helpful in chasing away predators (Baker, 1985). Yearling steers and even heifers are commonly grazed separately from cow-calf pairs to reduce disturbance as well as receive different management. In contrast to when sheep are free ranging, sheep present in large range bands under herding disturb other grazing animals, both livestock and big game. This results from the sheep functioning as a loose or tight band and from the activities of dogs, herders, and their horses (Nelson, 1984).

After reviewing studies in which interspecific dominance hierarachies have altered habitat use by free-ranging subordinate ungulates, Mosley (1999) integrated into the following dominance hierarchy the major rangeland ungulates in North America as follows: bison, horses, cattle, sheep, elk, mule deer, bighorn sheep, pronghorns, and white-tailed deer. Mosley concluded that when forage resources are plentiful, animal species commonly feed together, dominants move less and displace subordinates less frequently, and few agonistic encouters occur. In contrast, agonistic encounters should be expected more commonly when different animal species are clustered near scarce nutritional resources such as limited green pasture, hayfields, and disbursed hay or supplements.

In discussing behavioral interactions with mixed grazing, Nelson (1984) distinguished between **interference competition**—when one animal species takes aggressive defense of territory—and **disturbance competition**—when one animal species seemingly voluntarily leaves the vicinity of one or more other animal species. The latter is less commonly observed between large wild herbivore species than between large wild herbivores and domestic livestock. Wild herbivores, including white-tailed deer, mule deer, elk, bighorn sheep, and moose, all tend to vacate localized areas following introduction of livestock (Heitschmidt and Stuth, 1991); it was concluded this probably shows an innate social intolerance to livestock since these big game species tended to return to the vacated areas shortly after the livestock are removed.

It is often impossible to determine whether the exit by one species is resulting from interference or the mere presence of the prevailing species and/or from the latter's effects on the forage supply (Severson and Medina, 1983). Beck *et al.* (1996) found that elk distanced themselves from herded sheep but stayed in the same general area to use salt placed for the sheep. Interactions observed between elk, mule deer, and cattle in Arizona under free-ranging conditions were related primarily to the number of the respective species present, with large numbers of a particular species having a greater impact on the other species (Wallace and Krausman, 1987).

Aggressive behavior by grazing cattle against deer and elk apparently does not occur, and their total exclusion by the presence of cattle seems improbable (Mackie, 1976). On the other hand, Wisdom and Thomas (1996) found no North American studies that documented livestock aversion to wild ungulates. Shaw (1996) noted that bison are often reported to be intolerant of other ungulates and possibly should be grazed separately from them. However, he concluded that bison injury to other ungulates was found mostly at supplemental feeding areas or other concentrated and confined areas, but in large range units buffalo mostly grazed by themselves with little contact with other wild or domestic herbivores.

The domination of wild (feral) horses over cattle and pronghorn antelope under mixed grazing has been studied in the Red Desert of Wyoming (Miller, 1983; Plumb *et al.,* 1984; Sowell *et al.,* 1983). Aggressive behavior of the horses against the other grazing species occurred particularly around watering places, these conflicts materially increasing when the number of watering places was greatly restricted and/or when horse numbers were high. Both cattle and antelope generally would not water until the horses had left the watering place. When horses encountered cattle at the trough, several dominant horses would prompt the cattle to disperse from the tank through threatening actions such as biting and kicking. Only after the horses had left would the cattle return to watering.

When water in the Red Desert was scarce in one study and restricted to a single well, a herd of 200 horses took up to 5 hours to drink, and intraspecific as well as interspecific competition was at times intense (Miller, 1983). When horses were concentrated at the well, pronghorn would intermittently approach the water but leave each time before watering. Upon the approach of a large herd of horses, the antelope present at water invariably dispersed; however, the aggressive action of the horses towards cattle and antelope was primarily when water was in short supply. When there was enough room at a watering place for pronghorn to drink without getting closer than 10 feet to either cattle or horses, they drank mostly without interruption.

The interactions of cattle and white tailed deer have been observed in Texas at a single, central water facility under short-duration grazing (Prasad and Guthery, 1986). Deer appeared to partly avoid interaction with cattle by visiting the facilities earlier in the morning than the cattle, but the deer occasionally left the water either without drinking or before finishing upon the approach of cattle. Infrequent or no use of water was found by the deer at the cell center with increased cattle activity, and the problem was accentuated by more frequent visits and activity by humans and concentration of fencing at the cell center. It was concluded that livestock water provided only at the center of a short-duration grazing cell may be largely unavailable to deer as well as upland game birds, and regulated access or alternative watering sources were suggested.

Both elk and mule deer are frequently seen grazing and drinking in close proximity to cattle (Wallace and Krausman, 1987). Moose and cattle have been observed feeding within 10 feet of each other without apparent conflict (Dorn, 1970). There appears to be little or no problem of tolerance, stress, or behavioral characteristics between antelope and livestock since they graze together well (Yoakum, 1975). When sheep and antelope were confined in 240-acre range units in Wyoming's Red Desert, there was no evidence of any social stress placed on populations of either as a result of the presence of the other species. Antelope in western Utah tended to avoid areas currently or recently grazed by sheep, but this was attributed to dietary competition for black sagebrush (*Artemisia nova*) (Clary and Beale, 1983). No indications of social intolerance of cattle by bighorn sheep or other than minimal dietary overlap were found in an Arizona study (Dodd and Brady, 1988).

Both elk and deer appear to assume a subordinate role to cattle under range conditions, but Nelson (1984) cautioned it may be the result of either the presence of the cattle or their prior use of the standing forage. Wisdom and Thomas (1996) concluded that the aversion elk show to the presence of cattle may or may not restrict the grazing choices of the elk. The greatest effect of cattle on mule deer habitat selection on California mountain range occurred in late summer (Loft *et al.,* 1991); as cattle grazing pressure increased on the preferred sites, female deer shifted their use to sites less favored or avoided by cattle. However, many authors report diminishing numbers of elk and mule deer in areas frequented by cattle even under moderate cattle stocking levels (Severson and Medina, 1983; Wallace and Krausman 1987).

The lack of livestock grazing of plants highly palatable to big game animals in an adjoining area may be the attraction of big game into the alternative area. Lyon (1985) concluded it was confusing and unclear whether cattle tend to repel elk or elk have intolerance to the presence of cattle; possibly the answer is that elk are just more versatile. It was concluded that elk generally have greater mobility, tend to use steeper slopes, are not usually limited by drinking water, and have a natural tendency to graze lightly and move on; these factors tended to limit competition for space. These mobile characteristics apply similarly to mule deer (Nelson, 1984). Ragotzkie and Bailey (1991) concluded that mule deer preference for grazing units not currently being grazed by cattle on the Santa Rita Experimental Range

near Tucson may have been avoidance of cattle or more probably to a more attractive forage base in ungrazed pasture, or some combination.

The mobility of indigenous wildlife herbivores may give them a distinct advantage in finding high-quality forages when rangelands are temporal and/or spatially heterogeneous (Rittenhouse and Bailey, 1996). For example, big game often
have a much wider choice of habitats to graze, resulting both from innate choice
or management restrictions on livestock. Big game can generally utilize both
grazed and ungrazed (by livestock) units in a rotational grazing system while livestock are restricted to a specific unit. This permits the more mobile big game to remove high quality forage from a sward before livestock graze it and to graze any
regrowth before livestock are scheduled to return. However, in some situations the
mobility of the indigenous big game may be restricted because of other habitat requirements such as cover, and such advantages may not accrue.

Animal performance may be affected by changes in feeding behavior when social cohesion develops between individual animals of different species. This may
affect site selection for grazing and possibly plant species preference over time,
but the results cannot be fully anticipated. On improved pasture at Beltsville, MD,
a strong social relationship developed when one sheep and one steer grazed together. However, when multiples of both steers and sheep were in the same pasture, they grazed mostly as species groups (Bond *et al.,* 1976).

Cohesive tendencies were developed between kids and heifers in Texas studies
(Taylor *et al.,* 1988), but group affiliations appeared to prevail over individual kid-
heifer bondings. Yearling ewes were socially bonded to young cattle in New Mexico studies; this bonding provided protection to the sheep by the cattle when they
were threatened by predators (Anderson *et al.,* 1988). However, in a study of bonding of lambs and young cattle at the Jornada Range in New Mexico, the botanical
composition of diets in grasses and forbs differed between cattle and sheep regardless of bonding treatment. Diets of non-bonded lambs and lambs bonded to
cattle were similar in grasses, cacti, and shrubs, and liveweight changes were similar. Thus, it appears that cross-species bonding for the purpose of influencing diet
selection has minimal practical potential.

VI. COMPETITION UNDER MIXED GRAZING

Competition between grazing animals occurs only when there is a limited supply of one or more necessities of life. Competition can be for space, water, or cover but most commonly is for forage. Wisdom and Thomas (1996) concluded that
interspecific competition between ungulate herbivores must include the following
elements: (1) dietary overlap of both preferred plant species and space, (2) disturbance or displacement of one species by another, and (3) reduction in population
performance of one species by another. Also, failure to recognize that all grazing
animal species have surprising dietary flexibility can thwart evaluation of dietary
competition between animal species.

Forage competition can vary from intense to only moderate to virtually none; even negative competition (positive effects) is conceivable. In selected circumstances under mixed grazing, competition can actually be negative, i.e., synergistic, in that grazing by one species enhances the quality and/or quantity of grazing capacity for another. Opportunities for forage enhancement through manipulative grazing are discussed under "Manipulate Animal Habit by Grazing" in Chapter 16. Competition can also be materially reduced through the selective use of range and pasture development techniques, including plant control, prescribed burning, forage plant seeding, fertilization, etc. (Vallentine, 1989).

Interspecific forage competition, of course, cannot occur except under mixed grazing, although individual grazing animals of the same species can compete with each other. Dietary overlap between species is not sufficient evidence for exploitative competition (Hanley, 1982b; McInnis and Vavra, 1987); dietary overlap is important only if accompanied by spatial overlap, if shared foods are in short supply, or if one herbivore limits access of another to a preferred forage plant species in the absence of acceptable alternative forage plant species. Diet similarity studies are only a first step in assessing competitive interaction according to Thill and Martin (1986), and they further concluded that even high diet overlap is not sufficient evidence for competition in the absence of data showing diminished animal health or reproduction.

Various combinations of circumstances can cause severe competition between any two or among several kinds of animals grazing in common. The degree of forage competition between kinds of animals generally increases with (1) increasing similarity of diets, (2) increasing overlap of sites selected for grazing, (3) increased grazing pressures resulting from high stocking rates or low forage production, and (4) lack of alternatives beyond the most preferred forage plants and most preferred sites for grazing. Dietary overlap and the resulting extent of competition between animal species will be further determined by respective seasons of grazing, time of grazing relative to plant growth, severity of weather (especially drought), the vegetative species mix, and the kinds of grazing animals and their respective numbers in the animal mix.

Under mixed grazing, for example elk and cattle (Wisdom and Thomas, 1996), high stocking rates of either species can reduce animal performance through intraspecific as well as interspecific competition. Thus, beyond a few generalizations under extreme conditions, such as long-term overgrazing, it is difficult to formulate broadly based principles to guide managers of mixed grazing (Dwyer *et al.*, 1984). Competition interactions between grazing animal species is very complex and often multi-directional, and environmental factors conceivably can greatly modify both the direction and magnitude of forage competition. Severe competition may be highly seasonal, such as on limited, critical winter range where big game concentrate in winter if grazed simultaneously or previously by domestic livestock. For example, Wisdom and Thomas (1996) concluded that the potential for competition between elk and cattle in western U.S. was greater on winter and spring-fall range and lower on late spring and summer ranges, but prolonged seasonal drought may increase competition in late summer and early fall.

Rittenhouse and Bailey (1996) have generalized on the relationship between plant growth stage, nutritive levels, and competition. During initial growth stages, forages are green and growing, and the nutrient content of forage and diet quality is high. Forage quantity is the determining factor for competing herbivores when forage growth exceeds forage depletion. However, as the forage plants begin to mature, the abundance of high-quality forage and forage regrowth become the determinants of both animal competition and animal performance. Thus, competition among herbivores generally shifts from overall forage quantity to the quantity of less abundant high-quality forages. As the vegetation begins to mature, the rate of removal of the high-quality component of the landscape by each kind of animal becomes critical, and a reduction in grazing pressure or changing the mix of animals species at this time may be warranted.

Forage availability immediately prior to the most critical periods may be more important than during the critical period (Holechek, 1980). This is because most kinds and classes of grazing animals survive for long periods (30 days) with minimal forage intake if they have high initial body fat reserves. Where only winter grazing is limiting and little potential exists to provide natural forage, the practice of winter feeding of harvested roughages or even concentrates may have value for big game as well as domestic livestock (Urness, 1980).

Detailed observations or studies of forage selectivity and competition may be highly site or situation specific; severe competition between species of grazing animals is probably a more local than universal problem (Mackie, 1976, 1985). Even the potential or opportunity for exploitative competition among various species of large herbivores does not preclude that it will take place. Most wild ungulates appear to be fairly adaptable in their choice of food and habitat requirements (Mackie, 1976). Also, under typical rangeland situations, in which population densities of either or both competitors are at or close to their carrying capacities, it is difficult to distinguish the additional effects, if any, of interspecific competition from those of intraspecific competition.

Bryant and Taylor (1992) concluded that livestock management should be directed to maintaining a high floral diversity and suggested that white-tailed deer were more sensitive to changes in floral diversity and richness than were Spanish goats. Nevertheless, based on his own research and an extensive review of published studies, Walker (1994) has challenged the widely held hypothesis that dietary overlap invariably decreases as total biomass or plant species diversity increases. In contrast, he found that the dietary overlap between sheep and cattle tends to decrease as the available forage decreases. The apparent reason was that, when available forage becomes limiting, cattle shift their diet to the lower quality but more available forage resources but sheep were apparently capable of continuing to select their preferred diet. Walker (1994) was in agreement with the generally accepted opinion that sheep are more selective grazers than cattle and supported the hypothesis that a sympatric herbivore species (i.e., grazing in the same area with other species) should reduce competition by filling different food niches.

Cattle, sheep, and goats on mixed vegetation types in central Texas select a fairly large percentage of plants from the three major vegetation classes (grass, forbs, and browse) (Taylor, 1986a). As the available vegetation becomes limiting under full grazing, drought, or dormant growth periods, dietary overlap generally increases. Although significant variations occur among the diets of cattle, sheep, goats, and white-tailed deer, Bryant et al. (1979) observed a striking similarity in trends for the selection of grass, forbs, and browse. This suggested that competition among kinds of animals grazing yearlong together on the same range could perhaps best be summarized in terms of competition for green forage.

Cattle diets in central Texas generally show the least similarity with deer and goat diets, and deer and goat diets are the most similar, with sheep diets somewhat intermediate between the others (Merrill et al., 1957b; Rector and Huston, 1986). The diets of Spanish goats, because of higher browse consumption, are more similar to deer than Angora goats, while Angora goat diets are more similar to sheep than Spanish goats (Taylor, 1986a). As might be expected, deer numbers at the Sonora Station declined sharply under heavy goat grazing, were reduced even by moderate goat grazing, but were unaffected by moderate cattle grazing (Merrill et al., 1957b). Under combinations of drought, heavy grazing, and continuous grazing by livestock, deer production at the Kerr Wildlife Area in south central Texas was adversely affected through competition for food (McMahan and Ramsey, 1985; Teer, 1985). Deer mortality was most pronounced in fawns followed by summer death losses of does in the absence of green forage.

Diet overlap between white-tailed deer and cattle grazed in common on longleaf pine (Pinus palustris)-bluestem range in Louisiana averaged 21.5, 11.2, 19.6, and 30.9% during spring, summer, fall, and winter, respectively (Thill and Martin, 1986). Deer diets were affected more by burning than the presence or absence of cattle. It was concluded the generalist foraging behavior of deer should minimize the consequences of changes in the availability of certain forage species. Moderate grazing (40–50%) of pine-bluestem range from late spring through early fall had minimal negative impact on deer forage availability, but late fall and winter cattle grazing tended to reduce deer forage availability (Thill and Martin, 1989). While forage selectivity by deer was significantly reduced under moderate yearlong grazing by cattle relative to ungrazed conditions, deer diet quality was comparable to or better than under livestock-ungrazed conditions (Thill et al., 1995).

It is generally concluded that mule deer are well adapted to mid-successional communities that contain a good mix of grass, forb, and shrub species but are especially favored by browse communities in winter (Urness, 1976). Because of typically low levels of dietary and spatial overlap between cattle and mule deer, good management can largely prevent serious competition (Fig. 10.11). On mixed browse communities in the Great Basin, Austin and Urness (1986) concluded that the grazing effects of cattle on mule deer diets and nutrition in summer are minor when the intensity of cattle use is controlled such that cattle primarily use mostly

FIGURE 10.11   Good management can largely prevent serious competition between cattle and mule deer; in fact, spring cattle grazing when properly managed can be used to enhance deer winter range, as shown here in San Pete County, UT.

understory vegetation. Cattle management that leads to heavy browse utilization in late fall or winter, to concentration on critical deer winter-feeding areas, to heavy use of green herbage in late winter or early spring, or to high cattle stocking rates continued in prolonged drought can impact mule deer negatively.

Typically low levels of mule deer use of mountain meadows helps to minimize overlap for space or diets with cattle (Stuth and Winward, 1977). Further shifts were noted in deer use in the Sierra Nevadas from meadow-riparian and aspen areas to montane shrub habitat when cattle grazing began in early summer, but the change was greater under heavy grazing than under light grazing (Loft *et al.*, 1986; Kie *et al.*, 1988; Loft *et al.*, 1988). Heavy cattle grazing in this area also reduced the hiding cover for fawns, (2) increased the size of deer home ranges, and (3) increased deer feeding and travel time. These resulted from indirect effects of cattle on the habitat rather than the mere presence of the cattle; the effects were greater under heavy cattle grazing than light grazing but their significance was deemed uncertain.

In studies in mixed-browse communities in the Sheeprock Mountains of western Utah (Austin and Urness, 1986), few dietary or nutritional difference were determined for deer between areas simultaneously grazed or ungrazed by cattle. It was noted that deer preferred areas ungrazed by cattle when deer use was low. However, after deer use accumulated through the summer (to 99 deer days/acre), the selectivity by deer for areas ungrazed by cattle was eliminated.

Maintaining moderate stocking levels of both mule deer and cattle along with generally recommended cattle grazing management practices will minimize most serious competition problems between the two species. Within-year competition between livestock and deer could be eased where portions of the range are deferred or rested yearlong from cattle grazing but are continuously available to deer (Urness, 1976; Skovlin *et al.,* 1968). Avoiding concentrating cattle in winter protection areas used by deer is suggested (Bryant and Morrison, 1985). It is generally concluded that mule deer seldom have negative impacts on cattle either from spatial or forage considerations.

Greater opportunity exists for sheep than cattle to compete with mule deer because of range usage and dietary overlap; interspecific competition is almost assured if the supply of mutually preferred forage is inadequate to satisfy the demand of both animal populations (Mackie, 1976). Sheep in northern Utah utilized primarily herbs and made light use of shrubs during spring and early summer, thus having minimal impact on deer winter browsing (Jensen *et al.,* 1972). However, after July 15 and through the summer into fall, even moderate sheep use caused material utilization of shrubs preferred by mule deer in winter such as bitterbrush. Thus, sheep grazing in late fall through winter but not spring and early summer could be expected to impact deer during winter.

Numerous studies have shown a high degree of similarity in range usage and yearlong forage preferences between elk and cattle (Wisdom and Thomas, 1996) but to a lesser degree between elk and deer. Studies of elk and mule deer diets in the Blue Mountains of Oregon revealed that competition for forage between the two was not normally high, as elk consumed larger amounts of grass but less browse in their diets (Skovlin and Vavra, 1979). The preferred habitats of elk overlap those of cattle on the one hand and mule deer on the other. The competition of elk with mule deer can be expected to increase in the presence of cattle grazing because of resulting changes in vegetation type and forage selection by the elk in the presence of cattle (Mackie, 1976). When deer and elk occupy the same ranges and compete for foods, elk appear to be the better competitor; that is, they reach higher, are more versatile in their diets and choices of habitats, and are more adept at scraping snow off understory plants than are deer (Severson and Medina, 1983; Mackie, 1976).

It can be postulated that cattle have replaced bison of former years as the companion grazer to antelope in dry plains areas (Yoakum, 1975). Competition for forage or water is generally not a problem on range in good condition between cattle and antelope. Although both prefer low rolling terrain and grassland habitats, dietary overlap is minimal when a variety of grasses, forbs, and low shrubs are present. Overlap in diets of antelope and domestic sheep can be expected to be moderate or even high when a preferred grass component is not available for sheep diets. At the Desert Experimental Range in western Utah, pronghorn tended to avoid areas being grazed during the winter season by sheep and sought ungrazed, often steeper terrain. The prime factor in the avoidance of sheep-grazed areas was apparently dietary competition for black sagebrush (Smith and Beale, 1980). Diets of antelope and mule deer were similar in both summer and winter in the cold desert biome (Vavra and Sneva, 1978), but their spatial overlap is minimal in summer and minimal to only moderate in winter. Forage competition between moose and cattle is not expected to be significant because of widely different dietary composition (Dorn, 1970).

Horses, cattle, and bison have high dietary overlaps throughout the year; where their selected grazing areas are similar, competition can be expected to be high. On grass-shrub winter range, moderate to high dietary overlap can be expected between cattle and horses (Krysl et al., 1984; Sowell et al., 1983) and between sheep and cattle (Holechek et al., 1986; Vavra and Sneva, 1978). On salt-desert shrub and sagebrush-grass range in Oregon, the dietary overlap between horses and cattle was high each season of the year (62–78%, averaging 70%) (McInnis and Vavra, 1987). By contrast, the dietary overlap between these domestic herbivores and antelope ranged from 7–26%. Compared to the exploitative competition exhibited between the horses and cattle, the low levels of overlap between the domestic species and antelope provided a wider buffer for competitive coexistence.

The impact of forage competition in mixed grazing may be very different when the grazing occurs in different seasons rather than at the same time. For example, the temporal distribution may minimize or heighten the competition between elk and cattle (Wisdom and Thomas, 1996). When the same standing crop is grazed at different seasons of the year, the second species in time will be more subject to the grazing pressure and site and forage selectivity of the first, preceding species. The effects of prior grazing by a different ungulate herbivore (for example, elk prior to cattle) can be deleterious to the second species. However, in other scenarios, prior grazing by a different species may have minimal impact, or may even enhance the quality or quantity of forage available to the second species (see Chapter 16).

In a Rocky Mountain foothills area in Alberta, there was little contemporaneous spatial overlap of feral horses and cattle even though their summer diets showed 66% overlap (Salter and Hudson, 1980). However, over 90% of the sites utilized by cattle in summer had received prior use by horses—40% by horses in the spring and 50% used mostly the previous winter. Another example is the spring-summer grazing by cattle on Intermountain foothill range followed by mule deer grazing in fall and winter of the same area. If high cattle grazing pressures extend into the fall, thereby removing the browse that deer depend on during winter emergencies, the deer can be greatly impacted.

Elk and mule deer in the Intermountain Region during winter and early spring concentrate their grazing in intermixed timber and grass-steppe plant communities; the results of a study by Vavra and Sheehy (1987) indicated this grazing preceding domestic livestock grazing did not significantly impact the quantity and quality of the new standing crop. Concern has been expressed about the effect of late winter and early spring grazing of free-ranging elk on wet meadows and mountain range seedings on which access by cattle is prevented until late spring or early summer. Under this situation, "range readiness" will administratively determine when the cattle are permitted entry while elk generally closely follow the receding snowline. This permits elk to move into choice open parks and riparian areas before substantial plant growth has been made and before wet soils have become stabilized (Powell *et al.,* 1986). Heavy early concentration of elk, particularly when population numbers have become excessive, may seriously disadvantage subsequent cattle grazing by reducing the available forage.

11

GRAZING ANIMAL INTAKE AND EQUIVALENCE

I. FORAGE DRY MATTER INTAKE

Within the bounds of genetic potential, production by the grazing animal is primarily a function of quantity and quality of forage consumed. Both contribute directly to nutrient intake, the prime environmental basis of animal performance. Although diet quality is obviously important also, variation in voluntary forage intake has been deemed the most urgent factor determining level and efficiency of ruminant productivity (Demment and Van Soest, 1983). Data on diet quality without information on forage intake or ability to predict will poorly describe the nutritional status of grazing animals (Hakkila *et al.,* 1987).

Although the factors affecting voluntary feed intake are becoming better understood, precise quantitative intake projections under grazing conditions are difficult because of the numerous complex and often interacting factors affecting intake. Procedures used in the past for measuring intake by grazing animals, even under controlled experiment, have often been disappointing and somewhat unreliable (Cordova *et al.,* 1978). Nevertheless, the prediction or measurement of dry matter intake (i.e., feed consumption expressed on a dry matter basis) is a key component in (1) assessing free-choice nutrient intake, (2) directing needed dietary enhancements, (3) determining grazing capacity, and (4) applying appropriate management practices.

Control of feed intake is mostly indirect, except when high-nutrient-density rations that would exceed the animal's nutrient requirements if fed to appetite are limit fed (NRC, 1987). Increasing total forage dry matter intake is one way of correcting nutrient deficiencies. Conceptually, if an animal could eat enough it could satisfy its nutrient requirements from most low-quality forage. An understanding of the factors that restrict forage intake should suggest ways in which limitations may be overcome and the potential productivity of the animal more closely approached.

Forage intake by grazing animals is determined by a large number of animal (physical, physiological, and psychogenic), forage/dietary, weather, and management factors (Fig. 11.1). Those factors that increase forage dry matter intake or at least maintain high levels along with those that decrease forage intake are listed in

FIGURE 11.1   Forage intake by grazing animals is determined by a large number of animal and environmental factors; lactation will increase forage consumption by the range cow by an average of about 35%; ample quantities of palatable, highly digestible, readily available, and easily harvestable forage will help maintain high levels of forage intake (as shown on this ranch near Monte Vista, CO). (Forest Service Collection, National Agricultural Library.)

TABLE 11.1 Factors That Influence the Dry Matter Intake of Grazing Ruminants

Factors increasing/maintaining high (dry matter) intake	Factors decreasing forage (dry matter) intake
Animal physical factors	
Large body size (actual or metabolic)	Small body size (actual or metabolic)
Low body condition	Excessive body condition
Large reticulo-rumen capacity	Limited reticulo-rumen capacity
	Undeveloped rumen in young
	Distention of reticulo-rumen (fill)
Animal physiological/psychogenic factors	
High physiological energy demand	Low physiological energy demand
Lactation, advanced gestation, work, high rate of gain	Maintenance or early gestation only
	Temporary stress of estrus, rutting, or parturition
High milk production, suckling twins or triplets	Stress: disease, fever, parasites, grass tetany, bloat, fly attacks, weaning
Recovery from restricted feeding	Chemical factors contributing to satiety (in doubt)
Grazing experience (in doubt)	Lack of grazing experience (in doubt)
Forage/dietary factors	
High forage availability	Low forage availability, snow cover, drought
Low grazing pressure	High grazing pressure
High leaf:stem ratio, high green:dead ratio	Low leaf:stem ratio, low green:dead ratio
Forage rapidly harvestable	Forage difficult to harvest
High forage acceptability	Low forage acceptability
High forage palatability (in doubt)	Low forage palatability (in doubt)
High forage variety in diet (unproven)	Low forage variety in diet (unproven)
Forage succulent	Excessive forage water levels (in doubt)
Balanced diets (adequate N, P, Ca, Mg, NaCl, etc.)	Imbalanced diets (inadequate N, P, Ca, MG, NaCl, etc.)
High forage digestibility and rate of ingesta passage	Low forage digestibility and increased retention time (ruminants only)
	High selectivity for minimally occurring forage component
	Presence of toxicants or anti-palatability factors in forage
	Feeding time limited by excessive travel time to water, grazing disruption
Weather factors	
Temperature within zone of thermal neutrality	High temperature, extreme body heat load
	Extreme cold, strong winds, heavy precipitation
Management factors	
High protein supplements fed to balance deficiency	High energy supplements fed when forage unrestricted; substitution
Supplementation methods benefit grazing	Supplementation methods interrupt grazing
Free-choice water intake	Restricted water intake
Light-moderate stocking rates	Excessive stocking rates
Unlimited grazing time	Limit grazing
	Extensive travel time/distance

Table 11.1. The control of feed intake is apparently multifactorial since for any single treatment to suppress intake it has to be administered at an artificially high level (Forbes, 1986), and there is evidence that signals from the various receptors involved in negative feedback are interpreted by the central nervous system of the animal in an additive manner.

A. ANIMAL PHYSICAL FACTORS

Body size has a major effect on governing the level of voluntary feed intake (Allison, 1985; Freer, 1981), and fat-corrected body size of animals apparently adds further precision (Bailey *et al.,* 1996a). Feed intake or energy intake is commonly described in relation to $BW^{0.75}$ (body weight to the 0.75 power), the index for general metabolism, or more simply as a percent of body weight. Most estimates of intake for cattle and sheep grazing rangelands of the western U.S. fall within the range of 40–90 g of dry matter per kg $BW^{0.75}$ or from 1–2.8% of body weight (Cordova *et al.,* 1978). Based on summarizing the voluntary forage intake by sheep of 1215 different forages worldwide, Minson (1990) found that the mean voluntary intake was about 60 g per kg $BW^{0.75}$, with intake levels varying from 20–100 g per kg $BW^{0.75}$ and only 17% falling outside the 40–80 range.

Freer (1981) has cautioned against the use of $BW^{0.75}$ or any other fixed exponent of liveweight for comparing or predicting voluntary intake. He noted that voluntary intake usually must satisfy many other demands besides basal metabolism and that these may not be related to body weight in the same way. In studies on Montana winter range Adams *et al.* (1987) demonstrated that large cows had a higher absolute forage intake but a lower intake per unit of liveweight than small cows. In related Montana range studies, the larger 3/4 Simmental cows had greater voluntary intake than Hereford cattle under green forage conditions, but under conditions of low forage quantity and or quality the production potential of the 3/4 Simmental cattle was not achieved because of an inability to achieve adequate forage intake (Havstad and Doornbos, 1987).

Feed intake is controlled by physiological demand due to maintenance needs and production demands, but only up to the limits of the gastrointestinal capacity, and more particularly reticulo-rumen capacity in the ruminant (NRC, 1987). Forbes (1986) concluded that forage intake is controlled primarily by physical factors, while the intake of more concentrated diets is controlled mainly by the energy requirements. Animals in thin body condition generally consume more forage per unit of liveweight when other factors are not limiting (Allison, 1985). However, the body condition of beef cows grazing Northern Great Plains grassland in frigid winter weather had little effect on grazing time or on intake per unit of liveweight, suggesting other factors were having greater total impact on intake (Adams *et al.,* 1987).

Limited forage-holding capacity may be severe (1) in species with low rumen capacity:body size ratio (e.g., deer and pronghorn), (2) when the rumen is still developing in young offspring, and (3) during the last trimester of pregnancy. The latter probably results from rumen compression by the growing uterus and fetus and associated hormonal and discomfort factors (Wallace, 1984; Freer, 1981). Rumen compression from excessive abdominal fat seldom occurs in grazing ruminants but is theoretically possible (Forbes, 1986). Most young calves begin rumen function around 2–3 months of age. By 4 months of age, nursing calves may spend as much time grazing as their dams (Lauchbaugh et al., 1978), but nursing calves continue to consume less total dry matter than weaned calves (NRC, 1987).

Milk and forage intake by calves are negatively correlated, particularly in older calves, but augment each other in a nursing calf's diet. Calves on spring grassland range in Montana consumed 1 lb more forage for each 3.2 lb reduction in milk intake (Ansotegui et al., 1987). In subsequent studies, calves nursing low milk-producing cows were found to consume more forage than those nursing high milk-producing cows (Ansotegui et al., 1991); calves ate 0.3 lb more forage for each pound of reduction in fluid milk in July but 0.6 lb more in August and September, probably the result of advancing age and rumen development. When suckling, young calves (75 days of age) on low-quality shortgrass range in northeastern New Mexico were restricted in milk intake, they were not able to increase forage organic matter which resulted in decreased weaning weights, suggesting that forage intake by the young calves was limited by bulk fill (Sowell et al., 1996).

B. ANIMAL PHYSIOLOGICAL/PSYCHOGENIC FACTORS

The physiological status of the ruminant animal influences daily forage consumption. Forage intake of cows and ewes increases slightly during mid-gestation over maintenance alone, declines late in pregnancy (in spite of increasing energy needs), and is sharply reduced around parturition, but greatly increases during lactation (Forbes, 1986). In both cows and ewes, energy demand increases more rapidly than intake early in lactation, often requiring that body reserves be mobilized (NRC, 1987). For cattle, the initial postpartum lag in voluntary intake relative to increased energy requirements for lactation by 2–6 weeks apparently is caused by the time required for the rumen to increase in size and re-establish maximum volume (Heitschmidt and Stuth, 1991).

After the peak of lactation is reached, the level of voluntary intake often stays high while milk flow decreases and body reserves are replenished, then intake declines in late lactation (Forbes, 1986). Based on ad libitum feeding of cow and calf pairs on chopped native meadow hay, Hatfield et al. (1988) concluded that forage intake more closely paralleled differences in calf weight and milk production than cow weight alone; the largest variable in intake and corresponding estimates of grazing capacity was between high and medium milk production levels.

Lactating cows consume 35–50% more dry matter than gestating cows of the same weight and on the same diet under conditions of high feed availability (NRC, 1987). Forage intake values reported in the literature for lactating cows commonly range from 1.6–3.2% of body weight per day, with lactation-associated increases of 25–35% commonly reported (Kronberg et al., 1986). High-producing beef cows on Montana grassland summer range showed an increase of 53% forage intake for lactating over nonlactating cows (Havstad *et al.*, 1986a). Allison (1985) reported lactating 2-year-old heifers consumed 50% more forage than nonlactating animals of similar age.

Based on studies on grazing blue grama summer range in New Mexico, Rosiere *et al.* (1980) found that 2-year-old cows at 3 months postpartum (June 5–11) consumed 55% more forage than non-lactating heifers of similar age; at 5 months postpartum (August 6–13) the lactation increase in forage consumption had been reduced to 25% but did not consider forage intake by the calf. Cook *et al.* (1961) found that lactating ewes on Utah summer range consumed 26% more dry matter than dry ewes, not considering forage eaten by the lambs. Lactating cows and ewes rearing twins increase feed intake over those rearing singles. Feed intake averages approximately one-third higher when ewes are nursing one lamb (from birth to at least 10 weeks) and 50% more when nursing two lambs than from similar non-lactating ewes (NRC, 1987). Rosiere *et al.* (1980) reported that day-to-day variations of 20–25% in forage intake per animal are natural occurrences.

For a mature ewe weighing 154 lb the projected daily dry matter intake as a percentage of live weight has been given by NRC (1985) as follows: 1.7% for maintenance only; 2.0% for the first 15 weeks of gestation; 4.0% for the last 4 weeks of gestation (130–150% lamb crops expected); 5.5% for the first 6–8 weeks suckling singles; 6.2% for the first 6–8 weeks suckling twins; 4.0% for the last 4–6 weeks suckling singles; and 5.5% for the last 4–6 weeks suckling twins. However, the 1984 projections made for beef cattle (NRC, 1984) assumed no increased consumption associated with lactation over gestation (2 vs. 2%), thereby presumably depending upon an enriched ration to provide the additional nutrient requirements of lactation; but the latter seems improbable under most grazing conditions.

Intake differences between lactating cows of different biological types appear related to levels of milk production primarily and, to a lesser extent, to body weight (Havstad *et al.*, 1986a). Ferrell and Jenkins (1987) noted that maintenance accounts for 71–75% of the metabolizable energy required by the beef cow during the production cycle but also that maintenance appears to increase with increased potential for growth rate as well as with increased potential for milk production.

Martz *et al.* (1986) concluded that ruminal volatile fatty acids (VFA) do not appear to signal satiety in steers. His data indicated that voluntary intake of forages does not illicit changes in ruminal or blood VFA which correspond to the eating pattern of cattle and thus does not appear to be the signal to stop eating. Grovum (1987), however, concluded that chemical factors like VFA may affect the intake of both poor quality and moderate to good quality roughage by contributing to satiety. Freer (1981) concluded that within-day, short-term controls of feeding behavior seem more likely to be a response to gut distention than to changes in local or circulating levels of metabolites. The potential roles that VFAs, metabolites, hormones, and brain factors play in the control of feed intake have been reviewed by the National Research Council (1987); but the practical implications of these effects remain in doubt and are not discussed here.

Mertens (1987) recognized psychogenic mechanisms as affecting the animal's behavioral and metabolic intake responses to stimuli that are not related to physical capacity or energy demand. The effect of experience in grazing a particular forage plant species or mixture is uncertain as to magnitude and duration of effect on forage intake; some evidence suggests this may be substantial. In Oklahoma studies with early-weaned calves on wheat pasture, a period of adaptation was required to reach full intake rates (Paisley *et al.,* 1998); forage intake rate was 27% lower on the 20th day of grazing than on the 70th day.

The National Research Council (1987) concluded that both taste and smell can influence the selection and consumption of various foods for most animal species, that olfactory cues (smell) can influence whether or not a meal will be initiated, and that taste may affect the length of that meal.

C. FORAGE/DIETARY FACTORS: FORAGE AVAILABILITY

Environmental factors—principally forage availability and acceptability and weather—affect forage intake. Forage availability (i.e., herbage mass or total forage standing crop) is a major factor influencing intake by grazing animals (Ruyle and Rice, 1996; Dougherty *et al.,* 1992a). According to the National Research Council (1987), the quantity of available forage is the first limiting factor. Most studies show a tendency for reduced intake at high grazing intensities, but the relative effects of reduced forage quality as well as reduced forage quantity are often confounded. As the forage allowance decreases and the grazing pressure increases from the first through the last day of a short-duration grazing period, daily forage intake generally also decreases.

As grazing pressure increases and/or the plants mature, the animal is forced to consume plant parts with a slower rate and extent of digestion. Hunter (1991) concluded that when pasturage is abundant and of high nutritive value, daily feed intake may exceed 30 g dry matter per kg of liveweight and apparent digestibility of dry matter may exceed 65%. However, when only mature, senescent pasturage with low leaf content is available, intake can be as low as 10 g of dry matter per kg of liveweight and digestibility can be lower than 40%.

Heitschmidt and Stuth (1991) summarized that standing grass crops below 1000 kg per ha (891 lb per acre) on temperate native grasslands of North America restrict forage intake by sheep and cattle, but on improved pasture restrictive levels of standing crops should be anticipated at even higher levels, i.e., between 1000 and 4000 kg per ha (891 and 3564 lb per acre). Coleman (1992) noted that in most studies with cattle and sheep under continuous grazing, intake has increased as total herbage mass increased to about 1800 lb per acre and then remained constant. From an extensive review of literature, Malechek (1984) concluded that the scarcity of forage will conceivably limit intake at some undetermined level of availability but that forage intake on range is not universally reduced by heavy grazing.

Daily forage intake in Oklahoma studies has been shown to relate closely to herbage allowance. On tallgrass prairie, total daily forage intake of beef steers was reduced about 6% and daily digestible energy intake was reduced about 9% when forage allowance was reduced from 66–88 lb to 22–44 lb per animal unit day (McCollum *et al.*, 1990). The energy intake of beef steers on wheat pasture was also depressed by a shortage of forage; in one study (McCollum *et al.*, 1993), declines in both forage consumption and digestibility resulted from forage allowances less than 15–20 lb per 100 lb body weight daily, the breakpoint being at the level of 1000–1200 lb forage per acre. In a subsequent study (Redmon *et al.*, 1995), forage intake and forage organic matter digestibility declined at wheat forage allowance below 21–24 lb per 100 lb body weight daily and plateaued above level.

On grazing lands with abundant available forage, animals can selectively graze large mouthfuls of the most nutritious plant parts, usually leaves. As the quantity declines, the amount of intake per grazing bite declines. (Refer to "Rate of Ingestion" in Chapter 6.) This relationship also applies to herbivory on shrubs as well as herbaceous plants. When goats browsed on shrubs, Owens *et al.* (1992) found that intake rate was influenced by the size of bite the animal could obtain. Larger bites resulted in higher intake rates, but smaller bites were often more nutritious because of greater selectivity initiated. When grazing herbivores of all sizes can readily obtain large bites, forage intake rates are closely related to body size and can be quantitatively predicted from bite rates (Bailey *et al.*, 1996a).

Forage availability can be reduced by such factors as drought, ice or snow cover, other forage production factors, and high utilization levels. Ingestion rates by yearling heifers on crested wheatgrass in Utah were limited by forage availability less than 490 lb per acre (Olson *et al.*, 1986). Above that level, ingestion rate was controlled by plant physical and chemical properties, particularly crude protein content. Grazing time was relatively unresponsive to sward characteristics except at very low biomass levels. Most intensive grazing systems regulate intake through herbage allowance, often on a weekly or even daily or shorter basis (Dougherty *et al.*, 1992a). Under rotation grazing, bite size along with forage intake tend to fall from the first to the last day of each grazing period (Minson, 1990).

When free-ranging heifers in Montana (Havstad *et al.*, 1983) were grazed on mature crested wheatgrass during the summer when organic matter digestibility ranged from 33–43%, the results were somewhat different. Voluntary daily intake was not affected between forage availability of 800 down to 125 lb per acre and averaged about 1.25% of body weight daily. However, grazing time increased from 7.7–10.3 hr per day with declining availability. They concluded that under bulk-limiting conditions when forage quality remained constant, decreasing intake with decreasing forage availability would not be expected (i.e., within the normal ranges they studied). They further concluded that the decline in quality of available forage or stress caused by crowding is more apt than reduced levels of available forage to reduce voluntary intake under intensive grazing management. Reduced amounts of green material in the sward have further reduced forage intake below that caused by low herbage allowance (Baker *et al.*, 1981).

From studies on highly productive pasture in eastern U.S., Rayburn (1986) concluded that intake of cattle and sheep is maximum at forage availability of about 2000 lb per acre. The total range in dry matter intake relative to the 2000 lb per acre of standing forage crop was as follows: (1) no change but rather a plateau up to 3200 lb; (2) 97% at 1600 lb; (3) 93% at 1200 lb; (4) 83% at 800 lb; (5) 60% at 400 lb per acre; and (6) 35% at 200 lb per acre. Kothmann (1984) projected that intake of grazing animals on rangelands will be reduced by the following conditions: (1) feed is scarce, (2) feed is very abundant but of low bulk density, (3) animals graze very selectively (as for a small amount of green forage within a bulk of dry pasture), or (4) grazing time is managerially restricted.

Since the amount of leaf and the ratio of leaf to stem within harvest horizons generally determine the upper limit of intake, Heitschmidt and Stuth (1992) deemed it more appropriate to relate forage availability and intake to them than to the total forage standing crop. From cattle grazing studies on old world bluestems in Oklahoma, Forbes and Coleman (1993) found that intake per bite increased as the proportion of green leaf in the herbage mass increased; organic matter intake increased with increasing green leaf mass up to 1.07 Mg/ha (953 lb per acre) and then decreased. Noting that green mass, leaf:stem ratio, and live:dead ratio of herbage offered were generally more highly correlated with forage intake than was total herbage mass, Coleman (1992) concluded that grazing animals may, in fact, limit amount of forage intake in order to select those plant parts which more fully meet their needs.

While sward heterogeneity provides ruminants with the opportunity to graze selectively, Minson (1990) concluded that sward heterogeneity can affect the relation between forage intake and forage allowance as follows:

1. Reducing the leaf:stem ratio with increasing grazing pressure reduces intake of both grasses and shrubs.
2. Selection for green forage when the green:dead forage ratio goes down may greatly limit forage intake.
3. Forage mixes with wide differences between palatability of the plant species results in selective grazing and often a decrease in dry matter intake.
4. When forage soiling by feces reaches high levels, such as in dense swards and toward the end of the grazing period, both bite size and dry matter intake are reduced.

D. FORAGE/DIETARY FACTORS: FORAGE ACCEPTABILITY

Freer (1981) ranked (1) sward structure and the ability of the grazing animal to satisfy its appetite in a grazing day and (2) the rate of disappearance of digesta from the alimentary tract as highly relevant to the grazing animal. On mixed species forage stands, the grazing animal "has to select and harvest its diet from a mixed population of forage plants which vary, within and between individuals, not only in all those structural features that determine ease of harvesting and the disappearance of digesta from the gut, but in a range of other attributes that affect the acceptability of the material. . . . On the relatively homogenous swards of improved pastures, the effect of grazing selection on intake is through the rate of breakdown of the selected components rather than through any independent effect of palatability" (Freer, 1981).

The acceptability of forage plants, including palatability, strongly influences the grazing animal. It definitely affects forage selectivity when alternative choices are offered but may materially affect forage intake levels whether or not choices are offered (Arnold and Dudzinski, 1978). Forbes (1986) also suggested that smell, taste, and appearance of the feedstuff will have less effect on the level of dry matter intake when no choice is offered. Mayland (1986) and Walton (1983) have concluded that intake is high when the material on which the grazing animal is feeding is palatable, but may be very low if all alternative plant species are very low in palatability. Cattle on blue grama range in New Mexico consumed forage at 2.17% of body weight during active plant growth but only 1.49% when plants were dormant (Krysl *et al.*, 1987). Nevertheless, Wallace (1984) questioned whether palatability alone always has a consistent influence on either forage intake or animal performance.

Grovum (1987) ranked low forage palatability and an unfavorable protein:energy ratio (i.e., nitrogen status) over reticulo-rumen distention as the main factors limiting the intake of poor-quality roughage (overmature, weathered, low nutritive levels, etc.); with medium- and good-quality roughage, rumen distention was ranked as the priority factor. Walton (1983) provided rules of thumb for estimating daily forage intake by the ruminant animal based on forage quality (i.e., these factors considering both palatability and digestibility): (1) 2.5% of the animal's liveweight for a top-quality forage, (2) 2% for good-quality forage, and (3) only 1.5% for low-quality forage. However, these can be considered as only rough averages since they assume no effect from the many other factors known to affect forage intake.

Based on studies on acquired aversions to food ingested by ruminants, it has been suggested that providing forage from a variety of plant species, under grazing as well as in confinement, may increase food intake (Provenza, 1996a; Early and Provenza, 1998). A variety of forage species in the animal diet was concluded to circumvent aversion that builds up in high consumption of a single species, possibly the result of low intake levels of multiple aversive agents compared to the high intake level of a single aversive agent. "Offering different foods of similar nutritional value, offering foods of different nutritional value, and offering the same food in different flavors are all means of changing preference and potentially increasing intake" (Provenza, 1996a).

Freer (1981) concluded that in senescent herbage or straw with a digestibility of less than 40%, advanced maturity is commonly associated with levels of nitrogen and minerals that are low enough in the rumen to limit microbial activity, thus herbage intake. Deficits or imbalances of energy and/or protein can cause decreased feed intake in grazing herbivores, and phosphorus-deficient diets may be a cause of decreased intake with cattle, sheep, or goats (Howery *et al.,* 1998a). With roughage diets of high digestibility, possibly 65–80%, voluntary intake may be controlled less by physical factors than by the energy requirements of the animals (Freer, 1981). Provenza (1995) conceptualized that low concentrations of nutrients limit intake, intermediate concentrations cause intake to increase, but excessive rates and amounts of nutrient release cause intake to decrease.

Cows overwintered on range in Montana showed a high preference for the new growth on western wheatgrass during early spring following calving (Bellows and Thomas, 1976). However, the amount available was limited and the moisture content was high, 82% on April 21. Casual observations suggested these conditions caused the low dry matter intake experienced and were major factors contributing to daily weight losses of 2.7 lb.

Nitrogen deficiency can be a primary factor limiting feed intake, while also reducing net utilization of metabolizable energy (Wallace 1984) and thus animal performance (Judkins *et. al.,* 1987). Diet digestibility, and thus rate of passage, is reduced if the nitrogen requirements of rumen bacteria are not met (NRC, 1987). Within normal ranges of dietary protein content, voluntary intake probably is not affected by protein content. The critical protein level is lower in ruminants than monogastric species because the saliva of ruminants provides a substantial supply of urea for use in protein synthesis (Forbes, 1986). Although not adequately documented, deficiencies of salt (NaCl) and possibly other minerals, if severe, may also reduce forage intake.

The presence of toxicants or anti-palatability agents may reduce forage dry matter intake. (Refer to "Secondary Compounds as Anti-palatability Factors" in Chapter 9.) One example is the consumption of mountain big sagebrush (*Artemisia tridentata vaseyana*) by sheep. In Wyoming studies, this plant was not readily consumed when fed together with grass (Ngugi *et al.,* 1995). Levels as low as 10% of sagebrush in the diet adversely influenced intake and diet digestibility of wethers; wethers on a 30% sagebrush diet decreased daily dry matter intake from 8.8–2.3 lb per 100 lb of metabolic weight.

Grazing animals, particularly lactating females, may lose weight on green succulent grass in the spring, which has often been attributed to reduced dry matter intake brought on by the washy feeds and limited rumen capacity. However, Allison (1985) has concluded that high forage moisture levels—whether from high internal water content or rain water on the surface—seems not to affect forage drymatter intake. He noted that ruminants seem to have the ability to consume forages as high as 85% moisture without affecting dry matter intake, suggesting that excess water rapidly leaves the rumen and is subsequently voided.

However, Forbes (1986) concluded that excess water that is not in the free state but rather trapped inside cells, as in fresh grass or silage, may reduce dry matter intake, at least temporarily. "Voluntary free water intake plus water in the feeds consumed is approximately equal to the water requirements of cattle. Thus, dietary water concentration per se would not be expected to influence dry matter intake until total expected water intake per unit of dry matter is exceeded" (NRC, 1987). The latter may explain why adding water to silage fed to either sheep or steers by Phillips *et al.* (1991) substantially reduced forage dry matter intake.

High-moisture spring forages, as recorded by Pasha *et al.* (1994), were consumed by wethers in lesser quantities on a dry matter basis (12% less) and were less digestible than hay made from the same sward; the lower digestibility was explained by a faster rate of passage of the high-moisture forages compared to hay, but the associated decreased dry matter intake was not expected and was left unexplained. Minson (1990) found no evidence that voluntary intake differed substantially between low and moderate levels of water in forage but concluded that very high forage water levels may depress forage intake by requiring more time for ruminating, this probably associated with very wet forage being swallowed following only minimal chewing.

E. FORAGE/DIETARY FACTORS: INGESTA PASSAGE

Because high-quality forage has fast rates of digestion and passage through the gastrointestinal tract, ruminants grazing them are able to increase their grazing time and their herbage intake per day. Immature, highly digestible, slightly laxative forages will decrease retention time and rumen fill and thus stimulate intake. Providing the swards and herbage allowances are not limiting, high-quality forage may permit ruminants to reach daily consumption levels equivalent to as high as 5% of their liveweight (Dougherty, 1991). Voluntary intake is higher for legumes than for grasses and for temperate than for tropical forages, with legumes having a lower resistance to breakdown during chewing and rechewing (Minson, 1990).

Highly fibrous, slowly digestible forage (i.e., high in cell wall content) increases retention time; physical fill then becomes limiting, and intake is reduced. Since neutral detergent fiber (NDF) levels contribute directly to indigestibility, slow rate of passage, and fill, its inverse relationship to dry matter intake by ruminants causes it to be a major component of dry matter intake estimates. However, other factors that affect fill include particle size, chewing frequency and effectiveness, particle fragility, the indigestible NDF fraction, the rate of fermentation of the potentially digestible NDF fraction, and characteristics of reticular contractions (Allen, 1996). Wilson and Kennedy (1996) noted the existence of physical hindrances to the passage of even small particles from the rumen; they concluded this may be as limiting as is the time to break large particles to small particles on passage rate and voluntary feed intake.

The digestibility and rate of ingesta passage and its association with reticulorumen fill (distention) appear to be the primary mechanisms of forage intake regulation in large ruminants (Allison, 1985; Forbes, 1986; Freer, 1981; Galyean, 1987; NRC, 1987; Allen, 1996). The reticulo-rumen is generally regarded as the site where distention occurs, thereby limiting voluntary dry matter intake, but distention of the abomasum may also limit dry matter intake (Allen, 1996). Placing inert fill into the rumen often decreases dry matter intake but the experimental results have not been consistent (Allen, 1996). Since placing balls of different densities and number into the rumen failed to consistently affect dry matter intake, Schettini *et al.* (1999) concluded that factors other than distention of the gut have a large influence on voluntary intake of low-quality forage diets by ruminants. In contrast, partial or complete removal of rumen solids through a rumen fistula did not result in observable short-term effects on feeding behavior of steers (Maiga and Pfister, 1988).

The rate of breakdown of plant particles in the rumen of mule deer and elk was found by Spalinger *et al.* (1986) to be inversely related to lignin concentration and neutral detergent fiber concentration but even more to cell wall thickness. When modeling steer intake on rangeland, Johnson *et al.* (1986) projected that reticulo-rumen capacity restricted daily intake of non-digestible dry matter to 1.07% of animal weight based on the passage of non-digested feed from the digestive tract. Nevertheless, altering the digestibility and consequently rate of passage of a roughage can be expected to cause parallel changes in intake.

The proposition that ruminants increase their forage intake when digestibility goes down (Moen, 1984) cannot be accepted. When forage digestibility decreases with plant maturity, the grazing ruminant cannot compensate by eating more because the ingested material does not move through the intestinal tract fast enough. In studies with cattle on blue grama range in New Mexico (McCollum and Galyean, 1985), forage intake declined as the season progressed from early to late growing conditions (early August to late October). This decline was associated with reduced forage digestibility, higher levels of gastrointestinal tract fill, and longer residence time of particulate and fluid digesta phases in the rumen. In contrast, the cecal digestive system of the horse, because of rapid ingesta passage through the stomach and minimal restrictive/selective passage through the cecum, apparently does permit ingesting greater amounts of forage to compensate for lower quality diets (Janis, 1976).

Cattle grazing mature (40-day regrowth) coastal bermudagrass more completely masticated the forage during ingestion, resulting in smaller particles entering the rumen compared to when grazing immature forage (20-day regrowth) (Pond *et al.*, 1987). However, remastication and digestion in the rumen of the more mature regrowth were extended approximately 10 hr compared to when immature forage was ingested, and daily intake was reduced from 2.40–2.02 lb dry matter per 100 lb body weight. In his modeling of steer intake on rangeland, Johnson *et al.* (1986) accepted that intake was limited by physical factors such as rumen size up until forage digestibility of about 66% was reached; above this level, intake was based on (and also mostly limited to) the animal's physiological demand for energy.

F. WEATHER FACTORS

Temperature alone within the zone of thermal neutrality (14–68°F) has minimal effect on voluntary intake. However, "animals eat to keep warm and quit eating to prevent hyperthermia" (Forbes, 1986). Temperatures above 68°F increase body temperature and associated heat stress, which depress intake, particularly in the short run, with some acclimatization in the long run (Forbes, 1986). A reduction in feed intake represents a major cause of reduced productivity in heat-stressed ruminants (Robertshaw, 1987). The low productivity of research cattle on desert grassland ranges in southern New Mexico, in spite of good forage nutrient levels, was attributed to low intake resulting from high temperatures and associated reduced grazing time in summer but to low forage quality in late fall and winter (Hakkila *et al.*, 1987). Providing natural or artificial shade may reduce heat load on animals, thus allowing higher forage intake. Grazing at night provides an opportunity for compensatory intake while animals are acclimatizing to hot weather. High humidity will further increase the stress of high temperatures. Wind may reduce heat stress in hot environments, particularly when humidity is high.

Below the range of thermal neutrality (around 14°F), increased heat losses are compensated for by increasing the rate of heat production which will often increase feed intake if it is readily available (Forbes, 1986). In very cold environments, especially with strong winds, heavy precipitation, and muddy ground, livestock may seek shelter and intake will be reduced. Beverlin *et al.* (1989) found that daily forage intake of pregnant cows grazing Montana winter range was not affected by temperature fluctuations between 46 and 3°F. However, during inclement winter weather with snow cover, 600-lb steers on winter range in Nebraska reduced their intake to 4.4 lb dry matter daily compared to 8.4 lb during open weather (Clanton *et al.*, 1981). Both intake and digestibility of range forage were reduced on Montana winter range by adverse winter weather (Adams *et al.*, 1986; Adams, 1987). Providing grazing animals with shelter from inclement weather, when they are not grazing, should allow intake to increase during cold weather and thus help maintain adequate body temperature.

G. MANAGEMENT FACTORS

Many management factors—these operating at least indirectly as environmental factors—affect forage dry matter intake. Many management opportunities exist for increasing forage intake. These include ensuring ample forage availability and quality, moderate grazing intensity, and giving careful consideration to the best kind, amount, and method of supplementation. Growth-promoting implants tend to increase feed intake and weight gains, while monensin, a feed additive, typically decreases feed intake but enhances gains through greater metabolic efficiency (NRC, 1996).

The kind and amount of supplement fed can have a large influence on forage consumption of grazing animals. The goal of a supplementation program for grazing animals is commonly to maximize forage intake and utilization. Nevertheless, supplements that stimulate forage intake may require reducing stocking rates to enable an increased daily forage intake. However, there may be occasions when supplements that decrease forage intake may be desirable as a means of extending the forage supply or enabling more animals to be carried for a set time period on the grazing unit. This approach to supplementation often has much in common with complemental feeding.

Feeding small amounts (1–2 lb daily) of high-protein supplements to cattle on low- to medium-quality roughages or grazing on dormant native grass range usually increases forage dry matter intake (Lusby and Wagner, 1987). Increases in ruminal NH_3 and volatile fatty acid concentrations, forage digestibility, and ruminal digesta passage rates have also been noted (Caton et al., 1988). Each succeeding addition of protein supplement to a protein-deficient diet is accompanied by a diminishing increase in forage consumption until the optimal level of about 6–9% dietary crude protein is reached (Allden, 1981; Reeves et al., 1987; Allison, 1985). Supplementing ewes on Montana winter range with a 20% crude protein supplement has not reduced forage intake when fed at 0.2–0.3% of body weight (Thomas and Kott, 1995). The additions of shrubs to grass winter range has the potential to increase total dry matter intake, while enhancing limited dietary nutrient levels of phosphorus and carotene as well as protein.

The greater the protein deficiency in the diet, the greater the benefits of protein supplementation that can be expected on forage intake. Nevertheless, providing a protein supplement when dietary crude protein levels were apparently low has sometimes failed to affect forage intake, possibly resulting from assuming inflated protein requirements, underestimating the degree of selective grazing occurring, or other factors masking the supplementation effects. When protein is not limiting in the forage, the high protein supplement may function merely as a high-energy supplement (Kartchner, 1981; Cook and Harris, 1968b; Judkins et al., 1985; and Allden, 1981).

Grain-based supplements (high energy/low protein) tend to either maintain or more likely reduce forage intake, and 3 lb or more for cattle and 0.5–1 lb for sheep of such supplement may only cause substitution of forage by supplement (Allison, 1985; Lake et al., 1974; Rittenhouse et al., 1970; Bellows and Thomas, 1976; Cook and Harris, 1968b). Substitution has been greatest when herbage is abundant and livestock gains are good, and least when pasture has been less plentiful and gains are poor (Allden, 1981; Allison, 1985). Substitution of grain for forage may be desirable when the objective is to stretch the forage supply or enhance animal production levels by enriching the dietary energy levels but undesirable when maximizing utilization of forage is an objective (Horn and McCollum, 1987). However, when forage is sparse, feeding energy supplements may not materially affect the intake of grazed forage (Minson, 1990), unless grazing time is limited.

Forage intake reduction resulting from high-energy, low-protein supplements will be greater when the overall protein level in the forage and in the diet is low and when larger amounts of supplements are fed. Supplements high in nonstructural carbohydrates (i.e., high energy/low protein) often have a negative effect on the intake and digestibility of mature, low-quality forages but may stimulate fiber digestion by increasing microbial activity when fed in limited amounts (Bowman and Sanson, 1996). When high-energy/low-protein supplements are added to diets of medium- and high-quality forages, the effect on forage intake appears principally to be a one:one substitution. In general, it is seldom possible to make significant gains in the overall energy intake of cattle grazing winter grassland range by providing only a source of readily available starch (i.e., cereal grains characterized by high-energy/low-protein) (Kansas State University, 1995).

Supplementation methods that do not interfere with grazing time and patterns are also suggested for maximizing forage intake. In an inverse scenario, Dougherty *et al.* (1988) reported that supplementing beef steers with 3.5–10 lb of ground corn daily before grazing on stockpiled orchardgrass did not reduce forage dry matter intake rate, rate of biting, or bite size; this was probably the result of restricting grazing time in the study to 4 hr daily.

Restricting water consumption by livestock reduces dry matter intake, and any factor that reduces water consumption below 75% of free-choice consumption is apt to reduce forage intake (NRC, 1987; Forbes, 1986). Severe water restriction will severely limit feed intake and animal performance; less frequent watering can also reduce forage consumption (Musimba *et al.,* 1987). At least under high temperature conditions, adequate water for ruminant big game animals also appears critical. In a south Texas study with white-tailed deer, after four days of restricting water at 33 and 67% of *ad libitum* consumption, dry matter intake was reduced 16 and 52%, respectively, with some deer in the severe water restriction group refusing to eat (Lautier *et al.,* 1988). Both water and food intake are decreased when animals are forced to walk long distances in search of water (i.e., anything over one mile in Australia) (Squires, 1978). The data indicate a loss in food intake associated with increased walking despite the increasing energy cost.

II. ANIMAL EQUIVALENCE AND THE ANIMAL UNIT MONTH

A. BASIS OF ANIMAL EQUIVALENCE

Concepts of animal equivalence have been developed to express different kinds and classes of grazing animals in a common form. Animal equivalence, when quantified as animal unit equivalents (AUEs), provides a basis of summarizing grazing capacity needs and calculating stocking rates and other stocking variables. When used within a framework of animal unit equivalence, the animal unit month (AUM) can be used to quantify forage needs/forage supply relationships relative to grazing lands.

As previously discussed, dry matter intake of forages results from a complex interaction of many factors. These factors include not only animal factors but also forage, weather, and management factors. Consideration must be given to the fact that the grazing animal rather than the grazier primarily determines the upper limit of dry matter or nutrient intake. Thus, unless based only on animal factors, potential dietary intake can be anticipated to be highly variable and incapable of providing a constant around which animal equivalence can be based (Scarnecchia, 1985a).

Animal weight interpreted as liveweight was used by both Stoddart and Smith (1955) and Voisin (1959) as the single variable upon which animal equivalence was based. Stoddart and Smith adopted 1000 lb (453.6 kg) liveweight ("or roughly equivalent to a cow and a calf") as comprising one animal unit without attaching any intraspecific limitations. Thus, a 1000-lb cow or bull would comprise 1.0 animal unit; a 500-lb calf, or apparently even a 500-lb elk, would comprise 0.5 animal unit. This led to the use of direct exchange ratios (e.g., five sheep equal one cow). Voisin defined the animal unit as 500 kg (1102.5 lb) of intraspecific animal liveweight. Both definitions excluded herbage and other environmental variables, and neither differentiated between other animal factors such as lactation vs. non-lactation.

Since dry matter intake is more closely related to metabolic weight than liveweight, the following formula recognizes the latter relationship (Lewis et al., 1956):

$$\text{AUE} = \frac{W^{.75}}{1000^{.75}} \quad \text{where } W = \text{the average of monthly weight of the animal in question}$$

The Society for Range Management (1974) defined the **animal unit** as one mature (1000 lb) cow or the equivalent based upon average daily forage consumption of 26 lb dry matter per day. Scarnecchia (1985a) recognized this definition as being based both on animal weight and what was essentially a daily potential intake or animal demand, but this still failed to distinguish between a lactating cow with higher dry matter intake and a dry cow of equal weight.

Neely (1963) proposed alternative bases for an animal unit depending upon the stage of production. For classes of beef cattle where body maintenance is the goal, he suggested the mature, pregnant cow weighing 1000 lb. Where gain in body weight is an important factor he suggested that an 800-lb steer gaining 1.5 lb/day be used as the basis of the animal unit; however, this dual standard has neither been accepted nor used in actual practice. Scarnecchia (1986) has opted for the use of a single basis for expressing animal equivalence, since basing it vaguely on a combination of all of these variables simultaneously could not be the basis of good management.

Various bases have been considered for animal equivalence: (1) live weight, (2) metabolic weight, (3) energy needs/demand, (4) energy intake, (5) dry matter needs/demand, and (6) dry matter intake. Scarnecchia (1985a) concluded that only an animal's demand on pasturage (i.e., potential forage intake) allows consideration of this and other differences between classes of the same kind of grazing animal. Animal demand is concluded here to be the best single quantitative basis for animal unit equivalence, but other bases for animal unit equivalents have also been employed in some situations to provide greater versatility for its usage.

Animal demand is defined as potential forage dry matter intake of ungulate herbivores based solely on animal-related factors such as body size, body condition, stage of life cycle, production stage, etc. This definition excludes animal-pasture and animal-environment interactions that will affect intake on actual site-specific situations. It allows a given animal to have an animal unit equivalent independent of the kind and quality of herbage it is eating, the temperature of its environment, or even whether the animal is on pasture or in a drylot consuming primarily forage.

While simplifying an animal unit to a unit of animal demand equal to 12 kg (26.46 lb) of dry matter per day, Scarnecchia and Kothmann (1982) acknowledged this only approximated the demand of a mature, dry cow, without regard to losses to fouling or trampling. For further simplification while maintaining acceptable accuracy, 26 lb dry matter or 9 lb total digestible nutrients (TDN) per day of animal demand is utilized in this publication as a standard for the animal unit (equivalent to 2.5 lb dry matter or 0.9 lb TDN per 100 lb liveweight). Animal unit equivalence expressed as physiological energy needs including maintenance, growth, gestation, lactation, and change in body condition also permits comparing animal classes within species on the basis of energy requirements (Scarnecchia and Gaskins, 1987) (Note: See Scarnecchia and Gaskins [1987] for further discussion on describing animal demand in terms of energy, as well as on the relationships among units of animal demand, units of forage supply, units of intake, and animal units.)

B. ANIMAL UNIT EQUIVALENTS

Based on the preceding discussion, an **animal unit equivalent (AUE)** is defined as a number (either decimal or multiple) expressing the animal demand (for forage) of a particular kind and class of animal relative to that of an animal unit. It follows, then, that the **animal unit (AU)** is defined as a mature, non-lactating bovine weighing about 1000 lb or its equivalent in other classes or kinds of ungulate herbivores based on animal demand. Table 11.2 suggests animal unit equivalents for use with various kinds and classes of livestock and big game. (Note: defining the animal unit as an estimate of the total impact rate of a grazing animal on an ecosystem [such as in Perrier, 1996] is not only vague but compromises the traditional utility of the term and should be avoided [Scarnecchia, 1990].)

For the animal demand (i.e., potential forage dry matter intake) approach to animal equivalence to be most useful, it should be applied within an animal species. The equivalents provided in Table 11.2 can be used directly in determining mixed-species stocking rates or in converting from one kind of animal to another theoretically only when there is complete overlap in their diets (i.e., they are selecting the same proportions of the various plant species and plant parts in their diets).

Selectivity differences between classes of grazing animals can generally be ignored when determining animal unit equivalents. It is probable that the offspring of ruminant grazers, because of smaller mouth parts and only gradual transition into functional grazers themselves, will select forage somewhat different than their dams, but such differences can be expected to rapidly disappear by maturity (Smith *et al.,* 1986). However, differences between kinds of grazing animals can range

TABLE 11.2 Suggested Animal Unit Equivalents by Kind and Class of Herbivore Based on Animal Demand[a]

Kind and class of herbivore	Animal unit equivalents
Cattle	
Mature bull (24 months and over, 1700 lb average)	1.5
Young bull (18–24 months, 1150 lb average)	1.15
Cow and calf pair	1.35
Mature cow, non-lactating, 1000 lb	1.0
Pregnant heifer, non-lactating (18 months and over)	1.0
Yearlings (18–24 months, 875 lb average)	0.9
Yearlings (15–18 months, 750 lb average)	0.8
Yearlings (12–15 months, 625 lb average)	0.7
Calves (weaning to 12 months, 500 lb average)	0.6
Calves (weaning at 8 months, 450 lb average)	0.5
Sheep and goats	
Ewe and lamb pair	0.3
Doe and kid pair	0.24
Sheep, mature, non-lactating	0.2
Goats, mature, non-lactating	0.17
Weaned lambs and kids	0.14
Other animals	
Draft horse (mature)	1.5
Saddle horse (mature)	1.25
Bison (mature)	1.0
Moose (mature)	1.0
Elk (mature)	0.65
Deer, mule (mature)	0.23
Deer, whitetail (mature)	0.17
Antelope (mature)	0.17

[a]Consider all replacement heifers and young bulls 24 months and over, pregnant or lactating heifers over 18 months, and replacement sheep and goats 12 months and over as mature (adapted from Vallentine *et al.,* 1984). These equivalents are useful for converting between animal classes of the same species. However, interspecies conversions require the assumption of complete dietary uniformity and overlap, an assumption almost never true.

from high to very low levels of dietary overlap, and exchange ratios must be adjusted accordingly.

For growing cattle weighing 350–800 lb, an allowance of 0.11 AU/hundredweight of body weight, as suggested by Shultis and Strong (1955), is commonly used in developing animal unit equivalents. In areas or with breeds of livestock in which mature weights average significantly over or under the standard 1000 lb, greater accuracy can be obtained in figuring AUM requirements by adjusting the animal equivalents on a liveweight basis (or preferably metabolic weight, i.e., $W^{0.75}$).

Lactation by the dam and forage consumption by the suckling offspring increases animal demand substantially over maintenance or maintenance plus gestation of the dam alone. Based on a review of pertinent literature on the effects of lactation on forage intake and as a matter of convenience, a uniform AUE factor of 1.35 rather than a slightly inclining one has been assigned the cow-calf pair from calf birth to weaning. This is approximately equivalent to an AUE of 1.2 for a cow-calf pair through a production year. This is in general agreement with a yearling steer:cow-calf substitution ratio of 1.8:1 set by Forero *et al.* (1989); this weight-to-weight ratio provided equivalent utilization of shortgrass prairie and seeded sideoats grama in eastern Colorado. Rather than standardizing to the dry, mature cow, Waller *et al.* (1986) standardized to the lactating, mature cow and down-graded the AUEs for some but not all of the other classes of beef cattle.

Variability and lack of precision in defining an animal unit equivalent can limit if not destroy its effective use (Vallentine, 1965). A mere count of the number of breeding females should not be used to approximate the number of animal units in a cow-calf or ewe-lamb enterprise; many herds are comprised of a mix of animal classes. Continued use by some agencies and individuals of the mature cow "with or without calf" as the base animal unit, because of great difference in animal demand, results in a margin of error unacceptable in determining animal equivalents.

Historically, U.S. federal land management agencies counted all cattle 6 months of age and older as one animal unit, with suckling calves under 6 months of age being ignored for determining permit compliance. This oversimplification does not precisely record actual grazing pressures, discriminates against ranchers entering other than cow-calf pairs, and fails to foster conservation of the grazing resources. Filling permits with cow-calf pairs (1.35 AUE) would result in a 35% increase in animal demand over the standard non-lactating cow, while filling with weaned calves (0.5 AUE) would reduce the animal demand by 50%. These great discrepancies are gradually being corrected by assigning AUEs more appropriately to the specific animal classes being entered.

C. THE ANIMAL UNIT MONTH

An **animal unit month (AUM)** is the basic unit of grazing capacity and is defined as animal demand (i.e., potential forage dry matter intake) by one animal unit for one month (30 days). As a quantitative measure of carrying capacity, the animal unit month is further described as 780 lb of oven-dry forage (26 lb. dry matter daily × 30 days). For added convenience in making grazing capacity calculations, an **animal unit day (AUD)** is defined as 1/30 of an animal unit month. In some areas, the **animal unit year** (or **animal unit yearlong**) is equivalent to 12 AUMs when harvested by one animal unit over a continuous 12-month period is used.

The animal unit month concept was developed as a common measure of grazing capacity produced and available for utilization as well as for the grazing capacity needs of ungulate herbivores of specified numbers, kinds, and classes. This permits making grazing animal-forage balance projections and comparisons of current and historical grazing capacity needs and production (Fig. 11.2). The AUM is particularly useful with livestock production and growing enterprises using

FIGURE 11.2   The animal unit month provides the basis for making grazing animal-forage balance projections for the entire ranch or individual animal production or growing enterprises where grazing lands provide the principal source of carrying capacity.

range and other grazing lands as the principal sources of forage. However, under the wider scope of "carrying capacity," the AUM can be extended to include harvested roughages and even energy concentrates, particularly when fed in controlled amounts in conjunction with pasturage. Under this situation, harvested roughages and energy concentrates can be converted to quantitative AUMs on the basis of 270 lb of TDN (9 lb TDN per day × 30 days) or 540 therms (megacalories) of digestible energy being equivalent to one AUM (Table 11.3).

Since the grazier exerts only minimal direct control of forage consumption by grazing animals, free-choice consumption is assumed as the basis of the AUM. When harvested forages and limited amounts of energy concentrates are fed complementary to the pasturage but in controlled amounts, this can be calculated as added carrying capacity without seriously compromising the basic concept of the AUM. However, when harvested forages and concentrated energy feeds are fed with minimal intake restriction or free choice for more rapid weight gains, increased feed consumption will commonly increase energy intake by 50–100%. This suggests that under drylot conditions or when pasturage makes only minimal contribution to the daily ration, use of the AUM should be foregone and rations calculated on a nutrient weight basis.

The TDN equivalence of an AUM used by technicians in the past has been high-

TABLE 11.3   AUM Equivalents of Dry Roughages, Silages, and Grains for Maintenance or Minimal Weight Gains

| | | | AUM equivalents[a] | |
	Dry matter (%)	TDN (oven dry basis) (%)	Air dry basis[b] (AUM/T)	As fed basis (AUM/T)
Dry roughages				
Corn stover, mature	90	59	3.93	3.93
Sorghum fodder	90	58	3.87	3.87
Alfalfa hay	90	57	3.80	3.80
Sorghum stover	90	53	3.53	3.53
Bromegrass hay	90	53	3.53	3.53
Prairie hay	90	51	3.40	3.40
Meadow hay (sedge)	90	46	3.07	3.07
Wheat straw	90	44	2.93	2.93
Barley straw	90	41	2.73	2.73
Silages				
Corn fodder silage	28	70	4.67	1.45
Corn stover silage	27	58	3.86	1.16
Sorghum fodder silage	29	57	3.80	1.22
Alfalfa silage	29	57	3.80	1.22
Alfalfa-brome silage	30	55	3.67	1.22
Beet top silage	21	54	3.60	0.84
Energy feeds				
Corn grain	90	91	6.07	6.07
Milo grain	90	83	5.53	5.53
Barley grain	90	83	5.53	5.53
Oats grain	90	76	5.07	5.07

[a]Based on 270 lb of TDN or 540 megacalories (Mcal) or therms of digestible energy being equivalent to one AUM for maintenance or minimal weight gains (adapted from Vallentine et al., 1984).
[b]Assumed 90% oven dry.

ly variable, primarily as a result of variable definitions of the animal unit. Shultis and Strong (1955) considered 400 lb of TDN equivalent to one AUM, basing their conversion on an average requirement of 13.2 lb of TDN daily for cows nursing part of the year. Harris (1962) used 480 lb TDN as equivalent to an AUM based on the animal unit being a 1000-lb cow producing 25 lb of milk testing 4% fat daily and requiring about 16 lb TDN daily. Even when basing the animal unit on a mature beef cow in gestation, Sampson (1952) further defined an AUM as 300 lb of TDN. The TDN equivalence of an AUM must be standardized before being used. An equivalence of 270 lb of TDN equal to one AUM, as suggested and used in this publication, assumes that 9 lb TDN/day is typical of the energy requirement of the mature, 1000-lb, non-lactating bovine and further assumes only maintenance to typical weight gains on pasture for other animal classes.

AUMs produced or otherwise made available and required by associated animals can be totalled and compared on the basis of dry matter or energy values but not on a mixture of both. Hobbs *et al.* (1982), in working with big game animals, concluded that nutritionally based estimates of grazing capacity were satisfactory using either energy or protein, with a consideration of both being even more precise. However, estimates varied substantially between years since both quantity and quality of forage were involved. Wallmo *et al.* (1977) noted the problem of basing winter grazing capacity of mule deer in Colorado on a diet quality level capable of supplying maintenance energy requirements. Although grazing capacity of a particular range was calculated as zero, it was noted that many deer survived through the winter.

Kearl (1970) compared the utility of energy-based and standard dry matter-based carrying capacity; he concluded that the latter was more convenient in budgeting and planning work for livestock systems using pasturage but that the net energy system was preferred for drylot feeding situations. It is concluded that dry matter is the preferred basis for use with grazing because: (1) it relates directly to forage production, utilization, and disappearance, as well as grazing pressures; (2) TDN or digestible energy values are often unknown for pasturage, are difficult to determine precisely, and may not be fully meaningful for some forage sources; and (3) the effects of environment and activity factors on energy requirements of grazing animals are complex and not easily calculated.

Under the animal demand concept of animal equivalence, the AUM is only a quantitative measure of forage. Even when the AUM is based on energy content, still no consideration is given to mineral, protein, or vitamin content. Although AUMs from different forage sources have a similar gross carrying capacity, they often differ markedly in nutritive quality and ability to promote animal production. Since AUMs can differ greatly in quality, the sources of AUMs in relation to quality must be carefully recorded. The various sources of AUMs must then be utilized to most efficiently meet the nutrient requirements of each animal enterprise.

Both reasonable accuracy and simplicity are important in figuring carrying capacity needs in AUMs. Acceptable precision requires that a herd first be classified by age and class and then converted to animal units. For example, replacement heifers, weaned calves, yearlings, and bulls must be considered along with the breeding cows on a mixed-enterprise cattle ranch. Cow-calf enterprises typically require annually about 16 AUMs per breeding cow, while a mixed cow-calf-yearling enterprise may require 20 AUMs rather than only 12 AUMs per breeding cow. The actual carrying capacity in AUMs required per breeding cow on a cattle ranch will depend upon the management practices followed and, of course, will not be meaningful if the cow-calf enterprise is not found or plays an inconsequential role among the beef cattle enterprises on the ranch.

In addition to being the best measure for comparing the seasonal or yearlong carrying capacity needs with the sources of carrying capacity correspondingly allotted, the AUM also has other important uses. It provides a useful basis for leasing grazing capacity of private as well as publicly owned grazing lands. It can also be used as a prime indicator of ranch or grazing land value since the appraised market value and loan value depend primarily on the number of AUs it will support.

D. ANIMAL SUBSTITUTIONS

Substituting between different classes within animal species presents no problem in accounting for animal unit months, since differences in dietary and site preferences are generally minimal. However, when adjusting animal numbers disproportionally in animal species mixes or when converting from one species to another, great difficulty is met in predicting the effect on grazing capacity. Such changes modify the nature of the grazing capacity demand and, at the same time, the amount of grazing capacity in AUMs that can be provided.

An **animal substitution ratio** (synonymous with **animal conversion ratio** or **animal exchange ratio**) is a numerical ratio of numbers (or units or stocking levels) of one animal species to another, for use in partly or completely converting grazing use by one animal species to another or in partitioning grazing capacity between two animal species. For example, five sheep have been traditionally assumed equivalent in grazing capacity demand to one cow. However, it appears that an animal substitution ratio is universally site specific—and probably also management specific—since it is based on a unique set of environmental, forage, animal-herbage, and animal-area variables and depends on relative animal population levels (Scarnecchia, 1985a). It may be useful to relate AUMs produced and/or needed in terms of "sheep" AUMs, "cattle" AUMs, and "mule deer" AUMs.

Hobbs and Carpenter (1986) have concluded that an animal unit must be related to a specific kind of animal to be meaningful. As an example, Leckenby *et al.* (1982) suggested that allocation of forage to mule deer be based on "the conversion of available forage to deer unit months, a measure similar to animal unit months but also accounting for overlap of diet and season and area of range use." However, such a conversion must still remain essentially site or grazing-unit specific.

Conversion equivalents between animal species, even on a specified site, are not constant as sometimes believed but change with each shift in animal numbers from one kind of animal to another unless a complete conversion is made (Cook, 1954). A shift away from the optimum animal species mixture will reduce grazing capacity, while a shift from single use or a less optimum mixture towards the optimum mixture will increase grazing capacity. Cook (1954) concluded that conversion factors would be constant only at or near the point of optimum mix when changing from cattle to sheep or the reverse. Only then could shifts be made at the traditional ratio of 1:5 (cattle to sheep) or 5:1 (sheep to cattle).

In Table 11.4 hypothetical conversion factors have been utilized to explain the effects of partial to complete conversion between cattle and sheep on actual grazing capacity. The three factors considered are (1) advantage of mixed grazing, (2) advantage of animal species adaptation to site (principally botanical composition of the standing crop and the terrain), and (3) the relative numbers component in

TABLE 11.4   Direction and Relative Extent of Change in Grazing Capacity Anticipated from Making Partial to Complete Grazing Animal Conversions Using Hypothetical Situations and Assumed Exchange Ratios[a]

Order of exchange in hypothetical situations	Mixed grazing advantage[b]	Animal adaptation advantage[c]	Projected effect on grazing capacity	Assumed exchange ratios
Sheep to cattle (the trend of the recent past)				
Equally adapted range				
First 3rd	+	0	Increase	2–4:1
Middle 3rd or complete	0	0	Stay same	4–6:1
Last 3rd	−	0	Decrease	6–8:1
Browse-forb and/or steep topography range (better adapted to sheep)				
First 3rd	+	−	Stay same	4–6:1
Middle 3rd or complete	0	−	Decrease	7–9:1
Last 3rd	−	−	Greatly decrease	10–15:1
Grass range and/or gentle topography (better adapted to cattle)				
First 3rd	+	+	Greatly increase	1–3:1
Middle 3rd or complete	0	+	Increase	3–4:1
Last 3rd	−	+	Stay same	4–6:1
Cattle to sheep (reverse of recent past trend)				
Equally adapted range				
First 3rd	+	0	Increase	1:6–8
Middle 3rd or complete	0	0	Stay same	1:4–6
Last 3rd	−	0	Decrease	1:2–4
Browse-forb and/or steep topography (better adapted to sheep)				
First 3rd	+	+	Greatly increase	1:10–15
Middle 3rd or complete	0	+	Increase	1:7–9
Last 3rd	−	+	Stay same	1:4–6
Grass range and/or gentle topography (better adapted to cattle)				
First 3rd	+	−	Stay same	1:4–6
Middle 3rd or complete	0	−	Decrease	1:3–4
Last 3rd	−	−	Greatly decrease	1:1–3

[a]Assumes that all range areas included in examples have sufficient diversity available in forage species and terrain that maximum grazing capacity can be realized only under mixed grazing of cattle and sheep.
[b]Mixed grazing: +, entering; 0, neutral; −, leaving advantage.
[c]Animal adaptation to vegetation and/or terrain; +, more adapted; 0, neutral; −, less adapted.

order of exchange. In the example, it is assumed that all range areas have some diversity in forage species and topography, thereby maximizing grazing capacity only under mixed grazing. Thus, the marginal substitution ratio can be expected to be high initially if going from single use to multiple use but low initially if going from multiple use towards single use.

Hobbs and Carpenter (1986) have argued that animal unit equivalents should be weighted by approximations of dietary overlap. Investigators (Flinders and Conde, 1980; Botha et al., 1983) have attempted to calculate substitution ratios by adjusting animal unit equivalents based on metabolic weight for different animal species based on the average dietary overlaps of these animal species. These investigators proposed the following equation in calculating animal unit equivalents:

$$X = \frac{1000^{0.75}}{\text{liveweight}^{0.75} \times \% \text{ dietary overlap}}$$

In one example given by Flinders and Conde, sheep and cattle were given a 35% dietary overlap; rather than having 1 cow being equivalent to five sheep, 1 cow would then be equivalent to 14.3 sheep (5/0.35). In another example in which the dietary overlap between cattle and pronghorn was given as 8%, it was calculated that 1 cow would be equivalent to 68.7 pronghorn (metabolic weight equivalence of 5.5/0.08). However, left unclear are the exact circumstances under which these dietary overlaps would prevail. These conversion rates would be valid only when cattle under single grazing were totally converted to pronghorn grazing and then only up to the maximum grazing capacity of the habitat for pronghorns for single species grazing; dietary overlaps would also have to be determined for each case situation.

Scarnecchia (1985a, 1986) has effectively argued against this procedure and what amounts essentially to a redefinition of the animal unit equivalent concept as follows: (1) the dietary overlap of two animal species is highly dynamic rather than being constant, as required by that approach, as it is subject to many variables including season of use, stocking density, and stocking rate; and (2) incorporating an herbage-related factor greatly reduces, if not negates, the universality of the animal demand concept by making its application grazing unit-specific. It becomes impossible to compare stocking rates or other stocking variables between any two units if the animal unit equivalents in those stocking rates are specific to each grazing unit.

Severson *et al.* (1968), from a study on a big sagebrush range in Wyoming's Red Desert, reported highly variable equivalences between sheep and antelope depending upon the bases used in calculation. Using body weights for determining animal units, 1.03 sheep would be equivalent to (or used to replace) one antelope. Using total forage consumed, 1.13 sheep would be equivalent to one antelope. Based on shrub intake, 5.67 sheep were equivalent to one antelope, but when based on grass intake only 0.02 sheep was equivalent to one antelope. There was an 8.2% overlap in the diets of sheep and antelope in the study when considered on a year-long basis.

III. FORAGE ALLOCATION

A. THE NEEDS

Under mixed grazing, the manager must decide how many of each herbivore species to stock on a given grazing unit. Determining the optimum season, number, and mix of animal species must be based on the amounts of different forages produced in relation to their preference and acceptability by the different kinds of grazing animals. A determination of grazing capacity is meaningful only when consideration of the mix of animal species is included. In order to obtain the maximum allowable use of the total vegetation, efforts must be made to prevent overgrazing of important plant species or species groups and to maintain sufficient cover on the grazing lands to protect the soil from wind and water erosion. Animal numbers must be adjusted to meet both plant and animal needs in line with animal production objectives. Nevertheless, although empirical evidence indicates that multi-species grazing always increases grazing capacity, it is not possible to predict exactly how the interaction of several foraging species of herbivores will affect total forage demand (Walker, 1994).

A commonly expressed rationale is the need to determine grazing capacities when broken into component parts for the various herbivore species in the animal mix. **Forage allocation** is the partitioning of a standing forage crop or its associated grazing capacity between different kinds or classes of ungulate herbivores, or for the same animal species for different seasons of the year. However, forage allocation is a complex biological problem without a simple, objective solution (Fig. 11.3). For example, Leckenby et al. (1982) recommended that forage allocation for each animal species should be based on (1) season of use, (2) seasonal availability of the forage, (3) seasonal nutrient requirements of the grazing animals, (4) number of animals planned for by sex and class, and (5) conversion of available forage to animal months of grazing capacity for the respective animal species.

Forage allocation between domestic livestock and big game, generally under separate ownership in the U.S., has long been a controversial problem (Holechek, 1980). The most confounding and frustrating process facing public range managers, according to Wisdom and Thomas (1996), is that of stocking allocation among wild and domestic ungulates, the most controversial being that between elk and cattle. It has been commonly thought and promoted that a reduction in numbers of any herbivore species would improve forage conditions for the other (i.e., the antithesis of mixed grazing). There has often been public resistance to reducing big game animal populations even on rangelands overgrazed/overbrowsed by big game animals. Mossman (1975) noted that even under controlled game ranching there is a tendency to "stockpile" the wildlife and concluded that overstocking with big game animals just as with domestic animals is improper utilization and conservation of the grazing resources.

Van Dyne et al. (1984a) have reviewed the difficulty of developing useful and defensible forage allocation procedures, particularly when different animal species grazed together are under different ownerships or are being managed under objectives other than animal production. According to these authors, often "only livestock numbers can be reasonably manipulated. Does this mean, for example, that deer numbers are to be left unmanaged and changes in allocation taken from or added to livestock as wildlife numbers wax and wane? Or is that whole problem ignored and allocations simply set by an inflexible formula? Who should decide how many deer are enough on a given allotment, and what are the allocation

FIGURE 11.3   Forage allocation is a complex biological problem without simple, objective solution because of differences in dietary and site preferences; for example, to provide adequate elk grazing capacity after cattle grazing or vice versa. Forage allocations and substitution ratios should be considered site-specific and probably also management-specific.

criteria? If manipulation of numbers of some grazing animals is difficult, how are 'desired' allocations to be accomplished?"

Maintenance and improvement of perennial forage resources should always be a primary consideration when forage is allocated to different animal species. Vavra (1992) concluded that animal unit equivalencies can be used to make initial stocking estimates, but the vegetation must be the principal determiner in figuring proper animal numbers and mixes. In work on improved dryland subclover-perennial grass pastures, Bedell (1973) found that botanical composition of the available forage even in this simple forage plant mixture was influenced markedly by animal ratios. Sheep alone or in animal unit ratios of 2:1 or 1:1 with cattle resulted in a decline of the subclover both within and progressively among seasons. Grazing cattle alone or at a relative high ratio of cattle to sheep tended to hold the aggressive grasses in check while allowing more subclover to prevail longer into the grazing season and into succeeding years, thereby contributing to higher per acre animal performance. Similarly, grazing a 2:1 animal unit ratio between yearling steers and ewe-lamb pairs on irrigated Ranger alfalfa-orchardgrass pasture prevented excess pressure being placed on the alfalfa by the sheep component in the mix (Heinemann, 1969).

B. THE METHODS

The following information was listed by Holechek (1980) as necessary background information for allocating grazing capacity under mixed grazing: (1) the key forage plants for each animal species, (2) the degree of use on key plant species, (3) the ability (and opportunity) of each herbivore to switch from preferred to alternate forages, (4) the key areas where dual use occurs, and (5) the repeatability of dual use on key areas from year to year. Even with this information at hand there appears to be no firm formula which can be used to determine the precise proportion of different species of animals needed for an optimum animal mix; present forage conditions and past experience of the grazier must be included to obtain the optimal animal mix to improve both the vegetation and animal performance (Taylor, 1986a). To this should be added a continuing program of animal and vegetation monitoring to further refine the animal mix.

Several procedures are available for evaluating dietary similarity and overlap. The most common approach to allocating forage to different kinds of animals has been to maximize use based on different plant preferences of the different kinds of animals while typically including constraints on the degree of use, animal requirements, etc. (Rittenhouse and Bailey, 1996). Modeling to simulate dietary similarity and overlap can provide useful background information and predict if and how goals can be achieved but cannot be applied directly in making resource partitioning and optimal forage allocations in typically complex grazing situations (Hanley, 1982b; Van Dyne *et al.*, 1984b; Nelson, 1984). Rittenhouse and Bailey (1996) concluded that computer models developed to allocate forage have had major drawbacks in that (1) the grazing environment is represented by species composition, not nutrient content; and (2) they are not spatially or temporally explicit (i.e., are site-specific even when described in detail).

A forage allocation method based on a single key forage species was proposed by Smith (1965). The author's hypothesis was stated as: "Correct substitution rates of one grazing animal for another under common use are uniform, being governed at any point by the utilization standard of some single species. This key species may vary at different levels of animal combinations, thus changing the rate of substitution to another but still constant rate." His procedure was based on three assumptions: (1) sufficient forage of the key species is available so that animals are not compelled to adjust their normal forage preference to offset lack of forage; (2) common use does not alter the preference of either animal for the major forage species; and (3) the use factors for each animal species are proportional to its population on the grazing unit. The author suggested these three conditions may not always be precisely met but presumed the small deviations would be insignificant. However, it seems more likely that these three assumptions would seldom, if ever, be fully met and that this forage allocation method would be of doubtful validity.

Jensen (1984) proposed a "limiting factor method" as a more equitable forage allocation approach resulting from expanding the forage base: "In the limiting factor method the plant production available for allocation as grazable forage is aggregated by plant class, i.e., grasses, forbs, and shrubs. The pounds of forage consumed in a 30-day period for each animal in the management unit is multiplied by the dietary preference of each animal to determine the amount of vegetation each animal consumes by plant class. When the production of a plant class has been allocated, no additional animals may be grazed in the area even though forage is available from the other two plant classes." This approach also assumes no significant versatility in what animal species select under a maze of situations presented to them.

McInnis and Vavra (1987) utilized an index of dietary overlap that totalled the overlaps for all individual plant species. The percent overlap for each forage species shared by a pair of ungulate species was defined as the lesser percentage consumed by one ungulate species in its diet. However, Holechek *et al.* (1984) considered the results of such analyses not to be useful in making forage allocation decisions. Rather, they concluded it was the diet composition per se that must be utilized in quantitative decision making rather than overlap information described in an index. Since large herbivore species differ in their food habits and distribution across rangelands, animal equivalencies and forage allocation based solely on quantitative forage intake of each species do not give an accurate estimate of potential stocking rates for a mix of diverse herbivores (Vavra, 1992).

Wisdom and Thomas (1996; pages 167–171) suggested that stocking allocation between cattle and elk be based on the following steps: (1) calculation of the stocking rate to cattle as sole grazer, based on the biomass of key forage species, the allowable use of such forage, the forage demand of the animals, the length of grazing period, and the spatial distribution of animals within the planning area; (2) calculation of the stocking rate for elk as the sole grazer, based on these same parameters; and (3) making a *judgment decision* of stocking allocation between the two kinds of animals, based on the stocking rates calculated above, the extent of nonuniform spatial distribution of ungulates, and the potential for forage competition. They suggested the accuracy of stocking estimates would be improved if the nutritive supply of forage was considered but noted that such data are seldom readily available and are costly to obtain.

Wallace (1984) concluded generally that due to the complexities involved in obtaining meaningful preference values and to the very limited scope of their usefulness, they should not play a major role in forage allocation decisions. Hart (1980) questioned whether allowable use factors for the different species, considering the maze of environmental conditions under which they grow and are utilized, were adequate as the sole basis of forage allocation systems. He further noted that, even when accurately derived for a specific situation, they may prevent overuse of range but still not provide the best mix of animal species to harvest the entire forage complex.

A high-priority principle is that solutions for forage allocation problems will always depend on goals that have been set, and giving priority to one animal species in forage allocation is mostly arbitrary (Hart, 1980). This suggests that when forage allocations are made between different grazing animal species, particularly when under different ownerships, the solution will often be as much managerial or political as biological.

C. THE VARIABLES

One hypothesis is that livestock are better able to adjust their diets without detrimentally affecting their nutrition than big game animals (Dwyer *et al.,* 1984). This leads to the conclusion that when vegetation composition is altered through grazing, nutrition of competing wild animals declines in comparison with that of the more flexible domestic animals. This is a further extension of the concept that small ruminants (big game animals) with relatively lower reticulo-rumen capacity must be given priority in allowing them to select the most palatable and nutritious forage in order to tolerate large ruminant (livestock) competition, else they will be unable to compete successfully with the larger ruminants. An opposite hypothesis might be that since native herbivores have co-evolved with the vegetation, they are in fact more flexible than exotic livestock in being able to adjust their diets. However, either hypothesis being universally correct such that it must arbitrarily be adhered to is unlikely.

The wide-ranging review of ungulate herbivore diets made in Chapter 10 strongly suggests that probably all species (1) have surprisingly broad food habit adaptability, (2) are more flexible in their diets than generally believed, and (3) have an amazing ability to make limited if not substantial modifications in their diets when required to do so. While the capability of dietary flexibility seems assured in all ungulate herbivore species, its bounds are not well established (Fig. 11.4).

Urness (1986) has challenged "the myth that mule deer are obligate browsers" on winter range, while noting situations may arise when there are no alternatives; he has noted that this belief in obligatory deer browsing "dies hard despite many literature sources citing importance of green grass and forbs to their diets." In a study on big sagebrush range near Kremmling, CO, mule deer demonstrated versatility in moving from one or a few plant species to others as availability changed (Carpenter *et al.,* 1979). During a 30-day grazing trial beginning in mid-January in which deer were confined in a small range pasture, forbs and grasses comprised more than 50% of the diet but rapidly declined as forage supplies declined. During the same period, sagebrush went from 2–3% to about 30% of the diet.

Preoccupation with more preferred and important forage plants when assessing competition, and its extension to forage allocation, can be misleading. The result will lead to ignoring the vital roles of lesser plant species in the diets of grazing

FIGURE 11.4   The capability of dietary flexibility seems assured in all grazing animal species, both livestock and big game, but the bounds are not well established; photo shows a cow in Texas deviating from the typical diet.

animals and perhaps more subtle competitive relationships with respect to their use and availability (Mackie, 1976). Willms *et al.* (1980a), in studies on forested summer range in British Columbia, noted that when the availability of the forages of high preference to both deer and cattle was not limiting, the percent of diet overlap was high. As their availability declined, diet overlap decreased as both deer and cattle were forced into their individual food niches. In fact, the effect of declining availability of preferred forages on the dietary composition was less for deer than for cattle. They attributed the greater ability of deer to utilize their preferred forages despite reduced availability to the greater ability of deer to be selective and to the deer occupying a greater variety of habitats than did cattle.

Another problem with most forage allocation procedures is that they often fail to take into account the fact that the forage resource is highly variable from one year to another (Holechek, 1980). This suggests that animal numbers and mixes should optimally be adjusted to forage availability on a yearly or seasonal basis. Regardless of what method of forage allocation is used and how effectively implemented, continued monitoring of the results in terms of plant and animal response is required. Vegetation conditions can change rapidly, animal behavior and interface at the forage level may not be as expected, and even the best forage allocation methods will require fine tuning. Severe disturbances such as drought, cold, snow, or flood or a buildup in animal populations may induce animals to use forage or habitats not normally used. This may be a short-run solution to the problem unless the alternate habitats are marginal rather than merely less preferred and do not increase competition with animal species already present.

12

GRAZING CAPACITY INVENTORY

An inventory or estimate of how much forage is or will become available for grazing is the basis of projecting how many animals can be grazed and for how long. The animal unit month is the basic quantitative measure that permits comparison of the amount of grazing needed with the amount available and achieving a balance between the need and supply. Factors to consider and procedures to follow in calculating the demand side of grazing capacity were presented in Chapter 11; those for the yield side of grazing capacity are presented in this chapter.

Grazing capacity refers to the total number of AUMs produced and available for grazing per acre or from a specific grazing land unit, grazing allotment, the total ranch, or other specified land area. (The term can also be used to express a proper stocking rate.) Grazing capacity is necessarily limited to that which can be harvested by grazing animals from the **standing crop** (i.e., unharvested/ungrazed plant materials standing in place at a given time). The term **carrying capacity** can be used synonymously with grazing capacity, or it can play an even more useful role when differentiated to include all nutrient resources available on a given land area, including not only pasturage but also harvested forages and other feedstuffs used to complement the grazing resources, thereby providing a means of summarizing total capacity.

In ecological parlance, carrying capacity often refers to the animal population density when mortality equals recruitment, but this level is often associated with low animal performance, extensive die-offs, and damage to the habitat. Carrying capacity is here equated with optimal animal production on a long-term basis, variously set from animal maintenance to acceptable production, the same as with grazing capacity, and not for mere animal survival over the period of greatest stress. Thus, in grazing management usage, carrying capacity is well below levels that cause die-offs and destruction of habitat (Heady, 1994).

I. RELATIONSHIPS AMONG STOCKING VARIABLES

Suggested terminology along with definitions for describing the basic relationships between animal demand (for forage), forage mass (quantity), grazing land area, and time are given in Table 12.1. These terms express animal/area, animal/forage, and forage/animal relationships under both instantaneous and cumulative time frames (Scarnecchia and Kothmann, 1982; Scarnecchia, 1985b).

A. STOCKING RATE

Stocking rate is considered the most important variable in grazing management; unless it is near the proper level, regardless of other grazing practices, the objectives of grazing management will not be met (Walker, 1995). Stocking rate relates the total animal demand for forage to the total area available to provide it; it is defined as the animal demand that has been or will be made per unit of area over a period of time (i.e., AUD per acre or AUM per acre); it can also be expressed in terms of AUMs per grazing land unit. The stocking rate will greatly affect the quantity and/or quality of the available forage, which will directly affect animal response or production on both a per-head and a per-area basis (Fig. 12.1A). The stocking rate is one of the initial decisions required to utilize the grazing capacity provided by the standing crop. The process begins by setting the initial, recommended, or permitted stocking rate that is expected to achieve full proper use.

Since stocking rate is based on the amount and duration of the animal demand for forage, any changes from either projected animal numbers or length of **grazing period** (i.e., the length of time that grazing animals annually occupy a specific land area) will alter the stocking rate. Thus, the actual or realized stocking rate

TABLE 12.1 Summary of Terminology Derived from Basic Relationships between Animal Demand, Forage Quantity, Grazing Land Area, and Time[a]

Kind of relationship	Time instantaneous	Time cumulative
Animal–area	*Stocking density*—animal demand per unit area at any instant of time, i.e., AU/acre or AU/section of land.	*Stocking rate*—animal demand per unit area over a period of time, i.e., AUM/acre or AUD/acre or their reciprocals.
Animal–forage	*Grazing pressure* (*instantaneous*)— animal demand per unit weight of forage mass at any instant of time, i.e., AU/t.	*Grazing pressure* (*cumulative*)— animal demand per unit of forage mass over a period of time, i.e., AUM/t or AUD/t; synonym, *grazing pressure index*.
Forage–animal	*Forage allowance* (*instantaneous*)— weight of forage mass per unit of animal demand at any instant of time, i.e., t/AU; the reciprocal of grazing pressure.	*Forage allowance* (*cumulative*)— weight of forage mass per unit of animal demand over a period of time, i.e., t/AUM or lb/AUD; the reciprocal of cumulative grazing pressure.

[a]Adapted from Scarnecchia and Kothmann (1982) and Scarnecchia (1985b). AU, animal unit; AUM, animal unit month; AUD, animal unit day. Forage mass is expressed on a dry matter basis.

in terms of number of AUMs harvested or removed per unit of land area during the grazing season may differ substantially from the planned or initial stocking rate. It is the actual stocking rate that should become part of the historical record for each pasture unit for use in future planning. Stocking rate figures are reliable only to the extent that animal demand—involving an accurate count of kinds and classes of livestock and big game animals—and duration of grazing are known and properly recorded.

Stocking rate is interrelated with **stocking density** (i.e., the animal demand per unit area at any instant of time). The same stocking rate exists, but not necessarily accompanied by the same plant and animal response, when equivalent but inverse changes are made in stocking density and the length of grazing period. For example, the stocking rate of 50 AUMs per 100-acre unit (0.5 AUM per acre, or 2 acres per AUM) remains mathematically the same whether realized by: (1) 10 animal units over a 5-month uninterrupted grazing period, (2) 50 animal units during a single grazing period reduced to 30 days, or (3) 50 animal units that graze for 6 days during each of five distinct grazing periods during the same grazing season. Thus, the duration of grazing can be comprised of a single grazing period or the summation of several distinct grazing periods. Still another alternative for fulfilling the designated stocking rate during a set grazing period would be the (4) continual adjustment of stocking density in relation to forage growth rather than following a fixed stocking density.

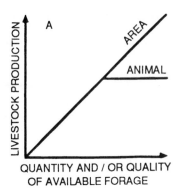

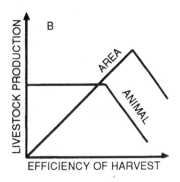

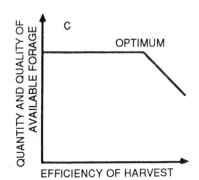

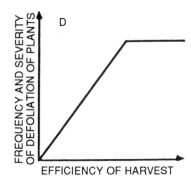

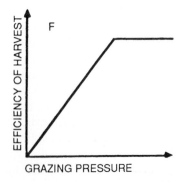

FIGURE 12.1   Relationships of selected components of grazing management (Heitschmidt, 1988).

B. GRAZING PRESSURE AND FORAGE ALLOWANCE

Grazing pressure relates the demand for forage by grazing animals to the amount of forage available to meet the demand, but the time period being referred to may be either a single instant of time or a period of time. Unless the time frame is otherwise clearly indicated, it is suggested that **grazing pressure (instantaneous)** be used to specify animal demand per unit weight of forage mass (dry matter basis) at any instant of time (i.e., AU/t) and **grazing pressure (cumulative)** be used to specify animal demand per unit weight of forage mass (dry matter basis) over a period of time (i.e., AUM/t or AUD/t). **Forage allowance** is the reciprocal of grazing pressure (i.e., weight of forage mass [dry matter basis] per unit of animal demand) but also is used in the alternative contexts of **instantaneous** (t/AU) or **cumulative** (t/AUM or t/AUD). Where a substantial component of the forage mass is rejected or is inaccessible, this fraction may preferably be excluded when estimating grazing pressure or forage allowance (Minson, 1990).

Grazing pressure (cumulative) (synonym **grazing pressure index**) is the major factor that affects the severity and frequency of defoliation of individual forage plants (Heitschmidt and Walker, 1983; Heitschmidt, 1988). As the grazing pressure (cumulative) increases, so does the frequency and intensity of defoliation and the efficiency of harvest, at least up to high levels before plateauing (Fig. 12.1E,F). Also, as grazing pressure increases over time, the most preferred plants will generally be defoliated more frequently and severely than the less preferred plants.

If the rate of depletion of the forage mass exceeds the rate of growth, the grazing animal under continuous grazing is offered a gradually diminishing forage allowance (i.e. the grazing pressure gradually increases). Under a rotation grazing system, such as a four-pasture, one-herd system, and assuming equivalent stocking rates, the forage allowance during each grazing period will be reduced by 75% while the grazing pressure will be increased fourfold. When following continuous grazing under ideal conditions, forage removal by grazing matches forage growth, and forage allowances (as well as grazing pressure) are constant (Dougherty *et al.*, 1992a; Heitschmidt and Stuth, 1991).

Grazing pressure has potential as a tool for monitoring availability of forage and predicting livestock performance. Only where forage is uniform in quality and is in sufficient quantity that it does not limit intake would grazing pressure not be expected to influence diet quality or animal production (Ralphs *et al.*, 1986b). In a study of cattle grazing on coastal bermudagrass during the growing season (Roth *et al.*, 1986b), a restriction in nutrient intake from forage or milk at the high levels of grazing pressure reduced average daily gains of stocker calves, lactating cows, and suckling calves.

In a related study, set grazing pressure levels were maintained over a 114-day grazing period by adjusting stocking densities through put-and-take variable stocking (Table 12.2) (Roth *et al.*, 1986a). Increasing grazing pressures reduced the standing forage crop available and the average daily gain of stocker calves,

TABLE 12.2   Relationships of Stocking Density, Grazing Pressure, and Weight Gains of Stocker Calves Summer Grazed on Coastal Bermudagrass, Overton, Texas, 1985[a]

	Grazing pressure level			
	High	Medium high	Medium low	Low
Stocking density (AU/acre, average for 114 days)	8.3	4.9	3.4	2.6
Standing crop available (average lb/acre dry matter)	720	1420	2420	4500
Actual grazing pressure (average AU/ton dry matter)	27.0	7.3	2.65	1.29
Average daily gain, stocker calves (lb)	0.29	1.05	1.28	1.39

[a]From Roth et al. (1986a).

with the biggest effect at medium high and high levels. At the high grazing pressure level of 27 AU per ton of herbage, the available standing crop averaged only 720 lb per acre and the average daily gain of stocker calves was held to 0.29 lb. Under the low grazing pressure level of 1.29 AU per ton, the available standing crop averaged 4500 lb per acre, and average daily gains were 1.39 lb. Under the high grazing pressure level, a younger mean age of leaf material and thus substantially lower fiber content was maintained, but animal performance was reduced as a result of restricted dry matter intake.

At given forage inventory levels, grazing pressure is directly related to stocking density; if one is doubled, so is the other. The same relationship exists between stocking rate and cumulative grazing pressure; if one is cut in half, the other is reduced by 50% also. The interaction of stocking density, grazing pressure, and the duration and number of the grazing and nongrazing periods is the basis of rotational grazing systems (Chapter 15). Grazing systems are management tools used to manipulate stocking density and grazing pressure. However, the grazing system does not set the total forage demand, as this is done by the stocking rate; the grazing system simply alters the distribution of the forage demand over time and space within the multi-grazing land unit complex.

C. GRAZING EFFICIENCY

Conversion of the aboveground net primary production of grazing lands into animal gains is not highly efficient. Heitschmidt (1984a) found that the ecological efficiency of grazing livestock on native grasslands in northern Texas—the efficiency between the primary producer (forage produced) and the primary consumer (weight gains by cattle)—was less than 3%. When compared under different grazing systems, the ecological efficiencies were 2.5% for heavy, continuous grazing;

1.2% for moderate, continuous grazing; and 2.2% for short-duration grazing with a moderately heavy stocking rate.

Grazing efficiency in forage harvesting will increase as the stocking rate and the associated cumulative grazing pressure increase (Heitschmidt, 1984b, 1988). Also, increasing efficiency of harvest will correspondingly increase the frequency and severity of defoliation of plants and, initially, animal production per unit area (Fig. 12.1D and B). Increasing efficiency of harvest will not initially affect the quantity and quality of the available forage and animal production per head but will reduce them after certain critical levels have been reached (Fig. 12.1C and B).

For a given amount of available forage, a higher proportion of plant energy and other nutrients will be channeled into the animal production cycle, and grazing animal production will increase as efficiency of harvest increases up to an optimal level; but declines in animal performance can be expected whenever the efficiency of harvest becomes excessive. This has been a nearly universal finding from grazing intensity studies covering the range of light and moderate to excessive grazing levels.

The cumulative grazing pressure (animal demand per ton) relates more directly to animal performance than stocking rate (animal demand per acre), and it may be a more useful basis of proper stocking since it automatically accounts for the typically large fluctuations in precipitation and thus forage production on rangelands (Hart, 1986). After evaluating the relationship between grazing pressure and the efficiency of forage harvest across several kinds of Texas grasslands, Kothmann et al. (1986a) concluded that proper grazing would fall within a ratio of animal demand to net forage production of 1:5 to 1:3. They concluded that harvest efficiency was maximized without restricting intake at a 1:3 forage demand-to-forage production ratio.

When grazing efficiency is based on the portion of the standing crop of forage consumed, it can be calculated as follows and can theoretically range from 0–100% (Scarnecchia, 1988):

$$\text{Grazing efficiency } (\%) = \frac{\text{Total forage intake (lb per acre)}}{\text{Total standing crop (lb per acre)}} \times 100 \quad \text{Formula 1}$$

when total standing crop of forage is calculated on the basis of either (1) standing crop at beginning of grazing period plus cumulative herbage growth, or (2) standing forage crop at the end of the grazing period plus cumulative herbage disappearance.

However, grazing animals never actually harvest 100% of the available forage, even when grazing annual forages or crop aftermath. Many plants are trampled down and soiled so that animals refuse to consume them, and some defoliated plant parts are discarded before being ingested. Thus, grazing efficiency, with a range of 0–100%, can also be based on relating forage consumption to total forage disappearance as follows:

$$\text{Grazing efficiency (\%)} = \frac{\text{Total forage intake (lb per acre)}}{\begin{array}{c}\text{Total cumulative forage}\\\text{disappearance (lb per acre)}\end{array}} \times 100 \quad \text{Formula 2}$$

Stocking rates, kinds and mixes of grazing animals, seasons of grazing, uniformity of utilization, and grazing systems all affect the efficiency of the grazing animal in consuming the forage produced in a stand, but perennial plant vigor and soil factors must also be considered.

Total herbivore consumption of the aboveground, annual biomass yield (Formula 1) has varied from as high as 40% in tallgrass ecosystems to 20% for a mixed prairie grassland, 15% for a shortgrass ecosystem, 5% in a desert grassland ecosystem, and even less under dense timber (Pieper, 1983). Of that actually consumed by ingestion, the amount consumed by insects, rodents, or other small organisms may be much greater than that consumed by the large herbivores (Van Dyne *et al.,* 1984a).

In a grazing intensity study on the Edwards Plateau of Texas, consumption accounted for almost half of the total annual net primary production at the heaviest stocking rate compared to only 22% at the lightest stocking rate (Ralphs *et al.,* 1986a). When grazing treatments on sandhills range in eastern Colorado utilized 30, 44, and 64% of the total standing crop, forage disappearance per steer day was 30, 20, and 17 1b, respectively (Sims *et al.,* 1976). On Louisiana bluestem range with pine overstory, 115, 100, and 70 lb of herbage daily, respectively, were available to sustain yearlong cattle stocking at light (26 acre per cow), medium (20 acre per cow), and heavy (13 acre per cow) rates (Pearson, 1975). When restricted to summer seasonal grazing under equivalent stocking rates, only 75 and 40 lb, respectively, were available for moderate and heavy stocking, suggesting greatly reduced levels of nonconsumptive losses under growing season compared to yearlong grazing.

In one grazing study with dairy cows on alfalfa, the waste of "usable forage" was reduced through daily strip grazing down to 11% by intensive management practices including clipping the pasture after each grazing cycle to remove the stemmy growth still standing (Porter and Skaggs, 1958). Efficient utilization is more difficult to achieve with coarse, stemmy species or after allowing better forage species to become coarse and stemmy. The forage refusal of dairy cows grazing irrigated sudangrass in one Nebraska study was reduced to only 51% by rotational grazing and to 34% under strip grazing (Rumery and Ramig, 1962).

Croplands used for grazing produce higher forage yields but sustain higher fixed costs as well as operating costs; therefore, grazing management must be geared to utilize as high a percentage of the forage produced as possible. Wedin and Klopfenstein (1985) proposed that forage consumption of aboveground biomass produced on perennial cropland pastures on the order of 65–70% is realistic. Wedin (1976) concluded that intensive grazing management was a means of increasing the true grazing capacity for a forage stand by more complete utilization. He estimated efficiency of grazing animals on tall, productive pasture mixtures in Iowa as follows:

Management	Percent dry matter lost	Calculated beef cow days per acre
Continuous grazing	50 or more	177
Rotational grazing	34	230
Daily strip grazing	25	266

Based on his calculations, daily strip grazing compared to continuous grazing could mean an additional 89 cow days per acre. Wedin (1976) further estimated that stored feeding (mechanically harvested, stored, and later fed) and green chopping could reduce the estimated loss of total dry matter produced (in the above example) to only 10 and 5%, respectively, with associated carrying capacity yields of 319 and 338 cow days per acre.

Grazing of improved pasture is seldom the most efficient way (even though it is often the cheapest way per acre) of utilizing forage (Fig. 12.2). **Green chopping** (also known as **zero grazing**) consists of mechanically harvesting forage and feeding it to animals while still fresh. Larsen and Johannes (1965), studying the utilization of alfalfa-smooth brome stands with dairy cows, reported that forage waste by cows on stored feeding (50:50 hay and silage) amounted to 8.5% of the dry matter of the forage fed, was reduced to 2% with green chopping, but increased to 33% of the forage dry matter under strip grazing. When dairy cows were grazed on al-

FIGURE 12.2   Grazing efficiency based on consumption of the standing crop is notably low, particularly on rangelands, but the alternative of zero grazing through mechanical forage harvesting is limited in scope by vegetation, terrain, and economic considerations. (Sperry-New Holland photo.)

falfa-smooth brome pasture during a 3-year study period, Van Keuren *et al.* (1966) found strip grazing increased cow days per acre by 14% over regular rotation, but green chopping increased cow days per acre 53% over regular rotation. The greater efficiency of green chopping over grazing has also been reported by others (Walton, 1983; Brown *et al.,* 1961; Hart *et al.,* 1976; Shaudys and Sitterley, 1959; Ittner *et al.,* 1954; and Hull *et al.,* 1961).

Most rangelands and many permanent pastures cannot physically be green chopped, and the cost of green chopping limits it as a substitute for grazing on many other forage producing lands even though the efficiency would be higher than for grazing. Nevertheless, the reasons for increased harvesting efficiency of green chopping over grazing suggest avenues in which grazing efficiency might be improved. Blaser *et al.* (1959) and Walton (1983) have reviewed the reasons for greater harvesting efficiency with green chopping as follows: (1) more uniform utilization, (2) less residue not utilized, (3) reduced losses from fouling and trampling, (4) less trampling damage of forage plants and the soil surface particularly where drainage is poor or irrigation is practiced, (5) reduced weed problems, (6) alternating growth and rest periods, and (7) harvesting at optimum growth stages for maximizing either dry matter or nutrient yield.

Forage disappearance and grazing efficiency of cattle have been compared under different stocking levels on a mixed grass prairie in north-central Texas (Allison *et al.,* 1982, 1983). When the cumulative forage allowance levels were at 22, 44, 88, and 110 lb per AUD (average of 66), total forage disappearance was 18.7, 26.5, 28.0, and 35.9 lb per AUD (average 27.3), respectively. When related to forage disappearance, the grazing efficiency (Formula 2) in consumption was 99, 78, 68, and 53% (average 74.5%), respectively.

These data suggest the possibility of reducing forage disappearance from other than grazing to very low levels by reducing the cumulative forage allowance. Forage intake did not vary greatly between treatments and averaged about 20 lb per AUD. In a study of stocking rates on Oklahoma prairie, grazing pressures were apparently too low for increased stocking rate to affect herbage disappearance per AUD (Brummer *et al.,* 1988).

II. UTILIZATION GUIDELINES

A. PROPER USE FACTORS

The conversion of the standing crop to grazing capacity is based on that portion that is not only edible and accessible but is also usable, i.e., **usable forage** that can be removed by grazing without damage to the forage plants. Usable forage in the standing crop is the summation of how fully each forage species can be defoliated and still maintain or improve in vigor. The **proper use factor** refers theoretically to the maximum degree of use by grazing (expressed as a percent), deemed to be physiologically correct from the standpoint of plant vigor, reproduction, longevity, and regrowth potential. However, while usable forage and proper use factors are helpful concepts, their determination and widespread application are virtually impossible because of the highly varied conditions under which forage plants grow and how they are grazed.

Attempts have been made to assess proper use (defoliation level) for each forage plant species by clipping and grazing studies and to extend and apply the findings to the maze of management situations that are met. However, the assignment of an average proper use factor or utilization standard to each species has variously been considered as useful to "a formidable undertaking" (Caldwell, 1984) to "hardly defensible" (Van Dyne *et al.,* 1984a) to an inviting theoretical concept which in practice is "almost useless" (Menke, 1987) to "should be buried immediately" (Scarnecchia, 1999).

There are many plant growth and environmental factors interacting at any given time that affect the degree of use a forage plant can tolerate. The impact of grazing use on forage plants varies by species of plant, season of use relative to plant phenology, duration of grazing periods and rest periods, competition from other plants, current and recent climatic factors, and when the impact is measured (while growing or post-growing) (Burkhardt, 1997). This makes the assignment of proper use factors, even within percentage ranges, quite arbitrary and open to many exceptions and variations. Some prefer the term **allowable use factor** as being more suggestive of its impreciseness and variability when taken beyond specific case-study situations.

A wide range of grazing intensity studies carried out on rangelands in western U.S. has been summarized by Holechek (1988) to reveal forage utilization levels associated with moderate stocking rates. These levels were then considered as best estimates of proper use factors, and the results have been adapted for presentation in Table 12.3. A positive relationship was found between average annual precipitation and apparent proper use factors; utilization rates under moderate stocking, assumed to be proper use, increased from lows in arid zones, intermediate in semi-arid and sub-humid zones, to highs in humid zones and those under irrigation.

There is general agreement that herbaceous forage plants can tolerate a higher level of defoliation after dormancy than before maturity. Table 12.3 suggests the higher suggested proper use factors for dormant season grazing compared to the lower factors for grazing during the growing season. McIlvain and Shoop (1961) found that mid and tall sandhills grasses in Oklahoma maintained their productivity in average rainfall years when 50% of the forage was removed by fall or when 75% use was made by the end of winter in yearlong grazing units. Another consideration is the conclusion by Holechek (1988) that ruminants consume an average of about 2% of their body weight per day in dry matter and may elevate their daily forage dry matter demand to 2.5% during active growth when forage is high in quality, but may lower their demand to only 1.5% during dormancy when the forage is of lower quality. This suggests that reduced demand per AUE and higher proper use factors for plants during dormancy may compensate for continued net dry matter losses in the standing crop during dormancy.

TABLE 12.3 Utilization Guidelines and Forage Conversion Factors for Selected Grazing Lands in the United States[a]

Suggested proper use factor (%)	Forage required per AUM (lb)[b]	Grazing land type
20–30	3900–2600	Alpine tundra
25–35	3120–2230	Southern desert shrublands
30–40	2600–1950	Northern desert shrublands Semi-desert grass and shrublands Sagebrush–grasslands Palouse prairie Oak woodland and chaparral
35–45	2230–1735	Western coniferous forest Western mountain grasslands Western mountain shrublands
40–50	1950–1560	Shortgrass prairie Northern mixed prairie Southern mixed prairie
45–60	1735–1300	Tallgrass prairie Southern pine forest Eastern deciduous forest
45–55	1735–1420	Cool-season pasture (western foothill and mountain)[c]
50–60	1560–1300	Irrigated perrenial pasture
50–60	1560–1300	California annual grassland
60–70	1300–1115	Crested wheatgrass Russian wildrye
—[d]	—[d]	Seeded annual pasture Crop aftermath
—[e]	—[e]	Hay meadow aftermath

[a]Adapted from Holechek (1988).

[b]Use the lower conversion factor (based on higher proper use factor) for vegetation in good condition and/or grazed during dormant season; the conversion factors were calculated as follows: 780 lb (i.e., usable forage/AUM) $\times$ 100 $\div$ proper use factor.

[c]Includes intermediate, pubescent, and tall wheatgrass; smooth brome; and orchardgrass.

[d]Can be as completely utilized by the end of the grazing season as acceptable animal performance will permit; divide pounds edible forage by 780 to determine AUMs.

[e]Post-maturity grazing must be regulated only to the extent of preventing mechanical damage to the root crowns of perennial plants or undue exposure to frost damage; exclude estimates of trampling losses and inedible residues from total herbage; divide pounds of edible forage by 780 to determine AUMs.

Proper use factors may have some use as management tools when these limitations are recognized and conclusions drawn with their use are considered suggestive rather than absolute. Their use in making initial recommendations are made more acceptable when followed by objective monitoring procedures. They do provide refinement over the "take half and leave half" criteria, which can be credited with having served a useful purpose in the past but is now known to be conservative in some situations but represent excessive utilization in others. Kothmann (1984) utilized a forage harvesting efficiency factor unique to the nature of the specific standing crop in place of a proper use factor to determine usable forage.

B. UTILIZATION METHODS

Utilization (synonym **degree of use**) refers to the proportion (usually percentage) of current year forage production that is consumed and/or destroyed by grazing animals. Utilization measurements have many uses in grazing management including that of making short-term and long-term adjustments in stocking rates (Fig. 12.3). Other uses include assessing physiological proper use, monitoring the adequacy of grazing distribution, determining key management areas and key species, determining the efficiency of forage-herbivore conversions, and evaluating grazing treatment effects.

Since details of alternative methods of determining forage plant utilization and their utility and adaptation to special situations are given in various reference manuals, with one exception they will not be repeated here. Although the ideal method of measuring utilization should be rapid, accurate, simple to use, and result in high precision among observers, Jasmer and Holechek (1984) have concluded that no method available meets all of these criteria. (Readers are particu-

FIGURE 12.3   The determination of forage utilization has many uses including that of making short-term and long-term adjustments in stocking rates; scene showing a Nebraska rancher monitoring utilization on his range.

larly referred to Cook and Stubbendieck, 1986, Chapter 5.) However, the following outline of utilization methods is provided to help in the utilization method selection process:

I. Descriptive/qualitative classes
    A. Descriptive only
    B. Descriptive plus percentage use categories (see following example)
II. Actual weight (clipping methods)
    A. Before and after grazing (utilizes plant "units," paired twigs, or total herbage)
    B. Caged comparisons
III. Weight (or volume) estimates
    A. Ocular estimate by plot
    B. Ocular estimate by plant (within plot)
IV. Indirect methods
    A. Counting (stems or plants grazed:ungrazed ratio)
    B. Height-weight methods
    C. Twig measurements (length or diameter reduction)
    D. Residue techniques (herbage or stubble height remaining)
    E. Photographic techniques (based on qualitative classes, growth form or appearance, height-weight data, or general appearance)
    F. Pellet group counts (gives only relative time spent)

One procedure for estimating utilization expresses results in relation to proper range use. It makes use of subjective utilization categories for the primary forage species (key species) including both descriptive and percent utilization ranges. (The example below assumes that 50% utilization is proper use; utilization ranges should be corrected for different proper use factors). The five categories of utilization are described as follows, with full use being the management objective:

Slight: 1 to 20% use of primary forage plants; practically undisturbed.

Moderate: 21–40% use of primary forage plants; most of accessible range grazed; little or no use of low-value plants.

Full: 41–60% use of primary forage plants; all of accessible range grazed; minimal or no use of low-value plants.

Close: 61–80% use of primary forage plants; all of the accessible range shows use and major areas are closely grazed; some use of low-value plants.

Severe: 81–100% use of primary forage plants; low-value plants carrying the grazing load.

C. USES AND LIMITATIONS

Scarnecchia (1999) has concluded that the transition of range utilization from a qualitative concept in the earlier years to a quantitative concept today cannot be justified conceptually or practically; furthermore, attempting to use utilization beyond that of a qualitative indicator is complicated by active herbage dynamics characteristic of most situations. One problem of basing grazing management on direct measurement of utilization is the difficulty of evaluating something already removed. Utilization is dependent upon knowing the current annual aboveground net primary production, which cannot be determined until the end of the growing season. Peak standing crop is often substantially less than total production, thereby resulting in a built-in bias of overestimating utilization compared to the standard definition (Frost *et al.*, 1994). If grazing occurs during the growth period rather than promptly after the cessation of growth, continuing growth tends to be confounded with utilization even when intensive utilization techniques are employed.

Another problem is that forage production on most non-irrigated grazing lands, particularly rangelands, varies from year to year, and a percentage take may be difficult to interpret. For example, 50% utilization during a high forage production year will probably have much less impact on the vegetation than even 25% in a severe drought year. Sharp *et al.* (1994) noted that utilization of crested wheatgrass at 80% in a good year would leave more residue on the ground than was produced in a poor year, based on a 5:1 ratio of forage production in a good compared to bad production year in Idaho. Furthermore, utilization of 50% of a key forage plant species may be achieved by 50% use on all the plants in the population or by 100% use on half of the plants and none on the rest; if the latter, then overgrazing may be serious without the utilization data showing it. Only when utilization guidelines are tailored for specific situations (i.e., time of use, what is measured, and how use is measured) might they be reliable indices for making management decisions (Frost *et al.*, 1994). Specifying proper use, percentage-wise, is a subjective evaluation specific to a certain site and set of conditions (Sharp *et al.*, 1994).

Using utilization standards alone to make grazing management decisions— particularly when applied to public lands as a regulatory tool—has more recently been equated with policing rather than managing the grazing resource (Sharp *et al.*, 1994; Burkhardt, 1997; McKinney, 1997). The general consensus among range scientists seems to be that conservative use levels cannot be the sole basis of grazing management (Burkhardt, 1996) and that simplistic guides such as utilization standards are not an acceptable substitute for experienced on-the-ground management, based on sound, long-term range trend information (Sharp *et al.*, 1994).

Basing the achievement of full proper use on plant residue or stubble heights rather than on utilization may be preferable because it is the residue levels left undefoliated that will have the greatest impact on plant health and on soil and watershed protection (Jasmer and Holechek, 1984). Residual dry matter levels (lb or t/acre) can be considered equal to:

$$\text{Standing crop} - \text{consumption} - \text{other dry matter losses}$$

However, this is more a monitoring tool than a predictive tool for achieving proper use because of the difficulty of predicting all of the above three required factors in advance of grazing.

Critical dry matter residue levels have been determined experimentally for some grazing land types. For example, Holechek (1988) summarized that 300 lb/acre of residue was satisfactory on Colorado shortgrass range, 160 lb per acre grass residue on Oregon big sagebrush ranges, and 250–1100 lb per acre in California annual grassland type, depending on the site. On the California annual grassland site studied by Hooper and Heady (1970), leaving 500 lb of herbage residue (mulch) at the time of the first rains in the fall was considered more appropriate than the previously accepted 2-in. stubble or 1000 lb mulch rule. Shoop and McIlvain (1971b) proposed that moderate grazing of Oklahoma sandhills grasslands left approximately $\frac{1}{3}$ of the average production of forage (about 350 lb per acre) at the end of the grazing year (just prior to spring regrowth). In contrast, on heavily grazed pastures, average forage residues of only 225 lb per acre were left, which was considered an unsatisfactory level.

A comprehensive evaluation of optimal amount of herbage residue to be left at the end of the summer grazing season on blue grama range in eastern Colorado was made by Bement (1969). It was concluded that residue levels, when related to average past stocking rates, could be used as cattle stocking rate guides since residue levels were related to gain per animal and gain per acre. Based on 19 years of data, Fig. 12.4 shows the calculated relationships between weight of ungrazed herbage, average stocking rates, and daily and per-acre gains.

According to data compiled by Bement (1969), leaving 350 lb of air-dry forage/acre at the end of the season, equivalent to 3.2 acres per yearling month, gave maximum daily summer gains/head of 1.45 lb; leaving 250 lb per acre, equivalent to 2.2 acres per yearling month, gave maximum animal gains per acre of 15 lb; but leaving 300 lb per acre, equivalent to 2.6 acres per yearling month, gave the max-

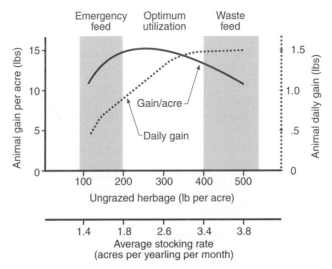

FIGURE 12.4   Stocking-rate guide for beef production on upland blue grama range grazed May 1 through October 31, eastern Colorado (Bement, 1969.)

imum dollar returns per acre from summer grazing with yearling cattle. Stocking recommendations were to leave 200 to 400 lb of ungrazed herbage per acre at the end of the 6-month summer grazing season; grazing down to 100 lb could provide additional emergency forage, but leaving more that 400 lb per acre of herbage was considered forage waste. It was deemed appropriate to remove down to 200 lb of herbage per acre in drought years if 400 lb was left in the best years.

The second alternative approach to setting guidelines for achieving full proper use is the use of minimum average stubble heights for grasses. Hall and Bryant (1995) considered that various stubble heights of the most palatable species were useful in predicting when unacceptable impacts—heavy use or trampling or both—were about to occur on riparian sites. Baker *et al.* (1981) concluded that residual sward height after grazing gave a better indication of the performance response of beef cows and suckling calves to vegetation conditions than actual herbage allowance.

Guidelines for grazing seeded grasses in the Ponderosa pine zone in Colorado were to leave a 2-in. average stubble height for crested wheatgrass and Russian wildrye, two plants tolerant of moderately close grazing, but a 4-in. stubble height for smooth brome and intermediate wheatgrass, these less tolerant of close grazing on the same site (Johnson, 1959). Grazing crested wheatgrass more lightly than a 2-in. stubble resulted in the development of wolf plants which was associated with stand depletion because of severe grazing of the grazed plants (Currie and Smith, 1970). Full-use stubble heights for crested wheatgrass were also set at 2 in. in southern Canada (Lodge *et al.,* 1972) but at 3 in. on semi-arid foothill range in central Utah (Frischknecht and Harris, 1968).

Utilization recommendations based on stubble heights for salt-desert shrub winter range in western Utah were 2–3 in. on Indian ricegrass (*Oryzopsis hymenoides*) and 1–1.5 in. on galleta (*Hilaria jamesii*); minimum remaining current twig growth was set at 1.5–2.5 in. on black sagebrush and winterfat (Hutchings and Stewart, 1953). On creeping bluestem range in Florida, leaving 12-in. stubble heights for this tallgrass was considered wasteful while grazing to an average 6-in. stubble height not only maintained plant vigor but provided more and better forage for grazing (Kalmbacher *et al.,* 1986).

III. SETTING INITIAL STOCKING RATES

Determining the grazing capacity of grazing land units is one of the most difficult tasks in grazing management. Because of the spatial variability of rangelands, in particular, the climatic variation, and the impact of grazing management practices, the proper stocking rate varies over both time and space while also being a function of management goals related to risk and catastrophe. Furthermore, the difficulty or even inability to accurately determine grazing capacity is a prime contributor to the technological problem associated with overgrazing (Walker, 1995).

Wilson (1996) has concluded that, in spite of the fact that research has led to much knowledge about grazing management, it has provided but minimal practical scientific advice on how to determine grazing capacity. There are few practical tools available to guide the estimation and implementation of sustaining grazing capacities. Most practitioners rely on some level of subjective judgment, knowledge of local factors, and experience in determining appropriate livestock numbers, particularly with the forages in question (Johnston *et al.,* 1996). Although grazing capacity must be determined on a case-by-case basis and this primarily on the ground, simulation models and decision support systems are now being developed and used in cumulating the effects of factors determining the grazing capacity of individual grazing units.

The objective may be to determine grazing capacity of the current forage crop or it may be to predict long-term average grazing capacity. If the latter, information will need to be accumulated over a period of several years. *Initial* stocking rates based on preliminary information must be replaced as soon as possible with *recommended* stocking rates as more data is accumulated on individual grazing units. Monitoring forage production and utilization throughout the grazing period will provide the basis of making further short-term adjustments in stocking rates as the grazing season progresses.

A. NATURAL VS. MANAGEMENT FACTORS

Grazing capacity is determined by a complex of plant production and usage factors. Estimates of the grazing capacity of any grazing unit in the short run will not be meaningful unless full consideration is given to management as well as natural site factors. No estimate of grazing capacity can be realistic without considering how the grazing land will be grazed and to what extent cultural treatments will be applied. Optimal stocking rates, and thus grazing capacity, become meaningful only when the objective and management details are specified; these management details consider such things as classes of animals, breeds of animals, timing of grazing, animal performance goals targeted, vegetation manipulation desired, and associated silvicultural systems (Scarnecchia, 1990).

According to Scarnecchia (1990), "The question 'What is the carrying capacity of this land?' is inherently limited in usefulness, because it (1) implies unstated objectives, (2) does not specify chosen management options, and (3) implies a single carrying capacity for a land area." If future management cannot be predicted, long-term grazing capacity will have to be based on natural characteristics inherent to the given tract of grazing land unless specified future management levels are assumed.

A listing of both natural (site) and management (including utilization) factors will be pertinent to determining the grazing capacity of all grazing lands:

Natural factors affecting grazing capacity	*Management factors affecting grazing capacity*
1. Climate and weather	1. Condition of forage stand resulting from past use
2. Height of water table	2. Adequacy of grazing distribution
3. Root zone depth	3. Meeting drinking water needs
4. Soil texture and structure (extremes)	4. Season of grazing

5. Natural soil fertility
6. Level of soil salinity
7. Physiography of area
8. Amount of vegetation
9. Quality of vegetation
10. Amount and distribution of drinking water

5. Kind/mix of grazing animals
6. Forage removal by other than assigned animals
7. Grazing methods used
8. Cultural treatments: weed control, fertilization, seeding, irrigation
9. Operational objectives and restrictions.

B. METHODS OF ESTIMATING STOCKING RATES

Several approaches are available for estimating grazing capacity and setting proper stocking rates. All depend more or less on trial-and-error estimates coupled with subsequent adjustments. The effectiveness of each method varies depending upon the kind of grazing land, but a combination of methods is generally required. Seven more or less distinct approaches to determining grazing capacity are recognized.

1. Initial Stocking Rate Tables

Initial stocking rate tables for native range and other kinds of grazing lands are available for many areas. Such standards or guidelines have been prepared variously by some federal and state land management agencies, state experiment stations or extension services, and consulting firms and private individuals. For example, Kirychuk and Tremblay (1995) prepared initial stocking rate (AUM per acre) tables for seeded dryland forages in Saskatchewan based on forage plant species, soil texture, soil zone, stand age, pasture condition, and nitrogen fertilizer application rate. However, such standards generally consider only average production years and average management practices, so they should be adapted and applied to individual grazing units with caution. Such standards are very useful in permitting the grazier to initiate grazing without having experience in a new geographical area and prior to making a more detailed inventory.

The former Soil Conservation Service (now Natural Resources Conservation Service) in many states prepared initial stocking rate tables for native rangelands based on range site (including precipitation zone) and range condition. These tables are still useful in that suggested initial stocking rates in acres per AUM (or AUMs per acre) are provided. These range site averages must be adjusted to local conditions affecting forage productivity. It should be determined if prepared stocking rate tables include a planned degree of undergrazing for improving range in lower than excellent condition. If so, consideration should be given to whether undergrazing is an effective range improvement practice for that range or whether full grazing combined with improved grazing systems, better distribution of grazing, or cultural treatments will be more effective.

2. Known Stocking Rates Plus Condition and Trend

This method is generally considered a reliable method on native range for adjusting initial stocking rates; it is also useful on other longer term grazing lands for which stand condition and trend criteria have been developed. It requires accurate stocking records by grazing unit for a period of years to develop a past history of stocking rates; the stand condition and direction of current/recent trend (i.e., improving, maintaining, or deteriorating) provide the basis for adjusting the stocking rates of the recent past. Since it is subjective rather than formula-derived, the utility of this method depends on good judgment in interpreting the evidence and applying it properly. This method combined with percent use of the key species is one of the methods used by USDA National Resources Conservation Service (1997) in establishing initial stocking rates.

An upward trend in the vegetation or vegetation plus soil indicators will provide evidence that the area is not being overgrazed, but it does not reveal per se whether it might be undergrazed. A downward trend suggests the stocking rate may be too heavy, but consideration should also be given to whether the season of grazing is incorrect, an unadapted kind or mix of animals is being grazed, or grazing is not being properly distributed. Consideration must be given to whether rainfall and other weather factors currently and recently have been unusually favorable or unfavorable.

Trend is best determined by accurately measuring range or pasture condition at the beginning and end of a definitive study period (5 years or more on native rangelands but less on medium-term perennial pasture), using identical procedures both times. When such before-and-after evaluations are unavailable, as frequently happens, reliance must be on general indicators of apparent trend that can be observed in a single survey. Such indicators must largely be based on (1) the vigor of the key forage species, (2) maintenance of the key forage species in the stand, (3) status of weedy species, (4) activity of local gullies, (5) soil stability, and (6) adequacy of protective ground cover.

3. Standing Crop of Usable Forage Converted to AUMs

This method is based on measuring or estimating forage mass, usually in green weight, and converting green weight to dry weight (oven-dry or possibly, more conveniently, air-dry weight based on 90% dry matter) (See Fig. 12.5.). Air-dry conversion factors can be determined by drying representative samples or using conversion tables based on forage species or species groups and stage of growth. Enough small plot clipping or estimates, and their stratification and subsampling if a mosaic of forage types is found in the grazing unit, must be made to accurately measure the standing crop of forage. Only current year's growth should normally be considered in measuring available forage. Biomass from plant species totally unacceptable or unavailable should be excluded from the standing crop, but a suggested proper use factor may be useful in further refining forage estimates for the remaining potential forage plant species.

The air-dry weight of the grazable portion of the standing forage crop is then summarized for the entire grazing unit after using appropriate slope and water adjustment factors on those acres where needed. The total weight of standing crop in the grazing unit is then divided by an appropriate conversion factor (from Table 12.3) for converting to grazing capacity in AUMs. (Note: the conversion

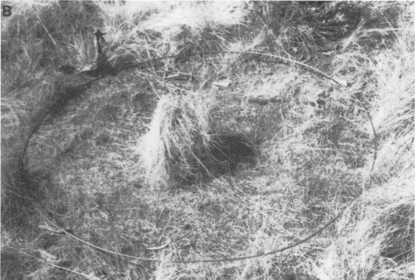

FIGURE 12.5 One method of estimating grazing capacity is by measuring or estimating the usable standing forage crop and converting to AUMs. (A) clipping the usable forage; (B) forage yield composited for weighing and converting to dry weight.

factor already incorporates the suggested proper use factor to derive usable forage.) An alternative procedure for complex native vegetation stands is to divide pounds of usable forage (summation of available forage for each plant species times its palatability factor by 780). Palatability factors (akin to harvest efficiency factors) can be determined by observing the percentage utilization made of each plant species under moderate or proper stocking rates. This approach permits greater attention to be given to kind of grazing animal and season of grazing. An adaptation of this method for stocking rate determinations is referred to by USDA National Resources Conservation Services as the "usable production method."

Holechek (1988) developed a procedure for calculating long-term stocking rates that included the following steps: (1) measuring or estimating the ungrazed standing crop of forage, (2) calculation of total usable forage by using suggested proper use factors (such as from Table 12.3), (3) calculation of forage demand, (4) calculation of unadjusted stocking rate, (5) adjustments for slope and travel distance, and (6) calculation of final stocking rate. Holechek and Pieper (1992) evaluated this procedure by comparing with other procedures and with actual long-term stocking rates on selected New Mexico experimental grazing units. When all forage species were included in the standing crop, the calculated stocking rates overestimated long-term stocking rates by 31%; when only perennial grass species were considered—equivalent to the four or five key forage species—the actual stocking rate was underestimated by only 10%. When not adjusted for slope and distance from water, all six stocking rate estimation procedures gave stocking rate estimates much higher than the range units actually carried.

4. Percent Utilization Method

This method has been used on seeded range and improved pasture, and also on native range if only key species are considered. The serious limitations of using only utilization standards to estimate grazing capacity were covered in the previous section.

This method is based on a comparison of actual utilization (degree of use) with the proper use factor, the latter based on experiment station data or other technical sources. In order to determine average annual grazing capacity by this method, three items must be known: (1) the average percent utilization of the key species by the end of the grazing season over a period of years, (2) average annual stocking levels on the pasture unit during these years, and (3) the proper use factor for the key species for the season in which grazed. Average annual grazing capacity in AUMs for the grazing unit is then computed as follows:

$$\frac{\text{Average AUMs removed annually} \times \text{proper use factor}}{\text{Average annual percent utilization}}$$

Determining AUMs removed annually and annual percent utilization of the major or key forage species requires that individual grazing unit records be kept annually over a period of years. (Refer to Fig. 12.4 for an example of a stocking rate guide based on weight of ungrazed herbage [1b per acre].)

5. Pasture Comparison Method

A direct estimate of grazing capacity can be made by comparing the kind of grazing land being inventoried with similar grazing land of demonstrated performance or by comparing with a mental ideal or standard. This method requires extensive experience and training in estimating grazing capacity and is subject to considerable personal bias. Experiment station grazing records on equivalent grazing lands and recommendations of experienced local managers should be considered. An initial best guess as to grazing capacity followed by monitoring to suggest subsequent adjustments is apt to prove a useful approach. This method can also be used as a common-sense governor when used to temper or modify grazing capacity estimates made using other estimate methods. This method is apt to be very unreliable when used on grazing land types with which the person is unfamiliar.

6. Energy-Based Methods

These methods are based on the premise that reliable evaluation of grazing capacity must treat forage amount and quality as integrated rather than distinct features of the habitat. (Readers are referred to Hanley and Rogers, 1989; Hobbs *et al.,* 1982, 1985); Moen, 1984; and McCall *et al.,* 1997, for further details]. Both animal equivalence and the AUM are utilized in this approach but expressed only in terms of energy (and/or protein). "Nutritional carrying capacity" has been described as the ratio of range nutrient supply divided by the nutrient demand of individual animals (Swartz and Hobbs, 1985). The two primary problems with this approach to grazing capacity has been (1) quantifying the functional relationship between the quality of forage and its biomass (i.e., energy levels), and (2) relating supportable animal numbers to quantitative animal nutritional requirements (i.e., energy demand).

Energy-based grazing capacity requires the following but often difficult-to-improbable assumptions: (1) the total useful energy in the standing crop (or habitat) can be quantified, (2) the grazing animals will mostly confine their energy intake to current physiological needs, (3) the array of animal and environmental factors that affect dry matter intake will result in no significant departures from energy needs, (4) animals will select high-energy forages over low-energy forages and other palatability and anti-palatability factors will not be overriding, and (5) high-energy forages will provide higher grazing capacity per unit of dry matter than low-energy forages. It is generally concluded that energy-based approaches to grazing capacity may be useful in conceptualizing and modeling to reveal animal-plant-habitat energy interrelations but too unwieldy and with too many variants and unknowns for widespread practical application (Wallace, 1984).

7. Forage-Density Method

This is a highly formularized procedure formerly used for estimating grazing capacity on native range but no longer considered valid. This method is seldom used today and should not be used, because it is laborious, requires a maze of subjective estimates that only assume objectivity, and is of unreliable accuracy. This method basically assumes that forage density (ground cover) and relative palatability, taken together, comprise a satisfactory index to the amount of forage available for grazing. This method is cited here for historical reasons and not with encouragement that it be given positive consideration.

C. ADJUSTMENTS FOR SLOPE AND WATER

Certain acres within a grazing unit will contribute reduced or no grazing capacity because of inaccessibility or for reasons other than the nature of the standing crop, i.e., distance from water, difficulty of access, unattractiveness of site to selected animal species, etc. (Fig. 12.6). Limited supplies of drinking water may limit realizable grazing capacity to levels considerably under forage limitations. These unrealizable portions of the standing crop must be excluded from estimates of grazing capacity. This problem will apply to all methods of calculating grazing capacity other than those based primarily on monitoring past stocking rates rather than being predictive.

Attempts have been made in the past to classify all areas within a grazing unit as either **suitable** or **unsuitable** for grazing; all usable forage on the suitable acres would be credited toward grazing capacity, while no contribution from the unsuitable acres would be allowed toward grazing capacity inventory. A classifica-

FIGURE 12.6   Adjustments must be made in estimating grazing capacity for herbage production that cannot be utilized because of inaccessibility, distance from water, or refusal of animals to use certain areas; mountain scene in Morgan County, UT, showing variable suitability for grazing by different animal species.

tion of unsuitable might result from one or a combination of two factors: (1) barrenness or inherent lack of forage or (2) acreage cannot or should not be grazed by specific ungulate herbivores because of inherently unstable soils, limited accessibility, or steep topography. However, strict adherence to the alternative designations of suitable or unsuitable, with most acreage falling somewhere in between, projects an "all-or-none" dilemma for each acre or smaller unit in the grazing unit.

A more acceptable approach to the problem has been to apply progressive adjustment factors (ranging from 0–100%) to the standing crop on all acres before being converted to usable forage. This appears to be a workable solution to the problem; adjustment factors can be applied to restricted acreages in the initial inventory of grazing capacity. However, further adjustments should be made based on careful observation of grazing patterns and noting which areas are being totally or partly avoided and to what extent; this consideration possibly should also be extended to excessive patchy grazing. It becomes readily apparent that such adjustment factors are greatly dependent on managerial ability to achieve full but not excessive utilization of the standing forage crop on each acre.

From a review of literature and personal observations, Holechek (1988) has suggested adjustment factors for slope and drinking water limitations. For cattle, he suggested the following slope adjustments: no reduction, 0–10% slopes; 30% reduction, slopes 11–30%; 60% reduction, 31–60% slopes; 100% reduction (considered ungrazable), slopes over 60%. No adjustment was suggested for sheep on slopes under 45%. A 50% reduction was suggested for cattle range acreage lying from 1–2 miles from drinking water and total exclusion of acreage lying over 2 miles from water, the latter assumed to be ungrazable by cattle. (Refer to Chapter 7 for a more complete discussion and documentation of factors affecting spatial patterns in use of grazing lands that affect grazing capacity.)

IV. DYNAMICS OF GRAZING CAPACITY

A. HERBAGE GROWTH, ACCUMULATION, AND DISAPPEARANCE

One of the best documented but poorly appreciated characteristics of the growth of forage plants is the rapid turnover of tissue that takes place (Parsons and Johnson, 1986). The standing forage crop is dynamic, with new herbage being added and existing herbage disappearing simultaneously (Scarnecchia, 1988). Forage produced into the standing crop does not enter a static storage situation. Dry matter flows continuously and rapidly through the standing crop of graminoids, forbs, and shrubs as new leaves are continually produced and old ones die. Herbage that is not removed by grazing or other defoliation will eventually die, decay, and disappear from the standing crop.

Net herbage accumulation or loss is, in fact, the difference between net herbage growth and net herbage disappearance during a specified time period (Scarnecchia and Kothmann, 1986). Loss of herbage from the standing crop results not only from consumption by grazing animals but also is associated with trampling and discarding of plant parts, removal by other animals, senescence and disease, weathering, and decay and return to mulch. Willms *et al.* (1996) reported biomass losses from senescence and weathering in the fescue grasslands during the over-wintering period at 27% for the grasses and 60% for the forbs. On bluestem range in Louisiana cattle intake across a combination of grazing intensities accounted for only 36–47% of the herbage disappearance, while more than 50% resulted from factors such as trampling, weather, and wildlife (Pearson, 1975). Similarly, on sandhills grasslands in Oklahoma under moderate grazing, 33–50% of the standing crop was consumed, 25–33% was lost to unknown causes, and 25–33% was considered left to maintain plant health and site stability (McIlvain and Shoop, 1961).

Annual and seasonal cycles in net accumulation and net loss of herbage from the standing crop will have a direct bearing on what is included in the inventory of grazing capacity since the peak standing crop lasts only a short time, and even the peak standing crop will underestimate total yield because of the continuing herbage disappearance. The botanical composition of the standing crop will be affected by differential growth and disappearance of individual plant species and plant parts over time. Also, during the dormant season, leaves and inflorescences of grasses deteriorate more rapidly than culms, this differential loss contributing to a deterioration of forage quality.

Ephemeral forbs and grasses add materially to the decline in herbage weight from the peak standing crop or some other point in time (Pieper *et al.,* 1974). During the growing season, animal production and grazing capacity are largely functions of the amount of green foliage that the animals can harvest; this suggests that utilization guidelines during the growing season would preferably be restricted to the green herbage. During plant dormancy, grazing capacity will depend upon retained acceptability and availability of forage in the standing crop. For example, on tobosa grasslands, Anderson (1988) concluded that grazing management should be based on "green" tobosa (new growth) and "brown" tobosa (current year's production that had senesced) but not on "gray" tobosa foliage (forage produced prior to the current year's growth but now considered only mulch).

Whether based on the current standing crop or on long-term averages or projections, deciding the timing of when to make the grazing capacity inventory as well as adjustments in stocking rates becomes critical. The relationships between forage production cycles and when grazing will occur must be considered. Will the grazing occur during the growing season, after the growing season has ended, yearlong, or some other combination of growing and dormancy periods? If during the growing season, will grazing be continuous, early or late, or in interrupted periods as with rotation grazing?

Annual and long-term grazing capacity is commonly determined at the point of peak standing crop. However, in the Southwest most decisions regarding adjustment in stocking rates on yearlong range are made at the end of the growing season in the fall (Holechek, 1988). At this time the standing crop is estimated and animal numbers are adjusted accordingly to carry through until new forage growth begins in spring or summer. Measuring at peak standing crop will overestimate herbage availability if grazing is subsequently delayed for an extended period.

On blue grama range in New Mexico, the standing crop peaked in September but declined 40% by November and still further by mid-winter (Pieper et al., 1974). This reduced herbage availability may not reduce grazing capacity during winter if plants can tolerate somewhat heavier utilization during dormancy and grazing animals reduce their consumption relative to body weight (the two latter effects being compensatory to a reduced standing crop). For example, Smith et al. (1986) have suggested that on a dry matter equivalent basis animals consuming an equivalent of 3% of their body weight when the standing forage is green may reduce consumption to only 2% of body weight when the foliage has matured.

When the standing crop is to be harvested by grazing during a short period of high animal density, short-term grazing capacity can be determined just prior to the beginning of grazing and forage produced during the grazing period mostly ignored. However, if both forage growth and harvest will take place simultaneously and over an extended grazing period, continuous monitoring of both forage production and utilization either by measurement or estimation are necessary. If the standing crop is not measured until after substantial grazing has occurred (unless representative plots have been protected from grazing by caging for comparison), grazing will have removed part of the forage produced prior to inventorying. Although this will provide information on forage remaining for current harvest, it will underestimate annual and long-term grazing capacity. (Utilizing cages to observe or measure the short-term interactions of forage production and utilization has been suggested as a useful management tool; small, fenced exclosures of from 0.25 up to 5 acres in size provide similar utility in observing or measuring annual forage production as well as the long-term effects of any grazing program.)

B. FLUCTUATIONS IN FORAGE PRODUCTION

While average forage production on perennial, dryland grazing lands suggests an average stocking rate, yearly variation in production determines the annual adjustments in stocking that may be required. In fact, the optimal grazing intensity at any given time is largely dependent upon the occurrence of future climatic conditions. Years ago Stoddart (1960) concluded that it must be "immediately evident that there is no single correct stocking rate for all years and that grazing capacity is not a constant feature of rangelands." Average or normal weather must be assumed in long-range ranch planning, but weather and thus the quantity and quality of vegetation produced are variable; therefore, seasonal and annual deviations from average grazing capacity must be incorporated into both the grazing inventory and the grazing plan.

McLeod (1997) concluded that the concept of carrying capacity can be applied to environments of low variance in annual forage production, but in environments characterized by a high degree of unpredictable variance, such as the shrubland he studied, the concept was considered not useful in describing plant-herbivore dynamics. He suggested that carrying capacity, as a function of resource availability, was not a measure of long-term equilibrium but rather of short-term potential. "The concept of safe capacity is close to the range manager's concept of carrying capacity when there is food limitation, such as a drought, [but] the notion that there is a sustainable herbivore density in all environments should be dispelled" (McLeod, 1997).

The concept of "safe" grazing capacity is sometimes equated with setting stocking rates so low that short-term grazing capacity would never be exceeded even in the worst season or year. However, basing stocking rates on such long-term rigidity ignores the opportunity and potential for making annual and seasonal adjustments in actual stocking density to accommodate forage production fluctuations, and the opportunity to "manage" grazing lands would be greatly restricted. Thus, it becomes apparent that any minimal carrying capacity (i.e., safe capacity) must include provision for constant monitoring to adjust even further downward in an extremely low production year or period of years or, alternatively, upwards during years of high forage yield.

Since rainfall and thus grazing capacity on non-irrigated, sub-humid to arid grazing lands are highly variable from year to year, emphasis should be given to forage production in worst-year and best-year scenarios in addition to the average year. There is a tendency to ignore or at least underestimate the magnitude of annual deviations in forage production from the long-time average in grazing capacity inventory and grazing management planning. Available data indicate that in the western U.S. weather can override grazing in its influence on annual vegetation production (Dwyer *et al.,* 1984). Heitschmidt and Walker (1996) concluded that the major challenge to grazing managers centers around capturing opportunities and avoiding pitfalls arising from deviations in forage production from the norms; managing solely for average forage production was considered a likely road to financial ruin in grazing enterprises.

Wide variations in annual herbage production on native and seeded rangelands are common in the western U.S. (Fig. 12.7). Annual grazing capacity on native range in north-central Texas during a 3-year study varied 22% above, 28% below, and 6% above the average of the 3 years (Kothmann *et al.,* 1986a). On shortgrass prairie in southeastern Alberta, average annual forage production during the 1930–1953 period averaged 317 lb per acre but varied from a low of 90 to a high of 825 lb (Smoliak, 1956). On low-elevation foothill range seeded to crested wheatgrass in central Utah, forage yields per acre varied from 420 to 935 lb during a 12-year study period (Frischknecht and Harris, 1968). On higher elevation seeded foothill range in Colorado, yearly forage production varied from 668 to 2457 lb per acre over a 10-year study period (Currie, 1970).

On mountain grasslands in western Montana, total herbage production was found in a 10-year study to differ as much as 200% between good and poor years (Mueggler, 1983). However, for two-thirds of the years it was between 80 and 85% of the long-term mean. Total production of graminoids was found to differ as much

FIGURE 12.7   Annual fluctuations in forage production resulting from weather conditions remain a major challenge to the manager of non-irrigated grazing lands in setting stocking rates; photos show the magnitude of differences in production and utilization of forage on this Edwards Plateau range (A) in a good year compared to (B) a severe drought year. (Texas Game and Fish Commission photos.)

as 275% between years, yet remain within 75% of their mean two-thirds of the time. It was also noted that a good year for grass production was not necessarily a good year for forb production. Also, the optimal year for grass production may not be the best year for browse production.

On semi-arid to arid grasslands of the Southwest, annual forage production fluctuates even more widely. On low-elevation, low-rainfall ranges at the Santa Rita Station near Tucson where annual grasses prevail, the production of grass herbage can drop from 655 lb in a good year to 3 lb in a bad year and back up to almost 900 lb in an exceptionally good summer growing season (Martin, 1975b). On higher elevation, higher rainfall range in the same general area, where perennial grasses dominate, the production of the perennial grasses is relatively stable, but total grass production can still vary from 500 lb in a poor year to 1300 lb in a very good year.

Prediction equations for forage yields have been developed for some range areas based on weather just prior to the growing season or early in the growing season. Such prediction equations have been most useful for areas where soil moisture at the beginning of forage growth is highly correlated with total growing season production. For example, it was concluded from a 14-year study at the Squaw Butte Station in southeastern Oregon that weather could be used to accurately predict crested wheatgrass production (Sneva, 1977). The highest correlations were between July–May precipitation plus March–May temperatures for predicting mature yields. Mean February temperature with March precipitation accounted for 83% of the variation in spring yield. Crested wheatgrass yields by May 15 varied from 67–437 lb and averaged 296 lb per acre. Similar procedures were used successfully for estimating annual herbage production on southwestern Idaho range (Hanson *et al.*, 1983).

On salt-desert shrub range in western Utah grazed only during the winter, the 1935–1947 average annual air-dry herbage production was 219 lb per acre, ranging from a high of 468 to a low of 75 lb per acre (Hutchings and Stewart, 1953). These researchers concluded that prediction equations could circumvent the need of sampling vegetation to estimate cumulative herbage production by the beginning of the dormancy grazing season. Their prediction equation utilized the previous 12-month precipitation to predict the October standing crop of forage and was found to be quite reliable during 4- to 11-in. rainfall years ($r = + 0.944$).

Where data are insufficient to estimate formula relationships between weather and forage production, subjective evaluations may be the best information that is currently available. However, only the locally experienced grazing managers are apt to be successful in this. Olson *et al.* (1989) noted that both dry and cold conditions in California can be particularly detrimental to forage growth but that either of these two conditions alone can be detrimental. On the other hand, average forage growth is usually expected when average degree-day conditions coincide with average or above-average moisture conditions, and high production is expected when warm degree-day conditions coincide with average or above-average moisture conditions.

C. ADJUSTING STOCKING RATES

Animal numbers may be managed under a practice of **set stocking** (numbers remain constant through most or all of the grazing season or grazing period) or **variable stocking** (numbers are varied to synchronize animal demand more closely with available forage). Using a fixed, season-long stocking rate based on average forage yields under set stocking is easier to manage by requiring fewer management decisions; a survey of Texas ranchers revealed that 45% indicated they made no periodic adjustments to rebalance animal numbers with fluctuating forage supplies (Hanselka *et al.,* 1990). However, set stocking can cause large differences in grazing pressure, may achieve only overstocking or understocking in the short term, can result in reduced forage intake and animal performance when forage supply falls sharply below animal demand, and fails to achieve drought preparedness.

Blaser *et al.* (1983) concluded that set stocking largely ignores the nutritional needs of ruminants and the dynamic characteristics of the standing crop during growth and grazing. In their study on improved pasture, average daily steer gains were improved under variable stocking compared to set stocking (1.72 vs. 1.57 lbs), and total steer gains per acre were increased by 61%. The higher animal performance under variable stocking was attributed to more of the forage being consumed when of high quality; under fixed stocking, the standing crop carried a higher proportion of stems and maturing and senescing plant tissue. During the rapid spring growth period, the stocking rate for variable stocking was approximately double that of fixed stocking but was proportionally reduced during the summer.

Managers of all grazing lands, and particularly rangelands, constantly face the problem of balancing animal demands with a fluctuating forage supply (Cox and Cadenhead, 1993). Since herbage growth, accumulation, and disappearance are variables that continuously modify standing crops of forage, monitoring the forage supply throughout the grazing period and periodically adjusting animal numbers (or length of grazing period), if needed, are an important management tool. Setting the stocking rate at the beginning of the grazing period is based only on an estimate of the ensuing forage supply, and only continuous monitoring will assure that the forage supply remains adequate to meet the forage demand through the end of the planned grazing period or that an undesirable surplus level does not develop.

The use of photoguides (picture references of known forage quantities for the various forage types) and/or exclosures, data about precipitation patterns and probabilities, plant growth cycles, and forage inventory procedures are recommended for even greater accuracy. This more detailed information is particularly useful and may be required before making adjustments in animal numbers. Conducting surveys of forage supplies at the end of forage production cycles, but also at strategic points within the plant growth period, will help estimate how long the accumulated forage supply will last at current stocking levels. Reliable techniques for making accurate quantitative estimates of forage supply and balancing forage demand with the supply are available from several printed sources (White and Richardson, 1991; Cox and Cadenhead, 1993; USDA-NRCS, 1997; Reynolds, 1998).

Utilization checks can also be used as the grazing period progresses to adjust short-term stocking rates. This permits adjustment of numbers of grazing animals or length of grazing period so that proper degree of use coincides with the end of the planned grazing period. A utilization check made during the grazing period can be used to estimate the number of AUMs remaining in the grazing unit as follows:

$$\frac{\% \text{ proper use} - \% \text{ actual use}}{\% \text{ actual use}} \times \text{AUMs removed by grazing thus far}$$

Utilization checks should be limited for practical purposes to one or a few key species on key areas only. Any additional forage growth made prior to the end of the grazing period should be credited to remaining grazing capacity.

The forage supply ideally should be monitored visually throughout the grazing season. Experience, visual appraisals of forage and livestock, and current weather conditions are useful and often the only methods used by graziers to evaluate forage availability. However, less than 20% of Texas ranchers were found to use quantifiable techniques to adjust stocking rates and to monitor the impact of their decisions on the forage resources (Hanselka *et al.,* 1990). While applying variable stocking in the short-term to meet fluctuating forage supplies is an optimal concept, it will prove challenging in practice. Too frequent, or possibly even any, short-term adjustments in animal numbers under variable stocking challenge managerial agility (Kansas State University, 1995). Can the manager come up with the additional grazing animals needed when unanticipated short-term forage surpluses arise, and can the extra animals be blended into existing animal enterprises? Can alternative forage sources be made available for the excess grazing animals during short-term periods of unpredicted forage deficits, or are there other practical solutions for handling the now excess animals?

13

GRAZING INTENSITY

Grazing intensity refers, in general, to the amount of quantitative animal demand for forage placed upon the standing crop forage or forage mass, and to the resulting level of defoliation made during grazing. Numerous grazing intensity studies have been completed on range and improved pasture in the U.S. and Canada. These studies have generally included at least three rates of stocking: a medium or moderate rate, at least in the beginning anticipated to be about optimum; a light stocking rate commonly allowing 35–50% more acreage per AUM than the medium rate; and a heavy stocking rate commonly allowing 35–50% less acreage per AUM than the medium rate. When 25 stocking rate studies conducted on native rangelands in North America were averaged, Holechek *et al.* (1999) found that heavy grazing made 57% use of the primary forage species, moderate grazing 43%, and light grazing 32%.

However, the labels of heavy grazing, moderate grazing, and light grazing have not always proven to be synonymous with overgrazing, proper grazing, and undergrazing. After several years of research some grazing intensity studies were found to include no stocking rate that resulted in overgrazing; in other studies, the stocking rate labeled as light proved to be the best stocking rate included in the study. Until a moderate stocking rate has proven to be the optimal rate or the term is clearly being used in that sense, it should be considered only as intermediate relative to other stocking rates or grazing intensities.

I. OVERGRAZING VS. OVERSTOCKING

Overgrazing refers to continued heavy grazing which exceeds the recovery capacity of the forage plants and creates deterioration of the grazing lands. Grazing is considered overgrazing only when it causes retrogressive vegetational and soil changes from a stated objective (Heady, 1994). Wilson and Macleod (1991) extended the concept to include deleterious effects on future animal production by defining overgrazing as a "concomitant vegetation change and loss of animal productivity arising from the grazing of land by herbivores." **Overstocking,** on the other hand, refers to placing so many animals on a grazing unit that overuse will result if continued to the end of the planned grazing period. Both terms imply stocking with too many animals; overgrazing has already caused the damage while overstocking presumes there is yet opportunity to correct stocking rates so as not to exceed full proper use of the key forage species by the end of grazing period.

Overgrazing and overstocking have their valid counterparts in **undergrazing** and **understocking**; both undergrazing and understocking imply only a partial use of the full grazing capacity and a probable waste of forage. Forage wasted by undergrazing cannot be recalled because of advanced deterioration; with understocking, time yet permits forage waste to be reduced by increasing animal density and thus demand during the remainder of the grazing period. Both overgrazing and undergrazing result from inappropriate managerial decisions. Overgrazing reduces potential animal production per unit area by limiting the amount of solar energy captured by plant species of high nutritive value by minimizing leaf area; undergrazing prevents maximized animal production per unit land area because plant species of high nutritive value are not fully utilized within the limits of sustainable production, and a large percentage of the herbage yield is incorporated into litter without being consumed by grazing animals (Heitschmidt and Stuth, 1991).

Proper use of the above terms permits the concept of "overgrazing with understocking" to be faced as a serious reality. "Overgrazing happens to individual plants [or patches of plants] not a plant species, a whole plant community, or a whole range. . . . Ranges are very seldom, if ever, overgrazed [but could be overstocked]. Only plants are overgrazed [or undergrazed] and the same thing is very seldom occurring to all of the plants at the same time" (Savory, 1987). The reality of the problem is that even on grazing lands deemed properly grazed, based on average level of defoliation, one will find overgrazed plants and undergrazed plants as well as properly defoliated plants of the same forage plant species.

The probability of finding undergrazed, overgrazed, and properly grazed plants of the same plant species on all grazing lands is high, particularly on heterogeneous rangelands. In grazing intensity studies on ponderosa pine-bunchgrass ranges at Manitou, CO, Johnson (1953) noted that even in the lightly grazed pastures, areas of moderate and heavy grazing occurred. Under the heavy grazing rates, more of the plants were grazed, and the stubble, on the average, was shorter. However, the stubble of the plants actually grazed in the lightly grazed pastures was about the same height as on the pastures that were moderately stocked, but fewer of the individual plants had any defoliation.

Many problems of differential defoliation along with at least partial solutions have been considered in previous chapters—severely defoliated and wolf plants found side by side, adjoining ungrazed and severely grazed patches, and differential grazing concentrations between larger adjoining areas. The conclusion drawn is that planning and monitoring of grazing practices affecting defoliation levels at the plant species level must consider not only average levels of defoliation but also the magnitude of the annual, seasonal, and even monthly variation within those averages.

It has been suggested that domestic livestock are mostly "severe selective grazers" in that they will severely graze some plants on the first day of grazing regardless of stocking density (Savory, 1987). While this observation is directed particularly to cattle, it seems to apply similarly to native grazers such as buffalo and elk. Observed differences in the effects of herbivory by domestic livestock and native herbivores probably results as much or more from the management applied to the two animal groups than to group differences per se. Native herbivores, except when managed under intensive game ranching, may vary widely relative to numbers, population trends, patterns of grazing, and animal concentrations. In contrast, domestic livestock can be maintained at consistently high and stable levels by supplemental feeding and watering and protection from natural predation and disease. Also, fences prevent migration to new areas when the abundance of preferred forage decreases. These factors combine to contribute to higher frequencies and intensities of defoliation and maintenance of grazing pressure by domestic livestock (Archer, 1994).

II. GRAZING INTENSITY EFFECTS ON VEGETATION AND SITE

A. DELETERIOUS EFFECTS OF OVERGRAZING

The deleterious effects of excessive levels of defoliation on plant morphology, physiology, reproduction, and growth have been discussed previously (see Chapter 5). As a result of numerous grazing intensity studies, clipping trials, and careful observation, overgrazing (i.e., excessive levels of defoliation over time) can be expected to have some or most of the following negative effects on the vegetation and site:

1. Reduced vigor of grazed plants and even kill them if defoliation is severe and prolonged (Fig. 13.1A).
2. Distorted growth patterns in plants otherwise resistant to overgrazing (e.g., grasses reduced to decumbent form or shrubs become hedged).
3. Reduced plant root system.
4. Reduced yield of the key forage species by reducing leaf area and photosynthetic capacity.
5. Delayed growth response of key forage plant species to favorable temperature and moisture.
6. Replacement of key plant species by less desirable or even worthless plants.
7. Accelerated brush and poisonous plant invasions.
8. Replacement of decreaser species by increaser species and eventually by invader species on rangelands and perennial pastures.
9. Replacement of midgrasses and tallgrasses in mixed grass vegetation by shortgrasses (Fig. 13.1B).
10. Reduced longevity of improved pasture stands.
11. Trampled and puddled soil when wet, decreased water infiltration, increased runoff, and increased severity of the microclimate (Hanson *et al.*, 1978).
12. Reduced amount of vegetation and subsequent amount of mulch covering the soil, thereby increasing the probability of wind and water erosion (Fig. 13.1C).
13. Reduced fuel load and possible elimination of the option of using prescribed burning as a brush control tool.

Aggressive non-forage plants on the site can be expected to magnify the deleterious effects of heavy grazing on the desirable forage plants because of differential defoliation levels and enhanced competition. Reduced grazing and even elimination of grazing on rangelands cannot be expected to reverse advanced trends towards brush or noxious perennial forbs without applying simultaneous plant control methods (Klipple and Bement, 1961; Dwyer *et al.*, 1984; Vallentine, 1989). The deleterious effects of extreme defoliation on bluebunch wheatgrass were found by Mueggler (1972) to be offset somewhat by an applied partial reduction of competing plants.

B. SHORT-TERM VS. LONG-TERM CONSIDERATIONS

The short-term effects of overgrazing are primarily those of excessive defoliation of forage plants in the standing forage crop, but continued overgrazing generally results in reducing range condition and a deterioration of botanical composition of the plant community. The degree of overgrazing that is accepted, if any, may relate to susceptibility of the particular forage stand to deterioration and to any short-term production advantages of overgrazing. Heavy grazing in the short

FIGURE 13.1   Fenceline contrasts often clearly demonstrate the results of heavy grazing: (A) heavy sheep grazing on the right in southern Utah; (B) heavy cattle grazing on left in grazing studies at Mandan, ND; and (C) heavy localized livestock grazing (left) compared to exclosure on desert winter range near Milford, UT.

FIGURE 13.1   Continued

term, implying overgrazing if continued over time, may have the following apparent advantages:

1. Results in more uniform grazing pressure on all plant species, thereby spreading grazing pressure to the less palatable species or individual ungrazed plants (Herbel and Pieper, 1991), (but even heavy grazing seldom eliminates selective grazing).
2. Increases efficiency of forage consumption (but decreases forage availability and intake at very high stocking levels).
3. May effectively delay seedstalk formation and induce vegetative regrowth during the active plant growth period (but only under ideal conditions of temperature, soil moisture, and soil fertility).

With improved pasture on tillable land, the projected/desired longevity of the forage stand, the interim potential for serious soil erosion, and the cost of plant stand restoration should be considered. However, on most rangelands the option of periodic site restoration is seldom practical and economical and often impossible as well.

Available forage generally decreases as grazing intensity increases. In the short term this decline occurs because the rate of forage depletion exceeds the rate of accumulation; in the long term this decline results from the interaction effects of both abiotic and biotic factors on plant growth and plant successional processes (Heitschmidt and Stuth, 1991). Grazing rangelands during the non-growing season or grazing the matured, residual foliage of improved pasture may somewhat reduce the deleterious effects of heavy grazing on herbaceous plants but often results in even greater reduction in animal response. Houston and Woodward (1966) found that low range condition correlated closely with heavy grazing on mixed grass range in Montana under summer grazing but less so under winter grazing.

Rainfall penetration and infiltration into the soil may be decreased and runoff increased by grazing, particularly by heavy grazing and trampling over long time periods. On shortgrass steppe near Cheyenne, WY, grazed for 12 years by cattle, there were no effects of grazing intensity on infiltration and runoff, and season-long and rotational grazing resulted in similar runoff rates (Frasier et al., 1996). However, on shortgrass range near Nunn, CO, grazed for 55 years by cattle, historic long-term animal grazing intensities did affect rainfall runoff, this attributed principally to impacts on soil physical properties. Runoff from rainfall simulator plots after 55 years of grazing intensity treatment at the Nunn study was 10, 30, and 50% under light, moderate, and heavy grazing intensities. Two years after removal of livestock, runoff was reduced to 5, 18, and 30%, respectively (Frasier et al., 1996).

Some range sites are much more tolerant of heavy grazing than others. Summer-long grazing of shortgrass range in eastern Colorado near Nunn at different grazing intensities for 32 years did not substantially change the plant composition of the shortgrass ecosystem nor cattle food habits and botanical composition of the diets (Hyder et al., 1966; Vavra et al., 1977). However, heavy grazing did substantially reduce herbage yields and increase grazing pressure. While unusual for many other natural plant communities, the productivity of shortgrass steppe is more sensitive to variability in precipitation than to differences in long-term grazing intensities (Milchunas et al., 1994). Shortgrass communities were found resistant to grazing and well adapted to drought through mechanisms of fast recovery and efficient utilization of precipitation.

On fine sands range sites in western Nebraska, 10 years of heavy grazing (74% utilization by end of summer grazing season) did not result in any major changes in vegetation or even livestock performance (Burzlaff and Harris, 1969). This tolerance of heavy grazing was attributed to slightly delayed turnout dates in the spring and to the resiliency of the vegetation stand during a period of average to favorable rainfall years. However, 2 years of severe drought followed the years reported in this Nebraska study. Although the mid- and tallgrasses were reduced by the severe drought under all grazing intensities, the deleterious effects of the drought on the forage stand, including wind erosion, were much more severe under the heaviest stocking rate.

The deleterious effects of heavy grazing (i.e., overgrazing) may merely be masked by a series of favorable rainfall years and remain mostly invisible until triggered by severe drought. Holechek et al. (1999), in reviewing 25 stocking rates studies over North America, found that the greatest benefit of light or conservative stocking rates in terms of forage production occurred in dry years. While average annual forage production across all years was 1597, 1473, and 1175 lb per acre, respectively, under light moderate, and heavy grazing intensity, in drought years corresponding figures were 1219, 986, and 820 lb per acre. While the cumulative effects of heavy grazing compounds the deleterious effects of drought on forage plants, conservative grazing enables greater tolerance of drought by the plants.

In grazing intensity studies on rough fescue grassland in Alberta, a very heavy rate of stocking supported from three to four times the recommended stocking level in the first 11 years of the study (Willms *et al.,* 1985, 1986a). Gains per acre in the short run favored the very heavy rate, but total forage production was reduced by 50% over the 35 years of the study under this rate. The subsequent loss in grazing capacity as the range declined in range condition often forced the removal of cattle before the end of the summer grazing season and required a reduction in stocking rate. The very heavy grazing rate also resulted in a loss of flexibility in managing cattle; how long the cattle could be kept on pasture became dependent on the variable available forage which was related to precipitation during the year. Under the very heavy rate, it was concluded the forage had to be utilized more like an annual crop without the benefit of potentially high production that such crops offer.

Herbage production of the predominant annual grasses on grass-woodland in California is not substantially influenced by intensities of grazing that are reasonable for livestock production (Rosiere, 1987). On coastal mountain ranges of northern California, the annual grass vegetation tolerated stocking rates even up to 2.5 times the rate considered moderate (Pitt and Heady, 1979).

After evaluating the long-term successional trends in salt-desert shrub vegetation at the Desert Range Station near Milford, UT, Norton (1978) found no evidence that heavy grazing affected the general trend in plant cover or species composition; the vegetation changes in dominant palatable and unpalatable species were apparently not a function of grazing pressure. Under both grazed and protected conditions, the least palatable shrub, shadscale, exhibited a short-term rise in total cover followed by a steady decline, and the more palatable co-dominant shrub, winterfat, consistently increased in cover. This led to the conclusion that inherent plant longevity, opportunity for plant replacement, and differential response to climatic pattern were more influential factors than grazing stress alone. While prolonged overgrazing of winterfat-dominated sites often fails to change the botanical composition, the vigor and resulting yields of winterfat are reduced by heavy grazing. Under 28 years of heavy grazing when combined with late winter use, the yield of winterfat was reduced to less than half of its original yield and budsage was even more severely reduced (Holmgren and Hutchings, 1972).

C. EFFECTS OF UNDERGRAZING

Undergrazing or nonuse does not generally result in damage to rangelands, although opportunities to favorably manipulate plant composition may be foregone. Most grazing systems incorporate short-term to yearlong rest; nonuse for range improvement may continue for a period of up 2 consecutive years. However, excessive buildup of plant mulch and debris and encouragement of the development of patch grazing and wolf plants in species such as crested wheatgrass by under-grazing should be considered.

Based on intensity of grazing studies in the Great Plains, Klipple and Bement (1961) recommended light grazing (i.e., incorporating a planned degree of under-grazing) for a period of years to increase the herbage-yielding ability of deterio-rated native shortgrass range. They suggested the following advantages with eco-nomic implications of using light grazing as a range improvement practice: (1) no additional fencing costs, (2) reduction in per head costs, (3) a continuing annual income during the interim years, (4) no additional capital outlay for cultural treat-ments, and (5) possible additional gains per head by grazing animals. It was rec-ommended that light grazing be applied to grasslands to achieve these objectives while they are still in fair or high poor condition, followed by a gradual return from light to moderate use.

Productive improved pasture is also less likely to be damaged by undergrazing than by overgrazing, but long-term manipulative grazing through light grazing is generally not appropriate because of planned return to tilled crops after a few years. However, undergrazing will result in excessive wastage of forage because of trampling, fouling, senescence, and shading of low-growing plant species (Roh-weder and Van Keuren, 1985); insect and rodent infestations are also encouraged by undergrazing, and foliage and root diseases often are more prevalent.

III. GRAZING INTENSITY EFFECTS ON ANIMALS AND ECONOMICS

A. GENERALIZED LIVESTOCK RESPONSE

The optimal grazing intensity to select depends not only on vegetation and soil response but also on livestock production goals and production economics. It is continually demonstrated that grazing intensity has a profound effect on animal performance, and the fact that improper grazing management, whatever its form, can greatly curtail livestock gains, livestock reproduction, and weight-for-age re-sponses is readily apparent. Grazing intensity has direct effects on livestock per-formance levels and on long-term economic returns to animal enterprises primar-ily based on grazing (Fig. 13.2). Although the long-term grazing intensity studies on rangeland have utilized livestock, there is every reason to believe that the ani-mal performance effects would be similar if applied to big game ungulate herbi-vores.

Controversy exists as to whether livestock condition and response are reliable as sole indicators of proper stocking rates or other grazing management practices (Launchbaugh et al., 1978). In some grazing studies, livestock response has been rapid enough to signal improper use of vegetation at early stages, but in other stud-ies the vegetation has been impacted long before animal response has been re-duced, particularly when masked by supplementation or other livestock manage-

FIGURE 13.2   Does overgrazing really matter? The answer to this question seems apparent from this scene in central New Mexico. (Forest Service Collection, National Agricultural Library.)

ment practices. The solution no doubt requires that both animal and plant indices be used in evaluating stocking rates.

The generalized livestock responses on long-term grazing lands—but with substantial application to short-term grazing lands, as well—to low grazing intensity vs. high grazing intensity (overgrazing) are summarized as follows:

Low grazing intensity	*High grazing intensity*
1. Higher gains per head	1. Lower gains per head
2. Lower gains per acre	2. Higher gains per acre
3. Higher wool and mohair yields per head	3. Reduced wool and mohair yields per head
4. Higher dry matter intake	4. Reduced dry matter intake
5. Less time and energy spent in grazing	5. More time and energy spent in grazing
6. Reduced forage harvest efficiency; forage may be wasted	6. Greater forage harvest efficiency; defoliation may be excessive
7. Higher nutritive quality of ingesta when forage plants are dormant	7. Lower nutritive quality of ingesta when forage plants are dormant
8. Greater nutritional adequacy of diet	8. Nutritional deficiencies of diet increased
9. Higher pregnancy rates	9. Lower pregnancy rates
10. Higher percent calf, lamb, and kid crops	10. Lower percent calf, lamb, and kid crops
11. Improved body condition	11. Reduced body condition

| 12. Supply of varying levels of drought emergency forage | 12. Minimal or no drought emergency forage |
| 13. Reduced losses from poisonous plants | 13. Increased losses from poisonous plants |

Gain per head and gain per acre from a large number of stocking rate studies have been found to rather consistently fit generalized curves (Harlan, 1958). These curves show gradual declines in gains per head as grazing intensity increases from light through moderate use but drop more sharply thereafter; however, heavy grazing rates persist in giving higher gains per acre even at grazing rates known to be detrimental to the vegetation and even individual animal performance. Typical animal growth curves for shortgrass range at Nunn, CO, and Cheyenne, WY, are found in Figs. 12.4 and 13.3, respectively.

B. LIVESTOCK PRODUCTION

The response of livestock gains to intensity of grazing are summarized for many individual range grazing studies in Table 13.1. Averaging the results of 25 North American stocking rate studies by Holechek *et al.* (1999) revealed that steer/calf gains per day were 2.3, 2.15, and 1.83 lb, respectively, under light, moderate, and heavy stocking rates; however, corresponding steer/calf gains per acre were 22.4, 33.8, and 40 lb, respectively.

Long-term rangeland studies have revealed that neither per-head livestock response (individual size, gain in weight, weaning weights, or calf weight weaned per cow) nor per-acre livestock response (per-acre livestock gains, number of calves produced per section of land) alone are reliable indicators of proper stock-

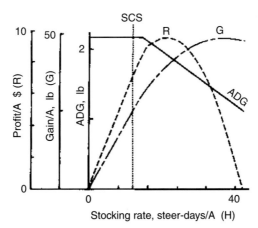

FIGURE 13.3   Response of average daily steer gain (ADG), gain per acre (G), and return per acre (R) to stocking rate on shortgrass range, Cheyenne, WY. Based on 150-day summer grazing season and 965 lb/acre of forage production; steer purchase and sale prices of $0.72 and $0.62/lb, respectively, and steer carrying costs of $0.70/day. SCS indicates the conservative stocking rate being recommended by the Soil Conservation Service. (Hart, 1986.)

TABLE 13.1 Livestock Gain Responses in Selected Studies of Grazing Intensity on Rangelands

Vegetation type (grazing period)	Location (study length)	Animal species	Grazing intensity[a]	Use level	Stocking level	Animal gain Pound/head for grazing period	Animal gain Pound/acre	Reference
West								
Pine–bunchgrass (June 1–Oct. 31)	Manitou, CO (6 years)	Cattle (yearling heifers)	H	58%		181	14.8	Johnson (1953)
			M	33%		222	16.0	
			L	16%		236	8.5	
Crested wheatgrass (Apr. 20–June 20)	Benmore, UT (11 years)	Cattle (mixed classes)	H	53%	20.2[b]	137 108 102[c]	39.7[d]	Frischknecht and Harris (1968)
			M	65%	16.6	153 163 109	43.4	
			L	80%	13.4	155 177 107	36.8	
Salt–desert shrub (Nov. 1–Apr. 30)	Milford, UT (7 years)	Sheep (ewes)	H	73% 52%[e]	17 acres/season	−0.1		Hutchings and Stewart (1953)
			M		14 acres/season	11.4		
			L		10 acres/season	8.5		
Crested wheatgrass (Apr. 20–May 25)	Ephraim, UT (7 years)	Sheep (ewes and lambs)	H	88%		20.3[f]	35.4[f]	Bleak and Plummer (1954)
			M	71%		20.3	32.0	
			L	59%		22.4	28.2	
Mountain grasslands (summer, 79 days average)	Bighorn Mtns., WY (7 years)	Cattle (yearling steers)	H	62%	2 acres/head	150	69.1	Beetle et al. (1961)
			M	44%	3 acres/head	174	56.6	
			L	17%	8.5 acres/head	190	22.4	
Pine–bunchgrass (summer, 4 months)	LaGrande, OR (10 years)	Cattle (cows and calves)	H	32%	20 acres/cow	168[a]	8.0[g]	Skovlin et al. (1976)
			M	26%	30 acres/cow	180	6.2	
			L	19%	40 acres/cow	194	5.1	
Great Plains								
Shortgrass (May 10–Nov. 10)	Nunn, CO (10 years)	Cattle (yearling heifers)	H	54%	9 acres/head	219	22	Klipple and Costello (1960)
			M	37%	15 acres/head	270	17	
			L	21%	23 acres/head	285	12	

Vegetation (dates)	Location (years)	Animal	Intensity	Grazing pressure/utilization	Stocking rate			Reference
Midgrass (May 1–Oct. 28)	Hays, KS (3 years)	Cattle (yearling steers)	H		2.0 acres/head	122	61	Launchbaugh (1957)
			M		3.4 acres/head	188	55	
			L		5.1 acres/head	217	43	
Mixed grass (May 15–Oct. 15)	Scottsbluff, NE (10 years)	Cattle (yearling steers)	H	74%	1.0 acre/steer month	246	49	Burzlaff and Harris (1969)
			M	58%	1.3 acres/steer month	248	38	
			L	53%	2.1 acres/steer month	249	23	
Mixed grass (yearlong)	Woodward, OK (9 years)	Cattle (cow–calf pairs)	H		12.0 acres/cow	314[h]	26[h]	McIlvain and Shoop (1961)
			M		17.4 acres/cow	424	24	
			L		22.4 acres/cow	437	20	
Mixed grass (May 1–Oct. 1)	Akron, CO (11 years)	Cattle (yearling steers)	H	64%	3.3 acres/head	198	58	Sims et al. (1976)
			M	44%	5.0 acres/head	228	44	
			L	30%	10.0 acres/head	237	24	
Shortgrass (May 20–Sept. 10)	Cheyenne, WY (10 years)	Sheep (ewe–lamb pairs)	H	0.6 in.[i]	74 day/acre	46.3[f]	35.8[f]	Lang et al. (1956)
			M	0.9 in.	51 day/acre	49.4	27.2	
			L	1.2 in.	31 day/acre	50.5	16.6	
Midgrass–shrubs (yearlong)	Miles City, MT (8 years)	Cattle (cow–calf pairs)	H		23.1 acres/cow	395[i]	14.0[i]	Reed and Peterson (1961)
			M		30.5 acres/cow	427	12.6	
			L		38.8 acres/cow	423	9.7	
Mixed grass (April–Dec.)	Manyberries, Alberta (19 years)	Sheep (ewes)	H	68%	7.5 acres/ewe	57.2 (126.6)[k]	16.2[l]	Smoliak (1974)
			M	53%	9.0 acres/ewe	58.1 (131.4)	13.6	
			L	45%	11.3 acres/ewe	58.0 (136.2)	11.0	
Fescue grassland (May 15–Nov. 15)	Lethbridge, Alberta (35 years)	Cattle (cow–calf pairs)	VH		2.0 AUMs/acre	226 (134)[m]	74.0 (43.5)[m]	Willms et al. (1986a)
			H		1.0 AUM/acre	302 (148)	49.0 (24.0)	
			M		0.65 AUM/acre	318 (188)	34.5 (20.5)	
			L		0.5 AUM/acre	304 (188)	25.0 (15.0)	
Mixed prairie (7 months summer)	Cottonwood, SD 10th–13th year	Cattle (cow–calf pairs)	H	69%	1.82 acres/AUM	316.7 (29.6)[n]	14.5[d]	Lewis et al. (1956)
			M	51%	2.85 acres/AUM	360.2 (80.9)	13.5	
			L	26%	3.78 acres/AUM	370.2 (135.8)	11.5	
Mixed prairie (yearlong)	Throckmorton, TX (8 years)	Cattle (cows and calf s)	H	abt. 75%	12.8 acres/AU	490[g]	34.4[g]	Kothman et al. (1970, 1971)
			M	abt. 45%	20.8 acres/AU	501	21.2	
			L	abt. 20%	28.4 acres/AU	506	16.4	

(continues)

TABLE 13.1 (continued)

Vegetation type (grazing period)	Location (study length)	Animal species	Grazing intensity[a]	Use level	Stocking level	Animal gain Pound/head for grazing period	Animal gain Pound/acre	Reference
Flint Hills bluestem (May 1–Oct. 1)	Manhattan, KS (17 years)	Cattle (yearling steers)	H		1.75 acres/head	215	122.6	Launchbaugh and Owensby (1978)
			M		3.30 acres/head	236	71.4	
			L		5.00 acres/head	228	45.6	
Grass–shrub savannah (yearlong)	Sonora, TX (21 years)	Cattle (growing)	H		13.3 acres/AUY	179	17.5	Taylor and Merrill (1986)
			M		20.0 acres/AUY	225	14.4	
			L		40.0 acres/AUY	262	8.6	
	Sonora, TX (21 years)	Sheep (growing)	H		13.3 acres/AUY	25.3 (8.9)[o]	13.5 (4.5)[o]	Taylor and Merrill (1986)
			M		20.0 acres/AUY	28.5 (9.3)	10.4 (3.3)	
			L		40.0 acres/AUY	33.4 (10.1)	6.4 (1.8)	
	Sonora, TX (15 years)	Angora goats (growing)	H		13.3 acres/AUY	18.2 (8.4)[p]	11.7 (5.4)[p]	Taylor and Merrill (1986)
			M		20.0 acres/AUY	20.3 (9.1)	8.8 (3.9)	
			L		40.0 acres/AUY	23.3 (9.5)	5.3 (2.0)	
Southeast								
Southern pine–bunchgrass (yearlong)	Palustris Expt. For., LA (11 years)	Cattle (cows and calves)	H	57%	13 acres/cow	421[j]	22.7[j]	Pearson and Whitaker (1974)
			M	49%	20 acres/cow	419	15.3	
			L	35%	26 acres/cow	444	14.0	
Pine–wiregrass (yearlong)	Charlotte Co., FL (10 years)	Cattle (cows and calves)	H	65–75%	15 acres/cow	294[j]	9.4[j]	Hughes (1974)
			M	45–55%	22 acres/cow	328	8.1	
			L	30–40%	36 acres/cow	355	4.9	

[a] VH, very heavy; H, heavy; M, moderate; L, light.   [b] Cattle days per acre.   [c] Left to right: yearlings, lactating cows, and calves.   [d] Total cattle gain per acre.   [e] Average use of 73% and 52% for Indian ricegrass and bud sagebrush, respectively; average use of all grasses was 40% and all shrubs 32%.   [f] Lamb gains only.   [g] Calf gains only.   [h] Calf weaning weight per cow.   [i] Remaining stubble height.   [j] Calf weaning weight.   [k] Summer gains of lambs (ewe weights at weaning).   [l] Total ewe and lamb gains in summer.   [m] Calf gain (cow gain).   [n] Adjusted weaning weight of calves (summer cow gains).   [o] Sheep gains (wool).   [p] Goat gains (mohair yield).

ing rates. Per-head response under light grazing is generally similar to or only slightly more than under moderate (proper) grazing intensity, but per-head response declines sharply under heavy stocking rates. On the other hand, production per acre rises almost linearly from zero grazing through light to moderate rates and continues upwards even after proper stocking rates have been considerably exceeded. However, a point is eventually reached at which livestock gains per acre are at a maximum but drop sharply as this point is exceeded.

In stocking rate studies on grass-shrub rangeland at Sonora, TX, gains per head typically decreased and gains per acre increased with increased grazing intensity when goats, sheep, and cattle grazed alone (Table 13.1) (Taylor and Merrill, 1986). However, the advantage of the heavy over the medium grazing intensity (representing a 50% increase in stocking rate) in livestock gains per acre was 21.5% for cattle, 31.4% for sheep, and 37.5% for goats. It was concluded that goats were less affected by high grazing intensity than sheep and cattle and that cattle were the most affected. Certain physical characteristics (i.e., small ratio of rumen volume to body weight, bipedal grazing stance, and prehensile tongue) apparently offered significant foraging advantage to the goat; sheep also evidenced prehensile grazing ability. Both sheep and goats apparently were also more willing to graze farther into the less palatable, "reserve" component of the vegetation, the goat being particularly noted for versatility in diet selection.

Average daily gain of yearling heifers grazed in an intensity of grazing study on the shortgrass steppe near Nunn, CO (Hart and Ashby, 1998), was shown to be inversely and linearly related to grazing pressure (cumulative); (i.e., grazing pressure index). Heifers initially weighing about 600 lb were grazed May to October in each of 55 consecutive years (1939–1994). The relationship of grazing pressure (cumulative) and herbage allowance (cumulative)—based on peak standing crop—to average daily gain during this long-term study were as follows:

Grazing pressure (cumulative) (heifer days/ton)	Herbage allowance (cumulative) (lb/AUD)	Average daily gain (lb)	Generalized stocking rate
25	114.3	1.54	Light
50	57.1	1.28	Moderate
100	28.6	0.84	Heavy
150	19.0	0.44	Very heavy

Long-term heavy grazing combines the effects of current grazing levels and range condition; heavy grazing over time leads to or maintains low range condition, while light or moderate grazing leads to or maintains higher range condition (Malechek, 1984). This is demonstrated by an evaluation of the results of a long-term grazing study at Miles City, MT (Woolfolk and Knapp, 1949; Reed and Peterson, 1961; Houston and Woodward, 1966) (see Table 13.2). At the end of the 25-year study, range condition on the summer range under heavy grazing was only 29% compared to 39% and 60% under moderate and light grazing, respectively. By contrast, the corresponding stocking levels applied over the years to the winter range caused only minor differences in range condition. Although the respective treatments were continued in the same pasture units, all stocking levels were correspondingly reduced during the last of the three study periods.

The first study period (1933–1936) in the Miles City study was terminated early because of the drought of the 1930s, while the last two periods were carried through the full productive cycles of the assigned cows. During the first lifetime production cycle, 1938–1945, the heavy grazing rate reduced average weaning weights about 30 lb compared to moderate and light grazing (Table 13.2). But, during the second lifetime cycle, 1950–1957, heavy grazing reduced average weaning weights by 60 lb over moderate grazing and 70 lb over light grazing. The percent calf crops weaned during years 7 through 14 were 82, 90, and 89 under heavy, medium, and light grazing. But, during the years 20 through 27 of the study, the corresponding average percent calf crop differences had widened to 63.0, 85.6, and 91.4, respectively.

It was concluded from yearlong grazing studies on native range at Woodward, OK, that calf condition and body weight reflected a shortage of forage under heavy grazing more than did changes in cow weights (McIlvain and Shoop, 1962a; USDA, Agricultural Research Service, 1961a). Lowered calf condition and declining weaning weights were considered useful in reflecting overgrazing. Calf performance was considered a useful measurement tool only if cattle got all of their roughage from rangeland and received supplemental protein only as needed in winter and salt all year, and calves were not creep fed. However, during 60-day spring grazing of crested wheatgrass in Utah, heavy grazing had minimal effect on calf gains but sharply reduced the gains of lactating cows (Frischknecht and Harris, 1968); the least effects of grazing intensity were on the yearling cattle.

Heavy grazing of rangelands over a period of years has commonly reduced actual weaning weights and summer gains of yearlings by 30 to 50 lb. Heavy grazing on rangelands has also generally reduced body weights and body condition of breeding females (cows, ewes, and does). This, in turn, has been related to lower conception rates and percent calf, lamb, and kid crops. With beef cows on rangeland, heavy year-round grazing has commonly reduced percent calf crop at weaning by 10 to 15% unless masked by heavy rates of supplemental feeding.

A survey of the effects of grazing intensity on pregnancy rates and body condition of spring calving cows in the Kansas Flint Hills was made by Sprott et al. (1981). Body condition at the time of pregnancy examination for cows under heavy stocking (less than 6 acres per AU) and light stocking (6 acres per AU or more) was 4.4 and 5.3, respectively. Heavy stocking rates also delayed conception; comparative pregnancy rates were as follows (heavy vs. light): after 20 days, 23 vs. 36%; after 40 days, 36 vs. 67%; and after 60 days (end of breeding), 76 vs. 84%. Lighter weaning weights under heavy grazing on south Florida range resulted from a combination of lower daily gains to weaning and about 10 days average younger age at weaning (Hughes, 1974). In a 10-year study on pine-bunchgrass range in Louisiana, cow weights at calf weaning time under heavy, moderate, and light grazing of 788, 778, and 819 lb, respectively, were associated with percent calf crops of 70, 73, and 82%, respectively (Pearson and Whitaker, 1974).

TABLE 13.2 Evaluation of Long-Term Stocking Rates on Mixed-Prairie Rangelands at the Fort Keogh Livestock and Range Research Laboratory, Miles City, Montana, 1932–1957[a]

Years	Grazing intensity	Stocking level[b]	Weaning weight (lb/calf)	Weaning weight (lb/acre)	Calf crop weaned (%)	Average hay fed cows (lb)	Weight of cows weaning calves, Nov. 1 (lb)	End of study Range condition (% climax) Summer units	Range condition (% climax) Winter units	Utilization key species (%) Summer units	Utilization key species (%) Winter units
1933–1936	Heavy	23.1 acres/cow	248	8.1	75.0		955				
	Moderate	30.5 acres/cow	297	7.7	79.0		1030				
	Light	38.8 acres/cow	300	6.1	79.0		1070				
1938–1945	Heavy	23.1 acres/cow	395	14.0	82.0	1068					
	Moderate	30.5 acres/cow	427	12.6	90.0	436					
	Light	38.8 acres/cow	423	9.7	89.0	470					
1950–1957	Heavy	29.3 acres/AUY	348	9.3	63.0		939	29	50	77	60
	Moderate	42.5 acres/AUY	408	8.2	85.6		1032	39	58	60	52
	Light	51.7 acres/AUY	420	7.4	91.7		1023	60	56	51	46

[a]Formerly U.S. Range Livestock Experiment Station; grazing treatments yearlong but on paired summer and winter grazing units; the three time periods represent different groups of Hereford cows; the 1933–1936 period was terminated because of drought, the latter two because of age of cows. Adapted from Woolfolk and Knapp (1949); Reed and Peterson (1961); Houston and Woodward (1966).

[b]Stocking levels indicate combined acreage of paired summer and winter units.

TABLE 13.3   Winter Sheep Grazing Study on Salt-Desert Shrub Range, Desert Range Station, Milford, Utah[a]

	Experimental pastures			Management pastures	
	Heavy	Moderate	Light	Heavy	Moderate
Stocking rate (sheep days/acre)	17	14	10	19	15
Sheep gains (lb)					
Nov. 15–Jan. 3	−1.8	2.5	2.0		
Jan. 4–Feb. 23	−0.3	3.5	0.5		
Feb. 24–Apr. 10	2.0	5.5	6.0		
Total winter	−0.1	11.5	8.5	1.1	9.3
Death loss (%)				8.1	3.1
Fleece weight (lb)				9.7	10.6
Lamb crop (%)				79.0	88.0
Lamb weaned per ewe (lb)				67.0	77.0
Net income per 3000 ewes ($; 1953 cost data)				5072.00	10,380.00
Net income per ewe ($)				1.69	3.45

[a]From Hutchings and Stewart (1953); 6.64 in. average annual precipitation; data 1938–1944.

Heavy grazing has also been shown to reduce wool and mohair yields per head while increasing yields per acre over light and moderate grazing (Table 13.1, under Taylor and Merrill, 1986). As shown in Table 13.3, moderate grazing by non-lactating ewes on Utah winter range not only increased fleece weight by about 1 lb per head but also decreased death loss (Figure 13.4). In studies at Manyberries, Alberta, heavy grazing reduced fleece weight by about one-half pound (10.0 vs. 9.4 lb) (Smoliak and Slen, 1972).

C. ECONOMIC RETURNS

The consideration of economic returns as well as plant and animal response has generally located the point of optimal stocking rate slightly beyond the initial drop in per-head response but much before maximum yield per acre has been reached. Hart (1986) has developed equations for various kinds of grazing lands in the Cheyenne, WY, area relating grazing pressure, average daily gain, and profitability. The equation developed for shortgrass range (with stocking rates from 0–40 steer days per acre) resulted in the profit curve (designated R) shown in Fig. 13.3. Maximum profit per acre (about $9.50) in this example occurred around the stocking rate at which the average daily gain and gain per acre curves crossed, i.e., after average daily gains had begun to decline but before gain per acre had peaked.

In a 9-year stocking rate study on Oklahoma sandhills range (Table 13.4), fixed land costs per cow were shown to be reduced under heavy grazing rates (over-

FIGURE 13.4   Grazing management including proper stocking rates maintained both the range and the sheep in good, productive condition at the Desert Experimental Range, Milford, UT.

stocking) compared to lighter rates (Fig. 13.5) (McIlvain and Shoop, 1961). However, the reduction in per-head livestock response under heavy grazing more than offset the advantage of reduced per-head land costs, resulting in a reduced residual return to management compared to moderate stocking (in this case, a negative return to management under heavy stocking). In contrast to land costs, other livestock costs (i.e., labor, veterinary care and drugs, feed supplements, interest and taxes on livestock, cow depreciation or replacement costs, and bull costs in the cow-calf enterprise) were nearly constant on a per-head basis. Wilson (1986) confirmed that in a developed economy variable costs are a significant part of total livestock production costs, and as grazing intensity increases toward the point of maximum production per acre, variable costs rise at a faster rate per acre than do gross returns.

From an ecological and conservation perspective, stocking rates should not exceed the grazing capacity of rangeland and medium-term grazing lands. A management strategy of systematically overgrazing and periodically applying cultural practices to restore the depleted forage stand may be applicable to short-term pasture, providing livestock performance remains acceptable. However, such a scenario will seldom, if ever, prove economically viable and ecologically sound on the longer term grazing lands. Excessive stocking rates on such grazing lands to take advantage of short-term gain potential commonly results in deterioration of the vegetation to the extent that annual net returns will be permanently lowered and left without practical restoration recourse. Torrell *et al.* (1991) concluded that,

TABLE 13.4   Analysis of Cow–Calf Stocking Rates on Mixed Grass Sandhill Range,
Woodward, Oklahoma, 1952–1960[a]

	Grazing intensity		
Item for comparison	Heavy	Moderate	Light
Acres per cow	12.0	17.4	22.4
Returns per cow			
Average weaning weights (lb)	388	461	491
Calf crop weaned (%)	81	92	89
Calf weaning weight/cow (lb)	314	424	437
Calf weaning weight/acre (lb)	26	24	20
Gross returns and costs per cow (1960 prices)			
Value of weaned calf ($)	87.80	101.87	107.95
Value of calf weaned per cow ($)	71.89	94.04	95.43
Land costs (per head) ($)	18.12	26.18	33.59
Other costs (per head) ($)[b]	56.57	59.06	58.98
Total costs ($)	74.69	85.24	92.57
Net returns (1960 prices) to:			
Land, labor, and management (per cow) ($)	29.62	50.88	53.12
Land, labor, and management (per acre) ($)	2.46	2.92	2.38
Management (per cow) ($)	−2.80	8.80	2.87
Management (per acre) ($)	−0.23	0.51	0.13

[a]From McIlvain and Shoop (1961).
[b]Other costs included veterinary costs, supplements, interest at 6% on cow and half of operating
costs, taxes, and depreciation on cow, cow death losses, and bull costs.

"Ranchers have no economic incentive as profit maximizers to overgraze continually . . . [Overgrazing] occurs in spite of the profit motive not because of it."

The grazing manager of intensive pasture may choose to maximize short-run annual net returns by stocking at somewhat higher than normal rates. Under management-intensive grazing in the Midwest, management emphasis has moved increasingly toward output per acre when profits are easier to generate, since cutting costs is not as crucial as increasing production in generating profit (Moore, 1999). When profit margins are tight, a shift is made towards optimizing individual animal performance; however, even in the short term, the optimal biological stocking rate (i.e., the level that maximizes animal performance) is consistently higher than the optimal economic rate. In a grazing study in Alabama using steers on grass and grass-legume pasture, the optimal biological stocking rate was found to be 2.22 steers per acre compared to 1.57 head per acre for the optimal economic stocking rate because of higher marginal costs under biological maximization (Olowolayemo et al., 1992).

Economists have verified that the planning horizon (number of years over which profit is to be maximized) has influenced the grazing intensity applied in the

FIGURE 13.5 Yearlong stocking rate studies with beef cattle at the USDA Southern Plains Experimental Range near Woodward, OK, demonstrated the added profits of moderate stocking over either overstocking or understocking.

past within livestock enterprises (Quigley *et al.,* 1984; Pope and McBryde, 1984); this relationship can be expected to continue. Many private landowners follow stocking rates greater than ecologically optimal to maximize short-term profits under a planning horizon of 1 to 10 years in order to pay mortgages, taxes, support the family, etc. Also, when the ratio of forage price ($/AUM) to the price of livestock ($/lb) decreases, heavier grazing is encouraged. However, a longer planning horizon (possibly 25–50 years) must include holding stocking rates to levels that will maximize and sustain high yield and profits over the long term.

Uncertainty of tenure of grazing and payment of grazing leases on an acreage basis promote a short-term planning horizon and reduced emphasis on sustainable grazing over the long term. The results of a state-wide survey of Texas ranchers revealed that stocking rates on short-term leases and/or payment made on a per-acre basis rather than per-head basis tend to promote excessive stocking rates (Rowan *et al.,* 1994). By contrast, moderate stocking rates are favored when lease contracts are long term, charge on a per-animal unit basis, and include some provision for seasonal adjustment and limitation of livestock numbers. Leases that provide flexibility, when kept within reasonable bounds, can provide greater incentive for conservative stocking. Hart (1980, 1986) has stressed the need to maintain flexibility in stocking rates to be able to adjust to short-term changes in economic situations.

An examination of opportunity costs by Hooper and Heady (1970) indicated that the economic loss from heavy grazing is several times that of light use. Grazing land managers who recommend and follow moderate or even light grazing are in effect advocating a small loss (opportunity cost of lighter grazing) as insurance against a large loss (opportunity cost of heavy grazing). Based on their study of range sheep enterprises, White and Morley (1977) recommended that risk avoidance, as indicated by the lowest gross margin or bank balance recorded over a long period, be the basis of the optimum stocking rate. They concluded the estimated optimal stocking rate, based upon risk avoidance, was slightly below that calculated to give maximum gross margins, but not so much below that gross incomes were seriously reduced.

From their grazing studies in Oklahoma, Shoop and McIlvain (1971b) concluded that moderate grazing was not only more profitable than heavy grazing in the long run but was more stable financially in the short run. They noted that heavy grazing was often profitable in non-drought years but disastrous in drought years. They likened overgrazing under their conditions of variable rainfall and forage supply—essentially universal to the western rangelands of North America and most of the world—as playing brinkmanship with the rangeland natural resources.

Management objectives may suggest some adjustments in stocking rates. High reproductive and growth performance with many individual animals making top performance or the general attractiveness in purebred herds may encourage lighter stocking rates, but full stocking rates may be suggested when only maintenance or low growth rates are acceptable.

D. NUTRIENT INTAKE

The significant effects that intensity of grazing have upon animal production result from differences in nutrient intake. As utilization becomes heavier and more even, the higher quality forages are readily consumed, the number of foraging choices declines, and performance generally declines. Heavy grazing deleteriously affects animal performance by reducing dry matter intake, nutrient composition in the ingesta, nutrient digestibility, or more commonly a combination of most or all of these factors. The specific causes of this reduction in nutrient intake are complex, seldom uniform, and vary from situation to situation (Malechek, 1984). Overgrazing promotes under-nutrition and even animal stress during winter and drought emergencies.

The greatest effect of heavy grazing on animal performance is apt to be in reducing total forage intake. The lower performance of the cows under heavy grazing at Miles City, MT, (Table 13.2), was attributed more to lack of sufficient volume of forage to consume than to specific nutrient deficiencies in the forage (Marsh *et al.,* 1959). Heavy grazing during winter increased the amount of hay that had to be fed, since the declining seasonal nutrient levels coincided with rising nutritional requirements of the pregnant cow. Cows in the lightly grazed pastures, where intake was not limited by the forage supply, were able to be carried through the winters with little or no hay feeding without showing nutrient deficiencies.

During seasons when most of the forage consists of growing plants with green leaves, grazing pressure may have little effect on nutrient intake until the point is reached that forage availability directly limits dry matter intake (Kothmann, 1980). During active plant growth periods, increased defoliation associated with heavy grazing can even result in increased forage quality, but it is doubtful that forage quality can be enhanced by any grazing practice during periods of plant dormancy (Heitschmidt and Walker, 1983). As the proportion of dry or dead forage increases in the standing crop towards the end of the growing season, sensitivity to grazing pressure appears to increase as the animals become more selective, and consumption may be reduced (Kothmann, 1980). Moreover, as opportunities for selective grazing are diminished, more of the mature herbage—lower in nutrient levels and in digestibility—will be ingested and nutrient intake further reduced. In contrast, during ample production years on north Texas grasslands, both animal performance and nutrient levels in forage ingested did not differ greatly between stocking rates (Pinchak et al., 1988).

Stocking levels also affect forage consumption on improved pasture. In a 4-year grazing study on orchardgrass-Ladino clover irrigated pastures at Davis, CA, stocking rates with yearling steers averaged 1.84, 3.64, and 5.46 steers per acre for the summer grazing season. Corresponding daily dry matter consumption per head was 18.1, 14.9, and 12.1 lb, while daily dry matter consumed per acre was 32.6, 53.3, and 65.1 lb. During the first year of the study, average daily gains were 1.81, 1.44, and 0.80 lb at the light, moderate, and heavy stocking rates (Hull et al., 1961). Calculated liveweight gain per acre and corresponding carcass gain per acre rose with increasing stocking rates to about 4 steers per acre but declined as stocking rates rose above this level. It was noted from the study that mechanically harvesting the forage as soilage (green chop) was less harmful to the forage plants, based on amount of regrowth, than the heaviest stocking rate; the heavy grazing removed forage irregularly from even below the height of the cut made by the harvester.

When grazing on dormant vegetation, such as on winter range, grazing animals select the more palatable and often more nutritious plants and plant parts. Without opportunity for remedial plant regrowth, continued grazing pressure will require more of the less palatable, more stemmy and woody, and often less nutritional forage to be consumed. The increased consumption of stems prompted by close grazing usually results in diets higher in low digestible fiber and lower in crude protein and useful energy. In studies with sheep on salt-desert shrub winter ranges in southwestern Utah, daily intake was less on poor condition than on good condition range (2.4 vs. 3.0 lb per day), and increased intensity of grazing reduced daily intake on both good condition (3.1–2.9 lb) and poor condition (2.6–2.2 lb) range (Cook et al., 1962). The combination of heavy grazing and poor condition range reduced metabolizable energy intake by about 50% (1752 vs. 894 kcal per day).

Further study revealed that both the content and digestibility of the nutrients in the diet decreased as animals were forced into higher utilization levels (Cook and Harris, 1968a).

Sheep on desert winter range in Utah were found to perform differently on pure than on mixed forage plant stands (Pieper *et al.*, 1959). On pure stands, protein, energy, and phosphorus tended to decrease as intensity of grazing increased. However, on some mixed vegetation types, the diet changed between plant categories of herbage and browse and between plant species as the grazing pressure increased, thereby maintaining or even occasionally increasing protein, phosphorus, and energy levels and/or their digestibility in the diet. In grazing studies with sheep on northern Utah mountain summer range, neither range condition nor grazing intensity significantly affected daily intake, largely due to substantial botanical dietary shifts as utilization increased (Cook *et al.*, 1965). Differences between years were highly significant, the higher intake rates being in years when the forage remained green longer and as a result was more palatable over a longer period.

Steers grazing post oak savannah near College Station, TX, were found to sustain nutrient intake during the first half of the growing season under heavy grazing; subsequent switching from grasses to dicots enabled the steers to sustain crude protein intake into late fall but digestible energy levels only into mid-summer (Stuth and Olson, 1986). Grazing animals are apparently able through selective grazing to maintain dietary protein levels but not energy levels until forage utilization reaches a critical level (Malechek, 1984). Grazing intensity had little impact on the nutritive quality of sheep diets on annual grass range in California at levels considered practical under yearlong use (Rosiere and Torell, 1985); this suggested that selective grazing by sheep on annual range was not as important a factor in sustaining adequate levels of nutrition as on other vegetation types.

Supplementation of grazing livestock is more critical at high stocking rates than at low stocking rates or grazing pressure. Protein and energy supplements can mask the low and declining animal production on overgrazed range. Livestock and big game can make good gains on overgrazed range for a few years if they are fed enough hay, grain, or protein supplement. This combination of overgrazing and extra supplements can even be profitable with livestock, and possibly with big game under game ranching, until the plant and soil resources are badly damaged or until a series of drought years is combined with low or dropping livestock prices (Shoop and McIlvain, 1971b). When the rate of removal of the high-quality component of the standing crop exceeds the rate of renewal, the need for supplementation must be considered or the consequences of undernutrition accepted (Rittenhouse and Bailey, 1996).

In north-central Texas studies with cattle handled under yearlong continuous grazing on mixed grasslands (Pinchak *et al.*, 1990), winter supplementation of range cows under heavy grazing did not stimulate forage intake and either replaced forage organic matter intake or substituted for insufficient forage availability. In southwest Texas studies on the Edwards Plateau (Huston *et al.*, 1993), low-level feeding of a high-protein supplement under heavy grazing increased intake of dormant range forage and decreased body weight loss, but high levels of low-protein supplements apparently increased nutrient status and reduced body weight loss primarily by providing supplemental nutrients.

Steers given extra supplements while grazing yearlong on a range overstocked for 3 years in Oklahoma gained 60 lb per head more annually than steers on moderately grazed range that were given only standard supplement (Shoop and McIlvain, 1971a). Winter supplementation (1.5 lb of high-protein supplement per day) in north Texas studies significantly increased production in a cow-calf enterprise in heavily grazed pastures and reduced animal production variability between years by about half (Heitschmidt *et al.,* 1982a; Whitson *et al.,* 1982). However, in some years winter supplementation improved weaning weights more under the moderate grazing than under the heavy grazing treatments, suggesting the supplement in the latter may have been required for body maintenance of the cows rather than contributing to milk production (Knight and Kothmann, 1986).

IV. MANAGING FORAGE PRODUCTION FLUCTUATIONS

A major problem faced by grazing managers is annual and seasonal fluctuations in forage production resulting from climate. On dryland range and pasture, even in sub-humid regions, forage production may vary by as much as 1:2 in consecutive years; in the arid Southwest, the range may be as great as 1:5. (Refer to Chapter 12, "Fluctuations in Forage Production".) Fluctuations in forage production are normal occurrences, and both severe droughts as well as bumper forage production years must be planned for. The challenge is to keep from overgrazing and damaging the forage resources in low production years but still realistically fully utilize the forage produced in high production years.

A. DROUGHT

The effects of drought are now less disastrous on the rangelands of particularly western North America than in the early days of the livestock industry. Many drought management measures are utilized today that were not then used. However, a better understanding of the relation of drought to grazing non-irrigated grazing lands is still needed to avoid permanent damage to long-term grazing lands and still maintain high levels of livestock production in the face of a highly variable forage supply.

Drought refers to prolonged dry weather during which time plants suffer from lack of soil moisture. Although anything less than average precipitation could be considered as drought, the term is more commonly used to refer to periods when precipitation is 75% or less than average for a considerable period of time (i.e., a few months to one or more years). Droughts occur in various lengths, cycles, and degrees of severity and cannot be adequately quantified or accurately predicted.

FIGURE 13.6   Black grama range in New Mexico severely depleted by multi-year drought, a natural phenomenon that can override, at least in the short term, the benefits of recent good grazing management.

Drought remains a major factor, if not the most difficult problem, in maintaining efficient, economical animal production enterprises on dryland pasture and range in subhumid to arid areas.

Severe drought reduces plant vigor and carbohydrate reserves in perennial plants, reduces both root growth and forage production, results in a shorter season of high-quality forage, and exposes more soil as a result of reduced plant cover. Even the more drought-hardy forage plant species—and also often the more potentially productive—are reduced in plant composition on the site (Fig. 13.6). Where the most drought-hardy plant species are also the least palatable or are worthless forage species, the competitive advantage provided them by drought may have serious long-term consequences on the botanical composition of the vegetation. Thus, the long-term competitive status of some plant species may be determined during even relatively short periods of drought (Caldwell, 1984). Also, poisonous plant problems almost invariably become more prominent during drought years because of the lack of alternative forage.

The forced sale of or severe reduction in the breeding herd may be the most serious effect of severe drought on livestock programs, but other negative consequences are (1) reduced weaning weights; (2) lower calf, lamb, or kid crop the following year; and (3) increased emergency feed costs. In cow-calf enterprises based primarily on rangelands, reductions in forage production may be as great as 60 to

80%, weaning weights reduced by 50–75 lb, and percent calf crop reduced by 15–40% the following year (Wallace and Foster, 1975). Lactating, 2-year-old heifers can be expected to suffer more from drought than any other single class of cattle. Fleece weight per ewe and percent lamb crop at the Texas Range Station at Barnhart were greatly affected by annual precipitation (Taylor *et al.,* 1986a). During a 10-year study period, both were severely depressed during two drought years; in the severe drought years, fleece weights at shearing in the study were reduced from 10.5–5.1 lb per ewe and percent lamb crop weaned from 130% to 74%.

Over the 4-year severe drought period of 1934–1937, the drought effects on the experimental range units at Miles City, MT, were sharply evident (Hurtt, 1951). During this period the six most important range forage species declined in ground cover to 8.5% of the 1933 level in all range units regardless of the planned rate of stocking. Drought reduced the forage supply so drastically that only 36 and 23% as much unsupplemented grazing use was obtained in 1934 and 1936 as was possible in 1933, the last year before the drought began. The effect of the drought was so severe on both the experimental range units and cattle that the study had to be temporarily halted.

The effects of drought on cattle production on native sandhills range near Woodward, OK, have been monitored under both moderate and heavy grazing intensity (Shoop and McIlvain, 1971b). During a 9-year study with cow-calf pairs, 5 years were considered non-drought years and 4 were drought years. Calf weaning weights favored moderate over heavy grazing by 50 lbs in non-drought years (490 vs. 440 lb) but by 100 lb in drought years (440 vs. 340 lb). The spread in percent calf crop weaned between moderate and heavy grazing was about 11% during both drought and non-drought years. Calf production per cow favored moderate grazing by 100 lb (460 vs. 360 lb) in non-drought years but by 125 lb (390 vs. 265 lb) in drought years.

The effects of drought and grazing intensity on steers grazed November 10 to October 1 were also included in the Oklahoma study (Shoop and McIlvain, 1971b). During the 12 non-drought years of this 17-year comparison, steers under moderate grazing gained an average of 25 lb more than under heavy grazing, i.e., overstocking (374 vs. 349 lb). In the 5 drought years steers under the moderate rate averaged 80 lb more than under the heavy rate (338 vs. 258 lb). As a result of the low production and the fewer acres allowed per head, each steer on the overstocked range in drought years had access to only 2600 lb of grazable forage compared to 6100 lb of grazable forage under moderate stocking. During these grazing studies, supplementation levels were held constant so that they did not mask the grazing intensity and year effects. High death loss of plants occurred under the combination of drought year and overgrazing; both wind and water erosion were greatly increased by heavy grazing and particularly in combination with drought.

Variations in annual rainfall and their effects on forage and cattle production on semi-desert grasslands on the Santa Rita research station near Tucson were monitored during the 1924–1950 period (Reynolds, 1954). Annual variations during this period were as follows: (1) June–September rainfall, 5–23 inches (9.5 in. average); (2) forage production, 35–680 lb per acre (370 lb average); (3) percent calf crop, 76–91%; and (4) weaning weights, 320 to 425 lb.

Special post-drought management is often required to assure perennial plant recovery following drought (Bedell and Ganskopp, 1980). Care must be taken not to restock to pre-drought levels too rapidly with the return of normal precipitation levels. Drought-stressed plants will be in low vigor and require some time to regain their vigor to resist normal grazing pressure. Reduced stocking for the dormancy period and subsequent growing season following return from drought should be considered in order to hasten range recovery. The rate of range recovery under conservative (light to moderate) grazing of semi-desert grasslands in the Southwest was shown to be about the same as under continuous protection from livestock grazing, provided the productivity of the range had not been greatly impaired during the drought (Rivers and Martin, 1980). However, following extended, severe drought and particularly on ranges with a history of heavy grazing, a growing season of deferred grazing or complete rest should be considered to aid recovery.

Weeds are another problem during the post-drought recovery period where the original perennial vegetation has been severely depleted. High rainfall following drought may stimulate high weed biomass production and severe competition with desirable, drought-damaged forage plants; fire hazards may develop upon the drying of flammable biomass such as cheatgrass. Excess moisture allows the germination and invasion of brush and other noxious perennials previously controlled by perennial forage plant competition. Thus, post-drought recovery may require or be greatly accelerated by applying prescribed weed control practices.

B. ACHIEVING FLEXIBILITY

Building flexibility into the livestock production and grazing management program is the initial step in meeting annual fluctuations in realizable grazing capacity. Thirty percent annual deviations in forage production either above or below average can probably be met through short-term management adjustments, but 50% deviations below the average will probably require severe measures (Holechek, 1988). Possibilities for building helpful flexibility in advance of future drought to help meet drought emergencies include:

1. Provide carrying capacity from a combination of range, temporary pasture, and harvested roughages. Including irrigated pasture (Nichols and Clanton, 1985) or subirrigated pasture or haylands helps to stabilize carrying capacity.
2. Organize forage programs for animal breeding enterprises to avoid having to make radical changes in animal numbers that would have serious genetic, economic, and managerial ramifications.
3. Maintain a portion of the cattle herd (possibly 25 to 35%) as steers or other livestock that can be increased, adjusted, or liquidated on relatively

short notice while maintaining the base cowherd. Where a cow-calf-year-ling operation is feasible, it is more flexible than a strictly cow-calf oper-ation.

4. Hold the breeding herd at a conservative level somewhat below the aver-age grazing capacity; this number can be set at 80% to 90% of grazing capacity in the median year.

5. Maintain long-term grazing lands in high condition because forage pro-duction is more stable than on low condition grazing lands.

6. Utilize any forage production prediction equations developed for the area.

7. Develop adequate fencing and stockwater developments to properly con-trol grazing in both good and drought years.

8. Maintain on hand a carryover supply of emergency feed such as hay, silage, or other harvested roughages.

9. Maintain a standing forage reserve by selecting grazing land each year to be left ungrazed or lightly grazed for emergency use. During severe drought years, moderately graze pasture units scheduled in a grazing sys-tem for light or no use during that year. (Note: this practice does not have the universal approval of range specialists since it may compromise the objectives of the special grazing system.)

10. Follow an orderly livestock marketing system rather than trying to out-guess the market; avoid situations as much as possible where emergency sale at any price is mandated.

11. Monitor stocking rates continually and adjust when needed and appropri-ate.

Continued stocking at near normal levels during moderate to severe drought is probably the greatest cause of range deterioration. Heavy grazing both before and during drought will cause even greater drought damage than under moderate graz-ing; the results may extend into several good years, and improvement in plant vig-or and production may be greatly delayed (Pieper and Donart, 1975). Reduced stocking rates during moderate to severe droughts should result in less drought damage to the vegetation and hasten its recovery following the drought.

Fixed, inflexible stocking rates on rangelands handicap making adjustments during drought-wet cycles and associated variation in forage production. Although nonuse can be taken or sometimes required on public grazing lands during drought years, the opportunity to increase stocking rates in bumper years is not generally provided. Administrative considerations additional to and often remote from those of maximizing livestock grazing affect stocking rates on public lands. Without ad-justable stocking, livestock numbers on public grazing lands must be set at a con-servative, below-average stocking rate to avoid serious damage to range. livestock. and even big game animals during drought and to provide for range recovery dur-ing good years.

The higher the stocking rate, the greater the requirement for flexibility in stocking rate, and the greater the risk. The managerial intent is commonly to maximize livestock production on a sustainable basis, but this magnifies the potential problems associated with major climatic variation (Heitschmidt and Stuth, 1991). As rate of stocking increases so does the effects and even frequency of catastrophic droughts. This requires that moderate stocking levels be maintained for sustained livestock production and that short-term stocking rates have some flexibility for adjustment to the extremes in forage production imposed by climate.

The use of high-density grazing systems is also geared to maximizing harvest efficiency, and this leads to sooner and more frequent encounters with reductions in forage supply associated with drought. A rapid response to drought becomes even more necessary, and a prompt decision to substantially destock, if necessary, will have better long-term expected economic return, with less variance, than taking a "wait-and-see" or hopeful inaction approach. This suggests that greater emphasis should be placed on minimizing climatic and financial risk in ranch management over attempting to maximize forage production and harvest efficiency in areas where drought is common (Thurow and Taylor, 1999).

Based on precipitation and forage production records kept at the Santa Rita research station near Tucson (Reynolds, 1954), it was concluded that reductions in livestock numbers were required in moderate droughts (41–70% below average rainfall) and severe droughts (over 70% below average), whereas slight droughts (0–40% below average rainfall) could be met by continuously conservative stocking levels. It was estimated that for the semi-desert grassland ranges, stocking should be reduced to 40% below the long-time average about 35% of the time (i.e., when droughts reach moderate and severe intensity).

A more detailed study of drought management strategies for the semi-desert ranges of the Southwest—but with obvious application to a much greater area—was made by Martin (1975a). He recommended limited flexible stocking or conservative constant stocking strategies over either constant stocking at average annual grazing capacity or flexible stocking strategies. Constant stocking at the average stocking level was considered impractical, if not impossible, because it resulted in overgrazing about half the time. The overstocking became increasingly severe if one dry year followed another, with mounting feed bills, declining range condition, and lowered animal productivity.

Martin found flexible stocking (90–140% of average) was too difficult to manage, the hazards of overgrazing were too great, and income varied greatly from year to year. Under flexible stocking, net sales were greatest in poor forage years when animal numbers were reduced and were lowest in good years when extra animals were added. Other limitations included difficulty in estimating variable forage production and adjusting animal numbers accordingly, costs of buying and selling extra animals, the possibility of introducing parasites and diseases through frequent additions to the herd, the natural reluctance to cull heavily enough in drought years, and problems of maintaining high animal genetic standards.

Limited flexible stocking, within the range of 70–110% of average, was considered a good system by Martin (1975a) if properly managed. It produced about the same income as constant stocking at 90% of average capacity, and with only moderate hazard of overstocking. Constant stocking at a conservative 90% of average proper stocking was the easiest approach to administer, produced relatively high income, and had a relatively low risk of overstocking but required being prepared to make some temporary stocking reductions in prolonged severe drought. This plan, based on data accumulated at the Santa Rita station near Tucson, resulted in moderate overstocking only 1 year in 3 and severe overstocking only 1 year in 15. Stocking at 80% of average grazing capacity further reduced chances of overstocking but was less profitable. However, to maintain animal quality it was considered necessary to retain a fixed number of replacement heifers each year under any drought management strategy.

A simulation study based on actual ranch properties but corrected to the same land/forage resource base was carried out by Foran and Smith (1991) to compare alternative drought management strategies on beef cattle ranches in the arid zone of central Australia. The three strategies were (1) "average-stock" (AS) 3000 AU strategy (ignores drought in hopes it will be of short duration), (2) "low-stock" (LS) 2000 AU strategy (avoids drought as much as possible by carrying only two-thirds the livestock numbers found under AS), and (3) "high-stock" (HS) 4000 AU strategy (stocked initially at 33% over AS; accepts the risk of drought but manages against it by quick destocking tactics of aggressively selling all non-breeding male stock and the older cows from the ranch at the first indications of drought).

The mean accumulated cash surpluses for the three strategies were $1.98 million for AS, $2.28 million for HS, and $1.34 for LS. AS and HS had highly variable financial returns, and both were projected to have gone broke during the drought decade of 1956–1965; LS maintained relatively constant financial returns and was projected to have survived the 1956–1965 drought with only a small deficit. Both AS and HS recovered quickly from 1-year droughts, but LS had higher annual returns when droughts lasted 2 years or longer. While HS gave higher average annual returns, the returns were more variable and risky; HS was projected to risk severe land degradation if errors were made in adjusting stocking rates and the destocking decisions were delayed. LS was projected to consistently have the highest livestock performance and an improving land/forage base and to make good returns even with the reduced animal numbers. The apparent advantage that the AS (i.e., "wait-and-see") manager had if the drought lasted 1 year were projected to be quickly lost it the drought extended to the second or third year.

C. MEETING THE LOWS

Preparing for the eventuality of drought is very important, but additional practices will be required to meet the emergencies presented at the onset and continuation of moderate to severe drought. Suggestions for balancing carrying capacity needs and availability during drought periods include:

1. Maximize grazing distribution over grazing land units and assure that stockwater and other distribution aids are fully used.

2. Sell yearling stocker cattle as soon as drought is indicated or move to drylot and feed growing-finishing rations.

3. "Tailor" the breeding herd a little closer than normal; sell dry cows, slow breeders, poor milkers, and older breeding animals and do it early in the season. However, this should be done with caution since brood cows must be maintained between years because major costs of frequently buying and selling brood stock will normally be prohibitive (Torell *et al.,* 1991) as well as genetically infeasible in a well-bred herd.

4. Consider curtailing the addition of replacement heifers into the cowherd for one year (Reece *et al.,* 1991).

5. Make greater use of temporary pasture, particularly where irrigation water is available to provide stable forage production, and of crop aftermath.

6. Utilize "junk" feeds not normally used; graze areas dominated by annual grasses and weeds early in the grazing season when available; utilize pricklypear or cholla cactus by singeing off spines before feeding, which provides forage that is acceptable but of low nutritive value (Correa *et al.,* 1987; Sawyer *et al.,* 1997) (Fig. 13.7).

7. Monitor closely to prevent excessive grazing pressure from developing in ongoing rotational grazing systems; delay the initiation of a new rotational grazing system until the drought ends (Reece *et al.,* 1991).

FIGURE 13.7   Full utilization of "junk" feeds and less palatable "forage reserves" may partly alleviate low carrying capacity during drought; photo shows the practice in south Texas of singeing spines off pricklypear to provide emergency forage. (Soil Conservation photo.)

8. Lease additional pasture or purchase additional harvested forages if the price is reasonable and economically justified; relocate livestock to non-drought areas.

9. Consider early weaning of calves (Herbel *et al.*, 1984) and lambs (Thomas *et al.*, 1987) as a means of reducing grazing capacity requirements—the dry cow will consume about 35% less forage than the lactating cow—enabling earlier culling and holding up subsequent reproductive performance.

10. Make a clear management decision, as early as possible, as to whether to maintain an animal herd through emergency feed purchase, by reducing the herd, or by selling the entire herd.

Supplemental feeding to overcome nutrient deficiencies in the remaining forage or providing additional carrying capacity through complemental or emergency feeding can mask much of the nutritional consequences of drought, but this must be based on the specific needs and objectives of the livestock program. The negative aspects of supplemental feeding during drought includes diminishing grazing distribution and encouraging a continuation or increase in grazing pressure and poor grazing management (Torell and Torell, 1996). Low feed costs, including governmentally subsidized supplemental feeds, promote livestock being left on grazing lands after the forage base is depleted, and supplements become the primary diet while deterioration of the forage base continues.

When the grazable forage has been fully utilized and subsequent carrying capacity must be provided with harvested feeds, strong consideration should be given to removing the animals from all grazing land units and placing them in drylot or holding traps. Continued access to pasture during drought emergencies may cause long-term damage from excessive grazing and trampling, expose the soil to wind and water erosion, prevent perennial plant species from taking advantage of any light rains, and accelerate poisonous plant problems. Concentrating animals on dryland pasture or range, either during drought emergencies or for routine winter feeding of harvested roughages, can seriously impair future forage production capability (Armbruster *et al.*, 1976).

Placing the livestock in drylot or small traps after the grazing capacity of pasture has been fully used will also reduce their maintenance requirements by reducing walking distances and time spent searching for forage. Livestock removed to drylot for emergency feeding should be divided into appropriate age, weight, and size groups to reduce competition for feed and to promote uniform feed consumption. Not only is part-year confinement of cows a useful tool in drought years, production levels per cow may be higher than under yearlong placement on rangeland in normal rainfall years (Herbel *et al.*, 1984).

The practice of "limit-grazing" has been used with temporary pasture to ration and extend the available forage supply by limiting animal access to a period of only 1 to 6 hours per day (Altom, 1978). However, this practice requires other feedstuffs, particularly other forage sources, to complement the reduced pasturage intake permitted. Limit-grazing has been proposed for the following uses: (1) continue livestock numbers during low to moderate drought or other poor growing season, (2) permit continued grazing of high numbers during fall and winter and then full-time grazing during the flush growth of spring, (3) use of the pasture as a high-protein supplement, and (4) in backgrounding growing animals for feedlot. The disadvantages of limit-grazing include increased labor costs, the alternative carrying capacity required, the need for good handling facilities, and careful regulation of stocking rate. Limit grazing will seldom be practical on extensive rangelands because of problems of frequent roundup, bunching, and removal of grazing animals.

D. MEETING THE HIGHS

The prospects of harvesting the excess forage produced in high production years is a much more pleasant challenge than meeting the dire consequences of severe drought. Some undergrazing of long-term pasture during bumper years may be justified and even desirable. This can be helpful in counter-balancing the effects of previous drought or overgrazing and in improving vegetation condition; carryover forage may also be valuable as a buffer against a slow spring greenup or drought the following year. However, to the extent that undergrazing would result only in wasted forage, suggestions for temporarily increasing the grazing demand include the following:

1. Hold stocker animals for later markets and increased gains.
2. Purchase additional growing animals for short-term gains, providing the price and economic picture are favorable, or lease out excess grazing capacity.
3. Put a greater portion of the forage production, primarily from improved pasture and meadow, into harvested forages.
4. Cull breeding animals later in the season if extra gains appear profitable.
5. Hold over slightly more replacement breeding stock, providing they are of high quality and are used to replace culls or could be marketed as bred females in drought emergency; but always remember that a drought year may well follow a high forage year.

14

GRAZING METHODS

I. INTRODUCTION TO SPECIAL GRAZING TECHNIQUES

Historically, an array of special grazing techniques has been collectively referred to as "grazing systems." These techniques range from simple to complex and are available to further fine tune the management of grazing. They exclude the selected "tools" employed in dispersing grazing and in achieving optimal stocking rates and optimal mixes of animal species; but many benefits attributed to special grazing techniques are the result of improved grazing distribution and improved stocking rates.

The difficulties met in naming, classifying, and describing these special grazing techniques have led to much confusion and misunderstanding. Differences in terminology are found among scientists, graziers, agencies, professional societies, and continents. Individual terms have often been given different meanings or inconsistently used; different terms have variously been distinct, partly overlapping, or nearly synonymous with other terms. A proliferation of terms has resulted from attempting to accommodate each variant of special grazing techniques.

Lewis (1983) proposed that "single grazing systems" be grouped by (1) continuous systems, (2) deferred systems, (3) rested systems, and (4) rotated systems; in addition, systems formed from two or more single systems were grouped as "combined grazing systems." In order to avoid "preconceived definitions as much as possible," Heady (1984) recommended that both the grazing treatments and the nongrazing treatments in specialized grazing techniques be treated as follows:

Grazing treatments	*Nongrazing treatments*
1. Season-long grazing	1. Season-long nongrazing
2. Late grazing	2. Late nongrazing
3. Early grazing	3. Early nongrazing
4. Short-term grazing	4. Short-term nongrazing
	5. Nonuse

These terms were related to the phenology of the key forage plants and the length of the forage year and avoided the quandary of fixed calendar dates. *Nonuse* was defined as not being grazed for many years (and not in rotation with grazing periods) because of ecological reasons or interference with other land uses.

Lacy and Van Poollen (1979) developed a dichotomous key for classifying "grazing systems" based on (1) full-year or part-year potential occupancy by grazing animals; (2) rotation used or not used; (3) length, timing, and frequency of nongrazing and grazing periods; and (4) scheduled vs. flexible application of treatments; unfortunately, this required coining new names for grazing techniques not in common use or generally accepted and a proliferation of terminology. In the first edition of *Grazing Management* (Vallentine, 1989), "grazing systems" were classified by (1) feasible grazing season (unrestricted, restricted to plant growing season, and restricted to plant dormancy season), and (2) whether or not rotation was incorporated into the system. While offering only a selected list, 45 "grazing systems" were included.

The Forage and Grazing Terminology Committee (FGTC) (1991, 1992), representing several disciplines and professional societies involved in grazing management, proposed that "a defined procedure or technique of grazing management designed to achieve a specific objective(s)" be referred to as a "grazing method." This approach has been approved and used by the Society for Range Management (1998) and has been incorporated into this second edition of *Grazing Management*. **Special grazing techniques** is a generic term employed to include both grazing methods and grazing systems.

A definition of **grazing method** has thus evolved to the following: a defined procedure or technique of grazing management based on a specified period of grazing and/or period of nongrazing which is designed to achieve a specific objective. By contrast, covered under "grazing system" are the more complex grazing management techniques that include one or more grazing methods in addition to rotation grazing. Most grazing methods require only a single grazing land unit to be operative, but grazing systems require multiple grazing land units (paddocks) and animal or pasture rotation. Individual grazing methods, as discussed later in this chapter, are grouped into (1) continuous grazing methods, (2) seasonal suitability grazing methods, (3) deferred/rested grazing methods, and (4) high-intensity grazing methods. The principal grazing systems are discussed in Chapter 15.

II. ROLE OF SPECIAL GRAZING TECHNIQUES

Special grazing techniques are not a panacea that can solve all problems met in managing grazing lands or reduce the importance of other aspects of good grazing land and livestock management (Shiflet and Heady, 1971) (Fig. 14.1). The hypothesis that special grazing techniques can largely negate the need for continuing and simultaneous focus on the basic principles of grazing management—optimal stocking rates, optimal kind or mix of animal species, optimal distribution of grazing, and optimal seasons of grazing—is untenable. Grazing methods and

FIGURE 14.1   Special grazing techniques are not a panacea for correcting all grazing problems, cannot substitute for the principles of grazing land management, and require greater rather than less management input to be successful.

systems can sometimes minimize but can seldom negate adverse plant response to grazing intensity, frequency, selectivity, and seasonality.

It almost appears in some cases that special grazing techniques have been introduced as substitutes for good livestock and forage management. Walker (1994) has noted the strong tendency toward implementing complex grazing techniques as the first step in improving grazing management rather than assessing current management in terms of the basic principles of grazing management. Heady (1974) has concluded that special grazing techniques have worked only when the graziers quit overgrazing. Heitschmidt (1988) has concluded that the impact of stocking rate in arid and semi-arid rangelands is of much greater magnitude than type of grazing "system" (i.e., special grazing technique).

Achieving the optimal grazing intensity might be considered the priority principle of grazing management, but the importance of timing, frequency, and selectivity of grazing animals also play important supporting roles. After reviewing numerous grazing studies, Van Poollen and Lacy (1979) concluded that adjustments in animal numbers have a greater effect on herbage production than do grazing "systems." Their summation was that livestock adjustments from heavy to moderate use accounted for 73% and use of specialized grazing "systems" only 27% of the total herbage response when both changes were implemented simultaneously.

The selection of a special grazing technique is sometimes based solely on personal choice or what is currently being promoted; however, the grazing method or grazing system selected must be adapted to the forage plant species being grazed, the grazing season, the physiography of the grazing land, the nutritional needs of the kind and class of livestock to be grazed, and management objectives. The grazing treatments required to improve grazing lands in poor condition may be very different from those needed to optimize forage conversion on grazing lands in good condition (Heady 1970). Platou and Tueller (1985) have noted that how native ungulates and plants in range ecosystems have co-evolved should be helpful in designing grazing systems for livestock production. This suggested to them that short-duration grazing should better fit the shortgrass plains and midgrass prairies, while the longer rest periods between grazing periods of the rest-rotation system should better fit the sagebrush steppe.

For a grazing method or grazing system to be effective and practical, the following characteristics are commonly suggested (adapted from Stoddart et al., 1975):

1. It is based on and suited to the physiological requirements and life history of the primary forage plants.
2. It will improve vegetation low in vigor or will maintain vegetation already in high condition.
3. It is adapted to existing soil conditions so erosion and puddling will not result from livestock trampling.
4. It will favor the desirable plants and promote high forage productivity.

5. It is not detrimental to animal performance and will hold animal disturbance at acceptable levels.
6. It is practical to implement and reasonably simple to operate.

The failure of the special grazing technique was deemed likely if any of these characteristics were missing.

The management objectives to be achieved must be given high priority in the
selection of a grazing method or grazing system. These objectives will generally
include one or more of the following:

Plant/site objectives

1. Restore forage plant vigor, raise vegetation
   condition.
2. Maintain high vegetation condition.
3. Attain more uniform distribution of grazing.
4. Reduce plant selectivity by the grazing animal.
5. Increase grazing capacity.
6. Provide a sustained and dependable forage supply.
7. Cope better with drought by carrying a deferred
   or rested unit for emergency use.

Animal/economic objectives

1. Provide acceptable if not maximum animal
   gains.
2. Maintain forage at high nutritional levels.
3. Meet animal nutrition needs during critical
   periods.
4. Fit into the total grazing management plan.

Special grazing techniques on federal grazing lands have often been looked
upon as a means of improving range conditions while still allowing grazing. Special grazing techniques have all too commonly been implemented on both public
and private rangelands for forage stand improvement and seldom for increased
livestock or wildlife production (Pieper, 1980). Unfortunately, the needs of livestock have not been adequately considered during the design and development of
such grazing systems (Kothmann, (1980). The rationale that special grazing techniques should be designed not for individual animal performance but to increase
gain per acre through improved carrying capacity seemingly ignores possible economic constraints on low animal performance (Launchbaugh *et al.,* 1978). Even
though some research suggests that rotation grazing can provide a higher grazing
capacity than continuous grazing in some situations, this advantage is often neu-
tralized by reduced animal performance (Bransby, 1991).

Special grazing techniques must provide an acceptably nutritious and abundant
forage supply for the grazing animals (Fig. 14.2). In general, livestock performance due to such techniques has been given much more attention on improved
pastures than on rangelands. "Nutrient intake is sensitive to grazing pressure index [i.e., cumulative grazing pressure]. Grazing systems that significantly increase
the grazing pressure index will probably reduce animal performance. This occurs
when animals are increased, area available is reduced, and/or time is increased"
(Kothmann, 1984). Special grazing techniques must avoid stress during critical periods in the livestock production cycle, such as special growth and development
requirements of replacement heifers (Launchbaugh *et al.,* 1978) and with cows,
especially first-calf heifers, shortly after calving (Malechek, 1984). "The key to
successful animal production is selective grazing by the animals; allowing animals

FIGURE 14.2   Special grazing techniques, including both grazing methods and grazing systems, must provide an acceptably nutritious and abundant forage supply for the grazing animals; showing cowherd on Nebraska Sandhills range (Soil Conservation Service photo).

to select the most nutritious parts of the total available forage increases nutrient intake and improves animal production. However, it is this selective grazing which creates the need for grazing systems" (Kothman, 1980).

III. ADAPTING SPECIAL GRAZING TECHNIQUES

Special grazing techniques have commonly been given more acclaim under practical application than under research, and at least some of the reasons for the greater apparent success of practice over research are apparent. A variety of improved practices are often initiated at the ranch level the same time the special grazing technique is begun, and the latter may erroneously be credited with the total added benefits rather than with its partial contribution. Care must be exercised against bias in assigning to a new specialized grazing technique the benefits that really accrue not directly to the grazing system but rather to building new fences, developing water, more salting or riding to improve distribution of grazing, and/or cultural grazing land treatments.

Hart *et al.* (1993) concluded that decreasing pasture size and reducing distance from water were more important for improving forage utilization patterns than implementing intensive rotational grazing systems. Laycock and Conrad (1981) concluded that equally good management applied to continuous summer grazing in the Intermountain West as used with rest-rotation grazing commonly removes the advantages sometimes assigned to rest-rotation grazing. Improved livestock genetics, health care, and nutrition when credited solely to specialized practices per se can bias the placement of credit for enhanced livestock performance. Nevertheless, initiation of specialized grazing techniques has often been accompanied by a renewed commitment to improve management on a broad scale, which is a fortunate result.

While rigid criteria have generally been applied in research studies, greater flexibility and greater attention to economic returns have been accorded under practice (Heady, 1970). The complex grazing system in practical use has been more willingly and frequently adjusted to better accommodate variable forage supply, weather fluctuations, inevitable drought, and changes in the objectives and economic pressures of the livestock operator. Tying grazing rotation to strict calendar dates rather than plant phenology is generally held to be counterproductive, as flexibility is considered the key for success. Jameson (1986) has suggested that both flexible scheduling and flexible stocking rates be incorporated into grazing systems. Also, if stocking rates under which experimental comparisons are made between continuous grazing and specialized rotation grazing systems are too low, forage availability may not be allowed to be a limiting factor in vegetation production (Dwyer *et al.,* 1984),

The selection of a special grazing technique should hinge upon the managerial skills and ability of the grazier to monitor forage plant and animal responses over time to make the best management decisions (Matches and Burns, 1985). Constraints such as competition for time with other farming or business operations may opt against a complex grazing system that demands continuous monitoring. It should be noted that continuous grazing, technically a grazing method and in some situations the special grazing technique, requires much less continual managerial input than does a complex grazing system such as short-duration grazing.

One Nebraska ranch owner (Salzman, 1983) has appropriately stated that, "The ranch manager is the key to success of a grazing system. . . . The operator can make a good grazing system fail and a poor system [almost?] work. . . . He should have a genuine interest in grass, a note at the bank, enough greed to want to make more money, willingness to take a chance, nerve enough to withstand criticisms of neighbors, willingness to accept and heed advice from technical people, and the time and inclination to observe and evaluate his program and to constantly update it. . . . The grazing system must fit the man, the ranch, the cattle, and be flexible to be successful."

The relative degree of animal concentration is an important characteristic of many grazing methods and all grazing systems. This can be expressed by using the **stocking density index,** defined as the ratio of the land area available in a single unit or among paddocks in a rotation grazing system to the land area available for grazing at any one time. The degree of animal concentration of special grazing techniques can be compared using the stocking density index as follows (when acre per head per season is the numerator and acre per head per instant of time is the denominator, the cumulative per-head acreage allowance per season is 12 acres, and the pasture units within each system are of equal size):

$\frac{12}{12} = 1$ (continuous grazing)

$\frac{12}{6} = 2$ (as found in two-unit alternate grazing)

$\frac{12}{2} = 6$ (as found in six-unit rotational grazing)

When expressed as the **grazing fraction** (i.e., the fraction of land in a single grazing unit or among paddocks in a rotation system which is being grazed at any given time), the level of animal concentration in the above examples would be $\frac{1}{1}$, $\frac{1}{2}$, and $\frac{1}{6}$, respectively.

IV. CONTINUOUS GRAZING METHODS

Continuous grazing, by definition a grazing method rather than a grazing system, allows animals unrestricted and uninterrupted access to a grazing unit for all or most of the grazing season. While **yearlong continuous grazing** extends essentially through the full 12 months, **growing season continuous grazing** and **dormant season continuous grazing** generally extend 2–6 months. From the beginning to the end of the grazing season, grazing animals under continuous grazing remain in the grazing unit. Growing season continuous grazing permits the use of either set stocking or variable stocking to synchronize more closely with quantity of available forage (see Chapter 12).

Continuous grazing has often been criticized as being historically detrimental to the vegetation; however, the cause of deterioration commonly has been due rather to heavy grazing and/or poor distribution of grazing. Under yearlong continuous grazing, particularly when available water is concentrated in an inadequate number of large facilities, heavily grazed areas around water remain large and of low productivity (Martin and Cable, 1974). Nevertheless, after an extensive review of grazing studies, Herbel (1974) found only limited success with any grazing system over continuous grazing on rangelands grazed only for a part of the year. It is generally concluded that continuous grazing is not an inherently inappropriate management procedure and may often be the best procedure, if proper tools are used to obtain uniform grazing distribution and if proper season of grazing and proper stocking rates are employed (Heady and Child, 1994; Laycock *et al.,* 1996)

Livestock often produce as well or better under continuous grazing of rangelands than under rotation. This is attributed to the least change in forage quality, to grazing animals having continuous opportunity for selecting their preferred plants, and to least disturbance of grazing under continuous grazing. This maximum opportunity for selectivity often improves nutrient intake but may result in undue intensity and frequency of defoliation on the better plants, which puts them at a disadvantage in plant competition and may cause undesirable vegetation changes (Heady, 1984).

Ely (1994) suggested that continuous grazing may permit excessive consumption of lower-yielding, highly palatable forage in the spring, leaving the higher yielding, less palatable forage to mature and become low quality during summer. Willms *et al.* (1992) noted that continuous grazing over the entire season minimized fencing costs and management requirements on Canadian foothills fescue prairie, but the cattle grazed selectively, avoiding certain plants or sites and preferring others, and the range became patchy grazed. However, these problems appear to be a problem in some vegetation types but less so in others.

Continuous grazing may be preferable over complex grazing systems on California grasslands, shortgrass plains, native sodgrasses, and other grasslands with few species of either extremely high or low palatability (Fig. 14.3). On California annual grasslands, Heady (1961) reported grazing animals were better able to search out ample forage when allowed to scatter under continuous grazing than when bunched under deferred-rotation grazing; he also noted that continuous yearlong grazing did, in fact, provide a partial deferment each spring when forage growth greatly exceeded forage utilization. Under yearlong moderate stocking of annual grasslands of the Sierra Nevada foothills, continuous yearlong grazing produced calves 55 lb heavier at weaning and provided greater grazing capacity than either sequence grazing or "rotated seasonal grazing" (Ratliff, 1986).

Under a year-round grazing program of cattle grazing on Oklahoma sandhills

FIGURE 14.3 Continuous grazing is generally preferable over complex grazing system for annual grasslands (shown here in California), shortgrass plains, and certain other grasslands.

grass range, no advantage was found in alternating from winter to summer range compared to yearlong continuous grazing (Shoop and Hyder, 1976). When grazed by yearling replacement heifers using moderate stocking rates, yearlong gains were slightly better under yearlong continuous (232 vs. 223 lb); heavy stocking rates were deleterious to cattle gains under either grazing technique. Continuous grazing during the growing season on similar range with yearling steers was considered to be a more efficient system of producing beef than rotation systems using 4- or 6-week grazing periods (McIlvain and Shoop, 1961).

Gray *et al.* (1982) has concluded that continuous grazing on rangelands will probably continue to be an accepted practice where adapted because of its lower management requirements, minimal livestock handling, and lower investment levels for improvements. However, because cattle are scattered over a wide area under continuous grazing, McCollum and Bidwell (1994) concluded that the amount of labor, bull power, and equipment costs are increased compared to a grazing system in which cattle are more concentrated.

V. SEASONAL SUITABILITY GRAZING METHODS

A. SEQUENCE GRAZING

The rationale for sequence grazing has been that many natural vegetation types evolved under intermittent grazing pressure from migrating herbivores (Heady and Pitt, 1979). The fact that migratory wild ungulate herbivores follow a distinct pattern from one vegetation type to another has led to the belief that grazing patterns might best be patterned to those under which the vegetation types evolved. However, "Mother Nature's system," as it has sometimes been referred to, seems to be more exemplary of how the use of different vegetation types might be correlated rather than how they must be used.

Valentine (1967) used the term "seasonal suitability grazing" to describe a grazing program for the Southwest consisting of grazing diverse range vegetation types in accord with the seasonal use requirements of and benefits to vegetation and livestock. This was based on each vegetation type having a most advantageous season of grazing even though seasonal use, particularly in the Southwest, is often not obligatory. It was concluded that on ranges or a combination of ranges with diverse vegetation types, this was a superior grazing program with respect to maintenance and improvement of the range, harvest of forage, and livestock production.

Holechek and Herbel (1982) further described "seasonal suitability grazing" as involving partitioning a ranch or grazing allotment into separate units on the basis of vegetation types. These vegetation types are then fenced and grazed separately. Terrain, ranch operation requirements, range condition, and range site differences are considered before the grazing land units are delineated and fenced. It was noted that the fenced units may not be contiguous and often vary as to size, location, and ownership. Their discussion of "Seasonal Suitability Grazing in the Western United States" (Holechek and Herbel, 1982) covers the major grazing regions of the West and is suggested for additional reading.

Sequence grazing, synonymous with but preferred to repeated seasonal grazing or seasonal suitability grazing, is a grazing method in which two or more grazing land units differing in forage species composition are generally grazed at the same time each year and sometimes for less than the full feasible grazing season. Sequence grazing takes advantage of differences among forage species and species combinations to extend the grazing season, enhance forage quality and/or quantity, or achieve some other management objective. Grazing units of dissimilar vegetation arranged and used in a series comprise a seasonal or year-round grazing plan. Sequence grazing can be applied to most kinds of grazing lands; the same basic principles apply to the use of complementary seeded range or improved pasture as well as native vegetation types. Grazing units of introduced forage species can readily be combined with those of native range into a sequential grazing plan.

Some vegetation types are environmentally unrestricted as to when they can be fit into a grazing plan, while others have a very short optimal or even feasible grazing period. Sequence grazing has been successfully extended to vegetation types having typically short optimal grazing seasons, such as crested wheatgrass, robust native vegetation of short palatability period, and cheatgrass or other annual grass stands where management towards a higher successional stage is deemed impractical. Where vegetation among different grazing land units is not substantially different (e.g., California annual grasslands) (Heady and Pitt, 1979), sequence grazing becomes essentially a rotational grazing system and probably will not be advantageous over full-season continuous grazing. Rotating seasonal grazing on black grama range in New Mexico did not improve forage production or cattle response over yearlong continuous grazing (Beck *et al.,* 1987); annual rainfall rather than grazing systems had the most influence on the native vegetation. Sequence grazing also provides for special uses year after year: breeding, calving, lambing, fattening, winter protection, and water accessibility in dry seasons.

B. FLEXIBLE SEQUENCE GRAZING

This grazing method, synonymous with best-pasture grazing, was designed for ranges where summer rainfall is usually spotty and varies greatly not only over time but from place to place (Martin, 1978a). It has been most commonly used in desert and semi-desert areas of the Southwest, where it permits shifting to local areas following showers that result in green forage or provide stockwater not otherwise available. It provides maximum flexibility since there are no scheduled times for moving livestock between range units. Instead, livestock are placed in the unit where rainfall has been the most favorable and forage is temporarily the best. After this range unit has received full utilization, an on-the-spot decision is then made as to the best range unit to be grazed next, and so forth.

Flexible sequence grazing is a restricted version of the pre-settlement natural grazing system, in which native herbivores followed the rains to the best forage areas. It tends to maximize animal gains, and, if the range is properly stocked, it appears to apply grazing pressure on the forage stands best able to withstand grazing because of favorable growing conditions (Martin 1978a). This special grazing technique can also incorporate aspects of selected deferment to further range improvement. It is sometimes used on other kinds of grazing lands by dividing up the ranch or grazing allotment into separate units and periodically making on-the-spot decisions as to which unit will be grazed next and for how long. Each decision is based on the quantity and quality of vegetation available for grazing or in some combination with management convenience.

On year-round black grama range in New Mexico, Beck *et al.* (1987) utilized the flexible sequence concept to determine the order in which the range units were grazed. Cattle were moved to units out of chronological order whenever thundershowers resulted in temporary green forage.

C. COMPLEMENTARY ROTATION GRAZING

Complementary rotation grazing involves several kinds of pasture or range (each of a separate plant species or mix) within a single fence in which animals rotate themselves as they see fit as the grazing season advances. In this regard, it is a one-unit variant of repeated seasonal grazing. Since direct management control of seasonal use is generally desired, the use of complementary rotation as a planned grazing method remains rare.

In Alberta, ewes during a 10-year comparison were reported to rotate themselves under a free-choice system at least as satisfactorily as under a scheduled rotation system using several units (Smoliak, 1968). Under the free-choice, complementary rotation the ewes typically preferred crested wheatgrass in the spring (late April to mid-June), then grazed Russian wildrye for about 3 weeks, shifted to native range during mid-July for a month, and then grazed Russian wildrye again until late October. In a followup study using yearling steers, 180-day summer gains favored the complementary rotation by 18 lb over the repeated seasonal grazing (267 vs. 249 lb) (Smoliak and Slen, 1974).

D. ROTATIONAL BURNING/GRAZING

Rotational burning/grazing is generally based on the grazing animals rotating themselves rather than by forcing such by cross-fencing a range unit into separate sub-units (Duvall, 1969). It is commonly used in areas of coarse range herbage such as in southeastern forest and salt marsh areas where prescribed fire is used to temporarily restore forage palatability. This special grazing technique is enabled by burning a proportional part of the total range unit every third or fourth year. Animals permitted access to the new growth on the area recently burned will graze there by preference.

Most of the grazing will be concentrated on the area most recently burned be-

FIGURE 14.4   Rotation burning-grazing provides not only the normal advantages of prescribed burning on this range area in the Southeast but also rotates cattle by attracting them to the palatable regrowth that develops on recently burned areas. (U.S. Forest Service photo.)

cause of the enhanced palatability; some grazing use will be made of the next oldest burn, but minimal or no use made of the oldest burn or burns (Fig. 14.4). For example, grass utilization under this program in one study on longleaf pine-bluestem range in central Louisiana averaged 78, 31, and 18% in the first, second, and third seasons after burning (Duvall and Whittaker, 1964). Close grazing of the new growth the first year kept the vegetation palatable and nutritious, while the lighter grazing the last 2 years of the cycle maintained or restored vigor in the grasses.

Rotational burning/grazing is better adapted to either growing season or year-long range rather than range grazed only during the dormant season. Improved forage value during summer and fall on southern forest ranges has been a major benefit of rotational burning. A further modification over prescribed burning the entire annual allotment portion of the range unit at one time is to burn at seasonal stages—part in late winter, part in spring, and part in midsummer (Duvall, 1969). This modification provides forage of relatively high quality over a longer period than making the annual burn on a single occasion. Schmutz (1978) has recommended the addition of rotational burning to established grazing systems for suppressing brush and controlling brush invasions into southwestern semi-desert grasslands. He suggested the addition of an additional grazing unit to the three-unit Santa Rita system, thereby permitting prescribed burning every fourth year following a growing-season rest to accumulate enough fuel to carry a fire.

E. INTENSIVE-EARLY STOCKING

The advantage in livestock gains resulting from concentrating grazing in the early part of the growing season when the forage is green and nutritious has been incorporated into **intensive-early stocking (IES)** (synonymous with **intensive-early grazing**) (Fig. 14.5). This one-unit grazing method has recently found favor on native grass ranges in the Midwest; it is based on high stocking density during rapid growth and, if discontinued at flowering, permitting plants to make regrowth and full plant recovery while not reducing the standing crop the next year. IES is commonly based on doubling stocking density (2×) for grazing during only the first half of the normal growing season, thereby maintaining the same stocking rate (AUMs per acre) as under seasonal continuous grazing.

Long-term comparisons between IES and seasonal continuous grazing using yearling steers have been made in Kansas on tallgrass prairie at Manhattan and on shortgrass range at Hays. On tallgrass prairie IES (2× stocking density) has markedly increased average daily gain and gain per acre as a result of eliminating late-season grazing (Table 14.1A); total gains per head have remained greater under seasonal continuous grazing because of the double length of grazing periods compared to IES (Smith and Owensby, 1978). Increasing stocking density on tallgrass prairie from 2× to 2.5× and 3×, with corresponding increases in stocking rates, has further increased gains per acre while not greatly affecting average daily gain during the restricted early season (Table 14.1B) (Owensby et al., 1988). Selectivity by animals for forage quality apparently does not become critical until late in the growing season when nutritive value drops.

Herbage production was sustained at all IES stocking rates during the 6-year study on tallgrass prairie at Manhattan. Percent composition and basal cover of the major dominants changed little during the study period. At the higher stocking rates, indiangrass appeared to be adversely affected, and Kentucky bluegrass was somewhat favored. The bluestems showed no negative response to IES; nongrazing from mid-July to frost apparently allows sufficient time for regrowth and storage of adequate food reserves for vigorous growth the following spring (Owensby et al., 1988). Although less standing forage remained at mid-season (July 15) under IES than continuous seasonal stocking (871 vs. 1301 lb dry matter per acre), more remained at the end of the growing season (1610 vs. 1334 lb) (Smith and Owensby, 1978). In Oklahoma the end-of-season standing crop of tallgrass residue

FIGURE 14.5   Intensive-early stocking concentrates grazing into the early part of the growing season when forage is green and nutritious; its application is largely limited to use with stocker yearlings on native grasslands of the Midwest; photo showing steers on Flint Hills range in Kansas. (Kansas State University photo.)

was similar between IES and continuous seasonal grazing because of late-season regrowth under IES and continued defoliation under continuous grazing (McCollum *et al.*, 1990).

Periodic late spring burning (about May 1) of tallgrass prairie has interacted beneficially with IES and has helped control many woody plants and herbaceous weeds, improved grazing distribution, and increased livestock production (Bernardo and McCollum, 1987; Launchbaugh and Owensby, 1978; Bock *et al.*, 1991). However, Kansas State University (1995) has suggested that carry-over forage produced during the last half of the growing season (i.e., following removal of cattle following IES) need not be wasted. Rather, it could be used from November 1 to the start of forage growth in the spring as a holding area for stockers as they were accumulated for another growing season of IES or used by the cowherd.

Eastern gamagrass is a large, warm-season bunchgrass native to the lower Midwest but also has high potential for producing growing beef cattle when seeded in pure stands. Because the grass is course but produces rapid regrowth, some form of rotational grazing has commonly been recommended for maintaining it in a rapid growth stage. While gains of growing cattle on eastern gamagrass have typically gained 2.5 lb daily during the early portion of the growing season, 0.4 to 1.3 lb daily have been typical later in the growing season. However, on seeded eastern gamagrass pasture in western Arkansas, IES has proven to be a highly benefi-

TABLE 14.1   Livestock Performance of Yearling Steers on Native Range in Kansas
Under Intensive-Early Stocking and Seasonal Continuous Grazing

A. Flint Hills Bluestem Range, Manhattan, 3 years[a]

	Intensive-early stocking	Seasonal continuous stocking		
	May 2– July 15 (75 days)	May 2– July 15 (75 days)	July 16– Oct. 3 (79 days)	May 2– Oct. 3 (154 days)
Stocking density	2× normal	normal	normal	normal
Stocking rate (acre/steer month)	0.68	0.68	0.68	0.68
Gain/steer (lb)	141	131	79	210
Average daily gain/steer (lb)	1.88	1.75	1.00	1.36
Gain/acre (lb)	83	39	23	62

[a]From Smith and Owensby (1978).

B. Flint Hills Bluestem Range, Manhattan, 6 years, 1982–1987[b]

	Intensive-early stocking		
	May 1–July 15		
Stocking density	3×	2.5×	2×[c]
Stocking rate (acre/steer month)	0.4	0.5	0.6
Gain/steer (lb)	165	163	167
Average daily gain/steer (lb)	2.17	2.14	2.20
Gain/acre (lb)	138	108	94

[b]From Owensby et al. (1988).
[c]Based on 3.1 acres/steer/5 month summer.

C. Shortgrass Range, Hays, 8 years, 1981–1988[d]

	Intensive-early stocking		Seasonal continuous	
	May 1–July 15		May 1–July 15	May 1–Oct. 1
Stocking density, early period	3×	2×	1×	[e]
Stocking rates (acre/steer month)	0.46	0.69	0.35	0.69
Gain/steer (lb)	88	110	116	204
Ave. daily gain/steer (lb)	1.21	1.50	1.59	1.39
Gain/acre (lb)	73	64	34	59

[d]From Olson and Launchbaugh (1989) and Olson et al. (1993).
[e]Base stocking rate was 3 acre/steer for the 5-month season.

cial grazing method (Gillen, 1999). When compared to continuous season-long grazing, IES increased stocking rates by about 50% while nearly doubling beef gain per acre without reducing gain per head.

On shortgrass range at Hays, KS, steer total gain and average daily gain for 2× intensive-early stocking and seasonal continuous grazing were not significantly different during the early period (May 1 to July 15; Table 14.1C) (Olson and Launchbaugh, 1989; Olson et al., 1993). Total steer gains and average daily gains were reduced by 3× IES compared to 2× IES, while gains per acre were increased. Under season-long continuous grazing, the percentage of total season gain during the early period ranged from 41 to 72% and averaged 57%. Late season (July 15 to Nov. 1) gains declined only slightly from early season gains under continuous grazing in contrast to the findings on tallgrass range at Manhattan. This was attributed to the forage quality on shortgrass prairie declining less as the season progressed than on tallgrass prairie. While it was noted during the early years of the study that 3× IES mostly prevented patch grazing (Ring et al., 1985), as the study progressed it became apparent that 3× IES was not sustainable on shortgrass vegetation.

The relatively greater importance of the cool-season midgrass component on shortgrass compared to tallgrass range became apparent. As the cool-season midgrass component (primarily western wheatgrass) at Hays declined during the study under 3× IES, a shift to the less productive warm-season grasses (principally blue grama and buffalograss) resulted. When determined on October 1, the herbage left was 2350 and 2180 lb, respectively, under 2× IES and seasonal continuous, but only 1530 lb under 3× IES. It was concluded that both 2× IES and season-long continuous were sustainable on shortgrass range. While 2× IES did not prove to be economically superior to continuous grazing in the study, two possible advantages were noted: (1) alternating 2× IES and seasonal continuous between two range units annually could reduce cattle marketing risks by allowing marketing twice per year, and (2) periods of nongrazing allow for integrating cultural range improvements (Olson et al., 1993).

Grazing yearling cattle on shortgrass range in eastern Colorado, Klipple (1964) compared grazing from May 10 to August 10 at 6 acres per head (intensive-early grazing) with grazing May 10 to November 10 at 13 acres per head (continuous seasonal grazing). Cattle gains under both treatments were similar during the first 3 months (about 1.9 lb per day), but cattle grazed season-long gained only 1.0 lb per day during the last 3 months. Concentrating grazing during the early, rapid-growth period resulted in 27.8 lb per acre annually compared to 21.1 lb per acre when grazing was extended through the full 6 months.

Greater forage-use efficiency early in the growing season under IES compared to late in the growing season is apparent (Launchbaugh, 1987). Under the higher stocking density of IES, more even distribution of grazing and greater use of the forb-shrub component is apparent by mid-season (Smith and Owensby, 1978; Brock and Owensby, 1999), but grazing patterns tended to even out as grazing continued under continuous season-long grazing until the end of the growing season.

However, the distribution of utilization was not improved by IES grazing on tall-grass prairie in Oklahoma (McCollum *et al.,* 1990). Another consideration is that IES does not require the additional fencing and handling facilities required for rotation grazing systems.

IES is suited best for weaned, growing livestock but is not practical for reproductive livestock (Launchbaugh and Owensby, 1978; Kansas State University, 1995). With stocker steers or heifers, IES has the advantages of reducing interest on livestock investment by 50% if cattle are owned only during the shortened early grazing period and of reaching favorable earlier markets, However, while gain per head during the forepart of the growing period and annual gain per acre are substantially increased by IES on tallgrass prairie, the reduced total gain per head under IES compared to total growing season can be a problem. Profitable ownership of steers may require total gains of 250–300 lb per head just to offset the negative margins on purchase to sale price. Consideration should be given to maintaining the advantage of high IES gains made on highly productive grassland range by providing alternative sources of high-quality forage during the remainder of the growing season, thereby extending the period of ownership while obtaining continued economical gains from forage.

Other alternatives for capitalizing on high IES gains include following improved winter gains in drylot (possibly 1.5 average daily gains) with IES as part of a finishing phase. Yet another alternative suggested by Kansas State University (1995), after stocking at 2× the normal seasonal rate for yearling animals to July 15, is to then move the livestock to a feedlot. Shortening the grazing period under IES to less than the first half of the growing season would limit gain per acre too much to be economical and would further limit total per head gains (Bernardo and McCollum, 1987); lengthening the grazing period much beyond the first half of the growing season will lower average daily gains but will increase total gains per head.

Utilizing the lush, nutritious, early spring growth of crested wheatgrass can substantially improve average daily gains. Yearling cattle were found to gain 3 lb daily during the first 28 days of spring growth but only 2 lb daily during the second 28-day growth period (Rhodes *et al.,* 1986). Although the gains of late grazing were not as high as those under early grazing and reduced total season rate of gain, the need to extend the grazing season under practical management must also be considered.

VI. DEFERRED/RESTED GRAZING METHODS

Deferment and other strategic periods of rest, the latter often extending for a full growing season or a full year or more, are based on providing nongrazing during periods that are expected to enhance the forage stand. These nongrazing treatments can be applied selectively or combined with rotation to enable being passed around systematically among multiple range units. A determination of the precise needs and anticipated benefits should precede the application of planned nongrazing. The benefits that can be expected from planned nongrazing depend upon when it is provided (Booysen and Tainton, 1978):

1. Early spring—provides relief when plants are drawing on their stored reserves and developing full leaf systems.
2. Spring—accelerates regrowth when potential is maximum.
3. Summer—benefits flowering and seed production.
4. Autumn—accelerate carbohydrate (TAC) buildup and storage.
5. Yearlong—enable seedlings to establish, preferred species to recover from very low vigor, or fine fuel to accumulate for subsequent prescribed burning.

Avoiding not being grazed every year at the most susceptible period of growth may greatly advantage many forage plant species and be an important ingredient of a special grazing technique.

A. DEFERMENT

Deferment provides for nongrazing from the breaking of plant dormancy until after seedset or equivalent stage of vegetative reproduction, accomplished by either delaying the beginning of spring grazing or discontinuing winter grazing early. It is best adapted to native or seeded rangelands where growth is seasonal. Current climatic patterns and plant phenological development are preferred over historic calendar dates for setting the beginning of deferment since these vary from year to year. Also, the calendar date when breaking of plant dormancy principally occurs will be later for warm-season grass swards than cool-season grass swards. In the Southwest, 90% of the growth of the semi-desert grasses occurs in July, August, and September, and these grasses require a different deferment period (Martin, 1978a). The deferment period in the Southwest is commonly extended to include both the lesser spring and the more prominent summer plant growth periods.

The objectives of deferment are to increase seed production, enhance seedling establishment, protect plants susceptible to trampling damage and defoliation in early spring, and to prevent overgrazing during low forage availability in early spring (Heady, 1984). Grazing after seed production, in addition to utilizing the forage crop, is suggested for trampling the seed into the soil while the coarse, standing plant materials are partially eaten or trampled to the soil surface. Deferment along with moderate grazing is apt to be more effective than light grazing alone or equal to 12-month rest in maintaining or improving the vigor of either seeded or native rangelands. Also, in contrast with most grazing systems incorporating 12-month rest, no annual forage production crop need be left unharvested.

Deferred grazing (synonym **selected deferment**) refers to the deferment of grazing on a grazing unit but not in systematic rotation with other units, and deferment is followed by grazing of the residual standing crop. Deferred grazing can be a one-time event or applied annually to the grazing unit where deemed most necessary. Deferred grazing is equally applicable to rangelands on which the grazing season is yearlong or limited to the growing season. Deferred-rotation grazing is a multi-unit grazing system in which deferment is systematically rotated among the respective grazing units in the system (see Chapter 15).

While deferred grazing is primarily adapted to native and seeded rangelands, animal performance in the short term ranges from nearly equivalent to continuous grazing to slightly reduced. Forage quality is seldom directly enhanced by deferment from grazing, although it may be indirectly enhanced if deferment induces a desirable qualitative change in species composition over time (Heitschmidt and Stuth, 1991). However, in order to maximize nutritive value of the forage consumed, deferment should not be applied to improved, intensively management pasture. Deferment is generally unnecessary to maintain vigor in improved pastures, shortens the green growth period, and reduces nutritive quality by advancing forage maturity.

For deferment to benefit the forage plants, stocking must be set at a moderate rather than heavy rate. A season-long stocking rate equivalent to moderate continuous grazing can be maintained under deferred grazing or deferred-rotation grazing. The same total area is available for grazing under deferred grazing as under continuous grazing during the grazing season since the deferred area is grazing after seedset, and stocking densities are not increased appreciably by deferring $\frac{1}{3}-\frac{1}{4}$ of the total area (Kothmann, 1984). When deferment is applied to a single range unit under deferred grazing, stocking density during the post-deferment grazing period can generally be increased by about 30% over stocking density under continuous stocking to take advantage of the shortened grazing period. Cool-season pasture—consisting of smooth brome, tall fescue, irrigated cool-season pasture, or winter cereals—has been suggested in Kansas as a means of providing alternative forage while deferment is being applied to native prairie range (Kansas State University, 1995).

Occasional deferment, as well as regularly scheduled deferment, is a technique that has practical application on such range types as midgrass, semi-desert bunchgrass, sagebrush-grass, and mountain grasslands to increase forage plant vigor, plant reproduction, carbohydrate root reserves, and general range condition (Fig. 14.6). Applying deferment every third or fourth year followed by grazing to about 50% of capacity in the fall effectively renovated crested wheatgrass in low vigor because of past heavy grazing (Sharp, 1970). However, since cattle prefer immature forbs to more mature growth, cattle grazing native prairie range in Kansas were considered more likely to suppress growth of "weedy" vegetation under conservative, season-long continuous stocking (or under IES) than under deferred grazing (Kansas State University, 1995).

Cattle browsing under continuous grazing in New Mexico has maintained fourwing saltbush in a hedged form and has resulted in relatively little leader growth (Pieper and Donart, 1978), but periodic deferment every third or fourth year during active growth has maintained productive, vigorous stands. Sixty-day spring deferment of fourwing saltbush has provided more leader growth than a full year

FIGURE 14.6 Deferment, either applied under deferred grazing or deferred-rotation grazing, has been an effective improvement tool on midgrass, semi-desert bunchgrass, sagebrush-grass, and mountain grassland range (shown here).

of rest, and rest beyond one year has provided more flower stalks but progressively less leader growth (Price *et al.,* 1989).

Deferment is apt to be the most beneficial on rangeland in high poor to low good condition; deferment may have little or no advantage on higher condition range and may not benefit range in very poor condition because of near total depletion of desirable plants (Launchbaugh and Owensby, 1978). Deferment is not required for perpetuation of annual grass range such as cheatgrass lands and California annual grasslands and may provide minimal advantage to any remaining perennial plants. Deferment of use on such annual grasses means almost complete loss of the current year's forage crop as well as increasing the fire hazard (Pechanec and Stewart, 1949).

A "semi-deferred" modification of deferred grazing was recommended by Canfield (1940) for semi-desert grassland ranges of New Mexico. The objective was to reduce summer grazing on black grama portions of the range to restore plant vigor and growth and save for winter grazing while encouraging summer grazing by livestock on other portions of the range such as tobosa flats. Partial deferment of the black grama areas was accomplished by placing salt and temporary water tanks in other areas during summer and partly herding livestock off the black grama areas. Since the "semi-deferred" system included no season of complete nonuse, many of the advantages of continuous grazing were reportedly maintained while accomplishing partial deferment of black grama areas by periodically reducing growing season use.

A partial deferment, providing that it includes the most critical plant growth period, may be nearly as effective as deferred grazing in restoring plant vigor while permitting grazing at less advanced plant growth stages (Heady, 1975). Herbel (1974) suggested that any nongrazing period should be as brief as is consistent with the vegetational objectives. On crested wheatgrass range, delaying the start of grazing under a partial deferment has improved maximum basal area and grass yields but somewhat at the expense of lower nutrient levels (Frischknecht and Harris, 1968).

B. REST

Rest is nongrazing of an area of grazing land for a specified period of time, ranging from a few days to a full year or more; it may be more fully described as **short-term, seasonal, yearlong,** etc. Historically, the term referred to nongrazing for a full year along with foregoing grazing on that year's complete forage crop, but the term now is commonly used to include any period of nongrazing excluding deferment. Thus, rest must be carefully described and interpreted in order to be meaningful.

Selected rest is a grazing method in which rest, typically for a full year, is selectively applied to the range unit deemed to be most in need; it is equally meaningful when applied to yearlong, growing season, or dormant season grazing lands. In some cases selected rest is continued through a second year or applied in alternate years. When combined with rotation grazing, and sometimes including additional grazing methods, rest is made a component of various grazing systems, including rotational rest, rest-rotation grazing, and Santa Rita grazing system (see Chapter 15).

Alternate-year rest and full grazing in intervening years has been studied in some vegetation types. Heavy grazing of crested wheatgrass in alternate years in Nevada studies produced a downward trend at two of three locations studied (Robertson *et al.,* 1970). Moderate grazing each year was concluded to be better than alternate-year rest after heavy use. One year's complete rest did not compensate for overgrazing in the previous year; however, forage production and plant vigor of heavily grazed crested wheatgrass in Idaho have been restored by letting it rest a year or two, or by deferred grazing, or just by alternating the timing of grazing during the growing season (Sharp, 1970). Alternate-year full grazing and rest of seeded crested wheatgrass and crested wheatgrass-brome grazing units on Diamond Mountain in northeastern Utah increased annual forage production (Laycock and Conrad, 1981). However, it was concluded that such high-producing stands could withstand full grazing more often than every other year.

Alternate yearlong rest (i.e., with total rest in alternate years) was compared to assigned summer-winter grazing on native switch cane range in the coastal plain of North Carolina (Hughes *et al.,* 1960). Cattle grazing had little effect on the vegetation when grazed in winter and generally was not detrimental to range grazed yearlong in alternate years, but it did hasten the decadence of switch cane grazed continuously in summer. Alternate-year deferred grazing was also effective in maintaining relatively high vigor and productivity of switch cane.

The use of yearlong rest as a general maintenance technique to be routinely included in grazing systems is now often challenged as being unnecessary as well as inefficient, and enthusiasm for its use has waned substantially. Its primary role should probably be directed to specific problem situations: severe drought emergencies, following reseeding or interseeding, providing fuel for prescribed burning, or in conjunction with critical site rehabilitation, particularly on mountain bunchgrass ranges. When rest is required, all livestock should be removed. Leaving only a few head on the grazing unit being rested, such as horses or even wild grazers such as elk and buffalo, will result in the more attractive areas such as near water and the most palatable plants continuing to receive the heaviest use. There is seldom justification for applying long-term rest to improved pasture.

An increasing consensus among graziers and range scientists has been voiced by Heady (1975): except in meeting special need situations, "resting of range pastures seems an extravagant use of herbage." Certain other problems have followed the use of rest as a routine management technique. Patchy grazing has commonly been accelerated by complete rest on productive sites; the development of old plant debris and wolf plants is encouraged in plant species such as crested wheatgrass (Heady, 1975). Favorable habitat for black grassbug populations has resulted from the higher levels of plant debris carryover into fall and winter (Haws *et al.,* 1973). The high cost of cross-fencing is ever present in all rotational grazing systems. Riding (drifting) and water control have reportedly permitted resting portions of winter range without the large costs of installing and maintaining fences (Zimmerman, 1980).

Discontinuing grazing when the key range species are flowering, resulting in rest (nongrazing) during seedset and maturation, may improve vigor of some forage species. This "late-season nongrazing" may enhance plant food storage in perennials and foster vigorous growth early the following year (Heady, 1984). Grazing is often terminated before livestock gains start dropping substantially late in the growing season. Advancing the date for terminating grazing under "removed 10-days early" improved average daily gains of cattle on crested wheatgrass in Utah (Frischknecht and Harris, 1968). Launchbaugh (1957) recommended that young cattle be removed from shortgrass range about September 1 to avoid weight losses when not being supplemented.

Seasonal continuous grazing (May 10 to November 10) of yearling beef cattle on shortgrass range in Colorado was compared to a two unit technique in which an equivalent herd was grazed early (May 10 to August 10) in one unit and late (August 10 to November 10) in the second unit (Klipple, 1964). When stocking densities were adjusted to achieve 40% utilization at the end of each grazing period, continuous grazing and the two-pasture grazing technique were similar in average daily gain (1.54 vs. 1.47 lb) and gain per acre (21.1 and 20.8 lb). However, vegetation under season-long continuous grazing maintained production better during the study than under either the early or late grazing.

VII. HIGH-INTENSITY GRAZING METHODS

A. CREEP GRAZING

Creep grazing permits juvenile herbivores to graze areas their dams or other mature animals cannot access at the same time, thus providing access by the juveniles to forage of higher quality and/or quantity. Creep grazing has some common goals with creep feeding but differs from creep feeding in that high-quality forage rather than concentrate feed is preferentially made available to juvenile offspring. Special creep pastures of inherently high-quality forage are provided adjoining the base pasture or pastures but are separated by a fence with special creep openings (slips) or creep gates adjustable to either calves or lambs. The juvenile offspring run with their dams in the base pasture, either continuously or rotationally grazed, but have continued access to the creep pasture as well. Once the first creep pasture has been topped, it can be opened temporarily to use by both the dams and the offspring. When full use or clean-up has been achieved, the first creep unit is closed and another is opened for access by the offspring. The progression continues in that order through the grazing season (Matches and Burns, 1985). In southern U.S., firebreaks seeded to low growing, highly palatable plants with long growing seasons have provided excellent creep pasture while the cowherd is in an adjoining forest-range unit (Linnartz and Carpenter, 1979).

Creep grazing provides an opportunity to increase calf weaning weights through the higher consumption of forage nutrients by the calf to supplement the milk obtained from the dam (Vicini *et al.*, 1982). Other advantages cited for creep grazing include (Rice *et al.*, 1987): (1) generally lower cost than creep-feeding grain, (2) less labor needed than creep feeding, (3) allows increased stocking rates in the base pasture, (4) reduces dependency on persistent milk production, and (5) justifies extra efforts in maintaining legumes in the creep pasture. Calves make best use of creep pasture after they are 3 to 4 months of age.

It has been noted that calves are unable to compete with their dams to maintain herbage intake at lower herbage allowances (Baker *et al.*, 1981). Also, the use of low grazing pressures in the base pasture to favor the calf from 3.5 months of age until weaning is impractical in that it causes wasted herbage and low calf production per land area (Blaser *et al.*, 1974). Based on their studies on hill pasture in

North Carolina, Harvey and Burns (1988) concluded that the greatest advantages from creep grazing accrued when the cows were stocked intensively and an adjoining creep pasture was available. However, when the base pasture retained high quality and lighter stocking rates allowed high forage availability, the relative advantage of creep grazing was greatly reduced. When the base pasture was comprised of tall fescue in West Virginia, regardless of forage allowance, the gains of suckling lambs were considered unacceptable (Adandedjan *et al.,* 1987). While reducing stocking rates below the traditional levels increased average daily gains of the lambs from 0.18 to 0.24 lb, providing adjoining alfalfa-smooth brome pasture for creep grazing increased average daily gains to 0.36 lb.

B. FIRST-LAST GRAZING

First-last grazing (also known as **topping-followup, leader-follower,** and **top-and-bottom grazing**) results from managing two groups of grazing animals, usually with different nutritional requirements. The favored group of animals is grazed first and followed by the less favored or trailing second group. This special grazing technique can be applied to individual grazing land units as a grazing method. It can also be applied to a series of paddocks grazed in rotation as a grazing system. The latter can be enabled by splitting each original paddock, thereby doubling the number of paddocks. First-last grazing has received only moderate use, in part because of increased management requirements, but has been recommended for high-yield mesic or irrigated pasture.

The objective of first-last grazing is to favor higher producing animals of the same class or animals of a different class or kind. The first herd is given the first choice of topping the standing crop and is benefitted by higher quality of forage and generally greater consumption, and the second herd is then required to "clean up" or fully use the residue. After the first herd has grazed off the higher quality forage, it can be moved into a third pasture and simultaneously be replaced by the second herd. In a grazing system, this process continues to the end of the grazing season (Matches and Burns, 1985). Minson (1990) noted that swards leniently grazed in the spring to ensure availability of forage late in the season may result in reduced nutrient intake as the forage supply becomes less leafy and more stemmy; he suggested this problem could be partly overcome by using first-last grazing throughout the grazing season.

An example of practical use of the first-last grazing is when a high-performance herd (growing-finishing livestock, dairy cattle, or a purebred herd) is followed by a reduced-performance herd (commercial beef cow-calf herd, ewe-lamb flock, the low milk producers in a dairy herd, or growing animals for which only light gains are acceptable). A dry breeding herd can be included in the second group, or it or other animals having only maintenance requirements could theoretically comprise a third herd of grazers. However, this further complicates the coordination and management required to operate the system successfully. Animal numbers and pasture size must be carefully balanced, and periodic adjustment in the size of each herd will often be required.

When yearling steers were grazed on improved pasture in West Virginia, the daily gains of first grazers and second grazers, respectively, were 1.36 lb and 0.92 lb compared to 1.13 lb by the "whole plant" control group (Blaser *et al.,* 1959). The advantage of the first grazers was attributed to opportunity for selective top grazing and associated increase in digestibility and quantity of herbage consumed. First-grazing was also found to substantially increase milk yield of lactating Holsteins over second-grazing—36.5 vs. 25.3 lb milk per day (Blaser *et al.,* 1969). When kikuyu (*Pennisetum clandestinum*) grass was grazed chronologically in 7-day periods by three groups of steers—leader, follower, and clean-up group—their respective average daily gains were 1.55, 1.25, and 0.99 lb (Campbell *et al.,* 1977).

Sanderson *et al.* (1987) compared first grazing by steers and second grazing by ewes with steers and ewes grazed continuously but independent of each other. First-grazer steers had a slight advantage over continuously grazed steers (1.88 vs. 1.77 lb average daily gains); the second-grazer ewes were somewhat disadvantaged compared to continuously grazed ewes (7.7 vs. 9.8 lb total gains). However, first-grazer steers grazed on buffalograss in Texas were not advantaged in daily gains or gains per acre over continuously grazed steers; this was attributed to maintenance of uniform quality in buffalograss and to its short, leafy growth that did not favor dietary selectivity (Mowrey *et al.,* 1986). Wedin *et al.* (1989) concluded that when both stocker steers and dry ewes were being grazed on Kentucky bluegrass-orchardgrass-birdsfoot trefoil pasture, first-last grazing of steers (followed by ewes) was more efficient in harvesting forage than was co-grazing but did favor steer gains over ewes.

C. FORWARD CREEP GRAZING

Forward creep grazing combines the advantages of both creep grazing and first-last grazing in providing higher quality forage to juvenile animals as a means of supplementing their milk intake once they are old enough to handle forages. It differs from creep grazing in that the offspring are permitted to "creep" ahead as first grazers and top each standing crop but are later followed by their dams as last grazers. Forward creep grazing can be applied to a single grazing land unit using temporary animal control measures or more commonly in rotation through a series of paddocks. Once a pasture rotation system is developed for use, the only additional requirement is to provide creep openings in the dividing fences.

Providing forward creep grazing in a Virginia improved pasture study benefited both the cows and the calves as follows (Blaser *et al.,* 1980): (1) calf daily gains were increased from 1.25–1.80 lb, (2) final calf weights were increased from 499.4–556.6 lb, and (3) cow weight losses during lactation were reduced from 0.42–0.02 lb daily. However, such large advantages in calf or lamb gains are apt to result only when high utilization levels are maintained in the base pasture (Jordan and Marten, 1970).

D. STRIP AND RATION GRAZING

Strip grazing is based on confining grazing animals to an area of grazing land under high stocking density and short grazing periods (typically $\frac{1}{2}$–3 days). It is most commonly applied to improved pasture using several grazing cycles during the growing season. Grazing and nongrazing periods are provided by manually advancing moveable fences (a grazing method) or rotation among multiple paddocks (a grazing system). **Ration grazing** is a variant of strip grazing in which animals are confined to areas that will provide one daily allowance (or, less commonly, a weekly or other short-period allowance) of forage per animal.

Strip grazing is generally limited to intensive, land-limited, high-production enterprises to justify the high degree of management and investment required (Matches and Burns, 1985). It is used primarily on short-term improved pasture, particularly with dairy cattle but is also adapted to pasture finishing of lambs or feeder cattle. When a new pasture area is provided daily, the quantity and quality of forage available to the grazing animals each day will be similar at the beginning of each grazing day rather than being cyclic. While strip grazing favors intensive utilization and land area production over continuous grazing, per-animal response will seldom exceed that under continuous grazing and may be less if moderately high grazing pressure is maintained under strip grazing.

When grazed under strip grazing, Van Keuren et al. (1966) reported a 14% increase in grazing capacity with dairy cattle on alfalfa-smooth brome pasture, and Hart et al. (1976) reported an 18% increase with beef steers on Coastal bermudagrass. The extra grazing capacity from strip grazing of dairy cows on alfalfa-brome pasture was found by Pratt and Davis (1962) to increase per-acre production of milk by 20–30% compared to continuous grazing while maintaining similar per-cow milk production. However, strip grazing does not always increase grazing capacity over regular rotation (USDA, Agricultural Research Service, 1960).

Grazing crossbred wether lambs during December and January on established alfalfa under irrigation in central California, Guerrero and Marble (1997) compared strip grazing (three subdivisions, each grazed 4–5 days)—considered to be a nonselective grazing alternative—with 14-day and 21-day set stocking. The result was a recommendation for the 21-day treatment over the other two since lambs gains were slightly more and required less moving and cross-fencing. The slightly greater average daily gains under 21-day set stocking was attributed to the lambs having more available leaves and a higher degree of selection over a longer time period.

E. FRONTAL GRAZING

This grazing method allocates forage within a pasture area by means of a sliding fence that cattle (and possibly other large herbivores) can advance at will to gain access to ungrazed forage. Cattle move the frontal control fence across a rectangular area to gain access to ungrazed forage by pushing on a cable along the feeding front with their heads; a back fence to concentrate the grazing is advanced manually. The fence mechanism, more fully described in Volesky (1990), consists of electric wires, sleds to carry the system, and a pace governor to assist in regulating the rate of advance.

Frontal grazing was developed in Argentina and has been only minimally used in the U.S. (Volesky, 1990). Its principal adaptation is to level terrain, improved pasture, and a landscape free of brush, trees, or other obstructions to the movement of the fence. Because of landscape limitations and requiring intensive management, frontal grazing appears impractical for use on range but provides a means of closely regulating defoliation intensity and duration and achieve high grazing efficiency on improved pasture.

Because livestock are confined under high stocking density and short grazing periods and ungrazed forage is continuously available in a narrow feeding front, forage loss due to trampling and defecation is minimized under frontal grazing, resulting in high pasture use efficiency. When frontal grazing was applied to Old World bluestem pasture in Oklahoma, nearly all plant tillers were routinely defoliated during each short grazing period, and the enhanced forage production allowed the maintenance of higher stocking rates (Volesky, 1994). The increased grazing capacity provided by frontal grazing over continuous grazing and rotation grazing was attributed to the unique timing and pattern of defoliation and subsequent opportunity for recovery (Volesky *et al.,* 1994). However, the greater grazing capacity under frontal grazing as applied in the Oklahoma study was not reflected in greater steer gains per acre; the lower daily steer gains under frontal grazing may have resulted from reduced forage intake and the time required for the steers to become proficient in moving the frontal fence.

F. MOB GRAZING

Mob grazing is the grazing of a grazing unit with a relatively large number of animals at a high stocking density for a short time period. It provides a means of forcing relatively more uniform grazing utilization even on the less preferred plant species (Gray *et al.,* 1982). High grazing pressure may permit its use for "grazing out" residual vegetation, to exert high levels of brush control, or to clean up during the non-growing season following less intensive grazing treatment during a prior grazing season (Booysen and Tainton, 1978). However, great care must be exercised to ensure that the use of this grazing method does not adversely affect desirable perennial forage plant stands; its prolonged application can also greatly reduce livestock performance. Counterparts of mob grazing are also found in the intensive application of high-intensity/low-frequency (HILF) grazing (see Chapter 15).

15

GRAZING SYSTEMS

A **grazing system** is defined as a specialization of grazing management based on rotating grazing animals among two or more grazing land units (paddocks) while defining systematically recurring periods of grazing and nongrazing. A grazing system will generally include one or more grazing methods in addition to rotation grazing. The Forage and Grazing Terminology Committee (FGTC; 1991), after differentiating grazing methods from grazing systems, further expanded the historical concept of "grazing system" to include "a defined, integrated combination of animal, plant, soil, and other environmental components," thereby approaching a grazing management plan. This expansion of "grazing system" has not been followed by the Society for Range Management (1998) and is not utilized here. Clear distinction must be made between the special grazing techniques incorporated into a grazing system and the day-to-day provision of forage from a wide variety of sources that comprise the forage-animal plan (Chapter 3).

Because of the maze of treatments, practices, and variants that can be incorporated into a grazing system, the name alone of a grazing system will seldom describe it adequately; also required for precision and clarification is a detailed description of the grazing system. The Society for Range Management (1998) has suggested that the first usage of the name of a grazing system in a publication always be followed by a description in the following prescribed order: number of grazing units, number of herds, and length of grazing periods and nongrazing periods for units within the system.

Additional criteria that are helpful in describing a grazing system include: (1) kind, class, and number of grazing animals in each herd; (2) number of grazing periods per year; (3) season of grazing in relation to plant and animal requirements; (4) stocking density index; and (5) grazing pressure (J. K. Lewis, 1983, 1984). Still other items ideally related include: (6) whether a systematic or flexible schedule is followed; (7) size, shape, and arrangement of grazing units; and (8) description of grazing methods or range improvement practices routinely included within the system. These factors are important not only in describing the grazing system but in initially designing the grazing system.

Numerous complex grazing systems can and, in fact, have been created from an assortment of grazing methods and other grazing management practices. Grazing systems have been designed to meet an array of management goals and to be applied to diverse vegetation types and conditions, often directed to special case situations and uniquely named. Thus, it is not surprising that there is only partial consensus as to the optimal role grazing systems can and should play as well as the comparative advantages between them. In attempting to simplify a comparison of grazing systems in this chapter, a generic approach has been taken in describing and discussing four categories of grazing systems: deferred grazing systems, rest grazing systems, HILF grazing systems, and short-duration grazing systems.

Decisions that must be made by the manager when designing a complex grazing system include the physical aspects of the system (items marked with an asterisk are interrelated with stocking rates): (1) the amount of land to include within the total system; (2) the number, size, shape, and arrangement of paddocks; (3) kind, number or extent, and location of fences and water developments; (4) kind (or mix), class, and number of animals (from 25 head of cattle optimum for small operations to 500 or more for large operations); (5) the number of herds (generally 1 to 3); (6) length of the grazing season; and (7) the grazing schedule (adapted from Kothmann, 1984).

The more intensive grazing systems (e.g., short-duration grazing and HILF) use one herd of livestock which rotates through several paddocks in an effort to exert maximum control over grazing and maximize range improvement. The less intensive grazing systems (e.g., Merrill grazing system) typically use one herd less than the number of paddocks in the system in an effort to realize many of the advantages of continuous grazing and maintain high livestock performance. In general, grazing systems on improved pasture have been designed to maximize animal production, those on rangeland to improve or maintain range condition (Taylor *et al.*, 1993c).

I. DEVELOPMENT OF GRAZING SYSTEMS

The enthusiasm for and promotion of individual grazing "systems"—often with great zeal—have been both cyclic and regional. This has led to the query as to whether Western-trained (American) graziers and scientists have not become "obsessed with implementing grazing rotations" (Sanford, 1983). Additional confusion has been generated by the strong advocacy roles taken historically throughout the 20th century by some individuals, groups, and agencies in supporting one or another kind of grazing system. The result has been a deluge of discussion, much practical application as well as misapplication, and considerable but still inadequate long-term experimentation on the various components of grazing systems.

Recommendations on which grazing system to use have gone through a series of crusades. Initially, the first crusade was the use of deferred-rotation grazing begun about the time of World War I and based on the work of Arthur W. Sampson. The second crusade, begun near the end of World War II, was that of rest-rotation grazing based on the work in California of August L. Hormay. The third crusade, that of short-duration grazing, came into vogue during the late 1970s and was recommended for widely different vegetation types in several countries by Allen Savory (Heady, 1999). Last, a consensus has been developing since about 1990—sometimes nearly approaching crusade level—that grazing systems may be more appropriate for special situations than for general use (Fig. 15.1). There is considerable evidence that some form of continuous grazing when combined with optimal stocking rates, kind or mix of animal species, season of grazing, and distribution of grazing will often negate the need of a complex grazing system, and still enable the selective application of one or more grazing methods as needed.

No single grazing system has universal adaptation to all forage types, climates, and management objectives and needs (Wilson, 1986; Dwyer *et al.*, 1984; Martin and Whitfield, 1973; Heady, 1999). Grazing systems vary widely in their potential for improving range conditions and/or improving livestock response. There is no perfect grazing system since they vary widely in their adaptation to different vegetation types, precipitation zones, terrain, and soil type as well as in their management and investment requirements. Wilson (1986) warned against extrapolation from one range type to another until more was known of the factors involved and rejected the wide advocacy of one system for all range types. Dwyer *et al.* (1984) cautioned against applying specialized grazing systems in the 8- to 12-in.

FIGURE 15.1   The consensus gaining in the 1990s seemed to be that complex grazing systems may be more appropriate for special situations rather than for general use, with some form of continuous grazing and selective application of one or more grazing methods an acceptable alternative; showing Angus heifers grazing crested wheatgrass in central Utah.

precipitation zone that minimize vegetational cover and thus hold a high potential for erosion.

Wilson (1986) concluded it was commonly accepted that a role of grazing systems on rangelands was to facilitate the maintenance of "non-resilient" range at stocking intensities that would otherwise cause vegetation deterioration. When stocking intensity was controlled at levels that achieved maximum animal production and economic return, he suggested that the resilience of many grazing lands will not be exceeded and no grazing system beyond continuous grazing will be required. When stocking intensity is above the threshold for deleterious change to the range under continuous grazing, he suggested that a change to a grazing system may have only marginal advantage in meliorating the effects. And, finally, he considered that successful grazing systems (i.e., using rotation or rest) are the exception rather than the rule.

It is generally agreed that (1) no grazing system will eliminate the need for proper stocking and practicing other techniques of good grazing management, (2) all grazing systems must be tailored to fit the needs of the specific case, (3) the manager is the key to the success of a specialized grazing system, and (4) flexibility must be maintained in adjusting to changing conditions. There have been outstanding results along with major failures in using grazing systems, leading Heady (1999) to conclude that, "Good managers can make any grazing system successful!" However, because of the potential for concentrating stress on both animals and plants if not properly and continually monitored, a complex grazing system requires even greater rather than less management attention compared to less intensive grazing management techniques.

One of the most commonly used approaches to evaluating individual grazing systems is their comparison against continuous grazing, probably the simplest and most common special grazing techniques. (Continuous grazing, defined as a grazing method rather than as a grazing system, was introduced along with its variants in Chapter 14.) For continuous grazing to serve as a valid base or standard against which comparisons in vegetation and livestock response can be made, it must be accompanied by optimal or equivalent levels of all aspects of good grazing management. When either continuous grazing or the grazing system being compared to continuous grazing has not received the full benefit of supporting good management, either through research or applied use, the comparison will invariably be biased. For example, intentionally handicapping the grazing method or system with a probable negative treatment such as heavier stocking for comparison with moderate continuous grazing has generally led to great difficulty in interpreting the cause-and-effect relationships of the results.

Comparisons of continuous grazing and rotational grazing have often been confounded by differences in size of grazing units. In a comparison of continuous grazing and short-duration grazing on midgrass prairie near Cheyenne, WY, Hart *et al.* (1989) found that the size of grazing unit affected cattle gains but the grazing system did not. When compared on 60-acre grazing units cattle gains were the same for continuous grazing as for short-duration grazing but were significantly less on a 512-acre continuously grazed unit. On the large unit cattle had to travel a maximum of 3.5 miles to water, and forage plant utilization averaged 41% but ranged from 60% near water to 30% at distances greater than 2.5 miles. By comparison, cattle had to travel only up to 1 mile on the 60-acre units, and plant utilization averaged 47% with no effect from distance from water. Only after separating the effects of subdivision from those of livestock rotation could a valid comparison be made between grazing treatments, and continuous grazing does not preclude subdividing grazing land into smaller grazing units.

Malechek (1984) emphasized that 10 years or more are apt to be required for range improvement to result from complex grazing systems in arid and semi-arid areas. Weather and soil moisture can obscure the treatment effects of a grazing system (Dwyer *et al.*, 1984). Martin (1978a) cautioned against short-term credit or fault of newly installed grazing systems since high rainfall or drought may be the principal effect; he recommended that a new grazing system be evaluated over a period of 6–12 years while it goes through several weather cycles. Herbel and Pieper (1991) have noted that favorable weather is often more effective than a specialized grazing system in providing a rapid improvement in species composition, at least in the short term. Comparisons of vegetation change in grazing systems studies have often reflected differences in climatic conditions rather than differences between grazing systems (Heitschmidt *et al.*, 1982b).

During an 8-year trial on semi-arid rangeland in South Africa, annual rainfall and stocking rate had a stronger effect on plant and animal production than did special grazing techniques—continuous grazing vs. six-paddock rotational grazing (Fourie and Bransby, 1988). Similarly, during an 8-year comparison of continuous and short-duration grazing on shinnery oak (*Quercus harvardii*) range in Texas, climate dictated all vegetation changes during the study, and the grazing system had no detectable influence (Dahl *et al.*, 1988). Based on studies at Manitou, CO, Currie (1976) found that fertilizer and/or 2,4-D spraying during a 5-year study period was effective in promoting improvement and recovery of pine-bunchgrass ranges, but the complex grazing systems he compared were not effective.

Many graziers consider that grazing units currently deferred, rested, or lightly grazed should provide a hedge against drought and be fully used during severe drought. However, Pieper and Donart (1975) held that it was important to retain a grazing system during a drought period even if confronted with the need for livestock reduction. They considered these treatment advantages as being even more important during drought to maintain the vigor of forage plants. Martin (1978a) recommended that return to the grazing schedule be made as soon as possible if a weather crisis forced a deviation from it.

II. ROTATION GRAZING

Rotation grazing is a generic term applied to moving grazing animals recurrently from one grazing unit (paddock) to another grazing unit in the same rotation series (group); in this regards, it is the opposite of continuous grazing. The following terms are pertinent for describing and discussing rotation grazing, by definition a basic component of each grazing system:

Paddock: One of the grazing units or subunits included in a rotation group (series)

Grazing period: One of a series of uninterrupted occupancies within a paddock

Nongrazing period: A period of rest (i.e., grazing animals are prevented access) that follows each grazing period within a paddock

Grazing period cycle: The sum of one grazing period and the following nongrazing period within a paddock

Grazing system cycle: The length of time required for all grazing methods and other treatments included in a grazing system to be passed through all paddocks within the series

Rotation grazing requires the combination of two or more paddocks into a common group or series to permit the scheduled transfer of grazing and nongrazing between paddocks. Combinations of grazing and nongrazing periods are primarily used to facilitate the rotation of grazing methods and other treatments among the respective grazing units; rotation grazing is less commonly the treatment per se.

Specific periods of grazing and nongrazing, stocking density, levels of plant defoliation, etc. are required for rotation grazing to make a transition into a grazing system.

Some special challenges met in applying rotation grazing, particularly to rangeland, include:

1. The necessity and cost of cross-fencing into grazing units (paddocks) and of additional stockwater provision and access roads
2. The requirement and probable additional cost of providing adequate livestock water in all grazing units
3. Utilizing grazing units of greatly different grazing capacity in the same series
4. Deciding how to handle the grazing system under prolonged drought—rigidly adhere to original schedule or impose temporary flexibility
5. Requirement for numerous short-term decisions because of the larger number of paddocks involved and the relatively rapid changes in the standing crop

A. THE PROS AND CONS

From a review of 15 different rangeland studies in which cattle were the grazing animals and some form of rotation grazing was compared with continuous grazing, Holechek *et al.* (1999) summarized their influences on rangeland vegetation. Across all studies forage production was 7% higher for rotation compared to continuous grazing. On semi-arid and desert range types, rotation showed no advantage, but on more humid range types, forage production averaged 20–30% higher under rotation grazing. Across these same 15 different rangeland studies, livestock performance favored continuous grazing over rotation grazing: 89.4 vs. 85.9% average calf crop and 504.6 vs. 494.1 lb average calf weaning weight for continuous and rotation grazing, respectively.

There is general agreement among range scientists that moving from continuous grazing to a complex grazing system will not enhance livestock performance in the short term and may reduce it (Herbel, 1974; Wilson, 1986; Pieper, 1980; Malechek, 1984; Rittenhouse, 1984). One exception may be the Merrill system developed in Texas which actually combines continuous grazing with scheduled deferment. From an extensive review of literature Herbel (1974) found there must generally be an improvement in range condition, and subsequently in range grazing capacity, to justify a rotation scheme on rangeland using livestock performance as a criterion.

Rittenhouse (1984) generally attributed any increase in animal performance from a complex grazing system as follows: (1) in the short term, to increased forage availability and decreased grazing pressure resulting from water development and fencing; and (2) in the long term, to improved range condition, which changes plant composition and increases productivity. Advantages of a grazing system for animal production arise only if the system favors (either promotes or maintains) the botanical composition that is the most productive or most nutritious for herbivores (Wilson, 1986).

There is generally a daily decline in quality of available forage and an increase in grazing pressure from the first to the last day of each short-term grazing period under rotation (Matches and Burns, 1985; Blaser *et al.,* 1974). Animals typically have access to the leafy, highest quality forage the first day of the grazing period, but the average quality of the remaining forage progressively declines over the next few days as the higher quality forage is removed (with each day's residue primarily becoming the next day's ration). As the short grazing period progresses, the leaf-to-stem and the dead-to-green plant material ratios increase, the legume component is reduced, and a greater proportion of the remaining forage becomes fouled.

A consistent finding has been that high-producing animals may suffer if they are compelled to eat low-quality forage under a fixed rotation schedule (USDA, Agricultural Research Service, 1960), and high-producing animals may lack opportunity to ingest sufficient high-quality forage to achieve their genetic potential. When cattle grazed seeded big bluestem pasture in Missouri, daily dry matter intake was estimated at 3.2, 2.9, and 2.4% of body weight, respectively, during the 1st, 3rd, and 7th day of the 7-day grazing period, compared to 2.7% under continuous grazing (Morrow *et al.,* 1994). Blaser *et al.* (1959) suggested that even short periods of forced heavy utilization may seriously lower either milk or meat production (Blaser *et al.,* 1959). Milk production of dairy cows has been noted to fluctuate from the first to the last day of each grazing period (Pratt and Davis, 1962).

Rotation grazing, once believed to have relevance in the control of internal parasites in sheep and possibly other grazing animal species, is now known to be ineffective (Morley, 1981; Tembely *et al.,* 1983). Free-living stages of internal parasites generally survive for periods far longer than the nongrazing periods incorporated into rotational grazing systems of either rangeland or improved pasture. Rotation must provide 3–6 months of rest for significant die-off of infective larvae, and such systems generally do not provide satisfactory forage utilization.

In Texas range studies all lambs became about equally infected regardless of whether grazed under continuous, switchback, Merrill four-pasture, or short-duration grazing (Tembely *et al.,* 1983). Simulation analysis in Texas indicated that nongrazing periods exceeding 150 days were necessary to minimize the rate and extent of spread of cattle fever ticks in rotation systems (Teel *et al.,* 1998). While concentrating animals in smaller areas provides greater opportunity to check livestock health, it also provides opportunity for diseases to spread more rapidly, and herd health should be checked often (Merrill, 1980; McCollum and Bidwell, 1994).

Rotation grazing provides a way to protect erect, highly productive plants sensitive to heavy and/or frequent defoliation, particularly legumes, and their reduction in grass-legume swards (USDA, Agricultural Research Service, 1960). Rotation grazing with short grazing periods and replenishment and regrowth periods of four weeks (nongrazing) have generally been found necessary for the persistence in grazed swards of tall-growing legumes including alfalfa, red clover, and birds-foot trefoil (Smith, 1970; Dougherty *et al.,* 1990; Hodgson, 1990; Bransby, 1991; Morley, 1981; Matches and Burns, 1985; Rohweder and Van Keuren, 1985; Leach and Clements, 1984). Such legumes are quite susceptible to severe defoliation during initial growth or regrowth, and they maintain fewer growing points below grazing height than do most grasses (Walton, 1983).

When grazed by dairy cows from May 10 to September 10 during a 3-year study, continuous grazing reduced alfalfa in an alfalfa-smooth brome forage stand from 60% to 10–15% (Pratt and Davis, 1962). In contrast, the composition of alfalfa under rotation grazing (5- to 7-day grazing, 35-day nongrazing) was not reduced below 50% of the stand. Van Keuren (1980) has suggested that grazing periods for tall-growing legumes such as alfalfa not begin until a height growth of 10 to 12 in. has been reached; he also suggested that nongrazing, recovery periods of at least 35 days for tall-growing legumes and 25 days for low-growing legumes be provided. For subtropical conditions Leach and Clements (1984) recommended nongrazing periods of 35–40 days for alfalfa. No difference was found in animal response between 24- and 36-day recovery periods when cattle grazed orchardgrass-trefoil or orchardgrass-Ladino clover pasture in California (Hull *et al.,* 1960).

Corah and Bartley (1985) have recommended a rotation grazing system with alfalfa in order to (1) obtain maximum forage production, (2) enable grazing at the ideal stage of growth (to improve average daily gain and achieve more uniform utilization of the forage plants), and (3) help in controlling bloat. During a 4-year study in Alberta with cattle on grass-alfalfa pasture, Walton *et al.* (1981) found that four-pasture rotation provided 40% more grazing capacity and doubled weight gains per acre during the last 2 years of the study, resulting in part from a decline of alfalfa under continuous grazing. When yearling steers in California were grazed on irrigated grass-legume pasture, gains per acre were higher for rotation grazing than for continuous grazing at a higher stocking rate but not at a medium stocking rate (Hull *et al.,* 1967); this resulted from the rotational grazing system providing more animal days of grazing per acre, but forage was sometimes short during the last 1 or 2 days of a 7-day grazing period.

The decline of less grazing-tolerant legumes in legume-grass stands may result from excessive stocking rates rather than from continuous grazing (Bransby, 1991). Allowing very heavy grazing pressures to develop by utilizing a fixed date of moving between paddocks can greatly damage alfalfa. Smith (1970) noted that under very heavy grazing pressures sheep even pawed the soil away from the alfalfa plant bases and ate the root crowns and exposed portions of the taproot. However, the alfalfa component in alfalfa-grass mixtures in a Manitoba study declined equally under continuous and short-duration grazing (from 70–50%) during a 4-year study under moderate stocking (Popp *et al.,* 1997). Consideration should be given to using in alfalfa-grass mixtures intended for grazing one of the several alfalfa cultivars developed for higher tolerance of grazing. Smith *et al.* (1989) concluded that selection for alfalfa grazing tolerance under continuous grazing from a broad-based population was, in fact, a method of improving the resulting germplasm while maintaining the potential for good forage yields.

The grazing regime experienced by plants (and probably animals as well) cannot be predicted solely from rate and intensity of herbivore rotation among paddocks (Coughenour, 1992). Even under continuous grazing, livestock may create smaller scale grazing systems of their own and rotate themselves within the confines of a single grazing unit. Also, under rotation grazing animals tend to be less settled and exert their own influences on management decisions (Hodgson, 1990). Rotation grazing does provide opportunity for organizing the sequential grazing and separation of groups of animals with different biological needs, i.e., breeding, isolation from breeding, calving and lambing, etc. (Hodgson, 1990).

B. DURATION OF NONGRAZING

Under rotation grazing "time" conceptually becomes a management variable for both the grazing interval and the sward regrowth interval in each grazing period cycle. A major objective of a rotational grazing system is to give the preferred plants growing on preferred grazing areas a chance to recover after each grazing period (Martin, 1978a). Thus, the length and timing of the nongrazing period is a critical part of any grazing system (Wilson, 1986). When nongrazing is applied during the growing season and where there is potential for regrowth available in multiple grazing period cycles, the duration of nongrazing in the grazing period cycle is considered the most important feature of rotation grazing. The nongrazing period in the cycle provides opportunity for the forage plant following defoliation to rebuild photosynthetic area, replenish carbohydrate reserves used in the early stages of regrowth, and maintain a vigorous root system (Walton, 1983).

Timely defoliation and recovery periods are of great importance in rotation grazing (Charette et al., 1969; Oesterheld and McNaughton, 1991). Following initial defoliation, the relative growth rate is rapid if the environmental resources are ample but gradually declines to lower levels. If the period of nongrazing is too short, forage plants will not have time to recover to high levels of forage supply and availability (i.e., tiller height and density, foliage density, leader growth) even though the regrowth will be highly nutritious. If the nongrazing period is too long, biomass recovery will be maximized but the foliage will be reduced in palatability and nutritive value before defoliation and ingestion occur. Since maximum green leaf production and harvest are major keys to animal performance, the nongrazing periods must be short enough to allow animals to maximize the harvest of green foliage.

If grazing pressure increases greatly near the end of the grazing period, animals will be forced to include in their diets coarse forage from previously ungrazed tillers. Also, as nongrazing periods are prolonged and plants advance toward maturity, regrowth response to subsequent defoliation may be diminished. Plants in a depressed state of vigor, or of a species less tolerant of grazing, or placed under high competitive pressure will require longer recovery periods (Caldwell, 1984).

Based on their work with switchgrass, Anderson *et al.* (1989) concluded that much longer regrowth periods were necessary following severe defoliation for stand maintenance and plant vigor.

Nongrazing periods of 10–20 days are commonly considered best during rapid growth periods of perennial grasses and legumes. During late spring and summer on irrigated pasture, where soil water and nutrients are not limiting factors, nongrazing periods of 20–30 days are commonly recommended. Following late spring and summer defoliation on dryland pasture and range, nongrazing periods of 40–60 days are common but may be extended through the remainder of the grazing season, depending upon utilization levels and potential and amount of regrowth. However, these nongrazing period lengths provide only broad guidelines; length of nongrazing and the rate of livestock advancement between pasture units in intensive rotation systems should theoretically be based on vegetation growth rather than on calendar days.

From their studies in the Nebraska Sandhills, Reece *et al.* (1996) found that multiple grazing periods initiated in June reduced total organic reserves by about $\frac{1}{3}$ in prairie sandreed and sand bluestem, both warm-season grasses. This led to the conclusion that rest periods following initial grazing in June should be longer than 60 days to avoid measurable reduction in total organic reserves and provide occasional deferment until mid-August. They suggested shifting the grazing sequence in multi-unit rotation systems each year by a sufficient number of paddocks to prevent the tallgrasses from being grazed at critical times for several consecutive years and concluded this would be more effective than a single pasture shift or reversing the sequence of grazing each year.

On improved mountain meadows Gomm (1979) found that 7-day nongrazing periods under rotation grazing failed to permit sufficient recovery from defoliation to sustain grazing through the growing season. Limiting recovery periods to 14 days compared to 21 days improved yearling cattle gains somewhat, particularly late in the season, resulted in greater uniformity of grazing, and apparently resulted in the utilization of higher quality forage. Regrowth periods in England on perennial ryegrass were recommended to be no less than 14 days to provide for adequate regrowth and not more than 28 days to prevent stemminess (Parsons and Penning, 1988). Optimal recovery periods on shrubs appear to be longer; under short-duration grazing in New Mexico, a 64-day rotation cycle but not 32-day rotation cycle increased leader growth over continuous grazing (Price *et al.*, 1989).

C. FREQUENCY OF DEFOLIATION

Optimal grazing management conceptually avoids repeated, severe defoliation of the forage plant or individual tillers without a recovery period. Theoretically, if the nongrazing periods are too short, the regrowth of plant tillers may be grazed before growth is optimized; if grazing periods are too long, tillers may be regrazed before the recovery interlude begins. While frequent defoliation under continuous grazing has tended to reduce desirable shrubs, rotation grazing has allowed shrubs to recover after browsing and maintain good vigor and survival (Teague, 1992).

Continuous grazing does not imply continuous defoliation of individual tillers of individual herbaceous plants, since intervals between defoliations often vary from 5 days to as long as 3 or 4 weeks on swards (Hodson, 1990). While it is commonly assumed that controlling patterns of defoliation and regrowth under rotation grazing will enhance herbage production, Hodgson (1990) concluded that the available evidence indicates that control even within quite wide levels is likely to have little impact on the amount of herbage produced and consumed per unit area. Thus, engineering rotation systems to control the frequency and intensity of defoliation of individual plants will probably remain mostly unsuccessful (Walker, 1995).

Frequency of defoliation is interrelated with intensity and selectivity of defoliation. Frequent defoliation generally results in more severe (intensive) defoliation. Kothmann (1984) suggested that frequency was the primary controlling factor in relation to intensity and selectivity of defoliation and could largely be controlled by regulating the length of the grazing period. Frequency of grazing was considered by Savory (1987) more a function of duration of grazing than even numbers of animals, this giving rise to his emphasis on minimizing the severity of grazing by reducing the length of the grazing periods. McKinney (1997) concluded that overgrazing results from herbivores regrazing the same plant and is a symptom of animals staying too long in the same spot.

Herbaceous plants that have been intensively defoliated but do not have regrowth tend to be avoided by grazing animals; this is also true of plants which have not been grazed but allowed to mature, but plants that have been intensively grazed and have then made regrowth are readily selected. Gammon and Roberts (1978) found previously defoliated tillers were selected for during periods of rapid growth but selected against during periods of slow or no growth. Frequent clipping of crested wheatgrass during the growing season produced herbage in the fall that was more leafy and more nutritious but with decreased yield of herbage (Cook *et al.*, 1958). A management dilemma results from animal gain per head and per acre being increased by high frequency and intensity of grazing during plant growth periods while low frequency and intensity increase range condition (Kothmann, 1984).

Grazing animals generally remove the top canopy of forage first, particularly in dense swards, even though the highest quality feed may be at the base of the standing crop (Kothmann, 1984). Seldom are plants completely defoliated the first time they are grazed. In studies with warm-season tallgrasses in Oklahoma (Jensen *et al.*, 1989), tillers were consistently only moderately defoliated the first time and more severely defoliated afterwards. Thus, it appears that high-intensity defoliation results primarily as a function of increments removed during successive defoliations. Each stocking rate under short-duration grazing was found by Hinant and Kothmann (1986) to have a characteristic mean stubble height to which the tillers of little bluestem and brownseed paspalum (*Paspalum plicatulum*) were grazed. Briske and Stuth (1986) found that tillers of brownseed paspalum were uniformly regrazed every third or fourth day under heavy grazing, but significant regrazing of tillers under moderate grazing did not occur until after 18 days of exposure to grazing.

It is generally held that repeated severe defoliation of desirable plants or areas can be reduced by increasing the stocking density and reducing the duration of grazing. Multi-unit grazing systems utilizing only one herd provide an opportunity to manipulate stocking density over a wide range with no change in stocking rate (Kothmann, 1984). Widely quoted has been Voisin's (1959) assertion that grazing periods should be sufficiently short to avoid regrazing immature regrowth, that is, until tolerance of defoliation has been restored by the forage plant upon achieving adequate leaf area and TAC storage levels.

However, repeated defoliations may not be as important as commonly believed if the intensity of defoliation is maintained at a reasonable level (Heady, 1984; Skovlin, 1987). As defoliation was found to be relatively infrequent under moderate continuous grazing, Gammon and Roberts (1978) found no great reduction in frequency of defoliation from the use of six-pasture rotation grazing over continuous grazing. From subsequent studies Gammon and Roberts (1980b) reported relatively few tillers were regrazed during 14-day grazing periods under moderate stocking rates.

Based on their cattle grazing studies on crested wheatgrass in Utah, Norton and Johnson (1986) concluded that any detrimental effect of continuous season-long grazing was unlikely to be caused by repeated defoliation during the same grazing season. At the moderate utilization level maintained during their study, only 17% of the plants were regrazed during a 6-week period in early summer. Any danger of pasture deterioration from continuous grazing was attributed more likely to come from repeated defoliation of the same plants or patches of plants year after year rather than from repeated defoliation over the short term. Even when sheep were heavily stocked in small paddocks on a foothill range in Utah during a 25-day period, plants were typically grazed only once or twice (Hodgkinson, 1980). The frequency of defoliation of individual tillers of Lehmann lovegrass increased slightly with stocking rate (Ruyle et al., 1986a), but more than 90% of the multiple defoliation events on individual, marked tillers occurred in the patches grazed the previous year.

When cattle were rotated more rapidly among eight paddocks in tallgrass prairie in Oklahoma (four rotation cycles compared to two rotation cycles), grazing schedule (1.3 and 1.8 times the recommended normal) had little effect on the height at which tillers of big and little bluestem were defoliated (Gillen et al., 1990). Increasing the number of grazing periods decreased the percentage of tillers defoliated each period but increased the cumulative defoliation frequency over the entire 152-day grazing season while not affecting the percentage of tillers ungrazed over the entire grazing season. Increasing the number of grazing periods at the same stocking rate appears not to promote secondary succession as effectively as fewer grazing periods with longer grazing and rest periods (Taylor et al., 1993a; Reece et al., 1996).

The studies of Jensen *et al.* (1990) on tallgrass prairie in Oklahoma provide assessment of the extent frequency and intensity of defoliation of individual tillers and selectivity between plant species can actually be altered within short grazing periods by manipulating herbage allowance (i.e., 22, 44, 66, and 88 lb per AUD). Their findings and conclusions are summarized as follows:

1. The maximum percentage of tillers grazed a single time during a given trial was commonly 50–80% but ranged as low as 20%.
2. Selectivity between species was reduced by decreasing herbage allowance, but this effect was not large until herbage allowance was below 44–50 lb per AUD, and selectivity was never completely removed. Similarly, Walker (1995) concluded that the use of traditional grazing systems does not appreciably affect selective grazing.
3. Grazing all tillers once in a grazing period, even within a plant species, is unlikely in a tallgrass community.
4. Leaf area removal was increasingly more severe as the number of defoliations increased per tiller.
5. The goal of grazing any individual tiller at no greater than moderate level within a grazing period is roughly equivalent to grazing any tiller no more than once; this would require many tillers to go ungrazed, but few or none would be severely grazed.
6. The concept of a single defoliation within a grazing period still has merit because that single defoliation would predict a moderate amount of leaf area removal for the defoliated tiller.

D. ALTERNATE GRAZING

Alternate grazing (synonym **switchback grazing**), utilizes two grazing units and one herd; grazing is typically during the growing season and is alternated between grazing units at intervals of $1-2\frac{1}{2}$-months. Its uses as well as its advantages on native rangeland have been rather limited. This simple grazing system, utilizing a 2-year grazing system cycle and involving consecutive grazing periods of March 15 to June 15, December 16 to March 15, and June 16 to December 15, was originally included in studies at the Texas Experimental Ranch (Kothmann *et al.,* 1971); however, lack of advantage over continuous grazing caused it to be dropped from the study. Alternate grazing has been used on seeded dryland spring pasture such as crested wheatgrass with some success; this use has entailed grazing one unit in early spring and the other in late spring and reversing the order of grazing in alternate years.

E. TWO-CROP VS. ONE-CROP GRAZING

Grazing crested wheatgrass from boot stage to anthesis has been referred to as **one-crop grazing** (Hyder and Sneva, 1963; Sharp, 1970). This grazing technique has permitted maximum root growth and harvesting maximum amounts of dry matter but has generally given no late spring regrowth for fall grazing. To meet the latter need **two-crop grazing** has been employed with dryland crested wheatgrass by grazing early (4- to 6-in leaf length) to boot stage and then again in late summer/early fall after curing. While it originated in Oregon (Hyder and Sneva, 1963), the two-crop grazing method has also been found effective with crested wheatgrass in Idaho (Sharp, 1970), in Utah (Harris *et al.,* 1968), and in Colorado (Currie, 1970).

The standing crop available in the fall under two-crop grazing has commonly consisted of a combination of spring residue and late spring and late summer regrowth. Grazing at a light to moderate level in the spring generally reduces seedstalk development and stimulates vegetative shoots and thus improves the quality of the fall-saved carryover forage; late-summer regrowth has further enhanced nutritive quality during years when it does occur. Two-crop management has been noted to (1) depress root growth slightly, (2) harvest a maximum of early forage, (3) reduce total herbage yield somewhat, and (4) permit high storage concentration of carbohydrates by late summer as well as (5) provide late summer and fall forage.

For maximum length of the spring grazing season or for spring-fall grazing, a rotational system combining two-crop and one-crop grazing and utilizing two pasture units (**two crop/one crop grazing**) is suggested, with the treatments being switched in alternate years. In a commercial ranching enterprise grazing yearling heifers, the combination of one-crop and two-crop systems maintained good daily gains over an extended grazing season, gave a forage-to-beef ratio of 10:1, kept forage reasonably nutritious, and prevented the development of wolf plants (Hedrick, 1967). An alternative is the use of a three-unit rotation in which two two-crop units are grazed in early spring, one one-crop unit grazed in mid-spring, and the two two-crop units grazed again in the late spring or in the fall, followed by a recombination of units the following year (Hyder and Sneva, 1963).

III. DEFERRED GRAZING SYSTEMS

A. ROTATIONAL DEFERMENT

The objectives of deferment and the application of deferred grazing (synonym selected deferment) as a grazing method have been discussed in Chapter 14. **Rotational deferment** consists of a multi-unit grazing system in which deferment is scheduled among the respective grazing units on a rotating basis. Although the number of herds under rotational deferment are variable, a low stocking density is maintained in all grazed units. Grazing of the standing crop follows deferment in the deferred unit but is continuous in the non-deferred grazing units. Sample schedules for rotational deferment are shown for growing season and for yearlong grazing application in Table 15.1 using three grazing units.

TABLE 15.1   Examples of Rotational Deferment

		Pasture units[a]		
Year	Period	A	B	C
A. Growing season application (6 months)				
1	May 1–June 30	ND	G	G
	July 1–August 31	ND	G	G
	September 1–October 31	G	G	G
2	May 1–June 30	G	ND	G
	July 1–August 31	G	ND	G
	September 1–October 31	G	G	G
3	May 1–June 30	G	G	ND
	July 1–August 31	G	G	ND
	September 1–October 31	G	G	G
B. Yearlong grazing application (12 months)				
1	May 1–August 31	ND	G	G
	September 1–December 31	G	G	G
	January 1–April 30	G	G	G
2	May 1–August 31	G	ND	G
	September 1–December 31	G	G	G
	January 1–April 30	G	G	G
3	May 1–August 31	G	G	ND
	September 1–December 31	G	G	G
	January 1–April 30	G	G	G

[a]Three grazing units, variable two or three herds or one herd with access to multiple grazing units.
G, grazing; ND, nongrazing constituting deferment.

B. MERRILL GRAZING SYSTEM

The Merrill grazing system was initiated at the Sonora Research Station in Texas by Leo Merrill in 1949 (Merrill, 1954). It utilizes four grazing units and three herds, which results in three units being grazed and one nongrazed at any given time (Table 15.2). Each unit is grazed for 12 months and then nongrazed for 4 months; this results in a grazing period cycle of 16 months and a grazing system cycle of 4 years. (Note: the Merrill grazing system has also been commonly referred to as "deferred-rotation" grazing but differs in that low stocking density is maintained by design.)

This system was developed for yearlong ranges of southwest Texas and adjoining areas; it combines the advantages of both continuous grazing and periodic nongrazing including deferment (Fig. 15.2). Since the herbaceous plant mixture and temperatures in the area permit plant growth during virtually any month of the year when moisture is adequate, each nongrazing period provides deferment for selected plant species. Even winter nongrazing is considered important in favor-

TABLE 15.2 The Merrill Grazing System Developed for Year-round Ranges of Texas and the Southwest

		Pasture units[a]			
Year	Period	A	B	C	D
1	March–June	NSP	G	G	G
	July–October	G	NSU	G	G
	November–February	G	G	NW	G
2	March–June	G	G	G	NSP
	July–October	NSU	G	G	G
	November–February	G	NW	G	G
3	March–June	G	G	NSP	G
	July–October	G	G	G	NSU
	November–February	NW	G	G	G
4	March–June	G	NSP	G	G
	July–October	G	G	NSU	G
	November–February	G	G	G	NW

[a]Four grazing units, three herds. G, grazing; NW, non-grazing during winter; NSU, non-grazing during summer; NSP, non-grazing during spring.

ing a balance of cool-season grasses and improving soil conditions including decreased bulk density and increased infiltration rates. The low stocking density reduces the impact of grazing during drought by spreading grazing continuously over $\frac{3}{4}$ of the total area within the system. Although the Merrill system at Sonora has produced higher animal performance and thus higher net returns than 21-day HILF, it has provided somewhat less grazing capacity and steady but slower range improvement rates (Merrill, 1980).

Livestock performance at the Sonora Research Station was found equal to or better than under yearlong continuous grazing (Merrill, 1969). The favorable cattle response to the Merrill system was attributed to low stocking density and grazing pressure (Kothmann, 1980; Merrill, 1980) and relatively infrequent movement of livestock among pastures (Herbel 1974). However, during a 10-year study at Sonora in which a 60:40 ratio of cattle and sheep was maintained under moderate stocking rates, the percent lamb crop and pounds of lamb weaned per acre were similar under yearlong continuous, the Merrill system, switchback (alternate grazing), and high-intensity/low-frequency rotation (Taylor *et al.,* 1986a). Under the 60:40 cattle-to-sheep ratio, the ease of management of continuous grazing was concluded to make continuous grazing only slightly less desirable than the Merrill system (Merrill, 1980). However, compared to the Merrill system, continuous grazing at Sonora left more vegetation unused and dormant, spot grazing was more of a problem, and the vegetation complex tended to cycle more.

The results of a study comparing the Merrill system with continuous yearlong grazing under moderate stocking rates on mixed grass range in northern Texas are provided in Table 15.3 (Kothmann *et al.,* 1971; Heitschmidt *et al.,* 1982a; and

FIGURE 15.2   The Merrill grazing system was developed at the Sonora Research Station (shown above) for similar areas in the Southwest and combines the advantages of periodic nongrazing including deferred grazing and continuous grazing. (Texas Agricultural Experiment Station photo by Robert Moen and Charles A. Taylor, Jr.)

Heitschmidt, 1986). Calf weaning weights and calf production per cow and per acre favored the Merrill grazing system. During the 4-year comparison (1982–1985) between continuous yearlong and the Merrill system, average calf weaning weights favored the Merrill system by 18 lb, but returns per acre slightly favored continuous moderate grazing because of reduced annual costs. The Merrill grazing system was concluded to be effective in increasing the stability of both plant and animal production in the north Texas study (Kothmann, 1984).

C. DEFERRED-ROTATION GRAZING

Deferred-rotation grazing differs from rotational deferment and the Merrill grazing system in that it is a one-herd grazing system (Table 15.4). (Note that the term "deferred-rotation grazing" is also used in Texas to refer to the Merrill grazing system, which tends to confuse these two very different grazing systems and requires that they be fully described when being referred to.) Deferred-rotation grazing combines characteristics of both deferred grazing and HILF grazing. While maintaining high stocking density, alternating periods of grazing and non-

TABLE 15.3   Average Cattle Performance Under the Merrill System and Continuous Yearlong Grazing at the Texas Experimental Ranch, Vernon, TX[a]

| | 1960–1968 | | 1960–1978 | |
	Merrill moderate	Continuous moderate	Merrill moderate	Continuous moderate
Average cow weights (lb)			986	966
Average calf weaning weights (lb)	521	501	478	467
Calf production/cow (lb)	487	441	445	417
Calf production/acre (lb)	24.6	21.2	24.8	22.5

| | 1982–1985 | |
	Merrill moderate	Continuous moderate
Average weaning weight (lb)	595	577
Calf production/cow (lb)	504	475
Calf production/acre (lb)	33	32
Net returns/cow ($)	131.24	133.59
Net returns/acre ($)	9.05	9.73

[a]Adapted from Kothmann et al. (1971); Heitschmidt et al. (1982a); Heitschmidt (1986).

grazing are assured and continuous grazing in any unit is ruled out. Thus, deferred-rotation grazing may reduce livestock response somewhat compared to rotational deferment or the Merrill system.

Moving livestock from vegetation maintained relatively immature by grazing to ungrazed mature vegetation should be avoided to prevent a sharp reduction in livestock gains (Kansas State University, 1995). Research near Manhattan, KS, has shown that concentrating livestock on two pastures to defer grazing on the third and grazing the deferred pasture heavily the latter part of the growing season reduced summer gains of yearling steers an average of 23 lb per head compared to continuous summer grazing (Launchbaugh and Owensby, 1978). Nevertheless, this increased the favored grass species and resulted in a 16% increase in grazing capacity. This suggests that deferred-rotation grazing should be applied to meet a specific need for range improvement or maintenance rather than universal use. Poor condition range should benefit more from deferred-rotation grazing at stocking rates commensurate with optimal livestock performance than does comparably stocked good to excellent condition range (Kansas State University, 1995).

Deferred-rotation grazing was recommended as an efficient grazing system for sheep on Intermountain sagebrush-grass, spring-fall sheep ranges (Pechanec and Stewart, 1949). It was suggested that variation between early and late spring and

TABLE 15.4 Examples of Deferred-Rotation Grazing

		Pasture units[a]		
Year	Period	A	B	C
A. Growing season application (6 months)				
1	May 1–June 30	ND	G	N
	July 1–August 31	ND	N	G
	September 1–October 31	G	N	N
2	May 1–June 30	N	ND	G
	July 1–August 31	G	ND	G
	September 1–October 31	N	G	N
3	May 1–June 30	G	N	ND
	July 1–August 31	N	G	ND
	September 1–October 31	N	N	G
B. Yearlong grazing application (12 months)				
1	May 1–August 31	ND	G	G
	September 1–December 31	G	N	G
	January 1–April 30	G	G	N
2	May 1–August 31	G	ND	G
	September 1–December 31	G	G	N
	January 1–April 30	N	G	G
3	May 1–August 31	G	G	ND
	September 1–December 31	N	G	G
	January 1–April 30	G	N	G
C. Dormant season application ($4\frac{1}{2}$ months)				
1	November 15–December 31	N	G	N
	January 1–February 15	G	N	N
	February 16–April 1	ND	ND	G
2	November 15–Decmeber 31	N	N	G
	January 1–February 15	N	G	N
	February 16–April 1	G	ND	ND
3	November 15–December 31	G	N	N
	January 1–February 15	N	N	G
	February 16–April 1	ND	G	ND

[a]Three grazing units, one herd. G, grazing; N, nongrazing; ND, nongrazing constituting deferment.

between spring and fall be incorporated to prevent grazing the native bunchgrasses always when most palatable. This concept was later incorporated into a spring rotation-fall continuous grazing system for study at the U.S. Sheep Experiment Station near Dubois, ID (Laycock, 1962). It was concluded that deferred-rotation grazing in the spring allowed sagebrush-grass range to be more heavily stocked without damage than did continuous grazing but that rotating grazing in the fall was unnecessary.

Deferred-rotation grazing was concluded to be a superior system for improving forage on forested pine-bunchgrass range in the Pacific Northwest and for restoring mountain watersheds while maintaining cattle production (Skovlin *et al.*, 1976). However, it was not recommended that deferred-rotation grazing invariably replace properly stocked continuous grazing on mountain summer range if good livestock distribution could be accomplished. Of note was the finding by Olson *et al.* (1999) on southern Utah forested rangeland grazed in common by cattle and sheep that deferred-rotation grazing appeared slightly superior for sheep performance but slightly inferior for calf performance. While deferred-rotation grazing improved ewe nutritional status shortly before breeding, average daily gains of calves were decreased during the last half of the summer grazing season.

Deferred-rotation grazing on high desert ranges of eastern Oregon was considered unsatisfactory since the damage to the forage plants caused when livestock grazing was concentrated in one grazing unit in early spring was not repaired by subsequent deferment (Hyder and Sawyer, 1951; McArthur, 1969). The deferred-rotation system compared to continuous grazing (May 1 to October 1) consisted of a three-pasture unit, one-herd system of three equal-length grazing periods, which provided a grazing sequence for each pasture of 2 years of spring use, 2 years of partial deferment, and 2 years of full deferment. Total forage production was greater under continuous grazing, but a more uniform increase over the range of bluebunch wheatgrass and Idaho fescue was obtained with deferred-rotation grazing. Deferred-rotation was costly in terms of feed quality but did improve distribution of grazing, something that might have been accomplished by water hauling under continuous grazing. A need to modify the deferred-rotation system to reduce heavy, early spring grazing was apparent; this can be accomplished by spreading early spring grazing over both pasture units not currently receiving deferment. Under deferred-rotation grazing livestock performance is more apt to be limited by nutrient levels than forage availability.

Rotating a deferment treatment has been suggested for use on Intermountain salt-desert shrub ranges (Table 15.4C) (Hutchings, 1954). Not only does this provide occasional protection against defoliation during the breaking of plant dormancy in late winter, but it also assures a forage supply saved for late winter (Fig. 15.3). Sheep under continuous season-long grazing commonly "top" range forage during early winter grazing, leaving themselves mostly dependent upon poorer quality "topped" plants by March. On native winter range, rotation grazing permitted a more uniform availability of forage and browse to the grazing animals throughout the winter.

Greater flexibility was provided on yearlong ranges in Arizona by modifying a one-herd, three-period, three-unit deferred-rotation grazing system (Schmutz and Durfee, 1980). This modification provided for substitution within certain guidelines of another range unit, or grazing was continued on the same unit for a longer

FIGURE 15.3   Deferred-rotation grazing, when applied to Intermountain salt-desert shrub winter ranges (shown above), provides occasional protection against defoliation during the breaking of plant dormancy in late winter and assures a forage supply saved for late winter. (U.S. Forest Service photo.)

period of time if it was capable of providing more usable forage due to more precipitation or other factors. When determining the best range unit, the next best unit, and so forth for each year, restrictions were adhered to in order to accomplish the objectives of deferred-rotation grazing. During each 3-year cycle: (1) no range unit was grazed more than twice during the same season, (2) each unit received seasonal nongrazing at least once during each of the spring and summer growing seasons typical of that area, (3) each unit was grazed sometime every year, and (4) cattle were moved to the next-best range unit when the present range unit was properly utilized.

IV. REST GRAZING SYSTEMS

A. ROTATIONAL REST

The objectives of rest and the application of **selected rest** have been discussed in Chapter 14. **Rotational rest** consists of a multi-unit grazing system in which rest (often specified for a 12-month period) is scheduled among the respective grazing units on a rotating basis (Table 15.5). When the number of grazing herds is one less than the number of grazing units, the non-rested units are normally continuously grazed.

B. REST-ROTATION GRAZING

Rest-rotation grazing is a more complex grazing system employing various combinations of yearlong rest, deferment, early season grazing, and full season

TABLE 15.5 Examples of Rotational Rest

| Year | Period | Pasture units[a] | | |
		A	B	C
A. Growing season application ($4\frac{1}{2}$ months)				
1	June 1–July 15	NR	G	G
	July 15–September 1	NR	G	G
	September 1–October 15	NR	G	G
2	June 1–July 15	G	NR	G
	July 15–September 1	G	NR	G
	September 1–October 15	G	NR	G
3	June 1–July 15	G	G	NR
	July 15–September 1	G	G	NR
	September 1–October 15	G	G	NR
B. Yearlong grazing application (12 months)				
1	May 1–August 31	NR	G	G
	September 1–December 31	NR	G	G
	January 1–April 30	NR	G	G
2	May 1–August 31	G	NR	G
	September 1–December 31	G	NR	G
	January 1–April 30	G	NR	G
3	May 1–August 31	G	G	NR
	September 1–December 31	G	G	NR
	January 1–April 30	G	G	NR

[a]Three grazing units, two herds. G, grazing; NR, nongrazing constituting 12-month rest.

grazing, commonly in a 3- to 5-year grazing system cycle. Rest-rotation grazing is a controversial grazing system in the West, yet widely promoted by some land management agencies and individuals even to the present day. Because its original chief designer and promoter was August L. (Gus) Hormay of the U.S. Forest Service, for many years it was also referred as the "Hormay grazing system." Based on preliminary observations at Harvey Valley in northeastern California, it was formally introduced as follows (Hormay and Talbot, 1961):

> Under rest-rotation grazing, stocking is based on the production and use of herbage from all the available forage species and not on the key species alone. Under this system, degree of use of plants does not have the same importance as under continuous seasonal grazing, because grazing is limited to a comparatively short time and is always followed by rest planned to be long enough to overcome the harmful effects of grazing. Further, under rest-rotation grazing little attention is paid to classifying plants according to palatability to determine forage production and stocking. . . . Plants are classified simply as forage or non-forage species, and stocking rate is calculated on the basis of production of all forage species.

In his training manual for "a course in range management," Hormay (1970) stated, "I have placed special emphasis on rest-rotation management of grazing." He further described periodical rest as "a powerful tool available to the range manager for increasing land productivity" and resolved that "only by this means can the main objectives of grazing management be realized." This grazing system was originally recommended for foothill and mountainous bunchgrass ranges with a 2- to 6-month grazing season, but attempts were later made to extend the system to virtually all types of range vegetation in western U.S. and Canada.

Inconsistencies in the original Harvey Valley study soon became apparent, and the results were found much less significant than originally believed (Vallentine, 1979; Laycock and Conrad, 1981). Based on data collected from this study through 1966, Ratliff *et al.* (1972) reported that "the response to rest-rotation has not been as dramatic at Harvey Valley as *in some other areas*" (presumably only observational at best) but that "it seems clear that rest-rotation grazing is ecologically superior to season-long grazing, and that range health at Harvey Valley relative to nearby allotments is better, and range condition trend upward, because of rest-rotation grazing." However, they acknowledged from their data that 7–20% greater livestock gains could be made from continuous grazing and that the cost-benefit of rest-rotation grazing was not completely satisfactory.

Based on data collected in the Harvey Valley study during 1965–1969, Ratliff and Reppert (1974) reported these somewhat startling conclusions: (1) "Continuous grazing appears to be more effective in controlling competing vegetation than it is damaging to Idaho fescue." (2) "The full [summer season] use treatments did not reduce nor did full-season rest improve Idaho fescue vigor on the Harvey Valley plots." (3) "It appears that range managers cannot key seed production into a set program of rest-rotation grazing." It was further reported that droughty springs rather than any grazing system per se caused the heaviest use of Idaho fescue and that grazing during drought was the most damaging effect. In springs favorable for plant growth, the actual grazing of Idaho fescue under continuous summer graz-ing was lighter and later, and the actual nongrazing periods were accompanied by reduced competition that allowed it to maintain its vigor and reproduce under mod-erate, continuous grazing. A previous observation (Ratliff, 1962) was that, al-though preferential grazing of range areas was continued under rest-rotation, some cattle were being forced into less preferred areas.

As originally outlined, rest-rotation grazing consisted of a 5-year cycle which included 2 years of rest (Table 15.6A) (Hormay and Talbot, 1961; Hormay, 1970). The consecutive annual treatments in the 5-year grazing system cycle were (1) continuous and heavy grazing, (2) rest, (3) deferred grazing, (4) rest, and (5) ear-ly nongrazing followed by full grazing. The system was designed for grazing sea-sons of 2 to 5 months in which grazing encompassed the green growing period. The number of grazing units available for grazing increased from 1–3 each year as the growing season progressed; each grazing season started with one herd con-centrated on a single grazing unit.

The original rest-rotation design, resting 40% of the pasture units each year, was advocated for improving range in low conditions rather than for maintaining range already in high condition. For maintaining range already in good condition, replacing one of the rest years by a second full grazing year was suggested (Hor-may, 1970); this reduced annual rest to 20% of the area and spread early spring grazing over two grazing units. From the original design many additional varia-tions have been derived. A 3-year, three-unit plan has included: (1) continuous and heavy grazing, (2) deferred grazing, and (3) rest, which results in resting $\frac{1}{3}$ of the range annually. A commonly used four-pasture unit, two-herd plan, shown in Table 15.6B, reduces annual rest to 25%. A plan using two herds and combining rota-tional rest on three spring units and rest-rotation on three summer units is shown in Table 15.6C; $\frac{1}{3}$ of both the spring units and the summer units is rested each year.

A major problem with rest-rotation grazing frequently reported is that forced heavy grazing one year in the cycle may cause more harm to the forage plants than combinations of rest and deferment can undo. Heady (1975) has calculated that 40% of the available land and perhaps 40–50% of the available forage are un-grazed every year. The remaining 60% has then received the increasing grazing pressure. This can be particularly damaging on both native and seeded forage species not tolerant of close grazing. Initiation of rest-rotation grazing will nearly always require immediate reductions in livestock numbers because of the inherent high forage waste and periodic high livestock density (Gray et al., 1982).

Success with rest-rotation grazing has largely been restricted to mountain bunchgrass range of steep, heterogenous terrain with minimal "suitable" range (Fig. 15.4). In the Pole Mountains of Wyoming, both rest-rotation and deferred-rotation grazing were superior to continuous summer grazing in spreading out the grazing more uniformly and reducing heavy utilization of meadows (Johnson, 1965). Where range conditions have been significantly damaged by improper graz-ing in the past, it has been commonly prescribed and sometimes been viewed as the only alternative to elimination of livestock grazing. It was suggested from al-

TABLE 15.6 Examples of Rest-Rotation Grazing

A. Original five-pasture plan[a]

Year	Treatment (unit A)	Pasture units[e] A	B	C	D	E
1	Graze full season	G	NDL	NR	ND	NR
2	Rest for vigor, litter	NR	G	NDL	NR	ND
3	Defer for seed, seed trampling, graze	ND	NR	G	NDL	NR
4	Rest for seedlings, vigor, litter	NR	ND	NR	G	NDL
5	Seedlings, graze after delay	NDL	NR	ND	NR	G

[a]Five grazing units, variable one to three herds. From Hormay (1970).

B. Plan for mountain ranges in western U.S.[b]

Year	Period	Pasture units[e] A	B	C	D
1	June15–August 15	G	G	ND	NR
	August 16–October 15	G	N	G	NR
2	June 15–August 15	NR	G	G	ND
	August 16–October 15	NR	G	N	G
3	June 15–August 15	ND	NR	G	G
	August 16–October 15	G	NR	G	N
4	June 15–August 15	G	ND	NR	G
	August 16–October 15	N	G	NR	G

[b]Four grazing units, two herds.

C. Combination plan for mountain spring and summer range[c]

Year	Period	Spring units[e] A	B	C	Summer units[e] D	E	F
1	June 15–July 15	NR	G	G			
	July 16–August 31				G	NR	ND
	September 1–October 15				N	NR	G
2	June 15–July 15	G	NR	G			
	July 16–August 31				ND	G	NR
	September 1–October 15				G	N	NR
3	June 15–July 15	G	G	NR			
	July 16–August 31				NR	ND	G
	September 1–October 15				NR	G	N

[c]Six grazing units (three spring units grazed with two herds under rotational rest; three summer units grazed with one herd under rest-rotation grazing).

TABLE 15.6   Continued

D. Northern desert shrub winter range plan[d]

Year	Period	Pasture units[e]			
		A	B	C	D
1	November 10–December 29 (50 days)	G	N	N	NR
	December 30–February 17 (50 days)	N	G	N	NR
	February 18–April 8 (50 days)	ND	ND	G	NR
2	November 10–December 29 (50 days)	NR	G	N	N
	December 30–February 17 (50 days)	NR	N	G	N
	February 18–April 8 (50 days)	NR	ND	ND	G
3	November 10–December 29 (50 days)	N	NR	G	N
	December 30–February 17 (50 days)	N	NR	N	G
	February 18–April 8 (50 days)	G	NR	ND	ND
4	November 10–December 29 (50 days)	N	N	NR	G
	December 30–February 17 (50 days)	G	N	NR	N
	February 18–April 8 (50 days)	ND	G	NR	ND

[d]Four grazing units, one herd.
[e]G, grazing; N, nongrazing; NDL, nongrazing constituting delayed grazing; ND, nongrazing constituting deferment; NR, nongrazing constituting 12-month rest.

lotment comparisons that rest-rotation would maintain vegetation cover and trend comparable to total livestock exclusion on breaks-type range in north-central Montana (Watts *et al.,* 1987).

A comprehensive 7-year comparison was made on Diamond Mountain in northeastern Utah between continuous summer-long grazing (July through September), alternate year summer-long, and 3-year rest-rotation (Laycock and Conrad, 1981). The study was conducted on fair to good condition range of intermingled mountain vegetation types and grazed at a moderate intensity. No differences between systems in plant cover, herbage production, or species composition were found during the study; no soil damage resulted. That there was a lack of differences and that continuous summer-long grazing maintained range just as productive as under rest-rotation were attributed to management; good distribution of water and salt and adequate riding to ensure uniform cattle distribution were provided in all units of all systems.

Daily gains of cows, calves, and yearlings were not significantly different between systems in the Utah study (Laycock and Conrad, 1981); gains per acre based on actual years grazed were similar but very different when based on all years whether grazed or not—11.3 lb per acre for rest-rotation, 9.5 lb per acre on alternate year grazing, and 18.3 lb per acre on summer-long continuous grazing every year. In addition to producing more beef, continuous summer grazing did not damage the soil or vegetation resource and did not require the additional fencing costs required for rest rotation-grazing.

FIGURE 15.4   Rest-rotation grazing has commonly been labeled an extravagant waster of forage; its application is now primarily directed to mountain bunchgrass range of steep, heterogenous terrain with minimal "suitable" range, such as shown above.

In a 5-year study on mountain rangeland in northeastern Oregon, grazing systems (rest-rotation, deferred-rotation, and season-long continuous) were compared under equivalent grazing pressures (AUMs per ton of forage) (Holechek *et al.,* 1987). The result was no differences in cattle weight gains, diet quality, or botanical composition of the cattle diets. No differences were found in crude protein, *in vitro* organic matter digestibility, and acid detergent fiber percentages in fistula samples used to indicate dietary quality. Cattle under rest-rotation and deferred-rotation were observed to use less accessible areas more than under continuous grazing, but rest-rotation grazing did not increase use of secondary forage species over deferred-rotation or season-long grazing. Equivalent livestock performance in the Utah and Oregon studies may have resulted, in part, from low grazing pressures and prolonged green growing periods on those favorable sites.

Based on desirable shrubs being disadvantaged by winter dormancy grazing compared to associated grasses, it has been suggested that about one-third of salt-desert shrub range should be rested each year (Blaisdell and Holmgren, 1984). This was predicted to allow palatable shrubs to increase in size and thereby develop a larger production base. Enhanced vigor and seed production have been attributed to permitting budsage and black sagebrush to re-establish and become important forage species. A four-pasture unit, one-herd plan that incorporates 1 year of rest

(25% of area) and 2 years of late nongrazing (delay or deferment on 50% of area) is shown in Table 15.6D; this system also assures that proportional fresh areas are available for grazing at the beginning of each of the three winter periods. Smith and Beale (1980) suggested leaving a portion of winter sheep allotments ungrazed each year to reduce pressure on black sagebrush and provide browse needed by antelope; an alternative suggestion was to alternate annual winter use by cattle and sheep, thereby mostly resting black sagebrush every other year without having to forego total forage use.

Two possible advantages of rest-rotation grazing are that the rest areas could provide emergency use in severe drought years and the scheduled rests can provide opportunities to insert any necessary cultural range improvement during the absence of livestock. Rest-rotation grazing has been concluded inappropriate for cattle grazing on riparian areas where early-seral shrubs dominate because the heavy grazing treatment will retard establishment and growth of willow-dominated communities (Vavra *et al.,* 1994). However, it was considered useful in maintaining late-seral shrub communities where shrubs have grown out of reach of grazing animals but is doubtful in inducing ongoing replacement of younger shrubs.

Attempts to extend the use of rest-rotation grazing into areas for which it was not designed have met with limited success and considerable failure. This includes grasslands of the Great Plains and prairies, seeded rangelands, and Southwest semi-arid rangelands. Heavy utilization of perennial desert grasses can be expected to be harmful even when followed with rest. Keeping the numbers of animals flexible to meet highly variable precipitation in desert areas and keeping levels of perennial grass utilization below 50% in all years were suggested by Hughes (1982). Nongrazing during dormancy per se, providing grazing pressures are not excessive, probably has little beneficial effect on herbaceous plants (Heady, 1984), at least in the short run.

In general, rest-rotation management of crested wheatgrass ranges is seldom warranted, but foothill or mountainous areas having a diverse mix of native and seeded vegetation types may be an exception (Austin *et al.,* 1983). Periodic heavy grazing during the growing season on sagebrush-grass range in Nevada resulted in restricted basal-area growth and lack of reproduction in the perennial grasses; the results were the same in the presence or absence of intense big sagebrush competition (Eckert and Spencer, 1986, 1987). Since rest-rotation grazing could only maintain early-seral vegetation in an unimproved condition over the 10-year study period, it was concluded that such sagebrush-grass areas were candidates for range improvement practices such as brush control and reseeding. The belief was expressed that rest-rotation had a better chance of success when the vegetation was mostly free of unpalatable species, such as sagebrush, which receive little grazing pressure and are very competitive with seedlings of native herbaceous forage species for soil water and nutrients.

Rest-rotation grazing was designed on the basis of plant responses with no serious consideration given to the nutritional needs of animals (Kothmann, 1984; Dwyer *et al.,* 1984). It greatly increases stocking density early in the spring and holds livestock on grazing units with advancing forage maturity for long grazing periods. When forage supplies are inadequate under high grazing pressures or when livestock are required to graze matured vegetation, livestock are apt to become nutritionally stressed. The peril point—the point beyond which continued grazing pressure will not only reduce production per head but also negate gain per acre—is easily passed in seasonal rest-rotations requiring high livestock concentrations, especially after mid-season when forage is dry (Launchbaugh *et al.* 1978). Because of reduced stocking rates required and unfavorable livestock response generally, Malechek (1984) suggested that informed livestock operators would not be inclined to willingly enter such programs. Changing from rest-rotation grazing to a grazing system with less grazing pressure has also reduced the consumption of locoweed on high mountain range and the associated levels of brisket disease in cattle (Ralphs *et al.,* 1984a).

In summary, rest-rotation grazing has not eliminated palatability differences between plant species or selective grazing, nor has it been as effective as anticipated in overcoming grazing distribution problems. Furthermore, the basic premise that complete absence of grazing is required for adequate seed production and seedling establishment seems doubtful (Kothmann, 1984). If important, it should be noted that nongrazing periods for grass seed production and seedling establishment are also incorporated into deferred grazing (Sindelar, 1988). A continuation of the grazing studies on Sierra Nevada meadows that initiated the rest-rotation movement led to the later conclusion that attention to management for proper residue levels and efforts to maintain good livestock distribution should minimize the need for a complex grazing system (Ratliff *et al.,* 1987).

C. SANTA RITA GRAZING SYSTEM

The Santa Rita grazing system was developed for the yearlong, semi-desert bunchgrass ranges of the Southwest (Fig. 15.5) where both precipitation and plant growth occur principally in mid-summer (Martin, 1978b). It comprises a one-herd, three-unit, three-year rotation cycle in which each grazing unit accumulates 24 months of nongrazing and 12 months of grazing during the 3-year grazing system cycle. The Santa Rita system is synchronized with the seasons of the year as follows:

Rest 12 months: November through October
Graze 4 months: November through February (graze first forage crop when dormant)
Rest 12 months: March through February
Graze 8 months: March through October (graze second forage crop when dormant and third crop when growing)

FIGURE 15.5 The Santa Rita grazing system was primarily designed for the yearlong, semi-desert bunchgrass ranges of the Southwest; although it includes rest treatments, all forage crops are utilized annually by grazing; photo shows good condition range at the Santa Rita Experimental Range in Arizona. (Forest Service Collection, National Agricultural Library.)

The Santa Rita grazing system was designed to provide a full year of rest before each grazing period but is unique in that each year's forage crop is utilized without the waste normally inherent under rest. It was observed that the full year's rest before each grazing period provided an accumulation of herbage that helped protect new growth from grazing, particularly in the early spring. The system was concluded to be advantageous in requiring only two livestock moves per year, enabling animals in a single range unit to be observed more closely and providing 12-month periods in which structural or other range improvements could be made that were not feasible when livestock were present.

However, a 13-year comparison of the Santa Rita grazing system with continuous grazing, both under moderate stocking rates, revealed no differences in grass, forb, or shrub densities or shrub cover by the end of the study (Martin and Severson, 1988). The lack of differences was attributed to near maximum initial plant densities and to moderate grazing during the study period. It was concluded that the Santa Rita system may not improve ranges that are already in good condition but should accelerate recovery of ranges in poor condition under situations of mid-to-late summer rainfall and forage production. The corollary may also be that continuous grazing initiated on such ranges in good condition, when properly managed, should not result in range deterioration.

V. HIGH-INTENSITY/LOW-FREQUENCY
(HILF) GRAZING SYSTEMS

HILF (also referred to as **slow-rotation grazing** or **high-utilization rotation grazing**) is the conventional rotation system employed in the past to sub-humid and semi-arid grazing lands; HILF was coined at the Sonora Research Station after initiating the system there in 1970 (Taylor, 1988) (Fig. 15.6). The grazing periods are generally over 2 weeks and often 30–45 days; the length of the nongrazing periods has varied but often has been between 30–90 days long (Table 15.7). Since one or two or sometimes more grazing period cycles per grazing season have been used, three to five pasture units have generally been required to operate HILF.

It is somewhat difficult to separate HILF from short-duration (or high-intensity/high-frequency, HIHF) grazing because they differ only by degree (i.e., there is a continuous gradient from one type to the other). Using the labels of "high-utilization grazing" and "high-performance grazing," respectively, Boysen and Tainton (1978) have contrasted HILF and short-duration grazing. They suggested that application of HILF was to regenerate depleted humid to sub-humid range-

FIGURE 15.6   "Conventional" rotation grazing, designated HILF by researchers at the Sonora Research Station (photo shown), has been characterized as favoring range improvement over livestock performance; showing livestock also being managed under multi-species grazing. (Texas Agricultural Experiment Station photo by Robert Moen and Charles A. Taylor, Jr.)

TABLE 15.7   High-Intensity, Low-Frequency Grazing System[a]

Year	Calendar period	Pasture units[b]			
		A	B	C	D
1	May 1–June 15	G	N	N	N
	June 15–August 1	N	G	N	N
	August 1–September 15	N	N	G	N
	September 15–October 31	N	N	N	G
2	May 1–June 15	N	G	N	N
	June 15–August 1	N	N	G	N
	August 1–September 15	N	N	N	G
	September 15–October 31	G	N	N	N

[a]Four-unit, one-herd, 4-year cycle with grazing during the growing season.
[b]G, grazing; N, nongrazing.

lands and thereby increase grazing capacity rather than to obtain high animal performance in the short term.

HILF is based upon obtaining forced utilization of vegetation by using high stocking densities and long grazing periods which make relatively long nongrazing periods necessary for plant recovery (Kothmann, 1984). Booysen and Tainton (1978) characterized HILF as applying high cumulative grazing pressure to maximize forage utilization by force grazing even the less preferred plant species. One suggested use was to clean up remaining forage during the non-growing season following less intensive grazing treatment during the prior growing season.

In a 10-year study on pinyon-juniper/blue grama range in New Mexico comparing continuous grazing and HILF, both under heavy grazing (i.e., 25% higher than recommended or 60–65% utilization), blue grama production was higher under HILF than continuous grazing treatment (Pieper et al., 1991). Although HILF compared to continuous grazing may result in more forage being carried through to the latter part of the growing season (Sims et al., 1976) or the dormant grazing period (Hutchings, 1954), this can be more a function of stocking rate than grazing system. Gray et al. (1982) also characterized HILF as offering much flexibility for brush management because of its high cumulative grazing pressure.

An initial hypothesis about HILF was that a high concentration of grazing animals would decrease selective grazing among plants and plant species (Kothmann, 1984). Cattle diets on the Texas coastal prairie were nutritionally similar on continuous grazing and deferred-rotation grazing, declining as the vegetation matured, but with HILF diet composition changed from highly preferred to less preferred species as the grazing period progressed (Drawe et al., 1988). However, later findings have suggested that a high density of grazing animals is unreliable in reducing selective grazing (Walker, 1995), even when forcing higher utilization.

While the desirable forage plants may respond well to HILF treatment, nutrient intake of grazing animals is generally reduced by the increased maturity of the vegetation and reduction of selectivity (Pfister *et al.*, 1984), and the forced utilization commonly lowers animal performance (Table 15.8). In a comparison with short-duration grazing (7-day grazing, 42-day nongrazing) in Texas, HILF (14-day graze, 84-day rest) was more effective in promoting succession from shortgrasses to midgrasses (Taylor *et al.*, 1993a) but was similar in diet quality under moderate stocking levels (Taylor *et al.*, 1997b). HILF can generally be recommended only where livestock performance is a secondary objective and vegetation manipulation through the use of animal impact is a high priority. Because of lower weaning weights, decreased calf crops, and reduced yearling gains under HILF, it has gradually been abandoned by Texas producers in favor of the Merrill or the short-duration systems or a return to continuous grazing (Bryant *et al.*, 1981b).

The high grazing pressure applied under the longer grazing periods of HILF often has a substantial effect on animal diets. Moving steers into a new environment (new grazing unit) and onto more mature forage during mid-season has consistently reduced their performance (Owensby *et al.*, 1973). Nearing the end of each grazing period, and more prominently nearing the end of the growing season or during dormancy, diets shift to more mature forage, and the lower nutritive forage has sharply reduced animal performance (Taylor *et al.*, 1980). Terminating grazing before advanced plant growth stages have been reached or applying reduced stocking rates has reduced the negative effects on animal performance but has reduced the benefits of more uniform grazing on vegetation composition.

Except during periods of ample, rapid growth, competition among cattle, sheep, goats, and white-tailed deer utilizing common range in Texas has been substantial under HILF (Taylor *et al.*, 1980). Based on grazing research at the Sonora Research Station, Taylor (1988) and Taylor *et al.* (1993c) have suggested a combination of HILF, short-duration grazing, and continuous grazing for year-round, 12-month application under multi-species grazing on the Edwards Plateau area as follows:

Mid-April to mid-September—Employ HILF during the active growing season to enhance both soil and vegetation, particularly increasing the warm-season bunchgrasses, using 100-day grazing period cycles.

Mid-September to mid-January—Employ short-duration grazing during the plant dormancy period to enhance livestock performance without damaging the warm-season grasses.

Mid-January to mid-April—Continue cattle in normal HILF rotation cycle while dispersing sheep and goats through all grazing units, thereby allowing continuous grazing to benefit lambing and kidding and utilize the annual forbs produced in late dormancy and early spring.

The use of HILF rotation has often benefited poor condition range while providing no advantage on good condition range when forage supplies are ample (Rogler, 1951; McIlvain and Savage, 1951; Smoliak, 1960; Pieper *et al.*, 1978; Frischknecht and Harris, 1968). A common conclusion from work ranging from switch cane in eastern North Carolina (Biswell, 1951) to Ponderosa pine-bunch-

TABLE 15.8 Comparisons of Rotation Grazing and Continuous Grazing

A. HILF, 30-day rotation, three-unit[a]

Treatment	Acres per head	Gains per head (lb)	Gain per acre (lb)
Continuous (heavy)	4.3	261	66
Rotation (heavy)	4.3	260	64
Continuous (moderate)	6.3	305	49
Rotation (moderate)	6.3	295	47

[a]April 15–October 4 average, yearling cattle, 7 years (1943–1949), sandhill range, Woodward, OK. From McIlvain and Savage (1951).

B. HILF, three-period ($1\frac{1}{2}$–3–$1\frac{1}{2}$ months), two-unit[b]

Treatment	Seasonal gains (lb)	Average daily gains (lb) by period			
		First	Second	Third	6-month
Rotation (moderate)	287	2.05	1.88	.31	1.58
Continuous (moderate)	301	2.05	2.02	.37	1.67

[b]May 1–October 1, yearling cattle, 9 years, mixed grass prairie, Manyberries, Alberta. From Smoliak (1960).

C. HILF, four-unit, one-herd[c]

Treatment	Average weaning weights (lb)	Weaning weight per acre (lb)
Yearlong continuous (moderate)	407	6.5
Yearlong continuous (heavy)	398	7.9
Rotation (heavy)	375	6.8

[c]Cows and calves, year-round, 6 years of data, shortages range, Fort Stanton, NM. From Pieper *et al.* (1978).

D. HILF, 10-day rotation, three-unit[d]

Treatment	Average daily gains (lb)			Gains per acre (lb)
	Yearlings	Calves	Cows	
Continuous (moderate)	2.47	1.73	2.43	39.0
Rotation (moderate)	2.37	1.70	2.40	42.9

[d]Sixty-day spring grazing season, 11 years of data, foothill crested wheatgrass, Benmore, UT. From Frischknecht and Harris (1968).

grass range in Colorado (Currie, 1978) has been that on good condition range the extra costs of labor, fencing, and developing watering facilities to initiate HILF have not been justified.

VI. SHORT-DURATION GRAZING (SDG) SYSTEM

A. DESIGNING THE SYSTEM

Although previously applied to improved pasture under various names, short-duration grazing began being applied to rangelands during the 1960s in Africa, mostly by trial and error, by Allan Savory and others. This system was begun as a departure from rigidly controlled nonselective grazing practices in an attempt to improve animal production. When introduced into the U.S. in the 1970s, it was labeled the "Savory grazing system," by which name it is still sometimes called. In more recent years, Savory has developed his grazing management concepts into a total management philosophy under the title of "Holistic Resource Management" (Savory, 1988) or some variant. As described at an early date by Savory (1978), short-duration grazing was based on grazing periods of 1 to 15 days followed by nongrazing periods of 20–60 days when utilizing 5–8 paddocks or preferably more.

Short-duration grazing (SDG) (also referred to as **high-intensity high-frequency [HIHF] grazing** or **rapid rotation grazing**) is now commonly designed (1) to employ one herd under high stocking density, (2) to include 5–12 paddocks in the system, (3) to have grazing periods of 3–10 days (less commonly 1–15 days), and (4) to employ two to several grazing cycles per year (Table 15.9). Grazing and nongrazing periods are either set or flexible under "time-control."

TABLE 15.9   Short-Duration Grazing System[a]

Cycle[b]	Approximate calendar[c]	Grazing period by grazing units (days)				
		A	B	C	D	E
1	May (last half)	3	3	3	3	3
2	June	6	6	6	6	6
3	July	6	6	6	6	6
4	August (plus first 10 days of September)	8	8	8	8	8

[a]Five-unit, one-herd, four cycles/year; for growing season grazing use on improved pasture or high-yield rangelands.

[b]Unit rotation in each cycle is units A through E.

[c]Beginning dates of each cycle and duration of each grazing period should be adjusted to plant growth rates; dates shown are approximate for Intermountain Region.

Although SDG utilizes high stocking densities, grazing pressures are reduced by shortening the grazing periods (Kothmann, 1984). The shorter grazing periods and moderate defoliation allow shorter rest periods and present animals with less mature forage, thereby potentially increasing diet quality over HILF. Stocking density, number of grazing units (or paddocks) in the system, and length of grazing period are all interrelated in SDG.

Determining when and how often to rotate livestock may be the most critical management decision with SDG next to establishing the stocking rate; yet, managers have minimal available information upon which to base rotation moves. A very large number of paddocks and extreme shortening of the grazing period to as few as 3 days, as recommended by Voisin (1959), compared to 10 days were concluded by Gammon and Roberts (1980b) to be unnecessary based on their African rangeland studies. From a comprehensive review of African studies, Skovlin (1987) found no evidence to justify the use of over 6–8 paddocks. The evidence is overwhelming that no criteria for moving can overcome the problems of overstocking.

The hypothesis that forage production in SDG will be increased when livestock density is increased by extensive subdivision and proliferation of number of paddocks was rejected by Heitschmidt et al. (1987a). It was concluded from studies in Texas that increasing the number of grazing units from 14–42 did not affect forage growth rate, forage production, harvest efficiency, or ground litter accumulation (Heitschmidt et al., 1986a), nor did it alter forage quality (Heitschmidt et al., 1987b). Seven or eight paddocks is the maximum number necessary to optimize SDG systems, and a further increase in number of paddocks and associated livestock density contributes to (1) additional stress associated with frequent moves, (2) disruption of grazing activity, and (3) restricted opportunity for forage selection (Taylor et al., 1993c).

The length of grazing periods for SDG has been commonly fixed according to the estimated time needed by key forage species to recover from grazing events. When both length of grazing period and length of recovery period are fixed, the number of paddocks needed can be determined by the following formula:

$$\text{No. paddocks} = 1 + \frac{\text{Recovery period in days}}{\text{Grazing period in days}}$$

For example, irrigated grass-legume pasture, requiring recovery of 25 days and utilizing a grazing period of 5 days, would require six paddocks. This formula also exhibits that as the number of paddocks increase and the number of days per grazing period remain constant, the length of the nongrazing recovery period increases.

The need to adjust the length of grazing cycles to changing climatic and vegetation conditions rather than following fixed-length cycles may have merit. This has led to the concept and application of "time-controlled grazing," relating the length of the grazing periods to the growth rate of the plants as has been emphasized by Savory (1983). This suggests that rotation of paddocks should be speeded up (1–3 days) during rapid growth periods and slowed down (7–14 days) during slow growth periods (Voisin, 1959; Savory, 1978). Voisin (1959) has suggested that a grazing unit should be grazed at the point where forage growth rate reaches a maximum, thus maintaining plants in the rapid accumulation phase. Savory (1987) has suggested that moves should be based on plant growth rate and that advancement into the next pasture unit should be made at the steepest part of the growth curve in the new unit.

Since total available forage was the only factor found to significantly change within grazing periods on midgrass range in Texas, Mosley and Dahl (1990) suggested it be used in timing rotation moves. Other suggestions proposed for basing moves between paddocks (mostly without verification) have been (1) correlation of use with phenological stage of a target species, (2) a given level of utilization of the initial standing crop, or (3) a minimum level of forage quality and quantity for a desired level of livestock performance. Dalrymple and Flatt (1993) have provided a list of indicators for use in making decisions as to rate and time of moving to new paddocks categorized and briefly discussed under (1) visual forage indicators, (2) measured or calculated forage indicators, and (3) visual and measured livestock indicators. However, the application of these indicators was not precisely treated and appears to rely primarily on observational and judgmental experience, possibly intentionally so, based on the explanation that "each forage in each situation has its own set of indicator ranges."

Data supporting the superiority of "time-controlled" flexible grazing cycles within SDG as well as precise criteria to be followed are both limited. Based on their research on sand shinnery oak range in Texas, Mosley and Dahl (1988) concluded that flexible rotation (grazing periods of less than 1 and up to 14 days) had no forage or animal advantages over fixed 7-day grazing periods; flexible rotation did not improve herbage crude protein levels, herbage digestibility, herbage moisture content, amount of available herbage, or animal gains (Mosley and Dahl, 1990). The use of flexible rotation (1–13-day grazing periods) compared to fixed 7-day grazing periods actually reduced heifer gains at the Sonora Research Station in spite of the intensive management it required (Taylor, 1988). When heifers and sheep were grazed together in a related study, flexible rotation did not significantly improve grazing distribution or enhance forage harvest efficiency (Taylor et al., 1993). Although not compared directly with fixed grazing periods, time-controlled grazing on mixed grass prairie near Cheyenne, WY, failed to improve average daily gain of steers, plant vigor or range condition, or forage production when compared to continuous grazing and rotational deferment (Manley et al., 1997).

The success of SDG may depend, in part, on the ability and desire of management personnel to properly manipulate grazing animals (Kothmann, 1984; Heitschmidt, 1986); it may require even a change in rancher lifestyle to cope with the need for making daily decisions and actions to ensure proper stocking control and needed changes (Quigley, 1987). Under time-controlled procedures, in particular, the manager must continually monitor forage growth and quality, forage utilization, rainfall and moisture conditions, and animal performance. The opportunities for mismanagement under intensively managed systems are much greater than under extensively managed systems, and even the best management cannot compensate fully for severe drought or setting excessive initial stocking rates (Heitschmidt, 1986). Attention must also be given to the fact that, because of high animal densities, larger quantities of water are required per grazing unit at any given time.

Livestock moved often, especially if forcefully moved and in poor fencing layouts, will be more likely to be stressed and suffer reduced individual performance. Forcefully moving livestock into a new grazing unit is generally undesirable and unnecessary, regardless of the grazing system being employed. Training them to move themselves in response to some signal that a new grazing unit is open will most likely accomplish the move (Savory, 1987). Any remaining stragglers can be drifted into the new paddock or can be intercepted by temporarily closing access to water where they will accumulate and can then be picked up. Care must be exercised that offspring are not separated from their dams during moves between paddocks; leaving the gate open between the old and new paddock for a couple of days will permit dams to go back for their offspring if separated (Kelton, 1982).

Experienced livestock are commonly found waiting at the gate when the time approaches to move into a new unit. Balph and Balph (1986) suggested basing moves to a new grazing unit on reaching a predeterminded, desired forage level rather than a predetermined time. They anticipated this might have training value with grazing animals to expedite making the moves as well as value in circumventing a decrease in foraging in anticipation by the livestock of the move.

SDG is not especially adaptable to small rangeland areas (Gray et al., 1982). A grazing system size of not less than 640 acres of rangeland but not more than 500 cows has been suggested (Kirby and Bultsma, 1984). A one-herd system such as short-duration grazing does not permit, except through associating several cell systems, maintaining the number of herds required for an intensive registered breeding or a cross-breeding program (Merrill, 1980). Nevertheless, combining cattle into a single, large herd may provide advantages of cattle being more docile and having fewer herds to look after (Bryant et al., 1988).

The wagon-wheel arrangement of paddocks in a **grazing cell** is often associated with SDG but is not required for the system (Figures 15.7 and 15.8). Advantages of the cell arrangement include providing a centralized watering source, equal access from all pastures, labor conservation, and the ease of handling livestock (Kirby and Bultsma, 1984). However, a continuing problem is that high livestock density and narrow triangular shape of the paddock cause excessive utilization and increased number and density of trails near the hub, particularly when water and supplement as well as access gates are all located at the hub.

Compared to continuous grazing and the Merrill system, trail density was increased four times under SDG at the Texas Experimental Ranch and was disproportionately concentrated near the cell center of the wagon-wheel design. The sac-

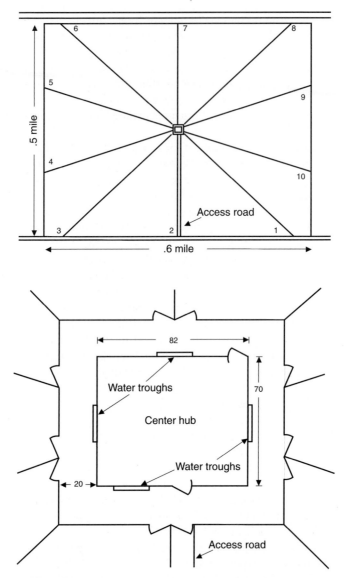

FIGURE 15.7 Diagram of cell system comprised of ten paddocks and a center hub, USU Tintic Valley Research Area near Eureka, UT (Malechek and Dwyer, 1983).

rifice area contained about 6% of the area (Walker and Heitschmidt, 1986b,c). This concentration of livestock is particularly serious in semi-arid and arid areas and on highly erosive soils. Thus, the wagon-wheel design appears more adapted to humid and sub-humid climates with gentle topography and uniform vegetation characteristics (Skovlin, 1987).

FIGURE 15.8   Photos of cell system at USU Tintic Valley Research Area, Eureka, UT: (A) corridor around the center hub into which all paddocks connect; (B) water troughs provided along the dividing fence between the corridor and the center hub livestock working area.

A fan design is similar to a wagon-wheel design but locates the hub on or near the perimeter rather than in the center of the paddock cluster (Smith *et al.,* 1986); it requires slightly more fencing but may fit better in a given situation. Block designs utilizing rectangular paddocks may be as efficient as the wagon-wheel design while reducing the narrow points at which excessive trampling is common (Voisin, 1959). Walker and Heitschmidt (1986a) calculated that using rectangular units and providing a 50-foot wide corridor in the middle to connect all paddocks at the Texas Experimental ranch rather than the wagon-wheel design would have required 17% less fence and the corridor would have contained less than 1% of the multi-unit system.

Ecological considerations may provide more natural subdivision by range types and meet their need for different management (Skovlin, 1987). Whatever the design, any centralized facility should be located on a firm, level site, and sandy, light-textured soils should be avoided unless a gravel blanket or other non-erosive ground cover is provided (Kirby and Bultsma, 1984). Minimizing shade at the central facility and moving stockwater and routine supplementation out into the pastures are suggested to discourage livestock from loafing at the center facility. Focusing on similar grazing capacity rather than uniform size in laying out the paddocks will solve many management problems and result in smoother operation of the grazing system. If paddocks within the grazing system are of substantially different size and grazing capacity, adjustments in the length of the grazing period in each paddock will be required.

Catch pens adjoining irrigated or mesic improved pasture units are sometimes employed to reduce treading impact on soil and plants. Five or six equal pie-shaped pasture units using electric fence and a catch pen about 200 feet in diameter at the pivot for supplemental feeding and watering are commonly utilized with center-pivot irrigation systems. When pasturing a unit with rectangular subdivisions under rotation, one approach is to subdivide into six equal paddocks with a catch pen at one end (about 30 feet wide and as long as the width of the four inside grazing units) (Corah and Bartley, 1985). Restricting alleyway usage to necessary animal movements will help minimize negative effects on alleyways such as erosion development and excess waste nutrient deposition (Gerrish, 1999).

Except on well-drained, sandy soils, irrigation water should generally be applied after livestock have been rotated from the pasture unit to prevent excessive trampling damage and then as needed through the regrowth period. For example, an irrigation schedule utilizing short-duration grazing (eight-unit, 3-day grazing, 21-day nongrazing) at the Sierra Foothill Range Field Station in California (Raguse et al., 1989) was required to meet the following criteria: (1) no irrigation while livestock were in the paddock, (2) no irrigation less than 3 days before livestock entry, (3) the non-irrigated interval not to exceed 7 days, and (4) a grazed paddock irrigated the evening of the same day livestock exited.

B. THE RESULTS ON IMPROVED PASTURE

Short-duration grazing, based on current-day nomenclature, is the rotation grazing system commonly used on mesic or irrigated pasture. Although there is no general agreement on the superiority of SDG over continuous grazing on improved

FIGURE 15.9   Although there is no complete consensus as to its relative merits when used on improved pasture, short-duration grazing is considered advantageous to the overall pasture productivity compared to continuous grazing; photo shows irrigated grass-legume pasture at the University of Nebraska North Platte Station.

pasture, there is some evidence that it is advantageous to overall pasture productivity (Nichols and Clanton, 1985; Popp *et al.,* 1997) (Fig. 15.9).

More uniform grazing, reducing selectivity between species and plant parts, maintaining alfalfa and upright clovers in the stand, allowing slightly heavier grazing without damage to the forage plants, and reducing bloat when legumes are in the plant mixture are often reported as advantages of SDG over continuous grazing on improved pasture. Since grazing animals are encouraged to eat coarse, more mature forage along with the bloat-causing immature, leafy portions of legumes, bloat is less likely to occur under short-duration grazing or strip grazing than under continuous grazing. Moving livestock to the next pasture in rotation when approximately 30–40% of the forage remains should both minimize excessive grazing pressure late in the grazing period while maintaining sufficient leaf area to keep the plants productive (Nichols and Clanton, 1985).

According to Matches and Burns (1985), three major advantages of rotation grazing over continuous grazing on improved pasture are (1) improved plant persistence, (2) opportunities to conserve (mechanically harvest) surplus forage, and (3) more timely thus more efficient utilization of forage. With rotation grazing, excess forage can be harvested as hay or silage for feeding during periods of low forage production; losses due to herbage trampling, fouling, and senescence are reduced by more timely utilization. On the other hand, continuous grazing has the advantage of lower input costs such as fencing and water facilities; also, management decisions are simplified because livestock are not being managed using high density and restricted area which require frequent moves from grazing unit to grazing unit (Matches and Burns, 1985).

When applied to improved pasture, SDG has sometimes increased grazing capacity and thus animal production per acre over continuous grazing but not per-head animal performance with cattle (Blaser *et al.,* 1959; Bransby, 1991, Bertelsen *et al.,* 1993) or with sheep (Thomas *et al.,* 1995). Per-head and per-acre gains of yearling beef steers on irrigated alfalfa-orchardgrass pastures in Washington were nearly the same under equal stocking rates (Heinemann, 1970); when stocking rates were increased in the rotation pastures to utilize the additional forage available, liveweight gains per acre increased 20% but gains per head remained similar to that under continuous grazing.

When designing a SDG study on irrigated pasture in California, Raguse *et al.* (1989) contrasted two utilization management approaches: (1) high accumulation-moderate utilization and (2) low accumulation-high utilization. Plant height was used as the basis of the treatments: at entry, 10–12 and 6–8 in., respectively; at exit, 4–6 and 3–4 in., respectively. Based on first-year data only, forage regrowth behavior and average daily gain per head were similar for both treatments but high accumulation-moderate utilization reduced animal gains per acre, the latter resulting from apparent lower transfer efficiency and higher fouling and trampling losses.

C. THE RESULTS ON RANGELANDS: PLANT AND SOIL RESPONSE

On rangelands, SDG is primarily adapted for use during the growing season, where the growing seasons are long, and on mesic sites where there is regrowth potential (Heady, 1984; Pieper and Heitschmidt, 1988). From mesic and sub-humid native and seeded ranges in the U.S., its use was subsequently expanded to more arid rangelands. However, the duration of optimum growing conditions on many rangelands tends to be relatively short, and by the time that leaf area has been replenished, optimum growing conditions may no longer exist (Reece, 1986). Its use during periods of dormancy when no new plant growth occurs is of doubtful advantage to the vegetation.

Short-duration grazing may increase the efficiency of forage harvesting on rangelands by applying a more uniform frequency and intensity of grazing (Hinant and Kothmann, 1986; Lundgren *et al.,* 1984). High stocking densities in a post oak savannah in Texas increased the percentage of leaves and tillers of the desirable grasses defoliated, but an inefficient harvest of available tillers remained (Briske and Stuth, 1982, 1986). Coughenour (1991) suggested that any increased grazing capacity resulting from SDG was likely a result of greater herd density leading to the fuller use of available tillers rather than a result of altered duration of grazing.

Nonuniform, patchy usage of tobosa grass in southern New Mexico has resulted from continuous grazing by cattle, leaving large amounts of forage ungrazed

(Senock *et al.,* 1993). While the percentage of tobosa tillers defoliated under continuous grazing was always less than 30%, percentage defoliation under SDG (11-paddock, 6-day average grazing periods) was always greater than 75%. While selective grazing may be reduced by high stocking density, such as where patch grazing or wolf plants are likely, achieving this only by increasing stocking rates frequently results in unstable situations (Kothmann, 1984). However, when the standing crop contains palatable as well as extremely unpalatable species, SDG alone at acceptable stocking rates cannot be expected to solve the problem of selective grazing (Heady, 1974; Araujo and Stuth, 1986). Cumulative grazing pressures resulting in maximizing nonselective grazing readily leads to a drop in dietary quality and can even shift livestock to the consumption of poisonous plants of otherwise low palatability.

SDG on rangeland is apt to be more effective in enhancing the competitive ability of the desirable forage plants than by forcing defoliation of the unpalatable species (Heitschmidt, 1984b). Because of the competitive advantage that can be afforded the more desirable plant species, this can be important in either maintaining or improving range condition (Heitschmidt, 1986). Herbel and Pieper (1991) concluded that even if SDG can result in improving range condition under proper stocking rates, return to continuous grazing once the desired range condition had been achieved should give the highest animal production.

SDG has the potential to distribute grazing animals more uniformly over the grazing unit and to utilize a greater proportion of the plant species in the standing crop (Malechek and Dwyer, 1983; Kothmann, 1980). Some improvement in grazing distribution and uniformity of use and thus grazing capacity was reported from spring-fall grazing of crested wheatgrass by cattle under short-duration grazing in Idaho (Sanders *et al.,* 1986). However, on mixed grass prairie in Texas (Walker *et al.,* 1987, 1989a; Scott *et al.,* 1995) and in North Dakota (Kirby *et al.,* 1986) SDG did not reduce site preference nor did it improve distribution of grazing by cattle.

Walker *et al.* (1989a) suggested that the grazing management strategy of forcing livestock to graze rangelands uniformly by using high grazing pressure may be ineffective; they proposed that grazing systems such as SDG may influence cattle preference for plant communities more by affecting the spatial availability of forage biomass than through increasing stocking density and grazing pressure per se. Bailey and Rittenhouse (1989) found no evidence that the relative consumption rate among patches or plant species in a grazing unit changed as a result of changes in animal density; that is, distribution of grazing was not immediately affected. However, observed changes in grazing patterns over time are more evident under high animal density because of higher rates of forage removal.

Where improved distribution of grazing does follow the initiation of SDG, the benefits may accrue from the cross-fencing and stockwater development as much as or more than from the grazing treatment per se. Short-duration grazing has not overridden the effects of distance from water on grazing utilization (Soltero *et al.,* 1989) or the attractiveness of watering points and shade trees as favored loafing sites (Walker *et al.,* 1989a). Reduced size of grazing units and closer distances to drinking water were concluded to be more important than rotation in determining either cattle gains or uniformity of utilization on shortgrass range (Hart *et al.,* 1988c).

Savory (1983) has proposed that, "As a general rule, the conventional or government-prescribed stocking rates can safely be doubled in the first year of operation [when converting to SDG] as time control is brought into the grazing handling." However, there seems to be no evidence that SDG regularly increases forage production in the short run, and in the long term increases presumably could result only through improving range condition (Bryant *et al.,* 1988; Heitschmidt, 1986; Heitschmidt *et al.,* 1986b; Pieper and Heitschmidt, 1988; Taylor, 1988; Anderson, 1988). While agreeing with the latter conclusion, McCollum and Bidwell (1994) suggested that SDG in tallgrass prairie may prevent further deterioration at higher than optimal stocking rates. During a 6-year grazing study on Nebraska tallgrass range comparing three stocking rates and three grazing systems, cattle gains per acre during the first 2 years were greater under continuous grazing, but gains per acre under SDG produced progressively higher gains per acre during the last 4 years of the study, especially as stocking rates increased (Anderson, 1999).

However, in a 10-year study at Sonora, TX, Taylor *et al.* (1997a) found that stocking rates of 1× (the recommended rate), 1.5×, 2×, or 2.5× under SDG all failed to sustain initial vegetation composition and prevent sideoats grama and other midgrasses from decreasing and curly mesquite from increasing. On crested wheatgrass-big sagebrush range near Burns, OR, Angell (1997) compared continuous grazing at a conventional stocking rate (1.5 AUM per acre) for 45 days in the spring with SDG at low (1.5 AUM per acre), medium (2.25 AUM per acre), and high (3.0 AUM per acre) stocking rates using two 3-day grazing periods per paddock separated by 27-day rest periods. Increasing stocking rate 50–100% greatly increased sagebrush seedlings and reduced crested wheatgrass tiller density; at the conventional stocking rates, both grazing techniques resulted in similar levels of crested wheatgrass tillering and yield and in big sagebrush density.

Thus, range scientists generally agree that any large increase in grazing capacity under SDG must generally result from a correction of a previous grazing inefficiency, such as a major distribution problem, substantial understocking, or undergrazing a substantial component of edible forage plants (Blackburn, 1983; Heitschmidt, 1986; Dahl *et al.,* 1992; Lewis and Volesky, 1988; Walker and Heitschmidt, 1986a), and any opportunity to double stocking rates will be quite rare. Based on an intensive review of research and practice with SDG in Africa, Skovlin (1987) concluded that claims of range improvement at double conventional stocking rates were not founded in fact.

It should be noted that when a grazing program is changed from continuous moderate to SDG heavy that any short-term increase in grazing capacity derived will result primarily from the heavier stocking rate rather than the change in grazing system per se. Such a practice might be supported for a few years but is apt eventually to lead to instability in the system and subsequent loss of grazing capacity through the effects of heavy grazing on the more productive grasses and reserve forages (Ralphs *et al.,* 1984b). There is no evidence that any grazing system can successfully counteract the deleterious effects of overstocking on a long-term basis (Pieper and Heitschmidt, 1988)

An increase in stocking rates of 20–30% from greater grazing efficiency under SDG may be possible without causing a decline in range condition in some situations, but Bryant *et al.* (1988) suggested that testimonials of increasing grazing capacity greatly in excess of this should be accepted with caution, and no increase in stocking rates should be made until range improvement is evident or greater grazing efficiency is clearly demonstrated. On mixed grass prairie in west central South Dakota grazed mid-May to mid-September, Volesky *et al.* (1990) compared SDG with continuous grazing using a mixture of heifer calves and ewe lambs. By basing stocking rates on end-of-season standing crop forage levels, they concluded that SDG permitted a modest increase in stocking rate (up to 25%). On mixed grass range in North Dakota, Kirby *et al.* (1996) reported being able to maintain range condition under SDG while increasing stocking rates up to 40% over continuous grazing.

Volesky *et al.* (1990) observed in South Dakota that mixing kinds of livestock in continuous grazing may permit a small stocking rate increase similar to SDG. However, Taylor *et al.* (1986b) cautioned against any expectation of increased grazing capacity in Texas due to moving from continuous grazing to SDG where the existing operation was already carrying a mixture of cattle, sheep, and goats and water was well distributed.

There is no evidence that SDG causes any hydrologic benefit from livestock trampling or "hoof action," infiltration rates mostly being reduced immediately after trampling, or seedling establishment in existing vegetation stands (Taylor *et al.,* 1993b; Balph and Malechek, 1985). (Refer to Chapter 5 for a more detailed discussion of SDG effects on soil.)

D. THE RESULTS ON RANGELANDS: ANIMAL RESPONSE

Cattle grazing crested wheatgrass under SDG were noted to be more dispersed and tended to graze less in synchrony; such behavior has the potential to promote more uniform use of the grazing unit (Balph and Balph, 1986). This grazing behavior was attributed to too many animals to form a single social unit and too little space for social units to be independent of one another. However, no significant differences in uniformity of utilization of crested wheatgrass in Utah were demonstrated with paddock size varying from 2.5 acres to 20 acres with herd sizes ranging from 3 to 24 head (Hacker *et al.,* 1988).

It was reported by Gammon and Roberts (1980a) that six-pasture SDG compared to continuous grazing increased daily travel distance, increased frequency of visits to each area but reduced the grazing time there, and increased frequency of drinking. Increased daily travel distance was also noted in Texas studies (Heitschmidt, 1986). From a review of grazing literature, Krysl and Hess (1993) concluded that as forage becomes limiting under high-density stocking, animals appear to reduce

TABLE 15.10   Average Cow–Calf Performance and Economic Returns at the Texas Experimental Ranch, Vernon, TX, 1982–1985, under Yearlong Grazing[a]

	Grazing treatments			
Item	Merrill moderate[b]	Continuous moderate	Continuous heavy	Short-duration heavy[c]
Acres/cow/year	14.5	14.5	12.5	8.9
Conception rate (%)	95	92	90	90
Calf crop weaned (%)	84	83	79	78
Average weaning weight (lb)	595	577	584	552
Production/cow (lb)	504	475	464	430
Production/acre (lb)	33	32	37	44
Net returns/cow ($)[d]	131.24	133.59	112.81	69.66
Net returns/acre ($)[d]	9.05	9.73	9.21	7.60

[a]From Heitschmidt (1986).
[b]Referred to as deferred-rotation grazing in the study.
[c]16-paddock, one-herd, cell-designed, flexible grazing periods (18 hours to 6 days) and nongrazing periods (30 to 65 days); referred to as rotation-grazing in the study.
[d]Over annual variable, ownership, and investment costs for cattle.

their grazing time, this probably a conditioned response of animals to the difficulty of prehension or anticipation of being moved to a new pasture.

SDG has failed to consistently improve individual animal performance compared to less intensive systems, particularly on arid and semi-arid rangelands (Heitschmidt, 1986; Pieper and Heitschmidt, 1988; Merrill, 1980). Gains of beef heifers were compared under different grazing systems during a 3-year study at the Sonora Research Station in Texas (Taylor, 1988). The average 12-month gains were as follows: SDG (7-day grazing period, 49-day cycle), 244 lb; HILF (14-day grazing period, 98-day cycle), 248 lb; and Merrill system, 244 lb. A fourth treatment, Merrill system plus brush control, produced 12-month gains of 302 lb, indicating that brush control had much greater effect on cattle gains than did grazing systems per se under the moderate stocking rates used in the study. Pregnancy rates of heifers grazing crested wheatgrass in Utah were lower under SDG than under continuous spring grazing (Olson and Malechek, 1988); the high density of animals or some other management practice may have interfered with breeding.

A comparison of animal performance and economic returns under four grazing treatments at the Texas Experimental Ranch, Vernon, are presented in Table 15.10. Production per cow was least under SDG because of lower conception rates, weaned calf crops, and calf weaning weights (Fig. 15.10). Because of the heavier stocking rate applied, SDG had the highest beef production per acre but still the lowest net return per acre and per head; it required larger per acre outlays for operating and cattle investment capital costs. Since the development and application

FIGURE 15.10 When compared with three other grazing treatments at the Texas Experimental Ranch, Vernon (typical range scene shown above), short-duration grazing at a heavy stocking rate produced the most beef per acre but was least profitable on both a per-cow and a per-acre basis (Heitschmidt, 1986). (Texas Agricultural Experiment Station photo by Rodney K. Heitschmidt.)

of SDG requires additional investment and management, increased livestock production is required to recover these greater expenses (Kothmann *et al.*, 1986b). Holechek *et al.* (2000), from a review of limited grazing studies in the Great Plains, concluded that short-duration grazing had no financial advantage over season-long continuous grazing.

SDG did not increase gain per head or per acre of sheep and goats at Sonora, TX (Taylor *et al.*, 1986b). Gains per head of yearling steers grazed during the 4-month summer season on shortgrass range near Cheyenne, WY, were the same (2.09 lb daily) under continuous, four-pasture rotational deferment, and eight-pasture short-duration (Hart, 1986); however, prior to making a change from rigid to flexible scheduling of rotations in the study, animal performance had suffered under SDG. No differences between systems were found in peak standing crop or in botanical composition or utilization of the standing crop. Eight-year average daily gains of stocker yearling steers on sand shinnery oak range in northern Texas under a moderate stocking rate (6.7 acres per steer) were 2.46, 2.01, and 1.61 lb daily under continuous, four-pasture rotation, and eight-pasture SDG, respectively (Dahl *et al.*, 1987). Forage species composition and forage production among

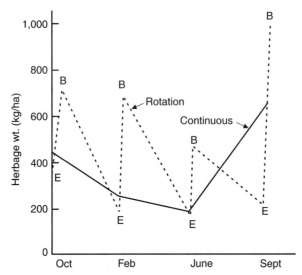

FIGURE 15.11   Herbage available under continuous and rotational grazing in New Mexico. B indicates herbage standing crop when cattle were shifted to an ungrazed pasture, and E indicates herbage standing crop when cattle left the pasture (Pieper, 1980).

treatments were similar. On crested wheatgrass in southern Idaho grazed during spring and fall, average daily gain of yearling cattle was similar between continuous and SDG even though stocking rates were gradually increased during the 5-year study under SDG to utilize the apparent greater grazing capacity (Sanders *et al.,* 1986).

Allison *et al.* (1983) compared daily cattle intake rates on Texas grasslands during 14-day, high-density grazing periods. Forage intake under very heavy and heavy stocking rates (22 and 44 lb per AUD cumulative forage allowance) decreased 22% and 14%, respectively, as herbage availability decreased near the end of each 14-day grazing period but remained relatively constant under the lighter grazing pressures. Minimal shifts occurred in botanical composition of cattle diets in the Texas studies from the first to the last day in each 7-day grazing period under SDG (Walker *et al.,* 1989b), which was attributed to minimal palatability differences among the plant species on offer. Mosley and Dahl (1989) reported that 7-day grazing periods were sufficiently short on tobosagrass range to prevent detectable changes in either diet botanical composition or forage availability or declines in diet nutrient quality.

SDG on rangeland, much the same as on improved pasture, greatly complicates the forage supply-demand situation for grazing animals. Diet selection is complicated by a relatively rapid rate of change in grazing pressure during the brief grazing period (Malechek, 1984; Dwyer *et al.,* 1984). Availability of forage at the beginning of each grazing period is at a high point when livestock enter a new pasture but declines rapidly to a low point when leaving, resulting in a sawtooth pattern of herbage availability when graphed (Fig. 15.11). Since forage availability in a SDG paddock is expected to decline as the grazing period progresses, managers must carefully monitor forage conditions and rotate animals to the next paddock before dietary intake becomes too low to maintain animal performance at a level similar to continuous grazing (Rittenhouse and Bailey, 1996).

The belief continues that under high animal density reducing the length of grazing periods to less than weekly may be beneficial in maintaining higher levels of selection for more nutritious forage and nutrient intake towards the end of the grazing period (Smith *et al.*, 1986). However, the quality and composition of cattle diets on rolling grasslands at the Texas Experimental Ranch were found to be no different between the first and last day of the grazing periods in the SDG paddocks (Walker *et al.*, 1989b). Increasing grazing pressure nearing the end of the 4-day grazing periods in a short-duration system at the Sonora Research Station had little effect on nutrient intake, diet selection, or dry-matter intake by Angora goats (Taylor and Kothmann, 1990).

A slight but nonsignificant trend of decreasing crude protein and dry-matter digestibility from the initiation to the termination of each grazing period was found by Kirby and Parman (1986) on North Dakota grasslands. However, in a continuation of grazing studies in this same area, Hirschfeld *et al.* (1996) found higher nitrogen content and forage digestibility in livestock diets under SDG than under continuous grazing, particularly late in the growing season. They attributed this difference to greater availability of cool-season grasses and sedges near the end of the season under SDG.

However, Ralphs *et al.* (1986b) reported significant reductions in quality of sheep diets between the first and last day of 3-day grazing periods and an even greater reduction with cattle. Also, the diet quality of beef heifers grazing crested wheatgrass during the spring declined significantly during even short 2- or 3-day grazing periods under SDG; decline in forage intake was also found but was less consistent (Olson *et al.*, 1989) Based on these findings, these researchers suggested that a proper rate of rotation was important and that grazing periods of no more than 2 days might maintain higher levels of nutrient intake than longer grazing periods. Nevertheless, advantages of reducing the length of grazing periods beyond 7 days will have to be large to justify the additional pasture units and management intensity required.

SDG has been designed to provide higher quality and availability of forage for improved animal performance over conventional rotation grazing (i.e., HILF), but it seems doubtful that SDG can consistently improve nutrient intake over continuous grazing. SDG applied to Texas grasslands did not consistently enhance diet quality or nutrient intake; it did appear to increase nutrient intake during rapid vegetation growth but not subsequently (Heitschmidt, 1986). Increasing the number of paddocks from 14–42 did not enhance nutrient intake in Texas studies (McKown *et al.*, 1991). When steers were grazed from late April to late September during a 6-year study on tallgrass prairie in Oklahoma, SDG compared to continuous grazing reduced both intake (11–20%) and diet nutrient composition, thereby explaining the lower weight gains and higher end-of-season residual standing vegetation when compared with continuous grazing (McCollum and Gillen, 1998; McCollum *et al.,* 1999). It was concluded that unless variable costs per steer could be reduced by 24–34% under SDG or the decline in gain per head reduced or eliminated, there was no economic incentive to implement SDG.

Stuth and Olson (1986) suggested that SDG systems be designed so that strategic access to nutritionally critical habitats is enabled, particularly late in the growing season. In spring grazing studies on crested wheatgrass in Utah, no differences in average diet quality of beef heifers were detected between SDG and continuous grazing, and SDG failed to extend the season of nutritious forage (Olson and Malechek, 1988).

After noting that short-duration grazing has failed to achieve many of the projected benefits expected of it, Holechek *et al.* (2000) concluded with the following: "short-duration grazing can facilitate improved management of livestock, and it gives ranchers more control over how specific parts of their ranch are grazed than continuous grazing. We believe it can be a useful grazing system for some ranchers if applied at conservative to moderate stocking rates."

VII. GRAZING SYSTEM EFFECTS ON GAME ANIMALS

Grazing systems can be expected to have varying effects on game animals depending not only on the animal species but also on the grazing system and under what stocking rates they are applied. Klebenow (1980) concluded that rotational grazing systems favor big game animals if they maintain or provide habitat diversity and adequate interspersion of food and cover. Payne and Bryant (1994) concluded that grazing systems are generally better for wildlife than continuous grazing but noted that the requirement of additional fencing may impede large animal movements. Brown (1986) noted that high fences provided for livestock control or predator exclusion could have serious effects on deer survival during drought. However, types of fences are available that control selected livestock species but do not materially affect large wild herbivores movements (Vallentine, 1989). Providing access to drinking water in each paddock or grazing unit other than at high livestock concentration areas such at cell centers is also recommended.

Elk, deer, and antelope frequent areas not currently being grazed by domestic livestock. Klebenow (1980) and many others consider this phenomenon to be partly a response away from high livestock density and partly in favor of the green growth in units not currently being frequented by livestock. (Refer to Chapter 10, "Interspecific Sociality," for further explanation and discussion.) Periodic nongrazing of domestic livestock during rest or deferment periods within grazing systems are attractive to wild herbivores because of greater forage selectivity and reduced competition for succulent regrowth when growing conditions are favorable.

Brown (1986) noted that white-tailed deer vacated each paddock under high-density grazing as livestock entered, but they returned shortly after cattle vacated, apparently in response to increased deer food availability following regrowth. In southeast Texas, white-tailed deer traveled 35% more in SDG than under continuous grazing in the summer, avoided SDG paddocks where cattle were present in the spring and early summer, but seemed to alternate between preferred habitats rather than following predictable paddock-to-paddock movements (Cohen *et al.,* 1989).

Heavy stocking rates combined with high livestock density in HILF and SDG systems are capable of damaging plants that are most valuable to white-tailed deer, including browse and forbs (Teer, 1996). While SDG and HILF may impact browse plants present in low density because of inadequate recovery periods provided following defoliation, these high-density grazing systems may be required to exert adequate control over brush species present in dense stands. It was suggested that forbs and shrubs should be monitored in conjunction with intensive grazing systems because of their primary concern to wildlife (Teer, 1985).

Ortega *et al.* (1997) concluded from work on mixed vegetation in the Coastal Bend region of southeast Texas that on high seral areas a continuous grazing system for cattle and white-tailed deer under moderate stocking rates may be the best solution because of fewer management decisions being required. The Merrill grazing system was considered acceptable to acheive range improvement goals; both continuous grazing and the Merrill system, a less intensive grazing system than SDG, induced somewhat greater forb consumption than SDG. This was in general agreement with Drawe *et al.* (1988) who had recommended moderate continuous grazing for livestock or moderate four-unit deferred-rotation grazing (Merrill system) with a variable livestock stocking rate which was increased in times of high forage production and decreased during times of drought.

From a 20-year study on the Edwards Plateau of Texas, it was concluded that the Merrill system produced the most desirable habitat for both livestock and wildlife; it increased the amount of decreaser plants over either livestock exclusion or deer/livestock exclusion (Reardon and Merrill, 1976). The Merrill system was recommended in Texas where concurrent production of white-tailed deer, bobwhites, and wild turkeys are desired (Bryant *et al.,* 1981a).

Rest-rotation grazing and deferred-rotation grazing are often preferred by big game managers in the Intermountain and Rocky Mountain West (Payne and Bryant, 1994; Wisdom and Thomas, 1996). Austin *et al.* (1983) suggested that resting one-third of foothill range units each year may provide a better grass-browse species mix for mule deer. Grazing systems that concentrate livestock affect mule deer distributions and diets, but if given the opportunity deer are apparently able to redistribute themselves within the same grazing unit or adjacent grazing unit to access available green regrowth (Peek and Krausman, 1996). Most grazing systems designed for livestock appear acceptable for pronghorns provided water is well distributed, fences enable passage by pronghorns while containing livestock, and ready access by pronghorns to preferred forbs and/or shrubs is enabled (Yoakum *et al.,* 1996).

Livestock management programs using specialized rotational grazing systems are usually not designed to accommodate significant increases of potentially competing wild herbivores. If the regrowth of vegetation on the areas recently vacated by livestock attract high numbers of wild herbivores, the benefits desired from livestock non-grazing could be negated and the intent of the grazing system thwarted (Dwyer *et al.*, 1984). For example, large herds of elk could quickly negate the anticipated benefits of deferred grazing or rest on mountain or foothill rangelands (Powell *et al.*, 1986). The concentration of elk in rested grazing units has been widely observed in Montana (Fristina, 1992), south central Utah (Werner and Urness, 1998), and eastern Idaho (Yeo *et al.*, 1993). Thus, it is imperative that elk as well as cattle populations be kept under adequate control from the standpoint of both numbers and distribution.

The effects of rotational grazing systems on bobwhite quail and wild turkey nesting were studied in Texas (Bareiss *et al.*, 1986). Coverage, density, and dispersion of suitable nest sites and loss rates of artificial nests were not affected by grazing treatment (e.g., SDG vs. continuous grazing). This confirmed that SDG did not increase trampling losses of ground nests. This was similar to the findings of Guthery *et al.* (1990) that SDG and continuous grazing had similar impacts on quail and turkey at similar stocking rates. Schulz and Guthery (1987) found that SDG had no negative effects on ideal wild turkey habitat—moderate brush cover, good soil depth and annual precipitation, and long growing season—but theorized that the high density used in SDG might be detrimental as sometimes reported in less ideal habitat.

SDG apparently favors quail by increasing herb species richness, diminishing plant litter accumulations, and providing habitat favored by quail during the fall and winter (Wilkins and Swan, 1992); by increasing bare ground and decreasing the height and coverage of grasses (Guthery *et al.*, 1990); and by increasing quail density associated with improved structure of herbaceous cover near the ground and the soil disturbance and trail forming associated with high stocking density (Schultz and Guthery, 1988). However, continuous grazing at light to moderate stocking rates has also been observed to create patchy environments favorable to upland game birds (Payne and Bryant, 1994).

Waterfowl in wetlands may be adversely affected by heavy grazing or high livestock density during breeding, nesting, and brood-rearing periods. Waterfowl in the northern Great Plains, in terms of breeding pairs, broods, and brood production, responded favorably to rest and deferred grazing treatments the previous year (Mundinger, 1976). However, responses were negative to heavy grazing pressure during the late summer and fall of the previous year and to spring grazing during the current season. The positive influences of special grazing systems appear to result from increased residual vegetation around the shoreline and maintaining desirable vegetation structure. However, Sedivec *et al.* (1990) found that duck nesting density was highest under nongrazing when compared with SDG and switchback grazing but percentage nesting success was lowest because of less safe nesting cover; total successful duck nests per unit area were similar across all three treatments during the 8-year study when grazing pressures were moderate.

16

GRAZING AS AN ENVIRONMENTAL TOOL

I. MANIPULATIVE BENEFITS OF ANIMAL GRAZING

It is widely recognized that grazing is a means of harvesting grazing resources, but it is less commonly known or acknowledged that grazing can also improve the forage resource. Attention to past and potential adverse effects of herbivory on native plant communities has tended to overshadow the positive influences of prescribed grazing on the forage and soil resources. However, for these benefits to be realized, or at least maximized, requires optimum grazing management in all of its aspects.

"Dual objective grazing management"—grazing for environmental enhancement while converting the forage crop—offers promise as a strategy for large animal herbivory in the future and could serve as the basis of grazing programs on both private and public grazing lands. Walker (1995) has concluded that dual objective grazing offers the solution to many public land grazing controversies and could legitimize the continuation of livestock grazing on such lands. He has projected that grazing as a tool for environmental enhancement of public grazing lands in the future will acquire priority equal to or greater than that for grazing as a means of forage conversion. However, it is seldom acceptable to relegate the performance of the livestock manipulator species to that of a secondary objective; livestock owners cannot be expected to utilize their animals to enhance wildlife habitat or other vegetation manipulation if livestock performance will be substantially reduced unless otherwise reimbursed for such under-performance (Mosley, 1994).

The benefits of large herbivore grazing—those natural effects of herbivory often capable of being maximized into manipulative tool effects—are summarized as follows:

1. Delay maturation and hold plants in a vegetative, forage producing stage (see Chapter 2).
2. Enhance nutritive value of available herbage by increasing the new growth:old growth ratio (Chapter 2).
3. Remove or prevent excessive thatch buildup, thereby enhancing attractiveness of the standing crop to grazing animals (Chapter 9).
4. Utilize selective grazing by one kind of grazing animal to enhance forage quantity, accessibility, palatability, and/or nutritive content for a another grazing animal species (see Chapter 10, also later in this chapter).
5. Stimulate compensatory growth or regrowth by pruning effect, such as tillering of wheat and other grasses or twig regrowth of bitterbrush (see Chapter 5).
6. Inoculate plant parts with saliva, which may stimulate plant growth (refer to a subsequent discussion of this theory).
7. Maintain optimal rather than excessive leaf area index in dense forage stands (see Chapter 5).
8. Reduce excess accumulations of standing dead vegetation and mulch that may chemically and physically inhibit new growth and delay soil warming in cool-climate areas.
9. Reduce excessive live foliage going into a harsh cold period and thereby reduce frost damage to plants.
10. Reduce plant water stress and conserve soil moisture by reducing transpiring surfaces, even though stomatal conductance per unit of leaf surface may be somewhat increased (Svejcar and Christiansen, 1987a).
11. Increase water yields by reducing transpiring surfaces and thus total evapotranspiration on a watershed.
12. Manipulate botanical plant composition through selective grazing for biological control of undesirable plants (refer to "Biological Plant Control by Grazing" later in this chapter).
13. Accelerate nutrient recycling in the ecosystem and make some nutrients more available (for example, nitrogen mineralization) (see Chapter 5).
14. Aid forage plant establishment through graze-out of resident vegetation, seed transport, and seed dispersal; provide hoof impact for microsite preparation and seed coverage; and suppress competing weeds during

plant establishment (refer to "Grazing To Aid Forage Plant Seeding" in this chapter).

15. Aid forest regeneration through selective grazing and suppressing competing understory vegetation (later in this chapter).

16. Manipulate flammable fine fuel by controlled cattle grazing (or sometimes sheep or horse grazing) for improved fire management: (a) use grazing to reduce flammable fine fuel levels in order to reduce the incidence and severity of wildfires and enhance the control of wildfires or (b) limit or exclude grazing for a growing season to accumulate fine fuel needed to carry prescribed burns (Pieper, 1994; Taylor, 1994).

17. Increase biodiversity at community and landscape levels through moderate grazing (see later in this chapter).

18. Provide trampling and puddling of the soil within a small reservoir basin prior to filling with water to reduce seepage losses.

Reardon *et al.* (1972) reported that the grazing animal causes plant growth stimulation by deposition of saliva on the plant tissue during grazing; they showed that animal saliva contained thiamine (Vitamin B_1) at concentrations previously reported to stimulate a growth response in plants. Johnston and Bailey (1972), taking saliva from the rumen of a fistulated cow rather than from the mouth, failed to find any stimulatory effect. In subsequent tests, Reardon *et al.* (1974) found a stimulatory response from hand application of thiamine but not saliva, but plants grazed by cattle sheep, or goats all had significantly taller foliage growth than mechanically clipped plants.

Additions of both thiamine and saliva in still later tests (Reardon and Merrill, 1978) resulted in significantly higher plant yields when grown in a sandy soil as opposed to those grown in a more fertile clay soil. They concluded that plants respond more to thiamine in the saliva under adverse conditions such as low soil fertility or excessive defoliation. Sideoats grama plants grown in sand and receiving bovine saliva, concentrated saliva, and thiamine applied directly to the plant yielded 79, 56, and 38% more dry matter, respectively, than control plants.

The benefits as well as even the concept and measurement of biodiversity remain uncertain and controversial. West (1993) notes that four basic reasons have been given by proponents on why biodiversity is important: (1) morality, (2) aesthetics, (3) economics, and (4) ecosystem services. Each of these benefit categories, however, remains somewhat intangible or even nebulous even in concept. While large animal herbivory continually affects ecosystem structure and function and can enhance plant species "richness" and even the retention of rare species, other natural phenomena such as climatic changes and biological invasions may soon overwhelm the influences of livestock in changing the biosphere (West, 1993).

Depending upon how livestock are managed, grazing can either increase or reduce biodiversity. Both ungrazed and heavily grazed areas are commonly less diverse in plant species than moderately grazed areas (Laycock *et al.*, 1996). At community and landscape levels, moderate grazing generally increases biodiversity because animals do not graze uniformly. Certain areas commonly remain ungrazed while others are grazed at various levels. This increases the patchiness of vegetation at the local level and produces a mosaic at the landscape level with an associated diversity in animal populations (Laycock *et al.,* 1996). Grazing at intermediate levels tends to increase overall plant species diversity by decreasing the competitive advantage of dominants and by creating gaps available for occupation by other plant species (Heitschmidt and Stuth, 1991).

II. BIOLOGICAL PLANT CONTROL BY GRAZING

Both common use and biological control of undesirable plants by grazing are based on selective grazing. In contrast to common use or mixed grazing (see Chapter 10), **biological control by grazing** is applied to achieve a desired directional succession in vegetation composition, leading to the elimination or substantial reduction of one or more undesirable or less desirable plant species. The vegetation can be manipulated to achieve a desired change by grazing a single kind of animal or an imbalanced mix of herbivore species but is effective only when the right kind or combination of grazing animals, season of grazing, and stocking density and rate are applied. Achieving the desired plant composition will often require a combination of plant control techniques, including but not limited to biological control by grazing, to reduce an undesirable noxious plant to minimal levels or help eliminate it.

In addition to achieving selective plant control, biological plant control by grazing often has advantages over other control methods as well, such as: (1) less effect on non-target plant species, (2) some natural fertility return, (3) reduced pesticide residues, (4) accepted as being environmentally friendly, (5) more sustained control, (6) lower direct costs, (7) conversion of weeds into animal protein (Popay and Field, 1996), and (8) creating positive net returns through added sales of meat and fiber (Walker, 1994). Since livestock can be used to meet desired levels of selective vegetation control or flammable fine fuel reduction, public or even private landowners may be willing to waive any grazing fees for these services or even pay for such grazing services in some situations to achieve the desired benefits (Davison, 1996). Using goats to selectively control brush species on California foothill brushlands or New England farmlands or subdivision property or cattle to reduce fine fuel loads on planted greenstrips and adjoining annual grass-sagebrush range in central Idaho are examples of potential livestock fees for services rendered (Davison, 1996).

Biological control by grazing is not universally effective; it requires the right combination of plant composition, animal acceptance of the offending plants, and season, intensity, and duration of grazing. Attempts to increase consumption of the targeted plant may be limited by inherent anatomical or morphological constraints in the animal, lack of experience with the plant, lack of an appropriate mentor animal, the presence of a strong anti-palatability agent in the plant, or slow adjustment of rumen microbial populations; feeding an anti-toxicant to absorb or bind plant allelochemicals present in many noxious plants has potential but may be unknown or unavailable (Olson, 1999). The availability of the targeted plant may also influence how much of the plant is grazed (Olson, 1999). If uncommon, individual plants may be heavily grazed as a novelty item but with minimal consumption per animal. If the plant comprises dense infestations and is common rather than novel, light utilization may result from animals seeking out more palatable alternatives. High consumption of the noxious plant per animal may also have negative effects on the animal.

To be most effective, biological control by grazing must often be combined in "integrated weed control" with conventional bio-control agents (e.g., insects or pathogens), herbicides, or mechanical or fire treatments. For example, effective control of large woody plant species often requires that they first be brought down within reach of the grazing animal manipulators by mechanical or chemical means or by burning before adequate grazing pressure can be achieved. While sheep and goats are principally used in grazing for biological control of broad-leaved plants in the U.S., camels have been promoted for shrub reduction in the desert Southwest with potential for economical brush utilization (Lewis and Volesky, 1988), but only minimal use of camels there has resulted.

Livestock grazing was determined to be ineffective in changing the within-year production or composition of California annual grasslands (Bartolome and McClaran, 1992). However, Kie and Loft (1990) found livestock grazing in late winter and spring useful in reducing the abundance of coarse annual grasses such as ripgut brome (*Bromus diandrus*) and wild oats (*Avena* spp.); this favored the growth of low-stature, spring-maturing forbs such as filaree (*Erodium* spp.) for mule deer use. Grazing was found generally not to be an effective tool for cheatgrass control in the Intermountain Region (Vallentine and Stevens, 1994); one possible exception was concentrating grazing of cheatgrass by sheep prior to seed production to reduce cheatgrass competition in mixed stands, providing the associated perennial grasses had opportunity to complete their life cycles. Mosley (1996) suggested spring grazing cheatgrass during two weekly periods separated by one to three weeks, each weekly grazing treatment being applied before the cheatgrass began to turn purple, with the goal of preventing the seed from reaching the soft dough stage.

A. BIOLOGICAL CONTROL BY SHEEP

Since sheep encourage many grass species by relatively heavier grazing of snowberry (*Symphoricarpos* spp.), geranium (*Geranium* spp.), dandelion (*Taraxacum officinale*), and butterweed (*Senecio serra*) on Intermountain summer range, moderate grazing by sheep tends to improve the range for cattle. Spring deferment and heavy fall grazing by sheep (60 sheep days per acre) on native sagebrush-grass range at Dubois, ID, improved range condition faster than total protection (Laycock, 1961, 1967). This grazing treatment increased grasses and forbs but decreased big sagebrush and raised range condition from poor to fair in 7 years. Heavy fall grazing following spring rest for 2 or more years in succession was recommended as a range improvement practice on sagebrush-grass range; however, it was noted that successful use of this treatment required that sufficient perennial grass be present to respond and that the range must not be grazed in the spring if heavily grazed in the fall.

Sheep grazing in late fall was also found by Frischknecht and Harris (1973) to effectively control big sagebrush on spring cattle range seeded to crested wheatgrass if applied before sagebrush became too dense. This treatment was effective when the density of big sagebrush was only about 1.5 plants per 100 square feet and primarily resulted in a decrease in sagebrush plant size and in limiting reproduction. Cattle were found ineffective in any season in controlling big sagebrush, even when big sagebrush was present in minimal amounts. Sheep on crested wheatgrass spring lambing range in New Mexico were observed to make comparatively heavy use of shrubs even under light stocking rates when the shrubs were present in lesser amounts (less than 10%), particularly when sheep movements were restricted in small paddocks (Springfield, 1961).

Sheep have inherent dietary preferences for forbs (Olson, 1999), and Olson and Lacey (1994) reported that sheep were effective agents to control perennial weeds and certain poisonous plants and currently were being used to control leafy spurge, spotted knapweed (*Centaurea maculosa*), tansy ragwort (*Senecio jacobaea*), and tall larkspur. Their guidelines for the effective use of sheep for biological weed control included:

1. Graze as a flock for more intensive defoliation.
2. Concentrate sheep in smaller areas by herding or by additional fencing to overcome patchy grazing.
3. Increase stocking density to (a) improve animal distribution spatially, and (b) reduce selective grazing. (While concentrating animals limits their ability to select, this does not always result in overcoming an animal's aversion to certain plant species [Olson, 1999].)
4. Time grazing to when weed species are most palatable and/or susceptible to defoliation and, if possible, when desirable species are least affected.
5. Alter time of grazing or adjust number of animals when necessary to prevent desirable plant species from being overgrazed.
6. End grazing period no later than when foliage on desirable plants reaches the minimum levels required for continued photosynthesis and rapid growth.
7. Repeat defoliation events when needed for adequate plant control, with adequate rest spaced in between for the benefit of the desirable species.
8. Use heavy weed infestations as bedgrounds.
9. Do not move sheep from noxious plant areas to weed-free areas until all weed seeds have passed through the digestive tracts, generally at least 5 days.

10. Use sheep that have had experience with grazing the plant species to be controlled. (Note that this may be unnecessary when time permits animals being trained on site; also, sheep having no previous experience grazing leafy spurge have learned to readily graze it within a 3–4 week period [Olson *et al.*, 1996].)

Cattle generally refuse to eat leafy spurge, partially or totally avoid leafy spurge-infested sites, and even develop aversions to it when mixed in grass hay or haylage, but sheep and goats will graze it (Kirby *et al.*, 1997; Heemstra *et al.*, 1999). Selective grazing of leafy spurge by sheep is considered an effective method for suppressing large infestations of leafy spurge, particularly in areas not accessible with ground equipment to apply alternative control (Lacey *et al.*, 1984). Sheep will consume significant amounts of leafy spurge in their diets (40–50% maximum) (Landgraf *et al.*, 1984). In crested wheatgrass stands grazed by sheep in Canada (Johnston and Peake, 1960), leafy spurge was reduced in basal area by 98% (from 3.14–0.22%) by 5 years of selective early spring grazing, while crested wheatgrass increased from 17.7–22.0%. However, sufficient leafy spurge plants remained to reinfest the area if the sheep grazing treatments were discontinued.

Sheep grazing will not eradicate leafy spurge but does materially suppress it and may keep it from spreading. Guidelines provided by Lacey *et al.* (1984) are as follows: (1) begin grazing in the spring when leafy spurge plants are several inches tall, and (2) schedule pasture rotations so that plants will be unable to set seed. After initial reduction of leafy spurge, sheep grazing as a continuing suppression measure will generally be needed. In Idaho studies, 3 years of repeated sheep grazing reduced leafy spurge seeds in the seedbank and reduced numbers of seedlings but had no effect on density of mature plants (Olson and Wallander, 1998). Maximum control levels of leafy spurge will require that sheep grazing be supplemented by herbicide application and/or bio-control by insects or pathogens. Three summers of repeated grazing negatively impacted spotted knapweed, but it was concluded that a long-term commitment to the treatment was required to slow the rate of increase of the noxious plant in native grass communities (Olson *et al.*, 1997).

Grazing sheep on traditional cattle range often reduces cattle losses from poisonous plants since several plants poisonous to cattle are safely consumed by sheep (Ely, 1994). Although the presence of plants poisonous principally to cattle may indicate greater suitability of the area for sheep than for cattle, converting from sheep to cattle may not be desirable or economically feasible. The alternative is to use sheep—or sometimes goats—to biologically control such plants and thereby reduce or even eliminate cattle losses.

On a body weight basis, about 20 times as much senecio (*Senecio longilobus* and *spatioides*) is required to poison a sheep as to poison a cow, and sheep will readily remove the plants by grazing, thereby reducing poisoning by cattle (Sperry *et al.*, 1964). Sheep grazing was found useful in California in reducing scattered stands of St. Johnswort (*Hypericum perforatum*) (Murphy *et al.*, 1954). To be effective, it was necessary to concentrate sheep grazing for short periods, avoiding long, continued heavy grazing. The combination of sheep grazing during March and April to reduce vigor by grazing off new shoots and again in early summer to remove flowering tops was suggested.

Sheep have also been effective in suppressing tansy ragwort (*Senecio jacobaea*), either alone or in conjunction with control by insects (Bedell *et al.*, 1981). It was suggested sheep be grazed from May to early June at an intensity that forced them to consume the foliage of tansy ragwort, this not only preventing the plants from flowering and setting seed but also reducing plant vigor by frequent defoliation (Sharrow and Mosher, 1982). Sheep can be grazed in advance of cattle to reduce tansy ragwort herbage to the extent that cattle will not be poisoned. Tansy ragwort provides useful forage for sheep and is not harmful to sheep as it is with horses, cattle, and goats.

Sheep are also more resistant to larkspur (*Delphinium* spp.) poisoning than are cattle and effectively reduce cattle losses when grazed prior to cattle turn-in or alternately grazed with cattle. Close herding and trailing sheep through tall larkspur patches or bedding them in patches greatly increases trampling of larkspur stalks and utilization of heads and leaves, which have the potential for reducing plant density. Sheep grazing at the bud or early flower stages is most effective, but forced grazing by close herding at other stages can also be effective. Reducing the acceptability of the tall larkspur is additive to defoliation in reducing cattle consumption. However, the use of sheep as a biological tool for reducing cattle losses from tall larkspur species has been variable, ranging from moderate to nearly elimination of cattle losses. Sheep grazing has generally been even more effective in reducing cattle losses from low larkspur species (Alexander and Taylor, 1986; Ralphs *et al.*, 1989; Ralphs *et al.*, 1991b; Manners *et al.*, 1996).

Sheep have long been a favorite for general suppression of weeds on ranchsteads, residential estates, and other building sites. Sheep were also effective in removing broad-leaved weeds from seedling alfalfa fields in the Imperial Valley of California (Bell and Guerrero, 1997; Guerrero *et al.*, 1999); 5-month old wether lambs readily consumed the weeds, often in preference to the alfalfa, and were as effective in general weed control as were herbicides. Lamb weight gains were similar whether grazing pure alfalfa stands, pure weed stands, or mixtures of alfalfa and weeds. Sheep have been used with variable results in removing from dairy pastures weeds that give off flavors to milk. However, sheep were ineffective in reducing Dyers woad (*Isatis tinctoria*) on Utah range (Farah and West, 1988); the sheep refused to graze the plants heavily enough and late enough into the spring to cause mortality or limit seed production.

B. BIOLOGICAL CONTROL BY GOATS

The most effective control of undesirable woody range plants by livestock grazing has generally been from the use of goats (Fig. 16.1). An FAO report (French,

FIGURE 16.1.   Goats being grazed on rangeland near Rocksprings, TX, following brush chaining for additional biological control of brush species. (Soil Conservation Service photo.)

1970) concluded that because of their general browsing tendencies, goats can be a potent factor in controlling woody plants and in preventing their return in areas of low and erratic precipitation throughout the world. Goats are well adapted to desolate, semi-arid, exhausted, poorly watered sites; they can often subsist after all other livestock have had to be removed and will travel long distances in search of preferred forage. However, only proper management of goats under such conditions permits them to be prominent in the solution rather than merely being a part of the problem.

Unless carefully managed, goats will also graze out desirable forage species. The goat can cope with a variety of dietary alternatives, but when the shrubs available for browsing are highly unpalatable and unacceptable or unavailable, the goat often readily shifts to herbaceous species and may only worsen an already deteriorated situation (Huss, 1972). In fact, the greatest use of brush by goats generally occurs when available green herbaceous forage becomes limiting; and the less palatable noxious weeds or brush may be avoided unless heavy grazing pressures force goats to consume plants they normally avoid (Taylor, 1992).

Goating has largely been restricted in the U.S. to Texas and the Southwest, but the potential exists in other southerly areas or seasonally in even more northern latitudes; for example, goats effectively controlled undesirable brush on rundown and abandoned pastures in Vermont (Wood, 1987). The goats were found capable even of destroying small trees and saplings by debarking, browsed higher than sheep by standing on their hind legs, and were not deterred by thorny vegetation.

It was suggested that the number of goats running with cattle or sheep be gradually reduced as the brush declined to prevent competitive foraging, but running a few goats with cattle and sheep was suggested to protect renovated pasture in Vermont from reinvasion by brush and some weedy forbs.

Goat management problems have restricted the use of goating in the U.S., particularly in more northerly latitudes. These management problems include the need for special goat-proof fences, sheds for protection from severe cold and rain, the shortage of winter forage where shrub species are deciduous, and predation by bobcats, coyotes, or dogs, particularly on the young kids (Rechenthin *et al.,* 1964). Goats can be highly competitive with deer and to a considerable extent with sheep, and long-term stocking rates must be adjusted accordingly.

Goats are commonly added on cattle range under mixed grazing in Texas to stabilize existing plant communities by exerting grazing pressure on forbs and browse plants. Since goats will utilize large amounts of browse in their diets, they have been widely used in Texas and adjacent Mexico to control low-growing brush and sprout regrowth. Goating is generally most effective when used to suppress brush sprouts following mechanical, herbicidal, or burning treatments because of greater acceptability and availability (Warren *et al.,* 1984b). Repeated defoliation of woody species by goats has been found to either control the plant growth and spread or kill the plants outright if continued long enough.

Goating has been effective on a rather wide range of shrubs in Texas including oaks (*Quercus* spp.), mesquite (*Prosopis glandulosa*), sumac (*Rhus* spp.), and hackberry (*Celtis* spp.). In east Texas, the yaupons (*Ilex* spp.), willow baccharis (*Baccharis salinca*), and greenbriar (*Smilax bona-nox*) are resistant to herbicidal control but are suppressed by goat browsing, since they are primary food sources when browsed by goats in winter (Lopes and Stuth, 1984). However, goats are not a universal solution to all brush problems; they were found quite ineffective in controlling creosotebush (*Larrea tridentata*) in New Mexico (Morrical *et al.,* 1984). Martin (1975b) concluded it was improbable that either goats or sheep could materially reduce the growth and spread of mesquite, juniper, catclaw acacia (*Acacia greggii*), or creosotebush on semi-desert ranges in Arizona.

Yearlong stocking rates commonly used to control woody sprouts in Texas are one goat for each 2 or 3 acres (Darrow and McCully, 1959; Norris, 1968; Rechenthin *et al.,* 1964); However, short-term grazing by larger numbers of goats, such as 5–8 goats per acre for a 30-day period, is more effective. The use of small pastures to increase grazing animal density also increases shrub control. Grazing should be carried to the point of leaf defoliation of the shrubs while assuring that other preferred forage plants are not excessively grazed. A high degree of brush control usually takes about 3 years of intensive defoliation. Subsequent light stocking with goats grazed in common with other kinds of livestock will provide continuing maintenance control of the brush. In order to adequately control sprouts in a follow-up treatment, goating should immediately follow mechanical or chemical clearing.

Goat enterprises have been added on many Texas ranches not only to control brush regrowth but also to provide additional ranch income. The economic returns from range goat enterprises have been similar to those from range cattle and sheep enterprises (Merrill and Taylor, 1976). The Angora goat produces mohair, and the Spanish goat is readily marketed for meat; however, the market for both of these products is quite variable. In a north-central Texas study, goat enterprises were added without reducing the original number of cattle, thereby approximately doubling the number of animal units maintained (Magee, 1957). The goat enterprises averaged 359 head in size and over a 5-year period paid for their original purchase price, the added fencing and shelter costs, all year-to-year costs incurred in handling the goats, and the cost of mechanically clearing an average of 518 acres of range.

When diets were compared on mixed vegetation range in south-central Texas, Spanish goats consumed a higher browse component than Angora goats, and Rambouillet, Karakul, and Barbado sheep consumed insufficient amounts of browse for effective brush suppression (Warren et al., 1981a, 1984a). Both the meat-type Spanish goat and the mohair-type Angora goat have been used successfully in biological brush control, but differences in physiology and nutritional requirements make the Spanish goat the more preferred for this use (Merrill and Taylor, 1976). The latter has been found to be more rangy and can browse to heights of 7 feet or more. It is readily available for purchase in the Southwest, is more prolific, is less vulnerable to extreme weather conditions, requires less management and labor, includes more browse in its diet than the Angora goat, and has less hair to become entangled in thorns or twigs.

Although the financial return from Spanish goats is less affected by heavy grazing pressure, the Angora goats may bring a greater return on investment when mohair prices are high (Taylor, 1986b). Spanish goats were clearly more willing to consume juniper than Angora goats in Texas studies (Pritz et al., 1997). However, about 25% of the Angora goats consumed more juniper than the average Spanish goat, suggesting that selecting for juniper consumption within an Angora goat population might be a viable alternative. Although goats were found capable of suppressing juniper over long-term grazing, the voluntary intake of juniper by goats is self-limited to only a partial but nevertheless important part of their diets (Riddle et al., 1999).

An important use of goats in brush control on California brushland has been to browse green shoots from brush stumps and roots when 6–12 in. high following initial mechanical control or fire (Spurlock et al., 1978). Goats browse a wide variety of chaparral shrubs but are least selective on young sprouts. Following a wildfire about 5 years previous, the diet of Spanish goats in California chaparral under summer grazing consisted of about 80% of scrub oak (*Quercus* spp.) and chamise (*Adenostema fasciculatum*), but little use was made of the manzanita (*Arctostaphylos* spp.) and ceanothus (*Ceanothus* spp.) under normal stocking rates (Sidamed et al., 1978). One suggested strategy has been to stock with enough goats to eat all leaves from all brush species two or three times per year (Green and Newell, 1982). A related strategy has been to stock two mature goats per acre the first year, one the second year, and one per 2 acres thereafter for continuing maintenance (Spurlock *et al.,* 1978).

Goats are a promising alternative to herbicides for controlling brush regrowth on fuelbreaks in southern California when concentrated o target areas (Green *et al.,* 1979; Green and Newell, 1982). Herding goats onto the selected areas concentrated goating effects sufficiently to achieve uniform browsing where brush seedlings and sprouting regrowth were less than a year old, but fencing and confinement grazing were necessary where brush regrowth was 5 years old.

The highest levels of goat stocking have reduced total shrub cover in Arizona chaparral stands, particularly when done in conjunction with an initial brush-crushing treatment or herbicide or burning treatment (Severson and Debano, 1991). Goats in Arizona chaparral utilized a much wider range of woody plants than other domestic livestock and resident big game, but the species they preferred were comparable (Knipe, 1983). When used in conjunction with site conversion, some form of intensive rotation grazing was suggested for concentrating the browsing and also protecting the area during critical periods of herbaceous plant establishment (Knipe, 1983).

Goat grazing of Gambel oak sprouts following mechanical treatment in Colorado was found to provide up to 95% sprout control (Davis *et al.,* 1975). Two defoliations of Gambel (*Quercus gambelii*) oak by concentrated goat grazing in short time periods under rotational grazing were found to be the most effective grazing treatment. No problems were observed from permitting the goats to mix freely with range cattle. A prior mechanical treatment was found necessary to allow the animals full access to all of the foliage, and roller chopping was suggested. In mechanically undisturbed oak the goats highlined the plants up to about 7 feet; this did not kill the larger oak plants but did aid in opening up the understory.

Goating of a Gambel oak community in northern Utah reduced both Gambel oak and serviceberry but increased the production of big sagebrush, apparently as a result of reduced competition from the taller shrubs (Riggs *et al.,* 1988; Riggs and Urness, 1989); both big sagebrush and big rabbitbrush were mostly avoided by the goats. It was concluded from this study that Spanish goats can be used periodically to manipulate the composition of oakbrush winter range in Utah and interrupt the successional trajectory that leads to increased oak at the expense of other shrubs associated on such sites (Urness, 1990; Riggs *et al.,* 1990). However, the scarcity of goats available locally was considered a limiting factor.

Walker *et al.* (1996) concluded that goats had a greater potential for biological control of leafy spurge than sheep as the goats demonstrated a consistently greater preference for the plant even in areas where sheep did not graze it readily. Spring grazing with goats in North Dakota followed by fall application of picloram plus 2,4-D has resulted in 98% control of leafy spurge (Lym, 1998). Leafy spurge-based diets fed to Angora goats nursing kids met nutritional requirements during the growing season and promoted good animal health and growth and production (Kirby *et al.,* 1997). While goats effectively grazed leafy spurge regardless of plant density and competed less with cattle in dietary selection than did sheep, goats were considered less marketable than sheep in the area (Lym, 1998).

C. BIOLOGICAL CONTROL BY CATTLE AND BIG GAME

Neither cattle nor big game animals are commonly employed to provide effective control of undesirable plants, although both groups play important roles in vegetation manipulation (see next section). However, small soapweed (*Yucca glauca*) has been controlled on Nebraska Sandhills grasslands by grazing with mature cattle during winter. Some ranchers report that winter feeding of protein supplements at more than the normal rate results in heavier grazing of soapweed. On the other hand, small soapweed in the Sandhills increases under continuous summer grazing by cattle. Following singeing off the spines of upright species of pricklypear in the Southwest, cattle readily consume the pads, resulting in reduction in stand (Fig. 16.2).

Following initial topkill by other control methods, heavy browsing by cattle in mid-summer has shown promise as an effective technique for controlling aspen suckers in aspen parkland of western Canada, where aspen is considered undesirable under intensive pasture management (Fitzgerald and Bailey, 1984). Cattle in August grazed aspen quite readily even when alternative forage was ample; however, defoliation was limited in the spring and early summer except after the herbaceous species had been consumed. In Oklahoma, cattle browsing effectively killed all American elm (*Ulmus americana*) sprouts in a 24-acre pasture following mechanical top removal (George and Powell, 1979).

FIGURE 16.2.   Cattle exerting biological control over pricklypear by eating the pads after the spines were burned off with a backpack burner; photo near Zapata, TX. (Soil Conservation Service photo.)

The opportunity to seasonally concentrate wild ungulate herbivores sufficiently to exert biological control of plant species is limited. In California chaparral, small burns of 4 to 5 acres in size attracted deer sufficiently to suppress sprouts and seedlings of chamise, oaks, and several other woody plants (Bentley, 1967; Biswell *et al.*, 1952). This localized heavy browsing by deer tends to maintain the herbaceous forage plants sought by cattle. On large burns, however, browse regrowth may be so abundant that brush control by resident deer is insufficient to be effective. On very large burns, sheep or goats may also be required to suppress sprouts. When grazed in mid-summer after grasses and forbs have dried, sheep and goats may materially retard the reinvasion of chaparral shrubs.

In browsing studies in the California foothills, deer helped maintain brush control, but browsing by both cattle and deer slowed even further the regrowth of brush and postponed the need for follow-up mechanical, chemical, or burning treatments (Johnson and Fitzhugh, 1990). In Sierra wet meadow communities in California, cattle grazing was found useful in forming tunnels and browseways in shrub vegetation, primarily after herbaceous forage had been consumed (Kie and Loft, 1990). The modification of structural diversity and opening of shrub stands were considered beneficial to some wildlife species such as mule deer. Waiting until after mule deer utilized the cover for giving birth to their young in early summer and then grazing by cattle to modify the vegetation structure by creating disturbances in the herbaceous and shrub vegetation was suggested.

III. MANIPULATING ANIMAL HABITAT BY GRAZING

There are numerous examples in grazing management literature showing or suggesting that increased availability, palatability, or production of forage plants for one animal species can result from the grazing or browsing activities of another or an assemblage of herbivore species; this has been referred to by Laycock *et al.* (1996) as "facilitative grazing." These induced favorable effects may accumulate over a period of years and result in lasting changes in botanical composition of the forage stand (i.e., an extension of biological control by grazing) by altering its structure by reducing heights of shrubs to levels that are available to shorter herbivores, and by increasing the diversity of the habitat. Or, these beneficial changes may be short term and result from altering the standing crop by increasing the productivity or accessibility of particular forage species or enhancing the nutritive quality or palatability of the available forage regrowth (Severson and Urness, 1994; Laycock *et al.*, 1996).

Severson and Urness (1994) concluded that, "Existing knowledge regarding the use of prescribed grazing by livestock as a tool to enhance wildlife habitats is meager, but enough information exists to warrant applications in some areas and investigate its applicability in others." While livestock can affect wildlife habitats negatively (arising generally from situations involving severely overgrazed grazing lands), opportunities exist for manipulating or directing livestock grazing to modify habitats more favorable to certain wild large herbivores. However, since it is impossible to maximize the habitat of all wildlife species at once, using livestock to enhance wildlife habitat requires firstly deciding which species' habitat is to be enhanced and just what vegetation change is sought (Mosley, 1994). Enhancing one herbivore's habitat may negatively affect another herbivore's habitat. For example, manipulating rangeland to provide more palatable browse for winter use by deer could be expected to unfavorably impact the optimal grassland habitat for bighorn sheep.

A. MANIPULATING VEGETATION COMPOSITION

Not all interactions between large herbivore species are competitive. It is well demonstrated that different species of grazing animals have different forage preferences. Since all grazing is selective relative to plant species, grazing by one species of animal tends to induce long-term vegetation trends in complex vegetation types away from the forage species preferred by that kind of animal. This gives competitive advantage to other plant species less palatable or unpalatable to the manipulating animal species. While such a long-term change may be disadvantageous to the animal species causing the trend, it can favor another animal species if the latter benefits from the trend. A point of emphasis is that opportunities to enhance production of a forage species are improved if there is limited overlap in forage selection between the manipulator animals and the animals to be benefited (Severson and Urness, 1994).

The complex interrelationships of grazing and big sagebrush led Laycock (1979) to conclude that grazing can maintain present levels of big sagebrush, cause its development into dense stands with minimal understory, or nearly eliminate it. Heavy grazing of the herbaceous understory, particularly by cattle, can give big sagebrush competitive advantage, particularly in the absence of fire. Frequent fire, on the other hand, coupled with heavy browsing by goats, sheep, or deer, can convert big sagebrush areas into annual/weedy grasslands. Spring and summer grazing by cattle often favors big sagebrush, while heavy fall grazing by sheep following rest during the spring, where the density of brush was initially not too great, has been found effective in materially reducing it. Complete protection from grazing, at least on sites where sagebrush is a natural part of the vegetation, is apt to be quite ineffective in reducing it because of its long life and competitive ability. And, finally, rangeland with a thick stand of big sagebrush and little or no perennial herbaceous production in the understory generally cannot be improved with grazing management alone.

Many, if not most, big game species are animals of seral stages somewhere below climax vegetation; here the mixed vegetation seems to provide the best habitat and highest carrying capacities (Teer, 1985). This suggests that on most range sites controlled grazing by livestock can be an important tool in the management of vegetation for wildlife. On many mule deer winter ranges in the Intermountain

Region livestock grazing shifted pre-settlement, grass-dominated communities to shrublands, partly as a result of grazing pressure on the grass understory and "fire-proofing" of many ranges by reducing the fine fuel, herbaceous understory; mule deer herds responded to the increase in available winter forage, particularly browse, and numbers of deer significantly increased (Austin *et al.,* 1986).

A strategy of managing critical big game range through the elimination of live-stock developed in the late 1930s; this proved mostly unrewarding and often seri-ously impacted rather than improved big game habitats. Smith (1949) formally documented on a northern Utah sagebrush-grass foothill range the decline of shrubs and the increase in herbaceous plants, particularly grasses where cattle had been removed for 11 years but heavy deer use had continued. When remeasure-ments were made in 1982, "the downward trend in shrub density and vigor on the 'deer range' continued to extinction and the former 'livestock range' (removed from livestock use about 1957) had the appearance of the deer range in 1948 (i.e., heavy impact on shrubs)" (Urness, 1990; Austin and Urness, 1998).

Beginning in the 1940s and 1950s state game and fish agencies in the Western states began to purchase operating cattle ranches deemed critical as elk or deer winter ranges. Livestock grazing was removed from these lands under the expec-tation that such would enhance the respective critical big game ranges. However, after 10–20 years it became apparent that the expected benefits were not being achieved and the big game use of these acquired ranches was shifting to adjacent private ranches where livestock were still grazed (Burkhardt, 1996). Eliminating the cattle grazing, when combined with continuing winter use by mule deer, re-sulted in altering the plant communities toward more grasses and fewer desirable shrubs.

Lack of grazing by a generalist herbivore (cattle) had allowed the vegetation to become rank and less palatable and nutritious. The grazing management strategy was then reversed and "managed" cattle grazing was reintroduced to these big game areas as part of the big game management program. Policies of managing state-owned big game range now commonly includes provision for controlled cat-tle grazing to promote a better balance of forage species for the big game. Live-stock grazing, particularly cattle and horses, plays a direct role in the establish-ment and maintenance of shrub-dominant plant communities for browsing by deer in winter (Austin and Urness, 1998), but unless land managers are able and will-ing to make that commitment, further declines in deer populations in the future can be expected (Urness, 1990).

Grazing by cattle can provide a favorable habitat for woody plant seedling es-tablishment by reducing the soil moisture utilized by grasses and by reducing fire frequency (Archer, 1994). Vigorous grass stands are also competitive with estab-lished shrubs. Hubbard and Sanderson (1961) in California found that grasses are competitive with bitterbrush, a highly palatable and productive browse plant for deer. The reduced vigor of bitterbrush and the increased vigor of the grasses ap-peared to be associated with reduced cattle grazing. When the herbaceous compe-tition was reduced by cattle grazing, bitterbrush previously in poor vigor increased average leader or twig length by 90% and average numbers of leaders by 223%.

FIGURE 16.3. Range reseeding in central Utah dominated by seeded grasses; cattle grazing for reducing competition from grasses needed for release of bitterbrush seedlings and enhancing grass regrowth for late winter and early spring use by mule deer (shown in background).

A balance of animal pressures has been concluded best for managing foothill ranges of the Intermountain area, such as winter use by big game and spring and early summer use by livestock (Smith and Doell, 1968). A vigorously growing stand of grass competes materially with palatable shrubs such as bitterbrush; late spring to early summer livestock grazing, particularly cattle, effectively increases vigor and production of browse plants valuable as winter forage for mule deer by utilization of the competing herbaceous understory and reducing its competitiveness with the bitterbrush (Fig. 16.3). The herbaceous forage should be grazed at the season and intensity that will maintain it but will not permit it to increase at the expense of the browse plants. Grazing into the fall, particularly during drought years or under heavy stocking rates, may result in a shift in cattle diets to the shrubs.

Cattle grazing 3- to 5-year old stands of bitterbrush near Burns, OR, increased both diameter and volume of the shrubs by their fifth growing season (Ganskopp *et al.*, 1999). However, since cattle began seeking out bitterbrush after the understory grasses cured, it was recommended that cattle be grazed before the most prominent grasses reached anthesis. During this period the shrubs were tolerant of any minimal use by cattle since there was generally ample opportunity for regrowth during the remaining growing season. These and similar observations in other areas have suggested it is advantageous to graze some cattle on range managed primarily for deer when cattle grazing is managed to make maximum use of grasses and minimum use of bitterbrush and other shrubs palatable to deer.

The spring-summer grazing of horses was found effective in manipulating sagebrush-grass range in studies in northern Utah to increase the seasonal production of bitterbrush (Reiner and Urness, 1982), big sagebrush, and true mountain mahogany (Austin *et al.,* 1994; Austin and Urness, 1995). Horses in these studies consistently exhibited high selectivity for grasses while grazing forbs only lightly and shrubs essentially never, thereby greatly increasing shrub volume and seedling recruitment, but horses were found mostly unavailable in sufficient numbers for effective vegetation manipulation (Urness, 1990). Spring-grazing sheep on Utah foothill range later winter-grazed by mule deer resulted in a 2.3-fold increase in total range grazing capacity over deer alone (Malechek, 1978). The spring sheep grazing apparently increased deer browse production when the sheep stocking rates were carefully regulated and the sheep were removed from the range prior to July.

However, from work also in northern Utah, it was suggested that spring grazing by sheep may be less effective than with cattle or horses in minimizing competition from herbaceous plants on shrubs because of the preference of sheep for forbs and reluctance to eat some grasses (Jensen *et al.,* 1972). Nevertheless, in subsequent studies in the same area winter diets of mule deer were higher in herbaceous components but lower in shrub components on range previously spring grazed by sheep than on range not previously grazed (Smith *et al.,* 1979). The spring sheep grazing treatment reduced total herbaceous plant material for winter but increased the proportion of green herbaceous material as a result of fall regrowth and increased bitterbrush yield, the latter resulting from release of moisture and nutrients accompanying the reduction of herbaceous herbage by the sheep. This procedure more than doubled the total annual stocking rate for mixed grazing without negatively affecting the quality of winter deer diets.

Alternatively, selective grazing by deer on cattle range may result in reduced or limited competition with cattle. In fact, where many shrubs occur that are palatable to deer but not to cattle, deer grazing on cattle range can beneficially control such shrubs and enhance herbaceous growth for spring and summer grazing by cattle. On chained and seeded foothill areas in central Utah, heavy winter and early spring grazing by mule deer, in combination with competition from the herbaceous understory, prevented even juniper and pinyon (*Pinus* spp.) from regaining dominance when present only as scattered small trees (Stevens *et al.,* 1975).

B. MANIPULATING ACCESSIBILITY, PALATABILITY, AND NUTRITIVE CONTENT

Preconditioning herbaceous forage, particularly bunchgrasses, by spring grazing (late vegetative to early reproductive stage) can be beneficial for subsequent winter grazing. The potential benefits of spring grazing by livestock for winter grazing by big game animals include (Clark *et al.,* 1998): (1) reducing standing litter accumulation, (2) reducing the number of reproductive culms (Ganskopp *et al.,* 1992), (3) delaying plant phenology, (4) increasing the accessibility of new green growth, (5) improving the nutritive quality of the forage, (6) decreasing foraging time and distance traveled per unit of nutrient ingested, and (7) decreasing the time big game animals are exposed to predators and harsh environmental conditions while foraging. However, a trade-off with spring conditioning of the forage is commonly a reduction in total forage quantity available during the dormant season, and realization of the potential benefits to forage quality is greatly dependent on plant growing conditions following spring defoliation.

The palatability and nutritional values of herbaceous forage may not be significant concerns during the green-forage growing season but become very important for big game or livestock when grazing animals must subsist on mature forage. The development of wolf plants and the buildup of thatch in cured grass may provide a physical barrier against deer grazing and even elk grazing of grass regrowth in late winter and early spring. Moderate or heavy fall defoliation by cattle grazing or burning was found by Willms *et al.* (1979, 1981) on big sagebrush-grass range in British Columbia to increase preference of bluebunch wheatgrass regrowth by deer in the spring. Deer preferred the burned plants over the grazed plants during spring grazing, probably because of lack of stubble, but readily grazed the fall-grazed plants after green growth exceeded the stubble height. Deer did not select forage from among standing litter of bluebunch wheatgrass when alternate sources were available. Deer were more affected by grass stubble than cattle, although deer were able to select closer to the height of short stubble and further below the height of long stubble (Willms *et al.*, 1980b). One proposed practice has been to "flash" graze bunchgrasses with cattle for a 2-week period early in late winter to remove rank, old growth in order to enhance the forthcoming regrowth for deer (Bryant and Morrison, 1985).

While heavy cattle grazing of grasses early in the spring may be competitive with deer, light or moderate use may be beneficial in prolonging the active growth period and reducing seedstalk production for later grazing by deer (Hanley, 1982b). Pitt (1986) concluded from a 2-year clipping study that spring defoliation of bluebunch wheatgrass can increase fall forage quality, but he pointed out that defoliation must not be excessive since the plant is susceptible to grazing damage at that time.

Austin *et al.* (1983) found that under snow-free and partial snow cover mule deer preferred areas of ungrazed crested wheatgrass to those previously grazed by cattle in spring. It appeared that the interference from cured growth in limiting green grass availability was more than compensated for by increased forage production. On foothill communities in northern Utah on which Gambel oak was reduced by goating, during winters when understory consumption was precluded by snow, the higher consumption of big sagebrush resulted in diets lower in fiber and tannins and more digestible (Urness, 1990; Riggs *et al.*, 1990). When snow cover was absent, deer shifted to grazing understory species which had not been materially altered by goating.

Spring-summer grazing of sheep on big game habitat in Oregon's Coast Range materially increased the quality (crude protein and dry matter digestibility) of most grasses and forbs for fall grazing by big game but reduced the quantity of total forage available (Rhodes and Sharrow, 1983, 1990). Grazing the previous year tended to increase the amount and earliness of new growth of graminoids for spring grazing by big game; the earlier spring growth on areas previously spring-summer grazed by sheep was attributed to increased soil temperatures from reduction of old growth and increased soil fertility from dung and urine deposition by the sheep. The reduction of the old, weathered material and stimulation of earlier graminoid growth in the spring was considered highly beneficial to both deer and elk.

On natural grasslands in northeastern Oregon a program of cattle grazing during the early portion of the growing season was employed to remove undesirable rank growth and encourage regrowth for ideal elk winter range (Anderson and Scherzinger, 1975). The objective was to stock cattle heavy enough to remove rank growth, maintain forage quality, make desirable forage more available, and reduce the incidence of wolf plants while still leaving an adequate volume of forage for the elk. It was concluded that this practice was one of the principal factors that resulted in a ten-fold increase in elk numbers during the study period: Vavra and Sheehy (1996) concluded that controlled spring livestock grazing in northeastern Oregon did enhance the quality of forage for wintering elk. It was projected that eliminating the livestock might well reduce the availability of high-quality forage for elk and induce them to migrate from traditional winter ranges in search of conditioned forage such as is found on private grazing lands.

Severson and Urness (1994) concurred that cattle grazing had potential in removing the old growth that accumulates around ungrazed plants, particularly grass, thereby enhancing access by elk: "These excessive amounts of leached residues are of poor nutritive quality and, during spring, can impede initiation of new growth by acting as an insulation layer. They also can make the new, nutritious growth less available to grazers when it does start growing." Jourdonnais and Bedunah (1990) compared the use of seasonal burning and fall cattle grazing as a means of reducing litter accumulations and increasing elk use of rough fescue on winter range in Montana. Elk use was greater with the cattle-grazed and burned treatments compared to the control, and the elk use was concentrated on rough fescue plants without heavy litter. While cattle grazing was less effective than burning in removing the plant litter, it created a mosaic of heavy to lightly grazed areas while maintaining some litter cover on the soil surface.

Elk forage quality was generally enhanced by light cattle grazing in southwestern Montana (Willard, 1989). Protein and phosphorus levels were increased in the grass herbage available to the elk; cattle removed more of the upper, more stemmy portions of the grasses and left more of the leafy material. Clark *et al.* (1998) applied three early-June defoliation treatments to bluebunch wheatgrass in the Blue Mountains of northeastern Oregon: (1) mid-boot stage, whole plant clipped; (2) mid-boot stage, half of plant clipped; and (3) inflorescence emergence stage, whole-plant clipped. All treatments increased the early November levels of crude protein and *in vitro* dry-matter digestibility, with the latter treatment having the greatest effect, but the control (no prior defoliation) retained higher dry matter levels over the defoliation treatments. Carefully managed, late spring sheep grazing was concluded to be an effective means for improving the winter forage quality of bluebunch wheatgrass and Idaho fescue for elk (Clark *et al.,* 2000); it was also projected as a means of improving the quality of grazed forage for livestock in winter and reducing supplementation costs.

In studies on bunchgrass foothill range in southeastern Washington, however, neither spring grazing by cattle nor fall burning to remove dead standing litter and enhance forage palatability provided increased winter use by elk (Skovlin *et al.,* 1983). Wambolt *et al.* (1997) found that spring grazing by cattle in a normal rest-rotation grazing system in southwestern Montana resulted in no significant increase in nutrient content of bluebunch wheatgrass for subsequent winter consumption by wildlife. While Brandyberry *et al.* (1993) found that preconditioning forage by spring grazing with cattle appeared to improve the nutritive quality of the available forage available for late fall grazing, the quality of diet selected by cattle grazing in November was not improved, possibly resulting from poor spring growing conditions or selective fall grazing. Westenskow-Wall *et al.* (1994) concluded that defoliation after the vegetative stage of plant growth for bluebunch wheatgrass is necessary to increase the winter quality of the forage, but this is dependent upon growing conditions after defoliation favoring regrowth. While the results are variable from using livestock to enhance nutritive value of forage for winter grazing by big game, further study appears warranted as forage quality is often more limiting than quantity at that time (Severson and Urness, 1994; Clark *et al.,* 1998).

The naturally spinescent growth form of blackbrush reduces the accessibility of its twigs for cattle that commonly browse blackbrush areas during fall and winter. Heavy goat browsing in southern Utah was found effective in removal of the spiny twig tips, and this stimulated sprouting from basal and axillary buds which resulted in large quantities of new growth accessible to cattle (Provenza, 1978; Provenza *et al.,* 1983a,b). However, the nutritious regrowth was found to be low in palatability because of increased levels of astringent tannins (Provenza and Malechek, 1984). Recommendations were to employ heavy goat browsing followed by one or two years of rest from browsing by cattle to allow tannin levels to decrease and an accumulation of palatable forage on the blackbrush plants to be made (Provenza *et al.,* 1983a,b). This provides forage of reduced tannin content while maintaining fair nutrient content and digestibility, while apparently also reducing the need for protein supplementation of the cattle.

IV. GRAZING TO AID FORAGE PLANT SEEDING

The transport and dispersal of seed and other plant propagules by grazing animals, either internally through the digestive systems or externally by temporary attachment to hair, fleece, or hooves, is always an ecological factor affecting a perennial forage stand, but the impact will range from favorable to deleterious depending upon the plant species and site being affected.

High-density grazing by livestock can be beneficial in seedbed preparation by graze-out and removing competing vegetation and soil treading to scarify the soil surface. Soil treading can also aid in planting naturally or artificially broadcasted seed. Benefits may be derived from the effects of hoof impact working the seed into the soil surface and compacting the soil around the seed. Livestock may also be employed for selective weed control during seed germination and seedling emergence of seeded perennials, but intensity and timing of grazing must be carefully regulated to minimize defoliation, trampling, and pull-up of establishing seedlings.

Livestock fed seeds of desirable forage species can serve as dispersal agents across the landscape through their dung, thereby hopefully creating patches of desirable plants (Archer and Pyke, 1991). This "fecal seeding" by livestock may be an effective, low-cost means of rangeland restoration (Ocumpaugh *et al.,* 1996), particularly when trying to reach sites of difficult access with machinery. Unfortunately, seed dispersal in dung per se does not provide seed-soil contact that is important in seed germination and seedling emergence (Barrow and Havstad, 1992). The greatest success can be expected from using selected forage species whose seed are capable of establishing from surface seeding. Nevertheless, when new forage stands are being established and the site is adapted to equipment for complete seedbed preparation, drilling is recommended as a more precise means of seed placement in the soil and for obtaining rapid and dependable stand establishment (Vallentine, 1989).

V. GRAZING AND FOREST REGENERATION

Agroforestry provides multiple-use and commercial opportunities to utilize forested lands in producing multiple products, such as timber and other wood products, livestock grazing, big game and upland game hunting, and other forms of recreation. The production of livestock and white-tailed deer under the trees on pine plantations in the Southeast has become a standard business practice (Fig. 16.4). In other areas transition has gradually been made from intolerance to tolerance of grazing in conjunction with timber production, and to more joint planning for both timber and grazing. There is opportunity to greatly increase cattle and sheep grazing on transitory range following timber cutting in the Pacific Northwest, and the conflict between grazing livestock and growing trees is more perceived than real (Mitchell *et al.,* 1982).

Much of the historical forestry literature has stressed that domestic livestock should be excluded from conifer regeneration sites, particularly until terminal leaders of the conifers are beyond the reach of animals. Although this may be a conservative safeguard, new methodology is becoming available for reducing livestock damage to conifer regeneration to inconsequential levels and even using livestock grazing as a manipulative tool in expediting conifer establishment. Although yet to be fully developed for many forested sites, prescription grazing for conifer

FIGURE 16.4. As an example of successful agroforestry, the production of livestock and white-tailed deer (not shown in photo) on pine plantations in the Southeast has become a standard business practice. (U.S. Forest Service photo.)

release by reducing competition from shrubs and herbaceous vegetation has had substantial application and warrants further study and refinement (Sharrow, 1994).

A. BROWSING AND MECHANICAL INJURY

The potential injury to conifer trees from ungulate herbivores is mostly limited to the seedling and sapling stages—within the first 3 years after planting on most productive sites—after which most commercial tree species avoid damage from ungulate herbivores. Damage to conifer seedlings can result either from defoliation and leader removal by browsing or from mechanical injury and breakage. Browsing is the primary cause of damage by big game animals, primarily deer. Mechanical damage is of more concern with domestic livestock, but even this can be minimized by careful control of grazing and by providing alternative sources of forage and browse.

The survival and growth of planted pines in the Southeast are not deleteriously affected by cattle grazing when proper grazing practices are applied (Tanner and Lewis, 1987). Most wild and domestic animals in the Southeast consume pine foliage only when they occupy areas with little alternative green foliage or with a limited variety of forages (Lewis, 1980). When animal stocking is based on the available forage, however, damage to planted pines resulting from livestock grazing has generally been minimal or inconsequential. Cattle grazing per se has had no negative influence on pine growth on wet-flatwoods in north Florida when proper stocking rates and other good grazing practices were followed (Tanner and Lewis, 1987).

Cattle occasionally damage pine seedlings in the Southeast by browsing, trampling, or by riding them during summer to remove insects (Lewis, 1984). The primary cause of severe damage and loss of planting stock is having too many animals where there is too little forage or placing minerals, water, and/or supplemental feed stations within young pine plantations. Cattle accustomed to pines and woodland grazing usually ignore pines while feeding. From studies on Ponderosa pine in Southwestern Oregon, Karl and Doescher (1998) speculated that cattle removal of terminal tissue was primarily attributable to a lack of salt and trace mineral provision.

Prescribed fires to control brush understory are generally excluded until longleaf pine and slash pine (*Pinus elliottii*) are 8 feet tall (Byrd, 1980). Because cattle tend to concentrate grazing on fresh burns, care must be taken that too many animals are not attracted onto areas prescribed burned. Prescribed burning of adjoining non-timbered areas can be used to attract animals away from sensitive tree regeneration areas, thereby preventing excess injury and achieving better distribution of livestock (Wade and Lewis, 1987).

When grasses are seeded during site preparation to provide site stability, competition with the tree seedlings should generally be minimized by controlled grazing. A recommended agroforestry approach for the Southeast has been to seed bahiagrass (*Paspalum notatum* and *media*) on the planting sites to provide improved, alternative forage during tree seedling establishment (Lewis *et al.*, 1985). However, Pearson *et al.* (1990a,b) concluded that Christmas tree production using slash pine and loblolly pine (*Pinus taeda*) was compatible and economical when associated with cattle grazing only when the trees were protected from grazing, trampling, and rubbing. Trampling damage from unrestricted grazing by cattle was most prominent during the first year after planting (about 8%) but virtually none occurred when protection was temporarily provided by an electric fence.

Some injury of seedlings of Southern conifers is acceptable because substantial to severe injury is required to affect long-term tree survival and growth or timber yield. Even heavy cattle grazing of pine regeneration in Louisiana up to 5 years of age, which killed 18% of the planted seedlings, left an adequate stand for timber production (Pearson *et al.*, 1971). Following the planting of loblolly pine in another Louisiana study, damage caused by cattle grazing during the first 3 years reduced seedling survival by 13% and reduced height growth of other seedlings but had minimal effects thereafter (Haywood, 1995).

Hughes (1976) simulated moderate to severe grazing damage on slash pine seedlings—defoliation, shoots girdled or removed, and stems bent. Combinations of types of damage increased plant losses and aggravated the stunting of trees, but simulated defoliation alone killed few trees. Without the other treatments, less than a full girdle of the stem was not detrimental. Damage to 30-month seedlings was much less than on 6- and 18-month seedlings. Girdling seedlings to simulate hoof scraping and bark eating showed that mortality was negligible except after complete girdling but that 75% girdling reduced seedling growth (Lewis, 1980). Since planted slash pine seedlings are relatively tolerant of removal of needles and clipping of the shoots, it was concluded that typical cattle browsing would have little effect on seedling survival or height growth.

Damage to pine seedlings by cattle grazing on Southern pine plantations occurs usually when seedlings are 1–3 feet tall, this being mostly mechanical rather than due to browsing, with minor damage when the trees are 3–5 feet tall. However, cattle need not be removed during the seedling planting season if cattle numbers are reduced to half the normal stocking rate (Byrd et al., 1984). Other precautions for limiting cattle damage during the first and possibly second establishment years are (1) minimize concentrations around feeding and watering areas or move these activities off the planting site, (2) prevent grazing during the winter months following planting when herbaceous forage matures and becomes scarce, and (3) delay nitrogen fertilization which may attract cattle until beyond the early tree seedling stage (Hughes, 1970; Byrd, 1980). However, these practices are largely precautionary as even moderate damage from livestock during establishment may have no significant effects 15–20 years later (Grelen et al., 1985).

Cattle damage to Ponderosa pine seedlings in Colorado under light to moderate stocking rates and good distribution of grazing resulted in damage to less than 1% of the seedling plants (Currie, 1978), far less than the damage caused by rodents and rabbits. Where livestock reduced the herbaceous ground cover by grazing in northern Idaho, the aboveground damage associated with rodents was also decreased (Kingery, 1987).

Browsing by mule deer and black-tailed deer in the Pacific Northwest has been a problem with Douglas fir regeneration (Crouch, 1968, 1980; Hines, 1973). Browsing on Douglas fir begins with the first snowfall and continues throughout the winter in the absence of more palatable browse; browsing often continues in subsequent years until seedlings grow out of reach of deer. Douglas fir seedlings are moderately palatable to deer, but browsing is reduced when other browse plants or green herbaceous forage is present. The great differences in palatability of different Douglas fir genotypes to deer suggest that genetic selection against palatability may be useful (Radwan and Crouch, 1978).

Suggestions for reducing deer browsing on Douglas fir seedlings have included: (1) reducing deer populations by heavier harvesting, (2) planting 5-foot saplings rather than seedlings, (3) using planting and handling techniques that encourage rapid growth beyond deer use, and (4) applying polyethylene bud envelopes for 1-year protection of small seedlings (Hines, 1973). In contrast with the numerous instances of elk and deer damage to Douglas fir seedlings, Hanley and Taber (1980) reported a study in which browsing actually increased Douglas fir regeneration, apparently as a result of decreased competition from shrubs which were browsed even more than the tree seedlings by big game animals.

Migratory deer have browsed up to 50% of the white fir seedlings in Sierran

mixed-conifer in California (Huntsinger and Bartolome, 1986; Kosco and Bartolome, 1983). Whereas big game were more likely to browse Douglas fir
seedlings, cattle damage resulted mostly from trampling rather than browsing
(Alejandro-Castro *et al.,* 1987). However, efforts made to prevent cattle concentrations and thus localize trampling damage can reduce damage by cattle to insignificant levels. In mixed conifer clearcuts in central California, no significant
trampling damage occurred to Douglas fir or white fir (*Abies concolor*) seedlings,
and the browsing damage was primarily caused by deer (Allen and Bartolome,
1989).

High white-tailed deer populations in northern hardwood forests of New York
(Tierson *et al.,* 1966) and Pennsylvania (Marquis and Brenneman, 1981) have
sometimes been a major limiting factor in tree reproduction. Artificial regeneration has been even more difficult to obtain than natural regeneration when deer
populations are excessive. Since fencing and other attempts to protect seedlings
from deer have been either ineffective or prohibitively expensive, the conclusion
has been that the only alternative is to bring the deer herds into better balance with
the habitat. Under more moderate deer numbers in the southern Appalachians, deer
browsing reduced the number of hardwood stems which exceeded the 4.5-foot level 7 years after cutting, but a sufficient number of seedlings escaped to provide adequate tree stocking (Harlow and Downing, 1970).

B. GRAZING TO AID TREE REGENERATION

Livestock grazing to promote the establishment and growth of conifer plantations or natural regeneration has been effectively used in both the Western and the
Southern coniferous forests of the U.S. (Fig. 16.5). First, high-density grazing pri-
or to tree planting has served as a site preparation technique by significantly reducing subsequent competition by herbaceous plants and shrubs. Second, following tree planting, manipulative livestock grazing must be carefully regulated to
maximize selective suppression of competing vegetation, thereby reducing moisture stress and enhancing establishment and growth of the pine seedlings. Careful
grazing will benefit tree seedling establishment by reducing competition of grasses and other herbaceous plants and by reducing fuels which could contribute to destructive wildfire.

Achieving these goals with livestock requires livestock numbers, distribution,
and timing of grazing to be carefully controlled (Doescher *et al.,* 1987; Krueger,
1985). Consideration of the following factors has been recommended when using
livestock grazing to suppress competing vegetation in conifer plantings (Doescher *et al.,* 1987):

1. Palatable forage must be available to minimize conifer damage; if necessary, plant palatable forage plant species in the understory.
2. In areas where moisture is limited during the growing season, vegetation
   should be grazed before stored soil moisture is depleted by competing
   herbaceous plants and shrubs.

FIGURE 16.5. Cattle being used on a pine plantation in southeastern U.S. to promote tree seedling establishment by suppressing and controlling competitive understory vegetation. (U.S. Forest Service photo.)

3. Animal numbers and their distribution must be controlled to reduce browsing and trampling damage and ensure uniform understory vegetation control.
4. Move animals when palatable plants have been properly grazed to avoid increased conifer seedling damage.

For grazing to be a useful tool in releasing newly planted conifer seedlings from competition from grass and other competing vegetation and enhancing tree seedling establishment, particularly on droughty sites, it must be applied reasonably soon after planting (Doescher *et al.,* 1987). Sheep grazing did not significantly improve Jeffrey pine (*Pinus jeffreyi*) seedling development in northeastern California, presumably because the competing vegetation was not manipulated soon enough or strong enough (McDonald *et al.,* 1996). If sheep were to be an effective silvicultural tool, it was recommended that sheep be (1) allowed to graze new plantations during the first growing season after planting, (2) brought to the plantation each spring thereafter to utilize the vegetation understory before carbohydrate reserves are replenished, and (3) concentrated in the area to achieve effective grazing pressure with precautions being taken that damage to conifer seedlings does not become intolerable.

Based on their studies of cattle grazing on replanted Douglas fir and Ponderosa pine in north-central Idaho, Kingery and Graham (1990) concluded that proper timing of grazing should include restriction of grazing during the first months after planting. Delaying grazing until after soils became drier and less subject to compaction and lateral displacement was suggested as a means of reducing the trampling effect; the delay would permit the seedlings to become more firmly established through new growth and expanded root systems. However, Karl and Doescher (1998) concluded that ponderosa pine seedlings in southwestern Oregon grazed early in the growing season were more tolerant of terminal tissue removal than later in the season, and understory vegetation control and soil water availability were best achieved then. Ponderosa pine seedlings sustaining tissue removal before winter bud set (April and May) were found more likely to recover by the end of the first year than seedlings sustaining tissue removal after bud set (August).

When begun 1 year after tree planting in southwestern Oregon, grazing by cattle during the next 3 years increased seedling volume significantly for both Ponderosa pine and Douglas fir (Doescher *et al.*, 1989). Improved water relations was one of the factors that enhanced the growth performance of the young conifer seedlings. While seedling survival was not stimulated by cattle grazing of Ponderosa pine seedlings planted following wildfire in northern California, the pine seedlings were taller and leaders were longer under season-long grazing than without grazing (Ratliff and Denton, 1995). Also, cattle grazing mixed conifer clearcuts in central California reduced shrub and herbaceous canopy cover to 8% by 6 years after timber harvest and conifer seedling planting, but tree seedlings showed no significant differences in height or basal diameter growth from grazing (Allen and Bartolome, 1989).

Grazing by cattle in an Oregon plantation study during the 20 years following tree planting enhanced the productivity of an assortment of coniferous species (Krueger and Vavra, 1984; Krueger, 1985). The added benefit in tree growth was attributed to improved moisture relations from grazing of the understory and the fertilizer effects from cattle urine and dung. Seeding the planting site to forage species had no effect on tree growth when grazed by cattle, and it enhanced the forage supply while making cattle management easier.

Controlled cattle grazing was also effective in minimizing mechanical damage while favorably manipulating planting site vegetation for tree regeneration with Engelmann spruce (*Picea engelmannii*) and lodgepole pine (McLean and Clark, 1980), Ponderosa pine (Rummell, 1951; Currie, 1978; Zimmerman and Neuenschwander, 1984), on mixed conifer sites in California (Allen, 1986; Huntsinger and Bartolome, 1986), and on Douglas fir plantations in western U.S. (Alejandro-Castro *et al.*, 1987). However, since Douglas fir seedlings are sensitive to trampling damage during the first year after planting, efforts must be made to prevent localized animal concentrations and the resulting trampling damage (Eissenstat *et al.*, 1982).

Sheep have proved effective in biological control of unwanted brush species in Douglas fir plantations and in accelerating tree regeneration and subsequent growth (Hedrick and Keniston, 1966; Leininger and Sharrow, 1983, 1987). One recommendation has been to concentrate sheep browsing in summer and fall when brush palatability is relatively high rather than in spring and winter when the relative palatability of Douglas fir seedlings is greatest. Although winter browsing by sheep should be avoided, light spring grazing is acceptable but requires preventing animal concentrations such as around bedgrounds and watering and salting places.

Sharrow (1994) concluded that sheep browsing of young conifers had relatively little impact upon growth unless the terminal leader or almost all of the current year's lateral branches were consumed; seedlings were found most likely to be browsed during the spring when lush new twig and needle growth was present or any time that other green feed became scarce. In north-central Idaho, sheep consumption of conifers was not affected by grazing season, and conifers averaged only 4% of sheep diets (Mbabaliye *et al.,* 1999). By contrast 58% of the sheep diets were graminoids; preference for graminoids was high in both early and late summer, with forbs and shrubs becoming more prominent in diets as environmental conditions caused graminoids to become drier and coarser than other forages in late summer.

APPENDIX

TABLE OF WEIGHTS, MEASURES, AND EQUIVALENTS

1. LENGTH MEASURE

1 mile = 5280 feet = 1760 yards = 320 rods = 80 chains = 1609.34 meters = 1.609 kilometers

1 chain = 66 feet = 22 yards = 100 links

1 rod = 16.5 feet = 5.5 yards

1 yard = 3 feet = 36 inches = 0.914 meter = 91.44 centimeters

1 foot = 12 inches = 0.305 meter = 30.48 centimeters

1 inch = 2.54 centimeters = 25.4 millimeters

1 kilometer = 1000 meters = 0.621 mile

1 meter = 10 decimeters = 39.37 inches = 3.281 feet = 1.094 yards

1 decimeter = 10 centimeters = 3.94 inches

1 centimeter = 10 millimeters = 0.394 inches

1 millimeter = 0.04 inch

2. AREA MEASURE

1 township = 36 sections = 36 square miles = 23,040 acres

1 section = 1 square mile = 640 acres = 2.59 square kilometers = 259.0 hectares

1 acre = 43,560 square feet = 160 square rods = 0.405 hectares = 4047 square meters = 10 square chains

1 square rod = 0.006 acres = 272.25 square feet = 25.293 square meters

1 square foot = 0.093 square meter

1 square inch = 6.451 square centimeters

1 square kilometer = 1,000,000 square meters = 100 hectares = 0.3861 square mile = 247.1 acres

1 hectare = 10,000 square meters = 2.471 acres

1 square centimeter = 0.155 square inch

3. CAPACITY MEASURE

1 gallon = 4 quarts = 8 pints = 16 cups = 256 tablespoons = 768 teaspoons
   = 231 cubic inches = 3.785 liters
1 quart = 2 pints = 4 cups = 64 tablespoons = 192 teaspoons = 0.946 liter
1 pint = 2 cups = 32 tablespoons = 0.473 liter = 473 milliliters
1 cup = 16 tablespoons
1 tablespoon = 3 teaspoons
1 liter = 1.057 quarts = 0.264 gallon = 61.02 cubic inches

4. WEIGHT MEASURE

1 ton = 2000 pounds = 907.18 kilograms
1 hundredweight (cwt.) = 100 pounds = 45.359 kilograms
1 pound = 16 ounces = 453.59 grams = 0.454 kilograms
1 ounce = 28.35 grams
1 kilogram = 2.205 pounds = 1000 grams = 1,000,000 milligrams
1 gram = 0.0022 pound = 0.0352 ounce = 1000 milligrams
1 gallon water = 8.346 pounds = 3785.655 grams
1 cubic foot of water = 62.43 pounds
1 cubic centimeter of water at 39.2°F = 1 gram

5. MISCELLANEOUS

1 pound per acre = 1.121 kilograms per hectare = 11.21 grams per square
   meter
1 kilogram per hectare = 0.891 pounds per acre
Degrees Fahrenheit = (9/5 × degrees Celsius) plus 32
Degrees Celsius = 5/9 × (degrees Fahrenheit minus 32)
Area of circle = 3.1416 × radius^2
Circumference of circle = 3.1416 × diameter
Grams per 96 square feet = pounds per acre

GLOSSARY[1]

Abomasum. The fourth compartment of the ruminant stomach, comprising the true stomach, in which occur digestive processes similar to those found in the nonruminant stomach.

Accessibility. (a) The ease with which an area can be reached or penetrated by large herbivores; (b) the ease with which large herbivores can reach and consume plants and plant parts. Accessibility, when related either to grazing area or forage plants, will vary depending upon kind or even class of grazing animal.

Accessible. (a) An area readily reached or penetrated by large herbivores; (b) plants or plant parts that are readily reached and consumed by large herbivores.

Agroforestry. The use and management of forested lands for integrating the production of timber and wood products with other agricultural crops, with or without animal production.

Alternate grazing. The repeated grazing and nongrazing (resting) of two grazing land units in sequence, typically during the growing season, commonly at 1- to $2\frac{1}{2}$-month intervals. Synonym, *switchback grazing*.

Animal conversion ratio. See *Animal substitution ratio*.

Animal day. One day's tenure upon grazing land by one animal; must be made specific as to kind and class of animal to be meaningful.

Animal demand (for forage). Potential forage dry matter intake of ungulate herbivores based solely on animal-related factors such as body size, body condition, stage of life cycle, production stage, etc.; the use of energy requirements of ungulate herbivores is an acceptable alternative basis for animal demand, particularly for feed concentrates and limit-fed harvested forages.

Animal month. One month's tenure upon grazing land by one animal; must be made specific as to kind and class of animal to be meaningful.

Animal substitution ratio. A numerical ratio of numbers of one animal species to another, considering management objectives and options, for use in partly or completely converting grazing capacity between two animal species. Such a ratio is site-specific since it is based on a unique set of environmental, forage, animal-herbage, and animal-area variables

---

[1]Terminology in this second edition of *Grazing Management* has been adapted to much but not all of the new and revised terminology found in *Terminology for Grazing Lands and Foraging Animals* (Forage and Grazing Terminology Comm. [FGTC], 1991; reprinted for wider access in FGTC, 1992) and in *Glossary of Terms Used in Range Management* (Society for Range Management [SRM], 1998; 4th ed.).

and requires knowledge of relative animal population levels at the time of conversion. Synonym, *animal exchange ratio* or *animal conversion ratio.*

Animal unit (AU). One mature, non-lactating bovine weighing about 1000 lb (454 kg) or its equivalent in other classes or kinds of ungulate herbivores based on animal demand or quantitative forage dry matter intake; assumes a standard daily forage intake of 26 lb (12 kg) on an oven-dry basis.

Animal unit conversion factor. See *Animal unit equivalent.*

Animal unit day (AUD). The $\frac{1}{30}$ of an animal unit month.

Animal unit equivalent (AUE). A number (either a decimal or multiple) expressing the animal demand (for forage) of a particular kind and class of animal relative to that of an animal unit. Such equivalents are satisfactory in respect to intraspecific potential forage intake but are less applicable for determining interspecific stocking rates or forage allocations between animal species because they do not account for degree of dietary and area use overlap. See *Animal substitution ratio.*

Animal unit month (AUM). (a) Animal demand, i.e., potential forage dry matter intake of one animal unit for a period of one month (30 days), this based on 780 lb of dry matter or less frequently on 540 therms of digestible energy or 270 lb of TDN. The AUM will be more meaningful when based on one specific animal species; this sometimes qualified as a "deer" or "sheep" AUM; an animal demand function. (b) (plural) Amount of grazing capacity when used as a common measure of pasturage, or carrying capacity of pasturage plus harvested roughages (and less frequently including feed concentrates also) produced on a specific land area or assembled at one place; a supply function that can be related to animal carrying capacity needs.

Animal unit year. Equivalent to 12 AUMs when harvested by one animal unit over a continuous 12-month grazing period. Synonym, *animal unit yearlong.*

Available forage. That portion of the *forage mass,* expressed as dry weight of forage per unit land area, that is accessible for consumption by a specified kind, class, sex, size, age, and physiological status of grazing animal. Synonym, *accessible forage.* See *Forage allowance.*

Aversion. A decrease in preference for food just eaten resulting from negative animal sensory responses and post-ingestive feedback to the food.

Balanced operation. A livestock or big game operation for which sufficient forage and other feed resources are available and provided during each season to promote continuous satisfactory maintenance and production of the animals.

Best-pasture grazing. See *Flexible sequence grazing.*

Biological control (by grazing). The application of selective grazing to achieve a desired directional succession in vegetation composition in order to achieve a predetermined objective.

Bite. Forage ingested as defined by a sequence of prehension, gripping, and severance motions; the smallest spatial foraging level or scale.

Bloat, legume. A condition in ruminants caused by the consumption of certain legume species and resulting in the retention of gas produced in normal rumen function as a frothy mass along with an inhibition of the eructation (belching) mechanism.

Browse. (n) Leaf and twig growth of shrubs, woody vines, and trees acceptable (edible) for animal consumption; (v) to consume browse. See *Graze.*

Browse line. A well-defined height up to which (or even down to which, in some cases) browse has been removed by herbivores. Synonym, *highline.*

Browser. An animal species that concentrates its plant consumption from woody plants rather than herbaceous plants.

Brush. Shrubs or small trees considered undesirable from the standpoint of planned use of the area; an undesirable noxious, woody plant. See *Shrub* and *Browse.*

C-3 plant. A plant employing the pentose phosphate pathway of carbon dioxide assimilation during photosynthesis; a cool-season plant.

C-4 plant. A plant employing the dicarboxylic acid pathway of carbon dioxide assimilation during photosynthesis; a warm-season plant.

Calf crop. The number of calves weaned from a given number of cows, usually expressed in percent; number of cows most commonly based on those exposed to breeding, or less commonly on those remaining in the herd at the end of the breeding season or at the beginning of calving.

Camp. A spatial foraging level defined as a set of feeding sites sharing a common foci for drinking, resting, and seeking cover.

Carrying capacity. All nutrient resources available on a given land area, including not only pasturage but also harvested forages and other feedstuffs used to complement the grazing resources, thereby providing a means of summarizing total ranch capacity or that allotted to a specific animal enterprise (best usage). Synonym, in part, *grazing capacity. Carrying capacity* includes *grazing* capacity but is not restricted to grazed storage.

Cecum. A blind sac forming the forepart of the large intestine in the horse and certain other non-ruminant herbivores; provides a medium for fermentation of fibrous materials.

Cell (or **grazing cell**). A grazing arrangement comprised of numerous subunits (i.e., paddocks), usually in a rotational grazing system, with a common central component provided with drinking water, animal handling facilities, and access between subunits.

Class of animal. Age, sex, or stage-of-production group within a kind (species) of animal.

Climax. The final or stable biotic community in a successional series; it is self-perpetuating and in equilibrium with the physical habitat.

Close herding. Handling a herd or flock of animals in a closely bunched manner, restricting the natural spread of the animals when grazing.

Common use. Synonym, *mixed grazing.*

Compensatory gains. Subsequent animal weight gains that are enhanced (or depressed) as a result of gains made during a prior period.

Competition. The general struggle for existence within a trophic level in which the living organisms (either plants or animals) compete for a limited supply of the necessities of life. See *Disturbance competition* and *Interference competition.*

Complementary pasture. A pasture of different kinds of forage, generally short-term (one season to 10 years) and of enhanced productivity, grazed chronologically (or sometimes simultaneously) with a primary forage resource base such as rangeland to provide additional or alternative grazing capacity. See *Supplementary pasture.*

Complementary rotation grazing. A variant of sequence grazing in which different areas of a seeded pasture or range unit are seeded to different species or mixes but not fenced separately, thereby permitting grazing animals to rotate themselves as they choose.

Concentrate feed. Grains or their products or other processed feeds that contain a high proportion of nutrients (over about 60% TDN) relative to bulk and are low in crude fiber (under about 20%).

Consumption. Dietary intake based on (1) amounts of specific forages and other feedstuffs or (2) amounts of specific nutrients.

Continuous grazing. Allowing animals unrestricted and uninterrupted access to a grazing land unit for all or most of the grazing season; includes *yearlong, growing season,* and *dormant season continuous grazing.* Synonym, *continuous stocking.*

Continuous stocking. See *Continuous grazing.*

Cool-season plant. A plant which generally makes the major portion of its growth during the late fall, winter, and early spring; these plants usually are C-3 plants.

Corral. A small enclosure for handling livestock. See *Paddock.*

Creep feeding. Supplemental feeding of young offspring in such a manner that the feed is not available to the dams or other mature livestock.

Creep grazing. Allowing juvenile animals to graze areas their dams or other mature animals cannot access at the same time, thus providing access by the juvenile animals to forage of higher quality and/or quantity.

Critical area. An area which must receive special management, treatment, or consideration because of inherent site factors, erosion potential, special values, or potential conflict of uses.

Crop aftermath. Regrowth of plants made after harvesting the primary crop, this available for grazing or additional mechanical harvesting; includes the pre-harvest excess foliage of small grain crops (i.e., *foremath*).

Crop aftermath/residue pasture. A kind of pasture in which grazing is provided as a secondary product of the land and is carried out after (or sometimes before) the primary crop is produced and harvested; consists of excess foliage of the primary crop, crop regrowth, stubble, crop residues, and weeds and volunteer growth.

Crop residue. Plant materials, with variable forage potential, remaining on the land as a consequence of harvesting seed, grain, or foliage of the primary crop; includes stubble, harvest refuse, weeds, and volunteer growth.

Crop-rotation pasture. Land on which grazing is maintained, generally for 3 to 10 years, in a designed crop rotation cycle, with perennial forage species mostly being utilized; replaces **tame pasture,** in part.

Cropland. Land devoted to the production of cultivated crops which may be used to produce *forage crops.*

Cropland pasture. Arable lands on which grazing is currently being realized but presumably under limited duration; includes crop-rotation pasture, temporary pasture, and crop aftermath/residue pasture.

Cumulative herbage disappearance. The amount of herbage that disappears from the standing crop because of grazing, senescence, or other causes over a period of time; unit: lb/acre or kg/ha.

Cured forage. Forage, either standing or mechanically harvested, that has been naturally or artificially dried and preserved for future use. See *Stockpiling.*

Decreaser. A plant species native to a site or plant community (or sometimes expanded to an intentionally introduced forage species) that will decrease in relative amount under continued heavy grazing.

Deferment. Nongrazing from the breaking of plant dormancy until after seed set or equivalent stage of vegetative reproduction, accomplished either by delaying the beginning of spring grazing or discontinuing winter grazing early.

Deferred grazing. The deferment of grazing on a grazing land unit but not in a systematic rotation with other units, followed by grazing of the residual standing crop. Synonym, *selected deferment.* See *Rotational deferment* and *Deferred-rotation grazing.*

Deferred-rotation grazing. A multi-unit, one-herd grazing system in which deferment is systematically rotated among the respective grazing land units; a combination of deferred grazing and low frequency rotation grazing. See *Merrill grazing system.*

Defoliation. The removal of plant leaves by grazing or browsing, cutting, chemical defoliant, or natural phenomena such as hail, fire, or frost.

Degree of use. See *Utilization.*

Density. The number of individuals per unit area and, by implication, the relative closeness of individuals to each other.

Desired plant community. Of the several plant communities that may occupy a site, the one that has been identified through a management plan to best meet the plan's objective for the site; it must protect the site as a minimum.

Digestibility, apparent. The balance of nutrients in the ingesta minus that in the feces.

Digestibility, true. Differs from apparent digestibility in requiring additionally that the metabolic products added back into the feces be accounted for and subtracted out of apparent nutrient losses through the feces.

Disturbance competition. Species behavioral interaction in which one animal species seemingly voluntarily leaves the vicinity of one or more other animal species. See *Interference competition.*

Dominant. (a) A plant species or species group which has considerable influence or control over associated plant species; (b) those individual animals which, by their aggressive behavior or otherwise, determine the behavior of one or more other animals, resulting in the establishment of a social hierarchy.

Drift fence. An open-ended fence used to retard or alter the natural movement of animals.

Drifting (or **drift**). The natural directional movement of animals; also the slow urging by the grazier of animals in a certain direction, utilizing the natural movement of the animals as much as possible.

Driving (or **drive**). Close control of livestock by the grazier when being moved towards a specific destination. See *Drifting.*

Drought. Prolonged dry weather, generally when precipitation is less that 3/4 of average for a considerable period of time; period during which plants suffer from lack of water.

Dual use. Mixed species grazing when only two kinds of grazing animals are involved.

Duration of grazing. See *Grazing period.*

Ease of prehension. The relative ease with which a plant or plant part can be approached and grasped in the mouth during grazing.

Ecological site. A kind of land with specific physical characteristics that differs from other kinds of land in its ability to produce distinctive kinds and amounts of vegetation and in its response to management; replaces *range site.*

Edge effect. (a) The influence of two adjoining plant communities on the vegetation and animals in the margin between them; also (b) the attraction the margin (edge) may have for certain animal species.

Edible. Acceptable for animal consumption; fit to be eaten.

Emergency feeding. Supplying feed to grazing animals when the standing crop is insufficient because of drought, deep snow or icing over, fires, or other emergencies. See *Maintenance feeding* and *Supplemental feeding.*

Extended feeding intervals. Providing supplemental feeds systematically but less frequently than daily.

Extensive grazing management. See *Grazing management.*

Feed. (n) Any non-injurious, edible material, including forage, having nutritive value for animals when ingested. Synonyms, *feedstuff* and *food.* (v) The act of providing feed to animals.

Feed ground. A designated area where livestock or big game are fed harvested forages and/or concentrates.

Feeding bout. See *Grazing bout.*

Feeding site. A spatial foraging level defined as a collection of patches in a contiguous area that animals graze during a foraging bout.

Feeding station. The area available in a half-circle shape in front of and to each side of the grazing animal while its front feet are temporarily stationary; a spatial foraging level defined as an array of plants available to a large herbivore without moving its front feet.

Feeding station interval. The pause exhibited by the grazing animal as it feeds at a given feeding station.

Fiber. The complex of forage components that are relatively resistant to digestion;

it is comprised principally of plant cell walls which are a composite of the structural polysaccharides cellulose and hemicellulose along with lignin, small amounts of protein, and cuticle waxes.

First-last grazing. Managing two groups of grazing animals, these usually with different nutritional requirements, in which the favored group is grazed first and followed by the less favored second group; applied to individual grazing land units or to a series of paddocks, the latter usually enabled by doubling the number of paddocks. Synonyms, *topping-followup, leader-follower,* and *top-and-bottom grazing.*

Flexible sequence grazing. A multi-unit grazing procedure based on periodic, on-the-spot decisions as to which unit is to be grazed next and for how long. Synonym, *best-pasture grazing.*

Flushing. Improving the nutrition of female breeding animals, usually by providing energy concentrates or lush pasture prior to and during the breeding season, as a means of stimulating estrus and ovulation.

Forage. (n) Edible parts of plants, other than separated grain, that can be consumed by grazing animals or mechanically harvested for feeding; includes the edible portion of herbage, also browse and mast. See *Forage mass* and *available forage.* (vt) To search for and consume forage.

Forage accumulation. The increase in forage mass per unit area over a specified period of time.

Forage allocation. The partitioning of the standing forage crop or its associated grazing capacity between different kinds or classes of ungulate herbivores.

Forage allowance (cumulative). Weight of forage dry matter (forage mass) per unit of animal demand over a period of time (i.e., T/AUM or lb/AUD; a forage-animal relationship; the reciprocal of *cumulative grazing pressure*.

Forage allowance (instantaneous). Weight of forage dry matter (forage mass) per unit of animal demand at any instant of time (i.e., T/AU); a forage/animal relationship; the reciprocal of *instantaneous grazing pressure*.

Forage-animal plans. Combined forage and management practices directed to meeting the nutritional needs of the grazing animals in specific production phases or throughout a production cycle.

Forage, available. See *Available forage*.

Forage crop. A crop generally consisting of cultivated plants and plant parts, other than separated grain, produced to be grazed or mechanically harvested for use as feed for animals.

Forage, grazed. See *Grazed forage*.

Forage, harvested. See *Harvested forage*.

Forage mass. The total dry weight of forage per unit area of land, usually above ground level and at a defined reference level. Forage mass may include forage not accessible to a particular kind of grazing animal. See *Available forage*.

Forage production. The weight of forage that is produced within a designated period of time on a given area; expressed as green, air-dry, or oven-dry weight and may be qualified as annual, current year's, or seasonal forage production.

Forage reserve. Standing forage specifically maintained for future or emergency use. See *Stockpiling*.

Forage value. A subjective evaluation of a forage plant species giving consideration to some or all of the following: palatability, length of palatable period, nutritive value, and productivity under grazing.

Foraging. The search for forage.

Foraging velocity. The rate at which large herbivores transit different portions of the landscape.

Forb. Any herbaceous plant other than grasses and grass-like plants (the latter including sedges and rushes).

Forward creep grazing. A combination of creep grazing and first-last grazing in which the juvenile offspring are allowed to "creep ahead" as first grazers but are later followed by their dams as last grazers; can be applied to a single grazing land unit or more commonly in rotation through a series of paddocks.

Free-ranging. The opportunity of grazing animals to roam or forage at will with no (or minimal) restriction from fencing or herding.

Frontal grazing. A grazing method that allocates forage within a pasture unit by means of a sliding fence that cattle can advance at will to gain access to ungrazed forage.

Full use. The maximum grazing use during a grazing season that can be made of range/pasture forage under a given grazing program without inducing a downward trend in vegetation condition.

Game cropping. Maintaining game animals in a wild state while harvesting them to keep populations in check and to reduce cycle extremes in numbers.

Game farming. A more intensified variant of game ranching in which animal management practices are similar to those applied to fully domesticated animals, often including the use of improved pastures, harvested forages, and feed concentrates.

Game ranching. Producing game animals under semi-domestication and intensive animal management to control breeding, health, nutrition, production, and marketing; a ranch earning enterprise.

Graminoids. True grasses and grass-like plants (e.g., sedges and rushes).

Grass. A member of the plant family *Poaceae.*

Grassland agriculture. A land management system emphasizing harvested forage crops, pasture, and rangelands for livestock production and soil stability.

Grass-like plants. Plants of the *Cyperaceae* (sedge) and *Juncaceae* (rush) families, which vegetatively resemble the true grasses.

Graze. (vt) The consumption of standing forage by ungulate herbivores (best usage); (vi) to put livestock to feed on standing forage.

Graze-out. Maximizing grazing of a standing crop prior to site conversion to another crop or alternative use.

Grazed forage. Forage consumed directly by the grazing animal from the standing crop. Synonym, *pasturage.* See *Harvested forage.*

Grazer. A grazing animal; (specific) an animal species that concentrates its plant consumption on standing herbaceous plants.

Grazier. A person who manages grazing animals.

Grazing. The act of eating forage from the standing crop comprised of (1) foraging, the search for forage; (2) defoliation, the removal of forage; and (3) ingesting the forage. Synonym, *herbivory.*

Grazing animal concentration index. See *Stocking density index.*

Grazing behavior. The foraging response elicited from an ungulate herbivore by its interaction with its surrounding environment.

Grazing bout. A period of concentrated grazing, typically lasting 1 to 4 hours with large herbivores, that is preceded and followed by non-grazing activities such as resting and ruminating. Synonyms, *foraging bout* or *feeding bout.*

Grazing capacity. (a) The optimal stocking rate that will achieve a target level of animal performance or other specific objective, while preventing deterioration of the ecosystem. Must consider both management objectives and management intensity to be accurate. (b) Total number of AUMs produced and available for grazing per acre or from a specific grazing land unit, a grazing allotment, the total ranch, or other specified land area. Synonym, in part, *carrying capacity* but with the latter term not being restricted to grazed forage.

Grazing cycle. The length of time between the beginning of one grazing period and the beginning of the next on the same paddock, where the forage is systematically grazed and rested; the sum of the grazing period and the following non-grazing (rest) period.

Grazing distribution. Dispersion of animals during grazing over a management unit or area.

Grazing efficiency. The percent of the total standing crop by weight that is ingested, or the percent of the total cumulative forage disappearance by weight that is ingested.

Grazing fraction. The fraction of land in a single grazing unit or among paddocks in a rotation system which is being grazed at any given time; expresses level of animal concentration; the inverse of the stocking density index.

Grazing intensity. A general term expressing (1) the amount of animal demand placed upon the standing crop or forage mass, and (2) the resulting level of plant defoliation made during grazing.

Grazing land. Any vegetated land area that is grazed or has the potential to be grazed by animals. Grazing lands, including both range and pasture, may be conveniently classified as *long term* (unlimited continuation of grazing projected), *medium term* (grazing tenure uncertain but projected at least 10 years), and *short term* (duration of grazing projected for 10 years or less).

Grazing land management. (a) The art and science of planning and directing the development, maintenance, and use of grazing lands to obtain optimum, sustained returns based on management objectives. (b) The manipulation of the soil-plant-animal complex of the grazing land in pursuit of a desired result (SRM).

Grazing management. The manipulation of animal grazing to achieve desired results based on animal, plant, land, or economic responses. Grazing management may be *extensive* (utilizes relatively large land area per animal and relatively low input levels of labor, resources, and capital) or *intensive* (utilizes relatively high input levels of labor, resources, and capital in attempting to increase quantity and/or quality of forage and thus animal production).

Grazing management plan. A comprehensive plan to secure the best practicable use of the grazing resources with grazing or browsing animals; the day-to-day provision of grazing capacity.

Grazing method. A defined procedure or technique of grazing management based on a specified period of grazing and/or period of nongrazing and is designed to achieve a specific objective. One or more grazing methods are utilized within a *grazing system.*

Grazing period. The length of time that grazing animals occupy a specific grazing area without interruption. The grazing period may consist of (1) the sole period of annual occupancy, (2) one of two or more irregular periods of occupancy during the year, or (3) one of several periods of occupancy characterizing each paddock within a rotation grazing series.

Grazing period cycle. The sum of one grazing period and the following nongrazing period within a paddock included in a rotation grazing series.

Grazing preference. The selection of certain plants or plant parts over others by grazing animals, this being animal response to the summation of plant factors affecting palatability. Synonym, *preference.*

Grazing pressure (cumulative). Animal demand per unit weight of forage dry matter (*forage mass*) available over a period of time (i.e., AUM/T or AUD/T); an animal/forage relationship. Synonym, *grazing pressure index.*

Grazing pressure index. Synonym, *grazing pressure (cumulative).*

Grazing pressure (instantaneous). Animal demand per unit weight of forage dry matter (*forage mass*) available at any instant of time (i.e., AU/T); an animal/forage relationship.

Grazing season. That time period during which grazing is normally possible and practical each year. The grazing season may be the whole year or a shorter time span and may be within or extend beyond the vegetation growing season.

Grazing system. A specialization of grazing management based on rotating grazing animals among two or more grazing land units (paddocks) while defining systematically recurring periods of grazing and nongrazing. A grazing system will generally include one or more grazing methods in addition to rotation grazing.

Grazing system cycle. The length of time required for all grazing methods and other treatments included in a grazing system to be passed through all paddocks within the series; the actual time required may extend from only part of a grazing season through a few years.

Grazing unit. A grazing unit, either range unit or pasture unit, is enclosed and separated from other land areas by fencing or other barriers; it may be a single area or one in a series of units (paddocks) in a rotation grazing system.

Green chop. Mechanically harvested forage fed to animals while still fresh. Synonym, *zero grazing.*

Growing season. That portion of the year when temperature and moisture typically enable plant growth.

Harvested forage. Forage mechanically harvested before being fed to animals in the form of hay, haylage, fodder, stover, silage, green chop, etc. See *Grazed forage* or *Pasturage.*

Heavy grazing. A comparative term which indicates that the stocking rate on an area is relatively greater than that on other similar areas; sometimes erroneously used to mean overgrazing.

Hedged. Describes the appearance of woody plants that have been repeatedly browsed so as to appear artificially pruned.

Hedging. The persistent browsing of terminal buds of browse species causing extensive lateral branching and a reduction in main stem growth.

Herbaceous. Vegetative growth with little or no woody component; non-woody vegetation such as graminoids and forbs.

Herbage. The biomass of herbaceous plants, generally above ground but may be specified to include edible roots and tubers. Herbage generally includes some plant material not edible or accessible to ungulate herbivores.

Herbage allowance. See *forage allowance.*

Herbivore. An animal that subsists principally or entirely on plants or plant materials.

Herbivory. See *grazing.*

Herd effect. A controversial concept that grazing under high animal concentration can be used to beneficially modify soil surface characteristics through hoof action.

Herd instinct. The natural intraspecies gregariousness expressed to different degrees by both wild and domestic grazing animal species.

Herding. The handling or tending of a herd or flock of grazing animals; (specifically) control exercised by the herder or grazier in keeping grazing animals together in a group. See *Close herding, Open herding,* and *Trailing.*

High-intensity/high-frequency (HIHF) grazing. Synonym, *short-duration grazing.*

High-intensity/low-frequency (HILF) grazing. A grazing system employing high to medium stocking density, commonly three to five grazing units, grazing periods generally over 2 weeks and often 30 to 45 days, nongrazing periods of 60 days or longer, and two to four (sometimes only one) grazing period cycles per year. Synonym, *slow rotation grazing* and *high utilization rotation grazing.*

Highline. Synonym, *browse line.*

Home range. The area over which an animal normally travels in search of food. A spatial foraging level defined as a collection of camps and confined by fences, barriers, extent of migration, or transhumance. Synonym, *landscape.*

Ice-cream plant. A plant of an exceptionally palatable plant species sought and grazed first by grazing and/or browsing animals. Plants of such species are usually over-utilized under proper stocking when present in mixed stands.

Improved pasture. Grazing lands to which have been applied substantial cultural treatments for developing and/or restoration as a means of improving quantity and quality of forage; a broad term generally implying enhanced productivity and potential to respond to intensive management; largely synonymous with *tame pasture,* which it replaces.

Increaser. A plant species native to a site or plant community (or sometimes extended to include an intentionally introduced species) that will increase in relative amount, at least for a time, under heavy use.

Ingesta. Nutritive materials consumed by an animal.

Intensive-early stocking (IES). A grazing method involving increased stocking density, often at about twice the normal level, during the first half of the growing season followed by nongrazing through the remainder of the growing season. Synonym, *intensive-early grazing (IEG).*

Intensive grazing management. See *Grazing management.*

Interference competition. Species behavioral interaction in which one animal species takes aggressive defense of territory against one or more other animal species. See *Disturbance competition.*

Intermediate feeders. An animal species that uses large amounts of grasses, forbs, and shrubs and has substantial capability to adjust their feeding habits to whatever kind of forage is available.

Intermittent grazing. Grazing for indefinite periods at irregular intervals.

Introduced species. A species not a part of the original fauna or flora of the area in question. See *Native species* and *Naturalized species.*

Invader. A plant species that was absent or only minimally present in the original vegetation and invades or rapidly increases following disturbance or continued heavy grazing.

Key area. A relatively small portion of a grazing land unit on which management of the entire unit is based; used as an indicator of proper grazing over the entire unit.

Key species. Forage species on which management of grazing is based; used as an indicator of proper grazing of the total vegetation.

Kid crop. The number of kids produced by a given number of does, usually expressed in percent kids weaned of does exposed to breeding.

Kind of animal. An animal species or species group of livestock or game animals.

Lamb crop. The number of lambs produced by a given number of ewes, usually expressed in percent of lambs weaned of ewes exposed to breeding.

Landscape. Synonym, *home range,* as a spatial foraging level.

Leaf area index (LAI). The ratio of the cumulative upper leaf surface area of the plant community to the corresponding ground area, expressed as a proportion. (LAI may exceed 1.)

Light grazing. A comparative term which indicates that the stocking rate on an area is relatively less than that on other similar areas; sometimes erroneously used to imply undergrazing.

Limit grazing. The practice of limiting the length of time daily that animals have access to grazable forage.

Loose running. Allowing grazing animals to graze without being herded. See *Free-ranging.*

Maintenance feeding. Supplying harvested forages and concentrates to provide minimal daily animal maintenance requirements, thereby partly or completely replacing grazing of the standing crop. See *Emergency feeding* and *Supplemental feeding.*

Malaise. A feeling of nausea or unpleasant

physical discomfort after ingesting a food or foods; a negative post-ingestive feedback.

Mast. Fruits and seed of shrubs, woody vines, trees, cacti, and other non-herbaceous plants acceptable (edible) for animal consumption.

Mastication, ingestive. Initial chewing prior to swallowing.

Mastication, ruminative. Chewing the cud after regurgitation.

Meadow. An area covered with grasses and/or legumes, often with the potential for either hay or grazing; openings in forests and grasslands of exceptional productivity, usually resulting from higher water content of the soil.

Merrill grazing system. A modification of rotational deferment consisting of a three-herd, four-paddock grazing system with a 4-year grazing system cycle which incorporates yearlong continuous grazing and periodic deferment and yearlong rest.

Mixed grazing. Grazing by two or more kinds of grazing animals on the same land unit, either at the same time or at different times of the year. Synonym, *common use.* See *Dual use.*

Mob grazing. Grazing a pasture or range unit with a relatively large number of animals at a high stocking density for a short time period.

Moderate grazing. A comparative term which indicates that the stocking rate of an area is intermediate between the rates of other similar areas; sometimes used to imply proper grazing.

Multi-species grazing. See *Mixed grazing.*

Native species. A species which is part of the original flora or fauna of the area in question. Synonym, *indigenous species.* See *Introduced species, Naturalized species,* and *Resident species.*

Naturalized species. A species not native to an area but which has adapted to that area and has established stable or expanding populations; does not require artificial inputs for survival and reproduction.

Nomadism. The habit of wandering from place to place, usually within a well-defined territory; characterizes the movement of people with their flocks and herds without a home base.

Nongrazing. The restriction or absence of grazing use on an area for a period of time, ranging from a short period of a few days to a year or more. Synonym, *rest* (in current usage).

Nongrazing period. The uninterrupted length of time that grazing animals are prevented access to a specific area. Synonym, *rest period* (in current usage).

Nonselective grazing. Utilization of standing forage by grazing animals with the objective of grazing all forage species, plants, and plant parts to a comparable degree; a conceptual objective seldom, if ever, achieved except under extreme overgrazing.

Nonuse. (a) Absence of use on the current year's forage production. (b) Lack of exercise, temporarily, of a grazing privilege on grazing lands or the authorization to do so without loss of preference for future considerations.

Noxious plant. A plant which is undesirable in light of planned land use or which is unwholesome to grazing lands or grazing animals. Synonym, *weed.*

Nutrient. Any food constituent or ingredient that is required for or aids in the support of life.

Nutrition. The ingestion, digestion, and assimilation of food by plants and animals.

Nutritional wisdom, theory of. The assumption that animals knowingly or instinctively select forage or other feedstuffs to meet their nutrient requirements.

Nutritive value. Relative capacity of a given forage or other feedstuff to furnish nutrition for animals; a general term unless reference is made to specific nutrients.

Open herding. A method of herding in which the individuals in a herd are allowed to spread naturally for grazing but are kept within a prescribed area of the grazing land unit.

Open range. See *Range, open.*

Opening date. The date on which an established grazing season begins or is legally permitted.

Optimal foraging, theory of. The assumption that grazing animals will optimize some objective function in their grazing; often that the animal will maximize its energy intake per unit of time or effort expended.

Overgrazing. Continued heavy grazing which exceeds the recovery capacity of the forage plants and creates deterioration of the grazing lands.

Overstocking. Placing so many animals on a grazing unit that overuse will result if continued unchanged to the end of the planned grazing period.

Paddock. (a) One of the multiple grazing units or subunits included in a rotation grazing series. (b) A relatively small enclosure used as an exercise or saddling area for horses.

Palatability. Summation of plant (or feed) characteristics that determine the relish with which a particular plant species or plant part (or feed) is consumed by an animal; partly synonymous with *acceptability.*

Palatability factor. An index to the grazing use (expressed in percent utilization) that is made of a specific forage species when the plant species mixture as a whole is deemed properly used; an expression of relative palatability. Synonym, *use factor;* formerly synonymous with *proper use factor.*

Pasturage. Synonym, *grazed forage.* See *Harvested forage.*

Pasture. (n) Land supporting mostly introduced forage plant species for grazing by ungulate herbivores. Rather than being managed as a natural ecosystem, pasture must generally be managed to arrest natural plant successional processes. See *Pasture unit, Paddock,* and *Pasturage;* similar terms not recommended as synonyms for pasture. The term "pastureland" is used as a land use for statistical and mapping purposes by USDA. (v) To graze.

Pasture, complementary. See *Complementary pasture.*

Pasture, crop aftermath/residue. See *Crop aftermath/residue pasture.*

Pasture, crop-rotation. See *Crop-rotation pasture.*

Pasture, cropland. See *Cropland pasture.*

Pasture, improved. See *Improved pasture.*

Pasture, perennial. See *Permanent pasture* and *crop rotation pasture.*

Pasture, permanent. See *Permanent pasture.*

Pasture/range, transitory. See *Transitory pasture/range.*

Pasture, supplementary. See *Supplementary pasture.*

Pasture, tame. See *Tame pasture.*

Pasture, temporary. See *Temporary pasture.*

Pasture unit. An area of pasture enclosed and separated from other land areas by fencing or other barriers. As defined, this term excludes *range unit* but is included within *grazing land unit*. See *Paddock*.

Patch. (a) A specific aggregation of forage plants. (b) A spatial foraging level defined as a cluster of feeding stations separated from other potential feeding stations; requires a break in foraging for transit from one patch to another patch.

Patch grazing. Close and often repeated grazing of small patches or even individual plants while adjacent similar patches or individual plants of the same species are lightly grazed or ungrazed.

Percent use. Removal by grazing of current's year plant growth when expressed as a percentage by weight of total plant growth.

Perennial pasture. See *Permanent pasture* and *Crop-rotation pasture*.

Permanent pasture. Medium-term grazing lands on which the forage stand is principally perennial grasses and legumes and/or self-seeding annuals and on which grazing tenure is indefinite but expected to exceed 10 years; replaces *tame pasture*, in part.

Phytomer. One modular unit of a plant consisting of a leaf, an internode, an axillary bud or potential bud, and a node.

Plant palatability. See *Palatability*.

Poisonous plant. A plant containing or producing substances that cause sickness, death, or deviation from normal state of health of animals.

Poloxalene. An anti-foaming agent fed to prevent legume bloat in ruminants.

Post-ingestive feedback. An animal's sensing of the nutritional or toxicological effects of recent food ingestion, either positive or negative, from which the animal accordingly adjusts its preference for the food by increasing or decreasing intake.

Preference. See *grazing preference*.

Preferred species. Plant species that are preferred by animals and are grazed first by choice.

Proper grazing. A general term for achieving continuously the proper use of a grazing area by grazing animals.

Proper stocking. Placing the number of animals on a given area that will result in proper use at the end of the planned grazing period.

Proper use. The degree and time of grazing use of current year's plant growth which, if continued, will either maintain or improve the condition of grazing lands.

Proper use factor. Maximum degree of use of a plant species by grazing, expressed as a percent, deemed to be physiologically correct from the standpoint of plant vigor, reproduction, longevity, and regrowth potential; may be applied to the standing crop as a whole. See *Palatability factor*.

Pseudostem. Concentrically arranged sheaths of fully expanded leaves that surround the immature growing leaves, tillers, and growing points.

Put-and-take stocking. See (*Variable stocking* and *Set stocking*.

Ranch. An establishment or firm with specific boundaries together with its lands and improvements, used for the grazing and production of livestock and/or game animals.

Ranch management. Manipulation of all ranch resources (financial, personnel, animal, and plant resources) to accomplish the specific management objectives set for the ranch.

Rancher. One who owns, leases, or manages a ranch.

Range. (n) Land supporting mostly native (indigenous) vegetation that either is grazed or has the potential to be grazed and is managed as a natural ecosystem. Range includes natural grasslands, shrublands, and grazable forestland. If part or all of the forage plant species are introduced, they are managed as a natural ecosystem. See *Rangeland*. (adj) Modifies resources, products, activities, practices, and phenomena as pertaining to range lands.

Range condition. Historically, the term has usually been defined in of two ways: (a) a generic term relating to present status of a unit of range in terms of specific values or potentials, or (b) the present state of vegetation of a range site in relation to the climax (natural potential) plant community for that site. This term is being phased out and replaced by preferred terms: *range similarity index* and *successional status*. See *Resource value rating (RVR)*.

Range management. Grazing land management as applied to native and seeded rangelands.

Range, open. Variously applied to mean (a) unfenced range, (b) range on which livestock grazing is permitted, (c) nontimbered range, or (d) range on which a livestock owner has unlimited access without benefit of land ownership or leasing.

Range/pasture inventory. (a) The systematic acquisition and analysis of resource information needed for planning and management of range and pasture lands; (b) the information acquired through range/pasture inventorying.

Range readiness. The stage of plant development at which grazing may begin without permanent damage to vegetation or soil; meaningful only when applied to certain seasonal ranges.

Range site. A less preferred term now being replaced by *ecological site*.

Range suitability. A subjective evaluation of the adaptability of a range area to grazing by a designated livestock or big game species.

Range trend. The direction of change in the plant community and the associated components of the site from the climax or other desired plant community.

Range unit. An area of rangeland enclosed and separated from other land areas by fencing or other barriers. As defined, this term excludes *pasture unit* but is included within the term *grazing unit*. See *Paddock*.

Rangeland. (a) (best use) Synonymous with *range;* (b) synonymous with *range* but excluding grazable woodland and grazable forestland (range without tree overstory) (SRM).

Rapid rotation grazing. See *Short-duration grazing*.

Ration grazing. Confining animals to an area of grazing land that will allot a predetermined amount of forage per animal on a daily basis (or less commonly a weekly or other short period).

Repeated seasonal grazing. See *Sequence grazing*.

Resident species. A species of plant or animal common to an area without distinction as to being native or introduced.

Residue. See *Crop residue*.

Resource value rating (RVR). The value of vegetation present on an ecological site for a particular use or benefit. RVRs may be established for each plant community capable of being produced on an ecological site, including exotic or cultivated plant species.

Rest. Nongrazing of an area of grazing land for a specified period of time (excluding *deferment* by definition), ranging from a few days to a full year or more; the duration of rest must be more fully described to be meaningful. Historically, the term referred to nongrazing for a full year along with foregoing grazing of that year's complete forage crop. See *Non-grazing.*

Rest period. The length of time that a specific land area is allowed to rest (remain ungrazed).

Rest-rotation system. A grazing system employing various combinations of yearlong rest, deferment, and full season grazing, commonly in a 3- to 5-year cycle. See *Rotational rest* and *Santa Rita grazing system.*

Reticulo-rumen. The anterior compartment of the ruminant stomach including the large rumen and the smaller reticulum.

Riparian. Pertaining to a zone between aquatic and terrestrial situations, such as bordering streams, rivers, and lakes, in which soil moisture is sufficiently in excess of that otherwise available locally so as to provide a more mesic habitat than that of contiguous uplands.

Rotation grazing. A generic term applied to moving grazing animals recurrently from one grazing unit (paddock) to another grazing unit in the same series (group); one of the basic components of a grazing system. See *Grazing system.*

Rotational burning/grazing. A grazing technique applied to grazing land under a single fence in which grazing animals are induced to rotate their grazing periodically among portions of the grazing land unit in accordance with enhanced palatability resulting from annually burning a portion of the area.

Rotational deferment. A multi-unit grazing system in which deferment is scheduled among the respective range units on a rotating basis. Although the number grazing herds is variable, commonly one less than the number of grazing units, grazing of the standing crop follows deferment in the deferred unit but is con-

tinuous in the other range units. See *Deferred-rotation grazing* and *Merrill grazing system.*

Rotational rest. A multi-range unit, usually multi-herd grazing system in which rest (often specified for a 12-month period) is scheduled among the respective range units on a rotating basis. See *Rest-rotation grazing* and *Santa Rita grazing system.*

Rough. The accumulation of mature living and dead vegetation—especially grasses and forbs—on forested range, marshland, or prairie.

Roughage. Plant materials and other feedstuffs high in fiber (20% or more) and low in total digestible nutrients (60% or less), usually bulky and coarse; synonymous with *forage* only in part.

Rumen. The large, first compartment of the stomach of a ruminant from which ingesta is regurgitated for rechewing and in which food is broken down (fermented) by symbiotic microbial action as an aid to digestion.

Ruminant. An even-toed, hoofed mammal that chews the cud and has a four-chambered stomach (*Ruminantia*).

Sacrifice area. A small portion of a grazing land unit that is willingly allowed to be overgrazed to obtain efficient overall use of the remaining majority of the unit.

Santa Rita grazing system. A variant of rest-rotation grazing utilizing three range units and one herd that accumulates 24 months of nongrazing and 12 months of grazing per 3-year grazing system cycle without foregoing grazing on any year's standing crop.

Satiety. Feeling of satisfaction "to the full" after ingesting adequate or optimal kinds and amounts of a food or foods; a positive post-ingestive feedback.

Savory grazing system. Formerly used synonymously with *short-duration grazing;* now formally extended into and replaced by "holistic resource management" or some variant.

Seasonal grazing. Grazing restricted to one season of the year, or possibly to more than one season of the year, but less than yearlong grazing.

Selected deferment. See *Deferred grazing.*

Selected rest. The practice of annually applying rest, typically for a full year, to the range unit where deemed most needed; equally meaningful when applied to yearlong, growing season, or dormant season grazing lands.

Selective grazing. The grazing of plant species, individual plants, or plant parts differently from random removal or from the average of what is available; a nearly universal phenomenon.

Selectivity ratio. The proportion in the animal diet of any plant species, species group, or plant part divided by the proportion it is found in the available herbage; an expression of relative preference. Synonym, *selectivity index.*

Sequence grazing. A grazing method in which two or more grazing units differing in forage species composition, are grazed in sequence, each unit generally grazed at the same time each year and sometimes for less than the full feasible grazing season. Sequence grazing takes advantage of differences among forage species and species combinations to extend the grazing season, enhance forage quality and/or quantity, or achieve some other management objective. Synonym, *repeated seasonal grazing.*

Set stocking. Keeping a fixed number of animals on a fixed area of grazing land during the time grazing is allowed. Synonym, *constant stocking* or *fixed stocking.* See *Variable stocking.*

Short-duration grazing. A rotational grazing system employing high stocking density, one herd, commonly 5 to 12 paddocks, grazing periods of 3 to 10 days (less commonly 1 to 15 days), and two to several grazing cycles per year; the common "rotation grazing" of improved pasture but has also been applied to range. Synonym, *rapid rotation grazing* or *high-intensity/high-frequency (HIHF) grazing.*

Spatial foraging levels. Scales of foraging attributed to large herbivores when functionally based on defined behaviors or characteristics: *home range* (or *landscape*), *camp, feeding site, patch, feeding station,* and *bite.*

Special grazing technique. A generic term including grazing methods and grazing systems; replaces the broader, historical definition of *grazing system.*

Spot grazing. See *Patch grazing,* a preferred term.

Standing crop. Unharvested/ungrazed plant material standing in place at a given time. Standing crop does not differentiate between edible and inedible nor between accessible and inaccessible plant material. The term can be qualified to refer only to forage, browse, mast, herbage, etc. See *Forage mass.*

Stocker. A beef animal being backgrounded (grown) prior to being finished for slaughter or entering the breeding herd.

Stocking density. Animal demand per unit area of land at any instant of time (i.e., AU/acre or AU/section of land); an animal/area ratio describing the relationship between number of animals and the corresponding area of land at any instant of time. See *Stocking rate.*

Stocking density index. The ratio of the land area available for grazing in a single unit or among paddocks in a grazing system to the land area available for grazing at any one time; expresses the level of animal concentration. Synonym, *Grazing animal concentration index*. See *Grazed fraction*.

Stocking intensity. A general term referring to animal demand-grazing land area relationships. See *Stocking density* and *Stocking rate*.

Stocking pressure. See *Grazing pressure*.

Stocking rate. Animal demand (for forage) per unit area of land over a period of time (i.e., AUM/acre or AUD/acre or their reciprocals); an animal/area ratio describing the relationship between number of animals and the corresponding land area being grazed over a specified period of time.

Stockpiling. Allowing standing forage to accumulate during rapid growth stages for grazing at a later period, often for fall and winter grazing during dormancy.

Strip grazing. Confining grazing animals to successive strips within a grazing unit, each strip being grazed under high stocking density for short periods (typically $\frac{1}{2}$ to 3 days); most commonly applied to improved pasture using several grazing period cycles during the growing season. Synonym, *Hohenstein system*. Grazing and nongrazing periods are provided by manually advancing moveable fences (a grazing method) or less commonly by rotation among multiple (strip) paddocks (a grazing system).

Stubble. The basal portion of herbaceous plants remaining after the top portion has been harvested either mechanically or by grazing animals.

Subunit. See *paddock*.

Suitable range. A subjective term describing an area of range that is accessible to a specific ungulate herbivore species and can be grazed by it on a sustained yield basis without damage to the resource.

Supplement. (n) A feedstuff high in specific nutrients such as protein, phosphorus,

sodium and calcium (salt), or energy and fed to remedy nutrient deficiencies of a range or pasture diet or other basal ration. Synonym, *supplemental feed*. (v) To feed a supplement.

Supplemental feeding. Supplying concentrates or harvested feed to correct deficiencies in the base diet. See *Emergency feeding* and *Maintenance feeding*.

Supplementary pasture. A pasture of higher quality forage grazed simultaneously and in conjunction with a base pasture or range unit; while the base unit provides the primary source of grazing capacity, the supplemental pasture of enhanced nutritive quality serves to correct nutrient deficiencies in the total animal diet. See *Complementary pasture*.

Sustained yield. The continuation of desired animal or forage production or yield of other related natural resources.

Sward. A plant community dominated by herbaceous species, such as grasses and legumes or other forbs, and characterized by a relatively short habit of growth and relatively continuous ground cover; comprised of natural or seeded, pure or mixed species stands.

Switchback grazing. See *Alternate grazing*.

Tame pasture. An archaic term referring to pasture on which the forage species are not native but have been intentionally introduced to the site. See *Pasture, Improved pasture*, and *Permanent pasture*.

Temporary pasture. A kind of pasture established on arable land for grazing during a single grazing season or shorter time period or planned for annual reestablishment.

Thatch. Accumulation of remnant flower stalks and old growth of herbaceous plants no longer attractive to grazing animals.

Tiller. A young vegetative lateral shoot of grasses growing upward within the enveloping leaf sheath.

Topping-followup grazing. See *First-last grazing.*

Total available carbohydrates (TAC). The pool of non-structural carbohydrates within a plant from which withdrawals can be made for maintenance, respiration, initial growth, and many other routine and emergency plant needs.

Trail. A well-defined path created by repeated passage of animals or vehicles.

Trail herding. See *driving.*

Trailing. The natural habit of livestock and big game animals to tread repeatedly along the same path or line; also controlled directional movement of livestock.

Trampling. Treading under foot; the damage to plants or soil resulting from the hoof impact of grazing animals.

Transhumance. Cyclical, annual movements of livestock between distinctive seasonal ranges, characterized by management-induced movement of livestock in company with or under the control of their graziers.

Transitory pasture/range. Land that provides grazing capacity during an interim period of uncertain duration, generally undeveloped but substantially modified from the original vegetation; includes go-back farmlands, timber clearings, pine plantations, and pre-development lands.

Two-crop grazing. A single-unit grazing method in which grazing is discontinued on grasses at boot stage and then fully grazed again around maturity.

Two-crop/one-crop grazing. A two-unit grazing system in which one grazing unit receives two-crop grazing and a second unit is grazed from boot stage to near maturity, with the treatments being switched in alternate years.

Undergrazing. Continued underuse, this often resulting in waste of forage.

Understocking. Placing so few animals on a given area that underuse will result when continued unchanged to the end of the planned grazing period.

Ungrazed. The status of grazing land and associated vegetation that is not grazed by ungulate herbivores.

Ungulate. A hoofed animal, including ruminants but also horses, tapirs, elephants, rhinoceroses, and swine.

Unsuitable range. A subjective term describing an area of range which (1) has no value for grazing because of barrenness or inherent lack of forage, or (2) should not be grazed by designated ungulate herbivores because of inherently unstable soils, limited accessibility, or steep topography.

Usable forage. That portion of the standing crop that can be removed by grazing without damage to the forage plants; may vary with season of use, plant species, and associated plant species.

Use factor. See *Palatability factor.*

Utilization. The proportion of current year's forage production (biomass) that is consumed and/or destroyed by grazing animals; may refer to a single plant species or to a portion or all of the vegetation. Synonym, *degree of use.*

Variable stocking. Placing a variable number of animals on a fixed area of grazing land during the time when grazing is allowed to achieve a management objective. Objectives of periodically adjusting animal numbers and thereby modifying the rate of forage removal include: (1) utilize forage at a rate similar to its growth rate, (2) achieve a desired level of utilization or grazing pressure, or (3) extend the period of forage availability or leave a forage surplus for a later grazing period. See *Set stocking*.

Vegetative. (a) Non-reproductive plant parts (leaf and stem) in contrast to reproductive plant parts (flower and seed) in developmental stages of plant growth. (b) Plant development stages prior to the sexual reproductive stage. Preventing the onset of sexual reproduction is generally associated with higher quantity and quality of forage production.

Voluntary intake. *Ad libitum* food intake achieved by an animal when an excess of forage or other feedstuffs is available for consumption.

Walkway. An earthen embankment constructed to improve the accessibility of marsh range.

Warm-season plant. A plant which makes most or all of its growth during the late spring to early fall period and is usually dormant in winter in temperate zones; these plants usually possess the C-4 photosynthetic pathway.

Wolf plant. An individual plant of a species generally considered palatable that remains ungrazed or mostly ungrazed when exposed to grazing.

Yearlong continuous grazing. A grazing method in which continuous grazing is applied to yearlong grazing lands.

Yearlong grazing. See *Yearlong continuous grazing*.

Yearlong range. Range that is or can be grazed yearlong.

Yearlong rest. Excluding grazing for a 12-month period, after which grazing of the corresponding forage crop may be foregone; synonymous with the historical use of the term *rest*.

Year-round grazing. A natural event or management scheme in which animals graze the entire year, enabled by yearlong range or a combination of grazing land units of different vegetation types or forage plant mixtures grazed sequentially.

Zero grazing. See *Green chop*.

LITERATURE CITED

Abaye, A. O., V. G. Allen, and J. P. Fontenot. (1994). Influence of Grazing Cattle and Sheep Together and Separately on Animal Performance and Forage Quality. *J. Anim. Sci.* **72**(4):1013–1022.

Abaye, A. O., V. G. Allen, and J. P. Fontenot. (1997). Grazing Sheep and Cattle Together or Separately: Effect on Soils and Plants. *Agron. J.* **89**(3):380–386.

Abdel-Magid, A. H., M. J. Trilica, and R. H. Hart. (1987a). Soil and Vegetation Responses to Simulated Trampling. *J. Range Mgt.* **40**(4):303–306.

Abdel-Magid, A. H., G. E. Schuman, and R. H. Hart. (1987b). Soil Bulk Density and Water Infiltration as Affected by Grazing Systems. *J. Range Mgt.* **40**(4):307–309.

Abouguendia, Z., and T. Dill. (1993). "Grazing Systems for Rangelands of Southern Saskatchewan." Sask. Agric. & Food, Regina, Sask. 12 pp.

Abu-Zanat, M., G. B. Ruyle, and R. W. Rice. (1988). Cattle Foraging Behavior in Grazed and Nongrazed Patches of Lehman Lovegrass. *Soc. Range Mgt. Abst. Papers* **41**:144.

Adams, D. C. (1985). Effect of Time of Supplementation on Performance, Forage Intake, and Grazing Behavior of Yearling Beef Steers Grazing Russian Wild Ryegrass in the Fall. *J. Anim. Sci.* **61**(5):1037–1042.

Adams, D. C. (1986). Getting the Most Out of Grain Supplements. *Rangelands* **8**(1):27–28.

Adams, D. C. (1987). Influence of Winter Weather on Range Livestock. *In* "Proceedings, Grazing Livestock Nutrition Conference, July 23–24, 1987, Jackson, Wyoming." Univ. Wyo., Laramie, Wyo., pp. 23–29.

Adams, D. C., and R. J. Kartchner. (1983). Effects of Time of Supplementation on Daily Gain, Forage Intake, and Behavior of Yearling Steers Grazing Fall Range. *Amer. Soc. Anim. Sci., West. Sect. Proc.* **34**:158–160.

Adams, D. C., and R. E. Short. (1988). The Role of Animal Nutrition on Productivity in a Range Environment. *In* R. S. White and R. E. Short (Eds.). "Achieving Efficient Use of Rangeland Resources, Fort Keogh Research Symposium, September 1987, Miles City, Mon." Mon. Agric. Expt. Sta., Bozeman, Mon., pp. 37–44.

Adams, D. C., T. C. Nelsen, W. L. Reynolds, and B. W. Knapp. (1986). Winter Grazing Activity and Forage Intake of Range Cows in the Northern Great Plains. *J. Anim. Sci.* **62**(5):1240–1246.

Adams, D. C., R. E. Short, and B. W. Knapp. (1987). Body Size and Body Condition Effects on Performance and Behavior of Grazing Beef Cows. *Nutrition Reports International* **35**(2):269–277.

Adams, D. C., R. B. Staigmiller, and B. W. Knapp. (1989). Beef Production from Native and Seeded Northern Great Plains Ranges. *J. Range Mgt.* **42**(3):243–247.

Adams, D. C., R. T. Clark, T. J. Klopfenstein, and J. D. Volesky. (1996). Matching the Cow with Forage Resources. *Rangelands* **18**(2):57–62.

Adandedjan, C. C., R. L. Reid, T. S. Ranney, and E. C. Townsend. (1987). Creep Grazing Lambs on Tall Fescue Pastures. *W. Va. Agric. & For. Expt. Sta. Bul.* **695**. 30 pp.

Alejandro-Castro, M., P. S. Doescher, and W. E. Drewien. (1987). Cattle Grazing as a Silvicultural Tool in Southwest Oregon. *Soc. Range Mgt. Abst. Papers* **40**:1.

Alexander, J. D. III, and J. E. Taylor. (1986). Sheep Utilization as a Control Method on Tall Larkspur Infested Cattle Range. *Soc. Range Mgt. Abst. Papers,* **39**:241.

Alexander, L. E., D. W. Uresk, and R. M. Hansen. (1983). Summer Food Habits of Domestic Sheep in Southeastern Montana. *J. Range Mgt.* **36**(3):307–308.

Ali, E., and S. H. Sharrow. (1994). Sheep Grazing Efficiency and Selectivity on Oregon Hill Pasture. *J. Range Mgt.* **47**(6):494–497.

Allden, W. G. (1981). Energy and Protein Supplements for Grazing Livestock. *In* F. H. W. Morley (Ed.), "Grazing Animals." Elsevier Sci. Pub. Co., Amsterdam, Neth., pp. 289–307.

Allen, B. (1986). Forest Grazing—Use of Livestock in Plantation Management. *Soc. Range Mgt. Abst. of Papers* **39**:229.

Allen, B. H., and J. W. Bartolome. (1989). Cattle Grazing Effects on Understory Cover and Tree Growth in Mixed Conifer Clearcuts. *Northwest Sci.* **63**(5):214–220.

Allen, E. O. (1968). Range Use, Foods, Condition, and Productivity of White-Tailed Deer in Montana. *J. Wildl. Mgt.* **32**(1):130–141.

Allen, G. C., and M. Devers. (1975). Livestock-Feed Relationships—National and State. *USDA Stat. Bul.* **530** (Supplement). 101 pp.

Allen, M. S. (1996). Physical Constraints on Voluntary Intake of Forages by Ruminants. *J. Anim. Sci.* **74**(12):3063–3075.

Allen, V. (1999). Anti-Quality Components in Forages: Overview, Significance, and Economic Impact. *Soc. Range Mgt. Abst. Papers* **52**:1.

Allen, V. G., D. D. Wolf, J. P. Fontenot, J. Cardina, and D. R. Notter. (1986a). Yield and Regrowth Characteristics of Alfalfa Grazed With Sheep. I. Spring Grazing. *Agron. J.* **78**(6):974–979.

Allen, V. G., L. A. Hamilton, D. D. Wolf, J. P. Fontenot, T. H. Terrill, and D. R. Notter. (1986b). Yield and Regrowth Characteristics of Alfalfa Grazed with Sheep. II. Summer Grazing. *Agron. J.* **78**(6):979–985.

Allison, C. D. (1985). Factors Affecting Forage Intake by Range Ruminants: A Review. *J. Range Mgt.* **38**(4):305–311.

Allison, C. D., M. M. Kothmann, and L. R. Rittenhouse. (1982). Efficiency of Forage Harvest by Grazing Cattle. *J. Range Mgt.* **35**(3):351–354.

Allison, C. D., M. M. Kothmann, and L. R. Rittenhouse. (1983). Forage Intake of Cattle as Affected by Grazing Pressure. *Proc. Intern. Grassland Cong.* **14**:670–672.

Altom, W. (abt. 1978). "Limit-Grazing of Small Grain Pastures." Noble Found., Ardmore, OK. 34 pp.

Ames, D. R. (1985). Energy Costs of Range Cows. *Amer. Soc. Anim. Sci., West. Sect. Proc.* **36**:4–5.

Ames, D. R., and D. E. Ray. (1983). Environmental Manipulation to Improve Animal Productivity. *J. Anim. Sci.* **57**(Supp. 2):209–220.

Anderson, B., A. G. Matches, and C. J. Nelson. (1989). Carbohydrate Reserves and Tillering of Switchgrass Following Clipping. *Agron. J.* **81**(1):13–16.

Anderson, B. (1999). Study Duration Can Affect Results and Conclusions of Grazing Studies. *Soc. Range Mgt. Abst. Papers* **52**:2.

Anderson, B. L., R. D. Pieper, and V. W. Howard, Jr. (1974). Growth Response and Deer Utilization of Fertilized Browse. *J. Wildl. Mgt.* **38**(3):525–530.

Anderson, D. M. (1987). Direct Measures of the Grazing Animal's Nutritional Status. *In* D. A. Jameson and J. Holechek (Eds.), "Monitoring Animal Performance and Production Symposium Proceedings, February 12, 1987, Boise, Idaho." Society for Range Management, Denver, CO, pp. 40–53.

Anderson, D. M. (1988). Seasonal Stocking of Tobosa Managed Under Continuous and Rotation Grazing. *J. Range Mgt.* **41**(1):78–83.

Anderson, D. M., and M. M. Kothmann. (1980). Relationship of Distance Traveled with Diet and Weather for Hereford Heifers. *J. Range Mgt.* **33**(3):217–220.

Anderson, D. M., C. V. Hulet, W. L. Shupe, and J. N. Smith (1988). Response of Banded and Unbanded Sheep to the Approach of a Trained Border Collie. *Soc. Range Mgt. Abst. Papers* **41**:187.

Anderson, E. W., and R. J. Scherzinger. (1975). Improving Quality of Winter Forage for Elk by Cattle Grazing. *J. Range Mgt.* **28**(2):120–125.

Anderson, K. L., E. F. Smith, and C. E. Owensby. (1970). Burning Bluestem Range. *J. Range Mgt.* **23**(2):81–92.

Anderson, M. S., H. W. Lakin, D. C. Beeson, F. F. Smith, and E. Thacker. (1961). Selenium in Agriculture. *USDA Agric. Handbook* **200**. 65 pp.

Anderson, V. J., and D. D. Briske. (1989). Species Replacement in Response to Herbivory: An Evaluation of Potential Mechanisms. *Soc. Range Mgt. Abst. Papers* **42**:13.

Angell, R. F. (1997). Crested Wheatgrass and Shrub Response to Continuous or Rotational Grazing. *J. Range Mgt.* **50**(2):160–164.

Angell, R. F., J. W. Stuth, and D. L. Drawe. (1986). Diets and Liveweight Changes of Cattle Grazing Fall Burned Gulf Cordgrass. *J. Range Mgt.* **39**(3):233–236.

Angell, R. F., H. A. Turner, and M. R. Haferkamp. (1987). Ramifications of Strip Grazing Harvested or Standing Flood Meadow Forages for Wintering Cattle. *Soc. Range Mgt. Abst. Papers* **40**:33.

Anonymous. (1985). Rangeland Resources and Management—A Report to the USDA Joint Council. *Rangelands* **7**(3):109–111.

Ansotegui, R. P., J. D. Wallace, K. M. Havstad, and D. M. Hallford. (1987). Effects of Milk Intake on Forage Intake and Performance of Suckling Range Calves. *Soc. Range Mgt. Abst. Papers* **40**:227.

Ansotegui, R. P., K. M. Havstad, J. D. Wallace, and D. M. Hallford. (1991). Effects of Milk Intake on Forage Intake and Performance of Suckling Range Calves. *J. Anim. Sci.* **69**(3):899–904.

Araujo, M. R., and J. W. Stuth. (1986). Stocking Rate Affects Diet Selection of Cattle in Rotational Grazing Systems. *Texas Agric. Expt. Sta. Prog. Rep.* **4431**. 2 pp.

Arave, C. W., P. H. Stewart, A. L. T. Hansen, and J. L. Walters. (1993). Primary Color Discrimination by Holstein Heifers. *Amer. Soc. Anim. Sci., West. Sect. Proc.* **44**:113–116.

Archer, S. (1994). Woody Plant Encroachment into Southwestern Grassland and Savannas: Rates, Patterns, and Proximate Causes. *In* M. Vavra, W. A. Laycock, and R. D. Pieper (Eds.), "Ecological Implications of Livestock Herbivory in the West." Soc. Range Mgt., Denver, CO, pp. 19–68.

Archer, S., and D. A. Pyke. (1991). Plant-Animal Interactions Affecting Plant Establishment and Persistence on Revegetated Rangeland. *J. Range Mgt.* **44**(6):558–565.

Archer, S., and L. L. Tiezen. (1986). Plant Responses to Defoliation: Hierarchial Considerations. *In* "Grazing Research at Northern Latitudes." Plenum Press, New York, pp. 45–59.

Ares, F. N. (1953). Better Cattle Distribution Through the Use of Meal Salt Mix. *J. Range Mgt.* **6**(5):341–346.

Ares, F. N. (1974). The Jornada Experimental Range. *Soc. for Range Mgt. (Denver, Colo.) Range Mono.* **1**. 74 pp.

Arias, J. E., C. T. Dougherty, N. W. Bradley, P. L. Cornelius, and L. M. Lauriault. (1990). Structure of Tall Fescue Swards and Intake of Grazing Cattle. *Agron. J.* **82**(3):545–548.

Armbruster, S., G. Horn, and K. Lusby. (1976). When Grass Is Short. *Okla. Agric. Expt. Sta. Res. Rep.* **742**, pp. 35–40.

Arnold, G. W. (1966). The Special Senses in Grazing Animals. II. Smell, Taste, and Touch and Dietary Habits in Sheep. *Australian J. Agric. Res.* **17**(4):531–542.

Arnold, G. W. (1985). Regulation of Forage Intake. *In* Robert J. Hudson and Robert G. White (Eds.). "Bioenergetics of Wild Herbivores." CRC Press, Boca Raton, FL, pp. 81–101.

Arnold, G. W., and M. L. Dudzinski. (1978). "Ethnology of Free Ranging Domestic Animals." Elsevier, N.Y. 198 pp.

Arnold, L. A., Jr., and D. Lynn Drawe. (1979). Seasonal Food Habits of White-Tailed Deer in the South Texas Plains. *J. Range Mgt.* **32**(3):175–178.

Arthun, D., S. Rafique, J. L. Holechek, J. D. Wallace, and M. L. Galyean. (1988). Effects of Forb and Shrub Diets on Ruminant Nitrogen Balance. II. Cattle Studies. *Amer. Soc. Anim. Sci., West. Sect. Proc.* **39:**204–207.

Artz, J. L., and E. I. Hackett. (1971). Wolf Plants in Crested Wheatgrass Seedings. *Amer. Soc. Range Mgt. Abst. Papers* **24:**17.

Asay, K. H., H. F. Mayland, and D. H. Clark. (1996). Response to Selection for Reduced Grass Tetany Potential in Crested Wheatgrass. *Crop Sci.* **36**(4):895–900.

Askins, G. D., and E. E. Turner. (1972). A Behavioral Study of Angora Goats on West Texas Range. *J. Range Mgt.* **25**(2):82–87.

Atwood, T. L., and R. F. Beck. (1987). Vegetation Parameters Inside and Outside Livestock Exclosures on Three Chihuahuan Desert Grassland Communities. *Soc. Range Mgt. Abst. Papers* **40:**34.

Auen, L. M., and C. E. Owensby. (1988). Effects of Dormant-Season Herbage Removal on Flint Hills Rangeland. *J. Range Mgt.* **41**(6):481–482.

Austin, D. D., and P. J. Urness. (1983). Overwinter Forage Selection by Mule Deer on Seeded Big Sagebrush-Grass Range. *J. Wildl. Mgt.* **47**(4):1203–1207.

Austin, D. D., and P. J. Urness. (1986). Effects of Cattle Grazing on Mule Deer Diet and Area Selection. *J. Range Mgt.* **39**(1):18–21.

Austin, D. D., and P. J. Urness. (1995). Effects of Horse Grazing in Spring on Survival, Recruitment, and Winter Injury Damage of Shrubs. *Great Basin Nat.* **55**(3):267–270.

Austin, D. D., and P. J. Urness. (1998). Vegetal Change on a Northern Utah Foothill Range in the Absence of Livestock Grazing Between 1948 and 1982. *Great Basin Nat.* **58**(2):188–191.

Austin, D. D., P. J. Urness, and L. C. Fierro. (1983). Spring Livestock Grazing Affects Crested Wheatgrass Regrowth and Winter Use by Mule Deer. *J. Range Mgt.* **36**(5):589–593.

Austin, D. D., P. J. Urness, and R. A. Riggs. (1986). Vegetal Change in the Absence of Livestock Grazing, Mountain Brush Zone, Utah. *J. Range Mgt.* **39**(6):514–517.

Austin, D. D., P. J. Urness, and S. L. Durham. (1994). Impacts of Mule Deer and Horse Grazing on Transplanted Shrubs for Revegetation. *J. Range Mgt.* **47**(1):8–11.

Bacon, C. W. (1995). Toxic Endophyte-Infected Tall Fescue and Range Grasses: Historic Perspectives. *J. Anim. Sci.* **73**(3):861–870.

Bailey, A. W. (1970). Barrier Effects of the Shrub *Elaeagnus commutata* on Grazing Cattle and Forage Production in Central Alberta. *J. Range Mgt.* **23**(4):248–251.

Bailey, C. B., and J. E. Lawson. (1987). Russian Wild Ryegrass Restricts the Formation of Siliceous Urinary Calculi in Range Calves. *Can. J. Anim. Sci.* **67**(4):1139–1141.

Bailey, C. B., R. Hironaka, and S. B. Slen. (1962). Effects of the Temperature of the Environment and the Drinking Water on the Body Temperature and Water Consumption of Sheep. *Can. J. Anim. Sci.* **42**(1):1–8.

Bailey, D. W. (1995). Daily Selection of Feeding Areas by Cattle in Homogeneous and Heterogeneous Environments. *Appl. Anim. Beh. Sci.* **45**(3–4):183–200.

Bailey, D. W. (1999). Influence of Species, Breed, and Type of Animal on Habitat Selection. *In* "Grazing Behavior of Livestock and Wildlife." *Idaho Forest, Wildl., and Range Expt. Sta. Bull.* **70,** pp. 101–108.

Bailey, D. W., and L. R. Rittenhouse. (1989). Management of Cattle Distribution. *Rangelands* **11**(4):159–161.

Bailey, D. W., and P. L. Sims. (1998). Association of Food Quality and Locations by Cattle. *J. Range Mgt.* **51**(1):2–8.

Bailey, D. W., and G. R. Welling. (1999). Modification of Cattle Grazing Distribution with Dehydrated Molasses Supplement. *J. Range Mgt.* **52**(6):575–582.

Bailey, D. W., J. W. Walker, and L. R. Rittenhouse. (1990). Sequential Analysis of Cattle Location: Day-to-Day Movement Patterns. *Applied Anim. Beh. Sci.* **25**(1–2):137–148.

Bailey, D. W., J. E. Gross, E. A. Laca, L. R. Rittenhouse *et al.* (1996a). Mechanisms That Result in Large Herbivore Grazing Distribution Patterns. *J. Range Mgt.* **49**(5):386–400.

Bailey, D. W., L. R. Rittenhouse, and D. M. Swift. (1996b). A Conceptual Model for Studying Grazing Distribution Patterns of Large Herbivores. *Proc. Internat. Rangeland Cong.* **5**(1):19–20.

Baker, D. L., and D. R. Hansen. (1985). Comparative Digestion of Grass in Mule Deer and Elk. *J. Wildl. Mgt.* **49**(1):77–79.

Baker, F. H. (1985). Multispecies Grazing: The State of the Science. *Rangelands* **7**(6):266–269.

Baker, M. J., E. C. Prigge, and W. B. Bryan. (1988). Herbage Production from Hay Fields Grazed by Cattle in Fall and Spring. *J. Prod. Agric.* **1**(3):275–279.

Baker, R. D., F. Alvarez, and Y. L. P. Le Du. (1981). The Effect of Herbage Allowance Upon the Herbage Intake and Performance of Suckler Cows and Calves. *Grass and Forage Sci.* **36**(3):189–199.

Balph, D. F., and M. H. Balph. (1986). The Application of Behavioral Concepts to Livestock Management. *Utah Sci.* **47**(3):78–85.

Balph, D. F., and J. C. Malechek. (1985). Cattle Trampling of Crested Wheatgrass Under Short-Duration Grazing. *J. Range Mgt.* **38**(3):226–227.

Balph, D. F., M. H. Balph, and J. C. Malechek. (1989). Cues Cattle Use To Avoid Stepping on Crested Wheatgrass Tussocks. *J. Range Mgt.* **42**(5):376–377.

Bareiss, L. J., P. Schulz, and F. S. Guthery. (1986). Effects of Short-Duration and Continuous-Grazing on Bobwhite and Wild Turkey Nesting. *J. Range Mgt.* **39**(3):259–260.

Barnes, R. F. (1982). Grassland Agriculture—Serving Mankind. *Rangelands* **4**(2):61–62.

Barrow, J. R., and K. M. Havstad. (1992). Recovery and Germination of Gelatin-Encapsulated Seeds Fed to Cattle. *J. Arid. Environ.* **22**(4):395–399.

Bartmann, R. M. (1983). Composition and Quality of Mule Deer Diets on Pinyon-Juniper Winter Range, Colorado. *J. Range Mgt.* **36**(4):534–541.

Bartmann, R. M., and L. H. Carpenter. (1982). Effects of Foraging Experience on Food Selectivity of Tame Mule Deer. *J. Wildl. Mgt.* **46**(3):813–818.

Bartmann, R. M., A. W. Alldredge, and P. H. Neil. (1982). Evaluation of Winter Food Choices by Tame Mule Deer. *J. Wildl. Mgt.* **46**(3):807–812.

Bartolome, J. W., and M. P. McClaran. (1992). Composition and Production of California Oak Savanna Seasonally Grazed by Sheep. *J. Range Mgt.* **45**(1):103–107.

Barton, R. K., L. J. Krysl, J. T. Broesder, S. A. Gunter, and M. B. Judkins. (1989). Effects of Protein Supplementation Time on Grazing Behavior, Nutrient Composition, Intake, Digesta Kinetics, and *In Situ* Rate of Digestion in Steers Grazing Dormant Tall Wheatgrass Pasture. *Amer. Soc. Anim. Sci., West. Sect. Proc.* **40**:349–352.

Barton, R. K., L. J. Krysl, M. B. Judkins, D. W. Holcombe *et al.* (1992). Time of Daily Supplementation for Steers Grazing Dormant Intermediate Wheatgrass Pasture. *J. Anim. Sci.* **70**(2):547–558.

Bawtree, A. H. (1989). Recognizing Range Readiness. *Rangelands* **10**(2):67–69.

Baxter, H. D., J. R. Owen, M. J. Montgomery, D. R. Waldo, and J. T. Miles. (1969). Pasturing Vs. Harvesting of a Grass-Legume Mixture. *Tenn. Agric. Expt. Sta. Bul.* **454**. 25 pp.

Bayoumi, M. A., and A. D. Smith. (1976). Response of Big Game Winter Range Vegetation to Fertilization. *J. Range Mgt.* **29**(1):44–48.

Beale, D. M., and A. D. Smith. (1970). Forage Use, Water Consumption, and Productivity of Pronghorn Antelope in Western Utah. *J. Wildl. Mgt.* **34**(3):570–582.

Beasom, S. L., and C. J. Scifres. (1977). Population Reactions of Selected Game Species to Aerial Herbicide Applications in South Texas. *J. Range Mgt.* **30**(2):138–142.

Beasom, S. L., L. LaPlant, and V. W. Howard, Jr. (1981). Similarity of Pronghorn, Cattle, and Sheep Diets in Southeastern New Mexico. *In* L. Nelson, Jr., and J. M. Peek (Eds.), "Proceedings of the Wildlife-Livestock Relationships Symposium, Coeur d'Alene, Idaho, April 20–22, 1981." Univ. Idaho, Moscow, Ida., pp. 565–572.

Beaver, J. M., and B. E. Olson. (1997). Winter Range Use by Cattle of Different Ages in Southwestern Montana. *Appl. Anim. Beh. Sci.* **51**(1–2):1–13.

Becerra, R. De Alba, J. Winder, J. L. Holechek, and M. Cardenas. (1998). Diets of 3 Cattle Breeds on Chihuahuan Desert Rangelands. *J. Range Mgt.* **51**(3):270–275.

Beck, J. L., J. T. Flinders, D. R. Nelson, C. L. Clyde *et al.* (1996). Elk and Domestic Sheep Interactions in a North-Central Utah Aspen Ecosystem. *USDA, For. Serv. Res. Paper* **INT-RP-491**. 114 pp.

Beck, R. F., R. P. McNeely, and H. E. Kiesling. (1987). Can Seasonal Grazing Improve Range and Cattle Production on Semidesert Areas? *Soc. Range Mgt. Abst. Papers* **40**:35.

Bedell, T. E. (1973). Botanical Composition of Subclover-Grass Pastures as Affected by Single and Dual Grazing by Cattle and Sheep. *Agron. J.* **65**(3):502–504.

Bedell, T. E., and D. C. Ganskopp. (1980). Rangelands in Dry Years: Drought Effects on Range, Cattle, and Management. *Pacific Northwest Coop. Ext. Pub.* **200**. 8 pp.

Bedell, T. E., R. E. Whitesides, and R. B. Hawkes. (1981). Pasture Management for Control of Tansy Ragwort. *Pacific Northwest Ext. Pub.* **210**. 6 pp.

Beetle, A. A., W. M. Johnson, R. L. Lang, M. May, and D. R. Smith. (1961). Effect of Grazing Intensity on Cattle Weights and Vegetation of the Bighorn Experimental Pastures. *Wyo. Agric. Expt. Sta. Bul.* **373**. 23 pp.

Behnke, R. J., and R. F. Raleigh. (1978). Grazing and the Riparian Zone: Impact and Management Perspectives. *USDA, For. Serv. Gen. Tech. Rep.* **WO-12**, pp. 263–267.

Beier, P., and D. R. McCullough. (1990). Factors Influencing White-Tailed Deer Activity Patterns and Habitat Use. *Wildl. Mono.* **109**. 51 pp.

Bell, C. E., and J. N. Guerrero. (1997). Sheep Grazing Effectively Controls Weeds in Seedling Alfalfa. *Calif. Agric.* **51**(2):19–23.

Bellows, R. A. (1985). Integration of Beef Cattle Reproduction and the Range Resource. *Amer. Soc. Anim. Sci., West. Sect. Proc.* **36**:1–3.

Bellows, R. A. (1988). Physiological Relationships That Limit Production of Range Beef Cattle. *In* R. S. White and R. E. Short (Eds.), "Achieving Efficient Use of Rangeland Resources, Fort Keogh Research Symposium, September 1987, Miles City, Mon." Mon. Agric. Expt. Sta., Bozeman, Mon., pp. 50–55.

Bellows, R. A., and O. O. Thomas. (1976). Some Effects of Supplemental Grain Feeding on Performance of Cows and Calves on Range Forage. *J. Range Mgt.* **29**(3):192–195.

Belsky, A. J., A. Matzke, and S. Uselman. (1999). Survey of Livestock Influences on Stream and Riparian Ecosystems in the Western United States. *J. Soil & Water Cons.* **54**(4):419–431.

Bement, R. E. (1968). Plains Pricklypear: Relation to Grazing Intensity and Blue Grama Yield on Central Great Plains. *J. Range Mgt.* **21**(2):83–86.

Bement, R. E. (1969). A Stocking-Rate Guide for Beef Production on Blue-Grama Range. *J. Range Mgt.* **22**(2):83–86.

Bennett, J. A., and R. D. Wiedmeier. (1992). Visual Body Condition Score as a Measure of Range Animal Response. *Rangelands* **14**(5):296–298.

Bentley, J. R. (1967). Conversion of Chaparral Areas to Grassland. *USDA Agric. Handbook* **328**. 35 pp.

Bentley, J. R., L. R. Green, and K. A. Wagnon. (1985). Herbage Production and Grazing Capacity on Annual-plant Range Pastures Fertilized with Sulfur. *J. Range Mgt.* **11**(3):133–140.

Berger, L. L., and D. C. Clanton. (1979). Rumensin Self-Fed on Pasture. *In* Nebraska Beef Cattle Report, 1979. *Neb. Agric. Ext. Cir.* **79–218**, pp. 16.

Bernardo, D. J., and F. T. McCollum. (1987). An Economic Analysis of Intensive-Early Stocking. *Okla. Agric. Expt. Sta. Res. Rep.* **P-887**. 35 pp.

Bertelsen, B. S., D. B. Faulkner, D. D. Buskirk, and J. W. Castree. (1993). Beef Cattle Performance and Forage Characteristics of Continuous, 6-Paddock, and 11-Paddock Grazing Systems. *J. Anim. Sci.* **71**(6):1381–1389.

Beverlein, S. K., K. M. Havstad, and M. K. Peterson. (1987). Winter Beef Cow Grazing Behavior as Affected by Environmental Stresses and Supplementation. *Soc. Range Mgt. Abst. Papers* **40**:166.

Beverlin, S. K., K. M. Havstad, E. L. Ayers, and M. K. Petersen. (1989). Forage Intake Responses to Winter Cold Exposure of Free-Ranging Beef Cows. *Appl. Anim. Beh. Sci.* **23**(1–2):75–85.

Bilbrough, C., and J. Richards. (1988). Morphological Constraints on Spring Growth Patterns Following Simulation of Winter Browsing on Sagebrush and Bitterbrush. *Soc. Range Mgt. Abst. Papers* **41**:61.

Bilbrough, C. J., and J. H. Richards. (1993). Growth of Sagebrush and Bitterbrush Following Simulated Winter Browsing: Mechanisms of Tolerance. *Ecology* **74**(2):481–492.

Biondini, M., R. D. Pettit, and V. Jones. (1986). Nutritive Value of Forages on Sandy Soils as Affected by Tebuthiuron. *J. Range Mgt.* **39**(5):396–300.

Biswell, H. H. (1951). Studies of Rotation Grazing in the Southeast. *J. Range Mgt.* **4**(1):52–55.

Biswell, H. H., R. D. Taber, D. W. Hedrick, and A. M. Schultz. (1952). Management of Chamise Brushlands for Game in the North Coast Region of California. *Calif. Fish & Game* **38**(4):453–484.

Bjugstad, A. J., and A. V. Dalrymple. (1968). Behavior of Beef Heifers on Ozark Ranges. *Mo. Agric. Expt. Sta. Bul.* **870.** 15 pp.

Blackburn, W. H. (1983). Livestock Grazing Impacts on Watersheds. *Rangelands* **5**(3):123–125.

Blackburn, W. H. (1984). Impacts of Grazing Intensity and Specialized Grazing Systems on Watershed Characteristics and Responses. *In* Natl. Res. Council/Natl. Acad. Sci. "Developing Strategies for Rangeland Management." Westview Press, Boulder, CO, pp. 927–983.

Blair, R. M., and L. E. Brunett. (1980). Seasonal Browse Selection by Deer in a Southern Pine-Hardwood Habitat. *J. Wildl. Mgt.* **44**(1):79–88.

Blaisdell, J. P., and R. C. Holmgren. (1984). Managing Intermountain Rangelands—Salt-Desert Shrub Ranges. *USDA, For. Serv. Gen. Tech. Rep.* **INT-163.** 53 pp.

Blaisdell, J. P., R. B. Murray, and E. D. McArthur. (1982). Managing Intermountain Rangelands—Sagebrush-Grass Ranges. *USDA, For. Serv. Gen. Tech. Rep.* **INT-134.** 41 pp.

Blankenship, J. O. (1969). Herderless Sheep Management on Mountain Ranges. *Amer. Soc. Range Mgt. Abst. Papers* **22:**19–20.

Blaser, R. E., H. T. Bryant, C. Y. Ward, R. C. Hammes, Jr., R. C. Carter, and N. H. MacLeod. (1959). Symposium on Forage Evaluation: VII. Animal Performance and Yields With Methods of Utilizing Pasturage. *Agron. J.* **51**(4):238–242.

Blaser, R. E., H. T. Bryant, R. C. Hammes, Jr., R. Bowman, J. P. Fontenot, and C. E. Polan. (1969). Managing Forages for Animal Production. *Va. Polytech. Inst., Res. Div. Bul.* **45.** 30 pp.

Blaser, R. E., E. Jahn, and R. C. Hammes, Jr. (1974). Evaluation of Forage and Animal Research. *In* R. W. VanKeuren (Ed.). "Systems Analysis in Forage Crops Production and Utilization." Crop Sci. Soc. Amer., Madison, Wisc., p. 1–27.

Blaser, R. E., J. T. Johnson, F. McClaugherty, J. P. Fontenot, R. C. Hammes, Jr., H. T. Bryant, D. D. Wolf, and D. A. Mays. (1983). Animal Production with Controlled and Fixed Stocking and Managed Stocking Rates. *Proc. Intern. Grassland Cong.* **14:**612–615.

Bleak, A. T., and A. P. Plummer. (1954). Grazing Crested Wheatgrass by Sheep. *J. Range Mgt.* **7**(2): 63–68.

Bock, B. J., S. M. Hannah, F. K. Brazle, L. R. Corah, and G. L. Kuhl. (1991). Stocker Cattle Management and Nutrition. *Kan. Agric. Ext. Cir.* **723.** 16 pp.

Boecker, E. L., V. E. Scott, H. G. Reynolds, and B. A. Donaldson. (1972). Seasonal Food Habits of Mule Deer in Southwestern New Mexico. *J. Wildl. Mgt.* **36**(1):56–63.

Bohn, C. C., and J. C. Buckhouse. (1985). Some Responses of Riparian Soils to Grazing Management in Northeastern Oregon. *J. Range Mgt.* **38**(4):378–381.

Bond, J., G. E. Carlson, C. Jackson, Jr., and W. A. Curry. (1967). Social Cohesion of Steers and Sheep as a Possible Variable in Grazing Studies. *Agron. J.* **59**(5):481–483.

Bond, J., T. S. Rumsey, and B. T. Weinland. (1976). Effect of Deprivation and Reintroduction of Feed and Water on the Feed and Water Intake Behavior of Beef Cattle. *J. Anim. Sci.* **43**(4):873–878.

Bonham, C. D., and K. Remington. (1991). Spatial Patterns of Forage and Grazing Intensity. *Soc. Range Mgt. Abst. Papers* **44:**14.

Booysen, P. de V., and N. M. Tainton. (1978). Grassland Management: Principles and Practice in South Africa. *Proc. Intern. Rangeland Cong.* **1:**551–554.

Boroski, B. B., and A. S. Mossman. (1996). Distribution of Mule Deer in Relation to Water Sources in Northern California. *J. Wildl. Mgt.* **60**(4):770–776.

Boroski, B. B., and A. S. Mossman. (1998). Water Use Patterns of Mule Deer (*Odocoileus hemionus*) and the Effects of Human Disturbance. *J. Arid Environ.* **38**(4):561–569.

Botha, P., C. D. Blom, E. Sykes, and A. S. J. Barnhoorn. (1983). A Comparison of the Diets of Small and Large Stock on Mixed Karoo Veld. *Proc. Grassland Soc. So. Africa* **18:**101–105.

Bowman, J. G. P., and D. W. Sanson. (1996). Starch- or Fiber-Based Energy Supplements for Grazing Ruminants. *Proc. Grazing Livestock Nutr. Conf.* **3:**118–135.

Bowman, J. G. P., and B. F. Sowell. (1997). Delivery Method and Supplement Consumption by Grazing Ruminants: A. Review. *J. Anim. Sci.* **75**(2):543–550.

Bowns, J. E. (1971). Sheep Behavior Under Unherded Conditions on Mountain Summer Ranges. *J. Range Mgt.* **24**(2):105–109.

Bowns, J. E. (1989). Common Use: Better for Cattle, Sheep, and Rangelands. *Utah Sci.* **50**(2):117–123.

Bowns, J. E., and C. F. Bagley. (1986). Vegetation Responses to Long-Term Sheep Grazing on Mountain Ranges. *J. Range Mgt.* **39**(5):431–434.

Bowns, J. E., and D. H. Matthews. (1983). Cattle Grazing With Sheep: A Plus for Rangelands and Production. *Utah Sci.* **44**(2):38–43.

Box, T. W., G. Brown, and J. Liles. (1965). Influence of Winter Supplemental Feeding of Cottonseed Cake on Activities of Beef Cows. *J. Range Mgt.* **18**(3):124–126.

Boyd, R. J. (1970). Elk of the White River Plateau, Colorado. *Colo. Div. Game, Fish, and Parks Tech. Pub.* **25.** 126 pp.

Brandyberry, S. D., T. DelCurto, and R. F. Angell. (1992). Physical Form and Frequency of Alfalfa Supplementation for Beef Cattle Winter Grazing Northern Great Basin Rangeland. *Amer. Soc. Anim. Sci., West. Sect. Proc.* **43**:47–50.

Brandyberry, S. D., T. DelCurto, R. K. Barton, K. J. Paintner, and K. H. Brandyberry. (1993). Effects of Early Spring Grazing of Rangelands Used in Winter Grazing Programs in the Northern Great Basin. *Amer. Soc. Anim. Sci., West. Sect. Proc.* **44**:223–226.

Bransby, D. I. (1991). Biological Implications of Rotational and Continuous Grazing: A Case for Continuous Grazing. *Forage & Grassland Conf.* **1991**:10–14.

Branson, F. A. (1953). Two New Factors Affecting Resistance of Grasses to Grazing. *J. Range Mgt.* **6**(3):165–171.

Bray, R. O., C. L. Wambolt, and R. G. Kelsey. (1987). Influence of Compounds Found in *Artemisia* on Mule Deer Preference. *Soc. Range Mgt., Abst. Papers* **40**:232.

Briske, D. D. (1986). Plant Response to Defoliation: Morphological Considerations and Allocation Priorities. *Proc. Internat. Rangeland Cong.* **2**:425–427.

Briske, D. D. (1991). Developmental Morphology and Physiology of Grasses. *In* "Grazing Management: An Ecological Perspective." Timber Press, Portland, OR, Chapter 6.

Briske, D. D. (1996). A Functional Interpretation of Herbivory Resistance in Mesic Grasslands. *Proc. Internat. Rangeland Cong.* **5**(1):68–69.

Briske, D. D., and V. J. Anderson. (1992). Competitive Ability of the Bunchgrass *Schizachyrium scoparium* as Affected by Grazing History and Defoliation. *Vegetatio* **103**(1):41–49.

Briske, D. D., and B. R. Brumfield. (1987). Grazing-Induced Modifications of Plant Structure and Tiller Demography of the Bunchgrass, *Schizachyrium scoparium:* A Simulation Approach. *Soc. Range Mgt. Abst. Papers* **40**:6.

Briske, D. D., and J. H. Richards. (1994). Physiological Responses of Individual Plants to Grazing: Current Status and Ecological Significance. *In* M. Vavra, W. Laycock, and R. D. Pieper (Eds.), "Ecological Implications of Livestock Herbivory in the West." Soc. Range Mgt., Denver, CO, pp. 147–176.

Briske, D. D., and J. W. Stuth. (1982). Tiller Defoliation in a Moderate and Heavy Grazing Regime. *J. Range Mgt.* **35**(4):511–514.

Briske, D. D., and J. W. Stuth. (1986). Tiller Defoliation in a Moderate and Heavy Grazing Regime. *Texas Agric. Expt. Sta. Prog. Rep.* **4420.** 1 p.

Briske, D. D., T. W. Boutton, and Z. Wang. (1996). Contribution of Flexible Allocation Priorities to Herbivory Tolerance in C_4 Perennial Grasses: An Evaluation with ^{13}C Labeling. *Oecologia* **105**(1):151–159.

Brock, B. L., and C. E. Owensby. (1999). Spatial and Temporal Cattle (*Bos taurus*) Grazing Patterns on Intensive-Early Stocked and Season-Long Stocked Pastures. *Soc. Range Mgt., Abst. Papers* **1999**:8.

Brown, J. R., and J. W. Stuth. (1986). Influence of Stocking Rate on Tiller Dynamics in a Rotational Grazing System. *Texas Agric. Expt. Sta. Prog. Rep.* **4416.** 1 p.

Brown, R. D. (Ed.). (1986). "Livestock and Wildlife Management During Drought." Caesar Kleberg Wildl. Res. Inst., Kingsville, Tex. 84 pp.

Brown, R. H., E. R. Beaty, R. A. McCreery, and John D. Powell. (1961). Coastal Bermudagrass Utilization: Soilage Vs. Continuous Grazing. *J. Range Mgt.* **14**(6):297–300.

Broyles, B. (1995). Desert Wildlife Water Developments: Questioning Use in the Southwest. *Wildlife Soc. Bul.* **23**(4):633–675.

Brummer, E. C., and J. H. Bouton. (1992). Physiological Traits Associated with Grazing-Tolerant Alfalfa. *Agron. J.* **84**(2):138–143.

Brummer, J. E., R. L. Gillen, and F. T. McCollum. (1988). Herbage Dynamics of Tallgrass Prairie Under Short Duration Grazing. *J. Range Mgt.* **41**(3):264–266.

Bryant, F. C., and B. Morrison. (1985). Managing Plains Mule Deer in Texas and Eastern New Mexico. *Texas Tech. Univ. Range and Wildl. Mgt. Note* **7.** 6 pp.

Bryant, F. C., and C. A. Taylor. (1992). Meat Goat/White-Tailed Deer Relationships on Rangelands. *In* "Proceedings of the International Conference on Meat Goat Production, Management, and Harvesting." Texas Agric. Ext. Serv., Corpus Christi, TX, pp. 157–165.

Bryant, F. C., M. M. Kothmann, and L. B. Merrill. (1979). Diets of Sheep, Angora Goats, Spanish Goats, and White-Tailed Deer Under Excellent Range Conditions. *J. Range Mgt.* **32**(6):412–417.

Bryant, F. C., C. A. Taylor, and L. B. Merrill. (1981a). White-tailed Deer Diets from Pastures in Excellent and Poor Range Condition. *J. Range Mgt.* **34**(3):193–200.

Bryant, F. C., F. S. Guthery, and W. M. Webb. (1981b). Grazing Management in Texas and Its Impact on Selected Wildlife. *In* L. Nelson, Jr. and J. M. Peek (Eds.), "Proceedings of the Wildlife-Livestock Relationships Symposium, Coeur d'Alene, Idaho, April 20–22, 1981." Univ. Idaho, Moscow.

Bryant, F. C., B. E. Dahl, R. D. Pettit, and C. M. Britton. (1988). Grazing Management: Research Experiences at Texas Tech University. *Texas Tech. Univ., Research Highlights, Noxious Brush and Weed Control; Range and Wildlife Management* **19**:7–13.

Bryant, F. C., B. E. Dahl, R. D. Pettit, and C. M. Britton. (1989). Does Short-Duration Grazing Work in Arid and Semiarid Regions? *J. Soil & Water Cons.* **44**(4):290–296.

Bryant, H. T., R. E. Balser, R. C. Hammes, Jr., and J. P. Fontenot. (1970). Effects of Grazing Management on Animal and Area Output. *J. Anim. Sci.* **30**(1):153–158.

Bryant, J. P., F. S. Chapin III, T. P. Clausen, and P. R. Reichardt. (1987). Effect of Resource Availability on Woody Plant-Mammal Interaction. *USDA, For. Serv. Gen. Tech. Rep.* **INT-222,** pp. 2–9.

Bryant, J. P., F. S. Chapin III, and D. R. Klein. (1983). Carbon/Nutrient Balance of Boreal Plants in Relation to Vertebrate Herbivory. *Oikos* **40**(3):357–368.

Bryant, L. D. (1982). Response of Livestock to Riparian Zone Exclusion. *J. Range Mgt.* **35**(6):780–785.

Buckhouse, J. C., and W. C. Krueger. (1981). What Caused Those Terracettes? *Rangelands* **3**(2):72–73.

Buckhouse, J. C., J. M. Skovlin, and Robert W. Knight. (1981) Streambank Erosion and Ungulate Grazing Relationships. *J. Range Mgt.* **34**(3):339–340.

Burke, D. J. (1987). The Role of Forages in the Horse Program. *Forage & Grassland Conf.* **1987:** 36–41.

Burkhardt, J. W. (1996). Herbivory in the Intermountain West: An Overview of Evolutionary History, Historic Cultural Impacts, and Lessons from the Past. *Idaho For., Wildl., & Range Expt. Sta. Bull.* **58.** 35 pp.

Burkhardt, J. W. (1997). Grazing Utilization Limits: An Ineffective Management Tool. *Rangelands* **19**(3):8–9.

Burleson, W. H., and W. C. Leininger. (1988). Intensive Grazing—Precautions. *Rangelands* **10**(4): 186–188.

Burlison, A. J., J. Hodgson, and A. W. Illius. (1991). Sward Canopy Structure and the Bite Dimensions and Bite Weight of Grazing Sheep. *Grass & Forage Sci.* **46**(1):29–38.

Burns, J. C. (1984). Managing Forage Availability for Animal Responses in Temperate-Species Grazing Systems. *Forage and Grassland Conf.* **1984:**386–393.

Burritt, E. A., and F. D. Provenza. (1989). Food Aversion Leaning: Conditioning Lambs to Avoid a Palatable Shrub (*Cercocarpus montanus*). *J. Anim. Sci.* **67**(3):650–653.

Burzlaff, D. F., and Lionel Harris. (1969). Yearling Steer Gains and Vegetation Changes of Western Nebraska Rangeland Under Three Rates of Stocking. *Neb. Agric. Expt. Sta. Bul.* **505.** 18 pp.

Busby, F. (1987). Go for the Gold. *J. Range Mgt.* **40**(2):98–99, 131.

Busso, C. A., J. H. Richards, and N. J. Chatterton. (1990). Nonstructural Carbohydrates and Spring Regrowth of Two Cool-Season Grasses: Interaction of Drought and Clipping. *J. Range Mgt.* **43**(4): 336–343.

Butcher, J. E. (1966). Snow as the Only Source of Water for Sheep. Amer. *Soc. Anim. Sci., West. Sect. Proc.* **17**:205–209.

Butcher, J. E., L. E. Harris, and R. J. Raleigh. (1959). Water Requirements for Beef Cattle. *Farm and Home Sci.* **20**(3):72–73.

Butler, J. L., and D. D. Briske. (1986). Tiller Dynamics in Response to Grazing. *Texas Agric. Expt. Sta. Prog. Rep.* **4417.** 1 pp.

Byrd, N. A. (Comp.). (1980 rev.). "Forestland Grazing: A Guide for Service Foresters in the South." USDA, For. Serv., Southeastern Area State and Private Forestry, Atlanta, GA. (For. Rep. SA-FRIO). 45 pp.

Byrd, N. A., C. E. Lewis, and H. A. Pearson. (1984). Management of Southern Pine Forest for Cattle Production. *USDA, For. Serv. Gen. Rep.* **R8-GR 4.** 22 pp.

Cable, D. R. (1982). Partial Defoliation Stimulates Growth of Arizona Cottontop. *J. Range Mgt.* **35**(5):591–593.

Cable, D. R., and R. P. Shumway. (1966). Crude Protein in Rumen Contents and in Forage. *J. Range Mgt.* **19**(3):124–128.

Caldwell, M. M. (1984). Plant Requirements for Prudent Grazing. *In* Natl. Res. Council/Natl. Acad. Sci. "Developing Strategies for Rangeland Management." Westview Press, Boulder, CO, pp. 117–152.

Caldwell, M. M., and J. H. Richards. (1986). Competitive Position of Species in Respect to Grazing Tolerance: Some Perspective on Ecophysiological Processess. *Proc. Internat. Rangeland Cong.* **2**:447–449.

Campbell, C. M., D. Reimer, J. C. Nolan, Jr., Y. N. Tamimi, and E. B. Ho-a. (1977). Grazing Management Systems for Fertilized Pastures. *Amer. Soc. Anim. Sci., West. Sect. Proc.* **28**:122–123.

Campbell, E. G., and R. L. Johnson. (1983). Food Habits of Mountain Goats, Mule Deer, and Cattle on Chopaka Mountain, Washington, 1977-1980. *J. Range Mgt.* **36**(4):488–491.

Campbell, J. B., E. Stringam, and P. Gervais. (1969). Pasture Activities of Cattle and Sheep. *Can. Dept. of Agric. Pub.* **1315.**, pp. 105–112 (Chap. 12).

Canfield, R. H. (1940 Rev.). Semi-deferred Grazing as a Restorative Measure for Black Grama Ranges. *USDA, SW For. and Range Expt. Sta. Res. Note* **80.** 4 pp.

Cardon, P. B., E. B. Stanley, W. J. Pistor, and J. C. Nesbitt. (1951). The Use of Salt as a Regulator of Supplemental Feed Intake and Its Effect on the Health of Range Livestock. *Ariz. Agric. Expt. Sta. Bul.* **239.** 15 pp.

Carman, J. G., and D. D. Briske. (1986). Does Long-Term Grazing Create Morphologic or Genetic Variation in Little Bluestem? *Texas Agric. Expt. Sta. Prog. Rep.* **4419.** 1 pp.

Carpenter, L. H., O. C. Wallmo, and R. B. Gill. (1979). Forage Diversity and Dietary Selection by Wintering Mule Deer. *J. Range Mgt.* **32**(3):226–229.

Caton, J. S., and D. V. Dhuyvetter. (1996). Manipulation of Maintenance Requirements with Supplementation. *Proc. Grazing Livestock Nutr. Conf.* **3**:72–82.

Caton, J. S., A. S. Freeman, and M. L. Galyean. (1988). Influence of Protein Supplementation on Forage Intake, In Situ Forage Disappearance, Ruminal Fermentation, and Digesta Passage Rates in Steers Grazing Dormant Blue Grama Rangelands. *J. Anim. Sci.* **66**(9):2262–2271.

Chai, K., L. P. Milligan, and G. W. Mathison. (1988). Effect of Muzzling on Rumination in Sheep. *Can. J. Anim. Sci.* **68**(2):387–397.

Chamrad, A. D., and T. W. Box. (1968). Food Habits of White-Tailed Deer in South Texas. *J. Range Mgt.* **21**(3):158–164.

Chaney, E., W. Elmore, and W. S. Platts. (1993). "Managing Change: Livestock Grazing on Western Riparian Areas." Environmental Protection Agency by the Northwest Resource Information Center, Eagle, ID. 31 pp.

Charette, L. A., V. S. Logan, and J. B. Campbell. (1969). Grazing Systems. *Can. Dept. Agric. Pub.* **1315**, pp. 113–121 (Chap. 13).

Choi, S. S., and J. E. Butcher. (1961). The Influence of Alternate Day Watering on Feed and Winter Consumption of Sheep Maintained Under Two Temperatures. *J. Anim. Sci.* **20**(3):678.

Christian, K. R. (1981). Simulation of Grazing Systems. *In* F. H. W. Morley (Ed.), "Grazing Animals." Elsevier Sci. Pub. Co., Amsterdam, Neth., p. 361–377.

Christopherson, R. J., and B. A. Young. (1972). Energy Cost of Activity in Cattle. *Univ. Alta. Feeders Day Report* **51**:40–41.

Church, S. B., M. J. McInerney, and K. M. Havstad. (1986). Growth Patterns of Beef Calves on Western Rangeland over Discrete Intervals in the Grazing Season. *Amer. Soc. Anim. Sci., West. Sect. Proc.* **37**:153–156.

Cid, M. S., and M. A. Brizuela. (1998). Heterogeneity in Tall Fescue Pastures Created and Sustained by Cattle Grazing. *J. Range Mgt.* **51**(6):644–649.

Clanton, D. C. (1979). Nonprotein Nitrogen in Range Supplements. *J. Anim. Sci.* **47**(4):765–779.

Clanton, D. C. (1982). Crude Protein System in Range Supplements. *Okla. Agric. Expt. Sta. Misc. Pub.* **109**, pp. 228–237.

Clanton, D. (1988). Grazing Cornstalks—A Review. *Neb. Agric. Res. Div. Misc. Pub.* **54**, pp. 11–15.

Clanton, D. C., and D. F. Burzlaff. (1966). Grazing Cattle on Subirrigated Meadows. *J. Range Mgt.* **19**(3):151–152.

Clanton, D. C., J. T. Nichols, and B. R. Somerhalder. (1971). Young Cows on Irrigated Pastures. *Neb. Agric. Ext. Cir.* **EC 71–218,** pp. 16–19.

Clanton, D. C., J. T. Nichols, and D. A. Yates. (1981). Continuous Grazing Affects on Steer Diet. *Neb. Agric. Ext. Cir.* **81–218,** pp. 17–18.

Clanton, D. C., R. L. Hildebrand, and L. E. Jones. (1971). Supplements for Yearling Cattle on Summer Range. *In* "Nebraska Beef Cattle Report." *Nebr. Ext. Cir.* **71–218,** pp. 7–9.

Clapperton, J. L. (1961). The Energy Expenditure of Sheep in Walking on the Level and on Gradients. *Proc. Nutr. Soc.* **20**:XXXI.

Clark, P. E., W. C. Krueger, L. D. Bryant, and D. R. Thomas. (1998). Spring Defoliation Effects on Bluebunch Wheatgrass: I. Winter Forage Quality. *J. Range Mgt.* **51**(5):519–525.

Clark, P. E., W. C. Krueger, L. D. Bryant, and D. R. Thomas. (2000). Livestock Grazing Effects on Forage Quality of Elk Winter Range. *J. Range Mgt.* **53**(1):97–105.

Clary, W. P. (1975). Range Management and Its Ecological Basis in the Ponderosa Pine Type of Arizona: The Status of Our Knowledge. *USDA, For. Serv. Res. Paper* **RM-158.** 35 pp.

Clary, W. P., N. L. Shaw, J. G. Dudley, V. A. Saab, *et al.* (1996). Response of a Depleted Sagebrush Steppe Riparian System to Grazing Control and Woody Plantings. USDA, For. Serv. Res. Paper Int-RP-492. 32pp.

Clary, W. P., and D. M. Beale. (1983). Pronghorn Reactions to Winter Sheep Grazing, Plant Communities, and Topography in the Great Basin. *J. Range Mgt.* **36**(6):749–752.

Clary, W. P., and G. D. Booth. (1993). Early Season Utilization of Mountain Meadow Riparian Pastures. *J. Range Mgt.* **46**(6):493–497.

Clary, W. P., and R. C. Holmgren. (1982). Desert Experimental Range: Establishment and Research Contribution. *Rangelands* **4**(6):261–264.

Clary, W. P., E. D. McArthur, D. Bedunah, and C. L. Wambolt (Comp.). (1992). Proceedings—Symposium on Ecology and Management of Riparian Shrub Communities. *USDA, For. Serv. Gen. Tech. Rep.* **INT-289.** 232 pp.

Clary. W. P. (1995). Vegetation and Soil Responses to Grazing Simulation on Riparian Meadows. *J. Range Mgt.* **48**(1):18–25.

Clawson, W. J., N. K. McDougald, and D. A. Duncan. (1982). Guidelines for Residue Management on Annual Range. *Univ. Calif., Div. Agric. Sci. Leaflet* **21327.** 4 pp.

Cluff, L. K., B. L. Welch, J. C. Pederson, and J. D. Brotherson. (1982). Concentration of Monoterpenoids in the Rumen Ingesta of Wild Mule Deer. *J. Range Mgt.* **35**(2):192–194.

Coblentz, B. E. (1977). Some Range Relationships of Feral Goats on Santa Catalina Island, California. *J. Range Mgt.* **30**(6):415–419.

Cochran, R. C., D. C. Adams, P. O. Currie, and B. W. Knapp. (1986). Cubed Alfalfa Hay or Cottonseed Meal-Barley as Supplements for Beef Cows Grazing Fall-Winter Range. *J. Range Mgt.* **39**(4):361–364.

Cohen, W. E., D. Lynn Drawe, F. C. Bryant, and L. C. Bradley. (1989). Observations on White-tailed Deer and Habitat Response to Livestock Grazing in South Texas. *J. Range Mgt.* **42**(5):361–365.

Cole, G. F. (1956). The Pronghorn Antelope: Its Range Use and Food Habits in Central Montana with Special Reference to Alfalfa. *Mon. Agric. Expt. Sta. Bul.* **516.** 63 pp.

Cole, S. (1999). Low Stress Livestock Handling. *In* "Grazing Behavior of Livestock and Wildlife." *Idaho Forest, Wildl., and Range Expt. Sta. Bull.* **70,** pp. 79–84.

Coleman, S. W. (1992). Plant-Animal Interface. *J. Prod. Agric.* **5**(1):7–13.

Coleman, S. W., D. C. Meyerhoeffer, and F. P. Horn. (1984). Semen Characteristics and Behavior of Grazing Bulls as Influenced by Shade. *J. Range Mgt.* **37**(3):243–247.

Collins, W. B., and P. J. Urness. (1982). Mule Deer and Elk Responses to Horsefly Attacks. *Northwest Sci.* **56**(4):299–302.

Collins, W. B., and P. J. Urness. (1983). Feeding Behavior and Habitat Selection of Mule Deer and Elk on Northern Utah Summer Range. *J. Wildl. Mgt.* **47**(3):646–663.

Commonwealth Scientific and Industrial Research Organization (CSIRO). (1990). "Feeding Standards for Australian Livestock: Ruminants." CSIRO Publications, East Melbourne, Victoria, Australia.

Conrad, B. E., and E. C. Holt. 1983. Year-Round Grazing of Warm-Season Perennial Pastures. *Texas Agric. Expt. Sta. Misc. Pub.* **1540.** 4 pp.

Cook, C. J., C. W. Cook, and L. E. Harris. (1948). Utilization of Northern Utah Summer Range Plants by Sheep. *J. For.* **46**(6):416–425.

Cook, C. W. (1954). Common Use of Summer Range by Sheep and Cattle. *J. Range Mgt.* **7**(1):10–13.

Cook, C. W. (1959). The Effect of Site on the Palatability and Nutritive Content of Seeded Wheat-grasses. *J. Range Mgt.* **12**(6):289–292.

Cook, C. W. (1962). An Evaluation of Some Common Factors Affecting Utilization of Desert Range Species. *J. Range Mgt.* **15**(6):333–338.

Cook, C. W. (1965). Plant and Livestock Responses to Fertilized Rangelands. *Utah Agric. Expt. Sta. Bul.* **455.** 35 pp.

Cook, C. W. (1966a). Development and Use of Foothill Ranges in Utah. *Utah Agric. Expt. Sta. Bul.* **461.** 47 pp.

Cook, C. W. (1966b). Factors Affecting Utilization of Mountain Slopes by Cattle. *J. Range Mgt.* **19**(4):200–204.

Cook, C. W. (1966c). The Role of Carbohydrate Reserves in Managing Range Plants. *Utah Agric. Expt. Sta. Mimeo. Ser.* **499.** 11 pp.

Cook, C. W. (1967). Increased Capacity Through Better Distribution on Mountain Ranges. *Utah Farm and Home Sci.* **28**(2):39–42.

Cook, C. W. (1970). Energy Budget of the Range and Range Livestock. *Colo. Agric. Expt. Sta. Bul.* **TB-109.** 28 pp.

Cook, C. W. (1971). Effects of Season and Intensity of Use on Desert Vegetation. *Utah Agric. Expt. Sta. Bul.* **483.** 57 pp.

Cook, C. W. (1972). Comparative Nutritive Values of Forbs, Grasses, and Shrubs. *USDA, For. Serv. Gen. Tech. Rep.* **INT-1,** pp. 303–310.

Cook, C. W. (1983). "Forbs" Need Proper Ecological Recognition. *Rangelands* **5**(5):217–220.

Cook, C. W., and L. E. Harris. (1950). The Nutritive Content of the Grazing Sheep's Diet on the Summer and Winter Ranges of Utah. *Utah Agric. Expt. Sta. Bul.* **342.** 66 pp.

Cook, C. W., and L. E. Harris. (1968a). Nutritive Value of Seasonal Ranges. *Utah Agric. Expt. Sta. Bul.* **472.** 55 pp.

Cook, C. W., and L. E. Harris. (1968b). Effect of Supplementation on Intake and Digestibility of Range Forage. *Utah Agric. Expt. Sta. Bul.* **475.** 38 pp.

Cook, C. W., and N. Jefferies. (1963). Better Distribution of Cattle on Mountain Ranges. *Utah Farm and Home Sci.* **24**(2):31, 48–49.

Cook, C. W., and L. A. Stoddart. (1963). The Effect of Intensity and Season of Use on the Vigor of Desert Range Plants. *J. Range Mgt.* **16**(6):315–317.

Cook, C. W., and J. Stubbendieck (Eds.). (1986). "Range Research: Basic Problems and Techniques." Soc. for Range Mgt., Denver, CO. 336 pp.

Cook, C. W., L. A. Stoddart, and F. E. Kinsinger. (1958). Responses of Crested Wheatgrass to Various Clipping Treatments. *Ecol. Monogr.* **28**(3):237–272.

Cook, C. W., J. E. Mattox, and L. E. Harris. (1961). Comparative Daily Consumption and Digestibility of Summer Range Forage by Wet and Dry Ewes. *J. Anim. Sci.* **20**(4):866–870.

Cook, C. W., K. Taylor, and L. E. Harris. (1962). The Effect of Range Condition and Intensity of Grazing upon Daily Intake and Nutritive Value of the Diet on Desert Ranges. *J. Range Mgt.* **15**(1):1–6.

Cook, C. W., M. Kothmann, and L. E. Harris. (1965). Effect of Range Condition and Utilization on Nutritive Intake of Sheep on Summer Ranges. *J. Range Mgt.* **18**(2):69–73.

Cook, C. W., L. E. Harris, and M. C. Young. (1967). Botanical and Nutritive Content of Diets of Cattle and Sheep Under Single and Common Use on Mountain Range. *J. Anim. Sci.* **26**(5):1169–1174.

Cook, C. W., J. K. Matsushima, and D. A. Cramer. (1981). Does Range Have a Place in Beef Production Systems of the Future? *Rangelands* **3**(4):143–144.

Cook, C. W., J. W. Walker, M. H. Ebberts, L. R. Rittenhouse, E. T. Bartlett, D. A. Cramer, P. T. Fagerlin, and M. C. McLean. (1983 rev.). Alternative Grass and Grain Feeding Systems for Beef Production. *Colo. Agric. Expt. Sta. Bul.* **579S**. 83 pp.

Cook, C. W., D. A. Cramer, and L. Rittenhouse. (1984). Acceptable Block Beef from Steers Grazing Range and Crop Forages. *J. Range Mgt.* **37**(2):122–126.

Cooper, C. S., R. R. Wheeler, and W. A. Sawyer. (1957). Meadow Grazing. I. A Comparison of Gains of Calves and Yearlings When Summering on Native Flood Meadows and Sagebrush-Bunchgrass Range. *J. Range Mgt.* **10**(4):172–174.

Cooper, S. M., and N. Owen-Smith. (1986). Effects of Plant Spinescence on Large Mammalian Herbivores. *Oecologia* **68**(3):446–455.

Cooperrider, A. Y., and J. A. Bailey. (1984). A Simulation Approach to Forage Allocation. *In* Natl. Res. Council/Natl. Acad. Sci. "Developing Strategies for Rangeland Management." Westview Press, Boulder, CO. pp. 525–559.

Coppedge, B. R., and J. H. Shaw. (1997). Effects of Horning and Rubbing Behavior by Bison (*Bison bison*) on Woody Vegetation in a Tallgrass Prairie Landscape. *Amer. Midl. Nat.* **138**(1): 189–196.

Coppedge, B. R., and J. H. Shaw. (1998). Bison Grazing Patterns on Seasonally Burned Tallgrass Prairie. *J. Range Mgt.* **51**(3):258–264.

Coppedge, B. R., D. M. Leslie, Jr., and J. H. Shaw. (1998). Botanical Composition of Bison Diets on Tallgrass Prairie in Oklahoma. *J. Range Mgt.* **51**(4):379–382.

Corah, L., and E. E. Bartley. (1985). Alfalfa Grazing—The Other Alternative. *Great Plains Extension* **GPE-1850**. 4 pp.

Cordova, F. J., J. D. Wallace, and R. D. Pieper. (1978). Forage Intake by Grazing Animals: A Review. *J. Range Mgt.* **31**(6):430–438.

Correa, A., D. M. Nixon, and C. Russel. (1987). An Economic and Nutritional evaluation of Pricklypear as an Emergency Forage Supplement. *Texas J. Agric. & Nat. Resources* **1**:41–44.

Cory, V. L. (1927). Activities of Livestock on the Range. *Texas Agric. Expt. Sta. Bul.* **367**. 47 pp.

Costello, D. F., and R. S. Driscoll. (1957). Hauling Water for Range Cattle. *USDA Leaflet* **419**. 6 pp.

Cotter, P. F., B. E. Dahl, and G. Scott. (1983). Utilization of a Short Duration Grazing System for Renovation and Management of Weeping Lovegrass (*Eragrostis curvula*) in the Rolling Plains of Texas. *Forage and Grassland Conf.* **1983**:24–30.

Coughenour, M. B. (1991). Spatial Components of Plant-Herbivore Interactions in Pastoral, Ranching, and Native Ungulate Ecosystems. *J. Range Mgt.* **44**(6):530–542.

Cox, J. R., and J. F. Cadenhead (Eds.). (1993). "Managing Livestock Stocking Rates on Rangelands: Proceedings of Symposia." Dept. Rangeland Ecol. & Mgt., Texas A&M Univ., College Station. 149 pp.

Cox, J. R., R. L. Gillen, and G. B. Ruyle. (1989). Big Sacaton Riparian Grassland Management: Seasonal Grazing Effects on Plant and Animal Production. *Applied Agric. Res.* **4**(2):127–134.

Cox, J. R., G. B. Ruyle, and B. A. Roundy. (1990). Lehmann Lovegrass in Southeastern Arizona: Biomass Production and Disappearance. *J. Range Mgt.* **43**(4):367–372.

Crane, K. K., M. A. Smith, and D. Reynolds. (1997). Habitat Selection Patterns of Feral Horses in Southcentral Wyoming. *J. Range Mgt.* **50**(4):374–380.

Crawford, R. J., Jr., M. D. Massie, D. A. Sleper, and H. F. Mayland. (1998). Use of an Experimental High-Magnesium Tall Fescue to Reduce Grass Tetany in Cattle. *J. Prod. Agric.* **11**(4):491–496.

Crop Sci. Soc. Amer. (1984). "Physiological Basis of Crop Growth and Development." Crops Sci. Soc. Amer., Madison, Wisc. 341 pp.

Crouch, G. L. (1968). Forage Availability in Relation to Browsing of Douglas-Fir Seedlings by Black-Tailed Deer. *J. Wildl. Mgt.* **32**(3):542–553.

Crouch, G. L. (1980). Postseason Hunting to Reduce Deer Damage to Douglas-Fir in Western Oregon. *USDA, For. Serv. Res. Note* **PNW-349.** 6 pp.

Culley, M. J. (1938). Grazing Habits of Range Cattle. *J. For.* **36**(7):715–717.

Currie, P. O. (1969). Use Seeded Ranges in Your Management. *J. Range Mgt.* **22**(6):432–434.

Currie, P. O. (1970). Influence of Spring, Fall, and Spring-Fall Grazing on Crested Wheatgrass Range. *J. Range Mgt.* **23**(2):103–108.

Currie, P. O. (1975). Grazing Management of Ponderosa Pine-Bunchgrass Ranges of the Central Rocky Mountains: The Status of Our Knowledge. *USDA, For. Serv. Res. Paper* **RM-159.** 24 pp.

Currie, P. O. (1976). Recovery of Ponderosa Bunchgrass Ranges Through Grazing and Herbicide or Fertilizer Treatments. *J. Range Mgt.* **29**(6):444–448.

Currie, P. O. (1978). Cattle Weight Gain Comparison Under Seasonlong and Rotation Grazing Systems. *Proc. Internat. Rangeland Cong.* **1:**579–580.

Currie, P. O. (1987). Herbage Yield and Cover Estimates as Guides for Predicting Livestock Management. *In* D. A. Jameson and J. Holechek (Eds.), "Monitoring Animal Performance and Production Symposium Proceedings, February 12, 1987, Boise, Idaho." Society for Range Management, Denver, CO. pp. 4–7.

Currie, P. O., and D. R. Smith. (1970). Response of Seeded Ranges to Different Grazing Intensities in the Ponderosa Pine Zone of Colorado. *USDA Prod. Res. Rep.* **112.** 41 pp.

Currie, P. O., D. W. Reichert, J. C. Malechek, and O. C. Wallmo. (1977). Forage Selection Comparison for Mule Deer and Cattle under Managed Ponderosa Pine. *J. Range Mgt.* **30**(5):352–256.

Currie, P. O., C. B. Edminster, and W. Knott. (1978). Effects of Cattle Grazing on Ponderosa Pine Regeneration in Central Colorado. *USDA, For. Serv. Res. Paper* **RM-201.** 7 pp.

Dahl, B. E., and D. N. Hyder. (1977). Developmental Morphology and Management Implications. *In* Ronald E. Sosebee (Ed.). Rangeland Plant Physiology. *Soc. Range Mgt. Range Sci. Ser.* **4**, pp. 257–290.

Dahl, B. E., J. C. Moseley, P. F. Cotter, and W. Matizha. (1987). Stocking Rate vs. Stock Density. *Texas Tech Univ., Research Highlights, Noxious Brush and Weed Control; Range and Wildlife Management* **18:**16–17.

Dahl, B. E., P. F. Cotter, R. L. Dickerson, and J. C. Mosley. (1988). Vegetation Response to Continuous Vs. Short Duration Grazing After 8 Years. *Soc. Range Mgt. Abst. Papers* **41:**139.

Dahl, B. E., P. F. Cotter, R. L. Dickerson, Jr., and J. C. Mosley. (1992). Vegetation Response to Continuous Versus Short Duration Grazing on Sandy Rangeland. *Texas J. Agric. & Nat. Resources* **5:**73–81.

Dalrymple, R. L., and B. Flatt. (1993). Using Objective Indicators To Help Make Rotational Grazing Decisions. *Forage and Grassland Conf.* **1993:**57–60.

Dana, A. C., J. Baden, and T. Blood. (1985). Ranching and Recreation: Covering Costs of Wildlife Production. *In* Jeff Powell (Ed.). "Holistic Ranch Management Workshop Proceedings, 23–30 May 1985, Casper, Wyoming." Wyo. Agric. Ext. Serv., Laramie, Wyo., pp. 61–76.

Daniels, T. K., J. G. P. Bowman, B. F. Sowell, E. E. Grings, and M. D. MacNeil. (1998). The Effects of Cow Age and Supplement Delivery Method on Forage and Liquid Supplement Intake. *Amer. Soc. Anim. Sci., West. Sect. Proc.* **49:**145–148.

Darrow, R. A., and W. G. McCully. (1959). Brush Control and Range Improvement in the Post Oak-Blackjack Oak Area of Texas. *Texas Agric. Expt. Sta. Bull.* **942.** 16 pp.

Daugherty, D. A., H. A. Turner, and C. M. Britton. (1979). Effects of Grazing Intensity on Steer Gains. *Amer. Soc. Anim. Sci., West. Sect. Proc.* **30:**139–142.

Daugherty, D. A., H. A. Turner, and C. M. Britton. (1979). Pasture Management for Increased Gains. *Ore. Agric. Expt. Sta. Spec. Rep.* **534,** p. 17–22.

Davis, G. G., L. E. Bartel, and C. W. Cook. (1975). Control of Gambel Oak Sprouts by Goats. *J. Range Mgt.* **28**(3):216–218.

Davison, J. (1996). Livestock Grazing in Wildland Fuel Management Programs. *Rangelands* **18**(6): 242–245.

Day, T. A., and J. K. Detling. (1990). Grassland Patch Dynamics and Herbivore Grazing Preference Following Urine Deposition. *Ecology* **71**(1):180–188.

DeFlon, J. G. (1986). The Case for Cheat Grass. *Rangelands* **8**(1):14–17.

Degen, A. A., and B. A. Young. (1981). Response of Lactating Ewes to Snow as a Source of Water. *Can. J. Anim. Sci.* **61**(1):73–79.

Degen, A. A., B. A. Young, and J. A. Francis. (1979). Snow Consumption by Beef Cattle. *Can. J. Anim. Sci.* **59**(4):826.

Delgiudice, G. D., and J. E. Rodiek. (1984). Do Elk Need Free Water in Arizona? *Wildl. Soc. Bul.* **12**(2):142–146.

Demarais, S., D. A. Osborn, and J. J. Jackley. (1990). Exotic Big Game: A Controversial Resource. *Rangelands* **12**(2):121–125.

Demment, M. W., and P. J. Van Soest. (1983). "Body Size, Digestive Capacity, and Feeding Strategies of Herbivores." Winrock International, Morrilton, Ark., 64 pp.

Dennis, R. E. (1981). Management of Perennial Irrigated Pastures. *Ariz. Agric. Ext.* **Q421.** 2 pp.

DeRamus, H. A., and C. P. Bagley. (1984). The Effect of Mefluidide Treatment on Bahia and Bermuda-grass and Nutrient Utilization. *Forage and Grassland Conf.* **1984:**152–156.

Dill, T. O., J. Menghini, S. S. Waller, and R. Case. (1983). Fee Hunting for Nebraska Big Game: A Possibility. *Rangelands* **5**(1):24–27.

Dodd, N. L., and W. W. Brady. (1988). Ecological Relationships of Sympatric Desert Bighorn Sheep and Cattle in Arizona. *Soc. Range Mgt. Abst. Papers* **41:**348.

Doescher, P. S., S. D. Tesch, and M. Alejandro-Castro. (1987). Livestock Grazing: A Silvicultural Tool for Plantation Establishment. *J. For.* 85(10):29–37.

Doescher, P. S., S. D. Tesch, and W. E. Drewien. (1989). Water Relations and Growth of Conifer Seedlings During Three Years of Cattle Grazing on a Southwest Oregon Plantation. *Northwest Sci.* **63**(5):232–240.

Dormaar, J. F., S. Smoliak, and W. D. Willms. (1989). Vegetation and Soil Responses to Short-Dura-tion Grazing on Fescue Grasslands. *J. Range Mgt.* **42**(3):252–256.

Dorn, R. D. (1970). Moose and Cattle Food Habits in Southwest Montana. *J. Wildl. Mgt.* **34**(3):559–564.

Dougherty, C. T. (1991). Influence of Ingestive Behavior on Nutrient Intake of Grazing Livestock. *Okla. Agric. Expt. Sta. Misc. Pub.* **MP-133,** pp. 74–82.

Dougherty, C. T., N. W. Bradley, P. L. Cornelius, and L. M. Lauriault. (1987). Herbage Intake Rates of Beef Cattle Grazing Alfalfa. *Agron. J.* **79**(6):1003–1008.

Dougherty, C. T., T. D. A. Forbes, P. L. Cornelius, L. M. Lauriault, N. W. Bradley, and E. M. Smith. (1988). Effects of Supplementation on the Ingestive Behaviour of Grazing Steers. *Grass & Forage Sci.* **43**(4):353–361.

Dougherty, C. T., N. W. Bradley, P. L. Cornelius, and L. M. Lauriault. (1989a). Short-Term Fasts and the Ingestive Behaviour of Grazing Cattle. *Grass and Forage Sci.* **44**(3):295–302.

Dougherty, C. T., N. W. Bradley, P. L. Cornelius, and L. M. Lauriault. (1989b). Accessibility of Herbage Allowance and Ingestive Behavior of Beef Cattle. *Applied Anim. Beh. Sci.* **23**(1–2):87–97.

Dougherty, C. T., M. Collins, N. W. Bradley, P. L. Cornelius, and L. M. Lauriault. (1990). Moderation of Ingestive Behavior of Beef Cattle by Grazing-Induced Changes in Lucerne Swards. *Grass and Forage Sci.* **45**(2):135–142.

Dougherty, C. T., N. W. Bradley, L. M. Lauriault, J. E. Arias, and P. L. Cornelius. (1992a). Allowance-Intake Relations of Cattle Grazing Vegetative Tall Fescue. *Grass and Forage Sci.* **47**(3):211–219.

Dougherty, C. T., M. Collins, N. W. Bradley, L. M. Lauriault, and P. L. Cornelius. (1992b). The Effects of Poloxalene on Ingestion by Cattle Grazing Lucerne. *Grass & Forage Sci.* **47**(2):180–188.

Dougherty, C. T., F. W. Knapp, P. B. Burrus, D. C. Willis, and N. W. Bradley (1993). Face Flies (*Musca autumnalis* De Geer) and the Behavior of Grazing Beef Cattle. *Applied Anim. Beh. Sci.* **35**(4):313–326.

Drawe, D. L., and T. W. Box. (1968). Forage Ratings for Deer and Cattle on the Welder Wildlife Refuge. *J. Range Mgt.* **21**(4):225–228.

Drawe, D. L., J. R. Frasure, and B. E. Dahl. (1988). Effect of Grazing Management on Cattle Diets and Nutrition in the Coastal Prairie. *Texas J. Agric. and Nutr. Res.* **2**:17–25.

Dunphy, D. J., M. E. McDaniel, and E. C. Holt. (1982). Effect of Forage Utilization on Wheat Grain Yield. *Crop Sci.* **22**(1):106–109.

Duvall, V. L. (1969). Grazing Systems for Pine Forest Ranges in the South. *Amer. Soc. Range Mgt. Abst. Papers* **22**:23.

Duvall, V. L., and L. B. Whitaker. (1964). Rotation Burning: A Forage Management System for Longleaf Pine-Bluestem Ranges. *J. Range Mgt.* **17**(6):322–326.

Duvall, V. L., and N. E. Linnartz. (1967). Influences of Grazing and Fire on Vegetation and Soil of Longleaf Pine-Bluestem Range. *J. Range Mgt.* **20**(4):241–247.

Dwyer, D. D. (1961). Activities and Grazing Preferences of Cows with Calves in Northern Osage County. *Okla. Agric. Expt. Sta. Bul.* **588.** 61 pp.

Dwyer, D. D., J. C. Buckhouse, and W. S. Huey. (1984). Impacts of Grazing Intensity and Specialized Grazing Systems on the Use and Value of Rangeland: Summary and Recommendations. *In* Natl. Res. Council/Natl. Acad. Sci.. "Developing Strategies for Rangeland Management." Westview Press, Boulder, Colo., pp. 867–884.

Dyksterhuis, E. J. (1986). Viewpoint: Crop or Range? *Rangelands* **8**(2):73–75.

Dziuk, P. J., and R. A. Bellows. (1983). Management of Reproduction of Beef Cattle, Sheep, and Pigs. *J. Anim. Sci.* **57**(Supp. 2):355–379.

Early, D., and F. D. Provenza. (1998). Food Flavor and Nutritional Characteristics Alter Dynamics of Food Preference in Lambs. *J. Anim. Sci.* **76**(3):728–734.

Eckert, R. E., Jr., and J. S. Spencer. (1986). Vegetation Response on Allotments Grazed Under Rest-Rotation Management. *J. Range Mgt.* **39**(20):166–174.

Eckert, R. E., Jr., and J. S. Spencer. (1987). Growth and Reproduction of Grasses Heavily Grazed Under Rest-Rotation Management. *J. Range Mgt.* **40**(2):156–159.

Eckert, R. E., Jr., F. F. Peterson, M. S. Meurisse, and J. L. Stephens. (1986). Effects of Soil-Surface Morphology on Emergence and Survival of Seedlings in Big Sagebrush Communities. *J. Range Mgt.* **39**(5):414–420.

Edgerton, P. J., and J. G. Smith. (1971). Seasonal Forage Use by Deer and Elk on the Starkey Experimental Forest and Range. Oregon. *USDA, For. Serv. Res. Paper* **PNW-112.** 12 pp.

Ehrenreich, J. H., and A. J. Bjugstad. (1966). Cattle Grazing Time Is Related to Temperature and Humidity. *J. Range Mgt.* **19**(3):141–142.

Eissenstat, D. M., J. E. Mitchell, and W. W. Pope. (1982). Trampling Damage by Cattle on Northern Idaho Forest Plantations. *J. Range Mgt.* **35**(6):715–716.

Eller, J. M. (1985). Biotechnology in Producing Range Livestock—Establishing Goals and Future Objectives. *In* "National Range Conference Proceedings, Oklahoma City, Oklahoma, November 6–8, 1985." U.S. Dept. Agric., Washington, D.C., pp. 112–113.

Ellis, J. E., and M. Travis. (1975). Comparative Aspects of Foraging Behaviour of Pronghorn Antelope and Cattle. *J. Appl. Ecol.* **12**(2):411–420.

Ellis, W. C., J. H. Matis, Carlos Lascano, Mehdi Mahloogi, and Kevin Pond. (1987). Size Reduction, Fermentation, and Passage of Forage Particles and Forage Intake by Cattle. *Okla. Agric. Expt. Sta. Misc. Pub.* **121,** p. 81–95.

Ely, D. G. (1994). The Role of Grazing Sheep in Sustainable Agriculture. *Sheep Res. J.* **10**(Spec. Issue):37–51.

Emerick, R. J. (1987). Answer to Silica Stones. *S. Dak. Farm & Home Res.* **38**:10–12.

Engle, D. M., and J. G. Schimmel. (1984). Repellent Effects on Distribution of Steers on Native Range. *J. Range Mgt.* **37**(2):140–141.

Erlinger, L. L., D. R. Tolleson, and C. J. Brown. (1990). Comparison of Bite Size, Biting Rate, and Grazing Time of Beef Heifers from Herds Distinguished by Mature Size and Rate of Maturity. *J. Anim. Sci.* **68**(11):3578–3587.

Farah, K. O., and N. E. West. (1988). Effects of Simulated and Sheep Grazing on Dyer's Woad. *Soc. Range Mgt., Abst. Papers* **41.**266.

Ferguson, R. B. (1972). Bitterbrush Topping: Shrub Response and Cost Factors. *USDA, For. Serv. Res. Paper* **INT-125**. 11 pp.

Ferrell, C. L., and T. G. Jenkins. (1987). Influence of Biological Types on Energy Requirements. *In* "Proceedings, Grazing Livestock Nutrition Conference, July 23–24, 1987, Jackson, Wyoming." Univ. Wyo., Laramie, Wyo., pp. 1–6.

Fisher, D .S., H. F. Mayland, and J. C. Burns. (1999). Variation in Ruminants' Preference for Tall Fescue Hays Cut Either at Sundown or at Sunup. *J. Anim. Sci.* **77**(3):762–768.

Fitzgerald, R. D., and A. W. Bailey. (1984). Control of Aspen Regrowth by Grazing with Cattle. *J. Range Mgt.* **37**(2):156–159.

Fitzhugh, H. A., J. J. Hodgson, O. J. Scoville, Thanh D. Nguyen, and T. C. Byerly. (1978). "The Role of Ruminants in Support of Man." Winrock International Livestock Research & Training Center, Morrilton, AR. 136 pp.

Flinders, J. T., and L. Conde. (1980). An Alternative Method of Calcuting Animal-Unit Equivalents Helps Ranchers and Wildlife. *Soc. Range Mgt. Abst. of Papers* **33**:48.

Flores, E., and J. C. Malechek. (1983). Applying the Concept of Feeding Stations to Heifers Grazing Variable Amounts of Available Forage. *Soc. Range Mgt. Abst. of Papers* **36**:95.

Flores, E. R., F. D. Provenza, D. F. Balph, and C. C. Parker. (1987). Effects of Experience in the Development of Foraging Skills of Range Sheep. *Soc. Range Mgt. Abst. Papers* **40**:170.

Flores, E. R., F. D. Provenza, and D. F. Balph. (1989a). Relationship Between Plant Maturity and Foraging Experience of Lambs Grazing Hycrest Crested Wheatgrass. *Applied Anim. Beh. Sci.* **23**(4):279–284.

Flores, E. R., F. D. Provenza, and D. F. Balph. (1989b). The Effect of Experience on the Foraging Skill of Lambs: Importance of Plant Form. *Applied Anim. Beh. Sci.* **23**(4):285–291.

Flores, E. R., E. A. Laca, T. C. Griggs, and M. W. Demment. (1993). Sward Height and Vertical Morphological Differentiation Determine Cattle Bite Dimensions. *Agron. J.* **85**(3):527–532.

Forage and Grazing Terminology Committee (Vivien G. Allen, Chair). (1991). "Terminology for Grazing Lands and Grazing Animals." Pocahontas Press, Blacksburg, VA. 38 pp.

Forage and Grazing Terminology Committee. (1992). Terminology for Grazing Lands and Grazing Animals. *J. Prod. Agric.* **5**(1):191–201.

Foran, B. D., and D. M. Stafford-Smith. (1991). Risk, Biology, and Drought Management Strategies for Cattle Stations in Central Australia. *J. Environ. Mgt.* **33**(1):17–33.

Forbes, J. M. (1986). "The Voluntary Food Intake of Farm Animals." Butterworth & Co., London, Eng. 206 pp.

Forbes, T. D. A. (1988). Researching the Plant-Animal Interface: The Investigation of Ingestive Behavior in Grazing Animals. *J. Anim. Sci.* **66**(9):2369–2379.

Forbes, T. D. A., and S. W. Coleman. (1985). Influence of Herbage Mass and Structure of Warm-Season Grass on Ingestive Behavior of Grazing Cattle. *Proc. Intern. Grassland Cong.* **15**:1123–1125.

Forbes, T. D. A., and S. W. Coleman. (1993). Forage Intake and Ingestive Behavior of Cattle Grazing Old World Bluestems. *Agron. J.* **85**(4):808–816.

Forbes, T. D. A., and J. Hodgson. (1985). The Reaction of Grazing Sheep and Cattle to the Presence of Dung From the Same or Other Species. *Grass and Forage Sci.* **40**(2):177–182.

Forbes, T. D. A., F. M. Rouquette, and J. W. Holloway. (1996). Grazing Behavior and Passage Kinetics of Tuli × Brahman, Brahman, Angus, and Angus × Brahman Cattle. *Proc. Internat. Rangeland Cong.* **5**(1):153–154.

Forbes, T. D. A., F. M. Rouquette, Jr., and J. W. Holloway. (1998). Comparisons Among Tuli-, Brahman-, and Angus-Sired Heifers: Intake, Digesta Kinetics, and Grazing Behavior. *J. Anim. Sci.* **76**(1):220–227.

Forero, L. C., and L. R. Rittenhouse. (1987). The Determination of a Yearling/Cow-Calf Substitution Ratio in Southeastern Colorado. *Soc. Range Mgt. Abst. Papers* **40**:217.

Forero, L., L. R. Rittenhouse, and J. E. Mitchell. (1989). A Cow-Calf vs. Yearling Substitution Ratio for Shortgrass Steppe. *J. Range Mgt.* **42**(4):343–345.

Forero, O., F. N. Owens, and K. S. Lusby. (1980). Evaluation of Slow-Release Urea for Winter Supplementation of Lactating Range Cows. *J. Anim. Sci.* **50**(3):532–538.

Forwood, J. R., A. G. Matches, and C. J. Nelson. (1988). Forage Yield, Nonstructural Carbohydrate Levels, and Quality Trends of Caucasian Bluestem. *Agron. J.* **80**(1):135–139.

Foster, L. (1982). Half a Century of Change. *Rangelands* **4**(2):70–71.

Foster, L., H. A. Turner, and R. J. Raleigh. (1971). Daily Versus Alternative Feeding of Range Supplements. *Ore. Agric. Expt. Sta. Spec. Rep.* **322**, pp. 1–4.

Fourie, J. H., and D. I. Bransby. (1988). Animal and Plant Responses to Continuous and Rotational Grazing at Four Stocking Rates on Semiarid Rangeland in South Africa. *Soc. Range Mgt. Abst. Papers* **41**:400.

Fox, D. G., C. J. Sniffen, and J. D. O'Connor. (1988). Adjusting Nutrient Requirements of Beef Cattle for Animal and Environmental Variations. *J. Anim. Sci.* **66**(6):1475–1495.

Frasier, G. W., R. H. Hart, and G. E. Schuman. (1996). Impact of Grazing Intensity on Infiltration/Runoff Characteristics of a Shortgrass Prairie. *Proc. Internat. Rangeland Cong.* **5**(1):159–160.

Frederickson, E., K. M. Havstad, and D. Doornbos. (1987). Vital Organ Mass and Reticulo-Rumen Volume of Different Types of Lactating Range Cows. *Soc. Range Mgt. Abst. Papers* **40**:228.

Freeman, P. K., and R. H. Hart. (1989). Satiety and Feeding Station Behavior of Grazing Steers. *Soc. Range Mgt. Abst. Papers* **42**:160.

Freer, M. (1981). The Control of Food Intake by Grazing Animals. *In* F. H. W. Morley (Ed.), "Grazing Animals." Elsevier Sci. Pub. Co., Amsterdam, Neth., pp. 105–124.

French, M. H. (1970). "Observations on the Goat." Food Agric. Organ., Rome, Italy. 204 pp.

Fribourg, H. A., and K. W. Bell. (1984). Yield and Composition of Tall Fescue Stockpiled for Different Periods. *Agron. J.* **76**(6):929–934.

Friedel, B. A., and R. J. Hudson. (1994). Productivity of Farmed Wapiti in Alberta. *Can. J. Anim. Sci.* **74**(2):297–303.

Frischknecht, N. C., and L. E. Harris. (1968). Grazing Intensities and Systems on Crested Wheatgrass in Central Utah: Response of Vegetation and Cattle. *USDA Tech. Bul.* **1388**. 47 pp.

Frischknecht, N. C., and L. E. Harris. (1973). Sheep Can Control Sagebrush on Seeded Range. *Utah Sci.* **34**(1):27–30.

Fristina, M. R. (1992). Elk Habitat Use within a Rest-Rotation Grazing System. *Rangelands* **14**(2): 93–96.

Fritz, J. O., K. J. Moore, and C. A. Roberts. (1987). Chemical Regulation of Quality and Botanical Composition of Alfalfa-Grass Mixtures. *Forage & Grassland Conf.* **1987**:166–170.

Frost, W. E., E. L. Smith, and P. R. Ogden. (1994). Utilization Guidelines. *Rangelands* **16**(6):256–259.

Fuls, E. R. (1992a). Semi-Arid and Arid Rangelands: A Resource Under Siege Due to Patch-Selective Grazing. *J. Arid Environ.* **22**(2):191–193.

Fuls, E. R. (1992b). Ecosystem Modification Created by Patch-Overgrazing in Semi-Arid Grassland. *J. Arid Environ.* **23**(1):59–69.

Funston, R. N., D. D. Kress, K. M. Havstad, and D. E. Doornbos. (1987). Grazing Behavior of Rangeland Beef Cattle Differing in Biological Type. *Soc. Range Mgt. Abst. Papers* **40**:229.

Gade, A. E. (Gene), and K. L. Johnson. (1986). Beefing Up the Winter Range; The J. C. Smith Ranch: A Demonstration of a Year-Round Grazing System. *In* K. L. Johnson (Ed.). "Crested Wheatgrass: Its Values, Problems, and Myths, Symposium Proceedings, Logan, Utah, October 3–7, 1983." Utah State Univ., Logan, Utah, pp. 339–342.

Gade, A. E., and F. D. Provenza. (1986). Nutrition of Sheep Grazing Crested Wheatgrass Versus Crested Wheatgrass-Shrub Pastures During Winter. *J. Range Mgt.* **39**(6):527–530. 1986.

Galt, H. D., B. Theurer, and S. C. Martin. (1982). Botanical Composition of Steer Diets on Mesquite and Mesquite-Free Desert Grassland. *J. Range Mgt.* **35**(3):320–325.

Galyean, M. L. (1987). Factors Influencing Digesta Flow in Grazing Ruminants. *In* "Proceedings, Grazing Livestock Nutrition Conference, July 23–24, 1987." Univ. Wyo., Laramie, Wyo., pp. 77–89.

Gammon, D. M., and B. R. Roberts. (1978). Patterns of Defoliation During Continuous and Rotational Grazing of the Matopos Sandveld of Rhodesia. 1. Selectivity of Grazing; 2. Severity of Defoliation; 3. Frequency of Defoliation. *Rhod. J. Agric. Res.* **16**(2):117–164.

Gammon, D. M., and B. R. Roberts. (1980a). Aspects of Defoliation During Short-Duration Grazing of the Matopos Sandveld of Zimbabwe. *J. Agric. Res.* **18**(1):29–38.

Gammon, D. M., and B. R. Roberts. (1980b). Grazing Behavior of Cattle During Continuous and Rotation Grazing of the Matopos Sandveld of Zimbabwe. *Zimbabwe J. Agric. Res.* **18**(1):13–27.

Gamougoun, N. D., H. Nascimento, R. Rice, R. P. Ogden, G. B. Ruyle, and E. L. Smith. (1987). Cattle Grazing Time as Affected by Topographic Differences, Stocking Rate, and Season in Southern Arizona. *Soc. Range Mgt. Abst. Papers* **40**:167.

Ganskopp, D., and R. Cruz. (1999). Efficiency of Beef Cattle in Rugged Terrain: GIS Analyses of Livestock Trails. *Soc. Range Mgt. Abst. Papers* **52**:22.

Ganskopp, D., and M. Vavra. (1986). Habitat Use by Feral Horses in the Northern Sagebrush Steppe. *J. Range Mgt.* **39**(3):207–212.

Ganskopp, D., and M. Vavra. (1987). Slope Use by Cattle, Feral Horses, Deer, and Bighorn Sheep. *Northwest Sci.* **61**(2):74–81.

Ganskopp, D., R. Angell, and J. Rose. (1992). Response of Cattle to Cured Reproductive Stems in a Caespitose Grass. *J. Range Mgt.* **45**(4):401–404.

Ganskopp, D., R. Angell, and J. Rose. (1993). Effect of Low Densities of Senescent Stems in Crested Wheatgrass on Plant Selection and Utilization by Beef Cattle. *Appl. Anim. Beh. Sci.* **38**(3–4):227–233.

Ganskopp, D., T. Svejcar, F. Taylor, J. Farstvedt, and K. Painter. (1999). Seasonal Cattle Management in 3- to 5-Year-Old Bitterbrush Stands. *J. Range Mgt.* **52**(2):166–173.

Garrison, G. A. (1953). Effects of Clipping on Some Range Shrubs. *J. Range Mgt.* **6**(5):309–317.

Garrison, G. A. (1972). Physiology of Shrubs: Carbohydrate Reserves and Response to Use. *USDA, For. Serv. Gen. Tech. Rep.* **INT-1,** pp. 271–278.

Garrott, R. A., G. C. White, R. M. Bartmann, L. H. Carpenter, and A. W. Alldredge. (1987). Movements of Female Mule Deer in Northwest Colorado. *J. Wildl. Mgt.* **51**(3):634–643.

Gatewood, R. G., L. R. Roath, and K. Oddie. (1987). The Influence of Management on Cow Condition and Its Interrelationship With Productivity. *Soc. Range Mgt. Abst. Papers* **40**:194.

Gdara, A. O., R. H. Hart, and J. G. Dean. (1991). Response of Tap- and Creeping-Rooted Alfalfas to Defoliation Patterns. *J. Range Mgt.* **44**(1):22–26.

Gee, C. K., L. A. Joyce, and A. G. Madsen. (1992). Factors Affecting the Demand for Grazed Forage in the United States. *USDA, For. Serv. Gen. Tech. Rep.* **RM-210.** 20 pp.

Gengelbach, G. P., D. E. Trokey, K. N. Grigsby, C. W. Peters *et al.* (1990). Comparison of Esophageal vs. Clipped Samples for Determination of Mineral Content of Three Cool-Season Grasses. *Amer. Soc. Anim. Sci., West. Sect. Proc.* **41**:241–244.

Genin, D., and A. Badan-Dangon. (1991). Goat Herbivory and Plant Phenology in a Mediterranean Shrubland of Northern Baja California. *J. Arid Environ.* **21**(1):113–121.

George, J. F., and J. Powell. (1979). Cattle Grazing Impacts on Small Cleared Areas in Dense American Elm Woodlands. *J. Range Mgt.* **32**(1):78–79.

George, J. R., and D. Obermann. (1989). Spring Defoliation To Improve Summer Supply and Quality of Switchgrass. *Agron J.* **81**(1):47–52.

Gerhart, W. A., and H. G. Fisser. (1977). An Evaluation of Three Grazing Management Systems on Northern Desert Shrub Vegetation of Southern Wyoming. *Soc. Range Mgt. Abst. Papers* **30**:52.

Gerrish, J. R., C. J. Nelson, R. E. Morrow, J. R. Forwood, W. C. Hires, V. E. Jacobs, and G. B. Garner. (1986). Calf Response to Creep Grazing. *Forage and Grassland Conf.* **1986**:197–200.

Gerrish, J. R., P. R. Peterson, and R. E. Morrow. (1995a). Distance Cattle Travel to Water Affects Pasture Utilization Rate. *Amer. Forage & Grassland Council Proc.* **1995**:61–65.

Gerrish, J. R., P. R. Peterson, and J. R. Brown. (1995b). Grazing Management Affects Soil Phosphorus and Potassium Levels. *Amer. Forage & Grassland Council Proc.* **1995**:175–179.

Gesshe, R. H., and P. D. Walton. (1981). Grazing Animal Preferences for Cultivated Forages in Canada. *J. Range Mgt.* **34**(1):42–45.

Gibb, M. J., R. D. Baker, and A. M. E. Sayer. (1989). The Impact of Grazing Severity on Perennial Ryegrass/White Clover Swards Stocked Continuously with Beef Cattle. *Grass & Forage Sci.* **44**(3):315–328.

Gibbens, R. P., and H. G. Fisser. (1975). Influence of Grazing Management Systems on Vegetation in the Red Desert Region of Wyoming. *Wyo. Agric. Expt. Sta. Sci. Monog.* **29.** 23 pp.

Gilbert, P. F., O. C. Wallmo, and R. B. Gill. (1970). Effect of Snow Depth on Mule Deer in Middle Park, Colorado. *J. Wildl. Mgt.* **34**(1):15–23.

Gillen, R. L. (1999). Grazing Management for Eastern Gamagrass. *Soc. Range Mgt. Abst. Papers* **1999**:23.

Gillen, R. L., W. C. Krueger, and R. F. Miller. (1984). Cattle Distribution on Mountain Rangeland in Northeastern Oregon. *J. Range Mgt.* **37**(6):549–553.

Gillen, R. L., F. T. McCollum, and J. E. Brummer. (1990). Tiller Defoliation Patterns under Short-Duration Grazing in Tallgrass Prairie. *J. Range Mgt.* **43**(2):95–99.

Glimp, H. A. (1988). Multi-Species Grazing and Marketing. *Rangelands* **10**(6):276–278.

Glimp, H. A., and S. R. Swanson. (1994). Sheep Grazing and Riparian and Watershed Management. *Sheep Res. J.* **10**(Spec. Issue):65–71.

Gluesing, E. A., and D. F. Balph. (1980). An Aspect of Feeding Behavior and Its Importance to Grazing Systems. *J. Range Mgt.* **33**(6):426–427.

Gold, W. G., and M. M. Caldwell. (1989). The Effects of the Spatial Pattern of Defoliation on Regrowth of a Tussock Grass. I. Growth Responses. *Oecologia* **80**(3):289–296.

Gold, W. G., and M. M. Caldwell. (1990). The Effects of the Spatial Pattern of Defoliation on Regrowth of a Tussock Grass. III. Photosynthesis, Canopy Structure, and Light Interception. *Oecologia* **82**(1):12–17.

Gomm, F. B. (1969). The Palatability of Range and Pasture Grasses in Montana. *Mon. Agric. Expt. Sta. Bul.* **619**. 29 pp.

Gomm, F. B. 1979. Grazing Management on Mountain Meadows. *In* Symposium Proceedings: Management of Intermountain Meadows. *Wyo. Agric. Expt. Sta. Res. J.* **141**, pp. 141–157.

Gomm, F. B., and H. A. Turner. (1976). Flexibility with Irrigated Pasture. *Rangeman's J.* **3**(2):37–38.

Goodson, N. J., D. R. Stevens, and J. A. Bailey. (1991). Effects of Snow on Foraging Ecology and Nutrition of Bighorn Sheep. *J. Wildl. Mgt.* **55**(2):214–222.

Goodwin, G. A. (1975). Seasonal Food Habits of Mule Deer in Southwestern Wyoming. *USDA, For. Serv. Res. Note* **RM-27**. 4 pp.

Gordon, I. J., and A. W. Illius. (1994). The Functional Significance of the Browser-Grazer Dichotomy in African Ruminants. *Oecologia* **98**:167–175.

Graham, K. T., L. Allen Torrell, and C. D. Allison. (1992). Costs and Benefits of Implementing Holistic Resource Management on New Mexico Ranches. *N. Mex. Agric. Expt. Sta. Bull.* **762**. 28 pp.

Gray, J. R., C. Steiger, Jr., and J. Fowler. (1982). Characteristics of Grazing Systems. *N. Mex. Agric. Expt. Sta. Res. Rep.* **467**. 16 pp.

Green, L. R., and L. A. Newell. (1982). Using Goats To Control Brush Regrowth on Fuelbreaks. *USDA, For. Serv. Gen. Tech. Rep.* **PSW-59**. 13 pp.

Green, L. R., C. L. Hughes, and W. L. Graves. (1978). Goat Control of Brush Regrowth on Southern California Fuelbreaks. *Proc. Intern. Rangeland Cong.* **1**:451–455.

Green, L. R., C. L. Hughes, and W. L. Graves. (1979). Goat Control of Brush Regrowth on Southern California Fuelbreaks. *Rangelands* **1**(3):117–119.

Greene, L. W. (1986). Grass Tetany in Beef Cattle: A Review. *Texas Agric. Expt. Sta. Prog. Rep.* **4478**. 3 pp.

Greene, L. W., R. K. Heitschmidt, and B. Pinchak. (1986). Mineral Composition of Forages in a Short-Duration Grazing System. *Texas Agric. Expt. Sta. Cons. Prog. Rep.* **4347**, pp 64–66.

Greenwood, G. B., and M. W. Demment. (1988). The Effect of Fasting on the Short-Term Cattle Grazing Behavior. *Grass and Forage Sci.* **43**(4):377–386.

Greenwood, P. T., and L. R. Rittenhouse. (1997). Feeding Area Selection: The Leader-Follower Phenomenon. *Amer. Soc. Anim. Sci., West. Sec. Proc.* **48**:267–269.

Grelen, H. E., H. A. Pearson, and R. E. Thill. (1985). Response of Slash Pines to Grazing from Regeneration to the First Pulpwood Thinning. *USDA, For. Serv. Gen. Tech. Rep.* **SO-54**, pp. 523–527.

Grigsby, K. N., F. M. Rouquette, Jr., M. J. Florence, R. P. Gillespie, W. C. Ellis, and D. P. Hutcheson. (1988). *Texas Agric. Expt. Sta. Cons. Prog. Rep.* **4593**, p. 16–22.

Grings, E. E., D. C. Adams, and R. E. Short. (1995). Diet Quality of Suckling Calves and Mature Steers on Northern Great Plains Rangelands. *J. Range Mgt.* **48**(5):438–441.

Grovum, W. L. (1987). A New Look at What Is Controlling Food Intake. *Okla. Agric. Expt. Sta. Misc. Pub.* **121**, pp. 1–40.

Grunes, D. L., and H. F. Mayland. (1984 Rev.). Controlling Grass Tetany. *USDA Leaflet* **561**. 8 pp.

Guerrero, J. N., and V. L. Marble. (1997). Lamb Grazing Strategies for the Irrigated Sonoran Desert. *Amer. Soc. Anim. Sci., West. Sec. Proc.* **48**:243–246.

Guerrero, J. N., M. I. Lopez, C. E. Bell, and B. Boutwell. (1999). Sheep Thrive on Weedy Alfalfa. *Calif. Agric.* **53**(5):29–32.

Guthery, F. S., C. A. DeYoung, F. C. Bryant, and D. L. Drawe. (1990). Using Short-Duration Grazing to Accomplish Wildlife Habitat Objectives. *In* Can Livestock Be Used to Enhance Wildlife Habitat? *USDA, For. Serv. Gen. Tech. Rep.* **RM-194**, pp. 41–55.

Hacker, R. B., B. E. Norton, M. K. Owens, and D. O. Frye. (1988). Grazing of Crested Wheatgrass, with Particular Reference to Effects of Pasture Size. *J. Range Mgt.* **41**(1):73–78.

Haferkamp, M. R., R. F. Miller, and F. A. Sneva. (1987). Mefluidide Effects on Forage Quality of Crested Wheatgrass. *Agron. J.* **79**(4):637–641.

Hakkila, M. D., J. L. Holechek, J. D. Wallace, D. M. Anderson, and M. Cardenas. (1987). Diet and Forage Intake of Cattle on Desert Grassland Range. *J. Range Mgt.* **40**(4):339–342.

Hall, F. C., and L. Bryant. (1995). Herbaceous Stubble Height as a Warning of Impending Cattle Grazing Damage to Riparian Areas. *USDA, For. Serv. Gen. Tech. Rep.* **PNW-GTR-362**. 9 pp.

Halloran, A. F., and O. V. Deming. (1956). Water Development for Desert Bighorn Sheep. *USDI, Fish & Wildl. Serv. Wildl. Mgt. Ser. Leaflet* **14**. 12 pp.

Halls, L. K. (Ed.). (1984). "White-Tailed Deer: Ecology and Management." Stackpole Books, Harrisburg, Pa. 870 pp.

Hamilton, R. G. (1999). Comparative Biodiversity Management Using Bison and Cattle. *Soc. Range Mgt. Abst. Papers* **52**:26.

Hanley, T. A. (1982a). Cervid Activity Patterns in Relation to Foraging Constraints: Western Washington. *Northwest Sci.* **56**(3):208–217.

Hanley, T. A. (1982b). The Nutritional Basis for Food Selection by Ungulates. *J. Range Mgt.* **35**(2):146–151.

Hanley, T. A., and K. A. Hanley. (1982). Food Resource Partitioning by Sympatric Ungulates on Great Basin Rangeland. *J. Range Mgt.* **35**(2):152–158.

Hanley, T. A., and J. J. Rogers. (1989). Estimating Carrying Capacity with Simultaneous Nutritional Constraints. *USDA, For. Serv. Res. Note* **PNW-RN-485**. 29 pp.

Hanley, T. A., and R. D. Taber. (1980). Selective Plant Species Inhibition by Elk and Deer in Three Conifer Communities in Western Washington. *For. Sci.* **26**(1):97–107.

Hanley, T. A., R. G. Cates, B. Van Horne, and J. D. McKendrick. (1987). Forest Stand-Age-Related Differences in Apparent Nutritional Quality of Forage for Deer in Southeastern Alaska. *USDA, For. Serv. Gen. Tech. Rep.* **INT-222**, pp. 9–17.

Hanselka, C. W., A. McGinty, B. S. Rector, R. C. Rowan, and L. D. White (1990). "Grazing and Brush Management on Texas Rangelands: An Analysis of Management Decisions." Texas Agric. Ext. Serv., College Station, TX. 22 pp.

Hansen, R. M. (1976). Foods of Free-Roaming Horses in Southern New Mexico. *J. Range Mgt.* **29**(4):347.

Hansen, R. M., and P. S. Martin. (1973). Ungulate Diets in the Lower Grand Canyon. *J. Range Mgt.* **26**(5):380–381.

Hansen, R. M., and L. D. Reid. (1975). Diet Overlap of Deer, Elk, and Cattle in Southern Colorado. *J. Range Mgt.* **28**(1):43–47.

Hansen, R. M., R. C. Clark, and W. Lawhorn. (1977). Foods of Wild Horses, Deer, and Cattle in the Douglas Mountain Area, Colorado. *J. Range Mgt.* **30**(2):116–118.

Hanson, C. L., A. R. Kuhlman, and J. K. Lewis. (1978). Effect of Grazing Intensity and Range Condition on Hydrology of Western South Dakota Ranges. *S. Dak. Agric. Expt. Sta. Bul.* **647**. 54 pp.

Hanson, C. L., R. P. Morris, and J. R. Wight. (1983). Using Precipitation to Predict Range Herbage Production in Southwestern Idaho. *J. Range Mgt.* **36**(6):766–770.

Harlan, J. R. (1958). Generalized Curves for Gain Per Head and Gain Per Acre in Rates of Grazing Studies. *J. Range Mgt.* **11**(3):140–147.

Harlow, R. F., and R. L. Downing. (1970). Deer Browsing and Hardwood Regeneration in the Southern Appalachians. *J. For.* **68**(5):298–300.

Harper, J. A., J. H. Harn, W. W. Bentley, and C. F. Yocom. (1967). The Status and Ecology of the Roosevelt Elk in California. *Wildl. Monogr.* **16**. 49 pp.

Harris, L. E. (1962). Measurement of the Energy Value of Pasture and Range Forage. Chap. 17 *in* "Pasture and Range Research Techniques." Comstock Pub. Co., Ithaca, NY.

Harris, L. E., M. L. Dew, and G. Q. Bateman. (1958). Irrigated Pastures—A Way to Maintain Beef Production. *Utah Farm and House Science* **19**(3):76–77, 80.

Harris, L., E., N. C. Frischknecht, and E. M. Sudweeks. (1968). Seasonal Grazing of Crested Wheatgrass by Cattle. *J. Range Mgt.* **21**(4):221–225.

Hart, R. H. (1978). Stocking Rate Theory and Its Application to Grazing on Rangelands. *Proc. Intern. Rangeland Cong.* **1**:547–550.

Hart, R. H. (1980). Determining a Proper Stocking Rate for a Grazing System. *In* Kirk C. McDaniel and Chris Allison (Eds.). "Grazing Management Systems for Southwest Rangelands: A Symposium." Range Impr. Task Force, N. Mex. State Univ., Las Cruces, N. Mex., pp. 49–64.

Hart, R. H. (1986). How Important Are Stocking Rates in Grazing Management? *In* P. E. Reece and J. T. Nichols (Eds.), "Proceedings, The Ranch Management Symposium, November 5–7, 1986, North Platte, Nebraska." Univ. Neb., Agric. Ext. Serv., Lincoln, Neb., pp. 77–87.

Hart, R. H., and M. M. Ashby. (1998). Grazing Intensities, Vegetation, and Heifer Gains: 55 Years on Shortgrass. *J. Range Mgt.* **51**(4):392–398.

Hart, R. H., W. H. Marchant, J. L. Butler, R. E. Hellwig, W. C. McCormick, B. L. Southwell, and G. W. Burton. (1976). Steer Gains Under Six Systems of Coastal Bermudagrass Utilization. *J. Range Mgt.* **29**(5):372–375.

Hart, R. H., M. J. Samuel, P. S. Test, and M. A. Smith. (1988a). Cattle, Vegetation, and Economic Responses to Grazing Systems and Grazing Pressure. *J. Range Mgt.* **41**(4):282–288.

Hart, R. H., J. W. Waggoner, Jr., T. G. Dunn, C. C. Kaltenbach, and L. D. Adams. (1988b). Optimal Stocking Rate for Cow-Calf Enterprises on Native Range and Complementary Improved Pastures. *J. Range Mgt.* **41**(5):435–441.

Hart, R. H., M. J. Samuel, and J. Bissio. (1988c). Role of Rotation and Distribution in Short-Duration Rotation Grazing on Range. *Soc. for Range Mgt. Abst. Papers* **41**:290.

Hart, R. H., M. J. Samuel, J. W. Waggoner, Jr., and M. A. Smith. (1989). Comparisons of Grazing Systems in Wyoming. *J. Soil & Water Cons.* **44**(4):344–347.

Hart, R. H., K. W. Hepworth, M. A. Smith, and J. W. Waggoner, Jr. (1991). Cattle Grazing Behavior on a Foothill Elk Winter Range in Southwestern Wyoming. *J. Range Mgt.* **44**(3):262–266.

Hart, R. H., J. Bissio, M. J. Samuel, and J. W. Waggoner, Jr. (1993). Grazing System, Pasture Size, and Cattle Grazing Behavior, Distribution, and Gains. *J. Range Mgt.* **46**(1):81–87.

Hart, S., and T. Sahlu. (1995). Limit-Grazing of Cool-Season Pastures for Wintering Angora Does. *Sheep & Goat Res. J.* **11**(1):1–3.

Harvey, R. W., and J. C. Burns. (1988). Creep Grazing and Early Weaning Effects on Cow and Calf Productivity. *J. Anim. Sci.* **66**(5):1109–1114.

Harwell, R. L., P. L. Strickland, and R. Jobes. (1976). Utilization of Winter Wheat Pasture in Oklahoma. *Okla. Agric. Expt. Sta. Res. Rep.* **743**. 22 pp.

Hatfield, P., D. Clanton, D. Sanson, and M. Nielsen. (1988). Forage Intake of Cows Differing in Milk Production. *Neb. Agric. Res. Div. Misc. Pub.* **54**, p. 7–9.

Hatfield, P. G., G. B. Donart, T. T. Ross, and M. L. Galyean. (1990). Sheep Grazing Behavior as Affected by Supplementation. *J. Range Mgt.* **43**(5):387–389.

Hatfield, R. D. (1989). Structural Polysaccharides in Forages and Their Digestibility. *Agron. J.* **81**(1):39–46.

Havstad, K. M., and D. E. Doornbos. (1987). Effect of Biological Type on Grazing Behavior and Energy Intake. *In* "Grazing Livestock Nutrition Conference." Univ. of Wyo., Laramie, pp. 9–15.

Havstad, K. M., and J. C. Malechek. (1982). Energy Expenditure by Heifers Grazing Crested Wheatgrass of Diminishing Availability. *J. Range Mgt.* **35**(4):447–450.

Havstad, K. M., A. S. Nastis, and J. C. Malechek. (1983). The Voluntary Forage Intake of Heifers Grazing a Diminishing Supply of Crested Wheatgrass. *J. Anim. Sci.* **56**(2):259–263.

Havstad, K. M., M. W. Wagner, S. L. Kronberg, D. E. Doornbos, and E. L. Ayers. (1986a). Forage

Intake of Different Beef Cattle Biological Types Under Range Conditions. *Mon. AgRes.* **3**(2): 17–19.

Havstad, K. M., W. J. Lathrop, E. L. Ayers, D. E. Doornbos, and D. D. Kres. (1986b). Grazing Behavior of Beef Cows Under Range Conditions. *Mon. AgRes.* **3**(2):20–21.

Haws, B. A., D. D. Dwyer, and M. G. Anderson. (1973). Problems with Range Grasses? Look for Black Grass Bugs. *Utah Sci.* **34**(1):3–9.

Haywood, J. D. (1995). Controlling Herbaceous Competition in Pasture Planted with Loblolly Pine Seedlings. *USDA, For. Serv. Res. Note* **SO-381**. 4 pp.

Hazam, J. E., and P. R. Krausman. (1988). Measuring Water Consumption of Desert Mule Deer. *J. Wildl. Mgt.* **52**(3):528–534.

Heady, H. F. (1961). Continuous Vs. Specialized Grazing Systems: A Review and Application to the California Annual Type. *J. Range Mgt.* **14**(4):182–193.

Heady, H. F. (1964). Palatability of Herbage and Animal Preference. *J. Range Mgt.* **17**(2):76–82.

Heady, H. F. (1970). Grazing Systems: Terms and Definitions. *J. Range Mgt.* **23**(1):59–61.

Heady, H. F. (1974). Theory of Seasonal Grazing. *Rangeman's J.* **1**(2):37–38.

Heady, H. F. (1975). "Rangeland Management." McGraw-Hill Book Co., New York. 460 pp.

Heady, H. F. (1984). Concepts and Principles Underlying Grazing Systems. *In* Natl. Res. Council/Natl. Acad. Sci. "Developing Strategies for Rangeland Management." Westview Press, Boulder, CO. pp. 885–902.

Heady, H. F. (1994). Summary: Ecological Implications of Livestock Herbivory in the West. *In* M. Vavra, W. Laycock, and R. D. Pieper (Eds.), "Ecological Implications of Livestock Herbivory in the West." Soc. Range Mgt., Denver, CO. pp. 289–297.

Heady, H. F. (1999). Perspectives on Rangeland Ecology and Management. *Rangelands* **21**(5):23–33.

Heady, H. F., and R. D. Child. (1994). "Rangeland Ecology and Management." Westview Press, Boulder, CO. 521 pp.

Heady, H. F., and M. D. Pitt. (1979). Seasonal Versus Continuous Grazing on Annual Vegetation of Northern California. *Rangelands* **1**(6):231–232.

Hedrick, D. W. (1967). Managing Crested Wheatgrass for Early Spring Use. *J. Range Mgt.* **20**(1): 53–54.

Hedrick, D. W., and R. F. Keniston. (1966). Grazing and Douglas-Fir Growth in the Oregon White-Oak Type. *J. For.* **64**(11):735–738.

Hedrick, D. W., W. M. Moser, A. L. Steninger, and R. A. Long. (1969). Animal Performance on Crested Wheatgrass Pastures During May and June, Fort Rock, Oregon. *J. Range Mgt.* **22**(4):277–280.

Heemstra, J. M., S. L. Kronberg, R. D. Neiger, and R. J. Pruitt. (1999). Behavioral, Nutritional, and Toxicological Responses of Cattle to Ensiled Leafy Spurge. *J. Anim. Sci.* **77**(3):600–610.

Hein, D. G., and S. D. Miller. (1992). Influence of Leafy Spurge on Forage Utilization by Cattle. *J. Range Mgt.* **45**(4):405–407.

Heinemann, W. W. (1969). Productivity of Irrigated Pastures Under Combination and Single Species Grazing. *Wash. Agric. Expt. Sta. Bul.* **717**. 6 pp.

Heinemann, W. W. (1970). Continuous and Rotation Grazing by Steers on Irrigated Pastures. *Wash. Agric. Expt. Sta. Bul.* **724**. 4 pp.

Heinemann, W. W., and T. S. Russell. (1969). Evaluation of Rotation Grazed Pastures from Esophageal and Hand Gathered Forage Samples. *Agron. J.* **61**(4):547–550.

Heitschmidt, R. K. (1984a). The Ecological Efficiency of Grazing Management. *Forage and Grassland Conf.* **1984**:378–385.

Heitschmidt, R. K. (1984b). Intensive Rangeland Grazing Systems: The Pros and Cons. Chapter 4. *In* "Symposium—Agriculture: The New Frontier, Brigham Young University, Provo, Utah, Mar. 27–30, 1984." Provo, UT.

Heitschmidt, R. (1986). Texas Perspective on Intensive Grazing. *In* Patrick E. Reece and James T. Nichols (Eds.). "Proceedings, The Ranch Management Symposium, November 5–7, 1986, North Platte, Nebraska." Univ. Neb., Agric. Ext. Serv., Lincoln, Neb., pp. 67–76.

Heitschmidt, R. (1988). Grazing Systems and Livestock Management. *In* Richard S. White and Robert E. Short (Eds.). "Achieving Efficient Use of Rangeland Resources, Fort Keogh Research

Symposium, September 1987, Miles City, Mon." Mon. Agric. Expt. Sta., Bozeman, Mon., pp. 101–106.

Heitschmidt, R. K., and J. W. Stuth (Eds.). (1991). "Grazing Management: An Ecological Perspective." Timber Press, Portland, OR. 259 pp.

Heitschmidt, R., and J. Walker. (1983). Short-Duration Grazing and the S ʼory Grazing Method in Perspective. *Rangelands* **5**(4):147–150.

Heitschmidt, R. K., and J. W. Walker. (1996). Grazing Management: Technology for Sustaining Rangeland Ecosystems. *Rangeland J.* **18**(2):194–215.

Heitschmidt, R. K., M. M. Kothmann, and W. J. Rawlins. (1982a). Cow-Calf Response to Stocking Rates, Grazing Systems, and Winter Supplementation at the Texas Experimental Ranch. *J. Range Mgt.* **35**(2):204–210.

Heitschmidt, R. K., D. L. Price, R. A. Gordon, and J. R. Frasure. (1982b). Short-Duration Grazing at the Texas Experimental Ranch: Effects on Aboveground Net Primary Production and Seasonal Growth Dynamics. *J. Range Mgt.* **35**(3):367–372.

Heitschmidt, R. K., R. A. Gordon, and J. S. Bluntzer. (1982c). Short-Duration Grazing at the Texas Experimental Ranch: Effects on Forage Quality. *J. Range Mgt.* **35**(3):372–374.

Heitschmidt, R. K., J. R. Frasure, D. L. Price, and L. R. Rittenhouse. (1982d). Short-Duration Grazing at the Texas Experimental Ranch: Weight Gains of Growing Heifers. *J. Range Mgt.* **35**(3):375–379.

Heitschmidt, R. K., S. L. Dowhower, and J. W. Walker. (1986a). Effect of Livestock Density (Numbers of Paddocks) in a Rotational Grazing System on Forage Production and Herbage Standing Crop. *Texas Agric. Expt. Sta. Prog. Rep.* **4421.** 2 pp.

Heitschmidt, R. K., S. L. Dowhower, and J. W. Walker. (1986b). Effect of Rotational Grazing on Forage and Litter Standing Crop. *Texas Agric. Expt. Sta. Prog. Rep.* **4422.** 2 pp.

Heitschmidt, R. K., S. L. Dowhower, and J. W. Walker. (1986c). Effect of Rotational Grazing on Protein Content of Available Forage. *Texas Agric. Expt. Sta. Prog. Rep.* **4428.** 1 p.

Heitschmidt, R. K., S. L. Dowhower, and J. W. Walker. (1987a). 14- Vs. 42-Paddock Rotational Grazing: Aboveground Biomass Dynamics, Forage Production, and Harvest Efficiency. *J. Range Mgt.* **40**(3):216–223.

Heitschmidt, R. K., S. L. Dowhower, and J. W. Walker. (1987b). 14-vs. 42-Paddock Rotational Grazing: Forage Quality. *J. Range Mgt.* **40**(4):315–317.

Heitschmidt, R. K., R. E. Short, and P. E. Grings. (1996). Ecosystems, Sustainability, and Animal Agriculture. *J. Anim. Sci.* **74**(6):1395–1405.

Hendrickson, J. R., L. E. Moser, K. J. Moore, and S. S. Waller. (1997). Leaf Nutritive Value Related to Tiller Development in Warm-Season Grasses. *J. Range Mgt.* **50**(2):116–122.

Henke, S. E., S. Demarais, and J. A. Pfister. (1988). Digestive Capacity and Diets of White-Tailed Deer and Exotic Ruminants. *J. Wildl. Mgt.* **52**(4):595–598.

Hepworth, K. W., P. S. Test, R. H. Hart, J. W. Waggoner, Jr., and M. A. Smith. (1991). Grazing Systems, Stocking Rates, and Cattle Behavior in Southeastern Wyoming. *J. Range Mgt.* **44**(3):259–262.

Herbel, C. H. (1974). A Review of Research Related to Development of Grazing Systems on Native Ranges of the Western United States. *USDA Misc. Pub.* **1271**, p. 138–149.

Herbel, C. H., and A. B. Nelson. (1966a). Activities of Hereford and Santa Gertrudis Cattle on a Southern New Mexico Range. *J. Range Mgt.* **19**(4):173–176.

Herbel, C. H., and A. B. Nelson. (1966b). Species Preference of Hereford and Santa Gertrudis Cattle on a Southern New Mexico Range. *J. Range Mgt.* **19**(4):177–181.

Herbel, C. H., and R. D. Pieper. (1991). Grazing Management. *In* J. Skujins (Ed.), "Semiarid Lands and Deserts: Soil Resource and Reclamation." Marcel Dekker, New York, pp. 361–385.

Herbel, C. H., F. N. Ares, and A. B. Nelson. (1967). Grazing Distribution Patterns of Hereford and Santa Gertrudis Cattle on a Southern New Mexico Range. *J. Range Mgt.* **20**(5):296–298.

Herbel, C. H., J. D. Wallace, M. D. Finkner, and C. C. Yarbrough. (1984). Early Weaning and Part-Year Confinement of Cattle on Arid Rangelands of the Southwest. *J. Range Mgt.* **37**(2):127–130.

Herd, D. B., and L. R. Sprott. (1986). Body Condition, Nutrition, and Reproduction of Beef Cows. *Texas Agric. Ext. Bul.* **1526.** 11 pp.

Hervert, J. J., and P. R. Krausman. (1986). Desert Mule Deer Use of Water Developments in Arizona. *J. Wildl. Mgt.* **50**(4):670–676.

Hess, B. W., K. K. Park, L. J. Krysl, M. B. Judkins *et al.* (1994). Supplemental Protein for Beef Cattle Grazing Dormant Intermediate Wheatgrass Pasture: Effects on Nutrient Quality, Forage Intake, Digesta Kinetics, Grazing Behavior, Ruminal Fermentation, and Digestion. *J. Anim. Sci.* **72**(8):2113–3123.

Hickey, W. C., Jr. (1961). Growth Form of Crested Wheatgrass as Affected by Site and Grazing. *Ecol.* **42**(1):173–176.

Hickey, W. C., Jr., and G. Garcia. (1964). Range Utilization Patterns as Affected by Fencing and Class of Livestock. *USDA, For. Serv. Res. Note* **RM-21.** 7 pp.

Hilton, J. E., and A. W. Bailey. (1972). Cattle Use of a Sprayed Aspen Parkland Range. *J. Range Mgt.* **25**(4):257–260.

Hines, W. W. (1973). Black-Tailed Deer Populations and Douglas-Fir Reforestation in the Tillamook Burn, Oregon. *Ore. State Game Comm. Game Res. Rep.* **3.** 59 pp.

Hinnant, R. T., and M. M. Kothmann. (1986). Frequency and Intensity of Defoliation Under Rotational Grazing. *Texas Agric. Expt. Sta. Prog. Rep.* **4424.** 1 p.

Hirschfeld, D. J., D. R. Kirby, J. S. Caton, S. S. Silcox, and K. C. Olson. (1996). Influence of Grazing Management on Intake and Composition of Cattle Diets. *J. Range Mgt.* **49**(3):257–263.

Hitz, A. C., and J. R. Russell. (1998). Potential of Stockpiled Perennial Forages in Winter Grazing Systems for Pregnant Beef Cows. *J. Anim. Sci.* **76**(2):404–415.

Hobbs, N. T. (1996). Modification of Ecosystems by Ungulates. *J. Wildl. Mgt.* **60**(4):695–713.

Hobbs, N. T., and L. H. Carpenter. (1986). Viewpoint: Animal-Unit Equivalents Should Be Weighted by Dietary Differences. *J. Range Mgt.* **39**(5):470.

Hobbs, N. T., and D. M. Swift. (1985). Estimates of Habitat Carrying Capacity Incorporating Explicit Nutritional Constraints. *J. Wildl. Mgt.* **49**(3):814–822.

Hobbs, N. T., D. L. Baker, J. E. Ellis, D. M. Swift, and R. A. Green. (1982). Energy- and Nitrogen-Based Estimates of Elk Winter-Range Carrying Capacity. *J. Wildl. Mgt.* **46**(1):12–21.

Hobbs, N. T., D. S. Schimel, C. E. Owensby, and D. S. Ojima. (1991). Fire and Grazing in the Tallgrass Prairie: Contingent Effects on Nitrogen Budgets. *Ecology* **72**(4):1374–1382.

Hodgkinson, K. C. (1980). Frequency and Extent of Defoliation of Herbaceous Plants by Sheep in a Foothill Range Community in Northern Utah. *J. Range Mgt.* **33**(3):164–168.

Hodgson, J. (1986). Grazing Behaviour and Herbage Intake. *In* J. Frame (Ed.), "Grazing." British Grassland Soc., Hurley, Berks, Eng., pp. 51–64.

Hodgson, J. (1990). "Grazing Management: Science into Practice." John Wiley & Sons, New York. 203 pp.

Hoehne, O. E., D. C. Clanton, and C. L. Streeter. (1968). Chemical Composition and In Vitro Digestibility of Forbs Consumed by Cattle Grazing Native Range. *J. Range Mgt.* **21**(1):5–7.

Hoffman, G. O., D. L. Huss, and B. J. Ragsdale. (1968). Grazing Systems for Profitable Ranching. *Tex. Agric. Ext. Misc. Pub.* **896.** 7 pp.

Hofmann, L., and R. E. Ries. (1988). Vegetation and Animal Production from Reclaimed Mined Land Pastures. *Agron. J.* **80**(1):40–44.

Hofmann, R. R. (1988). Anatomy of the Gastro-Intestinal Tract. *In* D. C. Church (Ed.). "The Ruminant Animal: Digestive Physiology and Nutrition." Prentice Hall, Englewood Cliffs, N.J. pp. 14–43.

Holechek, J. L. (1980). Concepts Concerning Forage Allocation to Livestock and Big Game. *Rangelands* **2**(4):158–159.

Holechek, J. L. (1981). Livestock Grazing Impacts on Public Lands: A Viewpoint. *J. Range Mgt.* **34**(3):251–254.

Holechek, J. L. (1982). Managing Rangelands for Mule Deer. *Rangelands* **4**(1):25–28.

Holechek, J. L. (1983). Considerations Concerning Grazing Systems. *Rangelands* **5**(5):208–211.

Holechek, J. L. (1984). Comparative Contribution of Grasses, Forbs, and Shrubs to the Nutrition of Range Ungulates. *Rangelands* **6**(6):261–263.

Holechek, J. L. (1988). An Approach for Setting the Stocking Rate. *Rangelands* **10**(1):10–14.

Holechek, J. L. (1991). Chihuahuan Desert Rangelands, Livestock Grazing, and Sustainability. *Rangelands* **13**(3):115–120.

Holechek, J. L., and C. H. Herbel. (1982). Seasonal Suitability Grazing in the Western United States. *Rangelands* **4**(6):252–255.

Holechek, J. L., and C. H. Herbel. (1986). Supplementing Range Livestock. *Rangelands* **8**(1):29–33.

Holechek, J. L., and R. D. Pieper. (1992). Estimation of Stocking Rate on New Mexico Rangelands. *J. Soil & Water Cons.* **47**(1):116–119.

Holechek, J. L., and T. Stephenson. (1983). Comparison of Big Sagebrush Vegetation in Northcentral New Mexico under Moderately Grazed and Grazing Excluded Conditions. *J. Range Mgt.* **35**(4): 455–456.

Holechek, J. L., and M. Vavra. (1983). Drought Effects on Diet and Weight Gains of Yearling Heifers in Northeastern Oregon. *J. Range Mgt.* **36**(2):227–231.

Holechek, J. L., M. Vavra, and J. Skovlin. (1982). Cattle Diet and Daily Gains on Mountain Riparian Meadow in Northeastern Oregon. *J. Range Mgt.* **35**(6):745–747.

Holechek, J. L., M. Vavra, and R. D. Pieper. (1982a). Botanical Composition Determination of Range Herbivore Diets: A Review. *J. Range Mgt.* **35**(3):309–315.

Holechek, J. L., M. Vavra, and R. D. Pieper. (1982b). Methods for Determining the Nutritive Quality of Range Ruminant Diets: A Review. *J. Anim. Sci.* **54**(2):363–376.

Holechek, J. L., M. Vavra, and R. D. Pieper. (1984). Methods for Determining the Botanical Composition, Similarity, and Overlap of Range Herbivore Diets. *In* Natl. Res. Council/Natl. Acad. Sci. "Developing Strategies for Rangeland Management." Westview Press, Boulder, Colo., pp. 425–471.

Holechek, J. L., J. Jeffers, T. Stephenson, C. B. Kuykendall, and S. A. Butler-Nance. (1986). Cattle and Sheep Diets on Low Elevation Winter Range in Northcentral New Mexico. *Amer. Soc. Anim. Sci., West. Sect. Proc.* **37**:243–248.

Holechek, J. L., T. J. Berry, and M. Vavra. (1987). Grazing System Influences on Cattle Performance on Mountain Range. *J. Range Mgt.* **40**(1):55–59.

Holechek, J. L., R. D. Pieper, and C. H. Herbel. (1989). "Range Management Principles and Practices." Prentice Hall, Englewood Cliffs, N. Jer. 501 pp.

Holechek, J. L., H. Gomez, F. Molinar, and D. Galt. (1999). Grazing Studies: What We've Learned. *Rangelands* **21**(2):12–16.

Holechek, J. L., H. Gomez, F. Molinar, D. Galt, and R. Valdez. (2000). Short-Duration Grazing: The Facts in 1999. *Rangelands* **22**(1):18–22.

Holland, C., W. Kezar, and Z. Quade (Eds.). (1990). "Pioneer Forage Manual—A Nutritional Guide." Pioneer Hi-Bred Internat., Des Moines, IA. 55 pp.

Hollingsworth-Jenkins, K. J., T. K. Klopfenstein, D. C. Adams, and J. B. Lamb. (1996). Ruminally Degradable Protein Requirement of Gestating Beef Cows Grazing Native Winter Sandhills Range. *J. Anim. Sci.* **74**(6):1343–1348.

Holmgren, R. C., and S. S. Hutchings. (1972). Salt Desert Shrub Response to Grazing Use. *USDA, For. Serv. Gen. Tech. Rep.* **INT-1**, p. 153–164.

Holt, G. A., and D. G. Wilson. (1961). The Effect of Commercial Fertilization on Forage Production and Utilization on a Desert Range Site. *J. Range Mgt.* **14**(5):252–256.

Hooper, J. F., and H. F. Heady. (1970). An Economic Analysis of Optimum Rates of Grazing in the California Annual-Type Grassland. *J. Range Mgt.* **23**(5):307–311.

Hooper, J. F., J. P. Workman, J. B. Grumbles, and C. W. Cook. (1969). Improved Livestock Distribution with Fertilizer—A Preliminary Economic Evaluation. *J. Range Mgt.* **22**(2):108–110.

Hormay, A. L. (1970). Principles of Rest-Rotation Grazing and Multiple-Use Land Management. *USDA, For. Serv. Training Text* **4**. 26 pp.

Hormay, A. L., and M. W. Talbot. (1961). Rest-Rotation Grazing—A New Management System for Perennial Bunchgrass Ranges. *USDA Prod. Res. Rep.* **51**. 43 pp.

Horn, B., and B. Anderson. (1988). Response of Warm-Season Tall Prairie Grasses to Defoliation by Clipping and Grazing. *Soc. Range Mgt. Abst. Papers* **41**:295.

Horn, G. W., and F. T. McCollum. (1987). Energy Supplementation of Grazing Ruminants. *In* "Proceedings, Grazing Livestock Nutrition Conference, July 23–24, 1987, Jackson, Wyoming." Univ. Wyo., Laramie, WY. pp. 125–136.

Horrocks, R. D., and J. F. Vallentine. (1999). "Harvested Forages." Academic Press, San Diego, CA. 384 pp.

Horrocks, R. D., and M. Zaifnejad. (1997). Late-Season Management of Alfalfa in Irrigated Valleys of the Intermountain West. *J. Prod. Agric.* **10**(1):96–101.

Horton, P. R., and A. W. Bailey. (1987). Vigour of Rough Fescue (*Festuca hallii*) Under Various Intensities and Seasons of Grazing in Central Alberta. *Soc. Range Mgt. Abst. Papers* **40**:186.

Houseal, G. A., and B. E. Olson. (1995). Cattle Use of Microclimates on a Northern Latitude Winter Range. *Can. J. Anim. Sci.* **75**(4):501–507.

Houseal, G. A., and B. E. Olson. (1996). Nutritive Value of Live and Dead Components of Two Bunchgrasses. *Can. J. Anim. Sci.* **76**(4):555–562.

Houston, W. R., and J. J. Urick. (1972). Improved Spring Pastures, Cow-Calf Production, and Stocking Rate Carryover in the Northern Great Plains. *USDA Tech. Bul.* **1451.** 21 pp.

Houston, W. R., and D. H. Van Der Sluijs. (1975). S-Triazine Herbicides Combined with Nitrogen Fertilizer for Increasing Protein on Shortgrass Range. *J. Range Mgt.* **28**(5):272–276.

Houston, W. R., and R. R. Woodward. (1966). Effects of Stocking Rates on Range Vegetation and Beef Cattle Production in the Northern Great Plains. *USDA Tech. Bul.* **1357.** 58 pp.

Hoveland, C. S., R. G. Durham, M. D. Richardson, and T. H. Terrill. (1990). Cutting Management of Endophyte-Free Tall Fescue. *Forage & Grassland Conf.* **1990**:125–128.

Howard, V. W., Jr., and D. G. DeLorenzo. (1975). Vegetation and Food Habits of Mexican Bighorn Sheep in the Game-Coin Enclosure near Red Rock. *N. Mex. Agric. Expt. Sta. Res. Rep.* **303.** 16 pp.

Howarth, R. E., R. K. Chaplin, K.-J. Cheng, B. P. Goplen *et al.* (1991). Bloat in Cattle. *Agric. Can. Pub.* **1858/E.** 34 pp.

Howery, L. D., F. D. Provenza, R. E. Banner, and C. B. Scott. (1996). Differences in Home Range and Habitat Use Among Individuals in a Cattle Herd. *Applied Anim. Beh. Sci.* **49**(3):305–320.

Howery, L. D., F. D. Provenza, G. B. Ruyle, and N. C. Jordan (1998a). How Do Animals Learn if Rangeland Plants Are Toxic or Nutritious? *Rangelands* **20**(6):4–9.

Howery, L. D., F. D. Provenza, R. E. Banner, and C. B. Scott. (1998b). Social and Environmental Factors Influence Cattle Distribution on Rangeland. *Applied Anim. Beh. Sci.* **55**(3–4):231–244.

Howery, L. D., D. W. Bailey, and G. W. Ruyle. (1999). Can Food Quality and Location Information Be Socially Transmitted by Cattle? *Soc. Range Mgt. Abst. Papers* **52**:34–35.

Hubbard, R. L., and H. R. Sanderson. (1961). Grass Reduces Bitterbrush Production. *Calif. Fish & Game* **47**(4):391–398.

Hudson, R. J., and S. Frank. (1987). Foraging Ecology of Bison in Aspen Boreal Habitats. *J. Range Mgt.* **40**(1):71–75.

Hudson, R. J., and M. T. Nietfeld. (1985). Effect of Forage Depletion on the Feeding Rate of Wapiti. *J. Range Mgt.* **38**(1):80–82.

Hudson, R. J., and R. G. White (Eds.). (1985). "Bioenergetics of Wild Herbivores." CRC Press, Boca Raton, FL. 314 pp.

Hughes, L. E. (1982). A Grazing System in the Mohave Desert. *Rangelands* **4**(6):256–257.

Hughes, L. E. (1983). Is No Grazing Really Better than Grazing? *Rangelands* **5**(4):159–161.

Hughes, R. H. (1970). Cattle Grazing Management on Pine-Wiregrass Range. *J. Range Mgt.* **23**(1): 71–72.

Hughes, R. H. (1974). Management and Utilization of Pineland Threeawn Range in South Florida. *J. Range Mgt.* **27**(3):186–192.

Hughes, R. H. (1976). Response of Planted South Florida Slash Pine to Simulated Cattle Damage. *J. Range Mgt.* **29**(3):198- 201.

Hughes, R. H., E. U. Dillard, and J. B. Hilmon. (1960). Vegetation and Cattle Response Under Two Systems of Grazing Cane Range in North Carolina. *N. Car. Agric. Expt. Sta. Bul.* **412.** 27 pp.

Hulet, C. V., D. M. Anderson, V. B. Nakamatsu, L. W. Murray, and R. D. Pieper. (1992). Diet Selection of Cattle and Bonded Small Ruminants Grazing Arid Rangeland. *Sheep Res. J.* **8**(1): 11–18.

Hull, J. L., J. H. Meyer, and G. P. Lofgreen. (1960). Effect of Recovery Interval of Irrigated Forage on the Performance of Grazing Steers. *J. Anim. Sci.* **19**(4):981–990.

Hull, J. L., J. H. Meyer, and R. Kromann. (1961). Influence of Stocking Rate on Animal and Forage Production from Irrigated Pasture. *J. Anim. Sci.* **20**(1):46–52.

Hull, J. L., J. H. Meyer, S. E. Bonilla, and W. Weltkamp. (1965). Further Studies on the Influence of Stocking Rate on Animal and Forage Production from Irrigated Pasture. *J. Anim. Sci.* **24**(3):697–704.

Hull, J. L., J. H. Meyer, and C. A. Raguse. (1967). Rotation and Continuous Grazing on Irrigated Pasture Using Beef Steers. *J. Anim. Sci.* **26**(5):1160–1164.

Hunt, L. J., N. E. Garza, Jr., C. A. Taylor, Jr., T. D. Brooks, and J. E. Huston. (1987). Influence of Supplemental Feeding on Voluntary Forage Intake and on Reproductive Efficiency in Angora Does. *SID Research Digest,* Summer 1987, pp. 18–20.

Hunter, R. A. (1991). Strategic Supplementation for Survival, Reproduction, and Growth of Cattle. *Okla. Agric. Expt. Sta. Misc. Pub.* **MP-133,** pp. 32–47.

Huntsinger, L., and J. W. Bartolome. (1986). Cattle as a Forester's Tool: Grazing Impacts on a Naturally Regenerating California Shelterwood. *Soc. Range Mgt. Abst. Papers* **39**:224.

Hurtt, L. C. (1951). Managing Northern Great Plains Cattle Ranges to Minimize Effects of Drought. *USDA Cir.* **865.** 24 pp.

Huss, D. L. (1972). Goat Response to Use of Shrubs as Forage. *USDA, For. Serv. Gen. Tech. Rep.* **INT-1,** pp. 331–338.

Huston, J. E. (1975). A Loud Shout for Sheep! *Rangeman's J.* **2**(5):136–137.

Huston, J. E., P. V. Thompson, and C. A. Taylor, Jr. (1993). Combined Effects of Stocking Rate and Supplemental Feeding Level on Adult Beef Cows Grazing Native Rangeland in Texas. *J. Anim. Sci.* **71**(12):3458–3465.

Huston, J. E., H. Lippke, T. D. A. Forbes, J. W. Holloway *et al.* (1997). Effects of Frequency of Supplementation of Adult Cows in Western Texas. *Amer. Soc. Anim. Sci., West. Sec. Proc.* **48**:236–238.

Huston, J. E., H. Lippke, T. D. A. Forbes, J. W. Holloway, and R. V. Machen. (1999). Effects of Supplemental Feeding Interval on Adult Cows in Western Texas. *J. Anim. Sci.* **77**(11):3057–3067.

Hutchings, S. S. (1954). Managing Winter Sheep Range for Greater Profit. *USDA Farm Bul.* **2067.** 46 pp.

Hutchings, S. S. (1958). Hauling Water to Sheep on Western Ranges. *USDA Leaflet* **423.** 8 pp.

Hutchings, S. S., and G. Stewart. (1953). Increasing Forage Yields and Sheep Production on Intermountain Winter Ranges. *USDA Cir.* **925.** 63 pp.

Hyder, D. N. (1972). Defoliation in Relation to Vegetative Growth. *In* "The Biology and Utilization of Grasses." Academic Press, Inc., New York and London, pp. 204–217.

Hyder, D. N., and R. E. Bement. (1964). Sixweeks Fescue as a Deterrent to Blue Grama Utilization. *J. Range Mgt.* **17**(5):261–264.

Hyder, D. N., and W. A. Sawyer. (1951). Rotation-Deferred Grazing as Compared to Season-long Grazing on Sagebrush-Bunchgrass Ranges in Oregon. *J. Range Mgt.* **4**(1):30–34.

Hyder, D. N., and F. A. Sneva. (1959). Growth and Carbohydrate Trends in Crested Wheatgrass. *J. Range Mgt.* **12**(6):271–276.

Hyder, D. N., and F. A. Sneva. (1963). Morphological and Physiological Factors Affecting the Grazing Management of Crested Wheatgrass. *Crop Sci.* **3**(3):267–271.

Hyder, D. N., R. E. Bement, E. E. Remmenga, and C. Terwilliger, Jr. (1966). Vegetation-Soils and Vegetation-Grazing Relations from Frequency Data. *J. Range Mgt.* **19**(1):11–17.

Hyder, D. N., R. E. Bement, E. E. Remmenga, and D. F. Hervey. (1975). Ecological Responses of Native Plants and Guidelines for Management of Shortgrass Range. *USDA Tech. Bul.* **1503.** 87 pp.

Irving, B. D., P. L. Rutledge, A. W. Bailey, M. A. Naeth, and D. S. Chanasyk. (1995). Grass Utilization and Grazing Distribution within Intensively Managed Fields in Central Alberta. *J. Range Mgt.* **48**(4):358–361.

Ittner, N. R., G. P. Lofgreen, and J. H. Meyer. (1954). A Study of Pasturing and Soiling Alfalfa with Beef Steers. *J. Anim. Sci.* **13**(1):37–43.

Ivins, J. D. (1955). The Palatability of Herbage. *Herb. Abstr.* **25**(2):75–79.

Jaindl, R. G., and S. H. Sharrow. (1987). Winter Grazing Effects on a Ryegrass-White Clover Pasture. *Soc. Range Mgt. Abst. Papers* **40**:187.

James, L. F. (1983a). Poisonous Plants: Aliphatic Nitro-containing Astragalus. *Rangelands* 5(6):256–257.

James, L. F. (1983b). Poisonous Plants: Locoweeds. *Rangelands* 5(5):224–225.

James, L. F., J. E. Butcher, and K. R. Van Kampen. (1970). Relationship Between *Halogeton glomeratus* Consumption and Water Intake by Sheep. *J. Range Mgt.* 23(2):123–127.

Jameson, D. A. (1963). Responses of Individual Plants to Harvesting. *Bot. Rev.* 29(4):532–594.

Jameson, D. A. (1986). What Shall We Do About Grazing Systems Studies. *Rangelands* 8(4):178–179.

Jameson, D. A. (1991). Effects of Single Season and Rotation Harvesting on Cool- and Warm-Season Grasses of a Mountain Grassland. *J. Range Mgt.* 44(4):327–329.

Jameson, D. A., and D. L. Huss. (1959). The Effect of Clipping Leaves and Stems on Number of Tillers, Herbage Weights, Root Weights, and Food Reserves of Little Bluestem. *J. Range Mgt.* 12(3):122–126.

Janis, C. (1976). The Evolutionary Strategy of the *Equidae* and the Origins of Rumen and Cecal Digestion. *Evolution* 30(4):757–774.

Jasmer, G. E., and J. L. Holechek. (1984). Determining Grazing Intensity on Rangeland. *J. Soil & Water Cons.* 39(1):32–35.

Jensen, C. H., A. D. Smith, and G. W. Scotter. (1972). Guidelines for Grazing Sheep on Rangelands Used by Big Game in Winter. *J. Range Mgt.* 25(5):346–352.

Jensen, E. H., R. R. Skivington, and V. R. Bohman. (1981). Dormant Season Grazing of Alfalfa. *Nev. Agric. Expt. Sta.* R-141. 4 pp.

Jensen, H. P., R. L. Gillen, and F. T. McCollum. (1989). Effects of Grazing Pressure on Defoliation Patterns of Tallgrass Prairie. *Soc. Range Mgt. Abst. Papers* 42:109.

Jensen, H. P., R. L. Gillen, and F. T. McCollum. (1990). Effects of Herbage Allowance on Defoliation Patterns of Tallgrass Prairie. *J. Range Mgt.* 43(5):401–406.

Jensen, J. C. (1984). Perspectives on BLM Forage Allocation: Calculations With Special Reference to the Limiting Factor Approach. *In* Natl. Res. Council/Natl. Acad. Sci. "Developing Strategies for Rangeland Management." Westview Press, Boulder, CO. pp. 473–524.

Jewiss, O. R. (1972). Tillering in Grasses—Its Significance and Control. *J. Brit. Grassland Soc.* 27(2):65–82.

Jiang, Z., and R. J. Hudson. (1994). Bite Characteristics of Wapiti (*Cervus elaphus*) in Seasonal *Bromus-Poa* Swards. *J. Range Mgt.* 47(2):127–132.

Johnson, D. A., K. H. Asay, and R. H. Skinner. (1988). Potential of Cool-Season Perennial Grasses for Late Fall and Winter Grazing on Intermountain Rangelands. *Soc. Range Mgt. Abst. Papers* 41:252.

Johnson, L. E., L. R. Albee, R. O. Smith, and A. L. Moxon. (1951). Cows, Calves, and Grass. *S. Dak. Agric. Expt. Sta. Bul.* 412. 39 pp.

Johnson, M. E., W. C. Russell, and J. D. Hanson. (1986). Effect of Environment on Intake of Steers Grazing Rangeland: A Model. *Amer. Soc. Anim. Sci., West. Sect. Proc.* 37:181–184.

Johnson, R. L. (1983). Mountain Goats and Mountain Sheep of Washington. *Wash. Dept. Game Biol. Bul.* 18. 196 pp.

Johnson, W. H., and E. L. Fitzhugh. (1990). Grazing Helps Maintain Brush Growth on Cleared Land. *Calif. Agric.* 44(5):31–32.

Johnson, W. M. (1953). Effect of Grazing Intensity upon Vegetation and Cattle Gains on Ponderosa Pine-Bunchgrass Ranges of the Front Range of Colorado. *USDA Cir.* 929. 36 pp.

Johnson, W. M. (1959). Grazing Intensity Trials on Seeded Ranges in the Ponderosa Pine Zone of Colorado. *J. Range Mgt.* 12(1):1–7.

Johnson, W. M. (1965). Rotation, Rest-Rotation, and Season-Long Grazing on a Mountain Range in Wyoming. *USDA, For. Serv. Res. Paper* RM-14. 16 pp.

Johnston, A., and C. B. Bailey. (1972). Influence of Bovine Saliva on Grass Regrowth in the Greenhouse. *Can. J. Anim. Sci.* 52(3):573–574.

Johnston, A., and R. W. Peake. (1960). Effect of Selective Grazing by Sheep on the Control of Leafy Spurge (*Euphorbia esula* L.). *J. Range Mgt.* 13(4):192–195.

Johnston, P. W., G. M. McKeon, and K. A. Day. (1996). Objective "Safe" Grazing Capacities for Southwest Queensland Australia: Development of a Model for Individual Properties. *Rangeland J.* 18(2):244–258.

Jones, T. A., and M. H. Ralphs. (1999). Preference of Endophyte-Free Robust Needlegrass by Cattle. *Soc. Range Mgt. Abst. Papers* **52**:38.

Jordan, L. A., and J. P. Workman. (1988). Fee Hunting Opportunities on Private Land in Utah. *Soc. Range Mgt. Abst. Papers* **41**:217.

Jordan, R. M., and G. C. Marten. (1970). Forward-Creep Grazing Vs. Conventional Grazing for Production of Suckling Lambs. *J. Anim. Sci.* **31**(3):598–600.

Jourdonnais, C. S., and D. J. Bedunah. (1990). Prescribed Fire and Cattle Grazing on an Elk Winter Range in Montana. *Wildl. Soc. Bull.* **18**(3):232–240.

Judkins, M. B., L. J. Kryst, J. D. Wallace, M. L. Galyean, K. D. Jones, and E. E. Parker. (1985). Intake and Diet Selection by Protein Supplemented Grazing Steers. *J. Range Mgt.* **38**(3):210–214.

Judkins, M. B., J. D. Wallace, M. L. Galyean, L. J. Krysl, and E. E. Parker. (1987). Passage Rates, Rumen Fermentation, and Weight Change in Protein Supplemented Grazing Cattle. *J. Range Mgt.* **40**(2):100–105.

Julander, O. (1966). How Mule Deer Use Mountain Rangeland in Utah. *Utah Academy Sci. Proc.* **43**(2):22–28.

Jung, H. J., and L. J. Koong. (1985). Effects of Hunger Satiation on Diet Quality by Grazing Sheep. *J. Range Mgt.* **38**(4):302–305.

Kalmbacher, R. S., K. R. Long, and F. G. Martin. (1984). Seasonal Mineral Concentration in Diets of Esophageally Fistulated Steers on Three Range Areas. *J. Range Mgt.* **37**(1):36–39.

Kalmbacher, R. S., F. G. Martin, and W. D. Pitman. (1986). Effect of Grazing Stubble Height and Season on Establishment, Persistence, and Quality of Creeping Bluestem. *J. Range Mgt.* **39**(3):223–227.

Kalmbacker, R. S., F. G. Martin. W. D. Pitman, and G. W. Tanner. (1994). South Florida Flatwood Range Vegetation Responses to Season of Deferment from Grazing. *J. Range Mgt.* **47**(1):43–47.

Kamil, A. C., and T. D. Sargent (Eds.). (1981). "Foraging Behavior: Ecological, Ethological, and Psychological Approaches." Garland STPM Press, New York. 534 pp.

Kansas State University, Depts. of Agron. and Anim. Sci. & Ind. (1995). "Fifty Years of Range Research Revisited with a Special Focus on Current Research." Kan. Agric. Ext. Serv., Manhattan. 118 pp.

Kanyama-Phiri, G., and B. E. Conrad. (1986). Influence of Grazing Frequency and Time of Grazing on Bite Rate. *In* Forage Research in Texas, 1985; *Texas Agric. Expt. Sta. Cons. Prog. Rep.* **4347,** pp. 15–16.

Karl, M. G., and P. S. Doescher. (1998). Ponderosa Pine Aboveground Growth after Cattle Removal of Terminal Tissue. *J. Range Mgt.* **51**(2):147–151.

Karn, J. F., and D. C. Clanton. (1977). Potassium in Range Supplements. *J. Anim. Sci.* **45**(6):1426–1434.

Kartchner, R. J. (1981). Effects of Protein and Energy Supplementation of Cows Grazing Native Winter Forage on Intake and Digestibility. *J. Anim. Sci.* **51**(2):432–438.

Kartchner, R. J., and D. C. Adams. (1982). Effects of Daily and Alternate Day Feeding of Grain Supplements to Cows Grazing Fall-Winter Range. *Amer. Soc. Anim. Sci., West. Sect. Proc.* **33**:308–311.

Kauffman, J. B., and W. C. Krueger. (1984). Livestock Impacts on Riparian Ecosystems and Streamside Management Implications: A Review. *J. Range Mgt.* **37**(5):430–438.

Kautz, J. E., and G. M. Van Dyne. (1978). Comparative Analyses of Diets of Bison, Cattle, Sheep, and Pronghorn Antelope on Shortgrass Prairie in Northeastern Colorado, USA. "Proceedings of the First International Rangeland Congress." Soc. Range Mgt., Denver, Colo., pp 438–443.

Kay, B. L., and D. T. Torell. (1970). Curing Standing Range Forage with Herbicides. *J. Range Mgt.* **23**(1):34–41.

Kearl, W. G. (1970). Comparison of Net Energy and Animal-Unit-Month Standards in Planning Livestock Feed and Forage Requirements. *Wyo. Agric. Expt. Sta. Res. J.* **35**. 20 pp.

Keeler, R. F. (1983). Deformed Calves from Poisonous Plants. *Rangelands* **5**(5):221–223.

Keeler, R. F., and W. A. Laycock. (1987). Use of Plant Toxin Information in Management Decisions. *Soc. Range Mgt. Abst. Papers* **40**:106.

Kelsall, J. P. (1968). "The Migratory Barren-Ground Caribou of Canada." Canada Wildl. Serv., Ottawa, Can. 340 pp.

Kelton, E. (1982). Rancher Boosts Grazing on Large Ranch Fenced in Cells. *Rangelands* **4**(6):258–260.

Kidunda, R. S., L. R. Rittenhouse, D. M. Swift, and R. W. Richards. (1993). Spatial Behavior of Free-Grazing Cattle: Movement from Patch to Patch. *Amer. Soc. Anim. Sci., West. Sect. Proc.* **44**:255–258.

Kie, J. G. (1987). Measures of Wild Ungulate Performance: Population Density and Condition of Individuals. *In* Donald A. Jameson and Jerry Holechek (Eds.). "Monitoring Animal Performance and Production Symposium Proceedings, February 12, 1987, Boise, Idaho." Society for Range Management, Denver, CO. pp. 23–36.

Kie, J. G., and E. R. Loft. (1990). Using Livestock To Manage Wildlife Habitat: Some Examples from California Annual Grassland and Wet Meadow Communities. *In* K. E. Severson (Tech. Coord.), "Can Livestock Be Used as a Tool to Enhance Wildlife Habitat?" *USDA, For. Serv. Gen. Tech. Rep.* **RM-194,** pp. 7–24.

Kie, J. G., C. J. Evans, E. R. Loft, and J. M. Menke. (1986). Cattle and Mule Deer Activity Patterns in the Sierra Nevada. *Soc. Range Mgt. Abst. Papers* **39**:338.

Kie, J. G., E. R. Loft, J. W. Menke, J. Winckel, and M. W. Demmet. (1988). Managing Mule Deer and Cattle on Sierra Nevada Summer Ranges in California. *Soc. Range Mgt., Abst. Papers* **41**:215.

Kingery, J. L. (1987). The Relationship of Animal Use to Tree Establishment on Forest Plantations in Northern Idaho. *Soc. Range Mgt. Abst. Papers* **40**:160.

Kingery, J. L., and R. T. Graham. (1990). Forest Land Grazing: Can Grazing and Reforestation Coexist? *West. Wildlands* **16**(2):20–22.

Kirby, D., and P. M. Bultsma. (1984). Implementing and Evaluating Short-Duration Grazing Systems: Preliminary Guidelines. *N. Dak. Agric. Ext.* **14 AGR-10.** 7 pp.

Kirby, D., and R. Lym. (1987). Grazing Behavior of Cattle in a Leafy Spurge Infested Pasture. *Soc. Range Mgt. Abst. Papers* **40**:168.

Kirby, D. R., and M. Parman. (1986). Botanical Composition and Diet Quality of Cattle Under a Short Duration Grazing System. *J. Range Mgt.* **39**(6):509–512.

Kirby, D. R., M. F. Pessin, and G. K. Clambey. (1986). Disappearance of Forage Under Short Duration and Seasonlong Grazing. *J. Range Mgt.* **39**(6):496–500.

Kirby, D., M. Parman, M. Pessin, and M. Humann. (1988). Dietary Overlap of Cattle and Sheep on Rotationally Grazed Rangeland. *SID Res. J.* **4**(3):6–11.

Kirby, D., W. Barker, and P. Nyren. (1996). Grazing Systems for the Mixed Grass Prairie of North Dakota. *Proc. Internat. Rangeland Cong.* **5**(1):291.

Kirby, D. R., T. P. Hanson, and C. Hull-Sieg. (1997). Diets of Angora Goats Grazing Leafy Spurge (*Euphorbia esula*)-Infested Rangeland. *Weed Tech.* **11**(4):734–738.

Kirychuk, B., and M. Tremblay. (1995). "Initial Stocking Rate Recommendations for Seeded Forages in Saskatchewan." Sask. Agric. & Food, Regina, Sask. 4 pp.

Kisserberth, W. C., W. B. Buck, M. E. Mansfield, and R. K. Manuel. (1986). Preferential Grazing by Cattle on Glyphosate-Treated Fescue Pastures. *Amer. J. Vet. Res.* **47**(3):696–698.

Klebenow, D. A. (1980). The Impacts of Grazing Systems on Wildlife. *In* K. C. McDaniel and C. Allison (Eds.). "Grazing Management Systems for Southwest Rangelands: A Symposium." Range Impr. Task Force, N. Mex. State Univ., Las Cruces, N. Mex., pp. 153–162.

Klein, L. (1994). "Winter Grazing and Alternate Feeds for Beef Cattle in Saskatchewan." Sask. Agric. & Food, Regina, Sask. 12 pp.

Klein, L. (1997). "Pasture and Forages for Wapiti." Grazing and Pasture Tech. Prog., Sask. Agric. & Food, Regina, Sask. 12 pp.

Klett, W. E., D. Hollingsworth, and J. L. Schuster. (1971). Increasing Utilization of Weeping Lovegrass by Burning. *J. Range Mgt.* **24**(1):22–24.

Klingsbiel, A.A., and P. H. Montgomery (1961). Land Capability Classification. *USDA Agric. Handbook* **210.** 21p.

Klipple, G. E. (1964). Early- and Late-Season Grazing Versus Season-Long Grazing of Shortgrass Vegetation on the Central Great Plains. *USDA, For. Serv. Res. Paper* **RM-11.** 16 pp.

Klipple, G. E., and R. E. Bement. (1961). Light Grazing—Is It Economically Feasible as a Range-Improvement Practice. *J. Range Mgt.* **14**(2):57–62.

Klipple, G. E., and D. F. Costello. (1960). Vegetation and Cattle Responses to Different Intensities of Grazing on Short-Grass Ranges on the Central Great Plains. *USDA Tech. Bul.* **1216.** 82 pp.

Knight, J. C., and M. M. Kothmann. (1986). Vegetation Trend and Cattle Performance in Several Grazing Systems in the Rolling Plains of Texas. *Texas Agric. Expt. Sta. Prog. Rep.* **4438.** 2 pp.

Knipe, O. D. (1983). Effects of Angora Goat Browsing on Burned-Over Arizona Chaparral. *Rangelands* **5**(6):252–255.

Knowlton, F. F. (1960). Food Habits, Movements, and Moose Populations in the Gravelly Mountains, Montana. *J. Wildl. Mgt.* **24**(2):162–170.

Korpela, E. J., and W. C. Krueger. (1987). Community Utilization and Preference by Livestock Grazing a Northeastern Oregon Riparian Zone. *Soc. Range Mgt. Abst. Papers* **40**:14.

Kosco, B. H., and J. W. Bartolome. (1983). Effects of Cattle and Deer on Regenerating Mixed Conifer Clearcuts. *J. Range Mgt.* **36**(2):265–268.

Kothmann, M. M. (1974). Grazing Management Terminology. *J. Range Mgt.* **27**(4):326–327.

Kothmann, M. M. (1980). Integrating Livestock Needs to the Grazing System. *In* K. C. McDaniel and C. Allison (Eds.), "Grazing Management Systems for Southwest Rangelands: A Symposium" Range Impr. Task Force, N. Mex. State Univ., Las Cruces, N. Mex., pp. 65–83.

Kothmann, M. M. (1984). Concepts and Principles Underlying Grazing Systems: A Discussant Paper. *In* Natl. Res. Council/Natl. Acad. Sci. "Developing Strategies for Rangeland Management." Westview Press, Boulder, CO. pp. 903–916.

Kothmann, M. M., and R. T. Hinnant. (1987). Direct Measures of the Nutritional Status of Grazing Animals. *In* D. A. Jameson and J. Holechek (Eds.). "Monitoring Animal Performance and Production Symposium Proceedings, February 12, 1987, Boise, Idaho." Society for Range Management, Denver, CO. pp. 17–22.

Kothmann, M. M., G. W. Mathis, P. T. Marion, and W. J. Waldrip. (1970). Livestock Production and Economic Returns from Grazing Treatments on the Texas Experimental Ranch. *Texas Agric. Expt. Sta. Bul.* **1100.** 39 pp.

Kothmann, M. M., G. W. Mathis, and Wm. J. Waldrip. (1971). Cow-Calf Response to Stocking Rates and Grazing Systems on Native Range. *J. Range Mgt.* **24**(2):100–105.

Kothmann, M. M., R. T. Hinnant, and J. F. Casco. (1986a). Vegetation Responses Under Rotational Grazing. *Texas Agric. Expt. Sta. Prog. Rep.* **4425.** 2 pp.

Kothmann, M. M., E. A. de Moraes, R. K. Heitschmidt, and J. W. Walker. (1986b). Botanical and Chemical Composition of Cattle Diets From Yearlong Continuous and Rotation Grazing Systems. *Texas Agric. Expt. Sta. Prog. Rep.* **4433.** 1 p.

Krausman, P. R. (Ed.). (1996). "Rangeland Wildlife." Soc. Range Mgt., Denver CO. 440 pp.

Krausman, P. R., A. J. Kuenzl, R. C. Etchberger, K. R. Rautenstrauch *et al.* (1997). Diets of Desert Mule Deer. *J. Range Mgt.* **50**(5):513–522.

Kreuter, U. P., R. C. Rowan, J. R. Conner, J. W. Stuth, and W. T. Hamilton. (1996). Decision Support Software for Estimating the Economic Efficiency of Grazingland Production. *J. Range Mgt.* **49**(5):464–469.

Kronberg, S. L., K. M. Havstad, E. L. Ayers, and D. E. Doornbos. (1986). Influence of Breed on Forage Intake of Range Beef Cows. *J. Range Mgt.* **39**(5):421–423.

Kronberg, S. L., R. B. Muntifering, E. L. Ayers, and C. B. Marlow. (1993). Cattle Avoidance of Leafy Spurge: A Case of Conditioned Aversion. *J. Range Mgt.* **46**(4):364–366.

Kropp, J. R., J. W. Holloway, D. F. Stephens, L. Knori, R. D. Morrison, and R. Totusek. (1973). Range Behavior of Hereford, Hereford X Holstein, and Holstein Non-Lactating Heifers. *J. Anim. Sci.* **36**(4):797–802.

Krueger, W. C. (1983). Cattle Grazing in Managed Forests. *In* B. F. Roche, Jr., and D. M. Baumgartner (Eds.). "Forestland Grazing, Proceedings of a Symposium Held February 23–25, 1983, Spokane, Washington." Wash. State Univ., Coop. Ext. Serv., Pullman, Wash., pp. 29–41.

Krueger, W. C. (1985). Grazing for Forest Weed Control. *In* D. M. Baumgartner, R. J. Boyd, D. W. Breuer, and D. Miller (Eds.), "Weed Control for Forest Productivity in the Interior West." Wash. State Univ., Conf. and Inst., Pullman, WA. pp. 83–88.

Krueger, W. C., and L. A. Sharp. (1978). Management Approaches to Reduce Livestock Losses from Poisonous Plants on Rangeland. *J. Range Mgt.* **31**(5):347–350.

Krueger, W. C., and M. Vavra. (1984). Twentieth-Year Results from a Plantation Grazing Study. *Ore. Agric. Expt. Sta. Spec. Rep.* **715**, pp. 20–24.

Krueger, W. C., W. A. Laycock, and D. A. Price. (1974). Relationships of Taste, Smell, Sight, and Touch to Forage Selection. *J. Range Mgt.* **27**(4):258–262.

Krysl, L. J., and B. W. Hess. (1993). Influence of Supplementation on Behavior of Grazing Cattle. *J. Anim. Sci.* **71**(9):2546–2555.

Krysl, L. J., M. E. Hubbert, B. F. Sowell, G. E. Plumb, T. K. Jewett, M. A. Smith, and J. W. Waggoner. (1984). Horses and Cattle Grazing in the Wyoming Red Dessert. I. Food Habits and Dietary Overlap. *J. Range Mgt.* **37**(1):72–76.

Krysl, L. J., M. L. Galyean, J. D. Wallace, F. T. McCollum, M. B. Judkins, M. E. Branine, and J. S. Caton. (1987). Cattle Nutrition on Blue Grama Rangeland in New Mexico. *N. Mex. Agric. Expt. Sta. Bul.* **727**. 35 pp.

Kufeld, R. C., D. C. Bowden, and D. L. Schrupp. (1988). Habitat Selection and Activity Patterns of Female Mule Deer in the Front Range, Colorado. *J. Range Mgt.* **41**(6):515–522.

Kunst, C. R., R. E. Sosebee, and M. J. Dumesnil. (1988). Herbicidal Control of Cholla. *Soc. Range Mgt. Abst. Papers* **41:**67.

Laca, E. A. (1998). Spatial Memory and Food Searching Mechanisms of Cattle. *J. Range Mgt.* **51**(4):370–378.

Laca, E. A., and I. M. Ortega. (1996). Integrating Foraging Mechanisms Across Spatial and Temporal Scales. *Proc. Internat. Rangeland Cong.* **5**(2):129–132.

Laca, E. A., E. D. Ungar, N. Seligman, and M. W. Demment. (1992). Effects of Sward Height and Bulk Density on Bite Dimensions of Cattle Grazing Homogeneous Swards. *Grass & Forage Sci.* **47**(1):91–102.

Laca, E. A., R. A. Distel, T. C. Griggs, and M. W. Demment. (1994). Effects of Canopy Structure on Patch Depression by Grazers. *Ecology* **75**(3):706–716.

Laca, E. A., L. Shipley, and E. Reid. (1999). Structural Anti-Quality Characteristics of Range and Forage Plants. *Soc. Range Mgt. Abst. Papers* **52:**41–42.

Lacey, J. R., and H. W. Van Poollen. (1979). Grazing System Identification. *J. Range Mgt.* **32**(2):38–39.

Lacey, J. R., and H. W. Van Poollen. (1981). Comparison of Herbage Production on Moderately Grazed and Ungrazed Western Ranges. *J. Range Mgt.* **34**(3):210–212.

Lacey, C. A., R. W. Kott, and P. K. Fay. (1984). Ranchers Control Leafy Spurge. *Rangelands* **6**(5):202–204.

Lachica, M., F. G. Barroso, and C. Prieto. (1997). Seasonal Variation of Locomotion and Energy Expenditure in Goats under Range Grazing Conditions. *J. Range Mgt.* **50**(3):234–236.

Lake, R. P., D. C. Clanton, and J. F. Karn. (1974). Intake, Digestibility, and Nitrogen Utilization of Steers Consuming Irrigated Pasture as Influenced by Limited Energy Supplementation. *J. Anim. Sci.* **38**(6):1291–1297.

Lamb, J. B., D. C. Adams, T. J. Klopfenstein, W. W. Stoup, and G. P. Lardy. (1997). Range or Meadow Regrowth and Weaning Effects on 2-Year-Old Cows. *J. Range Mgt.* **50**(1):16–19.

Lamb, S. H., and R. Pieper. (1971). Game Range Improvement in New Mexico. *N. Mex. Interagency Range Comm. Rep.* **9**. 28 pp.

Landgraf, B. K., P. K. Fay, and K. M. Havstad. (1984). Utilization of Leafy spurge (*Euphorbia esula*) by Sheep. *Weed Sci.* **32**(3):348–352.

Lane, M. A., M. H. Ralphs, J. D. Olsen, F. D. Provenza, and J. A. Pfister. (1990). Conditioned Taste Aversion: Potential for Reducing Cattle Loss to Larkspur. *J. Range Mgt.* **43**(2):127–131.

Lang, R., and L. Landers. (1960). (Beef) Production and Grazing Capacity from a Combination of Seeded Pastures Versus Native Range. *Wyo. Agric. Expt. Sta. Bul.* **370**. 12 pp.

Lang, R. L., O. K. Barnes, and F. Rauzi. (1956). Shortgrass Range—Grazing Effects on Vegetation and on Sheep Gains. *Wyo. Agric. Expt. Sta. Bul.* **343**. 32 pp.

Larsen, H. J., and R. F. Johannes. (1965). Summer Forage: Stored Feeding, Green Feeding, and Strip Grazing. *Wisc. Agric. Expt. Sta. Res. Bul.* **257**. 32 pp.

Lathrop, W. J., D. D. Kress, K. M. Havstad, D. E. Doornbos, and E. L. Ayers. (1985). Grazing Behavior of Rangeland Beef Cows Differing in Milk Production. *Amer. Soc. Anim. Sci., West. Sect. Proc.* **36**:66–69.

Lathrop, W. J., D. D. Kress, K. M. Havstad, D. E. Doornbos, and E. L. Ayers. (1988). Grazing Behavior of Rangeland Beef Cows Differing in Milk Production. *Applied Anim. Beh. Sci.* **21**(4):315–327.

Lauenroth, W. K., D. G. Milchunas, J. L. Dodd, R. H. Hart, R. K. Heitschmidt, and L. R. Rittenhouse. (1994). Effects of Grazing on Ecosystems of the Great Plains. *In* M. Vavra, W. Laycock, and R. D. Pieper (Eds.), "Ecological Implications of Livestock Herbivory in the West." Soc. Range Mgt., Denver, CO. pp. 69–100.

Launchbaugh, J. (1976). Graze First-Year Native Grass Plantings. *Kan. Agric. Expt. Sta.* **AES-23.** 2 pp.

Launchbaugh, J. L. (1957). The Effect of Stocking Rate on Cattle Gains and on Native Shortgrass Vegetation in West-Central Kansas. *Kan. Agric. Expt. Sta. Bul.* **394.** 29 pp.

Launchbaugh, J. L. (1986). Intensive-Early Season Stocking. *In* Patrick E. Reece and James T. Nichols (Eds.). "Proceedings, The Ranch Management Symposium, November 5–7, 1986, North Platte, Nebraska." Univ. Neb., Agric. Ext. Serv., Lincoln, Neb., pp. 52–56.

Launchbaugh, J. L. (1987.) The Use of Complementary Forages in a Reproductive Beef Cattle Operation. *Kan. Agric. Ext. Cir.* **681.** 14 pp.

Launchbaugh, J. L., and C. E. Owensby. (1978). Kansas Rangelands: Their Management Based on a Half Century of Research. *Kan. Agric. Expt. Sta. Bul.* **622.** 56 pp.

Launchbaugh, J. L., C. E. Owensby, F. L. Schwartz, and L. R. Corah. (1978). Grazing Management to Meet Nutritional and Functional Needs of Livestock. *Proc. Intern. Rangeland Cong.* **1**:541–546.

Launchbaugh, J. L., C. E. Owensby, J. R. Brethour, and E. F. Smith. (1983). Intensive-Early Stocking Studies on Kansas Ranges. *Kan. Agric. Expt. Sta. Rep. Prog.* **441.** 13 pp.

Launchbaugh, K. L., J. W. Stuth, and J. W. Holloway. (1990). Influence of Range Site on Diet Selection and Nutrient Intake of Cattle. *J. Range Mgt.* **43**(2):109–116.

Launchbaugh, K. L., J. W. Walker, and C. A. Taylor. (1999). Foraging Behavior: Experience or Inheritance? *In* "Grazing Behavior of Livestock and Wildlife." *Idaho Forest, Wildl., and Range Expt. Sta. Bull.* **70,** pp. 28–35.

Lautier, J. K., T. V. Dailey, and R. D. Brown. (1988). Effects of Water Restriction on Feed Intake of White-Tailed Deer. *J. Wildl. Mgt.* **52**(4):602–606.

Laycock, W. A. (1961). Improve Your Range by Heavy Fall Grazing. *Natl. Wool Grower* **51**(6):16, 30.

Laycock, W. A. (1962). Rotation Allows Heavier Grazing of Sagebrush-Grass Range. *Nat. Woolgrower* **52**(6):16–17.

Laycock, W. A. (1967). How Heavy Grazing and Protection Affect Sagebrush-Grass Ranges. *J. Range Mgt.* **20**(4):206–213.

Laycock, W. A. (1970). The Effects of Spring and Fall Grazing on Sagebrush-Grass Ranges in Eastern Idaho. *Intern. Grassland Cong. Proc.* **11**:52–54.

Laycock, W. A. (1978). Coevolution of Poisonous Plants and Large Herbivores on Rangelands. *J. Range Mgt.* **31**(5):335–342

Laycock, W. A. (1979). Management of Sagebrush. *Rangelands* **1**(5):207–210.

Laycock, W. A. (1983). Evaluation of Management as a Factor in the Success of Grazing Systems. *In* "Managing Intermountain Rangelands—Improvement of Range and Wildlife Habitats." *USDA, For. Serv. Gen. Tech. Rep.* **INT-157,** pp. 166–177.

Laycock, W. A., and P. W. Conrad. (1967). Effect of Grazing on Soil Compaction as Measured by Bulk Density on a High Elevation Cattle Range. *J. Range Mgt.* **20**(3):136–140.

Laycock, W. A., and P. W. Conrad. (1981). Responses of Vegetation and Cattle to Various Systems of Grazing on Seeded and Native Mountain Rangelands in Eastern Utah. *J. Range Mgt.* **34**(1):52–58.

Laycock, W. A., and R. O. Harniss. (1974). Trampling Damage on Native Forb-Grass Ranges Grazed by Sheep and Cattle. *Intern. Grassland Cong. Proc., Sect. Papers: Grassland Utilization* **12**:349–354.

Laycock, W. A., D. Loper, F. W. Obermiller, L. Smith *et al.* (1996). Grazing on Public Lands. *CAST Rep.* **129.** 70 pp.

Leach, G. J., and R. J. Clements. (1984). Ecology and Grazing Management of Alfalfa Pastures in the Subtropics. *Adv. Agron.* **37:**127–154.

Leaver, J. D. (Ed.). (1982). "Herbage Intake Handbook." Brit. Grassl. Soc., Hurley, Eng. 143 pp.

Leckenby, D. A., and A. W. Adams. (1986). A Weather Severity Index on a Mule Deer Winter Range. *J. Range Mgt.* **39**(3):244–248.

Leckenby, D. A., D. P. Sheehy, C. H. Nellis, R. J. Scherzinger, I. D. Lumen, W. Elmore, J. C. Lemos, L. Doughty, and C. E. Trainer. (1982). Wildlife Habitats in Managed Rangelands—The Great Basin of Southeastern Oregon: Mule Deer. *USDA, For. Serv. Gen. Tech. Rep.* **PNW-139.** 40 pp.

Leege, T. A., and W. O. Hickey. (1977). Elk-Snow-Habitat Relationships in the Pete King Drainage, Idaho. *Ida. Dept. Fish and Game Wildl. Bul.* **6.** 23 pp.

Leininger, W. C., and S. H. Sharrow. (1983). Sheep and Timber: Are They Compatible? *Ore. Agric. Expt. Sta. Spec. Rep.* **682,** p. 23–27.

Leininger, W. C., and S. H. Sharrow. (1987). Seasonal Diets of Herded Sheep Grazing Douglas-Fir Plantations. *J. Range Mgt.* **40**(6):551–555.

Leininger, W. C., B. D. Rhodes, and S. H. Sharrow. (1983). Herded Sheep as a Tool for Brush Control in Douglas-Fir Forest. *Soc. Range Mgt. Abst. Papers* **36:**229.

Lenssen, A. W., E. L. Sorensen, G. L. Posler, and L. H. Harbers. (1989). Sheep Preference for Perennial Glandular-Haired and Eglandular *Medicago* Populations. *Crop Sci.* **29**(1):65–68.

Leury, B. J., C. Stever-Kelly, K. L. Gatford, R. J. Simpson, and H. Dove. (1999). Spray-Topping Annual Grass Pasture with Glyphosate To Delay Loss of Feeding Value During Summer. IV. Diet Composition, Herbage Intake, and Performance in Grazing Sheep. *Austr. J. Agric. Res.* **50**(4):487–495.

Lewis, C. E. (1980). Simulated Cattle Injury to Planted Slash Pine: Girdling. *J. Range Mgt.* **33**(5):337–348.

Lewis, C. E. (1984). Warm Season Forage Under Pine and Related Cattle Damage to Young Pines. *In* Linnartz, Norwin E., and Mark K. Johnson (Eds.). "Agroforestry in the Southern United States." La. Agric. Expt. Sta., Baton Rouge, La., pp. 66–78.

Lewis, C. E., and W. C. McCormick. (1971). Supplementing Pine-Wiregrass Range with Improved Pasture in South Georgia. *J. Range Mgt.* **24**(5):334–339.

Lewis, C. E., and G. W. Tanner. (1987). Herbaceous Plant Responses to Pine Regeneration and Deferred-Rotation Grazing in North Florida. *Soc. Range Mgt. Abst. Papers* **40:**202.

Lewis, C. E., W. G. Monson, and R. J. Bonyata. (1985). Pensacola Bahiagrass Can Be Used to Improve the Forage Resource When Regenerating Southern Pines. *S. J. Appl. For.* **9**(10):254–259.

Lewis, J. K. (1983). Tentative Classificiation of Grazing Systems. *Soc. Range Mgt. Abst. Papers* **36:**231.

Lewis, J. K. (1984). Classification and Application of Grazing Systems. *Soc. Range Mgt. Abst. Papers* **37:**87.

Lewis, J. K. (1986). Practical Consideration of Range Ecosystems. *In* Patrick E. Reece and James T. Nichols (Eds.). "Proceedings, The Ranch Management Symposium, November 5–7, 1986, North Platte, Nebraska." Univ. Neb., Agric. Ext. Serv., Lincoln, Neb., pp. 9–18.

Lewis, J. K., and J. D. Volesky. (1988). Future Directions: Application of New Technology. *In* Richard S. White and Robert E. Short (Eds.). "Achieving Efficient Use of Rangeland Resources, Fort Keogh Research Symposium, September, 1987, Miles City, Mon." Mon. Agric. Expt. Sta., Bozeman, Mon., pp. 59–65.

Lewis, J. K., G. M. Van Dyne, L. R. Albee, and F. W. Whetzal. (1956). Intensity of Grazing: Its Effect on Livestock and Forage Production. *S. Dak. Agric. Expt. Sta. Bul.* **459.** 44 pp.

Lewis, M., T. Klopfenstein, and B. Anderson. (1988). Wintering Gain on Subsequent Grazing and Finishing Performance. Neb. Agric. Res. Div. Misc. Pub. 54, p. 34–35.

Linnartz, N. E., and J. C. Carpenter, Jr. (1979). Creep Pastures for Calves on Forest Range. *Rangelands* **1**(1):23.

Linnartz, N. E., and M. K. Johnson (Eds.). (1984). "Agroforestry in the Southern United States." La. Agric. Expt. Sta. Baton Rouge, La. 183 pp.

Lodge, R. W. (1970). Complementary Grazing Systems for the Northern Great Plains. *J. Range Mgt.* **23**(4):268–271.

Lodge, R. W., S. Smoliak, and A. Johnston. (1972). Managing Crested Wheatgrass Pastures. *Can. Dept. Agric. Pub.* **1473.** 20 pp.

Loft, E. R., J. W. Menke, and J. G. Kie. (1986). Mule Deer and Cattle Habitat Preferences in the Sierra Nevada Mountains. *Soc. Range Mgt. Abst. Papers* **39:**339.

Loft, E. R., J. W. Menke, and J. G. Kie. (1988). Competition Between Mule Deer and Cattle in Meadow–Riparian and Aspen Communities. *Soc. Range Mgt., Abst. Papers* **41:**72.

Loft, E. R., J. W. Menke, and J. G. Kie. (1991). Habitat Shifts by Mule Deer: The Influence of Cattle Grazing. *J. Wildl. Mgt.* **55**(1):16–26.

Long, K. R., R. S. Kalmbacher, and F. G. Martin. (1986). Effect of Season and Regrazing on Diet Quality of Burned Florida Range. *J. Range Mgt.* **39**(6):518–521.

Long, W. (1996). Free Market Wildlife Management: A Plus for Landowners, Hunters, and the Environment. *In* "Sharing Common Ground on Western Rangelands: Proceedings of a Livestock/Big Game Symposium." *USDA, For. Serv. Gen. Tech. Rep.* **INT-GTR-343,** pp. 9–10.

Longhurst, W. M., G. E. Connolly, B. M. Browning, and E. O. Garton. (1979). Food Interrelationships of Deer and Sheep in Parts of Mendocino and Lake Counties, California. *Hilgardia* **47**(6):191–247.

Lopes, E. A., and J. W. Stuth. (1984). Dietary Selection and Nutrition of Spanish Goats as Influenced by Brush Management. *J. Range Mgt.* **37**(6):554–560.

Love, R. M., and R. E. Eckert, Jr. (1985). Rangeland Ecosystems and Their Improvement. Chapter 51 *in* Maurice E. Heath, Robert F. Barnes, and Darrel S. Metcalfe. "Forages: The Science of Grassland Agriculture," Fourth Ed. Iowa State Univ. Press, Ames, Iowa.

Lowe, L. B. (1998). Prevention of Bloat in Pastured Cattle—Using Monensin Sodium Controlled Release Capsules (CRC). *Bovine Pract.* **32**(1):27–30.

Lugenja, M., R. Nicholson, and J. Launchbaugh. (1983). Grazing Behaviour of Steers on Seasonlong vs. Intensive Early Stocking. *Soc. Range Mgt. Abst. Papers* **36:**58.

Lundgren, G. K., J. R. Conner, and H. A. Pearson. (1984). An Economic Analysis of Five Forest-Grazing Management Systems in the Southeastern United States. *Tex. Agric. Expt. Sta. Misc. Pub.* **1551.** 8 pp.

Lusby, K. (Comm. Chm.). (1983). "Oklahoma Beef Cattle Manual." Okla. State Univ., Div. Agric., Stillwater, Okla. 179 pp.

Lusby, K. S., and D. G. Wagner. (1987). Effects of Supplements on Feed Intake. *Okla. Agric. Expt. Sta. Misc. Pub.* **121,** p. 173–181.

Lym, R. G. (1998). The Biology and Integrated Management of Leafy Spurge (*Euphorbia esula*) on North Dakota Rangelands. *Weed Tech.* **12**(2):367–373.

Lym, R. G., and D. R. Kirby. (1987). Cattle Foraging Behavior in Leafy Spurge (*Euphorbia esula* L.) Infested Rangeland. *Weed Tech.* **1:**314–318.

Lyon, L. J. (1979). Influences of Logging and Weather on Elk Distribution in Western Montana. *USDA, For. Serv. Res. Paper* **INT-236.** 11 pp.

Lyon, L. J. (1980). Coordinating Forestry and Elk Management. *Trans. N. Amer. Wildl. & Nat. Resources Conf.* **45:**278–287.

Lyon, L. J. (1985). Elk and Cattle on the National Forests: A Simple Question of Allocation . . . Or a Complex Management Problem? *West. Wildlands* **11**(1):16–19.

Lyon, L. J., and C. E. Jensen. (1980). Management Implications of Elk and Deer Use on Clear-Cuts in Montana. *J. Wildl. Mgt.* **44**(2):352–362.

MacCracken, J. G., and R. M. Hansen. (1981). Diets of Domestic Sheep and Other Large Herbivores in Southcentral Colorado. *J. Range Mgt.* **34**(3):242–243.

Mackie, R. J. (1970). Range Ecology and Relations of Mule Deer, Elk, and Cattle in the Missouri River Breaks, Montana. *Wildl. Monogr.* **20.** 79 pp.

Mackie, R. J. (1976). Interspecific Competition Between Mule Deer, Other Game Animals, and Livestock. *In* "Mule Deer Decline in the West: A Symposium." Utah State Univ., Logan, Utah, pp. 49–54.

Mackie, R. J. (1985). The Elk-Deer-Livestock Triangle. *In* Gar Workman (Ed.). "Western Elk Management Symposium." Utah State Univ., Logan, Utah, pp. 51–56.

Magee, A. C. (1957). Goats Pay for Clearing Grand Prairie Rangelands. *Texas Agric. Expt. Sta. Misc. Pub.* **206.** 8 pp.

Maiga, M. A., and J. A. Pfister. (1988). Influence of Rumen Fill and Feeding Behavior of Steers. *Soc. Range Mgt. Abst. Papers* **41:**276.

Majak, W., and T. A. McAllister. (1999). Pasture Management Systems that Reduce the Incidence of Frothy Bloat in Cattle. *Soc. Range Mgt. Abst. Papers* **52:**45.

Majak, W., J. W. Hall, and W. P. McCaughey. (1995). Pasture Management Strategies for Reducing the Risk of Legume Bloat in Cattle. *J. Anim. Sci.* **73**(5):1493–1498.

Majak, W., L. Stroesser, J. W. Hall, D. A. Quinton, and H. E. Douwes. (1996). Seasonal Grazing of Columbia Milkvetch by Cattle on Rangelands in British Columbia. *J. Range Mgt.* **49**(3):223–227.

Majerus, M. E. (1992). High-Stature Grasses for Winter Grazing. *J. Soil & Water Cons.* **47**(3):224–225.

Malechek, J. C. (1966). Cattle Diets on Native and Seeded Ranges in the Ponderosa Pine Zone of Colorado. *USDA, For. Serv. Res. Note* **RM-77.** 12 pp.

Malechek, J. C. (1978). Animal Production on Rangelands. *In* "Symposium: Agriculture, Everybody's Business." Brigham Young University, Provo, UT. pp. 1–18.

Malechek, J. C. (1984). Impacts of Grazing Intensity and Specialized Grazing Systems on Livestock Response. *In* Natl. Res. Council/Natl. Acad. Sci. "Developing Strategies for Rangeland Management." Westview Press, Boulder, Colo., pp. 1129–1158.

Malechek, J. C. (1986). Nutritional Limits of Crested Wheatgrass for Range Livestock Procution. *In* Kendall L. Johnson (Ed.). "Crested Wheatgrass: Its Values, Problems, and Myths; Symposium Proceedings, Logan, Utah, October 3–7, 1983." Utah State Univ., Logan, Utah, pp. 267–272.

Malechek, J. C., and D. D. Dwyer. (1983). Short-Duration Grazing Doubles Your Livestock? *Utah Sci.* **44**(2):32–37.

Malechek, J. C., and C. L. Leinweber. (1972). Forage Selectivity by Goats of Lightly and Heavily Grazed Ranges. *J. Range Mgt.* **25**(2):105–111.

Malechek, J. C., and B. M. Smith. (1974). Range Cow Behavior and Energy Conservation. *Utah Sci.* **35**(3):103–104.

Malechek, J. C., and B. M. Smith. (1976). Behavior of Range Cows in Response to Winter Weather. *J. Range Mgt.* **29**(1):9–12.

Manley, W. A., R. H. Hart, M. J. Samuel, M. A. Smith *et al.* (1997). Vegetation, Cattle, and Economic Responses to Grazing Strategies and Pressures. *J. Range Mgt.* **50**(6):638–646.

Manners, G. D., L. F. James, K. E. Panter, J. A. Pfister *et al.* (1996). Tall Larkspur Poisoning in Cattle Grazing Mountain Rangelands. *Proc. Internat. Rangeland Cong.* **5**(1):342–343.

Manske, L. L., and T. J. Conlon. (1986). Complementary Rotation Grazing System in Western North Dakota. *N. Dak. Farm Res.* **44**(2):6–10.

Marlow, C. B. (1985). Controlling Riparian Zone Damage with Little Forage Loss. *Mon. AgRes.* **2**(3):1–7.

Marlow, C. B., and J. Aspie. (1988). Channel Alterations under Time Control: Riparian Deferment and Traditional Defered-Rotation Grazing Strategies. *Soc. Range Mgt. Abst. Papers* **41:**73.

Marlow, C. B., and T. M. Pogacnik. (1986). Cattle Feeding and Restng Patterns in a Foothills Riparian Zone. *J. Range Mgt.* **39**(3):212–217.

Marquis, D. A., and R. Brenneman. (1981). The Impact of Deer on Forest Vegetation in Pennsylvania. *USDA, For. Serv. Gen. Tech. Rep.* **NE-65.** 7 pp.

Marsh, H., K. F. Swingle, R. R. Woodward, G. F. Payne, E. E. Frahm, L. H. Jonson, and J. C. Hide. (1959). Nutrition of Cattle on an Eastern Montana Range as Related to Weather, Soil, and Forage. *Mon. Agric. Expt. Sta. Bul.* **549.** 91 pp.

Marten, G. C. (1978). The Animal-Plant Complex in Forage Palatability Phenomena. *J. Anim. Sci.* **46**(5):1470–1477.

Marten, G. C., and J. D. Donker. (1964). Selective Grazing Induced by Animal Excreta. I. Evidence of Occurrence and Superficial Remedy; II. Investigation of a Causal Theory. *J. Dairy Sci.* **47**(7):773–776, 47(8):871–874.

Marten, G. C., C. C. Sheaffer, and D. L. Wyse. (1987). Forage Nutritive Value and Palatability of Perennial Weeds. *Agron. J.* **79**(6):980–986.

Martin, J. A. (Al), and D. L. Huss. (1981). Goats Much Maligned but Necessary. *Rangelands* **3**(5):199–201.

Martin, S. C. (1975a). Ecology and Management of Southwestern Semidesert Grass-Shrub Ranges: The Status of Our Knowledge. *USDA, For. Serv. Res. Paper* **RM-156.** 39 pp.

Martin, S. C. (1975b). Stocking Strategies and Net Cattle Sales on Semi-Desert Range. *USDA, For. Serv. Res. Paper* **RM-146.** 10 pp.

Martin, S. C. (1978). Grazing Systems—What Can They Accomplish? *Rangeman's J.* **5**(1):14–16.

Martin, S. C. (1978). The Santa Rita Grazing System. *Proc. Intern. Rangeland Cong.* **1:**573–575.

Martin, S. C., and D. R. Cable. (1974). Managing Semidesert Grass-Shrub Ranges. *USDA Tech. Bul.* **1480.** 45 pp.

Martin, S. C., and K. E. Severson. (1988). Vegetation Response to the Santa Rita Grazing System. *J. Range Mgt.* **41**(4):291–295.

Martin, S. C., and D. E. Ward. (1970). Rotating Access to Water to Improve Semidesert Cattle Range Near Water. *J. Range Mgt.* **23**(1):22–26.

Martin, S. C., and D. E. Ward. (1973). Salt and Meal-Balt Help Distribute Cattle Use on Semidesert Range. *J. Range Mgt.* **26**(2):94–97.

Martin, S. C., and C. R. Whitfield. (1973). "Grazing Systems for Arizona Ranges." Ariz. Interagency Range Comm., Tucson. 36 pp.

Martz, F. A., M. F. Weiss, and R. L. Belyea. (1986). Comparison of Ruminal Capacity and Volatile Fatty Acids as Regulators of Forage Intake. *Forage and Grassland Conf.* **1986:**90–96.

Mastel, K. L., J. W. Stuth, and H. A. Aljoe. (1987). Influence of Community Structure on Grazing Behavior of Cattle. *Soc. Range Mgt. Abst. Papers* **40:**169.

Masters, L., S. Swanson, and W. Burkhardt. (1996). Riparian Grazing Management That Worked (Parts I and II). *Rangelands* **18**(5):192–200.

Masters, R. A., and C. J. Scifres. (1984). Forage quality Responses of Selected Grasses to Tebuthiuron. *J. Range Mgt.* **37**(1):83–87.

Matches, A. G., and J. C. Burns. (1985). Systems of Grazing Management. 1985. Chapter 57 *in* Maruice E. Heath, Robert F. Barnes, and Darrel S. Metcalfe. "Forages: The Science of Grassland Agriculture," Fourth Ed. Iowa State Univ. Press, Ames, IA.

Matthews, D. H., and W. C. Foote. (1987). Multispecies Grazing on Public Lands: Impact on the Livestock. *Amer. Soc. Anim. Sci., West. Sect. Proc.* **38:**87–88.

May, G. J., L. W. Van Tassell, M. A. Smith, and J. W. Waggoner. (1999). Delayed Calving in Wyoming. *Rangelands* **21**(3):8–12.

Mayland, H. F. (1986). Factors Affecting Yield and Nutritional Quality of Crested Wheatgrass. *In* K. L. Johnson (Ed.), "Crested Wheatgrass: Its Values, Problems, and Myths; Symposium Proceedings, Logan, Utah, October 3–7, 1983." Utah State Univ., Logan, Utah, pp. 215–266.

Mayland, H. F., and C. F. Shewmaker. (1999a). Optimize Forage Quality by Afternoon Harvesting. *USDA, ARS-NWISRL Note* **99–01.** 2 pp.

Mayland, H. F., and G. E. Shewmaker. (1999b). Plant Attributes That Affect Livestock Selection and Intake. *In* "Grazing Behavior of Livestock and Wildlife." *Idaho Forest, Wildl., and Range Expt. Sta. Bul;.* **70,** pp. 70–74.

Mbabaliye, T., J. L. Kingery, and J. C. Mosley. (1999). Early Summer vs. Late Summer Diets of Sheep Grazing in a Conifer Plantation. *Sheep & Goat Res. J.* **15**(1):34–40.

McArthur, J. A. B. (1969). Grazing Systems in Eastern Oregon. *Amer. Soc. Range Mgt. Abst. Papers* **22:**24–25.

McCall, T. C., R. D. Brown, and L. C. Bender. (1997). Comparison of Techniques for Determining the Nutritional Carrying Capacity for White-Tailed Deer. *J. Range Mgt.* **50**(1):33–38.

McCalla, G. R., II, W. H. Blackburn, and L. B. Merrill. (1984a). Effects of Livestock Grazing on Infiltration Rates, Edwards Plateau of Texas. *J. Range Mgt.* **37**(3):265–269.

McCalla, G. R., II, W. H. Blackburn, and L. B. Merrill. (1984b). Effects of Livestock Grazing on Sediment Production, Edwards Plateau of Texas. *J. Range Mgt.* **37**(4):291–294.

McCaughey, W. P., and R. D. H. Cohen. (1990). Effect of Mefluidide on Yield and Chemical Composition of Crested Wheatgrass in East-central Saskatchewan. *Can. J. Plant Sci.* **70**(4):1081–1090.

McCollum, F. T.. III, and T. G. Bidwell. (1994). Grazing Management on Rangeland for Beef Production. *Okla. Agric. Ext. Cir.* **P-916.** 13 pp.

McCollum, F. T., and M. L. Galyean. (1985). Cattle Grazing Blue Grama Rangeland. II. Seasonal Forage Intake and Digesta Kinetics. *J. Range Mgt.* **38**(6):543–546.

McCollum, F. T., III, and R. L. Gillen. (1998). Grazing Management Affects Nutrient Intake by Steers Grazing Tallgrass Prairie. *J. Range Mgt.* **51**(1):69–72.

McCollum, F. T., III, R. L. Gillen, and H. P. Jensen. (1990). Forage Intake and Nutrient Intake by Steers Grazing Tallgrass Prairie at Different Forage Allowances. *Okla. Agric. Expt. Sta. Misc. Pub.* **129,** pp. 236–239.

McCollum, F. T., III, M. D. Cravey, S. A. Gunter, J. M. Mieres *et al.* (1993). Forage Availability Affects Wheat Forage Intake by Stocker Cattle. *Okla. Agric. Expt. Sta.* **P-933,** pp. 278–281.

McCollum, F. T., III, R. L. Gillen, B. R. Karges, and M. E. Hodges. (1999). Stocker Cattle Response to Grazing Management in Tallgrass Prairie. *J. Range Mgt.* **52**(2):120–126.

McConnell, B. R., and J. G. Smith. (1977). Influence of Grazing on Age-Yield Interactions in Bitterbrush. *J. Range Mgt.* **30**(2):91–93.

McCorquodale, S. M., K. J. Raedeke, and R. D. Taber. (1986). Elk Habitat Use Patterns in the Shrub-Steppe of Washington. *J. Wildl. Mgt.* **50**(4):664–669.

McCorquodale, S. M., K. J. Raedeke, and R. D. Taber. (1989). Home Ranges of Elk in an Arid Environment. Northwest Sci. 63(1):29–34.

McDaniel, A. H., and C. B. Roark. (1956). Performance and Grazing Habits of Hereford and Aberdeen-Angus Cows and Calves on Improved Pastures as Related to Types of Shade. *J. Anim. Sci.* **15**(1):59–63.

McDaniel, K. C., and J. A. Tiedeman. (1981). Sheep Use on Mountain Winter Range in New Mexico. *J. Range Mgt.* **34**(2):102–104.

McDonald, P. M., G. O. Fiddler, and P. W. Meyer. (1996). Vegetation Trends in a Young Conifer Plantation after Grazing, Grubbing, and Chemical Release. *USDA, For. Serv. Res. Paper* **PSW-RP-228.** 17 pp.

McDougald, N. K., Jr., W. E. Frost, N. K. McDougald, and D. E. Jones. (1989). Relationship of Supplemental Feeding Location to Cattle Use of Riparian Areas on Hardood Rangeland. *Soc. Range Mgt. Abst. Papers* **42:**161.

McGinnies, W. J., W. A. Laycock, T. Tsuchiya, C. M. Yonker, and D. A. Edmunds. (1988). Variability within a Native Stand of Blue Grama. *J. Range Mgt.* **41**(5):391–395.

McGinty, A. (1985). Poisonous Plant Management. *Texas Agric. Ext. Bul.* **1499.** 11 pp.

McGinty, W. A., F. E. Smeins, and L. B. Merrill. (1979). Influence of Soil, Vegetation, and Grazing Management on Infiltration Rate and Sediment production of Edwards Plateau Rangeland. *J. Range Mgt.* **32**(1):33–37.

McIlvain, E. H. (1976). Seeded Grasses and Temporary Pastures as a Complement to Native Rangeland for Beef Cattle Production. *In* "Proceedings for Symposium on Integration of Resources for Beef Cattle Production, February 16–20, 1976." Soc. Range Mgt., Denver, CO. pp. 20–31.

McIlvain, E. H., and D. A. Savage. (1951). Eight-Year Comparisons of Continuous and Rotational Grazing on the Southern Plains Experimental Range. *J. Range Mgt.* **4**(1):42–47.

McIlvain, E. H., and M. C. Shoop. (1961). "Stocking Rates and Grazing Systems for Producing Forage and Beef on Sand Sage Rangelands—A 20-Year Study." USDA, Agric. Res. Serv., Southern Great Plains Field Sta., Woodward, OK. Mimeo. (Preliminary Assembly). 90 pp.

McIlvain, E. H., and M. C. Shoop. (1962a). Calves Are Sensitive to Range Conditions. *West. Livestock J.* **40**(19):83–85.

McIlvain, E. H., and M. C. Shoop. (1962b). Daily Versus Every-Third-Day Versus Weekly Feeding of Cottonseed Cake to Beef Steers on Winter Range. *J. Range Mgt.* **15**(3):143–145.

McIlvain, E. H., and M. C. Shoop. (1971a). Moving and Mixing Range Steers. *J. Range Mgt.* **24**(4):270–272.

McIlvain, E. H., and M. C. Shoop. (1971b). Shade for Improving Cattle Gains and Rangeland Use. *J. Range Mgt.* **24**(3):181–184.

McIlvain, E. H., and M. C. Shoop. (1973). "Use of Farmed Forage and Tame Pasture to Complement

Native Range." Paper presented at Great Plains Beef Symposium, Lincoln, NE, May 29–31, 1973. 16 pp.

McIlvain, E. H., A. L. Baker, W. R. Kneebone, D. H. Gates, W. F. Lagrone, and E. A. Tucker. (1955). "Nineteen-Year Summary of Range Improvement Studies, 1937–1955." U.S. Southern Great Plains Field Station, Woodward, OK. 41 pp.

McInnis, M. L., and M. Vavra. (1987). Dietary Relationships Among Feral Horses, Cattle, and Pronghorn in Southeastern Oregon. *J. Range Mgt.* **40**(1):60–66.

McKinney, E. (1997). It May Be Utilization, But Is It Management? *Rangelands* **19**(3):4–5.

McKown, C. D., J. W. Walker, J. W. Stuth, and R. K. Heitschmidt. (1991). Nutrient Intake of Cattle on Rotational and Continuous Grazing Treatments. *J. Range Mgt.* **44**(6):596–601.

McLean, A., and M. B. Clark. (1980). Grass, Trees, and Cattle on Clearcut-Logged Area. *J. Range Mgt.* **33**(3):213–217.

McLean, A., and W. Willms. (1977). Cattle Diets and Distribution on Spring-Fall and Summer Ranges Near Kamloops, British Columbia. *Can. J. Anim. Sci.* **57**(1):81–92.

McLeod, M. N., and D. J. Minson. (1988). Large Particle Breakdown by Cattle Eating Ryegrass and Alfalfa. *J. Anim. Sci.* **66**(4):992–999.

McLeod, S. R. (1997). Is the Concept of Carrying Capacity Useful in Variable Environments? *Oikos* **79**(3):529–542.

McMahan, C. A. (1964). Comparative Food Habits of Deer and Three Classes of Livestock. *J. Wildl. Mgt.* **28**(4):798–808.

McMahan, C. A., and C. W. Ramsey. (1965). Response of Deer and Livestock to Controlled Grazing in Central Texas. *J. Range Mgt.* **18**(1):1–7.

McMahon, L. R., W. Majak. T. A. McAllister, J. W. Hall *et al.* (1999). Effect of Sainfoin on *In Vitro* Digestion of Fresh Alfalfa and Bloat in Steers. *Can. J. Anim. Sci.* **79**(2):203–212.

McMurphy, W. E., R. L. Gillen, D. M. Engle, and F. T. McCollum. (1990). The Philosophical Difference between Range and Pasture Management in Oklahoma. *Rangelands* **12**(4):197–200.

Mecke, M. B. (1979). Texotics—Introduced Ungulates on Texas Rangelands. *Soc. Range Mgt. Abst. Papers* **32**:39.

Melton, A. A., and J. K. Riggs. (1964). Frequency of Feeding Protein Supplement to Range Cattle. *Texas Agric. Expt. Sta. Bul.* **1025**. 8 pp.

Memmott, K. L., V. J. Anderson, and S. B. Monsen. (1998). Seasonal Grazing Impact on Cryptogamic Crusts in a Cold Desert Ecosystem. *J. Range Mgt.* **51**(5):547–550.

Menke, J. W. (1987). Indicators for Production Changes. *In* Donald A. Jameson and Jerry Holechek. "Monitoring Animal Performance and Production Symposium Proceedings, February 12, 1987, Boise, Idaho." Society for Range Management, Denver, CO. pp 12–16.

Menke, J. W., and M. H. Trlica. (1981). Carbohydrate Reserve, Phenology, and Growth Cycles of Nine Colorado Range Species. *J. Range Mgt.* **34**(4):269–277.

Menke, J. W., and M. J. Trlica. (1983). Effects of Single and Sequential Defoliations on the Carbohydrate Reserves of Four Range Species. *J. Range Mgt.* **36**(1):70–74.

Merrill, L. B. (1954). A Variation of Deferred-Rotation Grazing for Use Under Southwest Range Conditions. *J. Range Mgt.* **7**(4):152–154.

Merrill, L. B. (1969). Grazing Systems in the Edwards Plateau of Texas. *Amer. Soc. Range Mgt.* **22**:22–23.

Merrill, L. B. (1972). Selectivity of Shrubs by Various Kinds of Animals. *USDA, For. Serv. Gen. Tech. Rep.* **INT-1**, p. 339–342.

Merrill, L. B. (1980). Considerations Necessary in Selecting and Developing a Grazing System; What Are the Alternatives? *In* K. C. McDaniel and C. Allison (Eds.), "Grazing Management Systems for Southwest Rangelands: A Symposium." Range Impr. Task Force, N. Mex. State Univ., Las Cruces, N. Mex., pp. 29–35.

Merrill, L. B., and J. L. Schuster. (1978). Grazing Management Practices Affect Livestock Losses from Poisonous Plants. *J. Range Mgt.* **31**(5):351–354.

Merrill, L. B., and C. A. Taylor. (1976). Take Note of the Versatile Goat. *Rangeman's J.* **3**(3):74–76.

Merrill, L. B., and V. A. Young. (1954). Results of Grazing Single Classes of Livestock in Combination with Several Classes when Stocking Rates Are Constant. *Tex. Agric. Expt. Sta. Prog. Rep.* **1726.** 7 pp.

Merrill, L. B., J. G. Teer, and O. C. Wallmo. (1957). Reaction of Deer Populations to Grazing Practices. *Tex. Agric. Prog.* **3**(5):10–12.

Merrill, L. B., G. W. Thomas, W. T. Hardy, E. B. Keng, C. A. Rechenthin, D. C. Langford, T. A. Booker, Nolan Johnson, James G. Teer, Rudy J. Pederson, and J. E. Tatum. (1957a). Livestock and Deer Ratios for Texas Range Lands. *Tex. Agric. Expt. Sta. Misc. Pub.* **221.** 9 pp.

Mertens, D. R. (1987). Predicting Intake and Digestibility Using Mathematical Models of Ruminal Function. *J. Anim. Sci.* **64**(5):1548–1558.

Milchunas, D. G., and W. K. Lauenroth. (1988). Plant Community Relationships along a Multi-Disturbance Gradient: When the Lack of Grazing Is a Disturbance. *Soc. Range Mgt. Abst. Papers* **41**:76.

Milchunas, D. G., J. R. Forwood, and W. K. Lauenroth. (1994). Productivity of Long-Term Grazing Treatments in Response to Seasonal Precipitation. *J. Range Mgt.* **47**(2):133–138.

Miller, R. (1983). Habitat Use of Feral Horses and Cattle in Wyoming's Red Desert. *J. Range Mgt.* **36**(2):195–199.

Miller, R. F. (1989). Range Readiness. *The Grazier* (Ore. Agric. Ext. Serv.), No. **263**, pp. 3–5.

Miller, R. F., and G. B. Donart. (1981). Response of *Muhlenbergia porteri* Scribn. to Season of Defoliation. *J. Range Mgt.* **34**(2):91–94.

Miller, R. F., T. J. Svejcar, and N. E. West. (1994). Implications of Livestock Grazing in the Intermountain Sagebrush Region: Plant Composition. *In* M. Vavra, W. Laycock, and R. D. Pieper (Eds.), "Ecological Implications of Livestock Herbivory in the West." Soc. Range Mgt., Denver, CO. pp. 101–146.

Milton, A. A., and J. K. Riggs. (1964). Frequency of Feeding Protein Supplement to Range Cattle. *Texas Agric. Expt. Sta. Bul.* **1025.** 8 pp.

Miner, J. R., J. C. Buckhouse, and J. A. Moore. (1992). Will a Water Trough Reduce the Amount of Time Hayfed Livestock Spend in the Stream (and Therefore Improve Water Quality)? *Rangelands* **14**(1):35–38.

Minson, D. J. (1990). "Forage in Ruminant Nutrition." Academic Press, San Diego, CA. 483 pp.

Minyard, J. A. (1961). Selenium Poisoning in Beef Cattle. *S. Dak. Farm & Home Res.* **12**(1):1–2.

Mirza, S. N., and F. D. Provenza. (1990). Preference of the Mother Affects Selection and Avoidance of Foods by Lambs Differing in Age. *Appl. Anim. Behav. Sci.* **28**(3):255–263.

Mitchell, G. J., and S. Smoliak. (1971). Pronghorn Antelope Range Characteristics and Food Habits in Alberta. *J. Wildl. Mgt.* **35**(2):238–250.

Mitchell, J. E. (1983). Analysis of Forage Production for Assessments and Appraisals. *USDA, For. Serv. Gen. Tech. Rep.* **RM-98.** 26 pp.

Mitchell, J. E., and R. T. Rodgers. (1985). Food Habits and Distribution of Cattle on a Forest and Pasture Range in Northern Idaho. *J. Range Mgt.* **38**(3):214–220.

Mitchell, J. E., D. M. Eissenstat, and A. J. Irby. (1982). Forest Grazing—An Opportunity for Diplomacy. *Rangelands* **4**(4):172–174.

Mitchell, R. B., R. A. Masters, S. S. Waller, and K. J. Moore. (1991). Forage Quality of Big Bluestem in Response to Time of Burning, Fertilization, and Atrazine. *Forage & Grassland Conf.* **1991**:273–276.

Moen, A. N. (1984). Ecological Efficiencies and Forage Intake of Free-Ranging Animals. *In* Natl. Res. Council/Natl. Acad. Sci. "Developing Strategies for Rangeland Management." Westview Press, Boulder, CO. pp. 215–273.

Molyneux, R. J., and M. H. Ralphs. (1992). Plant Toxins and Palatability to Herbivores. *J. Range Mgt.* **45**(1):13–18.

Monsen, S. B., J. F. Vallentine, and K. H. Hoopes. (1990). Seasonal Dietary Selection by Beef Cattle Grazing Mixed Crested Wheatgrass-Forage Kochia Pastures. *Amer. Soc. Anim. Sci., West. Sect. Proc.* **41**:300–303.

Mooers, B. H. M., and G. R. B. Mooers. (1988). Crested Wheatgrass Seedling Establishment Under Different Animal Impact Intensities in Northeastern Montana. *Soc. Range Mgt. Abst. Papers* **41**:79.

Moore, K. C. (1999). Economics of Management-Intensive Grazing in Pasture Utilization. *In* "Grazing Land Economics and Policy Symposium, Omaha, Neb." Soc. Range Mgt., Denver, CO. pp. 20–25.

Moore, K. J., and J. J. G. Jung. (1999). Impact of Lignin on Fiber Digestion: Anti-Quality Considerations. *Soc. Range Mgt. Abst. Papers* **52**:52.

Moorefield, J. G., and H. H. Hopkins. (1951). Grazing Habits of Cattle in a Mixed-Prairie Pasture. *J. Range Mgt.* **4**(3):151–157.

Morgantini, L. E., and R. J. Hudson. (1985). Changes in Diets of Wapiti During a Hunting Season. *J. Range Mgt.* **38**(1):77–79.

Morley, F. H. W. (1981). Management of Grazing Systems. *In* F. H. W. Morley (Ed.). "Grazing Animals." Elsevier Sci. Pub. Co., Amsterdam, Neth., pp. 379–400.

Morrical, D. G., H. E. Kiesling, and R. F. Beck. (1984). Performance of Goats Grazing Creosotebush Dominated Rangeland and Their Affect on Canopy Cover. *Amer. Soc. Anim. Sci., West. Sect. Proc.* **35**:23–25.

Morrow, R. E., D. J. Quinlan, M. S. Kerley, J. R. Gerrish, and F. A. Martz. (1994). Influence of Grazing System on Intake When Cows Graze Big Bluestem Pastures. *Amer. Forage & Grassland Council Proc.* **1994**:229–232.

Moser, L. E. (1986). How Do Plants Respond to Grazing. *In* Patrick E. Reece and James T. Nichols (Eds.). "Proceedings, The Ranch Management Symposium, November 5–7, 1986, North Platte, Nebraska." Univ. Neb., Agric. Ext. Serv., Lincoln, NE. pp. 19–26.

Mosley, J. C. (1994). Prescribed Sheep Grazing to Enhance Wildlife Habitat on North American Rangelands. *Sheep Res. J.* **10**(Spec. Issue):79–91.

Mosley, J. C. (1996). Prescribed Sheep Grazing to Suppress Cheatgrass: A Review. *Sheep & Goat Res. J.* **12**(2):74–81.

Mosley, J. C. (1999). Influence of Social Dominance on Habitat Selection by Free-Ranging Ungulates. *In* Grazing Behavior of Livestock and Wildlife. *Idaho Forest, Wildl., and Range Expt. Sta. Bull.* **70**, pp. 109–118.

Mosley, J. C., and B. E. Dahl. (1988). Forage Availability as a Rotation Indicator for Short-Duration Grazing. *Soc. Range Mgt. Abst. Papers* **41**:395.

Mosley, J. C., and B. E. Dahl. (1989). Evaluation of 7-Day Grazing Periods for Short-Duration Grazing on Tobosagrass Rangeland. *Applied Agric. Res.* **4**(4):229–234.

Mosley, J. C., and B. E. Dahl. (1990). Evaluation of an Herbage-Based Method for Adjusting Short-Duration Grazing Periods. *Applied Agric. Res.* **5**(2):142–148.

Mosley, J. C., E. L. Smith, and P. R. Ogden. (1990). "Seven Popular Myths about Livestock Grazing on Public Lands." Idaho Forest, Wildl., and Range Expt. Sta., Univ. of Idaho, Moscow, ID. 18 pp.

Mossman, A. S. (1975). International Game Ranching Programs. *J. Anim. Sci.* **40**(5):993–999.

Mowrey, D. P., A. G. Matches, A. P. Martinez, and R. L. Preston. (1986). Steer Performance with Continuous Grazing or First-Last Grazing on Buffalograss. *Forage and Grassland Conf.* **1986**:47–51.

Mueggler, W. F. (1950). Effects of Spring and Fall Grazing by Sheep on Vegetation of the Upper Snake River Plains. *J. Range Mgt.* **3**(4):308–315.

Mueggler, W. F. (1965). Cattle Distribution on Steep Slopes. *J. Range Mgt.* **18**(5):255–257.

Mueggler, W. F. (1972). Influence of Competition on the Response of Bluebunch Wheatgrass to Clipping. *J. Range Mgt.* **25**(2):88–92.

Mueggler, W. F. (1983). Variation in Production and Seasonal Development of Mountain Grasslands in Western Montana. *USDA, For. Serv. Res. Paper* **INT-316.** 16 pp.

Mundinger, J. G. (1976). Waterfowl Response to Rest-Rotation Grazing. *J. Wildl. Mgt.* **40**(1):60–68.

Mungall, E. C., and W. J. Sheffield. (1994). "Exotics on the Range: The Texas Example." Texas A&M Univ. Press, College Station. 265 pp.

Murphy, Alfred H., R. Merton Love, and Lester J. Berry. (1954). Improving Klamath Weed Ranges. *Calif. Agric. Ext. Serv. Cir.* **437.** 16 pp.

Murphy, J. S., and D. D. Briske. (1992). Regulation of Tillering by Apical Dominance: Chronology, Interpretive Value, and Current Perspectives. *J. Range Mgt.* **45**(5):419–429.

Mushimba, N. K. R., R. D. Pieper, J. D. Wallace, and M. L. Galyean. (1987). Influence of Watering Frequency on Forage Consumption and Steer Performance in Southeastern Kenya. *J. Range Mgt.* **40**(5):412–415.

Nagy, J. G. (1979). Wildlife Nutrition and the Sagebrush Ecosystem. *In* "The Sagebrush Ecosystem: A Symposium." Utah State Univ., Logan, Utah, pp. 164–168.

Nagy, J. G., G. Vidacs, and G. M. Ward. (1967). Previous Diet of Deer, Cattle, and Sheep and Ability to Digest Alfalfa Hay. *J. Wildl. Mgt.* **31**(3):443–447.

Nascimento, H., N. Gamougoun, R. Rice, R. P. Ogden, G. B. Ruyle, and E. L. Smith. (1987). Herbage Biomass and Green Biomass of Grazed and Nongrazed Patches in Lehman Lovegrass Rangeland as Affected by Season and Stocking Rate. *Soc. Range Mgt. Abst. Papers* **40**:176.

Nascimento, H., R. W. Rice, G. B. Ruyle, and P. R. Ogden. (1989). Nutrient Composition and Utilization of Grazed and Ungrazed Patches of Lehmann Lovegrass Rangeland. *Soc. Range Mgt. Abst. Papers* **42**:164.

National Research Council. (1981a). "Effect of Environment on Nutrient Requirements of Domestic Animals." Natl. Acad. Press, Washington, D.C. 152 pp.

National Research Council. (1981b). "Nutrient Requirements of Goats: Angora, Dairy, and Meat Goats in Temperate and Tropical Countries." Natl. Acad. Press, Washington, D.C. 91 pp.

National Research Council. (1981c). "Nutritional Energetics of Domestic Animals and Glossary of Energy Terms," 2nd Rev. Ed. Natl. Acad. Press, Washington, D.C. 54 pp.

National Research Council. (1984). "Nutrient Requirements of Beef Cattle," 6th Rev. Ed. Natl. Acad. Press, Washington, D.C. 100 pp.

National Research Council. (1985). "Nutrient Requirements of Sheep," 6th ed. Natl. Acad. Sci., Washington, D.C. 112 pp.

National Research Council. (1987). "Predicting Feed Intake of Food-Producing Animals." Natl. Academy Press, Washington, D.C. 85 pp.

National Research Council. (1989). "Nutrient Requirements of Domestic Animals. Number 6. Nutrient Requirements of Horses," 5th Ed. Natl. Acad. Sci., Washington, D.C. 128 pp.

National Research Council (NRC). (1996, 7th rev. ed.). "Nutrient Requirements of Beef Cattle." National Academy Press, Washington, D.C. 242 pp.

Natl. Res. Council and Agric. Can. (1982). "United States-Canadian Tables of Feed Composition." 3rd Rev. Natl. Acad. Press, Washington, D.C. 148 pp.

Neely, W. V. (1963). A Management Tool for Range Evaluation. *Nev. Agric. Ext. Bul.* **111**. 14 pp.

Nelle, S. (1992). Exotics—At Home on the Range in Texas. *Rangelands* **14**(2):77–80.

Nelson, A. B., and C. H. Herbel. (1966). Activities and Species Preferences of Hereford and Santa Gertrudis Range Cows. *Amer. Soc. Anim. Sci., West. Sect. Proc.* **17**:403–408.

Nelson, J. R. (1984). A Modeling Approach to Large Herbivore Competition. *In* Natl. Res. Council/ Natl. Acad. Sci. "Developing Strategies for Rangeland Management." Westview Press, Boulder, Colo., pp. 491–524.

Ngugi, R. K., F. C. Hinds, and J. Powell. (1995). Mountain Big Sagebrush Browse Decreases Dry Matter Intake, Digestibility, and Nutritive Quality of Sheep Diets. *J. Range Mgt.* **48**(6):487–492.

Nichol, A. A. (1938). Experimental Feeding of Deer. *Ariz. Agric. Expt. Sta. Tech. Bul.* **75**. 39 pp.

Nichols, J. T., and D. C. Clanton. (1985). Irrigated Pastures. Chapter 54. *In* M. E. Heath, R. F. Barnes, and D. S. Metcalfe (Eds.), "Forages: The Science of Grassland Agriculture," 4th ed. Iowa State Univ. Press, Ames, IA.

Nichols, J. T., and D. C. Clanton. (1987). Complementary Forages with Range. *In* "Proceedings, Grazing Livestock Nutrition Conference, July 23–24, 1987, Jackson, Wyoming." Univ. of Wyo., Laramie, WY. pp. 137–144.

Nichols, J. T., and G. Lesoing. (1980). Irrigated Pasture and Native Range. Neb. Agric. Ext. Cir. 80–182, p. 27–29.

Nielsen, D. B., F. J. Wagstaff, and D. Lytle. (1986). Big-Game on Private Range. *Rangelands* **8**(1):36–38.

Nolan, T., and J. Connolly. (1977). Mixed Stocking by Sheep and Steers—A Review. *Herb. Abst.* **47**(11):367–374.

Norris, J. B. (1968). Biological Control of Oak. *Amer. Soc. Range Mgt. Abst. Papers* **21**:29.

Northup, B. K., and J. T. Nichols. (1998). Relationships between Physical and Chemical Characteristics of 3 Sandhills Grasses. *J. Range Mgt.* **51**(3):353–360.

Norton, B. E. (1978). The Impact of Sheep Grazing on Long-Term Successional Trends in Salt Desert Shrub Vegetation of Southwestern Utah. *Proc. Intern. Rangeland Cong.* **1**:610–613.

Norton, B. E., and P. S. Johnson. (1983). Pattern of Defoliation by Cattle Grazing Crested Wheatgrass Pastures. *Proc. Intern. Grassland Cong.* **14**:462–464.

Norton, B. E., and P. S. Johnson. (1986). Impact of Grazing on Crested Wheatgrass in Relation to Plant Size. *In* K. L. Johnson (Ed.), "Crested Wheatgrass: Its Values, Problems, and Myths; Symposium Proceedings, Logan, Utah, October 3–7, 1983." Utah State Univ., Logan, Utah, pp. 275–279.

Norton, B. E., P. S. Johnson, and M. K. Owens. (1983). Increasing Grazing Efficiency on Crested Wheatgrass. *Utah Sci.* **43**(4):110–113.

Nowak, R. S., and M. M. Caldwell. (1986). Photosynthetic Characteristics of Crested Wheatgrass and Bluebunch Wheatgrass. *J. Range Mgt.* **39**(5):443–450.

Noy-Meir, I. (1996). The Spatial Dimensions of Plant-Herbivore Interactions. *Proc. Internat. Rangeland Cong.* **5**(2):152–154.

O'Reagain, P. J. (1993). Plant Structure and the Acceptability of Different Grasses to Sheep. *J. Range Mgt.* **46**(3):232–236.

Ocumpaugh, W. R., and A. G. Matches. (1977). Autumn-Winter Yield and Quality of Tall Fescue. *Agron. J.* **69**(4):639–643.

Ocumpaugh, W. R., S. Archer, and J. W. Stuth. (1996). Switchgrass Recruitment from Broadcast Seed vs. Seed Fed to Cattle. *J. Range Mgt.* **49**(4):368–371.

Oesterheld, M., and S. J. McNaughton. (1991). Effect of Stress and Time for Recovery on the Amount of Compensatory Growth after Grazing. *Oecologia* **85**(3):305–313.

Ogden, P. R., R. W. Rice, E. L. Smith, and G. B. Ruyle. (1987). Cattle Performance as Influenced by Stocking Rate and Seasonal Use of Lehmann Lovegrass. *Soc. Range Mgt. Abst. Papers* **40**:193.

Olowolayemo, S. O., J. C. Reeves, N. R. Martin, Jr., R. R. Harris, and D. I. Bransby. (1992). Economics and Efficiency Factors for Optimal Stocking Rate. *Highlights Agric. Res.* (Ala.) **39**(2):12.

Olsen, J. D., and M. H. Ralphs. (1986). Feed Aversion Induced by Intraruminal Infusion with Larkspur Extract in Cattle. *Amer. J. Vet. Res.* **47**(8):1829–1833.

Olson, B. E. (1999). Manipulating Diet Selection to Control Weeds. *In* Grazing Behavior of Livestock and Wildlife. *Idaho Forest, Wildl., and Range Expt. Sta. Bull.* **70,** pp. 36–44.

Olson, B. E., and J. R. Lacey. (1994). Sheep: A Method for Controlling Rangeland Weeds. *Sheep Res. J.* **10**(Spec. Issue):105–112.

Olson, B. E., and J. H. Richards. (1987). Tiller Replacement in *Agropyron desertorum* Following Grazing: Implications for Bunchgrass Integrity. *Soc. Range Mgt. Abst. Papers* **40**:177.

Olson, B. E., and J. H. Richards. (1988). Timing and Sources of Regrowth in Crested Wheatgrass Following Grazing. *Soc. Range Mgt. Abst. Papers* **41**:249.

Olson, B. E., and J. H. Richards. (1989). Grazing Effects on Crested Wheatgrass Growth and Replacement in Central Utah. *Utah Agric. Expt. Sta. Bul.* **516.** 34 pp.

Olson, B. E., and R. T. Wallander. (1998). Effect of Sheep Grazing on a Leafy Spurge-Infested Idaho Fescue Community. *J. Range Mgt.* **51**(2):247–252.

Olson, B. E., R. T. Wallander, V. M. Thomas, and R. W. Kott. (1996). Effect of Previous Experience on Sheep Grazing Leafy Spurge. *Applied Anim. Beh. Sci.* **50**(2):161–176.

Olson, B. E., R. T. Wallander, and J. R. Lacey. (1997). Effects of Sheep Grazing on Spotted Knapweed Infested Idaho Fescue Community. *J. Range Mgt.* **50**(40):386–390.

Olson, K., M. George, and A. Murphy. (1989). Estimating Weather and Forage Relationships. *Univ. Calif. Range Sci. Rep.* **22.** 15 pp.

Olson, K. C., and J. L. Launchbaugh. (1989). Steer Growth and Nutritional Response to Intensive-Early Stocking of Shortgrass Rangeland. *Soc. Range Mgt. Abst. Papers* **42**:247.

Olson, K. C., and J. C. Malechek. (1988). Heifer Nutrition and Growth on Short Duration Grazed Crested Wheatgrass. *J. Range Mgt.* **41**(3):259–263.

Olson, K. C., R. L. Senft, and J. C. Malechek. (1986). A Predictive Model of Cattle Ingestive Behavior in Response to Sward Characteristics. *Amer. Soc. Anim. Sci. West. Sect. Proc.* **37**:259–262.

Olson, K. C., G. B. Rouse, and J. C. Malechek. (1989). Cattle Nutrition and Grazing Behavior during Short-Duration Grazing Periods on Crested Wheatgrass. *J. Range Mgt.* 42(2):153–158.

Olson, K. C., J. R. Brethour, and J. L. Launchbaugh. (1993). Shortgrass Range Vegetation and Steer Growth Response to Intensive-Early Stocking. *J. Range Mgt.* 46(2):127–132.

Olson, K. C., R. D. Wiedmeier, J. E. Bowns, and R. L. Hurst. (1999). Livestock Response to Multi-species and Deferred-Rotation Grazing on Forested Rangeland. *J. Range Mgt.* 52(5):462–470.

Olson, R., and A. M. Lewis. (1994). Winter Big Game Feeding: An Undesirable Wildlife Management Practice. *Wyo. Agric. Ext. Bull.* **1003.** 12 pp.

Orr, H. K. (1975). Recovery From Soil Compaction on Bluegrass Range in the Black Hills. *Trans. Amer. Soc. Agric. Eng.* **18:**1076–1081.

Ortega, I. M., S. Soltero-Gardea, F. C. Bryant, and D. L. Drawe. (1997). Evaluating Grazing Strategies for Cattle: Deer and Cattle Partitioning. *J. Range Mgt.* 50(6):622–630.

Ortega-Reyes, L., and F. D. Provenza. (1993). Experience with Blackbrush Affects Ingestion of Shrub Live Oak by Goats. *J. Anim. Sci.* 71(2):380–383.

Osko, T. J., R. T. Hardin, and B. A. Young. (1993). Chemical Repellents to Reduce Grazing Intensity on Reclaimed Sites. *J. Range Mgt.* 46(5):383–386.

Osuji, P. O. (1974). The Physiology of Eating and the Energy Expenditure of the Ruminant at Pasture. *J. Range Mgt.* 27(6):437–443.

Otsyina, R., C. M. Mckell, and G. Van Epps. (1982). Use of Range Shrubs to Meet Nutrient Require-ments of Sheep Grazing on Crested Wheatgrass During Fall and Early Winter. *J. Range Mgt.* **35**(6):751–753.

Owen-Smith, N., and S. M. Cooper. (1987). Classifying African Savanna Trees and Shrubs in Terms of Their Palatability for Browsing Ungulates. *USDA, For. Serv. Gen. Tech. Rep.* **INT-222,** pp. 43–47.

Owen-Smith, N., and P. Novellie. (1982). What Should a Clever Ungulate Eat? *Amer. Nat.* **119**(2):151–178.

Owens, F. N. (1988). Ruminal Fermentation. *In* D. C. Church (Ed.), "The Ruminant Animal: Diges-tive Physiology and Nutrition." Prentice Hall, Englewood Cliffs, NJ. pp. 145–171.

Owens, L. B., W. M. Edwards, and R. W. Van Keuren. (1997). Runoff and Sediment Losses Resulting from Winter Feeding on Pastures. *J. Soil & Water Cons.* 52(3):194–197.

Owens, M. K., K. L. Launchbaugh, and J. W. Holloway. (1991). Pasture Characteristics Affecting Spa-tial Distribution of Utilization by Cattle in Mixed Brush Communities. *J. Range Mgt.* 44(2):118–123.

Owens, M. K., D. E. Spalinger, and L. A. Newton. (1992). Foraging Behavior of Goats in Mixed Brush Communities of South Texas. *In* "Proceedings of the International Conference on Meat Goat Pro-duction, Management, and Harvesting." Texas Agric. Ext. Serv., Corpus Christi, TX, pp. 112–122.

Owensby, C. E., E. F. Smith, and K. L. Anderson. (1973). Deferred-Rotation Grazing with Steers in the Kansas Flint Hills. *J. Range Mgt.* 26(6):393–395.

Owensby, C. E., E. F. Smith, and J. R. Rains. (1977). Carbohydrate and Nitrogen Reserve Cycles for Continuous, Seasonlong and Intensive-Early Stocked Flint Hills Bluestem Range. *J. Range Mgt.* **30**(4):258–260.

Owensby, C. E., R. Cochran, and E. F. Smith. (1988). Stocking Rate Effects on Intensive-Early Stocked Flint Hills Bluestem Range. *J. Range Mgt.* 41(6):483–487.

Paisley, S. I., C. J. Ackerman, H. T. Puvis II, and G. W. Horn. (1998). Wheat Pasture Intake by Early Weaned Calves. *Okla. Agric. Expt. Sta.* **P-965,** pp. 202–207.

Parsons, A. J., and I. R. Johnson. (1986). The Physiology of Grass Growth Under Grazing. *In* J. Frame (Ed.). "Grazing." British Grassland Soc., Hurley, Berks, Eng., pp. 3–13.

Parsons, A. J., and P. D. Penning. (1988). The Effect of the Duration of Regrowth on Photosynthesis, Leaf Death, and the Average Rate of Growth in a Rotationally Grazed Sward. *Grass and Forage Sci.* 43(1):15–27.

Pasha, T. N., E. C. Prigge, R. W. Russell, and W. B. Bryan. (1994). Influence of Moisture Content of Forage Diets on Intake and Digestion by Sheep. *J. Anim. Sci.* 72(9):2455–2463.

Patton, W. W. (1971). "An Analysis of Cattle Grazing on Steep Slopes." M.S. Thesis, Brigham Young Univ., Provo, UT. 42 pp.

Payne, N. F., and F. C. Bryant. (1994). Rangeland Management Techniques: Controlled Grazing and Prescribed Burning. *In* N. F. Payne and F. C. Bryant (Eds.), "Techniques for Wildlife Habitat Management of Uplands." McGraw-Hill, New York, pp. 347–381.

Pearson, H. A. (1975). Herbage Disappearance and Grazing Capacity Determination of Southern Pine Bluestem Range. *J. Range Mgt.* **28**(1):71–73.

Pearson, H. A., and D. A. Rollins. (1986). Supplemental Winter Pasture for Southern Pine Native Range. *Forage and Grassland Conf.* **1986**:258–263.

Pearson, H. A., and L. B. Whitaker. (1972). Thrice-Weekly Supplementation Adequate for Cows on Pine-Bluestem Range. *J. Range Mgt.* **25**(4):315–316.

Pearson, H. A., and L. B. Whitaker. (1974). Forage and Cattle Responses to Different Grazing Intensities on Southern Pine Ridge. *J. Range Mgt.* **27**(6):444–446.

Pearson, H. A., L. B. Whitaker, and V. L. Duvall. (1971). Slash Pine Regeneration Under Regulated Grazing. *J. For.* **69**(10):744–746.

Pearson, H. A., V. C. Baldwin, and J. P. Barnett. (1990a). Cattle Grazing and Pine Survival and Growth in Subterranean Clover Pasture. *Agroforestry Syst.* **10**:161–168.

Pearson, H. A., T. E. Prince, Jr., and C. M. Todd, Jr. (1990b). Virginia Pines and Cattle Grazing—An Agroforestry Opportunity. *Southern J. Applied For.* **14**(2):55–59.

Pechanec, J. F., and G. Stewart. (1949). Grazing Spring-Fall Sheep Ranges of Southern Idaho. *USDA Cir.* **808**. 34 pp.

Peden, D. G., G. M. Van Dyne, R. W. Rice, and R. M. Hansen. (1974). The Trophic Ecology of *Bison bison* L. on Shortgrass Plains. *J. Applied Ecol.* **11**(2):489–497.

Pedersen, J. F., and D. A. Sleper. (1988). Considerations in Breeding Endophyte-Free Tall Fescue Forage Cultivars. *J. Prod. Agric.* **1**(2):127–132.

Pederson, J. C., and B. L. Welch. (1982). Effects of Monoterpenoid Exposure on Ability of Rumen Inocula to Digest a Set of Forages. *J. Range Mgt.* **35**(4):500–502.

Peek, J. M., and P. R. Krausman. (1996). Grazing and Mule Deer. *In* "Rangeland Wildlife." Soc. Range Mgt., Denver, CO. Chapter 11.

Perrier, G. K. (1996). The Animal Unit as an Ecological Concept. *Rangelands* **18**(1):30–31.

Petersen, R. G., W. W. Woodhouse, Jr., and H. Lucas. (1956). The Distribution of Excreta by Freely Grazing Cattle and Its Effects on Pasture Fertility: II. Effect of Returned Excreta on the Residual Concentration of Some Fertilizer Elements. *Agron. J.* **48**(10):444–449.

Petersen, R. G., H. L. Lucas, and G. O. Mott. (1965). Relationship Between Rate of Stocking and Per Animal and Per Acre Performance on Pasture. *Agron. J.* **57**(1):27–30.

Peterson, P. R., and J. R. Gerrish. (1995). Grazing Management Affects Manure Distribution by Beef Cattle. *Amer. Forage & Grassland Council Proc.* **1995**:170–174a.

Pfister, J. A. (1999). Behavioral Strategies for Coping with Poisonous Plants. *In* "Grazing Behavior of Livestock and Wildlife." *Idaho Forest, Wildl., and Range Expt. Sta. Bull.* **70,** pp. 45–59.

Pfister, J. A., and D. C. Adams. (1993). Factors Influencing Pine Needle Consumption by Grazing Cattle During Winter. *J. Range Mgt.* **46**(5):394–398.

Pfister, J. A., and D. R. Gardner. (1999). Consumption of Low Larkspur (*Delphinium nuttallianum*) by Cattle. *J. Range Mgt.* **52**(4):378–383.

Pfister, J. A., and K. W. Price. (1996). Lack of Maternal Influence on Lamb Consumption of Locoweed (*Oxytropis sericea*). *J. Anim. Sci.* **74**(2):340–344.

Pfister, J. A., G. B. Donart, R. D. Pieper, J. D. Wallace, and E. E. Parker. (1984). Cattle Diets Under Continuous and Four-Pasture, One-Herd Grazing Systems in Southcentral New Mexico. *J. Range Mgt.* **37**(1):50–54.

Pfister, J. A., J. Atto, and E. Nolte. (1987). Effects of Water Deprivation on Water Intake and Weight Responses of Sheep and Goats in the Northern Peruvian Desert. *Soc. Range Mgt. Abst. Papers* **40**:240.

Pfister, J. A., G. D. Manners, M. H. Ralphs, Z. X. Hong, and M. A. Lane. (1988). Effects of Phenology, Site, and Rumen Fill on Larkspur Consumption by Cattle. *J. Range Mgt.* **41**(6):509–514.

Pfister, J. A., M. H. Ralphs, G. D. Manners, D. R. Gardner *et al.* (1997a). Early Season Grazing by Cattle of Tall Larkspur (*Delphinium spp.*) Infested Rangeland. *J. Range Mgt.* **50**(4):391–398.

Pfister, J. A., D. R. Gardner, and K. W. Price. (1997b). Grazing Risk on Tall Larkspur-Infested Ranges. *Rangelands* **19**(5):12–15.

Pfister, J. A., K. E. Panter, and D. R. Gardner. (1998). Pine Needle Consumption by Cattle During Winter in South Dakota. *J. Range Mgt.* **51**(5):551–556.

Pfister, J., D. R. Gardner, B. L. Stegelmeier, K. E. Panter, and M. H. Ralphs. (1999). Alkaloids as Anti-Quality Factors in Plants on Western U.S. Rangelands. *Soc. Range Mgt. Abst Papers* **52**:63.

Phillips, C. J. C., J. K. Margerison, S. Azazi, A. G. Chamberlain, and H. Omed. (1991). The Effect of Adding Surface Water to Herbage on Its Digestion by Ruminants. *Grass & Forage Sci.* **46**(3):333–338.

Phillips, T. A. (1965). The Influence of Slope Gradient, Distance From Water, and Other Factors on Livestock Distribution on National Forest Cattle Allotments of the Intermountain Region. *USDA, For. Serv., Intermtn. Region Range Impr. Notes* **10**:9–19.

Pickford, G. D., and E. H. Reid. (1948). Forage Utilization on Summer Cattle Ranges in Eastern Oregon. *USDA Cir.* **796.** 27 pp.

Pieper, R. D. (1980). Impacts of Grazing Systems on Livestock. *In* K. C. McDaniel and C. Allison (Eds.), "Grazing Management Systems for Southwest Rangelands: A Symposium." Range Impr. Task Force, N. Mex. State Univ., Las Cruces, N. Mex., pp. 133–151.

Pieper, R. D. (1983). Consumption Rates of Desert Grassland Herbivores. *Proc. Intern. Grassland Cong.* **14**:465–467.

Pieper, R. D. (1994). Ecological Implications of Livestock Grazing. *In* M. Vavra, W. Laycock, and R. D. Pieper (Eds.), "Ecological Implications of Livestock Herbivory in the West." Soc. Range Mgt., Denver, CO. pp. 177–211.

Pieper, R. D., and R. F. Beck. (1980). Importance of Forbs on Southwestern Ranges. *Rangelands* **2**(1):35–36.

Pieper, R. D., and G. B. Donart. (1975). Drought on the Range: Drought and Southwestern Range Vegetation. *Rangeman's J.* **2**(6):176–178.

Pieper, R. D., and G. B. Donart. (1978). Response of Fourwing Saltbush to Periods of Protection. *J. Range Mgt.* **31**(4):314–315.

Pieper, R. D., and G. B. Donart. (1989). Vegetational Response to Grazing Management in South-Central New Mexico. *Soc. Range Mgt. Abst. Papers* **42**:200.

Pieper, R. D., and R. K. Heitschmidt. (1988). Is Short-Duration Grazing the Answer? *J. Soil & Water Cons.* **43**(2):133–137.

Pieper, R., C. W. Cook, and L. E. Harris. (1959). Effect of Intensity of Grazing upon Nutritive Content of the Diet. *J. Anim. Sci.* **18**(3):1031–1037.

Pieper, R. D., C. H. Herbel, D. D. Swyer, and R. E. Banner. (1974). Management Implications of Herbage Weight Changes on Native Rangeland. *J. Soil and Water Cons.* **29**(5):227–229.

Pieper, R. D., G. B. Donart, E. E. Parker, and J. D. Wallace. (1978). Livestock and Vegetational Response to Continuous and 4-Pasture, 1-Herd Grazing Systems in New Mexico. *Proc. Inter. Rangeland Cong.* **1**:560–562.

Pieper, R. D., E. E. Parker, G. B. Donart, J. D. Wallace, and J. D. Wright. (1991). Cattle and Vegetation Response to Four-Pasture Rotation and Continuous Grazing Systems. *N. Mex. Agric. Expt. Sta.* **Bull.** 756. 23 pp.

Pierson, F. B., and D. L. Scarnecchia. (1987). Defoliation of Intermediate Wheatgrass Under Seasonal and Short-Duration Grazing. *J. Range Mgt.* **40**(3):228–232.

Pinchak, W. E., S. K. Canon, R. K. Heitschmidt, and S. L. Dowhower. (1988). Longterm Effects of Heavy Grazing on Nutrient Intake Dynamics. *Soc. Range. Mgt. Abst. Papers* **41**:148.

Pinchak, W. E., S. K. Canon, R. K. Heitschmidt, and S. L. Dowhower. (1990). Effect of Long-Term, Yearlong Grazing at Moderate and Heavy Rates of Stocking on Diet Selection and Forage Intake Dynamics. *J. Range Mgt.* **43**(4):304–309.

Pinchak, W. E., M. A. Smith, R. H. Hart, and J. W. Waggoner, Jr. (1991). Beef Cattle Distribution Patterns on Foothill Range. *J. Range Mgt.* **44**(3):267–275.

Pitt, M. D. (1986). Assessment of Spring Defoliation to Improve Fall Forage Quality of Bluebunch Wheatgrass (*Agropyron spicatum*). *J. Range Mgt.* **39**(2):175–181.

Pitt, M. D., and H. F. Heady. (1979). The Effects of Grazing Intensity on Annual Vegetation. *J. Range Mgt.* **32**(2):109–114.

Platou, K. A., and P. T. Tueller. (1985). Evolutionary Implications for Grazing Management Systems. *Rangelands* **7**(2):57–61.

Platts, W. S. (1981a). Influence of Forest and Rangeland Management on Anadromous Fish Habitat in Western North America: Effects of Livestock Grazing. *USDA, For. Serv. Gen. Tech. Rep.* **PNW-124.** 25 pp.

Platts, W. S. (1981b). Sheep and Streams. *Rangelands* **3**(4):158–160.

Platts, W. S. (1986). Managing Fish and Livstock on Idaho Rangelands. *Rangelands* **8**(5):213–216.

Platts, W. S., and R. L. Nelson. (1985a). Streamside and Upland Vegetation Use by Cattle. *Rangelands* **7**(1):5–7.

Platts, W. S., and R. L. Nelson. (1985b). Will the Riparian Pasture Build Good Streams? *Rangelands* **7**(1):7–10.

Plice, M. J. (1952). Sugar Versus the Intuitive Choice of Foods by Livestock. *J. Range Mgt.* **5**(2):69–75.

Pluhar, J. J., R. W. Knight, and R. K. Heitschmidt. (1987). Infiltration Rates and Sediment Production as Influenced by Grazing Systems in the Texas Rolling Plains. *J. Range Mgt.* **40**(3):240–243.

Plumb, G. E., and J. L. Dodd. (1994). Foraging Ecology of Bison and Cattle. *Rangelands* **16**(3):107–109.

Plumb, G. E., L. J. Krysl, M. E. Hebbert, M. A. Smith, and J. W. Waggoner. (1984). Horses and Cattle Grazing on the Wyoming Red Desert, Ill. *J. Range Mgt.* **37**(2):130–132.

Plummer, A. P. (1975). Morphogenesis and Management of Woody Perennials in the United States. *USDA Misc. Pub.* **1271,** p. 72–80.

Pond, K. R., W. C. Ellis, and D. E. Akin. (1984). Ingestive Mastication and Fragmentation of Forages. *J. Anim. Sci.* **58**(6):1567–1574.

Pond, K. R., W. C. Ellis, C. E. Lascano, D. E. Akin, and Richard B. Russell. (1987). Fragmentation and Flow of Grazed Coastal Bermudagrass through the Digestive Tract of Cattle. *J. Anim. Sci.* **65**(2):609–618.

Popay, J., and R. Field. (1996) Grazing Animals as Weed Control Agents. *Weed Tech.* **10**(1):217–231.

Pope, C. A., III, and G. L. McBryde. (1984). Optimal Stocking of Rangeland for Livestock Production Within a Dynamic Framework. *West. J. Agric. Econ.* **9**(1):160–169.

Pope, L. S., A. B. Nelson, and W. D. Campbell. (1963). Feeding Protein Supplements to Range Beef Cows at 2, 4, and 6-Day Intervals. *Okla. Agric. Expt. Sta. Misc. Pub.* **70,** p. 49–51.

Popp, J. D., W. P. McCaughey, and R. D. H. Cohen. (1996). Effect of Grazing System, Stocking Rate, and Season of Use on Ingestive Behaviour of Stocker Cattle Grazing Alfalfa-Grass Pastures. *Amer. Soc. Anim. Sci., West. Sect. Proc.* **47:**170–173.

Popp, J. D., W. P. McCaughey, and R. D. H. Cohen. (1997). Grazing System and Stocking Rate Effects on the Productivity, Botanical Composition, and Soil Surface Characteristics of Alfalfa-Grass Pastures. *Can. J. Anim. Sci.* **77**(4):669–676.

Porter, R. M., and S. R. Skaggs. (1958). Forage and Milk Yields From Alfalfa Under Three Different Harvesting Systems. *N. Mex. Agric. Expt. Sta. Bul.* **421.** 9 pp.

Powell, J., G. Godbolt, and W. G. Hepworth. (1986). Questions About Livestock-Big Game Relations. *Rangelands* **8**(6):281–283.

Prasad, N. L. N. S., and Fred S. Guthery. (1986). Wildlife Use of Livestock Water Under Short-Duration and Continuous Grazing. *Wildl. Soc. Bul.* **14**(4):450–454.

Pratt, A. D., and R. R. Davis. (1962). Rotational Grazing and Green Chopping Compared. *Ohio Farm and Home Res.* **47**(3):38–39, 47.

Prescott, M. L., K. Olson-Rutz, K. M. Havstad, E. L. Ayers, and M. K. Petersen. (1989). Grazing Behavior Responses of Free-Ranging Beef Cows to Fluctuating Thermal Environments. *Amer. Soc. Anim. Sci., West. Sect. Proc.* **40:**458–460.

Prescott, M. L., K. M. Havstad, K. M. Olson-Rutz, E. L. Ayers, and M. K. Petersen. (1994). Grazing Behavior of Free-Ranging Beef Cows to Initial and Prolonged Exposure to Fluctuating Thermal Environments. *Applied Anim. Beh. Sci.* **39**(2):103–113.

Price, D. L., G. B. Donart, and G. M. Southward. (1989). Growth Dynamics of Fourwing Saltbush as Affected by Different Grazing Management Systems. *J. Range Mgt.* **42**(2):158–162.

Pritz, R. K., K. L. Launchbaugh, and C. A. Taylor, Jr. (1997). Effects of Breed and Dietary Experience on Juniper Consumption by Goats. *J. Range Mgt.* **50**(6):600–606.

Provenza, F. D. (1978). Getting the Most Out of Blackbrush. *Utah Sci.* **39**(4):144–146.

Provenza, F. D. (1990). Unraveling the Origins of Animal Appetites. *Utah Sci.* **51**(4):136–143.

Provenza, F. D. (1995). Post-ingestive Feedback as an Elementary Determinant of Food Preference and Intake in Ruminants. *J. Range Mgt.* **48**(1):2–17.

Provenza, F. D. (1996a). Acquired Aversions as the Basis for Varied Diets of Ruminants Foraging on Rangelands. *J. Anim. Sci.* **74**(8):2010–2020.

Provenza, F. D. (1996b). A Functional Explanation for Palatability. *Proc. Internat. Rangeland Cong.* **5**(2):123–125.

Provenza, F. D., and D. F. Balph. (1987). Diet Training, Behavioral Concepts, and Management Objectives. *USDA, For. Serv. Gen. Tech. Rep.* **INT-222,** pp. 132–136.

Provenza, F. D., and D. F. Balph. (1988). Development of Dietary Choice in Livestock on Rangelands and Its Implications for Management. *J. Anim. Sci.* **66**(9):2356–2368.

Provenza, F. D., and K. L. Launchbaugh. (1999a). Foraging on the Edge of Chaos. *In* "Grazing Behavior of Livestock and Wildlife." *Idaho Forest, Wildl., and Range Expt. Sta. Bull.* **70,** pp. 1–12.

Provenza, F. D., and K. L. Launchbaugh. (1999b). Herbivore Responses to Anti-Quality Factors in Forages. *Soc. Range Mgt. Abst. Papers* **52**:65.

Provenza, F. D., and J. C. Malechek. (1984). Diet Selection by Domestic Goats in Relation to Blackbrush Twig Chemistry. *J. Appl. Ecol.* **21**(3):831–841.

Provenza, F. D., J. E. Bowns, P. J. Urness, J. C. Malechek, and J. E. Butcher. (1983a). Biological Manipulation of Blackbrush by Goat Browsing. *J. Range Mgt.* **36**(4):513–518.

Provenza, F. D., J. C. Malechek, P. J. Urness, and J. E. Bowns. (1983b). Some Factors Affecting Twig Growth in Blackbrush. *J. Range Mgt.* **36**(4):518–520.

Provenza, F. D., D. Balph, J. Olsen, D. Dwyer, and M. Ralphs. (1987). Animal Grazing Behavior with Respect to Livestock Poisoning by Plants. *Soc. Range Mgt. Abst. Papers* **40**:110.

Provenza, F. D., D. F. Balph, J. D. Olsen, D. D. Dwyer, M. H. Ralphs, and J. A. Pfister. (1988). Toward Understanding the Behavioral Responses of Livestock to Poisonous Plants. *In* L. F. James, M. H. Ralphs, and D. B. Nielson (Eds.), "The Ecology and Economic Impact of Poisonous Plants on Livestock Production." Westview Press, Boulder, Colo., pp. 407–424.

Public Land Law Review Comm. (1970). "One Third of the Nation's Land." U.S. Govt. Print. Office, Washington, D.C. 342 pp.

Quigley, T. M. (1987). Short-Duration Grazing: An Economic Perspective. *Rangelands* **9**(4):173–175.

Quigley, T. M., J. M. Skovlin, and J. P. Workman. (1984). An Economic Analysis of Two Systems and Three Levels of Grazing on Ponderosa Pine-Bunchgrass Range. *J. Range Mgt.* **37**(4):309–312.

Quinn, J. A., and D. F. Hervey. (1970). Trampling Losses and Travel by Cattle on Sandhills Range. *J. Range Mgt.* **23**(1):50–55.

Radwan, M. A., and G. L. Crouch. (1978). Selected Chemical Constituents and Deer Browsing Preference of Douglas Fir. J. Chem. Ecol. 4(6):675–683.

Rafique, S., D. P. Arthun, M. L. Galyean, J. L. Holechek, and J. D. Wallace. (1988). Effects of Forb and Shrub Diets on Ruminant Nitrogen Balance. I. Sheep Studies. *Amer. Soc. Anim. Sci., West. Sect. Proc.* **39**:200–203.

Ragotzkie, K. E., and J. A. Bailey. (1991). Desert Mule Deer Use of Grazed and Ungrazed Habitats. *J. Range Mgt.* **44**(5):487–490.

Raguse, C. A., K. L. Taggard, J. L. Hull, C. A. Daley, and J. M. Connor. (1989). Short-Duration Grazing on Irrigated Pasture. *Calif. Agric.* **43**(4):4–7.

Raleigh, R. J. (1970). Symposium on Pasture Methods for Maximum Production in Beef Cattle: Manipulations of Both Livestock and Forage Management to Give Optimum Production. *J. Anim. Sci.* **30**(1):108–114.

Raleigh, R. J., and J. D. Wallace. (1965). Nutritive Value of Range Forage and its Effect on Animal Performance. *Ore. Agric. Expt. Sta. Spec. Rep.* **189.** 6 pp.

Ralphs, M. H. (1985). Poisonous Plants: The Snakeweeds. *Rangelands* **7**(2):63–65.

Ralphs, M. H. (1987). Cattle Grazing White Locoweed: Influence of Grazing Pressure and Palatability Associated with Phenological Growth Stage. *J. Range Mgt.* **40**(4):330–332.

Ralphs, M. H. (1992). Continued Food Aversion: Training Livestock To Avoid Eating Poisonous Plants. *J. Range Mgt.* **45**(1):46–51.

Ralphs, M. H., and J. D. Olsen. (1987). Alkaloids and Palatability of Poisonous Plants. *USDA, For. Serv. Gen. Tech. Rep.* **INT-222,** pp. 78–83.

Ralphs, M. H., and J. D. Olsen. (1989). Sheep Grazing Larkspur to Reduce Risk of Poisoning in Cattle. *Soc. Range Mgt. Abst. Papers* **42:**110.

Ralphs, M. H., and J. D. Olsen. (1990). Adverse Influence of Social Facilitation and Learning Context in Training Cattle to Avoid Eating Larkspur. *J. Anim. Sci.* **68**(7):1944–1952.

Ralphs, M. H., and J. A. Pfister. (1992). Cattle Diets in Tall Forb Communities on Mountain Rangelands. *J. Range Mgt.* **45**(6):534–537.

Ralphs, M. H., and L. A. Sharp. (1987). Management of Rangeland to Reduce Loss From Poisonous Plants. *Soc. Range Mgt. Abst. Papers* **40:**109.

Ralphs, M. H., L. F. James, D. B. Nielsen, and K. E. Panter. (1984a). Management Practices Reduce Cattle Loss to Locoweed on High Mountain Range. *Rangelands* **6**(4):175–177.

Ralphs, M., M. Kothmann, and L. Merrill. (1984b). Proper Stocking for Short-Duration Grazing. *Tex. Agric. Expt. Sta. Prog. Rep.* **4190.** 7 pp.

Ralphs, M. H., M. M. Kothmann, and C. A. Taylor, Jr. (1986a). Vegetation Response and Livestock Diet Selection With Increasing Stocking Rate Under Rotational Grazing. *Texas Agric. Expt. Sta. Prog. Rep.* **4435.** 2 pp.

Ralphs, M. H., M. M. Kothmann, and L. B. Merrill. (1986b). Cattle and Sheep Diets Under Short-Duration Grazing. **J. Range Mgt. 39**(3):217–223.

Ralphs, M. H., L. V. Mickelsen, and D. L. Turner. (1987). Cattle Grazing White Locoweed: Diet Selection Patterns of Native and Introduced Cattle. *J. Range Mgt.* **40**(4):333–335.

Ralphs, M. H., J. A. Pfister, J. D. Olsen, G. D. Manners, and D. B. Nielsen. (1989). Reducing Larkspur Poisoning in Cattle on Mountain Ranges. *Utah Sci.* **50**(2):109–115.

Ralphs, M. H., K. E. Panter, and L. F. James. (1990). Feed Preferences and Habituation of Sheep Poisoned by Locoweed. *J. Anim. Sci.* **68**(5):1354–1362.

Ralphs, M. H., K. E. Panter, and L. F. James. (1991a). Grazing Behavior and Forage Preferences of Sheep with Chronic Locoweed Toxicosis Suggest No Addiction. *J. Range Mgt.* **44**(3):208–209.

Ralphs, M. H., J. E. Bowns, and G. D. Manners. (1991b). Utilization of Larkspur by Sheep. *J. Range Mgt.* **44**(6):619–622.

Ralphs, M. H., D. Graham, R. J. Molyneux, and L. F. James. (1993). Seasonal Grazing of Locoweeds by Cattle in Northeastern New Mexico. *J. Range Mgt.* **46**(5):416–420.

Ralphs, M. H., D. T. Jensen, J. A. Pfister, D. B. Nielsen, and L. F. James. (1994a). Storms Influence Cattle to Graze Larkspur: An Observation. *J. Range Mgt.* **47**(4):275–278.

Ralphs, M. H., D. Graham, and L. F. James. (1994b). Social Facilitation Influences Cattle to Graze Locoweed. *J. Range Mgt.* **47**(2):123–126.

Ralphs, M. H., G. D. Manners, and D. R. Gardner (1998). Toxic Alkaloid Response to Herbicides Used to Control Tall Larkspur. *Weed Sci.* **46**(1):116–119.

Ratliff, R. D. (1962). Preferential Grazing Continues Under Rest-Rotation Management. *USDA, Pacific Southwest For. and Range Expt. Sta. Res. Note* **206.** 6 pp.

Ratliff, R. D. (1986). Cattle Responses to Continuous and Seasonal Grazing of California Annual Grassland. *J. Range Mgt.* **39**(6):482–485.

Ratliff, R. D., and R. G. Denton. (1995). Grazing on Regeneration Sites Encourages Pine Seedling Growth. *USDA, For. Serv. Res. Paper* **PSW-RP-223.** 11 pp.

Ratliff, R. D., and L. Rader. (1962). Drought Hurts Less with Rest-Rotation Management. *USDA, Pacific Southwest For. and Range Expt. Sta. Res. Note* **196.** 4 pp.

Ratliff, R. D., and J. N. Reppert. (1974). Vigor of Idaho Fescue Grazed Under Rest-Rotation and Continuous Grazing. *J. Range Mgt.* **27**(6):447–459.

Ratliff, R. D., J. N. Reppert, and R. J. McConnen. (1972). Rest-Rotation Grazing at Harvey Valley: Range Health, Cattle Gains, Costs. *USDA, For. Serv. Res. Paper* **PSW-77.** 24 pp.

Ratliff, R. D., M. R. George, and N. K. McDougald. (1987). Managing Livestock Grazing on Meadows of California's Sierra Nevada: A Manager-User Guide. *Univ. Calif., Div. Agric. & Nat. Resources Leaflet* **21421.** 9 pp.

Rauzi, R. J., and R. L. Lang. (1967). Effect of Grazing Intensity on Vegetation and Sheep Gains on Shortgrass Rangeland. *Wyo. Agric. Expt. Sta. Sci. Mono.* **4.** 11 pp.

Ray, M. L., A. E. Spooner, and R. W. Parham. (1969). Cow and Calf Nutrition and Management Under Different Grazing Pressures. *Ariz. Agric. Expt. Sta. Bul.* **749.** 25 pp.

Rayburn, E. B. (1986). "Quantitative Aspects of Pasture Management." Seneca Trail RC & D, Franklinville, NY.

Reardon, P. O., and L. B. Merrill. (1976). Vegetative Response Under Various Grazing Management Systems in the Edwards Plateau of Texas. *J. Range Mgt.* **29**(3):195–198.

Reardon, P. O., and L. B. Merrill. (1978). Response of Sideoats Grama Grown in Different Soils to Addition of Thiamine and Bovine Saliva. *Proc. Intern. Rangeland Cong.* **1**:396–397.

Reardon, P. O., C. L. Leinweber, and L. B. Merrill. (1972). The Effect of Bovine Saliva on Grasses. *Amer. Soc. Anim. Sci., West. Sect. Proc.* **23**:206–210.

Reardon, P. O., C. L. Leinweber, and L. B. Merrill. (1974). Response of Sideoats Grama to Animal Saliva and Thiamine. *J. Range Mgt.* **27**(5):400–401.

Reardon, P. O., L. B. Merrill, and C. A. Taylor, Jr. (1978). White-Tailed Deer Preferences and Hunter Success Under Various Grazing Systems. *J. Range Mgt.* **31**(1):40–42.

Reber, R. J. (1987). The Ruminant and Human Nutrition. *Forage and Grassland Conf.* **1987**:3–8.

Rechenthin, C. A. (1956). Elementary Morphology of Grass Growth and How It Affects Utilization. *J. Range Mgt.* **9**(4):167–170.

Rechenthin, C. A., H. M. Bell, R. J. Pederson, and D. B. Polk. (1964). "Grassland Restoration. II. Brush Control." USDA, Soil Cons. Serv., Temple, TX. 39 pp.

Rector, B. S., and J. E. Huston. (1986). Multispecies Grazing With Cattle, Sheep, and Goats. *Texas Agric. Expt. Sta. Prog. Rep.* **4448.** 3 pp.

Redmon, L. A., F. T. McCollum III, G. W. Horn, M. D. Carvey *et al.* (1995). Forage Intake by Beef Steers Grazing Winter Wheat with Varied Herbage Allowances. *J. Range Mgt.* **48**(3):198–201.

Redmon, L. A., E. G. Krenzer, Jr., D. J. Bernardo, and G. W. Horn. (1996). Effect of Wheat Morphological Stage at Grazing Termination on Economic Return. *Agron. J.* **88**(1):94–97.

Reece, P. E. (1986). The Long and Short of Short-Duration Grazing. *In* P. E. Reece and J. T. Nichols (Eds.), "Proceedings, The Ranch Management Syposium, November 5–7, 1986, North Platte, Nebraska." Univ. Neb., Agric. Ext. Serv., Lincoln, NE. pp. 37–51.

Reece, P. E., R. P. Bode, and S. S. Waller. (1988). Vigor of Needleandthread and Blue Grama after Short-Duration Grazing. *J. Range Mgt.* **41**(4):287–291.

Reece, P. E., J. D. Alexander III, and J. R. Johnson. (1991). Drought Management on Range and Pastureland: A Handbook for Nebraska and South Dakota. *Neb. Agric. Ext. Cir. EC* **91–123.** 23 pp.

Reece, P. E., J. E. Bummer, R. K. Engel, B. K. Northup, and J. T. Nichols. (1996). Grazing Date and Frequency Effects on Prairie Sandreed and Sand Bluestem. *J. Range Mgt.* **49**(2):112–116.

Reece, P. E., T. L. Holman, and K. J. Moore. (1999). Late-Summer Forage on Prairie Sandreed Dominated Rangeland after Spring Defoliation. *J. Range Mgt.* **52**(3):228–234.

Reed, J. D. (1999). Effects of Tannins (Proanthocyanidine) on the Digestion and Analysis of Fiber in Forages. *Soc. Range Mgt. Abst. Papers* **52**:66.

Reed, M. J., and R. A. Peterson. (1961). Vegetation, Soil, and Cattle Responses to Grazing on Northern Great Plains Range. *USDA Tech. Bul.* **1252.** 79 pp.

Reeves, G. W., J. A. Pfister, and M. A. Maiga. (1987). Influence of Protein Supplementation on Forage Intake and Dietary Selection of Grazing Steers. *Soc. Range Mgt. Abst. Papers* **40**:230.

Reichardt, P., T. Clausen, and J. Bryant. (1987). Plant Secondary Metabolites as Feeding Deterrents to Vertebrate Herbivores. *USDA, For. Serv. Gen. Tech. Rep.* **INT-222,** pp. 37–42.

Reid, R. L., and L. F. James. (1985). Forage-Animal Disorders. Chapter 46. *In* M. E. Heath, R. F. Barnes, and D. S. Metcalfe (Eds.), "Forages: The Science of Grassland Agriculture." Iowa State Univ. Press, Ames, Iowa; Fourth Ed.

Reiner, R. J., and P. J. Urness. (1982). Effect of Grazing Horses Managed as Manipulators of Big Game Winter Range. *J. Range Mgt.* **35**(5):567–571.

Renecker, L. A., and H. M. Kozak. (1987). Game Ranching in Western Canada. *Rangelands* **9**(5):213–216.

Reynolds, D. A. (Ed.). (1998). Rangeland Monitoring Manual: A Field Reference for Managers. *Wyo. Agric. Ext. Bull.* **1065.** 54 pp.

Reynolds, H. G. (1954). Meeting Drought on Southern Arizona Rangelands. *J. Range Mgt.* **7**(1): 33–40.

Reynolds, H. G. (1969). Aspen Grove Use by Deer, Elk, and Cattle in Southwestern Coniferous Forests. *USDA, For. Serv. Res. Note* **RM-138.** 4 pp.

Reynolds, H. W., R. M. Hansen, and D. G. Peden. (1978). Diets of the Slave River Lowland Bison Herd, Northwest Territories, Canada. *J. Wildl. Mgt.* **42**(3):581–590.

Reynolds, P. J., J. Bond, G. E. Carlson, C. Jackson, Jr., R. H. Hart, and I. L. Lindahl. (1971). Co-Grazing of Sheep and Cattle on an Orchardgrass Sward. *Agron. J.* **63**(4):533–536.

Rhodes, B. D., and S. H. Sharrow. (1983). Effect of Sheep Grazing on Big Game Habitat in Oregon's Coast Range. *Ore. Agric. Expt. Sta. Spec. Rep.* **682,** p. 28–31.

Rhodes, B. D., and S. H. Sharrow. (1990). Effect of Grazing by Sheep on the Quantity and Quality of Forage Available to Big Game in Oregon's Coast Range. *J. Range Mgt.* **43**(3):235–237.

Rhodes, L. A., V. R. Bohman, and A. Lesperance. (1986). Comparison of Grazing Systems for Crested Wheatgrass. *Amer. Soc. Anim. Sci., West. Sect. Proc.* **37:**227–230.

Rice, H. B., C. Absher, and L. Turner. (1987). Creep Grazing for Beef Calves. *Ky. Agric. Ext.* **ID-76.** 4 pp.

Rice, R. W., G. B. Ruyle, and K. S. Ramosketsi. (1990). The Selection Ability of Cattle Grazing Bunchgrass. *Amer. Soc. Anim. Sci., West. Sect. Proc.* **41:**293–294.

Rich, T. D., S. Armbruster, and D. R. Gill. (1976). Limiting Feed Intake with Salt. *Great Plains Extension* **GPE-1950.** 2 pp.

Richards, J. H. (1984). Root Growth Response to Defoliation in Two *Agropyron* Bunchgrasses: Field Observations with an Improved Root Periscope. *Oecologia* **64**(1):21–25.

Richards, J. H. (1986). Plant Response to Grazing: The Role of Photosynthetic Capacity and Stored Carbon Reserves. *Proc. Internat. Rangeland Cong.* **2:**428–430.

Richards, J. H., and M. M. Caldwell. (1985). Soluble Carbohydrates, Concurrent Photosynthesis, and Efficiency in Regrowth Following Defoliation: A Field Study with *Agropyron* Species. *J. Appl. Ecol.* **22**(3):907–920.

Richards, J. H., M. M. Caldwell, and B. E. Olson. (1987). Plant Production Following Grazing: Carbohydrates, Meristems, and Tiller Survival Over Winter. *In* D. A. Jameson and J. Holechek (Eds.), "Monitoring Animal Performance and Production Symposium Proceedings, February 12, 1987, Boise, Idaho." Society for Range Management, Denver, Colo., pp. 8–11.

Riddle, R. R., C. A. Taylor, Jr., J. E. Huston, and M. M. Kothmann. (1999). Intake of Ash Juniper and Live Oak by Angora Goats. *J. Range Mgt.* **52**(2):161–165.

Riewe, M. E. (1981). Expected Animal Response to Certain Grazing Strategies. *In* J. L. Wheeler and R. D. Moochrie (Eds.), "Forages Evaluation: Concepts and Techniques." CSIRO, E. Melbourne, Australia, pp. 341–355.

Riggs, R. A., and P. J. Urness. (1989). Effects of Goat Browsing on Gambel Oak Communities in Northern Utah. *J. Range Mgt.* **42**(5):354–360.

Riggs, R. A., P. J. Urness, and T. A. Hall. (1988). Diets and Weight Responses of Spanish Goats Used to Control Gambel Oak. *Small Rum. Res.* **1:**259–271.

Riggs, R. A., P. J. Urness, and D. D. Austin. (1988). Response of a Wasatch Oakbrush Community to Intensive Browsing by Spanish Goat. *Soc. Range Mgt. Abst. Papers* **41:**398.

Riggs, R. A., P. J. Urness, and K. A. Gonzalez. (1990). Effects of Domestic Goats on Deer Wintering in Utah Oakbrush. *J. Range Mgt.* **43**(3):229–234.

Ring, C. B., II, R. A. Nicholson, and J. L. Launchbaugh. (1985). Vegetational Traits of Patch-Grazed Rangeland in West-Central Kansas. *J. Range Mgt.* **38**(1):51–55.

Rittenhouse, L. R. (1984). "Impacts of Grazing Intensity and Specialized Grazing Systems on Livestock Response: A Discussant Paper." *In* "Developing Strategies for Rangeland Management." Westview Press, Boulder, CO. pp. 1159–1165.

Rittenhouse, L. R., and D. W. Bailey. (1996). Spatial and Temporal Distribution of Nutrients: Adaptive Significance to Free-Grazing Herbivores. *Proc. Grazing Livestock Nutr. Conf.* **3**:51–61.

Rittenhouse, L. R., and R. L. Senft. (1982). Effects of Daily Weather Fluctuations on the Grazing Behavior of Cattle. *Amer. Soc. Anim. Sci., West. Sect. Proc.* **33**:305–307.

Rittenhouse, L. R., D. C. Clanton, and C. L. Streeter. (1970). Intake and Digestibility of Winter-Range Forage by Cattle With and Without Supplements. *J. Anim. Sci.* **31**(6):1215–1221.

Rivers, W. J., and S. C. Martin. (1980). Perennial Grass Improves with Moderate Stocking. *Rangelands* **2**(3):105–106.

Roath, L. R., and W. C. Krueger. (1982a). Cattle Grazing Influence on a Mountain Riparian Zone. *J. Range Mgt.* **35**(1):100–103.

Roath, L. R., and W. C. Krueger. (1982b). Cattle Grazing and Behavior on a Forested Range. *J. Range Mgt.* **35**(3):332–338.

Robbins, C. T. (1983). "Wildlife Feeding and Nutrition." Academic Press, New York City. 343 pp.

Robbins, C. T., T. A. Hanley, A. E. Hagerman, O. Hjeljord, D. L. Baker, C. C. Schwartz, and W. W. Mautz. (1987a). Role of Tannins in Defending Plants Against Ruminants: Reduction in Protein Availability. *Ecology* **68**(1):98–107.

Robbins, C. T., S. Mole, A. E. Hagerman, and T. A. Hanley. (1987b). Role of Tannins in Defending Plants Against Ruminants: Reduction in Dry Matter Digestion. *Ecology* **68**(6):1606–1615.

Robbins, C. T., D. E. Spalinger, and W. van Hoven. (1995). Adaptation of Ruminants to Browse and Grass Diets: Are Anatomical-Based Browser-Grazer Interpretations Valid? *Oecologia* **103**(2):208–213.

Roberts, C. A., K. G. Moore, and J. O. Fritz. (1987). Chemical Regulation of Tall Fescue Growth and Quality. *Forage & Grassland Conf.* **1987**:162–165.

Roberts, W. P., Jr. (1961). Fencing Vs. Herding of Range Sheep. *Wyo. Agric. Expt. Sta. Mimeo. Cir.* **156**. 15 pp.

Robertshaw, D. (1987). Heat Stress. *In* "Proceedings, Grazing Livestock Nutrition Conference, July 23–24, 1987, Jackson, Wyoming." Univ. of Wyo., Laramie, WY. pp. 31–35.

Robertson, J. H., D. L. Neal, K. R. McAdams, and P. T. Tueller. (1970). Changes in Crested Wheatgrass Ranges Under Different Grazing Treatments. *J. Range Mgt.* **23**(1):27–34.

Robinson, D. L., L. C. Kappe, and J. A. Boling. (1989). Management Practices To Overcome the Incidence of Grass Tetany. *J. Anim. Sci.* **67**(12):3470–3484.

Rogler, G. A. (1944). Relative Palatability of Grasses Under Cultivation in the Northern Great Plains. *J. Amer. Soc. Agron.* **36**(6):487–496.

Rogler, G. A. (1951). A Twenty-five Year Comparison of Continuous and Rotation Grazing in the Northern Plains. *J. Range Mgt.* **4**(1):35–41.

Rohweder, D. A., and R. W. Van Keuren. (1985). Permanent Pastures. Chapter 52, *In* M. E. Heath, R. F. Barnes, and D. S. Metcalfe (Eds.), "Forages: The Science of Grassland Agriculture," 4th ed. Iowa State Univ. Press, Ames, IA.

Rollins, D. (1984). Determining Native Range Stocking Rates. *Okla. Agric. Ext. OSU Ext. Facts* **2855**. 2 pp.

Romo, J. T. (1994). Wolf Plant Effects on Water Relations, Growth, and Productivity in Crested Wheatgrass. *Can. J. Plant Sci.* **74**(4):767–771.

Romo, J. T., M. E. Tremblay, and D. Barber. (1997). Are There Economic Benefits of Accessing Forage in Wolf Plants of Crested Wheatgrass? *Can. J. Plant Sci.* **77**(3):367–371.

Rosiere, R. E. (1987). An Evaluation of Grazing Intensity Influences on California Annual Range. *J. Range Mgt.* **40**(2):160–165.

Rosiere, R. E., and D. T. Torell. (1985). Nutritive Value of Sheep Diets on Coastal California Annual Range. *Hilgardia* **53**(1):1–19.

Rosiere, R. E., R. F. Beck, and J. D. Wallace. (1975). Cattle Diets on Semidesert Grassland: Botanical Composition. *J. Range Mgt.* **28**(2):89–93.

Rosiere, R. E., J. D. Wallace, and R. D. Pieper. (1980). Forage Intake in Two-Year Old Cows and Heifers Grazing Blue Grama Summer Range. *J. Range Mgt.* **33**(1):71–73.

Roth, L. D., F. M. Rouquette, Jr., and W. C. Ellis. (1986a). Diet Selection and Nutritive Value of Coastal Bermudagrass as Influenced by Grazing Pressure. *Texas Agric. Expt. Sta. Cons. Prog. Rep.* **4347,** pp. 7–9.

Roth, L. D., F. M. Rouquette, Jr., and W. C. Ellis. (1986b). Influence of Grazing Pressure on Forage Digestibility, Intake, and Liveweight Gain. *Texas Agric. Expt. Sta. Cons. Prog. Rep.* **4347,** pp. 16–19.

Rothlisberger, J. A., W. W. Rowden, and J. E. Ingalls. (1962). Effect of Lengthened Feeding Interval on Winter and Summer Gains of Beef Calves. *Univ. Neb. Annual Feeders Day Progress Report* **50:**3.

Rouda, R. R., D. M. Anderson, L. W. Murray, and J. N. Smith. (1990). Distance Traveled by Free-Ranging Supplemented and Non-supplemented Lactating and Non-lactating Cows. *Appl. Anim. Beh. Sci.* **28**(3):221–232.

Rouda, R. R., D. M. Anderson, J. D. Wallace, and L. W. Murray. (1994). Free-Ranging Cattle Water Consumption in Southcentral New Mexico. *Appl. Anim. Beh. Sci.* **39**(1):29–38.

Roundtree, B. H., A. C. Matches, and F. A. Martz. (1974). Season Too Long for Your Grass Pasture? *Crops & Soil* **26**(7):7–10.

Roundy, B. A., and G. B. Ruyle. (1989). Effects of Herbivory on Twig Dynamics of a Sonoran Desert Shrub *Simmondsia chinensis* (Link) Schn. *J. Appl. Ecol.* **26**(2):701–710.

Roundy, B. A., G. B. Ruyle, A. E. Dobrenz, V. Wilson, and D. Floyd. (1987). Growth, Nutrient, and Water Status of Jojoba (*Simmondsia chinensis*) in Relation to Livestock Grazing. *USDA, For. Serv. Gen. Tech. Rep.* **INT-222,** pp. 146–153.

Rowan, R. C., L. D. White, and J. R. Conner. (1994). Understanding Cause/Effect Relationships in Stocking Rate Change Over Time. *J. Range Mgt.* **47**(5):349–354.

Rowland, M. E., and J. W. Stuth. (1989). Factors Affecting the Foraging Strategies of Cattle on Brush Managed Landscapes. *Soc. Range Mgt. Abst. Papers* **42:**165.

Rue, Leonard Lee, III. (1979). "The Deer of North America." Crown Pub., Inc. N. Y. 463 pp.

Rumery, M. G. A., and R. E. Ramig. (1962). Irrigated Sudangrass for Dairy Cows. *Neb. Agric. Expt. Sta. Bul.* **472.** 12 pp.

Rummell, R. S. (1951). Some Effects of Livestock Grazing on Ponderosa Pine Forest and Range in Central Washington. *Ecology* **32**(4):594–607.

Ruyle, G. B., and J. E. Bowns. (1985). Forage Use by Cattle and Sheep Grazing Separately and Together on Summer Range in Southwestern Utah. *J. Range Mgt.* **38**(4):299–302.

Ruyle, G. B., and D. D. Dwyer. (1985). Feeding Stations of Sheep as an Indicator of Diminished Forage Supply. *J. Anim. Sci.* **61**(2):349–353.

Ruyle, G. B., and R. W. Rice. (1996). Aspects of Forage Availability and Short-Term Intake Influencing Range Livestock Production. *Proc. Grazing Livestock Nutr. Conf.* **3:**40–50.

Ruyle, G. B., C. A. Wissler, and R. P. Young. (1986a). Cattle Utilization of Lehman Lovegrass: Patterns of Defoliation at Four Stocking Rates. *Soc. Range Mgt. Abst. Papers* **39:**396.

Ruyle, G. B., R. L. Grumbles, M. J. Murphy, and R. C. Cline. (1986b). Oak Consumption by Cattle in Arizona. *Rangelands* **8**(3):124–126.

Saliki, D. O., and B. E. Norton. (1987). Survival of Perennial Grass Seedlings under Intensive Grazing in Semi-Arid Rangelands. *J. Appl. Ecol.* **24**(1):145–151.

Salter, R. E., and R. J. Hudson. (1979). Feeding Ecology of Feral Horses in Western Alberta. *J. Range Mgt.* **32**(3):221–225.

Salter, R. E., and R. J. Hudson. (1980). Range Relationships of Feral Horses with Wild Ungulates and Cattle in Western Alberta. *J. Range Mgt.* **33**(4):266–271.

Salzman, S. E. (1983). Steps and Requirements in Establishment of Grazing Systems. *Rangelands* 5(5):212–213.

Sampson, A. W. (1952). "Range Management Principles and Practices." John Wiley & Sons, New York. 570 pp.

Sanders, K. D., L. A. Sharp, and M. A. Siebe. (1986). Short-Duration Grazing. *In* J. A. Tiedeman (Ed.), "Proceedings of the Short-Duration Grazing and Current Issues in Grazing Management Short-course Held January 21–23, 1986, at Kennewick, Washington." Wash. State Univ., Agric. Ext. Serv., Pullman, Wash., pp. 1–8.

Sanderson, M. A., W. F. Wedin, and D. G. Morrical. (1987). Sequential Grazing of Steers and Ewes. *Forage & Grassland Conf.* 1987:182–184.

Sanford, S. (1983). "Management of Pastoral Development in the Third World." John Wiley & Sons, Somerset, N. Jer. 316 pp.

Saunders, J. K., Jr. (1955). Food Habits and Range Use of the Rocky Mountain Goat in the Crazy Mountains, Montana. *J. Wildl. Mgt.* 19(4):429–437.

Savory, A. (1978). A Holistic Approach to Range Management Using Short-Duration Grazing. *Proc. Intern. Rangeland Cong.* 1:555–557.

Savory, A. (1983). The Savory Grazing Method of Holistic Resource Management. *Rangelands* 5(4):155–159.

Savory, A. (1987). "Center for Holistic Resource Management." Center for Holistic Resource Management, Albuquerque, N.M. Var. paged.

Savory, A. (1988). "Holistic Resource Management." Island Press, Washington, D.C. 564 pp.

Savory, A., and S. D. Parsons. (1980). The Savory Grazing Method. *Rangelands* 2(6):234–237.

Sawyer, J. E., L. A. Knox, G. B. Donart, and M. K. Petersen. (1997). Evaluation of the Effects of Spine Removal by Burning on the Nutritive Quality of Cholla Cactus for an Emergency Feed. *Amer. Soc. Anim. Sci., West. Sec. Proc.* 48:252–254.

Scarnecchia, D. L. (1985a). The Animal-Unit and Animal-Unit-Equivalent Concepts in Range Science. *J. Range Mgt.* 38(4):346–349.

Scarnecchia, D. L. (1985b). The Relationship of Stocking Intensity and Stocking Pressure to Other Stocking Variables. *J. Range Mgt.* 38(6):558–559.

Scarnecchia, D. L. (1986). Viewpoint: Animal-Unit Equivalents Cannot Be Meaningfully Weighted by Indices of Dietary Overlap. *J. Range Mgt.* 39(5):471.

Scarnecchia, D. L. (1988). Grazing, Stocking, and Production Efficiencies in Grazing Research. *J. Range Mgt.* 41(4):279–281.

Scarnecchia, D. L. (1990). Concepts of Carrying Capacity and Substitution Ratios: A Systems Viewpoint. *J. Range Mgt.* 43(6):553–555.

Scarnecchia, D. L. (1999). The Range Utilization Concept, Allocation Arrays, and Range Management Science. *J. Range Mgt.* 52(2):157–160.

Scarnecchia, D. L., and C. T. Gaskins. (1987). Modeling Animal-Unit-Equivalents for Beef Cattle. *Agric. Systems* 23(1):19–26.

Scarnecchia, D. L., and M. M. Kothmann. (1982). A Dynamic Approach to Grazing Management Terminology. *J. Range Mgt.* 35(2):262–264.

Scarnecchia, D. L., and M. M. Kothmann. (1986). Observations on Herbage Growth, Disappearance, and Accumulation Under Livestock Grazing. *J. Range Mgt.* 39(1):86–87.

Scarnecchia, D. L., A. S. Nastis, and J. C. Malechek. (1985). Effects of Forage Availability on Grazing Behavior of Heifers. *J. Range Mgt.* 38(2):177–180.

Schacht, W. H., A. J. Smart, B. E. Anderson, L. E. Moser, and R. Rasby. (1998). Growth Responses of Warm-Season Tallgrasses to Dormant Season Management. *J. Range Mgt.* 51(4):442–446.

Schake, L. M., and J. K. Riggs. (1972). Behavior of Beef Cattle in Confinement: A Technical Report. *Texas Agric. Expt. Sta. Tech. Rep.* 27. 12 pp.

Schettini, M. A., E. C. Prigge, and E. L. Nestor. (1999). Influences of Mass and Volume of Ruminal Contents on Voluntary Intake and Digesta Passage of a Forage Diet in Steers. *J. Anim. Sci.* 77(7):1896–1904.

Schmutz, E. M. (1978). Deferred-Rotation Grazing-Burning System for Southwestern Rangelands. *Ariz. Agric. Ext.* **Q178.** 2 pp.

Schmutz, E. M., and M. D. Durfee. (1980). The Next-Best Pasture Deferred-Rotation Grazing System. *Ariz. Agric. Ext.* **Q432.** 4 pp.

Schneider, N. R. (1999). Anti-Quality in Forages Associated with Nitrate and Cyanide (Prussic Acid). *Soc. Range Mgt. Abst. Papers* **52:**71–72.

Scholl, E. L., and R. J. Kinucan. (1996). Grazing Effects on Reproduction Characteristics of Common Curlymesquite (*Hilaria belangeri*). *Southwestern Nat.* **41**(3):251–256.

Schulz, P. A., and F. S. Guthery. (1987). Effects of Short-Duration Grazing on Wild Turkey Home Ranges. *Wildl. Soc. Bul.* **15**(2):239–241.

Schultz, P. A., and F. S. Guthery. (1988). Effects of Short-Duration Grazing on Northern Bobwhites: A Pilot Study. *Wildl. Soc. Bull.* **16**(1):18–24.

Schuster, J. L. (1984). The Importance of Rangeland and Range Conservation. *Rangelands* **6**(5):221–222.

Schwartz, C. C., and J. E. Ellis. (1981). Feeding Ecology and Niche Separation in Some Native and Domestic Ungulates on the Shortgrass Prairie. *J. Appl. Ecol.* **18**(2):343–353.

Schwartz, C. C., J. G. Nagy, and W. L. Regelin. (1980). Juniper Oil Yield, Terpenoid Concentration, and Antimicrobial Effects on Deer. *J. Wildl. Mgt.* **44**(1):107–113.

Scifres, C. J., J. R. Scifres, and M. M. Kothmann. (1983). Differential Grazing Use of Herbicide-Treated Areas by Cattle. *J. Range Mgt.* **36**(1):65–69.

Scott, C. B., F. D. Provenza, and R. E. Banner. (1995a). Dietary Habits and Social Interactions Affect Choice of Feeding Location by Sheep. *Appl. Anim. Beh. Sci.* **45**(3–4):225–237.

Scott, C. B., W. H. Schacht, C. S. McCown, and S. Hartmann. (1995b). Observations of Grass Community Dynamics in Short-Duration Grazing Systems in West Texas. *Texas J. Agric. & Nat. Resources.* **8**(1–12).

Scott, C. B., R. E. Banner, and F. D. Provenza. (1996). Observations of Sheep Foraging in Familiar and Unfamiliar Environments: Familiarity with the Environment Influences Diet Selections. *Appl. Anim. Beh. Sci.* **49**(2):165–171.

Scotter, G. W. (1980). Management of Wild Ungulate Habitat in the Western United States and Canada: A Review. *J. Range Mgt.* **33**(1):16–27.

Sedivec, K. K., T. A. Messmer, W. T. Barker, K. F. Higgins, and D. R. Hertel. (1990). Nesting Success of Upland Nesting Waterfowl and Sharp-Tailed Grouse in Specialized Grazing Systems in South-central North Dakota. *In* K. E. Severson (Tech. Coord.), "Can Livestock Be Used as a Tool to Enhance Wildlife Habitat?" *USDA, For. Serv. Gen. Tech. Rep.* **RM-194,** pp. 71–92.

Seegmiller, R. F., and R. D. Ohmart. (1981). Ecological Relationships of Feral Burros and Desert Bighorn Sheep. *Wildl. Monogr.* **78.** 58 pp.

Selting, J. P., and L. R. Irby. (1997). Agricultural Land Use Patterns of Native Ungulates in Southeastern Montana. *J. Range Mgt.* **50**(4):338–345.

Senft, R. L. (1986a). Cattle Spatial Use of Mountain Range: Effects of Grazing Systems and Landscape Pattern. *Amer. Soc. Anim. Sci., West. Sect. Proc.* **37:**231–234.

Senft, R. L. (1986b). Evaluating Short Duration Grazing on Mountain Range: A Modeling Study. *Amer. Soc. Anim. Sci., West. Sect. Proc.* **37:**249–251.

Senft, R. L., and J. C. Malechek. (1985). Short Duration Grazing Cell Parameters and Cattle Production: A Low Resolution Model. *Amer. Soc. Anim. Sci., West. Sect. Proc.* **36:**282–285.

Senft, R. L., and L. R. Rittenhouse. (1985). Effects of Day-to-Day Weather Fluctuations on Grazing Behavior of Horses. *Amer. Soc. Anim. Sci., West. Sect. Proc.* **36:**298–300.

Senft, R. L., L. R. Rittenhouse, and R. G. Woodmansee. (1982). Seasonal Patterns of Cattle Spatial Use of Shortgrass Prairie. *Amer. Soc. Anim. Sci., West. Sect. Proc.* **33:**291–293.

Senft, R. L., L. R. Rittenhouse, and R. G. Woodmansee. (1985a). Factors Influencing Patterns of Cattle Grazing Behavior on Shortgrass Steppe. *J. Range Mgt.* **38**(1):82–87.

Senft, R. L., L. R. Rittenhouse, and R. G. Woodmansee. (1985b). Factors Influencing Selection of Resting Sites by Cattle on Shortgrass Steppe. *J. Range Mgt.* **38**(4):295–299.

Senft, R. L., J. E. Bowns, and C. F. Bagley. (1986). Shifts in Cattle and Sheep Diets Under Various Grazing Systems on Mountain Pastures. *Amer. Soc. Anim. Sci., West. Sect. Proc.* **37**:252–254.

Senft, R. L., M. B. Coughenour, D. W. Bailey, L. R. Rittenhouse, O. E. Sala, and D. M. Swift. (1987). Large Herbivore Foraging and Ecological Hierarchies. *BioScience* **37**(11):789–799.

Senock, R. S., D. M. Anderson, L. W. Murray, and G. B. Donart. (1993). Tobosa Tiller Defoliation Patterns Under Rotational and Continuous Stocking. *J. Range Mgt.* **46**(6):500–505.

Severson, K. E. (Tech. Coord.). (1990). Can Livestock Be Used as a Tool to Enhance Wildlife Habitat? *USDA, For. Serv. Gen. Tech. Rep.* **RM-194.** 123 pp.

Severson, K. E., and L. F. Debano. (1991). Influence of Spanish Goats on Vegetation and Soils in Arizona Chaparral. *J. Range Mgt.* **44**(2):111–117.

Severson, K. E., and A. L. Medina. (1983). Deer and Elk Habitat Management in the Southwest. *J. Range Mgt. Mono.* **2.** 64.

Severson, K. E., and P. J. Urness. (1994). Livestock Grazing: A Tool to Improve Wildlife Habitat. *In* M. Vavra, W. Laycock, and R. D. Pieper (Eds.), "Ecological Implications of Livestock Herbivory in the West." Soc. Range Mgt., Denver, CO. pp. 232–249.

Severson, K., M. Morton, and W. Hepworth. (1968). Food Preferences, Carrying Capacities, and Forage Competition Between Antelope and Domestic Sheep in Wyoming's Red Desert. *Wyo. Agric. Expt. Sta. Sci. Mono.* **10.** 51 pp.

Sexson, M. L., J. R. Choate, and R. A. Nicholson. (1981). Diet of Pronghorn in Western Kansas. *J. Range Mgt.* **34**(6):489–493.

Sharp, L. A. (1970). Suggested Management Programs for Grazing Crested Wheatgrass. *Ida. For., Wildl. and Range Expt. Sta. Bul.* **4.** 19 pp.

Sharp, L., K. Sanders, and N. Rimbey. (1994). Management Decisions Based on Utilization—Is It Really Management? *Rangelands* **16**(1):38–40.

Sharrow, S. H. (1983a). Forage Standing Crop and Animal Diets Under Rotational Vs. Continuous Grazing. *J. Range Mgt.* **36**(4):447–450.

Sharrow, S. H. (1983b). Rotational Vs. Continuous Grazing Affects Animal Performance on Annual Grass-Subclover Pasture. *J. Range Mgt.* **36**(6):593–595.

Sharrow, S. H. (1994). Sheep as a Silvicultural Management Tool in Temperate Conifer Forest. *Sheep Res. J.* **10**(Spec. Issue):97–104.

Sharrow, S. H., and W. D. Mosher. (1982). Sheep as a Biological Control Agent for Tansy Ragwort. *J. Range Mgt.* **35**(4):480–482.

Sharrow, S. H., and I. Motazedian. (1987). Spring Grazing Effects on Components of Winter Wheat Yield. *Agron. J.* **79**(3):502–504.

Sharrow, S. H., W. C. Krueger, and F. O. Thetford, Jr. (1981). Effects of Stocking Rate on Sheep and Hill Pasture Performance. *J. Anim. Sci.* **52**(2):210–217.

Shaudys, E. T., and J. H. Sitterley. (1959). Green Chopping or Rotational Grazing? *Ohio Farm and Home Res.* **44**(320):68–69, 79.

Shaw, J. H. (1996). Bison. *In* "Rangeland Wildlife." Soc. Range Mgt., Denver, CO. Chapter 14.

Shaw, R. B., and J. D. Dodd. (1979). Cattle Activities and Preferences Following Strip Application of Herbicide. *J. Range Mgt.* **32**(6):449–452.

Sheffield, R. E., S. Mostaghimi, D. H. Vaughan, E. R. Collins, Jr., and V. G. Allen. (1997). Off-Stream Water Sources for Grazing Cattle as a Stream Bank Stabilization and Water Quality BMP. *Trans. Amer. Soc. Agric. Eng.* **40**(3):595–604.

Sheppard, A. J., R. E. Blaser, and C. M. Kincaid. (1957). The Grazing Habits of Beef Cattle on Pasture. *J. Anim. Sci.* **16**(3):681–687.

Shiflet, T. N., and H. F. Heady. (1971). Specialized Grazing Systems—Their Place in Range Management. *USDA, SCS* **TP-152.** 13 pp.

Shipley, L. A. (1999). Grazers and Browsers: How Digestive Morphology Affects Diet Selection. *In* Grazing Behavior of Livestock and Wildlife. *Idaho Forest, Wildl., and Range Expt. Sta. Bull.* **70,** pp. 20–27.

Sholar, J. R., J. F. Stritzke, J. L. Caddel, and R. C. Berberet. (1988). Response of Four Alfalfa Cultivars to Fall Harvesting in the Southern Plains. *J. Prod. Agric.* **1**(3):266–270.

Shoop, M. C., and D. N. Hyder. (1976). Growth of Replacement Heifers on Shortgrass Ranges of Colorado. *J. Range Mgt.* **29**(1):4–8.

Shoop, M. C., and E. H. McIlvain. (1971a). Efficiency of Combining Improvement Practices that Increase Steer Gains. *J. Range Mgt.* **24**(2):113–116.

Shoop, M. C., and E. H. McIlvain. (1971b). Why Some Cattlemen Overgraze—And Some Don't. *J. Range Mgt.* **24**(4):252–257.

Shoop, M. C., and E. H. McIlvain. (1972). Heavy Grazing in September-October Injures Blue Grama and Weeping Lovegrass. *Soc. Range Mgt. Abst. Papers* **25**:28.

Shoop, M. C., R. C. Clark, W. A. Laycock, and R. M. Hansen. (1985). Cattle Diets on Shortgrass Ranges with Different Amounts of Fourwing Saltbush. *J. Range Mgt.* **38**(5):443–449.

Short, H. L., W. Evans, and E. L. Boeker. (1977). The Use of Natural and Modified Pinyon Pine-Juniper Woodlands by Deer and Elk. *J. Wildl. Mgt.* **41**(3):543–559.

Shultis, A., and H. T. Strong. (1955). "Choosing Profitable Beef Production." Univ. Calif., Agric. Ext. Serv., Berkeley, Calif. 29 pp.

Sidamed, A. E., S. R. Radosevich, J. G. Morris, and W. L. Graves. (1978). An Assessment of Goat Grazing in Chaparral. *Calif. Agric.* **32**(10):12–13.

Sievers, E. (1985). Burning and Grazing Florida Flatwoods. *Rangelands* **7**(5):208–209.

Simpson, J. R., and T. H. Stobbs. (1981). Nitrogen Supply and Animal Production From Pastures. *In* F. H. W. Morley (Ed.), "Grazing Animals." Elsevier Sci. Pub. Co., Amsterdam, Neth., pp. 261–287.

Sims, P. L. (1989). Efficiency of Cow Types on Native Range and Integrated Forage Systems. *Soc. Range Mgt. Abst. Papers* **42**:198.

Sims, P. L. (1993). Cow Weights and Reproduction on Native Rangeland and Native Rangeland-Complementary Forage Systems. *J. Anim. Sci.* **71**(7):1704–1711.

Sims, P. L., R. K. Lang'at, and D. N. Hyder. (1973). Developmental Morphology of Blue Grama and Sand Bluestem. *J. Range Mgt.* **26**(5):340–344.

Sims, P. L., B. E. Dahl, and A. H. Durham. (1976). Vegetation and Livestock Response at Three Grazing Intensities on Sandhill Rangeland in Eastern Colorado. *Colo. Agric. Expt. Sta. Tech. Bul.* **130**. 48 pp.

Sindelar, B. W. (1987). Plant Population Response to Time Control Grazing in Southwestern Montana. *Soc. Range Mgt. Abst. Papers* **40**:179.

Sindelar, B. W. (1988). Opportunities for Improving Grass Reproductive Efficiency on Rangelands. *In* R. S. White and R. E. Short (Eds.), "Achieving Efficient Use of Rangeland Resources, Fort Keogh Research Symposium, September 1987, Miles City, Mon." Mon. Agric. Expt. Sta., Bozeman, Mon., pp. 77–85.

Skiles, J. W. (1984). A Review of Animal Preference. *In* Natl. Res. Council/Natl. Acad. Sci. "Developing Strategies for Rangeland Management." Westview Press, Boulder, CO. pp. 153–213.

Skovlin, J. M. (1957). Range Riding—The Key to Range Management. *J. Range Mgt.* **10**(6):269–271.

Skovlin, J. M. (1965). Improving Cattle Distribution on Western Mountain Rangelands. *USDA Farm Bul.* **2212**. 14 pp.

Skovlin, J. M. (1984). Impacts of Grazing on Wetlands and Riparian Habitat: A Review of Our Knowledge. *In* Natl. Res. Council/Natl. Acad. Sci. "Developing Strategies for Rangeland Management." Westview Press, Boulder, CO. pp. 1001–1103.

Skovlin, J. (1987). Southern Africa's Experience with Intensive Short-Duration Grazing. *Rangelands* **9**(4):162–167.

Skovlin, J., and M. Vavra. (1979). Winter Diets of Elk and Deer in the Blue Mountains, Oregon. *USDA, For. Serv. Res. Paper* **PNW-260**. 21 pp.

Skovlin, J. M., P. J. Edgerton, and R. W. Harris. (1968). The Influence of Cattle Management on Deer and Elk. *Trans. N. Amer. Wildl. and Nat. Res. Conf.* **33**:169–181.

Skovlin, J. M., R. W. Harris, G. S. Strickler, and G. A. Garrison. (1976). Effects of Cattle Grazing Methods on Ponderosa Pine-Bunchgrass Range in the Pacific Northwest. *USDA Tech. Bul.* **1531**. 40 pp.

Skovlin, J. M., P. J. Edgerton, and B. R. McConnell. (1983). Elk Use of Winter Range as Affected by Cattle Grazing, Fertilizing, and Burning in Southeastern Washington. *J. Range Mgt.* **36**(2):184–189.

Stephenson, T. E., J. L. Holechek, and C. B. Kuykendall. (1985). Diets of Four Wild Ungulates on Winter Range in Northcentral New Mexico. *Southwestern Nat.* **30**(3):437–441.

Stephenson, T. R., M. R. Vaughan, and D. E. Andersen. (1996). Mule Deer Movements in Response to Military Activity in Southeast Colorado. *J. Wildl. Mgt.* **60**(4):777–787.

Steuter, A. A. (1999). Comparative Ecology of Bison and Cattle on Mixed Grass Prairie. *Soc. Range Mgt. Abst. Papers* **52**:99.

Stevens, D. R. (1974). Rocky Mountain Elk-Shiras Moose Range Relationships. *Naturaliste Can.* **101**(3–4):505–516.

Stevens, R., B. C. Giunta, and A. P. Plummer. (1975). Some Aspects in the Biological Control of Juniper and Pinyon. *In* "Proceedings of Pinyon-Juniper Symposium." Utah State Univ., Logan, UT. pp. 77–82.

Stewart, G., and I. Clark. (1944). Effect of Prolonged Spring Grazing on the Yield and Quality of Forage from Wild-Hay Meadows. *J. Amer. Soc. Agron.* **36**(3):238–248.

Stine, K. (1998). GLA (Grazing Lands Application), What Is It? Can I Use It? *Rangelands* **20**(4):28–30.

Stoddart, L. A. (1960). Determining Correct Stocking Rate on Range Land. *J. Range Mgt.* **13**(5):251–255.

Stoddart, L. A., and A. D. Smith. (1955). "Range Management," 2nd Ed. McGraw-Hill Book Co., NY. 433 pp.

Stoddart, L. A., A. D. Smith, and T. W. Box. (1975). "Range Management," 3rd ed. McGraw-Hill Book Co., NY. 532 pp.

Stout, D. G., and B. Brooke. (1985a). Quantification of Tiller Pull-Up During Grazing of Pinegrass. *Can. J. Plant Sci.* **65**(4):943–950.

Stout, D. G., and B. Brooke. (1985b). Rhizomes and Roots Below Clipped Pinegrass Tillers Have a Higher Percent Carbohydrate when Attached to Other Nonclipped Tillers. *J. Range Mgt.* **38**(3):276–277.

Stout, D. G., A. McLean, B. Brooke, and J. Hall. (1980). Influence of Simulated Grazing (Clipping) on Pinegrass Growth. *J. Range Mgt.* **33**(4):286–291.

Stuedemann, J. A., and C. S. Hoveland. (1988). Fescue Endophyte: History and Impact on Animal Agriculture. *J. Prod. Agric.* **1**(1):39–44.

Stuth, J. W. (1991). Foraging Behavior. *In* (R. K. Heitschmidt and J. W. Stuth, Eds), "Grazing Management: An Ecological Approach." Timber Press, Portland, OR. pp. 65–83.

Stuth, J. W., and B. H. Lyons (Eds.). (1993). "Decision Support Systems for the Management of Grazing Lands: Emerging Issues." Parthenon Publishing Group, Pearl River, NY. 301 pp.

Stuth, J. W., and P. D. Olson. (1986). Influence of Stocking Rate on Nutrient Intake of Steers in a Simulated Rotational Grazing System. *Texas Agric. Expt. Sta. Prog. Rep.* **4436**. 2 pp.

Stuth, J. W., and M. S. Smith. (1993). Decision Support for Grazing Lands: An Overview. *In* J. W. Stuth and B. G. Lyons (Eds.), "Decision Support Systems for the Management of Grazing Lands: Emerging Issues." Parthenon Publishing Group, Pearl River, NY. pp. 1–35.

Stuth, J. W., and A. H. Winward. (1977). Livestock-Deer Relations in the Lodgepole Pine-Pumice Region of Central Oregon. *J. Range Mgt.* **30**(2):110–116.

Svejcar, T., and S. Christiansen. (1987a). Grazing Effects on Water Relations of Caucasian Bluestem. *J. Range Mgt.* **40**(1):15–18.

Svejcar, T., and S. Christiansen. (1987b). The Influence of Grazing Pressure on Rooting Dynamics of Caucasion Bluestem. *J. Range Mgt.* **40**(3):224–227.

Swanson, A. F., and K. Anderson. (1951). Winter Wheat for Pasture in Kansas. *Kan. Agric. Expt. Sta. Bul.* **345**. 32 pp.

Swartz, C. C., and N. T. Hobbs. (1985). Forage and Range Evaluation. *In* Robert J. Hudson and Robert G. White (Eds.). "Bioenergetics of Wild Herbivores." CRC Press, Boca Raton, Fla., pp. 25–51.

Tanner, C. B., and C. P. Mamaril. (1959). Pasture Soil Compaction by Animal Traffic. *Agron. J.* **51**(6):329–331.

Tanner, G. W., and C. E. Lewis. (1987). Woody Plant Responses to Pine Regeneration and Deferred-Rotation Grazing in North Florida. *Soc. Range Mgt. Abst. Papers* **40**:203.

Tanner, G. W., J. M. Inglis, and L. H. Blankenship. (1978). Acute Impact of Herbicide Strip Treatment on Mixed-Brush White-Tailed Deer Habitat on the Northern Rio Grande Plain. *J. Range Mgt.* **31**(5):386–391.

Tanner, G. W., L. D. Sandoval, and F. G. Martin. (1984). Cattle Behavior on a South Florida Range. *J. Range Mgt.* **37**(3):248–251.

Taylor, C. A., Jr. (1986a). Multispecies Grazing—Vegetation Manipulation. *Texas Agric. Expt. Sta. Prog. Rep.* **4426.** 2 pp.

Taylor, C. A., Jr. (1986b). Multispecies Grazing—Forage Selection. *Texas Agric. Expt. Sta. Prog. Rep.* **4427.** 2 pp.

Taylor, C. A., Jr. (1986c). Angora Goats and Rotational Grazing. *Texas Agric. Expt. Sta. Prog. Rep.* **4446.** 2 pp.

Taylor, C. (1988). Grazing Systems: Are They Worth the Trouble? *In* "Grazing Management Field Day, October 18, 1988." Texas Agric. Expt. Sta. Tech. Rep. 88–1, pp. 10–16.

Taylor, C. A., Jr. (1992). Brush Management Considerations with Goats. *In* "Proceedings of the International Conference on Meat Goat Production, Management, and Harvesting." Texas Agric. Ext. Serv., Corpus Christi, TX. pp. 166–176.

Taylor, C. A., Jr. (1994). Sheep Grazing as a Brush and Fine Fire Fuel Management Tool. *Sheep Res. J.* **10**(Spec. Issue):92–96.

Taylor, C. A., Jr., and M. M. Kothmann. (1990). Diet Composition of Angora Goats in a Short-Duration Grazing System. *J. Range Mgt.* **43**(2):123–126.

Taylor, C. A., Jr., and L. B. Merrill. (1986). Cattle, Sheep, and Goat Production From Fixed Yearlong Stocking Rates. *Texas Agric. Expt. Sta. Prog. Rep.* **4441.** 2 pp.

Taylor, C. A., Jr., and M. H. Ralphs. (1992). Reducing Livestock Losses from Poisonous Plants through Grazing Management. *J. Range Mgt.* **45**(1):9–12.

Taylor, C. A., M. M. Kothmann, L. B. Merrill, and D. Elledge. (1980). Diet Selection by Cattle Under High-Intensity Low-Frequency, Short-Duration, and Merrill Grazing Systems. *J. Range Mgt.* **33**(6):428–434.

Taylor, C. A., Jr., N. Garza, and L. B. Merrill. (1986a). Rambouillet Ewe Response to Grazing Systems at the Texas Range Station. *Texas Agric. Expt. Sta. Prog. Rep.* **4440.** 2 pp.

Taylor, C. A., Jr., T. Brooks, and N. Garza. (1986b). Livestock Production From Two Four-Pasture Deferred-Rotation and Two Rotational Grazing Systems. *Texas Agric. Expt. Sta. Prog. Rep.* **4449.** 4 pp.

Taylor, C. A., Jr., J. E. Bowns, and M. H. Ralphs. (1987). The Importance of Poisonous Plants as Forages. *Soc. Range Mgt. Abst. Papers* **40**:107.

Taylor, C. A., Jr., N. E. Garza, Jr., D. M. Anderson, C. V. Hulet, J. N. Smith, and W. L. Shupe. (1988). Bonding of Kids to Heifers. *Soc. Range Mgt. Abst. Papers* **41**:311.

Taylor, C. A., Jr., T. D. Brooks, and N. E. Garza. (1993a). Effects of Short-Duration and High-Intensity, Low-Frequency Grazing Systems on Forage Production and Composition. *J. Range Mgt.* **46**(2):118–121.

Taylor, C. A., Jr., N. E. Garza, Jr., and T. D. Brooks. (1993b). Grazing Systems on the Edwards Plateau of Texas: Are They Worth the Trouble? I. Soil and Vegetation Response. *Rangelands* **15**(2):53–57.

Taylor, C. A., Jr., N. E. Garza, Jr., and T. D. Brooks. (1993c). Grazing Systems on the Edwards Plateau of Texas: Are They Worth the Trouble? II. Livestock Response. *Rangelands* **15**(2):57–60.

Taylor, C. A., Jr., M. H. Ralphs, and M. M. Kothmann. (1997a). Vegetation Response to Increasing Stocking Rate under Rotational Stocking. *J. Range Mgt.* **50**(4):439–442.

Taylor, C. A., Jr., C. L. Robinson-Hicks, M. M. Kothmann, and J. E. Huston (1997b). Sheep Diets and Performance from Two Rotational Grazing Methods. *Sheep & Goat Res. J.* **13**(3):135–141.

Taylor, T. H., and W. C. Templeton, Jr. (1976). Stockpiling Kentucky Bluegrass and Tall Fescue Forage for Winter Pasturage. *Agron. J.* **68**(2):235–239.

Teague, W. R. (1992). Effects of Intensity and Frequency of Defoliation on Shrubs. *In* "Proceedings of the International Conference on Meat Goat Production, Management, and Harvesting." Texas Agric. Ext. Serv., Corpus Christi, TX. pp. 141–156.

Teel, F. D., W. E. Grant, S. L. Marin, and J. W. Stuth. (1998). Simulated Cattle Fever Tick Infestations in Rotational Grazing Systems. *J. Range Mgt.* **51**(5):501–508.

Teer, J. G. (1975). Commercial Uses of Game Animals on Rangelands of Texas. *J. Anim. Sci.* **40**(5):1000–1008.

Teer, J. G. (1985). Strategies and Techniques for Production of Wildlife and Livestock on Western Rangelands. *In* J. Powell (Ed.). "Holistic Ranch Management Workshop Proceedings, 28–30 May 1985, Casper, Wyoming." Wyo. Agric. Ext. Serv., Laramie, WY. pp. 91–99.

Teer, J. G. (1996). The White-Tailed Deer: Natural History and Management. *In* "Rangeland Wildlife." Soc. Range Mgt., Denver, CO. Chapter 12.

Telfer, E. S. (1978a). Cervid Distribution, Browse, and Snow Cover in Alberta. *J. Wildl. Mgt.* **42**(2): 352–361.

Telfer, E. S. (1978b). Habitat Requirements of Moose—The Principal Taiga Range Animal. *Proc. Intern. Rangeland Cong.* **1**:462–465.

Telford, J. P., W. C. Ellis, and K. R. Pond. (1983). Effect of Stocking Rate and Plant Maturity on Intake, Digestibility, Fill, and Flow in Cattle Grazing Annual Ryegrass. *Tex. Agric. Expt. Sta. Prog. Rep.* **4091.** 5 pp.

Tembely, S., T. M. Craig, L. B. Merrill, and R. Dusik. (1983). The Effect of Various Grazing Systems on Gastrointestinal Helminths of Sheep on Native Range. *Texas Agric. Expt. Sta. Cons. Prog. Rep.* **4171,** p. 139–141.

Terrill, C. E. (1975). Game Animals and Agriculture. *J. Anim. Sci.* **40**(5):1020–1022.

Tesar, M. B., and J. L. Yager. (1985). Fall Cutting of Alfalfa in the North Central USA. *Agron. J.* **77**(5):774–778.

Thetford, F. O., R. D. Pieper, and A. B. Nelson. (1971). Botanical and Chemical Composition of Cattle and Sheep Diets on Pinyon-Juniper Grassland Range. *J. Range Mgt.* **24**(6):425–431.

Theuer, C. B., A. L. Lesperance, and J. D. Wallace. (1976). Botanical Composition of the Diet of Livestock Grazing Native Ranges. *Ariz. Agric. Expt. Sta. Tech. Bul.* **233.** 19 pp.

Thilenius, J. F. (1975). Alpine Range Management in the Western United States—Principles, Practices, and Problems: The Status of Our Knowledge. *USDA, For. Serv. Res. Paper* **RM-157.** 32 pp.

Thilenius, J. F., and G. R. Brown. (1987). Herded Vs. Unherded Sheep Grazing Systems on an Alpine Range in Wyoming. *USDA, For. Serv. Gen. Tech. Rep.* **RM-147.** 8 pp.

Thill, R. E. (1984). Deer and Cattle Diets on Louisiana Pine-Hardwood Sites. *J. Wildl. Mgt.* **48**(3):788–798.

Thill, R. E., and A. Martin, Jr. (1986). Deer and Cattle Diet Overlap on Louisiana Pine-Bluestem Range. *J. Wildl. Mgt.* **50**(4):707–713.

Thill, R. E., and A. Martin, Jr. (1989). Deer and Cattle Diets on Heavily Grazed Pine-Bluestem Range. *J. Wildl. Mgt.* **53**(3):540–548.

Thill, R. E., A. Martin, Jr., H. F. Morris, Jr., and A. T. Harrel. (1995). Effects of Prescribed Burning and Cattle Grazing on Deer Diets in Louisiana. *USDA, For. Serv. Res. Paper* **SO-289.** 13 pp.

Thomas, J. R., H. R. Cosper, and W. Bever. (1964). Effects of Fertilizers on the Growth of Grass and Its Use by Deer in the Black Hills of South Dakota. *Agron. J.* **56**(2):223–226.

Thomas, J. W., and D. E. Toweill (Comp. & Eds.). (1982). "Elk of North America: Ecology and Management." Stackpole Books, Harrisburg, Pa. 698 pp.

Thomas, J. W., C. Maser, and J. E. Rodiek. (1979). Wildlife Habitats in Managed Rangelands—The Great Basin of Southeastern Oregon: Edges. *USDA, For. Serv. Gen. Tech. Rep.* **PNW-85.** 17 pp.

Thomas, V. M., and R. W. Kott. (1995). A Review of Montana Winter Range Ewe Nutrition Research. *Sheep & Goat Res. J.* **11**(1):17–27.

Thomas, V. M., E. Ayers, and R. W. Kott. (1987). Influence of Early Weaning of Range Lambs on Ewe Weight Change and Lamb Performance During a Drought. *Mon. AgRes.* **4**(1):14–16.

Thomas, V. M., C. M. Hoaglund, and R. W. Kott. (1992). Influence of Daily Versus Alternate-Day Supplementation on the Production of Gestating Ewes Grazing Winter Range. *Sheep Res. J.* **8**(3):85–90.

Thomas, V. M., R. W. Kott, and R. W. Ditterline. (1995). Sheep Production Response to Continuous and Rotational Stocking on Dryland Alfalfa/Grass Pasture. *Sheep & Goat Res. J.* **11**(3):122–126.

Thompson, D. C., and K. H. McCourt. (1981). Seasonal Diets of the Porcupine Caribou Herd. *Amer. Midl. Nat.* **105**(1):70–76.

Thompson, D. J., D. A. Quinton, and K. Boersma. (1999). Seasonal Grazing Regimes on a Mid-Grassland Plant Community. *Soc. Range Mgt. Abst. Papers* **52:**79.

Thompson, J. R. (1968). Effect of Grazing on Infiltration in a Western Watershed. *J. Soil and Water Cons.* **23**(2):63–65.

Thorhallsdottir, A. G., F. D. Provenza, and D. F. Balph. (1987). Role of Social Models in the Development of Dietary Habits in Lambs. *Soc. Range Mgt. Abst. Papers* **40:**165.

Thorhallsdottir, A. G., F. D. Provenza, and D. F. Balph. (1990). Ability of Lambs to Learn about Novel Foods While Observing or Participating with Social Models. *Appl. Anim. Beh. Sci.* **25**(1–2): 25–33.

Thurow, T. L., and C. A. Taylor, Jr. (1999). The Role of Drought in Range Management: Viewpoint. *J. Range Mgt.* **52**(5):413–419.

Thurow, T. L., W. H. Blackburn, and C. A. Taylor, Jr. (1988). Infiltration and Interill Erosion Responses to Livestock Grazing Strategies, Edwards Plateau, Texas. *J. Range Mgt.* **41**(4):296–302.

Tierson, W. C., E. F. Patric, and D. F. Behrend. (1966). Influence of White-Tailed Deer on the Logged Northern Hardwood Forest. *J. For.* **64**(12):801–805.

Todd, J. W. (1975). Foods of Rocky Mountain Sheep in Southern Colorado. *J. Wildl. Mgt.* **39**(1):108–111.

Torell, D. T., I. D. Hume, and W. C. Weir. (1972). Flushing of Range Ewes by Supplementation, Drylot Feeding, or Grazing of Improved Pasture. *J. Range Mgt.* **25**(5):357–360.

Torrell, L. A., K. S. Lyon, and E. B. Godfrey. (1991). Long-Run Versus Short-Run Planning Horizons and the Rangeland Stocking Rate Decision. *Amer. J. Agric. Econ.* **73**(3):795–807.

Torrell, L. A., and R. C. Torrell. (1996). Evaluating the Economics of Supplementation Practices. *Proc. Grazing Livestock Nutr. Conf.* **3:**62–71.

Torell, P. J., and L. C. Erickson. (1967). Reseeding Medusahead-Infested Ranges. *Idaho Agric. Expt. Sta. Bul.* **489.** 17 pp.

Trammell, M. A., and J. L. Butler. (1995). Effects of Exotic Plants on Native Ungulate Use of Habitat. *J. Wildl. Mgt.* **59**(4):808–816.

Tribe, D. E. (1950). The Behaviour of the Grazing Animal: A Critical Review of Present Knowledge. *J. Brit. Grassland Soc.* **5**(3):209–224.

Trlica, M. J. (1977). Distribution and Utilization of Carbohydrate Reserves in Range Plants. *In* Ronald E. Sosebee (Ed.). "Rangeland Plant Physiology." *Soc. Range Mgt. Range Sci. Ser.* **4,** pp. 73–96.

Trlica, M. J., M. Buwai, and J. W. Menke. (1977). Effects of Rest Following Defoliations on the Recovery of Several Range Species. *J. Range Mgt.* **30**(1):21–27.

Truscott, D. R., and P. O. Currie. (1983). The Role of Sight in the Selective Grazing Process of Cattle. *Soc. Range Mgt. Abst. Papers* **36:**51.

Tueller, P. T. (1979). Food Habits and Nutrition of Mule Deer on Nevada Ranges. *Nev. Agric. Expt. Sta. Rep.* **128.** 104 pp.

Tueller, P. T., and J. D. Tower. (1979). Vegetation Stagnation in Three-Phase Big Game Exclosures. *J. Range Mgt.* **32**(4):258–263.

Turner, K. E., D. P. Belsky, and J. M. Fedders. (1996). Canopy Management Influences on Cool-Season Grass Quality and Simulated Livestock Performance. *Agron. J.* **88**(2):199–205.

Turner, K. E., J. A. Paterson, M. S. Kerley, and J. R. Forwood. (1990). Mefluidide Treatment of Tall Fescue Pastures: Intake and Animal Performance. *J. Anim. Sci.* **68**(10):3399–3405.

Twidwell, E. K., K. D. Johnson, J. H. Cherney, and J. J. Volenec. (1987). Changes in Forage Quality of Switchgrass Morphological Components with Maturation. *Forage and Grassland Conf.* **1987:**81–85.

Ueckert, D. N., M. W. Wagner, J. L. Petersen, and J. E. Huston. (1988). Performance of Sheep Grazing Fourwing Saltbush During Winter. *Texas Agric. Expt. Sta. Prog. Rep.* **4579.** 2 pp.

Ueckert, D. N., J. L. Peterson, J. E. Huston, and M. W. Wagner. (1990). Relative Value of Fourwing Saltbush as a Source of Supplemental Protein for Yearling Ewes During Winter. *Texas Agric. Expt. Sta. Prog. Rep.* **4786.** 4 pp.

Urness, P. J. (1976). Mule Deer Habitat Changes Resulting from Livestock Practices. "Mule Deer Decline in the West: A Symposium." Utah State Univ., Logan, p. 21–35.

Urness, P. J. (1980). "Supplemental Feeding of Big Game in Utah." *Utah Div. Wildl. Resources Pub.* **80–88.** 12 pp.

Urness, P. J. (1986). Value of Crested Wheatgrass for Big Game. *In* K. L. Johnson (Ed.), "Crested Wheatgrass: Its Values, Problems, and Myths; Symposium Proceedings, Logan, Utah, October 3–7, 1983." Utah State Univ., Logan, Utah, pp. 147–153.

Urness, P. J. (1990). Livestock as Manipulators of Mule Deer Winter Habitats in Northern Utah. *In* K. E. Severson (Tech. Coord.), "Can Livestock Be Used as a Tool to Enhance Wildlife Habitat?" *USDA, For. Serv. Gen. Tech. Rep.* **RM-194,** pp. 25–40.

Urness, P. J., D. D. Austin, and L. C. Fierro. (1983). Nutritional Value of Crested Wheatgrass for Wintering Mule Deer. *J. Range Mgt.* **36**(2):225–226.

USDA, Agric. Res. Serv. (1960). Utilizing Forage from Improved Pastures. *USDA, ARS* **22–53.** 9 pp.

USDA, Agric. Res. Serv. (1961a). Calf Weights: Stocking Guide for Cattlemen. *Agric. Res.* **10**(1): 12–13.

USDA, Agric. Res. Serv. (1961b). Crested Wheatgrass. *Agric. Res.* **10**(2):3–4.

USDA, Agric. Res. Serv. (1968). 22 Plants Poisonous to Livestock in the Western States. *USDA Agric. Info. Bul.* **327.** 64 pp.

USDA, Agric. Res. Serv. (1987). Ranchers Battle Leafy Spurge. *Agric. Res.* **35**(6):6–9.

USDA, Extension Service. (1986). "Grazing Lands and People: A National Program Statement and Guidelines for the Cooperative Extension Service." USDA, Ext. Serv., Washington, D.C. 17 pp.

USDA, For. Serv. (1969). "Structural Range Improvement Handbook." U.S. Forest Serv., Intermountain Region, Ogden, UT. Not paged.

USDA, For. Serv. (1972). The Nation's Range Resources—A Forest-Range Environmental Study. *USDA For. Resource Rep.* **19.** 147 pp.

USDA, For. Serv. (1981). An Assessment of the Forest and Range Land Situation in the United States. *USDA For. Resource Rep.* **22.** 352 pp.

USDA, Inter-Agency Work Group on Range Production. (1974). "Opportunities to Increase Red Meat Production from Ranges of the United States, Phase I-Non-Research." USDA, Washington, D.C. 100 pp.

USDA, Natl. Resources Cons. Serv. (1997). "National Range and Pasture Handbook." USDA, Natl. Resources Cons. Serv., Washington, D.C.

USDA, Soil Cons. Serv. (1976). "National Range Handbook." USDA, Soil Cons. Serv., Washington, D.C. Not paged.

Valentine, K. A. (1947). Distance from Water as a Factor in Grazing Capacity of Rangeland. *J. For.* **45**(10):749–754.

Valentine, K. A. (1967). Seasonal Suitability, a Grazing System for Ranges of Diverse Vegetation Types and Condition Classes. *J. Range Mgt.* **20**(6):395–397.

Valentine, K. A. (1970). Influence of Grazing Intensity on Improvement of Deteriorated Black Grama Range. *N. Mex. Agric. Expt. Sta. Bul.* **553.** 21 pp.

Vallentine, J. F. (1963). Water for Range Livestock. *Neb. Agric. Ext. Serv. Cir.* **63–156.** 16 pp.

Vallentine, J. F. (1965). An Improved AUM for Range Cattle. *J. Range Mgt.* **18**(6):346–348.

Vallentine, J. F. (1978). More Pasture or Just Range for Rangemen? *Rangeman's Journal* **5**(2):37–38.

Vallentine, J. F. (1979). Grazing Systems as a Management Tool. *In* "The Sagebrush Ecosystem: A Symposium." Utah State Univ., Logan, Utah, pp. 214–219.

Vallentine, J. F. (1988). Grazing Lands: An Integrative Common Denominator. *Rangelands* 10(5):218–220.

Vallentine, J. F. (1989). "Range Development and Improvements," 3rd Ed. Academic Press, San Diego, CA. 524 pp.

Vallentine, J. F., and A. R. Stevens. (1994). Use of Livestock To Control Cheatgrass—A Review. *In* S. B. Monsen and S. G. Kitchen (Comps.). Proceedings—Ecology and Management of Annual Rangelands. *USDA, For. Serv. Gen. Tech. Rep.* **INT-GTR-313,** pp. 202–206.

Vallentine, J. F., C. W. Cook, and L. A. Stoddart. (1963). Range Seeding in Utah. *Utah Agric. Ext. Serv. Cir.* **307.** 20 pp.

Vallentine, J. F., R. P. Shumway, and S. C. James. (1984). "Cattle Ranch Planning Manual." Brigham Young Univ. Pub., Provo, Utah. 315 pp.

Van Dyne, G. M., and H. F. Heady (1965). Botanical Composition of Sheep and Cattle Diets on a Mature Annual Range. *Hilgardia* **36**(13):465–492.

Van Dyne, G. M., W. Burch, S. K. Fairfax, and W. Huey. (1984a). Forage Allocation on Arid and Semiarid Public Grazing Lands: Summary and Recommendations. *In* Natl. Res. Council/Natl. Acad. Sci. "Developing Strategies for Rangeland Management." Westview Press, Boulder, CO. pp. 1–25.

Van Dyne, G. M., P. T. Kortopates, and F. M. Smith (Eds.). (1984b). Quantitative Frameworks for Forage Allocation. *In* Natl. Res. Council/Natl. Acad. Sci. "Developing Strategies for Rangeland Management." Westview Press, Boulder, CO. pp. 289–416.

Van Keuren, R. W. (1980). Managing Established Pastures. *Forage and Grassland Conf.* **1980:** 165–175.

Van Keuren, R. W., A. D. Pratt, H. R. Conrad, and R. R. Davis. (1966). Utilization of Alfalfa-Bromegrass as Soilage, Strip Grazing, and Rotational Grazing for Dairy Cattle. *Ohio Agric. Expt. Sta. Res. Bul.* **989.** 20 pp.

Van Poollen, H. W., and J. R. Lacey. (1979). Herbage Response to Grazing Systems and Stocking Intensities. *J. Range Mgt.* **32**(4):250–253.

Van Soest, P. J. (1967). Development of a Comprehensive System of Feed Analysis and Its Application to Forage. *J. Anim. Sci.* **26**(1):119–128.

Van Soest, P. J. (1982). "Nutritional Ecology of the Ruminant." O & B Books, Corvallis, OR. 373 pp.

Van Tassell, L. W., C. Phillips, and W. G. Hepworth. (1995). Livestock to Wildlife Is Not a Simple Conversion. *Rangelands* **17**(6):191–193.

Van Vuren, D. (1984). Summer Diets of Bison and Cattle in Southern Utah. *J. Range Mgt.* **37**(3):260–261.

Varner, L. W., and L. H. Blankenship. (1987). Southern Texas Shrubs—Nutritive Value and Utilization by Herbivores. *USDA, For. Serv. Gen. Tech. Rep.* **INT-222,** pp. 108–112.

Varner, L. W., L. H. Blankenship, and S. C. Heineman. (1986). Nutritional Value of South Texas Deer Food Plants. *Texas Agric. Expt. Sta. Cons. Prog. Rep.* **4347,** p. 66–75.

Vartha, E. W., A. G. Matches, and G. B. Thompson. (1977). Yield and Quality Trends of Tall Fescue Grazed with Different Subdivisions of Pasture. *Agron. J.* **69**(6):1027–1029.

Vavra, M. (1992). Livestock and Big Game Forage Relationships. *Rangelands* **14**(2):57–59.

Vavra, M., and R. J. Raleigh. (1976). Coordinating Beef Cattle Management With the Range Forage Resource. *J. Range Mgt.* **29**(6):449–452.

Vavra, M., and D. P. Sheehy. (1987). Grazing Relationships of Three Large Herbivores on Seasonal Rangelands. *Soc. Range Mgt. Abst. Papers* **40:**162.

Vavra, M., and D. P. Sheehy. (1996). Improving Elk Habitat Characteristics with Livestock Grazing. *Rangelands* **18**(5):182–185.

Vavra, M., and F. Sneva. (1978). Seasonal Diets of Five Ungulates Grazing the Cold Desert Biome. *Proc. Intern. Rangeland Cong.* **1:**435–437.

Vavra, M., R. W. Rice, and R. E. Bement. (1973). Chemical Composition of the Diet, Intake, and Gain of Yearling Cattle on Different Grazing Intensities. *J. Anim. Sci.* **36**(2):411–414.

Vavra, M., R. W. Rice, R. M. Hansen, and P. L. Sims. (1977). Food Habits of Cattle on Shortgrass Range in Northeastern Colorado. *J. Range Mgt.* **30**(4):261–263.

Vavra, M., W. A. Laycock, and R. D. Pieper (Eds.). (1994). "Ecological Implications of Livestock Herbivory in the West." Soc. Range Mgt., Denver, CO. 297 pp.

Velez, E., H. Tejada, H. E. Kiesling, R. F. Beck, and G. M. Southward. (1991). Grazing Behavior of Angora Goats Confined to a Semidesert Mesquite Brush Dominated Rangeland. *Amer. Soc. Anim. Sci., West. Sect. Proc.* **42:**177–179.

Vicini, J. L., E. C. Prigge, W. B. Bryan, and G. A. Varga. (1982). Influence of Forage Species and Creep Grazing on a Cow-Calf System. I. Intake and Digestibility. II. Calf Production. *J. Anim. Sci.* **55**(4):752–764.

Vickery, P. J. 1981. Pasture Growth Under Grazing. *In* F. H. W. Morley (Ed.), "Grazing Animals." Elsevier Sci. Pub. Co., Amsterdam, Neth., pp. 55–57.

Villalobos, G., D. C. Adams, T. J. Klopfenstein, and J. B. Lamb. (1992). Evaluation of High Quality Meadow Hay as a Winter Supplement for Gestating Beef Cows in the Sandhills of Nebraska. *Amer. Soc. Anim. Sci., West. Sect. Proc.* **43**:367–370.

Villalobos, G., D. C. Adams, T. J. Klopfenstein, J. T. Nichols, and J. B. Lamb. (1997). Grass Hay as a Supplement for Grazing Cattle. I. Animal Performance. *J. Range Mgt.* **50**(4):351–356.

Vogel, K. P., H. J. Gorz, and F. A. Haskins. (1985). Viewpoint: Forage and Range Research Needs in the Central Great Plains. *J. Range Mgt.* **38**(5):477–479.

Voisin, A. (1959). "Grass Productivity." Philosophical Library, New York. 349 pp.

Volesky, J. D. (1990). Frontal Grazing: Forage Harvesting of the Future? *Rangelands* **12**(3):177–181.

Volesky, J. D. (1994). Tiller Defoliation Patterns under Frontal, Continuous, and Rotation Grazing. *J. Range Mgt.* **47**(3):215–219.

Volesky, J. D., J. K. Lewis, and C. H. Butterfield. (1990). High-Performance Short-Duration and Repeated-Seasonal Grazing Systems: Effect on Diets and Performance of Calves and Lambs. *J. Range Mgt.* **43**(4):310–315.

Volesky, J. D., F. De Achaval O'Farrell, W. C. Ellis, M. M. Kothmann *et al.* (1994). A Comparison of Frontal, Continuous, and Rotation Grazing Systems. *J. Range Mgt.* **47**(3):210–214.

Wade, D. D., and C. E. Lewis. (1987). Managing Southern Grazing Ecosystems with Fire. *Rangelands* **9**(3):115–119.

Wagner, F. H. (1978). Livestock Grazing and the Livestock Industry. *In* H. P. Brokaw (Ed.), "Wildlife and America." Council on Environmental Quality, Washington, D.C., pp. 121–145.

Wagnon, K. A. (1963). Behavior of Beef Cows on a California Range. *Calif. Agric. Expt. Sta. Bul.* **799.** 58 pp.

Wagnon, K. A. (1965). Social Dominance in Range Cows and Its Effect on Supplemental Feeding. *Calif. Agric. Expt. Sta. Bul.* **819.** 32 pp.

Wagnon, K. A. (1968). Use of Different Classes of Range Land by Cattle. *Calif. Agric. Expt. Sta. Bul.* **838.** 16 pp.

Wagnon, K. A., and H. Goss. (1961). The Use of Molasses to Increase the Utilization of Rank, Dry Forage and Molasses-Urea as a Supplement for Weaner Calves. *J. Range Mgt.* **14**(1):5–9.

Wagnon, K. A., H. R. Guilbert, and G. H. Hart. (1959). Effects of Various Methods of Supplemental Feeding. *Calif. Agric. Expt. Sta. Bul.* **765,** p. 30–32.

Wairimu, S., and R. J. Hudson. (1993). Foraging Dynamics of Wapiti (*Cervus elaphus*) Stags During Compensatory Growth. *Appl. Anim. Beh. Sci.* **36**(1):65–79.

Walker, J. W. (1994). Multispecies Grazing: The Ecological Advantage. *Sheep Res. J.* **10**(Spec. Issue):52–64.

Walker, J. W. (1995). Grazing Management and Research Now and in the Next Millennium: A Viewpoint. *J. Range Mgt.* **48**(4):350–357.

Walker, J. W., and R. K. Heitschmidt. (1986a). Grazing Behavior of Cattle in Moderate, Continuous, and Rotational Grazed Systems. *Texas Agric. Expt. Sta. Prog. Rep.* **4429.** 2 pp.

Walker, J. W., and R. K. Heitschmidt. (1986b). Effect of Various Grazing Systems on Type and Density of Cattle Trails. *Texas Agric. Expt. Sta. Prog. Rep.* **4430.** 2 pp.

Walker, J. W., and R. K. Heitschmidt. (1986c). Effect of Various Grazing Systems on Type and Density of Cattle Trails. *J. Range Mgt.* **39**(5):428–431.

Walker, J. W., R. M. Hansen, and L. R. Rittenhouse. (1981). Diet Selection of Hereford, Angus × Hereford, and Charolais × Hereford Cows and Calves. *J. Range Mgt.* **34**(3):243–245.

Walker, J. W., R. K. Heitschmidt, and S. L. Dowhower. (1985). Evaluation of Pedometers for Measuring Distance Traveled by Cattle on Two Grazing Systems. *J. Range Mgt.* **38**(1):90–93.

Walker, J. W., R. K. Heitschmidt, and J. W. Stuth. (1987). The Effect of Grazing Systems on Livestock Distribution. *Soc. Range Mgt. Abst. Papers* **40**:211.

Walker, J. W., R. K. Heitschmidt, and S. L. Dowhower. (1989a). Some Effects of a Rotational Grazing Treatment on Cattle Preference for Plant Communities. *J. Range Mgt.* **42**(2):143–148.

Walker, J. W., R. K. Heitschmidt, E. A. De Moraes, M. M. Kothmann, and S. L. Dowhower. (1989b). Quality and Botanical Composition of Cattle Diets under Rotational and Continuous Grazing Treatments. *J. Range Mgt.* **42**(3):239–242.

Walker, J. W., S L. Kronberg, S. L. Al-Rowaily, and N. E. West. (1996). Managing Noxious Weeds with Livestock: Studies on Leafy Spurge. *Proc. Internat. Rangeland Cong.* **5**(1):586–587.

Wallace, J. D. (1984). Some Comments and Questions on Animal Preferences, Ecological Efficiencies, and Forage Intake. *In* Natl. Res. Council/Natl. Acad. Sci. "Developing Strategies for Rangeland Mangement." Westview Press, Boulder, CO. pp. 275–287.

Wallace, J. D., and L. Foster. (1975). Drought on the Range: Drought and Range Cattle Performance. *Rangeman's J.* **2**(6):178–180.

Wallace, J. D., J. C. Free, and A. H. Denham. (1972). Seasonal Changes in Herbage and Cattle Diets on Sandhill Grassland. *J. Range Mgt.* **25**(2):100–104.

Wallace, J. D., S. Rafique, and E. E. Parker. (1988). Response of Yearling Range Heifers to Type and Frequency of Supplemental Feeding. *Amer. Soc. Anim. Sci., West. Sect. Proc.* **39**:226–229.

Wallace, M. C., and P. R. Krausman. (1987). Elk, Mule Deer, and Cattle Habitats in Central Arizona. *J. Range Mgt.* **40**(1):80–83.

Waller, S. S. (1986). Seasonal Forage Production Patterns of Grasses. *In* Patrick E. Reece and James T. Nichols (Eds.). "Proceedings, The Ranch Management Symposium, November 5–7, 1986, North Platte, Nebraska." Univ. Neb., Agric. Ext. Serv., Lincoln, Neb., pp. 27–34.

Waller, S. S., P. E. Reece, L. E. Moser, and G. A. Gates. (1985). "Understanding Grass Growth: The Key to Profitable Livestock Production." Trabon Print. Co., Kansas City, Mo. 20 pp.

Waller, S. S., L. E. Moser, and B. Anderson. (1986). A Guide for Planning and Analyzing a Year-Round Forage Program. *Neb. Agric. Ext. Cir.* **86–113**. 19 pp.

Wallmo, O. C. (1969). Response of Deer to Alternate-Strip Clearcutting of Lodgepole Pine and Spruce-Fir Timber in Colorado. *USDA, For. Serv. Res. Note* **RM-141**. 4 pp.

Wallmo, O. C., L. H. Carpenter, W. L. Regelin, R. B. Gill, and D. L. Baker. (1977). Evaluation of Deer Habitat on a Nutritional Basis. *J. Range Mgt.* **30**(2):122–127.

Walton, P. D. (1983). "Production and Management of Cultivated Forages." Reston Pub. Co., Reston, VA. 336 pp.

Walton, P. D., R. Martinez, and A. W. Bailey. (1981). A Comparison of Continuous and Rotational Grazing. *J. Range Mgt.* **34**(1):19–21.

Wambolt, C. L. (1996). Mule Deer and Elk Foraging Preference for 4 Sagebrush Taxa. *J. Range Mgt.* **49**(6):499–503.

Wambolt, C. L., and A. F. McNeal. (1987). Selection of Winter Foraging Sites by Elk and Mule Deer. *J. Environ. Mgt.* **25**(3):285–291.

Wambolt, C. L., R. G. Kelsey, T. L. Personius, K. D. Striby, A. F. McNeal, and K. M. Havstad. (1987). Preference and Digestibility of Three Big Sagebrush Subspecies and Black Sagebrush as Related to Crude Terpenoid Chemistry. *USDA, For. Serv. Gen. Tech. Rep.* **INT-222**, pp. 71–73.

Wambolt, C. L., M. R. Frisina, K. S. Douglass, and H. W. Sherwood. (1997). Grazing Effects on Nutritional Quality of Bluebunch Wheatgrass for Elk. *J. Range Mgt.* **50**(5):503–506.

Ward, A. L. (1976). Elk Behavior in Relation to Timber Harvest Operations and Traffic on the Medicine Bow Range in South-Central Wyoming. *In* "Elk-Logging-Roads Symp. Proc." Univ. Idaho, Moscow, ID. p. 32–43.

Warren, L. E., D. Ueckert, and M. Shelton. (1981a). Dietary Choices of Selected Breeds of Sheep and Goats Grazing in West Texas. *Tex. Agric. Expt. Sta. Prog. Rep.* **3907**. 8 pp.

Warren, L., M. Shelton, D. N. Ueckert, and A. D. Chamrad. (1981b). Forage Selectivity of Goats in South Texas Plains. *Texas Agric. Expt. Sta. Prog. Rep.* **3908**. 8 pp.

Warren, L. E., D. N. Ueckert, and J. M. Shelton. (1984a). Comparative Diets of Rambouillet, Barbado, and Karakul Sheep and Spanish and Angora Goats. *J. Range Mgt.* **37**(2):172–180.

Warren, L. E., D. N. Ueckert, M. Shelton, and A. D. Chamrad. (1984b). Spanish Goat Diets on Mixed-Brush Rangeland in the South Texas Plains. *J. Range Mgt.* **37**(4):340–342.

Warren, S. D., W. H. Blackburn, and C. A. Taylor, Jr. (1986a). Effects of Season and Stage of Rotation Cycle on Hydrologic Condition of Rangeland Under Intensive Rotation Grazing. *J. Range Mgt.* **39**(6):486–491.

Warren, S. D., W. H. Blackburn, and C. A. Taylor, Jr. (1986b). Soil Hydrologic Response to Number of Pastures and Stocking Density Under Intensive Rotation Grazing. *J. Range Mgt.* **39**(6):500–504.

Watts, C. R., L. C. Eichorn, and R. J. Mackie. (1987). Vegetation Trends Within Rest-Rotation and Season-Long Grazing Systems in the Missouri River Breaks, Montana. *J. Range Mgt.* **40**(5):393–396.

Weaver, J. E., and G. W. Tomanek. (1951). Ecological Studies in a Midwestern Range: The Vegetation and Effects of Cattle on Its Composition and Distribution. *Univ. Neb., Cons. & Survey Div. Neb. Cons. Bul.* **31.** 82 pp.

Weaver, R. A., F. Vernoy, and B. Craig. (1959). Game Water Development on the Desert. *Calif. Fish & Game* **45**(4):333–342.

Wedin, W. F. (1976). Integration of Forage Resources for Beef Cattle Production in the Western Cornbelt. *In* "Proceedings for Symposium on Integration of Resources for Beef Cattle Production, February 16–20, 1976." Society for Range Management, Denver, CO. pp. 4–19.

Wedin, W. F., and T. J. Klopfenstein. (1985). Cropland Pastures and Crop Residues. Chapter 53. *In* M. E. Heath, R. F. Barnes, and D. S. Metcalf (Eds.), "Forages: The Science of Grassland Agriculture," 4th Ed. Iowa State Univ. Press, Ames, Iowa.

Wedin, W. F., M. A. Sanderson, C. Ohlsonn, and D. G. Morrical. (1989). Multispecies Grazing Systems in Iowa, Midwestern USA. *Proc. Internat. Grassland Cong.* **16**:1125–1126.

Weigel, J. R., and C. M. Britton. (1987). Effects of Short-Duration Grazing Trampling on Seedling Emergence and Soil Strength. *Soc. Range Mgt. Abst. Papers* **40**:180.

Welch, B. L., and E. D. McArthur. (1979). Variation in Winter Levels of Crude Protein Among *Artemisia tridentata* Subspecies Grown in a Uniform Garden. *J. Range Mgt.* **32**(6):467–469.

Welch, B. L., E. D. McArthur, and J. N. Davis. (1983). Mule Deer Preference and Monoterpenoids (Essential Oils). *J. Range Mgt.* **36**(4):485–487.

Welch, B. L., E. D. McArthur, and R. L. Rodriguez. (1987). Variation in Utilization of Big Sagebrush Accessions by Wintering Sheep. *J. Range Mgt.* **40**(2):113–115.

Welker, J. M., and D. D. Briske. (1986). Carbon and Nitrogen Sharing Among Bunchgrass Tillers in Response to Defoliation. *Texas Agric. Expt. Sta. Prog. Rep.* **4418.** 1 p.

Welty, L. E., R. L. Ditterline, and L. S. Prestbye. (1988). Fall Management of Alfalfa. *Mon. AgRes.* **5**(2):16–19.

Weltz, M., and M. K. Wood. (1986). Short-Duration Grazing in Central New Mexico: Effects on Infiltration Rates. *J. Range Mgt.* **39**(4):365–368.

Werner, S. J., and P. J. Urness. (1998). Elk Forage Utilization within Rested Units of Rest-Rotation Grazing Systems. *J. Range Mgt.* **51**(1):14–18.

West, N. E. (1993). Biodiversity of Rangelands. *J. Range Mgt.* **44**(1):2–13.

Westenskow-Wall, K. J., W. C. Krueger, L. D. Bryant, and D. R. Thomas. (1994). Nutrient Quality of Bluebunch Wheatgrass Regrowth on Elk Winter Range in Relation to Defoliation. *J. Range Mgt.* **47**(3):240–244.

Whisenant, S. G., and F. J. Wagstaff. (1991). Successional Trajectories of a Grazed Salt-Desert Shrubland. *Vegetatio* **94**(2):133–140.

White, D. H., and F. H. W. Morley. (1977). Estimation of Optimal Stocking Rate of Merino Sheep. *Agric. Systems* **2**:289–304.

White, L. D. (1988). Technology Transfer and Total Ranch Management. *In* Richard S. White and Robert E. Short (Eds.). "Achieving Efficient Use of Rangeland Resources, Fort Keogh Research Symposium, September, 1987, Miles City, Montana." Mon. Agric. Expt. Sta., Bozeman, Mon., pp. 125–128.

White, L. D., and C. Richardson. (1989). How Much Forage Do You Have? *Texas Agric. Ext. Bull.* **1646.** 8 pp.

White, L. M. (1973). Carbohydrate Reserves of Grasses: A Review. *J. Range Mgt.* **26**(1):13–18.

White, L. M. (1977). Perenniality and Development of Shoots of 12 Forage Species in Montana. *J. Range Mgt.* **30**(2):107–110.

White, L. M. (1989). Growth Regulators' Effect on Crested Wheatgrass Forage Yield and Quality. *J. Range Mgt.* **42**(1):46–50.

White, L. M. (1990). Mefluidide Effect on Caucasian Bluestem Leaves, Stems, Forage Yield, and Quality. *J. Range Mgt.* **43**(3):190–194.

White, L. M. (1991). Mefluidide Effect on Weeping Lovegrass Heading, Forage Yield, and Quality. *J. Range Mgt.* **44**(5):501–508.

White, L. M., and J. R. Wight. (1973). Development and Growth of Several Forage Species in the Northern Great Plains. *Soc. Range Mgt. Abst. Papers* **26**:25.

White, L. M., and J. R. Wight. (1980). Optimal Time to Harvest 9 Forages for Maximum Livestock Weight Gains per Hectare. *Soc. Range Mgt. Abstr. Papers* **33**:19.

White, R. J. (1987). "Big Game Ranching in the United States." Wild Sheep and Goat International, Mesilla, N. Mex. 355 pp.

Whitson, R. E., R. K. Heitschmidt, M. M. Kothmann, and G. K. Lundgren. (1982). The Impact of Grazing Systems on the Magnitude and Stability of Ranch Income in the Rolling Plains of Texas. *J. Range Mgt.* **35**(4):526–532.

Whyte, R. J., and B. W. Cain. (1981). Wildlife Habitat on Grazed or Ungrazed Small Pond Shorelines. *J. Range Mgt.* **34**(1):64–68.

Wiggers, E. P., D. D. Wilcox, and F. C. Bryant. (1984). Cultivated Cereal Grains as Supplemental Forages for Mule Deer in the Texas Panhandle. *Wildl. Soc. Bul.* **12**(3):240–245.

Wilkins, R. J., and E. A. Garwood. (1986). Effects of Treading, Poaching, and Fouling on Grassland Production and Utilization. *In* J. Frame (Ed.), "Grazing." British Grassland Soc., Hurley, Berks, Eng., pp. 19–31.

Wilkins, R. N., and W. G. Swank. (1992). Bobwhite Habitat Use Under Short-Duration and Deferred-Rotation Grazing. *J. Range Mgt.* **45**(6):549–553.

Willard, E. E. (1989). Cattle Grazing Influence on Forage Quality of Elk Summer Range. *Soc. Range Mgt. Abst. Papers* **42**:213.

Williams, N. C., and E. H. Cronin. (1996). Five Poisonous Range Weeds—When and Why They Are Poisonous. *J. Range Mgt.* **19**(5):274–279.

Williams, O. B. 1981. Evaluation of Grazing Systems. *In* F. H. W. Morley (Ed.). "Grazing Animals." Elsevier Sci. Pub. Co., Amsterdam, Neth., pp. 1–12.

Williams, R. M., R. T. Clark, and A. R. Patton. (1942). Wintering Steers on Crested Wheatgrass. *Mon. Agric. Expt. Sta. Bul.* **407**. 18 pp.

Willms, W. (1978). Forage Strategy of Ruminants. *Rangeman's J.* **5**(3):72–74.

Willms, W. D. (1989). Distribution of Cattle on Slope without Water Restrictions. *Soc. Range Mgt. Abst. Papers* **42**:111.

Willms, W. D. (1992). Influence of Summer Cutting and Fertilizer Application on Altai Wild Rye in Winter. *Can. J. Plant Sci.* **72**(1):173–179.

Willms, W. D., and K. A. Beauchemin. (1991). Cutting Frequency and Cutting Height Effects on Forage Quality of Rough Fescue and Parry Oat Grass. *Can. J. Anim. Sci.* **71**(1):87–96.

Willms, W. D., and L. M. Rode. (1998). Forage Selection by Cattle on Fescue Prairie in Summer or Winter. *J. Range Mgt.* **51**(5):496–500.

Willms, W., A. McLean, R. Tucker, and R. Ritcey. (1979). Interactions Between Mule Deer and Cattle on Big Sagebrush Range in British Columbia. *J. Range Mgt.* **32**(4):299–304.

Willms, W., A. McLean, R. Tucker, and R. Ritcey. (1980a). Deer and Cattle Diets on Summer Range in British Columbia. *J. Range Mgt.* **33**(1):55–59.

Willms, W., A. W. Bailey, A. McLean, and R. Tucker. (1980b). The Effects of Fall Grazing or Burning Bluebunch Wheatgrass Range on Forage Selection by Deer and Cattle in the Spring. *Can. J. Anim. Sci.* **60**(1):113–122.

Willms, W., A. W. Bailey, A. McLean, and R. Tucker. (1981). The Effects of Fall Defoliation on the Utilization of Bluebunch Wheatgrass and Its Influence on the Distribution of Deer in Spring. *J. Range Mgt.* **34**(1):16–18.

Willms, W. D., S. Smoliak, and J. F. Dormaar. (1985). Effects of Stocking Rate on a Rough Fescue Grassland Vegetation. *J. Range Mgt.* **38**(3):220–225.

Willms, W. D., S. Smoliak, and G. B. Schaalje. (1986a). Cattle Weight Gains in Relation to Stocking Rate on Rough Fescue Grassland. *J. Range Mgt.* **39**(2):182–187.

Willms, W. D., S. Smoliak, and A. W. Bailey. (1986b). Herbage Production Following Litter Removal on Alberta Native Grasslands. *J. Range Mgt.* **39**(6):536–540.

Willms, W. D., J. F. Dormaar, and G. B. Schaalje. (1988). Stability of Grazed Patches on Rough Fescue Grasslands. *J. Range Mgt.* **41**(6):503–508.

Willms, W. D., B. W. Adams, and J. F. Dormaar. (1992). Grazing Management of Native Grasslands. Handbook 1. Foothills Fescue Prairie. *Agric. Can. Pub.* **1883/E.** 33 pp.

Willms, W. D., B. W. Adams, and J. F. Dormaar. (1996). Seasonal Changes of Herbage Biomass on the Fescue Prairie. *J. Range Mgt.* **49**(2):100–104.

Willms, W. D., J. King, and J. F. Dormaar. (1998). Weathering Losses of Forage Species on the Fescue Grassland in Southwestern Alberta. *Can. J. Plant Sci.* **78**(2):265–272.

Wilson, A. D. (1986). Principles of Grazing Management Systems. *Proc. Internat. Rangeland Cong.* **2:**221-225.

Wilson, A. D., and N. D. Macleod. (1991). Overgrazing: Present or Absent? *J. Range Mgt.* **44**(5):475–482.

Wilson, J. R. (1983). Effects of Water Stress on Herbage Quality. *Proc. Intern. Grassland Cong.* **14:**470–472.

Wilson, J. R., and P. M. Kennedy. (1996). Plant and Animal Constraints to Voluntary Feed Intake Associated with Fibre Characteristics and Particle Breakdown and Passage in Ruminants. *Austr. J. Agric. Res.* **47**(2):199–225.

Wilson, P. N., D. E. Ray, and G. B. Ruyle. (1987). A Model for Assessing Investments in Intensive Grazing Technology. *J. Range Mgt.* **40**(5):401–404.

Wiltbank, J. N. (1964). Reasons for Poor Reproductive Performance. *In* "Fort Robinson Beef Cattle Research Station Field Day Report, April 30, 1964." USDA, ARS, Crawford, NE.

Wimer, S. K., J. K. Ward, B. E. Anderson, and S. S. Waller. (1986). Mefluidide Effects on Smooth Brome Composition and Grazing Cow-Calf Performance. *J. Anim. Sci.* **63**(4):1054–1062.

Winchester, C. F., and M. J. Morris. (1956). Water Intake Rates of Cattle. *J. Anim. Sci.* **15**(3):722–740.

Winder, J. A., D. A. Walker, and C. C. Bailey. (1995). Genetic Aspects of Diet Selection in the Chihuahuan Desert. *J. Range Mgt.* **48**(6):549–553.

Winder, J. A., D. A. Walker, and C. C. Bailey. (1996). Effect of Breed on Botanical Composition of Cattle Diets on Chihuahuan Desert Range. *J. Range Mgt.* **49**(3):209–214.

Winter, S. R. (1994). Managing Wheat for Grazing and Grain. *Texas Agric. Expt. Sta. Misc. Pub.* **1754.** 7 pp.

Winter, S. R., and E. K. Thompson. (1987). Grazing Duration Effects on Wheat Growth and Grain Yield. *Agron. J.* **79**(1):110–114.

Wisdom, M. J., and J. W. Thomas. (1996). Elk. *In* "Rangeland Wildlife." Soc. Range Mgt., Denver, CO. Chapter 10.

Wolf, D. D., and R. E. Blaser. (1981). Flexible Alfalfa Management: Early Spring Utilization. *Crop Sci.* **21**(1):90–93.

Wood, G. M. (1987). Animals for Biological Brush Control. *Agron. J.* **79**(2):319–321.

Wood, M. K., and W. H. Blackburn. (1981). Grazing Systems: Their Influence on Infiltration Rates in the Rolling Plains of Texas. *J. Range Mgt.* **34**(4):331–335.

Woolfolk, E. J., and B. Knapp, Jr. (1949). Weight and Gain of Range Calves as Affected by Rate of Stocking. *Mon. Agric. Expt. Sta. Bul.* **463.** 26 pp.

Wright, H. A. (1970). Response of Big Sagebrush and Three-tip Sagebrush to Season of Clipping. *J. Range Mgt.* **23**(1):20–22.

Wydeven, A. P., and R. B. Dahlgren. (1983). Food Habits of Elk in the Northern Great Plains. *J. Wildl. Mgt.* **47**(4):916–923.

Yabann, W. K., E. A. Burritt, and J. C. Malechek. (1987). Sagebrush (*Artemisia tridentata*) Monoterpenoid Concentrations as Factors in Diet Selection by Free-Grazing Sheep. *USDA, For. Serv. Gen. Tech. Rep.* **INT-222,** pp. 64–70.

Yates, D. A. (1980). "Nutrition Management of the Productive Mature Beef Cow." Wyo. Agric. Ext. Serv., Laramie, WY. 3 pp.

Yelich, J. V., D. N. Schutz, and K. G. Odde. (1988). Effect of Time of Supplementation on Performance and Grazing Behavior of Beef Cows Grazing Fall Native Range. *Amer. Soc. Anim. Sci., West. Sect. Proc.* **39:**58–60.

Yeo, J. J., J. M. Peek, W. T. Wittinger, and C. T. Kvale. (1993). Influence of Rest-Rotation Cattle Graz-
    ing on Mule Deer and Elk Habitat Use in East-Central Idaho. *J. Range Mgt.* **46**(3):245–250.
Yoakum, J. D. (1975). Antelope and Livestock on Rangelands. *J. Anim. Sci.* **40**(5):985–992.
Yoakum, J. D. (1978). Managing Rangelands for the American Pronghorn Antelope. *Proc. Intern.*
    Rangeland Cong. **1:**584–587.
Yoakum, J. D., B. W. O'Gara, and V. W. Howard, Jr. (1996). Pronghorn on Western Rangelands. *In*
    "Rangeland Wildlife." Soc. Range Mgt., Denver, CO. Chapter 13.
Young, B. A., and A. A. Degen. (1991). Effect of Snow as a Water Source on Beef Cows and Their Calf
    Production. *Can. J. Anim. Sci.* **71**(2):585–588.
Young, J. A., and R. A. Evans. (1984). Historical Aspects of Winter Grazing. *Rangelands* **6**(5):206–
    209.
Younger, V. B., and C. M. McKell. (1972). "The Biology and Utilization of Grasses." Academic Press,
    N. Y. 426 pp.
Zhang, J., and J. T. Romo. (1995). Impacts of Defoliation on Tiller Production and Survival in North-
    ern Wheatgrass. *J. Range Mgt.* **48**(2):115–120.
Zimmerman, E. A. (1980). Desert Ranching in Central Nevada. *Rangelands* **2**(5):184–186.
Zimmerman, G. T., and L. F. Neuenschwander. (1984). Livestock Grazing Influences on Community
    Structure, Fire Intensity, and Fire Frequency Within the Douglas-fir/Ninebark Habitat Type.
    J. Range Mgt. **37**(2):104–110.

INDEX